Mechanical Performance of Sustainable Bio-Based Compounds

Mechanical Performance of Sustainable Bio-Based Compounds

Editors

Domenico Acierno
Antonella Patti

MDPI • Basel • Beijing • Wuhan • Barcelona • Belgrade • Manchester • Tokyo • Cluj • Tianjin

Editors
Domenico Acierno
CRdC Tecnologie Scarl
Regional Center of
Competence New
Technologies for Productive
Activities Scarl
Naples
Italy

Antonella Patti
Department of Civil
Engineering and Architecture
(DICAr)
University of Catania
Catania
Italy

Editorial Office
MDPI
St. Alban-Anlage 66
4052 Basel, Switzerland

This is a reprint of articles from the Special Issue published online in the open access journal *Polymers* (ISSN 2073-4360) (available at: www.mdpi.com/journal/polymers/special_issues/mech_perform_bio_compd).

For citation purposes, cite each article independently as indicated on the article page online and as indicated below:

LastName, A.A.; LastName, B.B.; LastName, C.C. Article Title. *Journal Name* **Year**, *Volume Number*, Page Range.

ISBN 978-3-0365-6685-6 (Hbk)
ISBN 978-3-0365-6684-9 (PDF)

Contents

polymers

Editorial

Special Issue "Mechanical Performance of Sustainable Bio-Based Compounds"

Antonella Patti [1,*] and Domenico Acierno [2,*]

[1] Department of Civil Engineering and Architecture (DICAr), University of Catania, Viale Andrea Doria 6, 95125 Catania, Italy
[2] Regional Center of Competence New Technologies for Productive Activities Scarl, Via Nuova Agnano 11, 80125 Naples, Italy
* Correspondence: antonella.patti@unict.it (A.P.); acierno@crdctecnologie.it (D.A.)

Citation: Patti, A.; Acierno, D. Special Issue "Mechanical Performance of Sustainable Bio-Based Compounds". *Polymers* **2022**, *14*, 4832. https://doi.org/10.3390/polym14224832

Received: 8 November 2022
Accepted: 9 November 2022
Published: 10 November 2022

Publisher's Note: MDPI stays neutral with regard to jurisdictional claims in published maps and institutional affiliations.

The global production of plastic is increasing, and plastic represents one of the most popular materials, widespread in countless applications in commercial and industrial fields and everyday life. However, the main drawback to consider in this widespread employment is the difficulty of properly disposing of plastic at the end of its life cycle, and its consequent accumulation in nature. Plastic wastes, including microplastics (1 to 5000 μm particles), accumulate in the oceans, seriously altering the marine environment and the lives of its inhabitants. Furthermore, millions of tons of CO_2 are emitted into the atmosphere, generated from both the production processes and the final incineration at the end of the products' life cycles. The huge emission of greenhouse gases such as CO_2 further contributes to global warming by inducing dramatic and irreversible weather changes [1].

In this context, scientific research should focus on alternative solutions for the sustainable development of production processes, including biomass waste valorization [2] and the use of green technologies, avoiding risks for human health, and limiting the environmental impact. For example, fruit and vegetable wastes have high water contents and high concentrations of biodegradable organic substances (e.g., carbohydrates, lipids, and organic acids). These can be efficiently recovered and valorized as sources of polymer materials for active packaging [3]. Innovative technologies, involving both thermal and non-thermal methods, have been applied for the extraction and fortification of sensitive bioactive compounds, as reported in a recent review by Basri et al. [3].

The purpose of this Special Issue was to collect original contributions, both research papers and reviews, showing recent results and/or positive advances in the behavior of new sustainable biomaterials under applied mechanical stress, in both static and dynamic modes, and evaluating the characteristics of resistance, moduli and/or viscoelasticity in view of potential applications in biotechnology and biomedicine [4], tissue engineering [5], medical devices [6,7], surgical or dental implants [8], agriculture [9], packaging [3,10,11], textiles [1], automotives [12–14], green buildings and construction [15], radiation dosimetry [16], and 3D printing [17–19].

Wound-care products have been prepared by using gellan gum and virgin coconut oil (VCO) and developing microemulsion-based hydrogels [7]. The effects of drying methods on carboxymethyl cellulose and citric acid coating layers on cotton threads have been examined in [6], with regard to wound-dressing applications.

Arboblend V2 Nature biopolymer (which features lignin as a basic matrix; a significant amount of polylactic acid; and small amounts of natural additives such as resins, waxes, and shellac) was covered with three ceramic powders: Amdry 6420 (Cr_2O_3), Metco 143 (ZrO_2 $18TiO_2$ $10Y_2O_3$), and Metco 136F (Cr_2O_3-$xSiO_2$-$yTiO_2$). These systems have been considered suitable for operating in harsh conditions, such as in the automotive industry, to replace plastic materials, in light of their adhesion strength; microindentation hardness; and thermal, structural, and morphological properties [12].

Bio-based fibers can be efficiently applied in packaging when the mechanical properties of the corresponding fiber materials exceed those of conventional paperboard. Hot-pressing has been considered an efficient method for improving both the wet and dry strength of lignin-containing paper webs. Different pressing conditions for webs formed with thermomechanical pulp have been studied by Mattsson et al. [11].

Lignin and soy flour have been proposed as binders in the fabrication of *Rhizophora* spp. particleboard, for use as a phantom material in radiation dosimetry applications [16]. An increased internal bond strength was confirmed in samples with binders, which may indicate better structural integrity and physicomechanical strength.

A new sustainable food system based on rice bran biopolymers, in light of the potential application of ultrasound technology for enhancing the production of resistant starch, was proposed in [20].

An overview of recent potential applications of biopolymers in textiles was provided by Patti and Acierno [1]. Here, the use of biopolymers in various textile processes, from spinning processes to dyeing and finishing treatment, is proposed as a possible solution for reducing the environmental impact of the textile industry.

Bio-based materials have been considered a broad class of organic components ranging from agro-food renewable resources [3] (e.g., polysaccharides and proteins), bacterial activities (e.g., polyhydroxyalkanoates [21]), the conventional synthesis of bio-derived monomers (e.g., polylactides [19] and polyglycolides), or synthetic monomers (e.g., polycaprolactones, polyesteramides, bio-polyester, etc.).

These systems allow current needs to be met without dangerously burdening the future from ecological, economic, and human points of view (sustainability).

Biomaterials can be converted into chemical elements through the action of biological or physical agents (biodegradability) and can also be transformed into natural fertilizers for agriculture, once degraded (compostability).

A schematization of four types of plastics (fossil-based and non-biodegradable plastics, fossil-based and biodegradable, natural-based and non-biodegradable, and natural-based and biodegradable) is provided in [1].

However, applications of biodegradable polymers to replace conventional fossil-based plastics remain limited by the former's poor water resistance, high-dimensional stability, and processability [22]. Most biodegradable systems have been made from polyester. The main drawback of these biopolymers is their absorption of moisture, which gives rise to dangerous degradation phenomena. In this perspective, Titone et al. [9] investigated the effect of moisture on the processing and mechanical properties of a biodegradable polyester used for injection molding. The mechanical properties of sheep wool fibers under the impact of humid air and UV irradiation have been analyzed in [23].

The addition of different fillers can effectively improve the performance of biopolymers by allowing the functional properties and degradation rate to be controlled [22]. There have been various attempts to improve the tensile, flexural, hardness, and impact strength of bioplastics by incorporating reinforcement materials, such as inorganic and lignocellulosic fibers [24]. PLA/boehmite composites were prepared using a twin-screw extruder by investigating the effects of the modification of polylactide using dicumyl peroxide as a crosslinker and Joncryl (a patented, multifunctional, reactive polymer with improved thermal stability/chain extenders for specific food-contact applications) as a chain extender on a boehmite distribution [19]. Hernandez et al. investigated the effects of microspheres, carbon fibers, or polyethylene glycol added to polyhydroxybutyrate (PHB) resin on the thermal and mechanical properties of the final compounds [21]. A polymer film based on poly(lactic acid) (PLA) as a matrix and polyaniline (PAni) as a conductive filler was prepared by Wong et al. [10], to obtain antistatic properties. Polyaniline was chosen due to to its good biocompatibility and conductivity and excellent performance in combination with a PLA matrix. The crazing technique has been applied to enhance the polymer-degrading activity of enzymes.

For specific applications, it is important to prevent alterations upon contact with biological fluids, reactions with the human body, and the release of harmful species (biocompatibility). Zirconium oxide nanoparticles (ZrO_2 NPs) have been recognized for their high biocompatibility and resistance to wear and corrosion. El-Tamimi et al. [8] investigated the effects of adding salinized nano ZrO_2 particles on the microstructure, hardness, and wear resistance of acrylic denture teeth.

Sustainable compounds can also be obtained from natural lignocellulosic fibers [14,17,18,25] or plant residues [15,26]. Birch (*Betula pendula* Roth.) and beech (*Fagus sylvatica* L.) solid wood and plywood were overmolded with polyamide 6 and polypropylene (PP) to investigate the mechanical properties and interfacial adhesion in [14]. Wood flour and wood pellets manufactured from secondary processing mill residues, originating from white cedar, white pine, spruce-fir and red maple, were added to polypropylene, in the presence and absence of the coupling agent maleic, to evaluate the tensile, impact, and flexural strength [25]. Wood flour, supplied from a wood-processing factory, was added to recycled polyethylene that was made by regenerating films from greenhouse covers, to investigate the rheological responses of the final materials [18]. Additionally, 3D-printed objects have been prepared by depositing a PLA-fused filament onto jute fabrics and mainly characterized in terms of their tensile and flame-retardancy properties [17]. Pandanus amaryllifolius (a member of the Pandanaceae family, abundant in south-east Asian countries) fibers were extracted via a water-retting-extraction process and investigated as a potential fiber reinforcer in a polymer composite [26].

However, the main challenge in the full development of these bio-based compounds, as replacements for traditional-plastic-based formulations, is the lack of good mechanical characteristics, resulting from poor interfacial adhesion between the filler and matrix [13,27,28] or weak miscibility between the two phases in blend preparations [29]. Different strategies should be studied in order to adapt these systems to our needs and improve their efficiency. An alkaline and benzoyl chloride treatment of sugar palm fibers was proposed by Sherwani et al. [13]. The combination of treated sugar palm fiber/glass fiber reinforced PLA hybrid composites resulted in good physical and mechanical properties, especially after the fiber pretreatment. This composite was designed to replace acrylonitrile butadiene styrene (ABS) plastic in motorcycle batteries [13]. A new modifying agent based on polyacrylic-acid-grafted organosolv lignin was synthesized via free-radical copolymerization and used in combination with chitosan fiber in polylactide acid in the study by Tanjung et al. [27]. Urea-treated halloysite nanotubes have been utilized to reinforce epoxidized natural rubber and to improve the overall properties of composites [28]. Additives based on vegetable oil (sunflower, rapeseed, and castor oil) have been synthesized and introduced in PLA/starch blends to improve their miscibility [29].

Funding: This research received no external funding.

Acknowledgments: A. Patti and D. Acierno wish to sincerely thank all participants to this Special Issue, authors, reviewers, and members of the Editorial Office.

Conflicts of Interest: The authors declare no conflict of interest.

References

1. Patti, A.; Acierno, D. Towards the Sustainability of the Plastic Industry through Biopolymers: Properties and Potential Applications to the Textiles World. *Polymers* **2022**, *14*, 692. [CrossRef] [PubMed]
2. Phojaroen, J.; Jiradechakorn, T.; Kirdponpattara, S.; Sriariyanun, M.; Junthip, J.; Chuetor, S. Performance Evaluation of Combined Hydrothermal-Mechanical Pretreatment of Lignocellulosic Biomass for Enzymatic Enhancement. *Polymers* **2022**, *14*, 2313. [CrossRef] [PubMed]
3. Basri, M.S.M.; Shah, N.N.A.K.; Sulaiman, A.; Tawakkal, I.S.M.A.; Nor, M.Z.M.; Ariffin, S.H.; Ghani, N.H.A.; Salleh, F.S.M. Progress in the Valorization of Fruit and Vegetable Wastes: Active Packaging, Biocomposites, By-Products, and Innovative Technologies Used for Bioactive Compound Extraction. *Polymers* **2021**, *13*, 3503. [CrossRef] [PubMed]
4. Makarova, A.O.; Derkach, S.R.; Kadyirov, A.I.; Ziganshina, S.A.; Kazantseva, M.A.; Zueva, O.S.; Gubaidullin, A.T.; Zuev, Y.F. Supramolecular Structure and Mechanical Performance of κ-Carrageenan-Gelatin Gel. *Polymers* **2022**, *14*, 4347. [CrossRef] [PubMed]

5. Rodríguez-Soto, M.A.; Polanía-Sandoval, C.A.; Aragón-Rivera, A.M.; Buitrago, D.; Ayala-Velásquez, M.; Velandia-Sánchez, A.; Peralta Peluffo, G.; Cruz, J.C.; Muñoz Camargo, C.; Camacho-Mackenzie, J.; et al. Small Diameter Cell-Free Tissue-Engineered Vascular Grafts: Biomaterials and Manufacture Techniques to Reach Suitable Mechanical Properties. *Polymers* **2022**, *14*, 3440. [CrossRef]
6. Khalid Khairunnisa-Atiqah, M.; Mohd Salleh, K.; Ainul Hafiza, A.H.; Syazwani, N.; Mazlan, N.; Mostapha, M.; Zakaria, S. Impact of Drying Regimes and Different Coating Layers on Carboxymethyl Cellulose Cross-Linked with Citric Acid on Cotton Thread Fibers for Wound Dressing Modification. *Polymers* **2022**, *14*, 1217. [CrossRef]
7. Muktar, M.Z.; Bakar, M.A.A.; Amin, K.A.M.; Rose, L.C.; Ismail, W.I.W.; Razali, M.H.; Razak, S.I.A.; Panhuis, M. In Het Gellan Gum Hydrogels Filled Edible Oil Microemulsion for Biomedical Materials: Phase Diagram, Mechanical Behavior, and In Vivo Studies. *Polymers* **2021**, *13*, 3281. [CrossRef]
8. El-Tamimi, K.M.; Bayoumi, D.A.; Ahmed, M.M.Z.; Albaijan, I.; El-Sayed, M.E. The Effect of Salinized Nano ZrO_2 Particles on the Microstructure, Hardness, and Wear Behavior of Acrylic Denture Tooth Nanocomposite. *Polymers* **2022**, *14*, 302. [CrossRef]
9. Titone, V.; Correnti, A.; La Mantia, F.P. Effect of Moisture Content on the Processing and Mechanical Properties of a Biodegradable Polyester. *Polymers* **2021**, *13*, 1616. [CrossRef]
10. Wong, P.Y.; Takeno, A.; Takahashi, S.; Phang, S.W.; Baharum, A. Crazing Effect on the Bio-Based Conducting Polymer Film. *Polymers* **2021**, *13*, 3425. [CrossRef]
11. Mattsson, A.; Joelsson, T.; Miettinen, A.; Ketoja, J.A.; Pettersson, G.; Engstrand, P. Lignin Inter-Diffusion Underlying Improved Mechanical Performance of Hot-Pressed Paper Webs. *Polymers* **2021**, *13*, 2485. [CrossRef] [PubMed]
12. Mazurchevici, S.N.; Marguta, A.; Istrate, B.; Benchea, M.; Boca, M.; Nedelcu, D. Improvements of Arboblend V2 Nature Characteristics through Depositing Thin Ceramic Layers. *Polymers* **2021**, *13*, 3765. [CrossRef] [PubMed]
13. Sherwani, S.F.K.; Zainudin, E.S.; Sapuan, S.M.; Leman, Z.; Khalina, A. Physical, Mechanical, and Morphological Properties of Treated Sugar Palm/Glass Reinforced Poly (Lactic Acid) Hybrid Composites. *Polymers* **2021**, *13*, 3620. [CrossRef] [PubMed]
14. Stadlmann, A.; Mautner, A.; Pramreiter, M.; Bismarck, A.; Müller, U. Interfacial Adhesion and Mechanical Properties of Wood-Polymer Hybrid Composites Prepared by Injection Molding. *Polymers* **2021**, *13*, 2849. [CrossRef] [PubMed]
15. Ferrandez-García, A.A.; Ortuño, T.G.; Ferrandez-Villena, M.; Ferrandez-Garcia, A.; Ferrandez-García, M.T. Evaluation of Particleboards Made from Giant Reed (*Arundo donax* L.) Bonded with Cement and Potato Starch. *Polymers* **2021**, *14*, 111. [CrossRef] [PubMed]
16. Zuber, S.H.; Hashikin, N.A.A.; Yusof, M.F.M.; Aziz, M.Z.A.; Hashim, R. Influence of Different Percentages of Binders on the Physico-Mechanical Properties of Rhizophora spp. Particleboard as Natural-Based Tissue-Equivalent Phantom for Radiation Dosimetry Applications. *Polymers* **2021**, *13*, 1868. [CrossRef]
17. Franco-Urquiza, E.A.; Escamilla, Y.R.; Llanas, P.I.A. Characterization of 3D Printing on Jute Fabrics. *Polymers* **2021**, *13*, 3202. [CrossRef]
18. Patti, A.; Cicala, G.; Acierno, S. Rotational Rheology of Wood Flour Composites Based on Recycled Polyethylene. *Polymers* **2021**, *13*, 2226. [CrossRef]
19. Makwakwa, D.; Ojijo, V.; Bandyopadhyay, J.; Sinha Ray, S. Flow Characteristics, Mechanical, Thermal, and Thermomechanical Properties, and 3D Printability of Biodegradable Polylactide Containing Boehmite at Different Loadings. *Polymers* **2021**, *13*, 2019. [CrossRef]
20. Vaitkeviciene, R.; Bendoraitiene, J.; Degutyte, R.; Svazas, M.; Zadeike, D. Optimization of the Sustainable Production of Resistant Starch in Rice Bran and Evaluation of Its Physicochemical and Technological Properties. *Polymers* **2022**, *14*, 3662. [CrossRef]
21. Hernandez, M.M.; Gupta, N.S.; Lee, K.S.; Dumont, J.H.; Pital, A.C.; Marrone, B.L.; Iverson, C.N. Characterization of Polyhydroxybutyrate-Based Composites Prepared by Injection Molding. *Polymers* **2021**, *13*, 4444. [CrossRef] [PubMed]
22. Glaskova-Kuzmina, T.; Starkova, O.; Gaidukovs, S.; Platnieks, O.; Gaidukova, G. Durability of Biodegradable Polymer Nanocomposites. *Polymers* **2021**, *13*, 3375. [CrossRef] [PubMed]
23. Starkova, O.; Sabalina, A.; Voikiva, V.; Osite, A. Environmental Effects on Strength and Failure Strain Distributions of Sheep Wool Fibers. *Polymers* **2022**, *14*, 2651. [CrossRef] [PubMed]
24. Boey, J.Y.; Lee, C.K.; Tay, G.S. Factors Affecting Mechanical Properties of Reinforced Bioplastics: A Review. *Polymers* **2022**, *14*, 3737. [CrossRef]
25. Pokhrel, G.; Gardner, D.J.; Han, Y. Properties of Wood-Plastic Composites Manufactured from Two Different Wood Feedstocks: Wood Flour and Wood Pellets. *Polymers* **2021**, *13*, 2769. [CrossRef]
26. Diyana, Z.N.; Jumaidin, R.; Selamat, M.Z.; Alamjuri, R.H.; Yusof, F.A.M. Extraction and Characterization of Natural Cellulosic Fiber from Pandanus amaryllifolius Leaves. *Polymers* **2021**, *13*, 4171. [CrossRef]
27. Tanjung, F.A.; Arifin, Y.; Kuswardani, R.A. Influence of Newly Organosolv Lignin-Based Interface Modifier on Mechanical and Thermal Properties, and Enzymatic Degradation of Polylactic Acid/Chitosan Biocomposites. *Polymers* **2021**, *13*, 3355. [CrossRef]
28. Surya, I.; Waesateh, K.; Saiwari, S.; Ismail, H.; Othman, N.; Hayeemasae, N. Potency of Urea-Treated Halloysite Nanotubes for the Simultaneous Boosting of Mechanical Properties and Crystallization of Epoxidized Natural Rubber Composites. *Polymers* **2021**, *13*, 3068. [CrossRef]
29. Nagy, B.; Miskolczi, N.; Eller, Z. Improving Mechanical Properties of PLA/Starch Blends Using Masterbatch Containing Vegetable Oil Based Active Ingredients. *Polymers* **2021**, *13*, 2981. [CrossRef]

polymers

MDPI

Review

Towards the Sustainability of the Plastic Industry through Biopolymers: Properties and Potential Applications to the Textiles World

Antonella Patti [1,*] and Domenico Acierno [2,*]

1 Department of Civil Engineering and Architecture (DICAr), University of Catania, Viale Andrea Doria 6, 95125 Catania, Italy
2 CRdC Nuove Tecnologie per le Attività Produttive Scarl, Via Nuova Agnano 11, 80125 Naples, Italy
* Correspondence: antonella.patti@unict.it (A.P.); acierno@crdctecnologie.it (D.A.)

Abstract: This study aims to provide an overview of the latest research studies on the use of biopolymers in various textile processes, from spinning processes to dyeing and finishing treatment, proposed as a possible solution to reduce the environmental impact of the textile industry. Recently, awareness of various polluting aspects of textile production, based on petroleum derivatives, has grown significantly. Environmental issues resulting from greenhouse gas emissions, and waste accumulation in nature and landfills, have pushed research activities toward more sustainable, low-impact alternatives. Polymers derived from renewable resources and/or with biodegradable characteristics were investigated as follows: (i) as constituent materials in yarn production, in view of their superior ability to be decomposed compared with common synthetic petroleum-derived plastics, positive antibacterial activities, good breathability, and mechanical properties; (ii) in textile finishing to act as biological catalysts; (iii) to impart specific functional properties to treated textiles; (iv) in 3D printing technologies on fabric surfaces to replace traditionally more pollutive dye-based and inkjet printing; and (v) in the implants for the treatment of dye-contaminated water. Finally, current projects led by well-known companies on the development of new materials for the textile market are presented.

Keywords: textiles; sustainability; biopolymers

Citation: Patti, A.; Acierno, D. Towards the Sustainability of the Plastic Industry through Biopolymers: Properties and Potential Applications to the Textiles World. *Polymers* **2022**, *14*, 692. https://doi.org/10.3390/polym14040692

Academic Editor: Helena Felgueiras

Received: 1 December 2021
Accepted: 7 February 2022
Published: 11 February 2022

1. Introduction

Currently, textile production is constantly expanding by catering to fashion, style, and marketing needs, as well as increasingly competitive technical challenges [1]. The prospect of innovation and development contrasts, however, with the high pollution of the environment caused by this industrial activity [2]. Between 1975 and 2018, the production of chemical and textile fibers nearly quadrupled, from 23.94 million metric tons to 105.6 million metric tons: a more than fourfold increase in less than 40 years [3]. Concerns about the environmental impact of fiber production and subsequent disposal operations have grown in prominence as demand for fabrics has increased. Several recent studies show that the textile industry endangers freshwater and atmosphere micro-systems due to the use of industrially harmful and toxic chemicals during the manufacturing process, as well as the release of these chemicals [4–6]. Inadequate collection and careless disposal of solid waste pollutes the land and air, endangering human health and the environment. As a result, countries should devote a significant portion of their effort to waste management [7].

In this context, emphasis was placed on the development of alternative eco-friendly solutions, green technologies, and sustainable manufacturing processes in order to avoid risks to human health while also limiting environmental issues.

In our previous work [8], attention was paid to recent scientific research activities promoting the sustainability requirements of textile production and circular economy actions through the recovery of textile waste products and recycling fibers into composite materials.

However, bio-based materials, i.e., a broad class of organic constituents produced from renewable resources endowed with specific structural and functional characteristics such as biodegradability, composability, or biocompatibility, could offer a potential solution to replace the conventional plastics (primarily made from polyester (PET) [9,10], polyamide (PA), and polypropylene (PP) [11–13]) involved in fabric and yarn production [14]. Biopolymers can be used in the textile industry as the constituent base material for the production of filaments and yarns, or to replace harmful chemicals in pre-treatment and finishing operations by providing textile substates with various potential functionalities, such as antibacterial and flame-retardant activity, UV protection, electric conductivity, and hydrophobicity [15]. In this study, we aimed to collect the most recent studies on prospective biopolymers in the development of textile production by demonstrating the potential of these materials in replacing common oil-derived plastics.

2. Background

Several applications of plastic-based materials exist today, involving not only commercial and industrial fields, but also everyday life products. The global production of plastics is continuously increasing due to different intrinsic properties, such as lightweight, low-cost, durable, and chemical resistant properties. In addition, the increase in the human population, together with rapid economic growth, continuous urbanization, and lifestyle changes are other factors considered responsible for the growing demand for plastic products [2].

However, the main drawback that should be considered in the widespread employment of plastic consists in the difficulty of properly disposing of plastic at the end of its life cycle, and its consequent accumulation in nature. It has been predicted that millions of tons of plastics will be released into the surroundings in the coming years [16]. Every year, tons of plastic waste accumulate in the oceans, including microplastics (1 to 5000 μm particles), seriously altering the marine environment and the lives of its inhabitants [17].

According to Jambeck et al. (2015) [18], 4.8 to 12.7 million tons of plastic debris enter the ocean each year, with cumulative inputs expected to increase tenfold by 2025.

Furthermore, millions of tons of CO_2 will be emitted in the atmosphere generated both from production processes and also from the final incineration at the end of the products' life cycle [19]. The huge emission of greenhouse gases such as CO_2 further contribute to global warming by leading to dramatic and irreversible weather changes. As a result of the industrial revolution, the CO_2 concentration rose to over 390 ppm, and the global temperature rose by 0.9 °C. In 2019, the former rose to over 400 ppm, while the latter rose by 1.0 °C. Based on current trends, it is possible to imagine these levels exceeding 450 ppm and 1.5 °C in the next 20–40 years [20]. Environmental impact deriving from production, use, consumption, and end of life of plastics, is illustrated in Figure 1.

At the end of 2019, with the spread of the severe acute respiratory infection syndrome coronavirus 2 (SARS-CoV-2) pandemic, also known as COVID-19, which has affected millions of people worldwide, hundreds of countries, and thousands of victims, the global economy was influenced in three ways: directly, through production and demand, or supply chain and market disruption; and indirectly, through its financial impact on firms and financial markets [21]. As illustrated in a very recent paper by Mittal et al. [22], with the closing of shops and restaurants during lockdowns, people's living conditions have been changed, causing concerns over waste generation routines. Plastic wastes have increased due to the extensive use of face masks, vaccine containers, packaging from food, medical industries, and online purchases. In order to reduce the cumulative effects, the substitution of petroleum-based plastics with bio-based plastics has also been presented. Data and hypothetical calculations on the amplified amount of plastic waste generated during COVID-19 have been collected in the work of Shams et al. [23].

The textile, fashion, and style industries were hit particularly hard by the pandemic's spread and the associated containment measures. Shahbandeh (2021) [24] reported a drastic decrease in European textile production (equal to 26.9%) compared to the same period in 2019 (April and June 2020), when coronavirus cases reached a global peak. However, a remarkable increase in extra-EU textile trade of 154.2% was also confirmed. This was attributed to the import of personal protective equipment (PPE) that was required during Europe's coronavirus crisis. Face masks and isolation gowns were identified as the first weapons of contagion protection in 2020. Authorities all over the world have encouraged, and in some cases mandated, the use of face masks in public places, and every adult on the planet has worn one.

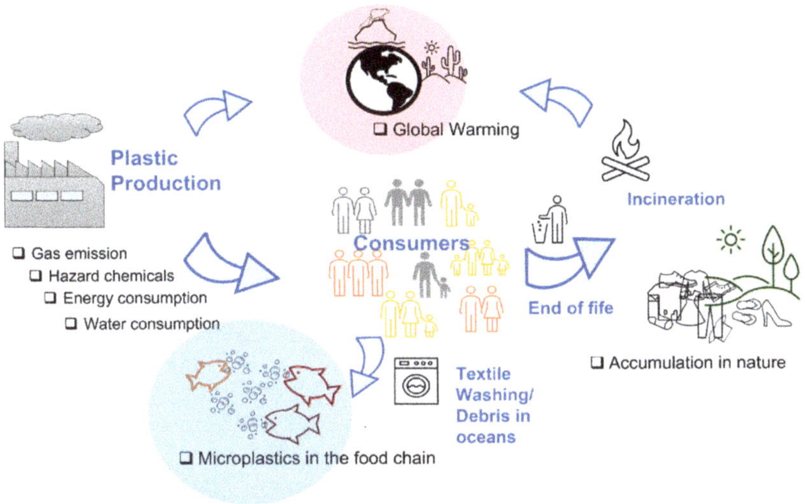

Figure 1. Environmental impact of plastics (black points) and corresponding sources (blue elements).

Face masks and gowns are mostly made from polypropylene given the versatile nature of this polymer, which allows thermal processes in different ways and forms, and the quick production on a large scale from monomers via chain-growth polymerization [25].

However, the unpredictability of the pandemic left the municipal solid waste (MSW) sector unprepared, even in countries with a longer tradition of sustainable waste management. Contrary to expectations, single-use masks and gloves have little impact on waste management, accounting for only 1% of the residual municipal solid waste collected annually. On the other hand, the dispersal of abandoned masks and gloves outside of indoor environments is causing environmental issues [26].

3. Biopolymers

Biopolymers can be produced from vegetal or animal-derived polysaccharides (i.e., starch, cellulose, lignin), proteins (i.e., chitosan, collagen) and lipids (i.e., wax, fatty acids), bacterial activities (i.e., polyhydroxyalkanoates (PHA)), the conventional synthesis of bio-derived monomers (i.e., polylactides (PLA) and polyglycolide (PLG)) or synthetic monomers (i.e., polycaprolactone (PCL)s, polyester-amides (PEA), an aliphatic or aromatic co-polyester (PBSA, PBAT)) [27].

The term "bio-based" does not necessarily imply "biodegradable" [28]. According to ASTM D6400-99 [29], a biodegradable plastic is a degradable plastic in which the degradation results from the action of naturally occurring microorganisms, such as bacteria, fungi, and algae. The ability to be decomposed by microorganisms depends on the chemical structure rather than the origin: petroleum-derived plastics can also be biodegradable [28]. A compostable plastic is a plastic that undergoes degradation through biological processes during composting to yield CO_2, H_2O, inorganic compounds, and biomass at a rate that is consistent with other compostable materials, and leave no visible, distinguishable or toxic residue [29]. Composting is a biological process in which the organic material is decomposed primarily by microorganisms, to produce a soil-like substance, called humus. The composting conditions are affected by environmental factors, such as water or moisture content, temperature, acidity, enzyme specificity; or polymer factors, such as polymer structure and chain flexibility, crystallinity, and molecular weight, as well as copolymer compositions, size, and shape [30].

Plastics can be broadly classified into the four categories listed below (Figure 2): (i) fossil-based and non-biodegradable plastics (PET, PP; PE; PA, thermosets); (ii) fossil-based and biodegradable (PCL, PBAT), (iii) natural-based and non-biodegradable (bio-PP, bio-PE, bio-PET), and (iv) natural-based and biodegradable (PLA, PHA, lignocellulose, starch, proteins) [31,32].

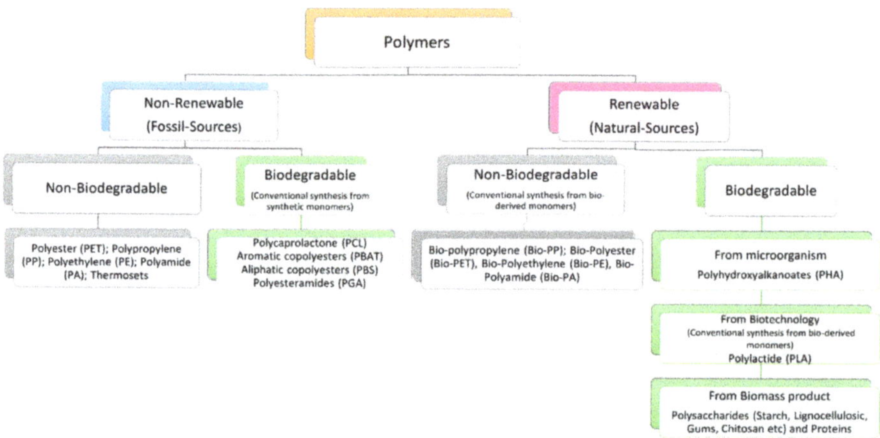

Figure 2. Schematization of biodegradable and non-biodegradable plastics [31,32].

A transition to more sustainable plastics requires the development of systems not derived from fossil resources and towards carbon-neutral blocks, with specific functional and recyclable properties. In general, there are three methods to synthesize biopolymers from renewable sources: to make use of natural polymers that can be modified but remain similar to a large extent (such as starch), to produce monomers from fermentation that follow polymerization, and to produce biopolymers directly through microorganisms [33]. The following biodegradable products are currently on the market (Figure 3):

- Poly(lactic acid) (PLA) is one of the most commercially successful bio-plastics. The monomer (lactide acid), is produced through fermentation from renewable sources such as starch or sugar. Other carbohydrate resources include: leaves, stems and stalks from corn fiber, corn stover, sugarcane bagasse, rice hulls, woody crops, and forest residues [34]. Jem and Tan (2020) estimated the global production volume of PLA in 2019 to be around 190,000 tons [20]. However, their modest thermal, mechanical and rheological characteristics, as well as their incompatibility with the process and recycling technologies [35], limit the practical application of PLAs, making necessary their copolymerization or blending with other polymers;
- Polyhydroxyalkanoates (PHA) accumulate inside the cells of microorganisms as granules for energy storage under restrictive nutrient conditions and high concentrations of carbon [28];
- Poly(3-hydroxybutyrate-co-3-hydroxyvalerate)(PHBV) is obtained by the biosynthesis of diverse bacteria in the presence of specific nutrients (e.g., nitrogen, phosphorous, or sulfur) and an abundance of carbon sources [36];
- Polycaprolactone (PCL) is obtained through the condensation of 6-hydroxycaproic (6-hydroxyhexanoic) acid or ring-opening polymerization (ROP) of ε-caprolactone (ε-CL). Industrially ε-CL is produced from the oxidation of cyclohexanone. The products

ε-CL and 6-hydroxyhexanoic acid are also intermediary products in the oxidation process using microorganisms [37];

- Poly(butylene-adipate-co-terephthalate) (PBAT) is produced through the polycondensation reaction of butanediol (BDO), adipic acid (AA), and terephthalic acid (PTA), using general polyester manufacturing technology [38];
- Poly(butylene succinate) (PBS) is produced through the polycondensation reaction of succinic acid (or dimethyl succinate) and butanediol (BDO). The monomers can be derived from fossil-based or renewable resources. Succinic acid can be obtained through fermentation of the microorganism; renewable feedstocks, such as glucose, starch, or xylose; or by chemical process via the hydrogenation of maleic anhydride [39];
- Poly(glycolic acid) (PGA) is attained by the direct poly-condensation polymerization of glycolic acid, via the solid-state polycondensation of halogen acetates, by reacting formaldehyde with carbon monoxide, or via the ring opening polymerization of glycolide, the cyclic dimer of glycolic acid (current industrial route) [40];
- Biomass: polysaccharides (starch, lignocellulose, chitosan, gums) and proteins (silk, collagen, soy, zein).

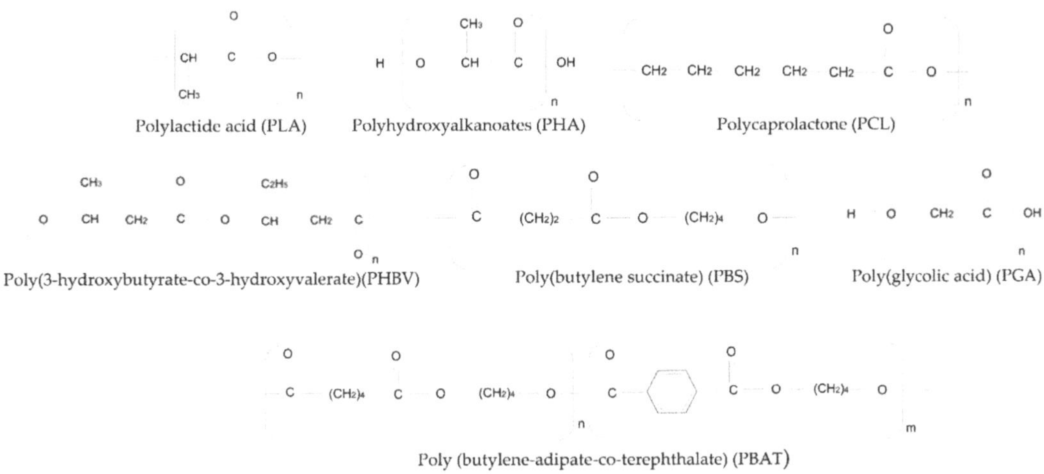

Figure 3. Chemical structure of main biodegradable polymers.

In recent years, there has been a growing trend towards the replacement of fossil-based conventional plastics with identical molecules derived from renewable sources [41]:

- Bio-poly(ethylene) (bio-PE) is produced from the dehydration of ethanol from sugar to produce ethylene, which, upon purification and polymerization. produces polyethylene [42];
- Bio-polypropylene (bio-PP). Biomass can be converted into bio-based propylene by multiple processes, such as cracking, gasification, fermentation, metathesis, and dehydrogenation. Methanol, isopropanol, ethanol, butanol, and glycerin, are good candidates for propylene synthesis [43];
- Bio-poly(ethylene terephthalate) (bio-PET) is obtained through bio-based ethylene glycol as the monomer. Its precursors, terephthalic acid (TA) and ethylene glycol (EG), are both produced from biomass (forest residues, corn stover). Bis(hydroxyethyl)terephthalate (BHET) is synthesized from ethylene glycol (EG) via a transesterification reaction with dimethyl terephthalate (DMT) or an esterification reaction with terephthalic acid (TA). Next, BHET pre-polymerization and melt polycondensation form low-Mw PET or solid-state polymerization to produce high-Mw PET [44];
- Bio-poly(trimethylene terephthalate) (bio-PTT) is produced from biobased 1,3-propanediol (1,3-PDO) [41];

- Bio-polyamide (bio-PA). The main source of bio-based monomers (11-aminoundecanoic acid, 1,8-octanedicarboxylic acid, 1,10-decanediamine) is castor oil derived from oil crops. Other biomass (sugar, starch, lignocellulose)-derived monomers are adipic acid, caprolactam, and 1,4-butanediamine [45].

Therefore, in light of sustainable green manufacturing, future developing fabrics should be biodegradable. Fabrics made with similar fiber characteristics, but derived from renewable resources, such as bio-PET and bio-PA, could be used as alternatives to current fabrics made with synthetic fibers. However, the debate over the potential pollution induced during the latter's life cycle remains open, as the latter, not being biodegradable, can equally cause an aggravation of problems already highlighted with the use of fossil-based filaments during the life cycle and at the end of their use [46].

Nevertheless, the step forward in reducing the environmental impact during the production of fibers from renewable sources cannot be neglected. The lack of biodegradability could not be considered an issue as durable bio-plastics have identical performance compared to petroleum-based ones, and could be directly applied in existing recycling systems. Furthermore, during the production of durable bio-plastics, although these are still more expensive, a large reduction in greenhouse gas (GHG) imprint emission is obtained compared to petro-equivalents [47]. Life-cycle assessments (LCAs) for bio-PE and PLA bioplastics were compared to those of two fossil-derived plastics, high-density polyethylene (HDPE) and low-density polyethylene (LDPE), in terms of greenhouse emissions (GHG) and fossil fuel consumption (FFC). The results demonstrated the benefits of bio-based plastic pathways over fossil-based pathways by showing GHG equal to −1.0 and 1.7 kg CO_2e per kg for bio-PE and PLA with no biodegradation, compared with 2.6 and 2.9 kg CO_2e per kg for LDPE and HDPE; and FFC equal to 29 and 46 MJ per kg of bio-PE and PLA, compared with 73 and 79 MJ per kg of LDPE and HDPE. However, despite the benefits of biogenic carbon uptake, at the end-of-life, PLA emissions were increased from 16% to 163%, passing from composting to landfill because less CH_4 was emitted in the composting gas [48].

A comparison between the properties of bio-based and traditional plastics is presented in Table 1.

Table 1. Main properties of common bio-based and traditional plastic materials.

	Monomers and Sources	Tensile Strength [MPa]	Biodegradability		Water Vapor Permeability (g·mil/m²·Day·kPa)	Density (g/cm³)
			Environment	Biodegradability (%)		
PLA	Lactide acid from starch or sugar cane corn stover, sugarcane bagasse, rice hulls, woody crops and forest residues [34]	37−66 (at yield) [49]	Compost (30 days)	60–70 [49]	400 [50]	1.21–1.25 [51]
		61.6–49.6 [52]	Soil (120 days)	0 [49]		
		21–60 [51]	Marine (180 days)	3–4 [49]		
PHA	Hydroxyalkanoates accumulated inside the cells of microorganisms [28]	17–104 [52]	Compost (180 days)	94 [53]	/	1.25 [54]
			Soil (90 days)	100 [53]		
			Marine (365 days)	52–82 [53]		

Table 1. *Cont.*

	Monomers and Sources	Tensile Strength [MPa]	Biodegradability		Water Vapor Permeability (g·mil/m^2·Day·kPa)	Density (g/cm^3)
			Environment	Biodegradability (%)		
PHBV	Hydroxybutyric monomers and hydroxyvalerate from carbon sources [36]	20−40 (at yield) [49]	Compost (28 days)	80 [49]	30 [50]	1.18–1.262 [51]
		40 [51]	Soil (280 days)	48.5 [49]		
			Marine (180 days)	38–45 [49]		
PCL	6-hydroxycaproic acid, ε-caprolactone [37]	25−33 (at yield) [49]	Compost (91 days)	100 [55]	200 [50]	1.11–1.146 [51]
		20.7–4 [51]	Soil (280 days)	50 [55]		
			Marine (365 days)	30 [56]		
PBAT	Butanediol (BDO), adipic acid (AA) and terephthalic acid (PTA) [38]	13−15 (at yield) [49]	Compost (45 days)	34–67 [49]	1200 [50]	1.26 [57]
			Soil (120 days)	6.6 [49]		
			Marine (28 days)	1–1.4 [49]		
PBS	Succinic acid from glucose, starch, xylose, or oil sources [39]	30−35 (at yield) [49]	Compost (160 days)	90 [49]	90 [50]	1.26 [58]
		24.8 [52]	Soil (28 days)	1 [49]		
			Marine (28 days)	1 [49]		
Bio-PE	Ethanol from sugar [42]	Similar to petroleum-based [49] 17.9–33.1 [52]	Very slow [52]/Not biodegradable [59]		3 [50]	Similar to petroleum-based [60]
Bio-PP	Methanol, isopropanol, ethanol, butanol, and glycerin from biomass [43]	29.3–38.6 [52]	Very slow [52]		/	Similar to petroleum-based [60]
Bio-PET	Terephthalic acid and ethylene glycol from biomass (forest residues, corn stover) [44]	Similar to petroleum-based [49] 50 [52]	Very slow [52]/Not biodegradable [59]		/	Similar to petroleum-based [60]
Bio-PA	11-aminoundecanoic acid, 1,8-octanedicarboxylic acid, 1,10-decanediamine) from castor oil [45]	37−41 (at yield) [49]	Not biodegradable		/	Similar to petroleum-based [60]
PE	Ethylene from fossil sources	10–32 (at yield) [49]	Not biodegradable		1.7–8.7 [61]	1.3–1.8 [60]
PP	Propylene from fossil sources	15–27.5 (at yield) [49]	Not biodegradable		4 [50]	~0.95 [60]
PET	Terephthalic acid from fossil sources	55−79 (at yield) [49]	Not biodegradable		20 [50]	1.33–1.48 [60]
PA	Caprolactam(nylon–6), Adipic acid (nylon–6,6) from fossil sources	49–69 [61]	Not biodegradable		350 [50]	1.07–1.24 [60]

4. Traditional Textile Materials

One of the most basic human needs since the Prehistoric Age has been to cover and protect the body from cold and/or bad weather. Many anthropological and archeological studies have demonstrated the practical and symbolic importance of fabrics throughout human history, particularly in rituals, events, and ceremonies [62]. The first woven fibers were discovered in Egypt during the Neolithic era: they were made of very fine-quality hemp, i.e., linen, in a complex pattern ("brocaded") with fringes at the edges [63]. There is evidence of a wool trade in Iran, and there were woolen garments in Babylon. Around 3000 BC, Egypt and India had well-developed spinning and weaving of linen and cotton. During the Bronze Age (2000 BC), discoveries of animal fibers for burial were made in Northern Europe. Turkey became skilled in carpet manufacturing during the Middle Ages; Palermo (Sicily) became famous for the production of elaborate silk and gold fabrics, and Lyon (France) became the most important silk manufacturing center in Europe. Prior to the industrial revolution, the textile industry grew more as a trade in fine ornamental yarns and beautiful textiles than as a result of technological advancement. However, W. Lee of Woodborough invented the knitting machine in England at the end of the 1500s, followed by cotton gin in 1793, the flying shuttle for weaving, invented by John Kay, and the mechanized loom, invented by Edmund Cartwright [64]. Textile production, based on both natural and synthetic fibers, has grown significantly in tandem with population growth, as well as with the economic and commercial aspects of fashion and style.

Textile fibers can be divided into two main categories: natural and synthetic fibers [65]. The former originate from animals, vegetables/plants, or mineral sources. Examples of animal-based fibers are wool, silk, and hair (alpaca, goat, horse), whereas plant-based fibers are constituted by seed, bast, leaf, wood; finally, mineral fibers include asbestos, fibrous brucite, and wollastonite [66]. As confirmed in a recent study by Uddin [67], natural fibers better meet human requirements in view of comfort and aesthetics. Cotton, silk, and wool are the three most common natural textile fibers used in clothing production. Recently, the market has witnessed the introduction of "organic" cotton, which is cotton grown without the use of chemicals, pesticides, or fertilizers. This has shifted production toward more sustainable development. Organic cotton is primarily produced in countries such as Turkey, China, India, and United States [68]. Wool fiber is well known for its warmth and is commonly used in winter clothing, with renewability and recyclability characteristics that contribute to its popularity in this industry. Silk fiber is well-known for its unique softness and low linear density, despite being produced in much smaller quantities than cotton and wool.

Synthetic fibers, also known as man-made fibers, are typically made from synthetic materials, such as petrochemicals (polyamide, polyester, polypropylene) [65]. However, some synthetic fibers, such as rayon, modal, and the more recently developed Lyocell (a regenerated cellulose fiber made from dissolving pulp), are made from natural cellulose (bleached wood pulp). Synthetic fibers can often be produced at a lower cost and in greater quantities than natural fibers. For clothing, natural fibers can provide some advantages, such as comfort, over their man-made counterparts [69].

In 2019, more than 60% of the world consumption of natural and man-made fibers belonged to synthetic fibers, primarily made from polyester, followed by cotton, cellulosic, and wool [70]. Data on global fiber production are presented in Figure 4 [70]. In 2019, China was the main consumer of natural and synthetic fibers, employing more than half of the total, followed by other Asian, North American, and European countries [71].

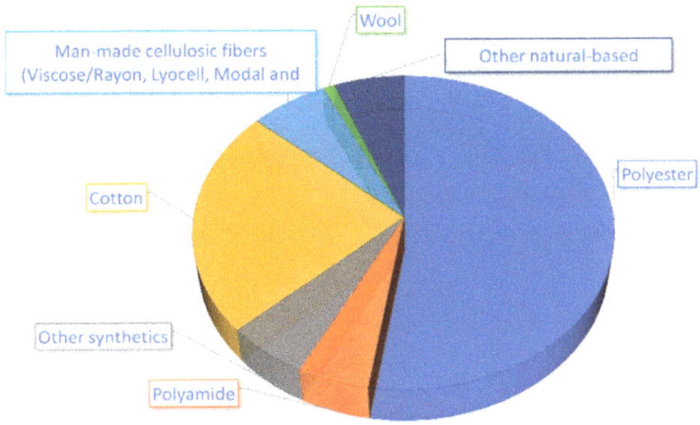

Figure 4. Global fiber production in 2019 (as reported in [70]).

5. Polluting Aspects by Textile Industry

The manufacturing process of textiles consists of four primary stages: yarn production, fabric production, textile production, and finishing treatment.

Pesticides, insecticides, and fertilizers are used in raw material treatments for cellulose- and protein-based natural fibers. After the first step, which is the removal of impurities, a series of continuous operations is followed. This series consists of blending, mixing, cleaning, carding, drawing, roving, and spinning; these processes are mainly mechanical, not requiring chemical applications [67]. By contrast, monomers, chemical agents, precursors, catalysts, and a variety of auxiliary chemicals are used in the preparation of synthetic polymer-based yarns [67].

Three general stages can be distinguished in man-made manufacturing yarn processes: the preparation of the spinning fluid from solid polymers or monomers; fiber spinning by extrusion; and, subsequently, solidification, mechanical, thermal and chemical treatment to improve properties [72].

Several processes are applied during the production of fabrics, including: sizing chemicals to improve absorbent capacity; oxidizing agents to improve strength, hygroscopicity, dye absorbency, and brightness; the use of pigment and chromophore agents, as well as dyes, to improve color [8].

The ultimate appearance and aesthetic properties of textile materials are determined by textile finishing, which has the potential to change various physical and chemical properties of textile materials in response to consumer demands [73]. Specific finishing treatments have been used to impart water and oil repellency, antibacterial resistance, and flame resistance [8]. Textile finishing, including washing, bleaching, dyeing, and coating, is applied to bulk textiles or garments after weaving and/or synthetic material production [74]. These processes are conventionally performed through typical pad-dry-cure techniques, consuming intensive amounts of energy and massive quantities of water, usually discharged as effluent [74]. The pollutants involved in each operation of the textile industry are summarized in Figure 5 [75].

With the growth of the world's population and rising living standards, large amounts of production and post-consumer fiber waste have accumulated in nature or in landfill [76]. Fibers, yarns, fabric scraps, and apparel cuttings are examples of production waste generated by fiber producers, textile mills, and fabric and apparel manufacturers. Pre-consumer waste are products with mistakes in design, faults, or incorrect colors, while post-consumer waste refers to garments or household items that the owner no longer requires and chooses to discard [8].

The cleaning of clothing made of synthetic fabrics appears to be a significant source of contamination for sewage due to the production of micro- and nano-sized plastic fragments released in washing water. Recently, microplastics have been revealed as new sources of pollution in the oceans. Because microplastics are not visible to human eye, and do not degrade in an aqueous environment, they can become food for plankton and other organisms that live in the oceans by entering the food chain [77].

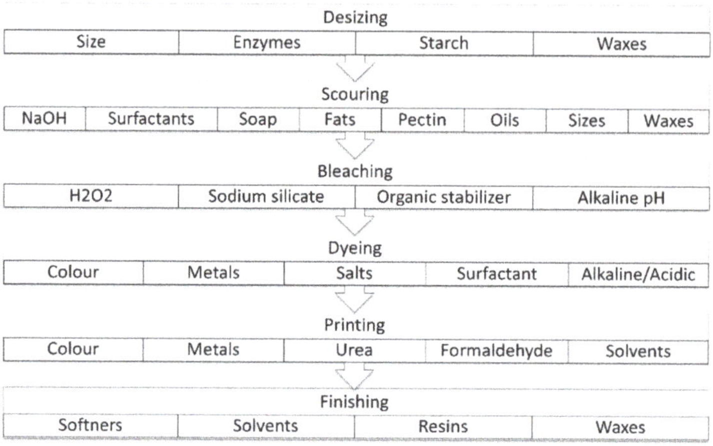

Figure 5. Schematization of main pollutants employed in stages of textile wet processing. Reproduced from [75]. Copyright (2016), with permission from Elsevier Ltd.

6. Properties of Bio-Based Fibers and Fabrics

Choosing an appropriate fiber among various renewable fibers derived from biomass necessitates an awareness of sustainability based on the product's GHG imprint and the energy requirements of its production and distribution [46]. The selection of material should be based on the product's value chain, previous use in clothing, and a comparison of its physical parameters with the most commonly used synthetic fibers derived from PET (terylene, dacron, etc.) [46].

6.1. Biodegradability

Polymer degradation occurs as an effect of different mechanisms (photodegradation, thermo-oxidative degradation, and biodegradation) throughout chemical reactions such as chain scission, crosslinking, side group elimination, chemical structure modification [78]. Biodegradation consists in the breakage of organic matter by means of microorganisms through two main pathways developed under aerobic (in the presence of oxygen) and anaerobic (in the absence of oxygen) conditions [79] (Figure 6). The final products of aerobic biodegradation are carbon dioxide, water, biomass, and oxidation products of nitrogen and sulfur; whereas, in the case of anaerobic biodegradation, hydrocarbons, methane, carbon dioxide, biomass, and reduction products of nitrogen and sulfur are released [80].

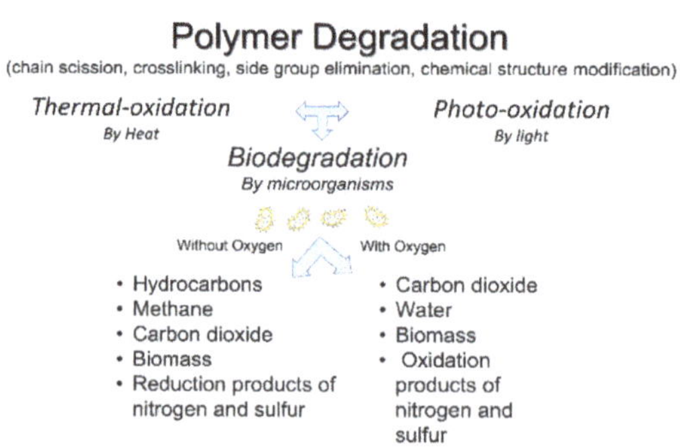

Figure 6. Products of polymer biodegradation under aerobic or anaerobic conditions.

Environmental biodegradation occurs in the case of uncontrolled conditions (temperature, moisture, pH, nutrients, oxygen level, presence of organisms, and composition of the waste), without human involvement. If conditions are inadequate, the biodegradation process leads to waste accumulation [79].

Once discarded in the environment, polylactide is hydrolyzed into low-molecular-weight oligomers, and then converted into CO_2 and H_2O by microorganisms present in the environment [81]. However, microorganisms able to degrade PLA are not widespread in the soil, making PLA biodegradation less feasible than that of other polyesters, such as PHB, PCL, and PBS [82].

Aliphatic bio-polyesters (PLA; PGA, PCL, etc.) have been commonly used in biodegradable products. The hydrolytic and/or enzymatic chain cleavage of these materials leads to αhydroxyacids, which could be assimilated by the human body or in composts [83]. However, poor mechanical properties and degradation time restrict the applications of these biopolymers. Copolymerization or blending have been designed to increase the performance enabling a range of mechanical properties and degradation rates [83]. The fibers of PLA-PCL, PGA-PCL, PDO (polydioxone) and PGA, with two different diameters (150 μm and 400 μm), were characterized in terms of degradation rate under three different environments (water, NaCl and PBS). After immersion in an aqueous medium, the first phenomenon to occur was the penetration of water from the surface to the center due to the negative gradient of water concentration (pure diffusion), which was faster compared with degradation. Therefore, it was considered that the hydrolysis of ester bonds started homogeneously. Despite the usual low degradation of PCL, PGA-PCL, the corresponding fibers were the fastest in terms of weight loss, followed by PGA, PDO, and the less biodegradable PLA-PCL [83].

Cotton fibers are mostly made of cellulose, which is biodegraded by microorganisms that secrete enzymes called cellulases. Cellulases catalyze the hydrolysis and oxidation of the cellulose molecular chain into cellobiose units, which then degrade into glucose and glucose derivatives, both of which are non-hazardous to the environment [84]. The biodegradation of cotton fabric is affected by textile finishing (silicone softener, durable press, water repellent, and a blue reactive dye). The rate of degradability of textile microfibers during laundering decreased with durable press- and water-repellant finishing treatment. The presence of crosslinking (durable press) and hydrophobicity (water repellent) on the surface slows down the initial adsorption of enzymes excreted by microorganisms in the inoculum. Therefore, the use of manufacturing techniques such as derivatization, blending, and coating to improve bio-based material performance in practical applications affects the environmental impact of the final product [85].

Experimental evidence of the aquatic biodegradation of cotton, rayon, and polyester-based fabrics has demonstrated that during, laundering cellulose-based fabrics releases more microfibers (0.2–4 mg/g fabric) than synthetic textiles (0.1–1 mg/g fabric). However, cotton and rayon fibers degraded in aquatic conditions faster than polyester fibers, which, by contrast, persisted in the environment for a long time [86].

The biodegradation of wool (natural keratin fiber), cotton (a natural cellulose fiber), and fiber of PLA was evaluated at 35 °C for 42 days to determine the time-dependent changes in weight loss, strength loss, and morphology under natural soil and aqueous medium conditions. The results made it possible to determine that the degradation rates in natural soil were higher than in aqueous medium, listed in the following order: cotton > wool > PLA fiber. This led to the conclusion that natural fibers degrade more easily than man-made biodegradable PLA fibers [87].

Based on a study by Egan and Salmon [79], the percentage biodegradation of textile fibers in various environments can be summarized as follows: (i) cotton fibers are biodegradable in soil (180 days) and anerobic conditions (30 days), and semi-biodegradable in compost (45 days); (ii) PLA-based fibers are biodegradable in compost (45 days), semi-biodegradable under anerobic conditions (30 days) and seawater (90 days), and completely non-biodegradable in soil (180 days); (iii) PET fibers are never biodegradable.

6.2. Mechanical Performance

Polylactide acid (PLA) is one of the most promising biopolymers in textiles, since it possesses similar characteristics to synthetic fibers, with superior biodegradability compared to other biopolymers. It is soft to the touch, features a silky sheen, and offers good durability. However, the breaking strength of pure PLA was very low, making it necessary to set specific parameters for the production and processing of PLA fibers [88], or for blending it with other polymers [89].

Persoon et al. [90] analyzed the effect of draw ratio and temperature on the tensile and thermal properties of melt-spun monofilament and multifilament fibers from PLA. The fibers were produced by melt spinning and subsequent solid-state, by choosing two matrices, different in viscosity and molecular weight. The final results demonstrated that the melt draw ratio, solid-state draw ratio, and drawing temperature strongly affected the mechanical properties of PLA fibers. By increasing the solid-state ratio, the orientation and crystallinity were increased. The authors concluded that a higher solid-state draw ratio was preferable when manufacturing PLA fibers with higher orientation, crystallinity, and tenacity, and that as a result, the melt draw ratio should be reduced accordingly.

Weft-knitted single-jersey fabrics were prepared by combining 100% PLA yarn or PLA/lyocell blended yarns. By increasing the proportion of the lyocell fibers, the bursting strength of corresponding fabrics was increased (bursting strength of basic PLA of 300 kPa), although it remained lower than that of PET/cotton blended fabric [91].

Endowed with strong antimicrobial activity and excellent biocompatible and biodegradable properties, PHBV can be represented as a potential candidate for replacing petroleum-derived polymers. Unfortunately, the mechanical strength, water sorption and diffusion, and electrical and/or thermal properties are all lacking, necessitating its use in conjunction with other polymers [92]. In this regard, PLA/PHBV blends (100% bio-based and fully degradable) have been investigated in terms of spinnability, as well as their mechanical and thermal characteristics. The results made it possible to demonstrate the promising application of these fibers to textiles: the addition of PHBV was shown to be compatible with PLA and increased the flexibility of the blend. The highest tenacity and shrinkage were recorded with PLA/PHBV blends of 95/5 and 90/10. At these concentrations, the fibers were subjected to knitting to prepare knitting socks [93]. A further investigation of dyeing processes of textiles made from PLA/PHBV was also performed by determining the effects of different dyes, dyeing temperatures, dyeing times, pH and liquor ratio on the dye exhaustion in the dyeing process, color fastness, and mechanical properties. The final results confirmed commercially acceptable levels of the measured characteristics, and the feasibility of the designed dyeing process. The calculated amount of energy required for the whole dyeing was comparable with that of polyethylene terephthalate [94]. Other studies went further in testing the antibacterial ingredient, polyhydroxy butyrate (PHB), of the PHBV material as a potential candidate for antimicrobial activity, in textile fabrics made from poly (hydroxybutyrate-co-hydroxy valerate)/polylactide acid (PHBV/PLA) with natural cotton fibers [95]. The effects of blend yarn, fabric structures, and distributions of fibers on the antimicrobial properties of the knitted fabrics were analyzed.

A comparison of the mechanical, thermal, and surface properties, as well as the anti-bacterial behavior, of five different types of synthetic multi-filament yarn and the corresponding knitted fabrics was presented in the work of Huang et al. [96]. Three bio-based materials (PLA/PHBV, PLA, Cupro) and two petroleum-based materials (PET, PA6) were considered. The final results made it possible to attest that the bursting strength, extension, and recovery of jersey-knitted fabrics from PLA/PHBV yarns satisfy industrial requirements. For these fabrics, a bursting strength of 375.8 kPa was measured, which was higher than that of PLA fabrics, similar to that of PA 6, and significantly lower than PET fabrics. Dyed PLA/PHBV knitted fabrics displayed the worst abrasion resistance (2250 rubs) compared with other fabric samples (the highest: PET fabrics with abrasion resistance > 50,000 rubs), promising tearing property, and good air permeability, as well as excellent antibacterial performance against staphylococcus aureus, klebsiella pneumoniae, and candida albicans.

6.3. Breathability and Comfort

"Clothing is comfortable when we are unaware of it" [97]. Factors that increase the awareness of wearing a garment, such as its pressure on the skin, its temperature, and its wetness, decrease the comfort [98]. Depending on personal choices, discomfort arises when the pressure approaches that of diastolic blood pressure (80 mmHg), three-quarters of the body surface is covered by liquid sweat, and a constant core temperature is maintained at different work rates for the body [97]. This means that during human activity, the generated heat should be transferred to the environment. The main mechanism of heat transport, especially during physical activities, is water evaporation and its transport to the surroundings [98].

Polylactide acid has a higher natural hydrophilicity than most other thermoplastic polymers, including polypropylene, nylon, and PET, because water molecules can enter in PLA macromolecules through polar oxygen linkages [99]. Although PLA fibers are not as wettable as cotton, the improved wettability characteristics of these fibers determine a larger moisture vapor transmission compared to PET or nylon fiber-based fabrics. This aspect allows the greater "breathability" of garments such as shirts, dresses, underwear, and shoes composed of PLA-based fibers when these fibers are used in place of fibers such as PET or nylon [99].

6.4. Antibacterial Properties

Resistance against microbial attack is one of the important aspects of the textile industry. A good contact area and the absorption of moisture are the two main causes of microbial growth, which leads to unpleasant odors, dermal infections, allergic reactions, and fabric deterioration. Thus, the incorporation of antimicrobial agents on textile products capable of overcoming these issues is critical [100]. Chitin and chitosan are highly versatile biomaterials that have gained widespread attention due to their unique properties, such as nontoxicity, biocompatibility, biodegradability, low allergenicity, biological activity, low cost, and so on [101]. One of the most common applications of chitosan in the textile industry is as an antimicrobial agent, given its ability to provide protection against allergies and infectious diseases, as well as moisture retention and wound-healing capabilities [102]. Because of the presence of reactive amino and hydroxyl groups along the backbone, chitosan has some intriguing properties for use in textile dyeing and finishing. However, low water solubility at neutral pH and poor durability on textile surfaces limit the widespread use of chitosan [103]. Experiments have been conducted on the use of crosslinking and graft polymerization to chemically modify chitosan to produce water-soluble bioactive derivatives. For example, glutaric dialdehyde was chosen as the crosslinking agent to chemically bond chitosan to cotton fabric by verifying the antibacterial behavior of chitosan against Escherichia coli and the Hay bacillus. It was found that 0.3 g/L chitosan solution effectively inhibited W. Escherichia coli, whereas 0.5 g/L chitosan solution was required against Hay bacillus [104].

The worldwide epidemic caused by the coronavirus 2 (SARS-CoV-2) pandemic has focused attention on rules, attitudes, and best practices for virus disease prevention. As a result, there has been a surge of interest in the development of surfaces that inhibit or prevent the adhesion of these microorganisms [105]. The health risks posed by the SARS-CoV-2 viruses and the need to protect the environment have shifted the focus to safe and durable antiviral and antibacterial textiles made from bio-based, renewable materials. New fibers and environmentally friendly advanced technologies to improve the mechanical properties and the antimicrobial behavior of textiles have been examined in [105]. Chemical treatment to modify natural and synthetic fibers is one of the viable and promising methods used in textile coating and finishing to improve their antibacterial, ultraviolet-protection, flame-retardant, and anti-static properties [106]. However, the involvement of hazard chemicals during chemical modification treatments raised environmental concerns by encouraging the use of a variety of natural and biodegradable coatings for wet chemical surface modification [107]. A series of novel chitosan-based water-dispersible polyurethanes have been experimentally assessed for upgrading the antibacterial activity of polyester/cotton textiles [108]. Nano-chitosan-polyurethane dispersions (NCS-PUs) were prepared via polymerization processes and applied on polyester cotton fabric (PCF) as a finishing agent, using the pad dry cure technique. Antibacterial activity, ultraviolet protection factor, tear, and tensile strength were measured for both dyed and printed textiles, demonstrating the NCS-PUs treatment's usefulness and effectiveness in increasing the desired properties of coated samples. The gradual increase in concentration of chitosan into the PU backbone significantly enhanced resistance against *B. subtilus, S. aureus* (gram-positive) and *E. coli* (gram-negative).

6.5. Flammability

A comparison of the flammability properties of PLA and PET fibers showed low flammability and less smoke generation than PET [109]. However, the high flammability of PLA is well known, which can result in the production of toxic gases during combustion in vitiated atmospheres [110]. As a result, PLA's flame-retardant properties have become an issue, limiting the scope of its applications [111]. Although most systems have not yet been commercialized, phosphorus-based PLA formulations have proven to be quite effective at increasing the flame-retardant features of corresponding PLA-based materials [110]. In this regard, intumescent formulations mainly composed of acid and carbon sources was analyzed in the work of Cayla [112]. Lignin from wood waste was selected as the carbon source, and added in different PLA-based preparations, to ammonium polyphosphate (AP) by melt extrusion, and then hot-pressed into sheets. The spinnability, thermal behavior, and flame retardancy of the developed PLA formulations containing lignin and/or ammonium polyphosphate were explored. MFI results confirmed that not all the formulations were spinnable and endowed with an increase in thermal flame-retardant properties. Depending on the concentration, the presence of ammonium polyphosphate in blends may not only degrade the macromolecular PLA chains, but also change interactions between all of the compounds. This may result in an increase in free volume of the PLA matrix and lignin agglomeration. A type of regenerated cellulose fiber that is widely used as bio-derived material in textiles, public transportation, electrical equipment, and building industries is viscose fibers. It offers numerous advantages, such as biodegradability, excellent air permeability, non-toxicity, low cost, and environmental advantage, with the main disadvantage being its high susceptibility to ignition. In order to overcame this drawback, alginate fibers were added to viscose yarns, and the effect of the content on the thermal stability, flame retardancy, and combustion behaviors of the prepared viscose/alginate-blended nonwoven fabrics, were investigated. The final results confirmed changes in the thermal stability of developed nonwoven fabrics by decreasing flammable gas emissions, thus improving the flame retardancy and inhibiting the smoke release [113].

The tenancy values of common natural, man-made, and bio-based fibers, are summarized in Table 2. The melting point and indications of moisture absorption in terms of moisture regain (the weight percentage of water in a material versus the material's dry weight) or overall moisture management capacity (an index between 0–1 expressing the ability of the fabric to manage the transport of liquid moisture, with 0–0.2 being poor, 0.2–0.4 being fair, 0.4–0.6 being good, 0.6–0.8 being very good, and 0.8–1 being excellent) are also shown.

Table 2. Main properties of most common natural, synthetic, and bio-based fibers.

	Yarn Production Process	Tenacity (cN/dtex)	Melting Point (°C)	Moisture Absorption	
				Moisture Regain (%) -	Overall Moisture Management Capacity
PLA	Melt spinning at 210 to 235 °C [93]	0.90–1.09 [93] 4–5.5 [114] 3.2–5.5 [115]	162.4–162.7 [93] 175–180 [114]	0.5 [114] 0.4–0.6 [115]	/
PLA/PHBV 90/10 (*w/w*)	Melt spinning at 210 to 230 °C [93]	0.46–1.09 [93]	164.3–164.9 [93]	/	/
PLA/PHBV 80/20 (*w/w*)	Melt spinning at 210 to 230 °C [93]	0.59–0.64 [93]	164.4–164.8 [93]	/	/
Cotton	/	1.8 [116] 3.1 [117] 1.9–3.1 [115]	/	7.5 [88] 8.5 [115]	0.63 [116]
Viscose	/	2 [116]	/	/	0.62 [116]
Viscose/Modal 50/50	Ring spinning technique [116]	2 [116]	/	/	0.6 [116]
Viscose/Cotton 50/50	Ring spinning technique [116]	1.5 [116]	/	/	0.64 [116]
Cotton/Silk (85/15)	Ring and siro spinning systems [117]	1.6–2.2 [117]	/	/	0.58–0.62 [117]
Cotton/Silk (55/45)	Ring and siro spinning systems [117]	2.42–2.52 [117]	/	/	0.57–0.66 [117]
Wool	/	1–1.4 [115]	/	14–18 [88] 14.5 [115]	/
Silk	/	4.7 [117] 1.9–5.1 [115]	/	10.5 [118] 30 [117]	/
Polyester (PET)	Melt spinning at 280–290 °C [114]	3.5–5 [114] 5.6 [119]	265 [114]	0.2–0.4 [88] 0.4 [114]	/
Polyamide	Melt spinning	3.5–5 [114] 6.6 [119]	214 [114]	4.1 [88] 4.5 [114]	/
Polypropylene	Melt spinning	3.5–5 [114] 6.5 [119]	175 [114]	0 [114]	/

7. Recent Applications of Biopolymers to Textiles

7.1. Textile Finishing and Treatment

Conventional textile wet processing poses a significant challenge to the development of eco-friendly processes and green products given the negative effects on human health and the environment due to the requirement of a significant amount of energy for heating, drying, and/or steaming, in addition to the machinery, resulting in an increase in greenhouse gas emissions and carbon footprint [120]. Recently, there has been an increasing awareness of hygienic lifestyle, concern over carbon and water footprint, and a desire to meet consumer demands at the same time, ensuring the sustainability of our eco-systems. The application of green chemistry principles and aspects of cleaner production have been considered for preserving the environment, economy, and society [121]. The use of enzymes in the textile pre-treatment of natural fibers, such as depilling, desizing, scouring, and so on, has already proven to be a highly profitable and sustainable alternative to the harsh toxic chemicals used in the textile industry [122]. Enzymes are proteins made up of amino acids, responsible for thousands of metabolic processes that sustain life. They act as catalysts, accelerating chemical reactions in highly specific and efficient ways while not altering or being consumed [121]. Different enzymes are already used in experiments for the textile industry, and they are able to impart specific functional features to treated textiles: chitosan [103], cyclodextrin [123], alginate [124], or plant-based bioactive materials (aloe vera) [125], essential oils (jasmine) [126], natural dyes extracted from different parts of plants such as bark, leaf, root, and flowers containing common coloring materials such as tannin, flavonoids, and quinonoids [127].

In order to impart shrink resistance to wool fabric, an eco-friendly treatment involving a sequential combination of enzymes followed by polysaccharide-based bio-polymers was proposed by Kadam et al. [128]. For the experiments, wool fabric was considered and treated firstly with laccase and protease enzymes, and then coated with three polysaccharide biopolymers (chitosan, wheat starch, and gum arabic) using the pad-dry-cure technique. On the developed samples, a characterization based on microscopy and IR spectroscopy, tensile, frictional, and bending analysis was performed. Except for friction, shrink-resistance treatment has no effect on tensile or bending properties. The combination of treatments preserved the whiteness of the wool fabric while reducing its yellowness. The result obtained in the presence of chitosan treatment was found to reduce the wool fabric shrinkage to <4%, which is deemed comparable to traditional chlorine-based methods.

UV-B radiation (wavelengths from 280 to 315 nm) results in harm to humans, particularly in terms of its effects on the eyes and skin [129]. It can cause damage to DNA and RNA structures in humans, leading to helix distortion and alterations in transcriptional programs [130]. In this regard, Dominguez-Pacheco et al. [131] examined the effect of adding natural pigments, extracted from the cooking of commercial white corn kernels, on the improvement of fabric protection against UV radiation. The optical absorption spectra of natural pigments and treated textiles were obtained using photoacoustic spectroscopy. During the cooking of agricultural grains, several components, such as phenols and flavonoids, contained in the superficial layers were released. Maceration and microwave oven-assisted extraction were the two adopted methods to capture the natural pigments from the samples. The experimental results confirmed the higher optical absorption coefficient of textiles treated with bio-based additives, compared to textiles to which a chemical anti-UV agent had been added, and a wider band in the optical penetration length. This indicated that the previous systems possessed greater opacity and less penetration by UV light.

7.2. Printing on Textiles by Fused Deposition Modeling (FDM)

A recent development in 3D printing processes increased the potential of this technology by promoting its accessibility and providing a new platform for design, customization, and innovation [132]. Many fashion designers are taking advantage of this innovation by producing textiles, clothing, jewelry, notions, or shoes [133]. Sophistication and fidelity to style have been added to the printed textiles [134], but functional characteristics such as rigidity [135] and abrasion resistance [136] have also been imparted. In the work of Singh et al. [137], cotton-knitted fabric and tulle net fabric based on nylon were selected for better deposition of the fused plastic material inside the fabric. The effect of the infill percentage on the adhesion property was investigated. The authors concluded that the fiber direction in the fabric and the first layer played a larger role in affecting the adhesion properties than the platform temperature.

To improve the environmental footprint of the production process for smart and functional textiles by avoiding unnecessary use of water, energy, and chemicals while minimizing waste, Hashemi Sanatgar et al. [138] proposed 3D printing as a more cost-effective textile functionalization process than conventional printing processes such as screen and inkjet printing. They investigated the adhesion characteristics, depending on the fabric and filler type, of polymers and nanocomposite layers, printed directly onto the textile fabrics by using the FDM method. Nylon was printed on polyamide 66 (PA66) fabrics, polylactic acid on PA66 and PLA fabrics, and finally nanocomposites of PLA and carbon black or multi-wall carbon nanotubes on PLA fabric. The results demonstrated that different

3D printing process variables, such as the extruder temperature, platform temperature, and printing speed could all have a significant impact on the adhesion force. The breaking strength of 3D printed layers was reduced by increasing the extruder temperature. This was interpreted as a sign of increased brittleness caused by higher processing temperatures. Next, as the printing speed increased, the adhesive force decreased. In fact, improving the printing speed reduced macromolecule penetration into the fabric, resulting in cohesive forces, which are greater than adhesive ones. No significant linear effect on the adhesion force of the platform temperature (chosen close to the glass transition temperature) was verified. This was attributed to the poor mobility of macromolecules, which reduced the diffusion of the printed polymer in the fabric structure by leading to negligible adhesion. The platform temperature was found to play an important role in increasing the adhesion when the value was higher than the glass point. The highest value of the adhesive force was verified for samples made from PLA-based nanocomposites applied to PLA fabrics.

By contrast, Mpofu et al. [139] concentrated their study on fabric properties affecting the adhesion of 3D printed PLA polymer onto selected fabrics (acrylic, cotton, polyester/cotton blend and polyester) (Figure 7). The effect of fabric areal density, warp and weft count, fabric thickness and roughness, ends/inch and picks/inch, was analyzed through a regression model. The former four parameters were discovered to have a positive effect on the investigated property, while the last were revealed to exert a negative effect. In terms of fiber types, acrylic-based fabrics had the highest adhesion force to PLA, while polyester-based fabrics had the lowest.

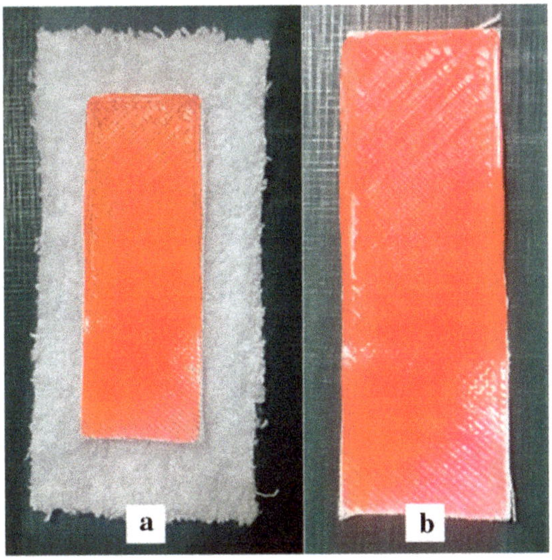

Figure 7. Example of PLA printed on cotton fabric (**a**) before cutting (**b**) after cutting the edges. Reprinted [139].

Many clothing companies have begun to use 3D printing to create accessories and soles. However, due to a lack of flexible materials, it is still not possible to produce ready-to-wear clothing. In this regard, Uysal and Stubbs [140] developed a new method for printing textile-like surfaces. Different flexible structures were made of layers by combining different materials, PLA and lay-foom (made from a rubber-elastomeric polymer and a PVA-component), numbers of layers, and repeating patterns (polygon, rectangles, floral). Sewing patterns were printed and assembled into a three-dimensional garment using a single printing step instead of typical production steps in the textile supply chain, such as fabric production, dyeing, colour printing, cutting, and the application of other components, such as inner linings.

7.3. Treatment of Dye-Contaminant Water

One of the aspects of textile production that causes the most pollution is the contamination of fresh water, particularly from the dye treatment [141]. This reduces the quality of water and renders it unfit for consumption by inhabitants. Some dyes are more resistant and difficult to degrade completely by using photolysis, biological and chemical decomposition, and other ordinary approaches [142]. Advanced treatment technologies, such as adsorption processes, advanced oxidation processes (AOPs), and membrane processes have been shown to be promising alternatives for micropollutant removal; however, high operating costs and the formation of by-products and concentrated residues limit their application [143]. From an environmental point of view, it is also essential to develop new technologies for the wastewater treatment and recycling of dye-contaminated water. A recent review by Sirajudheen et al. [144] analyzed the applications of chitin and chitosan to the adsorption of textile dyes from water. Chitin and chitosan are among the most abundant natural biopolymers on the planet. They are endowed with distinct chemical, mechanical, optical, and physical properties as a result of their structural characteristics, such as high porosity, low density, renewability, and biodegradability. Unmodified and modified chitin and chitosan, as well as their various derivatives, are used in applications such as dye adsorption [145], air pollution [146], and heavy metal adsorption [147]. However, their main disadvantages include their low absorption capacity, poor mechanochemical stability, and low surface area. In order to improve these features, various approaches have been experimentally adopted: imprinting with metallic species [148] or minerals [149], the modification of the biomaterial surface through the crosslinker [150], and grafting [151], resulting in more active sites that produced more reliable dye–adsorbent interactions.

7.4. Composites

Numerous studies on the use of natural fibers of different types (hemp [152], kenaf [153], jute [154], flax [155], curaua [156]) and architectures (short-fiber [157], non-woven mat [158], and woven fabrics [159]) as reinforcements for PLA polymer have been conducted in order to improve the biodegradability, mechanical properties, thermal properties, and flame retardancy of corresponding composites [160]. By interlacing 3D-braided yarn produced by the solid braiding method, a plain woven fabric in flax material was produced. The obtained system was then combined with PLA polymer to create a sheet using the solution casting technique. Finally, composite laminates were prepared by film stacking and compression molding, and their tensile, flexural, and impact properties were studied in relation to the number of layers of fabric and loading along the warp and weft directions. The final results confirmed that fax fibers worked as effective reinforcing agents for the PLA polymer, improving its thermal and mechanical properties [161].

Polymers and composites can absorb varying amounts of water depending on their chemical nature, formulation, and environmental conditions of humidity and temperature [162]. Composite applications, ranging from civil structures to medical implants, necessitate long-term studies in moist environments [163]. The dominant mechanism in the phenomenon of moisture penetration is the diffusion of water molecules into the matrix, and also into the fibers, which is enabled by a capillary flow along the fiber–matrix interface, followed by diffusion from the interface into the bulk resin and transport via micro cracks [164]. Often, the surface damage and cracks caused by absorption facilitate the entry of water into the composite [165]. Liquid swelling is an important experiment to understand the composites' performance in wet environments. The mechanical and swelling behavior of a fully biodegradable "green" textile composite made from Ecoflex polymer and ramie fabric by using the hot compression molding technique was analyzed in the work of Kumar et al. [166]. The tensile strength, tensile modulus, elongation at break, and diffusion characteristics of the composites in water, naphthenic oil, and diesel were measured. From the values, the tensile strength and Young's modulus of Ecoflex/ramie mat composite were more than that of the neat polymer. The mechanism of the diffusion followed the classical Fick law of mass transport with good approximation. The polar molecules of water easily penetrated in the cellulosic polar fiber (ramie); thus, the Ecoflex/ramie fabric composite absorbed more water than diesel and lubricating oil.

7.5. Personal Thermal Management

Personal thermal management is currently receiving significant attention and interest because it can keep people comfortable while saving energy at the same time. This technology aims to heat or cool the human body locally, without wasting energy used for heating, ventilation, and air conditioning. Nonetheless, previous studies highlighted weak points, such as limited working temperature, poor comfort, and low textile reliability. In a study by Wu et al. [167], a skin-friendly personal insulation textile and a thermoregulation textile capable of performing both passive heating and cooling with a single piece of textile and zero energy input was developed. A freeze-spinning process was used to create a micro-structured biomaterial from breathable and antibacterial silk fibroin, resulting in good thermal insulation, low thermal emissivity, and good dyeability. Next, the obtained microstructure fibers were filled with biocompatible phase-change materials (poly(ethylene glycol), PEG) through the impregnation and coated with polydimethylsiloxane (PDMS) to enhance the hydrophobic and mechanical properties and prevent material leakage (Figure 8). As a result, the insulation textile was transformed into a thermoregulation textile with good water hydrophobicity, mechanical robustness, and working stability.

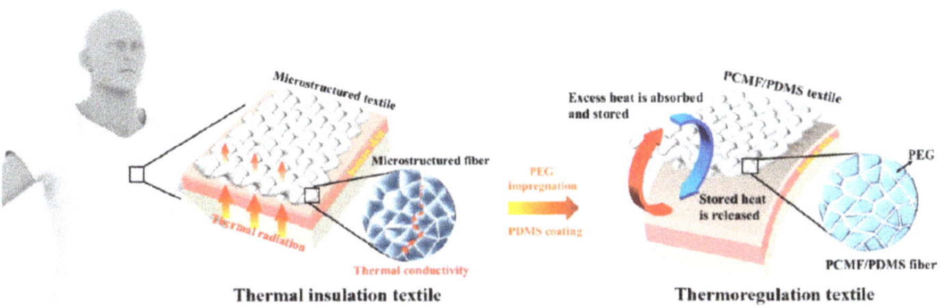

Figure 8. Schematization of thermal insulation textile made from microstructure fibers impregnated with biocompatible PEG and coated with PDMS. Reprinted [167].

7.6. Counterfeiting Sector

Silk luminescence was achieved in the work of Zhang et al. [168] through the chemical coating of the surfaces of natural fibers with luminescent gold nanoclusters (AuNCs). The synthesis was achieved through an easy, eco-friendly, and highly reproducible method. The silk was immersed in an alkaline aqueous solution containing hydrogen tetrachloroaurate (III) hydrate ($HAuCl_4$), and a redox reaction between the protein-based silk and an Au salt precursor occurred. From the experimental data, the good optical properties of the luminescent silk coated with AuNCs were established by its relatively long wavelength emission, high quantum yields, long fluorescent lifetime, and photostability. Compared to the pristine fibers, golden silk possessed superior mechanical properties, a good ability to inhibit the penetration of UV radiation, and lower toxicity in vitro. The authors proposed nanocluster-coated silk fibers as potential candidates for the commercial silk textile industry, tissue engineering, cell adhesion, antibacterial materials, biosensors and, particularly, the anti-counterfeiting sector.

7.7. Hospital Clothing

With the purpose of developing disposable hospital clothing and reducing the social costs resulting from the harmful effects of pollutants, Reza Saffari et al. [169] studied the improvement of antibacterial characteristics of nonwoven biodegradable textiles, made from polylactide, by using titanium dioxide (TiO_2) coating. The coating treatment involved a clean and environmentally benign process, i.e., the low-temperature plasma technique.

Through contact with cold plasma, several concurrent processes caused chemical and physical changes to the fabric surface's characteristics. To activate the plasma surface, gases such as oxygen, nitrogen, hydrogen, and ammonia were used. The interaction of these gases with the surface resulted in the formation of various chemical functional groups. The solvents, surfactants, and drying ovens typically used in the pretreatment and finishing of textile fabrics are not required in the case of plasma, by making this coating technique clean and environmental safety. The final results made it possible to determine a reduction in both bacteria, *S. aureus* and E. coli, in the treated textiles as a function of the coating processing time.

8. Introducing Bio-Sustainable Textile Materials to the Market

In 2018, Aquafil and Genomatica announced a multi-year collaboration to create a commercially viable bioprocess for producing caprolactam from plant-based renewable ingredients. This process was applied to the production of 100% sustainable nylon by combining the skills of the first (Aquafil) in the production of polyamide 6 with those of the second (Genomatica) process technologies to produce chemicals from alternative feedstocks in a more cost-effective, sustainable, and performance-oriented manner [170].

In 2019, in Trentino (Italy), a vegan and coated fabric named VEGEA was developed by a company of the same name. The name originated from the union of VEG (Vegan) and GEA (Mother Earth). It was chosen as an alternative material totally based on oil- and animal-derived sources, characterized by a high content of vegetal/recycled raw materials such as vegetal oils and natural fibers from agroindustry. In fact, VEGEA is a plant-based technical fabric, obtained from the treatment of the fibers and oils of marc, a natural derivative of wine production, including grape skins, seeds and stalks. Its production can also be considered entirely sustainable and "vegan-friendly"), since it does not use oil or pollutants, and does not consume water or animal derivatives [171].

The North Face and Spiber released the "Moon Parka" in 2019. This silk-like yarn was made from bio-fabricated brewed proteins and produced through a fermentation process involving sugars and microbes. Vivo Barefoot developed a bio-based vegan version of their popular performance shoes, the Primus Lite, made with 30% biobased materials. Flexible and stretchy characteristics were produced by a combination of an algae-based natural foam, Vietnamese natural rubber, and biosynthetic material derived from corn [172].

Eastman and DuPont Biomaterials promoted a fabric collection made from sustainable, bio-based materials by combining Eastman NaiaTM and DuPontTM Sorona fibers to create garments with exceptional stretch and recovery, luxurious drape, and a smooth, soft feel (September 2020). Eastman Naia is a cellulosic fiber applied in womenswear, whereas DuPont™ Sorona is a high-performance polymer certified as a bio-based product. In June 2021, a new fabric collection made with sustainable, bio-based materials was launched by Dupont and JayaShree Textiles [173].

DuPont Biomaterials and Welspun India announced a new, bio-based home textile collection, including bath towels and bedsheets (October 2021). Bolt Threads, a material solutions company, created a revolutionary, certified bio-material, called Mylo, by drawing inspiration from nature. It is created by engineering mycelium, which is composed primarily of renewable natural ingredients. It is marketed as a sustainable alternative to leather that will be made available to the public in 2022 by a consortium of companies (Adidas, Kering, Lululemon, and Stella McCartney) [174].

9. Conclusions

This work offers an overview of recent potential applications of bio-materials in the textile industry. The majority of textile fibers used around the world are made from petroleum-derived plastics (polyester, polyamide, polypropylene). However, in addition to all the benefits of fossil-based plastics (such as their light weight, low cost, durability, and chemical resistance), the negative environmental consequences of their production and uses have recently received significant attention. Textile production is one of the most pollutive industries due to its use of hazardous chemicals, consumption of water and energy, and high gas emissions. Other sources of contamination arise from the end-of-life of these products due to waste accumulation in nature, in oceans, or in landfills, or due to further gas discharges during incineration. Given the significant contamination of our ecosystem caused by synthetic fibers involved in the textile industry, biomaterials derived from renewable resources or endowed with biodegradability characteristics have been proposed as a possible green solution for reducing the environmental impact of fabric production. The use of polymers derived from renewable sources (both biodegradable and non-biodegradable) would result in reduced greenhouses emissions (GHG) and fossil fuel consumption (FFC) when compared to common fossil-based, non-biodegradable polymers. Although less biodegradable compared to natural-based fibers (wool, cotton), aliphatic polyester bio-based fibers are biodegraded more quickly compared to PET fibers. Furthermore, the larger moisture vapor transmission of bio-based polymers compared to PET, nylon and PP materials, allows greater breathability by corresponding fabrics. Bio-based fibers are also endowed with good mechanical resistance and antibacterial properties. Research studies confirmed the applicability of biopolymers in blend formulations to the production of antibacterial fabrics, to increasing fabric resistance against harmful UV radiation, to the thermoregulation of passive heating and cooling, to composites, or in the anti-counterfeiting sector. The application of biopolymers was also found to be useful in the pre-treatment or finishing of fabrics, in 3D printing to replace traditional printing methods involving hazards chemicals, and in the purification of dye-contaminant water. Finally, several suggestions from major brands to shift production toward environmentally friendly materials and green technologies were also presented.

Author Contributions: Conceptualization, A.P. and D.A.; writing—original draft preparation, A.P.; writing—review and editing, D.A. All authors have read and agreed to the published version of the manuscript.

Funding: This research received no external funding.

Institutional Review Board Statement: Not applicable.

Informed Consent Statement: Not applicable.

Data Availability Statement: The data presented in this study are available on request from the corresponding author.

Acknowledgments: A. Patti wishes to thank the Italian Ministry of Education, Universities and Research (MIUR) in the framework of Action 1.2 "Researcher Mobility" of The Axis I of PON R&I 2014–2020 under the call "AIM-Attrazione e Mobilità Internazionale".

Conflicts of Interest: The authors declare no conflict of interest.

References

1. Patti, A.; Cicala, G.; Acierno, D. Textile Production. Encyclopedia. Available online: https://encyclopedia.pub/7964 (accessed on 2 November 2021).
2. Zhang, H.; Themelis, N.J.; Bourtsalas, A. Environmental impact assessment of emissions from non-recycled plastic-to-energy processes. *Waste Dispos. Sustain. Energy* **2021**, *1*, 3. [CrossRef]
3. Garside, M. Chemical and Textile Fibers Production Worldwide 1975–2018. Statista. Available online: https://www.statista.com/statistics/263154/worldwide-production-volume-of-textile-fibers-since-1975/#statisticContainer (accessed on 15 May 2020).
4. Roy Choudhury, A.K. Environmental Impacts of the Textile Industry and Its Assessment Through Life Cycle Assessment. In *Roadmap to Sustainable Textiles and Clothing. Textile Science and Clothing*; Muthu, S., Ed.; Springer: Singapore, 2014; pp. 1–39.
5. Cabrera, F.C. Eco-friendly polymer composites: A review of suitable methods for waste management. *Polym. Compos.* **2021**, *42*, 2653–2677. [CrossRef]
6. Okafor, C.C.; Madu, C.N.; Ajaero, C.C.; Ibekwe, J.C.; Nzekwe, C.A.; Okafor, C.C.; Madu, C.N.; Ajaero, C.C.; Ibekwe, J.C.; Nzekwe, C.A. Sustainable management of textile and clothing. *Clean Technol. Recycl.* **2021**, *1*, 70–87. [CrossRef]
7. Yalcin-Enis, I.; Kucukali-Ozturk, M.; Sezgin, H. Risks and Management of Textile Waste. In *Nanoscience and Biotechnology for Environmental Applications*; Gothandam, K., Ranjan, S., Dasgupta, N., Lichtfouse, E., Eds.; Springer: Cham, Switzerland, 2019; Volume 22, pp. 29–53.

8. Patti, A.; Cicala, G.; Acierno, D. Eco-Sustainability of the Textile Production: Waste Recovery and Current Recycling in the Composites World. *Polymers* **2020**, *13*, 134. [CrossRef] [PubMed]
9. Patti, A.; Acierno, D. The Puncture and Water Resistance of Polyurethane- Impregnated Fabrics after UV Weathering. *Polymers* **2019**, *12*, 15. [CrossRef] [PubMed]
10. Patti, A.; Costa, F.; Perrotti, M.; Barbarino, D.; Acierno, D. Careful Use of Silica Nanoparticles in the Textile Treatment for Potential Large-scale Production. *Chem. Eng. Trans.* **2021**, *84*, 91–96. [CrossRef]
11. Patti, A.; Costa, F.; Perrotti, M.; Barbarino, D.; Acierno, D. Polyurethane Impregnation for Improving the Mechanical and the Water Resistance of Polypropylene-Based Textiles. *Materials* **2021**, *14*, 1951. [CrossRef]
12. Patti, A.; Acierno, D. The effect of silica/polyurethane waterborne dispersion on the perforating features of impregnated polypropylene-based fabric. *Text. Res. J.* **2020**, *90*, 1201–1211. [CrossRef]
13. Patti, A.; Acierno, D. Effect of waterborne polyurethane on mechanical properties of impregnated fabrics. In *AIP Conference Proceedings*; AIP Publishing LLC: Melville, NY, USA, 2018; Volume 1981, p. 020139.
14. Acquavia, M.A.; Pascale, R.; Martelli, G.; Bondoni, M.; Bianco, G. Natural polymeric materials: A solution to plastic pollution from the agro-food sector. *Polymers* **2021**, *13*, 158. [CrossRef]
15. Abdellatif, F.H.H.; Abdellatif, M.M. Utilization of sustainable biopolymers in textile processing. In *Green Chemistry for Sustainable Textiles*; Woodhead Publishing: Cambridge, UK, 2021; pp. 453–469.
16. Geyer, R. Production, use, and fate of synthetic polymers. In *Plastic Waste and Recycling*; Academic Press: Cambridge, MA, USA, 2020; pp. 13–32.
17. Lebreton, L.; Slat, B.; Ferrari, F.; Sainte-Rose, B.; Aitken, J.; Marthouse, R.; Hajbane, S.; Cunsolo, S.; Schwarz, A.; Levivier, A.; et al. Evidence that the Great Pacific Garbage Patch is rapidly accumulating plastic. *Sci. Rep.* **2018**, *8*, 4666. [CrossRef]
18. Jambeck, J.R.; Geyer, R.; Wilcox, C.; Siegler, T.R.; Perryman, M.; Andrady, A.; Narayan, R.; Law, K.L. Plastic waste inputs from land into the ocean. *Science (80-)* **2015**, *347*, 768–771. [CrossRef] [PubMed]
19. Nicholson, S.R.; Rorrer, N.A.; Carpenter, A.C.; Beckham, G.T. Manufacturing energy and greenhouse gas emissions associated with plastics consumption. *Joule* **2021**, *5*, 673–686. [CrossRef]
20. Jem, K.J.; Tan, B. The development and challenges of poly (lactic acid) and poly (glycolic acid). *Adv. Ind. Eng. Polym. Res.* **2020**, *3*, 60–70. [CrossRef]
21. Fiber Market Statistics 2021: Growth, Statistics, Trends, Share, Revenue and Forecast Outlook till 2027 with Prominent Regions and Countries Data—Market Watch. Available online: https://www.marketwatch.com/press-release/fiber-market-statistics-2021-growth-statistics-trends-share-revenue-and-forecast-outlook-till-2027-with-prominent-regions-and-countries-data-2021-10-06 (accessed on 3 November 2021).
22. Mittal, M.; Mittal, D.; Aggarwal, N.K. Plastic accumulation during COVID-19: Call for another pandemic; bioplastic a step towards this challenge? *Environ. Sci. Pollut. Res.* **2022**, *29*, 11039–11053. [CrossRef]
23. Shams, M.; Alam, I.; Mahbub, M.S. Plastic pollution during COVID-19: Plastic waste directives and its long-term impact on the environment. *Environ. Adv.* **2021**, *5*, 100119. [CrossRef]
24. Coronavirus Impact on Textile Industry in Europe 2020. Statista. Available online: https://www.statista.com/statistics/1131136/coronavirus-impact-on-textile-industry-europe/ (accessed on 3 November 2021).
25. Nghiem, L.D.; Iqbal, H.M.N.; Zdarta, J. The shadow pandemic of single use personal protective equipment plastic waste: A blue print for suppression and eradication. *Case Stud. Chem. Environ. Eng.* **2021**, *4*, 100125. [CrossRef]
26. Ragazzi, M.; Rada, E.C.; Schiavon, M. Municipal solid waste management during the SARS-COV-2 outbreak and lockdown ease: Lessons from Italy. *Sci. Total Environ.* **2020**, *745*, 141159. [CrossRef]
27. Ramos, Ó.L.; Pereira, R.N.; Cerqueira, M.A.; Martins, J.R.; Teixeira, J.A.; Malcata, F.X.; Vicente, A.A. Bio-Based Nanocomposites for Food Packaging and Their Effect in Food Quality and Safety. In *Food Packaging and Preservation*; Elsevier: Amsterdam, The Netherlands, 2018; pp. 271–306.
28. Sathya, A.B.; Sivasubramanian, V.; Santhiagu, A.; Sebastian, C.; Sivashankar, R. Production of Polyhydroxyalkanoates from Renewable Sources Using Bacteria. *J. Polym. Environ.* **2018**, *26*, 3995–4012. [CrossRef]
29. Astm, D. 6400—99e1 Pdf Free Download. Available online: https://civileng.co/astm-d-6400-99e1-pdf-free-download/ (accessed on 27 January 2022).
30. Kale, G.; Kijchavengkul, T.; Auras, R.; Rubino, M.; Selke, S.E.; Singh, S.P. Compostability of bioplastic packaging materials: An overview. *Macromol. Biosci.* **2007**, *7*, 255–277. [CrossRef]
31. Gedde, U.W.; Hedenqvist, M.S.; Hakkarainen, M.; Nilsson, F.; Das, O. Plastics and Sustainability. In *Applied Polymer Science*; Springer: Cham, Switzerland, 2021; pp. 489–504.
32. Anne, B. Environmental-Friendly Biodegradable Polymers and Composites. In *Integrated Waste Management*; IntechOpen: London, UK, 2011; pp. 341–364.
33. Ebnesajjad, S. *Handbook of Biopolymers and Biodegradable Plastics: Properties, Processing, and Applications*; William Andrew: Norwich, NY, USA, 2012; ISBN 1455728349.
34. Singhvi, M.; Gokhale, D. Biomass to biodegradable polymer (PLA). *RSC Adv.* **2013**, *3*, 13558–13568. [CrossRef]
35. Patti, A.; Acierno, D.; Latteri, A.; Tosto, C.; Pergolizzi, E.; Recca, G.; Cristaudo, M.; Cicala, G. Influence of the processing conditions on the mechanical performance of sustainable bio-based PLA compounds. *Polymers* **2020**, *12*, 2197. [CrossRef] [PubMed]

36. Antonio, G.P.; Fabbricino, P.M.; Policastro, G.; Fabbricino, Á.M.; Fabbricino, M.; Panico, A. Improving biological production of poly(3-hydroxybutyrate-co-3-hydroxyvalerate) (PHBV) co-polymer: A critical review. *Rev. Environ. Sci. Bio/Technol.* **2021**, *20*, 479–513. [CrossRef]
37. Labet, M.; Thielemans, W. Synthesis of polycaprolactone: A review. *Chem. Soc. Rev.* **2009**, *38*, 3484–3504. [CrossRef] [PubMed]
38. Jian, J.; Xiangbin, Z.; Xianbo, H. An overview on synthesis, properties and applications of poly(butylene-adipate-co-terephthalate)–PBAT. *Adv. Ind. Eng. Polym. Res.* **2020**, *3*, 19–26. [CrossRef]
39. Xu, J.; Guo, B.H. Poly(butylene succinate) and its copolymers: Research, development and industrialization. *Biotechnol. J.* **2010**, *5*, 1149–1163. [CrossRef]
40. Samantaray, P.K.; Little, A.; Haddleton, D.M.; McNally, T.; Tan, B.; Sun, Z.; Huang, W.; Ji, Y.; Wan, C. Poly(glycolic acid) (PGA): A versatile building block expanding high performance and sustainable bioplastic applications. *Green Chem.* **2020**, *22*, 4055–4081. [CrossRef]
41. Hatti-Kaul, R.; Nilsson, L.J.; Zhang, B.; Rehnberg, N.; Lundmark, S. Designing Biobased Recyclable Polymers for Plastics. *Trends Biotechnol.* **2020**, *38*, 50–67. [CrossRef]
42. Mohsenzadeh, A.; Zamani, A.; Taherzadeh, M.J. Bioethylene Production from Ethanol: A Review and Techno-economical Evaluation. *ChemBioEng Rev.* **2017**, *4*, 75–91. [CrossRef]
43. Wang, Z.; Ganewatta, M.S.; Tang, C. Sustainable polymers from biomass: Bridging chemistry with materials and processing. *Prog. Polym. Sci.* **2020**, *101*, 101197. [CrossRef]
44. Lamberti, F.M.; Román-Ramírez, L.A.; Wood, J. Recycling of Bioplastics: Routes and Benefits. *J. Polym. Environ.* **2020**, *28*, 2551–2571. [CrossRef]
45. Radzik, P.; Leszczyńska, A.; Pielichowski, K. Modern biopolyamide-based materials: Synthesis and modification. *Polym. Bull.* **2020**, *77*, 501–528. [CrossRef]
46. Norddahl, B. Chemical Product Design in a Sustainable Environment. *Chem. Eng. Trans.* **2021**, *86*, 1339–1344. [CrossRef]
47. Philp, J.C.; Bartsev, A.; Ritchie, R.J.; Baucher, M.A.; Guy, K. Bioplastics science from a policy vantage point. *New Biotechnol.* **2013**, *30*, 635–646. [CrossRef] [PubMed]
48. Benavides, P.T.; Lee, U.; Zarè-Mehrjerdi, O. Life cycle greenhouse gas emissions and energy use of polylactic acid, bio-derived polyethylene, and fossil-derived polyethylene. *J. Clean. Prod.* **2020**, *277*, 124010. [CrossRef]
49. Zhao, X.; Cornish, K.; Vodovotz, Y. Narrowing the Gap for Bioplastic Use in Food Packaging: An Update. *Environ. Sci. Technol.* **2020**, *54*, 4712–4732. [CrossRef]
50. Wu, F.; Misra, M.; Mohanty, A.K. Challenges and new opportunities on barrier performance of biodegradable polymers for sustainable packaging. *Prog. Polym. Sci.* **2021**, *117*, 101395. [CrossRef]
51. Farah, S.; Anderson, D.G.; Langer, R. Physical and mechanical properties of PLA, and their functions in widespread applications—A comprehensive review. *Adv. Drug Deliv. Rev.* **2016**, *107*, 367–392. [CrossRef]
52. Chen, G.Q.; Patel, M.K. Plastics derived from biological sources: Present and future: A technical and environmental review. *Chem. Rev.* **2012**, *112*, 2082–2099. [CrossRef]
53. Meereboer, K.W.; Misra, M.; Mohanty, A.K. Review of recent advances in the biodegradability of polyhydroxyalkanoate (PHA) bioplastics and their composites. *Green Chem.* **2020**, *22*, 5519–5558. [CrossRef]
54. Vandi, L.J.; Chan, C.M.; Werker, A.; Richardson, D.; Laycock, B.; Pratt, S. Wood-PHA Composites: Mapping Opportunities. *Polymers* **2018**, *10*, 751. [CrossRef]
55. Al Hosni, A.S.; Pittman, J.K.; Robson, G.D. Microbial degradation of four biodegradable polymers in soil and compost demonstrating polycaprolactone as an ideal compostable plastic. *Waste Manag.* **2019**, *97*, 105–114. [CrossRef] [PubMed]
56. Lu, B.; Wang, G.X.; Huang, D.; Ren, Z.L.; Wang, X.W.; Wang, P.L.; Zhen, Z.C.; Zhang, W.; Ji, J.H. Comparison of PCL degradation in different aquatic environments: Effects of bacteria and inorganic salts. *Polym. Degrad. Stab.* **2018**, *150*, 133–139. [CrossRef]
57. Pietrosanto, A.; Scarfato, P.; Di Maio, L.; Incarnato, L. Development of Eco-Sustainable PBAT-Based Blown Films and Performance Analysis for Food Packaging Applications. *Materials* **2020**, *13*, 5395. [CrossRef] [PubMed]
58. Saffian, H.A.; Yamaguchi, M.; Ariffin, H.; Abdan, K.; Kassim, N.K.; Lee, S.H.; Lee, C.H.; Shafi, A.R.; Alias, A.H. Thermal, Physical and Mechanical Properties of Poly(Butylene Succinate)/Kenaf Core Fibers Composites Reinforced with Esterified Lignin. *Polymers* **2021**, *13*, 2359. [CrossRef] [PubMed]
59. Iwata, T. Biodegradable and bio-based polymers: Future prospects of eco-friendly plastics. *Angew. Chem.-Int. Ed.* **2015**, *54*, 3210–3215. [CrossRef] [PubMed]
60. Rahman, M.H.; Bhoi, P.R. An overview of non-biodegradable bioplastics. *J. Clean. Prod.* **2021**, *294*, 126218. [CrossRef]
61. Bastarrachea, L.; Dhawan, S.; Sablani, S.S. Engineering Properties of Polymeric-Based Antimicrobial Films for Food Packaging. *Food Eng. Rev.* **2011**, *3*, 79–93. [CrossRef]
62. Nosch, M.-L.; Michel, C.; Harlow, M. *Prehistoric, Ancient Near Eastern & Aegean Textiles and Dress*; Oxbow Books: Oxford, UK, 2014.
63. Barber, E.J.W. *Prehistoric Textiles: The Development of Cloth in the Neolithic and Bronze*; Princeton University Press: Princeton, NJ, USA, 1991.
64. Fisher, C.H. History of Natural Fibers. *J. Macromol. Sci. Part A-Chem.* **1981**, *15*, 1345–1375. [CrossRef]
65. Yusuf, M. *Handbook of Renewable Materials for Coloration and Finishing*; John Wiley & Sons: Hoboken, NJ, USA, 2018; ISBN 1119407842.

66. Ngo, T.-D. Natural Fibers for Sustainable Bio-Composites. In *Natural and Artificial Fiber-Reinforced Composites as Renewable Sources*; IntechOpen: London, UK, 2017; ISBN 978-1-78923-061-1.
67. Uddin, F. Introductory Chapter: Textile Manufacturing Processes. In *Textile Manufacturing Processes*; IntechOpen: London, UK, 2019; ISBN 978-1-78985-106-9.
68. Günaydin, G.K.; Yavas, A.; Avinc, O.; Soydan, A.S.; Palamutcu, S.; Şimşek, M.K.; Dündar, H.; Demirtaş, M.; Özkan, N.; Kıvılcım, N.M. Organic Cotton and Cotton Fiber Production in Turkey, Recent Developments. In *Organic Cotton*; Springer: Singapore, 2019; pp. 101–125. [CrossRef]
69. El Nemr, A. From natural to synthetic fibers. In *Textiles: Types, Uses and Production Methods*; El Nemr, A., Ed.; Nova Science Publishers: Hauppauge, NY, USA, 2012; pp. 1–152. ISBN 9781621002390.
70. Preferred Fiber and Materials Market Report (PFMR) Released!—Textile Exchange. 2020. Available online: https://textileexchange.org/2020-preferred-fiber-and-materials-market-report-pfmr-released-2/ (accessed on 3 November 2021).
71. Natural and Man-Made Fibers Overview—Chemical Economics Handbook (CEH). IHS Markit. Available online: https://ihsmarkit.com/products/fibers-chemical-economics-handbook.html (accessed on 3 November 2021).
72. Kaplan, D.; McGrath, K. *Protein-Based Materials*; Springer Science & Business Media: Boston, MA, USA, 2012; ISBN 1461240948.
73. Choudhury, A.K.R. Sustainable chemical technologies for textile production. In *Sustainable Fibres and Textiles*; Elsevier: Amsterdam, The Netherlands, 2017; pp. 267–322. ISBN 9780081020425.
74. Reay, D.; Ramshaw, C.; Harvey, A. Application Areas—Miscellaneous Process Industries. In *Process Intensification*; Butterworth-Heinemann: Oxford, UK, 2013; pp. 349–391. ISBN 978-0-08-098304-2.
75. Holkar, C.R.; Jadhav, A.J.; Pinjari, D.V.; Mahamuni, N.M.; Pandit, A.B. A critical review on textile wastewater treatments: Possible approaches. *J. Environ. Manag.* **2016**, *182*, 351–366. [CrossRef]
76. Ütebay, B.; Çelik, P.; Çay, A. Textile Wastes: Status and Perspectives. In *Waste in Textile and Leather Sectors*; IntechOpen: London, UK, 2020; pp. 39–52. ISBN 978-1-78985-244-8.
77. De Falco, F.; Di Pace, E.; Cocca, M.; Avella, M. The contribution of washing processes of synthetic clothes to microplastic pollution. *Sci. Rep.* **2019**, *9*, 6633. [CrossRef]
78. Vasile, C. *Handbook of Polyolefins*, 2nd ed.; CRC Press: Boca Raton, FL, USA, 2000; ISBN 0203908716.
79. Egan, J.; Salmon, S. Strategies and progress in synthetic textile fiber biodegradability. *SN Appl. Sci.* **2021**, *4*, 22. [CrossRef]
80. Karamanlioglu, M.; Preziosi, R.; Robson, G.D. Abiotic and biotic environmental degradation of the bioplastic polymer poly(lactic acid): A review. *Polym. Degrad. Stab.* **2017**, *137*, 122–130. [CrossRef]
81. Tokiwa, Y.; Calabia, B.P. Biodegradability and biodegradation of poly(lactide). *Appl. Microbiol. Biotechnol.* **2006**, *72*, 244–251. [CrossRef] [PubMed]
82. Suyama, T.; Tokiwa, Y.; Ouichanpagdee, P.; Kanagawa, T.; Kamagata, Y. Phylogenetic affiliation of soil bacteria that degrade aliphatic polyesters available commercially as biodegradable plastics. *Appl. Environ. Microbiol.* **1998**, *64*, 5008–5011. [CrossRef]
83. Vieira, A.C.; Vieira, J.C.; Guedes, R.M.; Marques, A.T. Degradation and Viscoelastic Properties of PLA-PCL, PGA-PCL, PDO and PGA Fibres. *Mater. Sci. Forum* **2010**, *636–637*, 825–832. [CrossRef]
84. Goynes, W.R.; Moreau, J.P.; Delucca, A.J.; Ingber, B.F. Biodeterioration of Nonwoven Fabrics. *Text. Res. J.* **1995**, *65*, 489–494. [CrossRef]
85. Zambrano, M.C.; Pawlak, J.J.; Daystar, J.; Ankeny, M.; Venditti, R.A. Impact of dyes and finishes on the aquatic biodegradability of cotton textile fibers and microfibers released on laundering clothes: Correlations between enzyme adsorption and activity and biodegradation rates. *Mar. Pollut. Bull.* **2021**, *165*, 112030. [CrossRef]
86. Zambrano, M.C.; Pawlak, J.J.; Daystar, J.; Ankeny, M.; Cheng, J.J.; Venditti, R.A. Microfibers generated from the laundering of cotton, rayon and polyester based fabrics and their aquatic biodegradation. *Mar. Pollut. Bull.* **2019**, *142*, 394–407. [CrossRef]
87. Sun, Y.; Luo, J.; Ni, A.; Bi, Y.; Yu, W. Study on Biodegradability of Wool and PLA Fibers in Natural Soil and Aqueous Medium. *Adv. Mater. Res.* **2013**, *641–642*, 82–86. [CrossRef]
88. Yang, Y.; Zhang, M.; Ju, Z.; Tam, P.Y.; Hua, T.; Younas, M.W.; Kamrul, H.; Hu, H. Poly(lactic acid) fibers, yarns and fabrics: Manufacturing, properties and applications. *Text. Res. J.* **2020**, *91*, 1641–1669. [CrossRef]
89. Yang, Q.; Dou, Y. The Strength and Elongation of PLA Fiber Yarn. *Adv. Mater. Res.* **2012**, *460*, 271–274. [CrossRef]
90. Persson, M.; Cho, S.-W.; Skrifvars, M. The effect of process variables on the properties of melt-spun poly(lactic acid) fibres for potential use as scaffold matrix materials. *J. Mater. Sci.* **2013**, *48*, 3055–3066. [CrossRef]
91. Jabbar, A.; Tausif, M.; Tahir, H.R.; Basit, A.; Bhatti, M.R.A.; Abbas, G. Polylactic acid/lyocell fibre as an eco-friendly alternative to polyethylene terephthalate/cotton fibre blended yarns and knitted fabrics. *J. Text. Inst.* **2019**, *111*, 129–138. [CrossRef]
92. Rivera-Briso, A.L.; Serrano-Aroca, Á. Poly(3-Hydroxybutyrate-co-3-Hydroxyvalerate): Enhancement Strategies for Advanced Applications. *Polymers* **2018**, *10*, 732. [CrossRef]
93. Pivsa-Art, S.; Srisawat, N.; O-Charoen, N.; Pavasupree, S.; Pivsa-Art, W. Preparation of Knitting Socks from Poly (Lactic Acid) and Poly [(R)-3-Hydroxybutyrate-co-(R)-3-Hydroxyvalerate] (PHBV) Blends for Textile Industrials. *Energy Procedia* **2011**, *9*, 589–597. [CrossRef]
94. Zhang, Z.H.; Xu, Z.Q.; Huang, X.X.; Tao, X.M. Dyeing processes of 100% bio-based and degradable polylactide/poly (hydroxybutyrate-co-hydroxyvalerate) textiles. *Text. Res. J.* **2016**, *87*, 2066–2075. [CrossRef]

95. Liu, S.; Ma, L.; Ding, X.; Wong, K.C.; Tao, X.M. Antimicrobial behavior, low-stress mechanical properties, and comfort of knitted fabrics made from poly (hydroxybutyrate-co-hydroxyvalerate)/polylactide acid filaments and cotton yarns. *Text. Res. J.* **2021**, *92*, 284–295. [CrossRef]
96. Huang, X.X.; Tao, X.M.; Zhang, Z.H.; Chen, P. Properties and performances of fabrics made from bio-based and degradable polylactide acid/poly (hydroxybutyrate- co-hydroxyvalerate) (PLA/PHBV) filament yarns. *Text. Res. J.* **2016**, *87*, 2464–2474. [CrossRef]
97. Keighley, J.H. Breathable Fabrics and Comfort in Clothing. *J. Ind. Text.* **1995**, *15*, 89–104. [CrossRef]
98. Gebben, B. A water vapor-permeable membrane from block copolymers of poly(butylene terephthalate) and polyethylene oxide. *J. Memb. Sci.* **1996**, *113*, 323–329. [CrossRef]
99. Dugan, J.S. Novel Properties of PLA Fibers. *Int. Nonwovens J.* **2001**, *33*, 5–10. [CrossRef]
100. Pinho, E.; Magalhães, L.; Henriques, M.; Oliveira, R. Antimicrobial activity assessment of textiles: Standard methods comparison. *Ann. Microbiol.* **2010**, *61*, 493–498. [CrossRef]
101. Younes, I.; Rinaudo, M.; Harding, D.; Sashiwa, H. Chitin and Chitosan Preparation from Marine Sources. Structure, Properties and Applications. *Mar. Drugs* **2015**, *13*, 1133–1174. [CrossRef] [PubMed]
102. Shirvan, A.R.; Shakeri, M.; Bashari, A. Recent advances in application of chitosan and its derivatives in functional finishing of textiles. In *The Impact and Prospects of Green Chemistry for Textile Technology;* Woodhead Publishing: Sawston, UK, 2018; pp. 107–133. ISBN 9780081024911.
103. Shahid-ul-Islam; Butola, B.S. Recent advances in chitosan polysaccharide and its derivatives in antimicrobial modification of textile materials. *Int. J. Biol. Macromol.* **2019**, *121*, 905–912. [CrossRef] [PubMed]
104. Zhang, Z.; Chen, L.; Ji, J.; Huang, Y.; Chen, D. Antibacterial Properties of Cotton Fabrics Treated with Chitosan. *Text. Res. J.* **2016**, *73*, 1103–1106. [CrossRef]
105. Balasubramaniam, B.; Prateek; Ranjan, S.; Saraf, M.; Kar, P.; Singh, S.P.; Thakur, V.K.; Singh, A.; Gupta, R.K. Antibacterial and Antiviral Functional Materials:Chemistry and Biological Activity toward Tackling COVID-19-like Pandemics. *ACS Pharmacol. Transl. Sci.* **2021**, *4*, 8. [CrossRef]
106. El-Sayed, E.; Hassabo, A.G. Recent advances in the application of plasma in textile finishing (A Review). *J. Text. Color. Polym. Sci.* **2021**, *18*, 33–43. [CrossRef]
107. Pan, H.; Zhao, T.; Xu, L.; Shen, Y.; Wang, L.; Ding, Y. Preparation of novel chitosan derivatives and applications in functional finishing of textiles. *Int. J. Biol. Macromol.* **2020**, *153*, 971–976. [CrossRef]
108. Muzaffar, S.; Abbas, M.; Siddiqua, U.H.; Arshad, M.; Tufail, A.; Ahsan, M.; Alissa, S.A.; Abubshait, S.A.; Abubshait, H.A.; Iqbal, M. Enhanced mechanical, UV protection and antimicrobial properties of cotton fabric employing nanochitosan and polyurethane based finishing. *J. Mater. Res. Technol.* **2021**, *11*, 946–956. [CrossRef]
109. Parmar, M.S.; Singh, M.; Tiwari, R.K.; Saran, S. Study on flame retardant properties of poly(lactic acid) fibre fabrics. *Indian J. Fibre Text. Res.* **2014**, *39*, 268–273.
110. Tawiah, B.; Yu, B.; Fei, B. Advances in Flame Retardant Poly(Lactic Acid). *Polymers* **2018**, *10*, 876. [CrossRef] [PubMed]
111. Bourbigot, S.; Fontaine, G. Flame retardancy of polylactide: An overview. *Polym. Chem.* **2010**, *1*, 1413–1422. [CrossRef]
112. Cayla, A.; Rault, F.; Giraud, S.; Salaün, F.; Fierro, V.; Celzard, A. PLA with Intumescent System Containing Lignin and Ammonium Polyphosphate for Flame Retardant Textile. *Polymers* **2016**, *8*, 331. [CrossRef] [PubMed]
113. Liu, Y.; Tao, Y.; Wang, B.; Li, P.; Xu, Y.J.; Jiang, Z.M.; Dong, C.H.; Zhu, P. Fully bio-based fire-safety viscose/alginate blended nonwoven fabrics: Thermal degradation behavior, flammability, and smoke suppression. *Cellulose* **2020**, *27*, 6037–6053. [CrossRef]
114. Perepelkin, K.E. Polymeric materials of the future based on renewable plant resources and biotechnologies: Fibres, films, plastics. *Fibre Chem.* **2005**, *37*, 417–430. [CrossRef]
115. Zupin, Z.; Dimitrovski, K. Mechanical Properties of Fabrics Made from Cotton and Biodegradable Yarns Bamboo, SPF, PLA in Weft. In *Woven Fabric Engineering;* IntechOpen: London, UK, 2010; ISBN 978-953-307-194-7.
116. Basit, A.; Latif, W.; Baig, S.A.; Rehman, A.; Hashim, M.; Zia Ur Rehman, M. The mechanical and comfort properties of viscose with cotton and regenerated fibers blended woven fabrics. *Mater. Sci.* **2018**, *24*, 230–235. [CrossRef]
117. Uzumcu, M.B.; Sari, B.; Oglakcioglu, N.; Kadoglu, H. An Investigation on Comfort Properties of Dyed Mulberry Silk/Cotton Blended Fabrics. *Fibers Polym.* **2019**, *20*, 2342–2347. [CrossRef]
118. Ramadan, A.M.; Mosleh, S.; Gawish, S.M. Weighting and improvement of silk properties. *J. Appl. Polym. Sci.* **2004**, *93*, 1743–1747. [CrossRef]
119. De Araújo, M. Natural and man-made fibres: Physical and mechanical properties. In *Fibrous and Composite Materials for Civil Engineering Applications;* Woodhead Publishing: Sawston, UK, 2011; pp. 3–28.
120. Saxena, S.; Raja, A.S.M.; Arputharaj, A.; Saxena, S.; Arputharaj, Á.A. Challenges in Sustainable Wet Processing of Textiles. In *Textiles and Clothing Sustainability;* Springer: Singapore, 2017; pp. 43–79. ISBN 10.1007/9789811.
121. Eid, B.M.; Ibrahim, N.A. Recent developments in sustainable finishing of cellulosic textiles employing biotechnology. *J. Clean. Prod.* **2021**, *284*, 124701. [CrossRef]
122. Radhakrishnan, S. Application of Biotechnology in the Processing of Textile Fabrics. In *Roadmap to Sustainable Textiles and Clothing. Textile Science and Clothing Technology;* Springer: Singapore, 2014; pp. 277–325.
123. Bezerra, F.M.; Lis, M.J.; Firmino, H.B.; Da Silva, J.G.D.; Valle, R.D.C.S.C.; Valle, J.A.B.; Scacchetti, F.A.P.; Tessaro, A.L. The Role of β-Cyclodextrin in the Textile Industry—Review. *Molecules* **2020**, *25*, 3624. [CrossRef]

124. Kim, H.W.; Kim, B.R.; Rhee, Y.H. Imparting durable antimicrobial properties to cotton fabrics using alginate–quaternary ammonium complex nanoparticles. *Carbohydr. Polym.* **2010**, *79*, 1057–1062. [CrossRef]
125. Mondal, M.I.H.; Saha, J.; Rahman, M.A. Functional Applications of Aloe vera on Textiles: A Review. *J. Polym. Environ.* **2021**, *29*, 993–1009. [CrossRef]
126. Liu, C.; Liang, B.; Wang, Y.; Li, Y.; Shi, G. Core–shell nanocapsules containing essential oil for textile application. *J. Appl. Polym. Sci.* **2018**, *135*, 45695. [CrossRef]
127. Kasiri, M.B.; Safapour, S. Natural dyes and antimicrobials for green treatment of textiles. *Environ. Chem. Lett.* **2014**, *12*, 1–13. [CrossRef]
128. Kadam, V.; Rani, S.; Jose, S.; Shakyawar, D.B.; Shanmugam, N. Biomaterial based shrink resist treatment of wool fabric: A sustainable technology. *Sustain. Mater. Technol.* **2021**, *29*, e00298. [CrossRef]
129. Solomon, K.R. Effects of ozone depletion and UV-B radiation on humans and the environment. *Atmosphere-Ocean* **2010**, *46*, 185–202. [CrossRef]
130. Berens, P.J.T.; Molinier, J. Formation and Recognition of UV-Induced DNA Damage within Genome Complexity. *Int. J. Mol. Sci.* **2020**, *21*, 6689. [CrossRef]
131. Dominguez-Pacheco, A.; Gomez-Uzeta, B.; Hernandez-Aguilar, C.; Cruz-Orea, A.; Cruz-San Martín, V. Optical properties of textile materials added with UV protective biomaterials. *SN Appl. Sci.* **2020**, *2*, 93. [CrossRef]
132. Tanij Tania, R. 3D Printing Technology: The Surface of Future Fashion. *Int. J. Comput. Appl.* **2017**, *157*, 975–8887. [CrossRef]
133. Pasricha, A.; Greeninger, R. Exploration of 3D printing to create zero-waste sustainable fashion notions and jewelry. *Fash. Text.* **2018**, *5*, 30. [CrossRef]
134. Sitotaw, D.B.; Ahrendt, D.; Kyosev, Y.; Kabish, A.K. Additive Manufacturing and Textiles—State-of-the-Art. *Appl. Sci.* **2020**, *10*, 5033. [CrossRef]
135. Meyer, P.; Döpke, C.; Ehrmann, A. Improving adhesion of three-dimensional printed objects on textile fabrics by polymer coating. *J. Eng. Fiber. Fabr.* **2019**, *14*, 1558925019895257. [CrossRef]
136. Eutionnat-Diffo, P.A.; Chen, Y.; Guan, J.; Cayla, A.; Campagne, C.; Nierstrasz, V. Study of the Wear Resistance of Conductive Poly Lactic Acid Monofilament 3D Printed onto Polyethylene Terephthalate Woven Materials. *Materials* **2020**, *13*, 2334. [CrossRef] [PubMed]
137. Singh, A.; Yadav, P.K.; Singh, K.; Bhaskar, J.; Kumar, A. Design of 3D Printed Fabric for Fashion and Functional Applications. In *Lecture Notes in Mechanical Engineering*; Springer: Singapore, 2021; pp. 729–735. ISBN 9789811585418.
138. Hashemi Sanatgar, R.; Campagne, C.; Nierstrasz, V. Investigation of the adhesion properties of direct 3D printing of polymers and nanocomposites on textiles: Effect of FDM printing process parameters. *Appl. Surf. Sci.* **2017**, *403*, 551–563. [CrossRef]
139. Mpofu, N.S.; Mwasiagi, J.I.; Nkiwane, L.C.; Njuguna, D. Use of regression to study the effect of fabric parameters on the adhesion of 3D printed PLA polymer onto woven fabrics. *Fash. Text.* **2019**, *6*, 24. [CrossRef]
140. Uysal, R.; Stubbs, J.B. A new method of printing multi-material textiles by fused deposition modelling (FDM). *Tekstilec* **2019**, *62*, 248–257. [CrossRef]
141. Jo, W.K.; Tayade, R.J. Recent developments in photocatalytic dye degradation upon irradiation with energy-efficient light emitting diodes. *Chinese J. Catal.* **2014**, *35*, 1781–1792. [CrossRef]
142. Javaid, R.; Qazi, U.Y. Catalytic Oxidation Process for the Degradation of Synthetic Dyes: An Overview. *Int. J. Environ. Res. Public Health* **2019**, *16*, 2066. [CrossRef]
143. Luo, Y.; Guo, W.; Ngo, H.H.; Nghiem, L.D.; Hai, F.I.; Zhang, J.; Liang, S.; Wang, X.C. A review on the occurrence of micropollutants in the aquatic environment and their fate and removal during wastewater treatment. *Sci. Total Environ.* **2014**, *473–474*, 619–641. [CrossRef]
144. Sirajudheen, P.; Poovathumkuzhi, N.C.; Vigneshwaran, S.; Chelaveettil, B.M.; Meenakshi, S. Applications of chitin and chitosan based biomaterials for the adsorptive removal of textile dyes from water—A comprehensive review. *Carbohydr. Polym.* **2021**, *273*, 118604. [CrossRef]
145. Sakkayawong, N.; Thiravetyan, P.; Nakbanpote, W. Adsorption mechanism of synthetic reactive dye wastewater by chitosan. *J. Colloid Interface Sci.* **2005**, *286*, 36–42. [CrossRef] [PubMed]
146. Mohraz, M.H.; Golbabaei, F.; Yu, I.J.; Mansournia, M.A.; Zadeh, A.S.; Dehghan, S.F. Preparation and optimization of multifunctional electrospun polyurethane/chitosan nanofibers for air pollution control applications. *Int. J. Environ. Sci. Technol.* **2019**, *16*, 681–694. [CrossRef]
147. Vakili, M.; Deng, S.; Cagnetta, G.; Wang, W.; Meng, P.; Liu, D.; Yu, G. Regeneration of chitosan-based adsorbents used in heavy metal adsorption: A review. *Sep. Purif. Technol.* **2019**, *224*, 373–387. [CrossRef]
148. Tianwei, T.; Xiaojing, H.; Weixia, D. Adsorption behaviour of metal ions on imprinted chitosan resin. *J. Chem. Technol. Biotechnol.* **2001**, *76*, 191–195. [CrossRef]
149. Shi, Y.; Zhang, Q.; Feng, L.; Xiong, Q.; Chen, J. Preparation and adsorption characters of Cu(II)-imprinted chitosan/attapulgite polymer. *Korean J. Chem. Eng.* **2014**, *31*, 821–827. [CrossRef]
150. Shawky, H.A. Synthesis of ion-imprinting chitosan/PVA crosslinked membrane for selective removal of Ag(I). *J. Appl. Polym. Sci.* **2009**, *114*, 2608–2615. [CrossRef]
151. Jiang, X.; Sun, Y.; Liu, L.; Wang, S.; Tian, X. Adsorption of C.I. Reactive Blue 19 from aqueous solutions by porous particles of the grafted chitosan. *Chem. Eng. J.* **2014**, *235*, 151–157. [CrossRef]

152. Hu, R.; Lim, J.K. Fabrication and Mechanical Properties of Completely Biodegradable Hemp Fiber Reinforced Polylactic Acid Composites. *J. Compos. Mater.* **2016**, *41*, 1655–1669. [CrossRef]
153. Yussuf, A.A.; Massoumi, I.; Hassan, A. Comparison of polylactic Acid/Kenaf and polylactic Acid/Rise husk composites: The influence of the natural fibers on the mechanical, thermal and biodegradability properties. *J. Polym. Environ.* **2010**, *18*, 422–429. [CrossRef]
154. Gibeop, N.; Lee, D.W.; Prasad, C.V.; Toru, F.; Kim, B.S.; Song, J. Il Effect of plasma treatment on mechanical properties of jute fiber/poly (lactic acid) biodegradable composites. *Adv. Compos. Mater.* **2013**, *22*, 389–399. [CrossRef]
155. Sanivada, U.K.; Mármol, G.; Brito, F.P.; Fangueiro, R. PLA Composites Reinforced with Flax and Jute Fibers—A Review of Recent Trends, Processing Parameters and Mechanical Properties. *Polymers* **2020**, *12*, 2373. [CrossRef] [PubMed]
156. Kandemir, A.; Longana, M.L.; Panzera, T.H.; del Pino, G.G.; Hamerton, I.; Eichhorn, S.J. Natural Fibres as a Sustainable Reinforcement Constituent in Aligned Discontinuous Polymer Composites Produced by the HiPerDiF Method. *Materials* **2021**, *14*, 1885. [CrossRef] [PubMed]
157. Gunti, R.; Ratna Prasad, A.V.; Gupta, A.V.S.S.K.S. Preparation and properties of successive alkali treated completely biodegradable short jute fiber reinforced PLA composites. *Polym. Compos.* **2016**, *37*, 2160–2170. [CrossRef]
158. Moyo, M.; Kanny, K.; Velmurugan, R. Performance of Kenaf Non-woven Mat/PLA Biocomposites under Medium Velocity Impact. *Fibers Polym.* **2020**, *21*, 2642–2651. [CrossRef]
159. Porras, A.; Maranon, A. Development and characterization of a laminate composite material from polylactic acid (PLA) and woven bamboo fabric. *Compos. Part B Eng.* **2012**, *43*, 2782–2788. [CrossRef]
160. Siakeng, R.; Jawaid, M.; Ariffin, H.; Sapuan, S.M.; Asim, M.; Saba, N. Natural fiber reinforced polylactic acid composites: A review. *Polym. Compos.* **2019**, *40*, 446–463. [CrossRef]
161. Kanakannavar, S.; Pitchaimani, J. Fabrication and mechanical properties of braided flax fabric polylactic acid bio-composites. *J. Text. Inst.* **2021**, 1–13. [CrossRef]
162. Chamas, A.; Moon, H.; Zheng, J.; Qiu, Y.; Tabassum, T.; Jang, J.H.; Abu-Omar, M.; Scott, S.L.; Suh, S. Degradation Rates of Plastics in the Environment. *ACS Sustain. Chem. Eng.* **2020**, *8*, 3494–3511. [CrossRef]
163. Geethamma, V.G.; Thomas, S. Diffusion of water and artificial seawater through coir fiber reinforced natural rubber composites. *Polym. Compos.* **2005**, *26*, 136–143. [CrossRef]
164. Pavlidou, S.; Papaspyrides, C.D. The effect of hygrothermal history on water sorption and interlaminar shear strength of glass/polyester composites with different interfacial strength. *Compos. Part A Appl. Sci. Manuf.* **2003**, *34*, 1117–1124. [CrossRef]
165. Srivastava, V.K. Influence of water immersion on mechanical properties of quasi-isotropic glass fibre reinforced epoxy vinylester resin composites. *Mater. Sci. Eng. A* **1999**, *263*, 56–63. [CrossRef]
166. Kumar, K.A.A.; Sreekala, M.S.; Arun, S.; Kumar, K.A.A.; Sreekala, M.S.; Arun, S. Studies on Properties of Bio-Composites from Ecoflex/Ramie Fabric-Mechanical and Barrier Properties. *J. Biomater. Nanobiotechnol.* **2012**, *3*, 396–404. [CrossRef]
167. Wu, J.; Hu, R.; Zeng, S.; Xi, W.; Huang, S.; Deng, J.; Tao, G. Flexible and Robust Biomaterial Microstructured Colored Textiles for Personal Thermoregulation. *ACS Appl. Mater. Interfaces* **2020**, *12*, 19015–19022. [CrossRef] [PubMed]
168. Zhang, P.; Lan, J.; Wang, Y.; Xiong, Z.H.; Huang, C.Z. Luminescent golden silk and fabric through in situ chemically coating pristine-silk with gold nanoclusters. *Biomaterials* **2015**, *36*, 26–32. [CrossRef] [PubMed]
169. Saffari, M.R.; Kamali Miab, R. Antibacterial property of PLA textiles coated by nano-TiO_2 through eco-friendly low-temperature plasma. *Int. J. Cloth. Sci. Technol.* **2016**, *28*, 830–840. [CrossRef]
170. Aquafil. Aquafil and Genomatica Join Forces for Bio-Nylon—Target More Sustainable Apparel, Carpets and Fibers. Available online: https://www.aquafil.com/newsmedia/aquafil-and-genomatica-join-forces-for-bio-nylon-target-more-sustainable-apparel-carpets-and-fibers/ (accessed on 23 November 2021).
171. VEGEA: Innovative Biomaterials for fashion & Design. Available online: https://www.vegeacompany.com/ (accessed on 23 November 2021).
172. Biomaterials: Sportswears Leads the Way into the New Era of Textiles. Available online: https://thesustainablemag.com/fashion/biomaterials-sportswears-leads-the-way-into-the-new-era-of-textiles/ (accessed on 23 November 2021).
173. Eastman Naia^{TM} and DuPont Biomaterials Collaborate to Launch New Sustainable Fabric Collection. Available online: https://www.eastman.com/Company/News_Center/2020/Pages/Eastman-Naia-and-DuPont-biomaterials-collaborate-sustainable-fabric.aspx (accessed on 23 November 2021).
174. Bolt Threads—News. Available online: https://boltthreads.com/news/ (accessed on 23 November 2021).

polymers

MDPI

Article

Performance Evaluation of Combined Hydrothermal-Mechanical Pretreatment of Lignocellulosic Biomass for Enzymatic Enhancement

Jiraporn Phojaroen [1], Thitirat Jiradechakorn [1], Suchata Kirdponpattara [1,2], Malinee Sriariyanun [2,3], Jatupol Junthip [4] and Santi Chuetor [1,2,*]

1 Department of Chemical Engineering, Faculty of Engineering, King Mongkut's University of Technology North Bangkok (KMUTNB), Bangkok 10800, Thailand; s6301031810028@email.kmutnb.ac.th (J.P.); s6301031820066@email.kmutnb.ac.th (T.J.); suchata.k@eng.kmutnb.ac.th (S.K.)

2 Biorefinery and Process Automation Engineering Centre (BPAEC), King Mongkut's University of Technology North Bangkok (KMUTNB), Bangkok 10800, Thailand; macintous@gmail.com

3 The Sirindhorn International Thai-German Graduate School of Engineering, King Mongkut's University of Technology North Bangkok (KMUTNB), Bangkok 10800, Thailand

4 Faculty of Science and Technology, Nakhon Ratchasima Rajabhat University, Nakhon Ratchasima 30000, Thailand; jatupol.j@nrru.ac.th

* Correspondence: santi.c@eng.kmutnb.ac.th

Abstract: Pretreatment is a crucial process in a lignocellulosic biorefinery. Corncob is typically considered as a natural renewable carbon source to produce various bio-based products. This study aimed to evaluate the performance of the hydrothermal-mechanical pretreatment of corncob for biofuels and biochemical production. Corncob was first pretreated by liquid hot water (LHW) at different temperatures (140–180 °C) and duration (30, 60 min) and then subjected to centrifugal milling to produce bio-powders. To evaluate the performance of this combined pretreatment, the energy efficiency and waste generation were investigated. The results indicated that the maximum fermentable sugars (FS) were 0.488 g/g biomass obtained by LHW at 180 °C, 30 min. In order to evaluate the performance of this combined pretreatment, the energy efficiency and waste generation were 28.3 g of FS/kWh and 7.21 kg of waste/kg FS, respectively. These obtained results indicate that the combined hydrothermal-mechanical pretreatment was an effective pretreatment process to provide high energy efficiency and low waste generation to produce biofuels. In addition, the energy efficiency and waste generation will be useful indicators for process scaling-up into the industrial scale. This combined pretreatment could be a promising pretreatment technology for the production of biofuels and biochemicals from lignocellulosic valorization.

Keywords: hydrothermal-mechanical pretreatment; agricultural wastes; energy efficiency; waste generation

Citation: Phojaroen, J.; Jiradechakorn, T.; Kirdponpattara, S.; Sriariyanun, M.; Junthip, J.; Chuetor, S. Performance Evaluation of Combined Hydrothermal-Mechanical Pretreatment of Lignocellulosic Biomass for Enzymatic Enhancement. *Polymers* **2022**, *14*, 2313. https://doi.org/10.3390/polym14122313

Academic Editor: Domenico Acierno

Received: 11 May 2022
Accepted: 3 June 2022
Published: 8 June 2022

1. Introduction

The amount of energy requirement has increased as part of a rapidly growing population, which directly causes several problems in terms of environmental sustainability and an economic crisis. The deficiency in conventional energy resources such as petroleum and natural gases has been a struggle for the sustainable development goals.

Several proven alternative energy resources have been developed to partially replace the main energy resources. Lignocellulosic biomass represents its potential for producing various value-added products including biofuels, biochemicals, and biomaterials via a biorefinery [1–4]. The utilization of lignocellulosic material as a renewable carbon source for high value-added products has been developed over the last decade [5]. The lignocellulosic biomass is typically composed of three major natural biomolecules including cellulose hemicellulose and lignin, which are associated together with complex chemical bonding to form a recalcitrant structure [6,7]. This structural complexity and recalcitrance make the proper lignocellulosic valorization become more difficult and have a low conversion yield [8–10]. However, the scientific breakthrough and research on lignocellulosic valorization have been emphasized to be more applicable on a large scale in order to reduce the environmental concerns. To maximize the use of lignocellulosic biomass, one of the most important processes in a biorefinery is the pretreatment process, which is responsible for altering and deconstructing the complex structure for the release of active components to be converted into active component biomolecules for further applications. The pretreatment process mostly affects the physical and chemical barriers of the lignocellulosic structure for its enhancement [11,12].

The existing pretreatment technologies have been divided into four categories including physical, chemical, physicochemical, and biological pretreatment. The physical pretreatment consists of alleviating the physical properties (crystallinity, degree of polymerization, and porosity) of the material by using several technologies such as mechanical size reduction and irradiation. The chemical pretreatment is mostly the use of acid and alkaline pretreatments, which are typically responsible for breaking down the chemical bonding in the lignocellulosic structure. Moreover, ionic liquid and deep eutectic solvents (DESs) have recently been developed for lignocellulosic pretreatment, which significantly enhances the delignification. The physicochemical pretreatment mostly consists of the steam explosion and liquid hot water, and the biological pretreatment consists of mainly using the enzymes of fungi for the removal of lignin [1,7].

However, these pretreatment technologies possess drawbacks such as high energy consumption, undesired waste generation, high installation costs, and long duration. The pretreatment process basically counts for 40% of the total capital cost of lignocellulosic valorization because of the high energy consumption, high chemical usage, and undesired waste generation [2,13]. Even though the existing pretreatment technologies lead to the enhanced utilization of lignocellulosic material, the single pretreatment process is inadequate to cover all of the biorefinery processes in terms of the technological, economic, and environmental considerations. Moreover, the existing biorefinery process is still developing because of the high investment cost, in particular, the pretreatment process, which generally counts for 30–40% of the total investment cost.

The recent challenges in developing the proper biorefinery processes include the reduction in investment costs, no undesired waste generation, no inhibitors for biological conversion, and a comparable production yield. Several studies have revealed that the combination of pretreatment has been effective in various aspects such as high active component recovery, less waste generation, economic feasibility, and environmentally-friendly processes.

Hydrothermal pretreatment presents an industrial potential for integration into the biorefinery process. It consists of hemicellulose solubilization and lignin removal via the autohydrolysis reaction of water at high temperature and pressure, and in particular, this process does not require the addition and recovery of chemicals different from water, and it can be said that hydrothermal processes are environmentally-friendly fragmentation processes [14,15]. Autohydrolysis is used to pretreat the lignocellulosic material to disrupt the linkage between cellulose-hemicellulose and lignin, which could enhance the enzymatic hydrolysis. Several cellulosic enzymes such as cellulase provide an effective saccharification, in particular, cellulose [16].

The hydrothermal pretreatment consists of using the liquid state of water at high pressure and temperature to form an H_3O^+ for hemicellulose solubilization and partial lignin removal [17]. The hydrothermal pretreatment is typically operated at an elevated temperature between 140–200 °C and high pressure in the range of 10–20 bar in a Parr reactor [18,19]. A number of studies on hydrothermal pretreatment indicated its efficiency to solubilize the hemicellulose component by approximately 60–80% compared to the untreated samples [20,21]. Furthermore, the hydrothermal pretreatment, along with the chemical catalysts, could achieve a high lignin removal for further enzymatic accessibility.

On the other hand, one of the most efficient physical pretreatments is mechanical size reduction, which can tackle the physical properties including cellulose crystallinity and the degree of polymerization of lignocellulosic material for further processes [22–24]. However, the mechanical size reduction process typically needs a high energy input to break down the internal structure, and this technology has its own limitations with regard to integration into the biorefinery process.

To overcome these aforementioned limitations of pretreatment technology, a combination of hydrothermal and mechanical pretreatment is an effective and promising strategy in an integrated biorefinery for the production of biofuels, biochemicals, and biomaterials [25]. The hydrothermal pretreatment, followed by mechanical size reduction, leads to the solubilization of the hemicellulose and partial lignin removal for enzymatic enhancement as well as making the material softer and more porous. Moreover, the combined process affects the crystallinity of the cellulose to be more amorphous, thus increasing the reactivity of the particles in subsequent processing [26]. This current study investigated the effect of a combined hydrothermal-mechanical pretreatment of corncob on its performance on the production of fermentable sugars. The combined process was evaluated in terms of energy efficiency and waste generation compared to its efficiency in enhancing the production yield of fermentable sugars. Two established indicators, energy efficiency and waste generation, can be used as the crucial factor for the further techno-economic analysis of the process. The process integration could define the holistic bioethanol production from corncob material. This research could lead to further development, application management, and valorization of agricultural byproducts to create a bio-circular economy and to decrease CO_2 emissions, which would consequently slow down global warming.

2. Materials and Methods

2.1. Raw Material Preparation

Corncob (CC) was obtained from farmers in Maetha District, Lampang Province, Thailand. Samples were oven-dried until the moisture content in the samples was approximately 8–10%. Samples were coarsely ground by knife milling to obtain the particle size of approximately 2–4 mm and subjected to a sieve with a screen size of 1 mm. The samples were stocked in a Ziploc plastic bag at room temperature for the further pretreatment process.

2.2. Hydrothermal Pretreatment

CC was pretreated by using a hydrothermal technique without a chemical catalyst. A sample of 30 g CC was placed in a stainless-steel Parr reactor 1 L. This pretreatment was conducted by using distilled water boiled under a pressure of 20 bar and a temperature between 140–180 °C for two different durations (30 and 60 min) and a rotation speed of 200 rpm [26]. The solid fraction was collected and washed with distilled water for neutralization and oven-dried at 60 °C to obtain a constant weight. The dried CC was stocked at room temperature for the mechanical size reduction process.

2.3. Mechanical Pretreatment

In this study, the CC samples hydrothermally pretreated at three different temperatures and two different durations were used to conduct the mechanical size reduction. The CC samples were ground by using centrifugal milling using a 0.25 mm screen size. The centrifugal milling was equipped with a digital wattmeter to measure the intensity and voltage during the grinding process.

In the case of the specific energy consumption calculation due to the mechanical size reduction, this can be solved with Equation (1)

$$E_{sp} = \int_{t_0}^{t} (P_t - P_{t_0})dt/m \quad (1)$$

where E_{sp} is the total specific energy consumption (kWh/kg); P_t is the power in watt consumed at time t; P_{t0} is the average power consumption in Watt under empty conditions (without biomass); and m is the mass in kg of material to be ground.

2.4. Monosaccharide Concentration Analysis

The monosaccharide concentration including glucose, xylose, and arabinose of the biomass was determined using the standard method for biomass analysis provided by the National Renewable Energy Laboratory (NREL), Golden, CO, USA [27]. The concentration of monosaccharides such as glucose and xylose in the soluble fraction was measured using high-performance liquid chromatography (CTO-10AS VP, Shimadzu, Kyoto, Japan) equipped with an Aminex HPX-87 H column (Bio-Rad Laboratories, Inc., Hercules, CA, USA). The column temperature was 65 °C with 0.005 M of sulfuric acid as the mobile phase at a flow rate of 0.5 mL/min. Monosaccharides of an analytical grade with a known concentration were used as the standards [12].

2.5. Enzymatic Saccharification

Enzymatic hydrolysis of the untreated and pretreated CC was performed by using commercial enzymes Cellulast CTec2 (Novozymes, Belgrave, Denmark). The reaction (5 mL total volume) contained 5% of the solid biomass sample (on a dry weight basis) with 10 FPU/g enzyme loading in 50 mM of sodium acetate buffer, pH 5 adjusted with acetic acid. Sodium azide was added at the end of the experiment to inhibit microbial growth. The reaction was incubated at 50 °C for 72 h with 200 rpm agitation [12]. The experiment was performed in triplicate. The amount of monosaccharide concentration was quantified by HPLC as described above. Moreover, the enzymatic efficiency was calculated as:

$$\text{Enzymatic Efficiency } (\%) = \frac{The\ amount\ of\ glucose\ (\text{kg})}{The\ amount\ of\ cellulose\ (\text{kg})} * 100 \quad (2)$$

2.6. Energy Efficiency and Waste Generation Evaluation

Energy efficiency was used to evaluate the performance of an integrated hydrothermal-mechanical pretreatment. It was defined by the ratio of output and input, where the input is the total energy consumption and the output is the total fermentable sugars released by enzymatic hydrolysis. The energy efficiency was calculated according to Chuetor et al. [12] as:

$$\text{Energy Efficiency}\left(\frac{\text{kg}}{\text{kWh}}\right) = \frac{Total\ amount\ of\ fermetable\ sugars\ (\text{kg})}{Total\ energy\ consumption\ (\text{kWh})} \quad (3)$$

Waste generation was used to investigate the undesired waste generated during the pretreatment process, which was defined by the ratio of total waste generation and the total amount of fermentable sugars obtained from enzymatic hydrolysis. The waste generation was calculated as:

$$\text{Waste generation}\left(\frac{\text{kg}}{\text{kg}}\right) = \frac{Total\ amount\ of waste\ generation\ (\text{kg})}{Total\ amount\ of\ fermentable\ sugars\ (\text{kg})} \quad (4)$$

where the total waste is calculated by the difference between the total reactant (biomass + water) and the total fermentable sugars.

3. Results and Discussion

3.1. Effect of Temperature and Duration on Enzymatic Hydrolysis Efficiency

Hydrothermal pretreatment of CC was conducted at different temperatures ranging from 140–180 °C with two different durations (30 and 60 min). Figure 1 illustrates the variation of the biochemical compositions, cellulose, hemicellulose, lignin, and others of each sample. It was noticed that the cellulose content relatively increased with the increasing temperature whereas the hemicellulose content decreased in a solid fraction. The maximum cellulose content was 52.90%, obtained by CC 180 °C at 60 min, which corresponded to a 100% increase compared to the native CC. Furthermore, Figure 1 indicates that the hemicellulose content trended to decrease significantly because of its solubilization during hydrothermal pretreatment. Imman et al. revealed that the hydrothermal pretreatment of corncob was effectively caused by the solubilization of hemicellulose, which was due to the hydronium ions being responsible for breaking down the cellulose–hemicellulose–lignin linkages [28,29]. Concerning the lignin content, it still remained in the solid fraction and was hard to remove during hydrothermal pretreatment, which would certainly restrict the enzymatic hydrolysis of CC due to the lignin content and its derived inhibition impact [30,31].

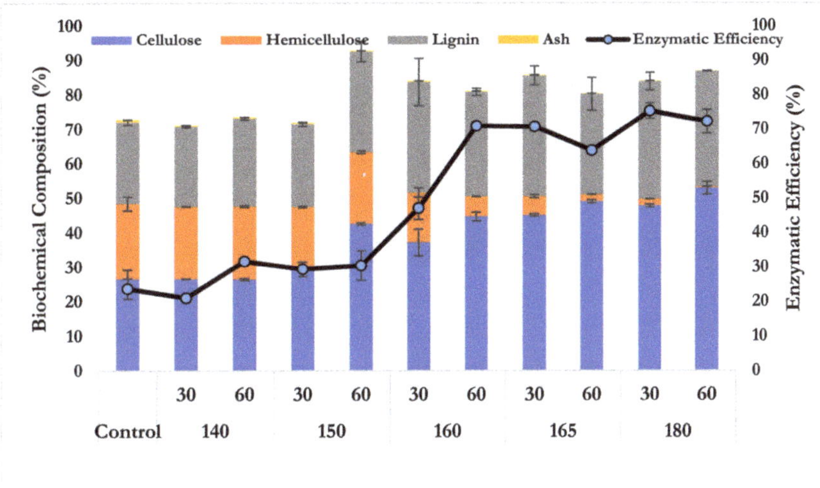

Figure 1. The biochemical composition and enzymatic hydrolysis efficiency of corncob pretreated by the hydrothermal pretreatment.

From the aforementioned results in the biochemical composition, it interestingly noticed that the hydrothermal pretreatment was effective for hemicellulose solubilization and cellulose enrichment, which was subsequently favorable for further enzymatic hydrolysis [32,33]. Figure 1 also shows that the enzymatic efficiency increased with the temperature. The highest enzymatic efficiency was 75.30% obtained by CC at 180 °C at 30 min, which corresponded to a 216% increase compared to a native CC. This large augmentation of enzymatic hydrolysis efficiency was due to the hemicellulose solubilization during hydrothermal pretreatment [34].

3.2. Evolution of Energy Consumption during the Combined Pretreatment Process

In this study, the CC was pretreated hydrothermally in the Parr reactor at three different temperatures (145, 165, and 180 °C) and two different durations (30 and 60 min), followed by mechanical size refining, and evaluated for energy consumption and enzymatic hydrolysis. The combined hydrothermal-mechanical pretreatment led to a decrease in the particle size, which subsequently increased the specific surface area for further enzymatic digestibility. Table 1 shows the amount of energy used in each process of the combined pretreatment method. The total energy consumption during the hydrothermal pretreatment increased when the temperature increased, while the energy consumption during the grinding process decreased because the CC structure had been destroyed due to the hydrothermal pretreatment process before.

Table 1. The energy consumption of corncob pretreated by combined pretreatment at different conditions.

Temperature (°C)	Time (min)	E Mechanical (kWh/kg)	E LHW (kWh/kg)	E Drying (kWh/kg)
140	30	0.532	6.628	0.649
	60	0.321	6.534	0.650
165	30	0.126	9.964	0.650
	60	0.102	9.972	0.652
180	30	0.033	16.567	0.659
	60	0.025	16.639	0.660

The highest total energy consumption was obtained from CC pretreated for 30 min at 180 °C, which corresponded to 17.8 kWh/kg of biomass. It was approximately 66% of the addition of energy compared to the lowest total energy consumption that was obtained by CC pretreated for 60 min at 165 °C, which corresponded to 10.7 kWh/kg of biomass. Interestingly, the essential energy consumption was due to the energy-consuming during hydrothermal pretreatment, which was in the range of 6.5–16.5 kWh/kg biomass. This indicates that the energy consumption of the Parr reactor increased with the increased temperature. Concerning the energy consumption due to the mechanical size reduction, the energy consumption decreased with the increased temperature that caused the structural changes during pretreatment [35,36].

3.3. Effect of Pretreatments on Enzymatic Hydrolysis

The effect of pretreatment duration on the enzymatic hydrolysis was evaluated. In this study, three different temperatures and two different durations were used to investigate the evolution of fermentable sugar concentration after pretreatment. The total fermentable sugar content was calculated to evaluate the performance of a combined hydrothermal-mechanical process. Figure 2 illustrates that when the temperature increased, the amount of fermentable sugar increased ($T_{180\,°C} > T_{165\,°C} > T_{140\,°C}$), respectively. In the case of pretreatment duration, the longer the duration in the reactor, the more fermentable sugars were released ($D_{60\ min} > D_{30\ min}$). These indicate that the temperature and duration of the pretreatment affected the internal structural deconstruction of the corncob to enzyme accessibility.

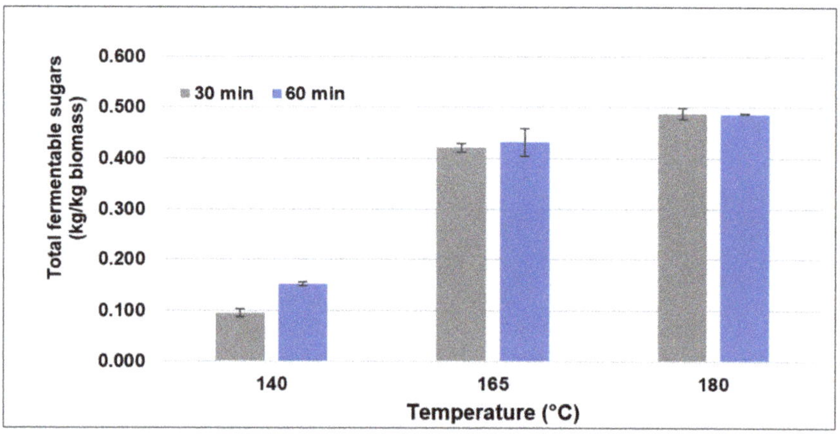

Figure 2. The total fermentable sugars from the enzymatic hydrolysis of different pretreatment conditions.

The highest fermentable sugar was 0.488 kg/kg of biomass obtained from CC pretreated at 180 °C for 30 min, which corresponded to a 356.07% increase in the fermentable sugar compared to the control (0.107 kg/kg of biomass). This increase in fermentable sugar was due to the structural deconstruction of the biomass and the removal of the hemicellulose and lignin. The high amount of fermentable sugars obtained by enzymatic hydrolysis was related to the enzymatic efficiency, as seen in Figure 1. Several studies have suggested that the augmentation of fermentable sugar concentration was due to the structural alteration through the combined pretreatment process, which is typically an important stage for lignocellulosic valorization.

3.4. *Evaluation of Energy Efficiency and Waste Generation of Combined Hydrothermal-Mechanical Pretreatment*

To evaluate the performance of the combined hydrothermal-mechanical pretreatment, the relationship between the total energy consumption and total fermentable sugars was evaluated through the estimation of the energy efficiency. The energy efficiency was used to evaluate the performance of this developed combined pretreatment to provide fermentable sugars. Energy efficiency is defined as the ratio of the total product (fermentable sugars) to the total energy consumption consumed during pretreatment. Table 2 indicates that the highest energy efficiency was obtained by the CC pretreated at 165 °C for 60 min, which corresponded to 0.041 kg of product/kWh. The effect of temperature on energy efficiency showed that the elevated temperature provided the high energy efficiency that was due to the high fermentable sugars obtained from enzymatic hydrolysis. On the other hand, the effect of pretreatment duration revealed that the longer duration released more fermentable sugars. These obtained results were relatively associated with the obtained results of the biochemical compositions and enzymatic efficiency, as seen in Table 1.

Table 2. The comparison of the energy efficiency and waste generation at different pretreatment conditions.

Temperature (°C)	Reducing Sugars (kg/kg Biomass)		Total Energy Consumption (kWh/kg Biomass)		Energy Efficiency (kg FS/kWh)		Waste Generation (kg Waste/kg Product)	
	30 min	60 min	30 min	60 min	30 min	60 min	30 min	60 min
140	0.095	0.152	17.808	17.505	0.0055	0.0088	26.71	15.34
165	0.421	0.432	10.740	10.726	0.0396	0.0406	8.42	8.9
180	0.488	0.487	17.259	17.324	0.0283	0.0282	7.21	7.23

Beyond sustainability, the combined hydrothermal-mechanical pretreatment has also been investigated for waste generation during pretreatment. The waste generation represents the total amount of waste generated during the pretreatment process, which is calculated from the ratio of the total waste generated and total fermentable sugars. The obtained results showed that the minimum waste generation was 7.21 kg of waste/kg product obtained from the CC pretreated at 180 °C at 30 min, as seen in Table 2. The waste generation was the total waste generated in solid and liquid form in the process. To compare different pretreatment conditions, the waste generation was varied between 7.2–26.7 kg of waste/kg of fermentable sugars.

4. Conclusions

The development of an economically environmentally-friendly pretreatment technology for integration into a biorefinery is important for industrial-scale production. The combined hydrothermal-mechanical pretreatment is potentially an alternative solution process for lignocellulosic valorization. The integrated hydrothermal and mechanical pretreatment could achieve the structural alteration of lignocellulosic material for enzymatic enhancement. The combined pretreatment provided a high fermentable sugar concentration compared to the untreated material. The maximum fermentable sugar was 0.488 kg/kg of biomass obtained at 180 °C at 30 min, which was due to the hemicellulose solubilization via the hydrothermal pretreatment and the increase in the specific surface area via the mechanical size reduction. The energy efficiency was used to evaluate the performance of the combined pretreatment, which could be used for further economic analysis. The highest energy efficiency was 0.0283 kg of fermentable sugars/kWh. On the other hand, the waste generation was investigated in terms of the environmental impacts of the process. The lowest waste generation was 7.21 kg of waste/kg product obtained at 180 °C at 30 min. The aforementioned results showed that the combined hydrothermal-mechanical pretreatment provided a high fermentable sugar concentration, which consequently enhanced further biofuel production.

These obtained results suggest that the combination of hydrothermal-mechanical pretreatment could be a promising pretreatment technology in terms of energy efficiency and an environmentally-friendly process for lignocellulosic valorization.

Author Contributions: Investigation, resources, formal analysis, J.P.; Investigation, resources, formal analysis, T.J.; Writing—review and editing, S.K.; Writing—review and editing, M.S.; Writing—review and editing, J.J.; Conceptualization, methodology, validation, investigation, writing—original draft preparation, writing—review and editing, visualization, supervision, funding acquisition, S.C. All authors have read and agreed to the published version of the manuscript.

Funding: This research obtained financial support from the A New Researcher Scholarship of CSTS, NSTDA, Thailand under contract NO. JRA-CO-2565-16402-TH.

Institutional Review Board Statement: Not applicable.

Informed Consent Statement: Not applicable.

Data Availability Statement: This study did not report any data.

Acknowledgments: The authors also thank the Department of Chemical Engineering, KMUTNB, and TGGS for the facilities support. We would also like to thank the BIOTEC-JGSEE Integrative Biorefinery Laboratory (IBL) for the supporting equipment used in the experiments.

Conflicts of Interest: The authors declare no conflict of interest.

References

1. Ashokkumar, V.; Venkatkarthick, R.; Jayashree, S.; Chuetor, S.; Dharmaraj, S.; Kumar, G.; Chen, W.-H.; Ngamcharussrivichai, C. Recent advances in lignocellulosic biomass for biofuels and value-added bioproducts—A critical review. *Bioresour. Technol.* **2021**, *344*, 126195. [CrossRef] [PubMed]
2. Zhao, L.; Sun, Z.F.; Zhang, C.C.; Nan, J.; Ren, N.Q.; Lee, D.J.; Chen, C. Advances in pretreatment of lignocellulosic biomass for bioenergy production: Challenges and perspectives. *Bioresour. Technol.* **2022**, *343*, 126123. [CrossRef]
3. Patel, A.; Shah, A.R. Integrated lignocellulosic biorefinery: Gateway for production of second generation ethanol and value added products. *J. Bioresour. Bioprod.* **2021**, *6*, 108–128. [CrossRef]
4. Abuhay, A.; Mengie, W.; Tesfaye, T.; Gebino, G.; Ayele, M.; Haile, A.; Yillie, D. Opportunities for New Biorefinery Products from Ethiopian Ginning Industry By-products: Current Status and Prospects. *J. Bioresour. Bioprod.* **2021**, *6*, 195–214. [CrossRef]
5. Li, H.; Liang, Y.; Li, P.; He, C. Conversion of biomass lignin to high-value polyurethane: A review. *J. Bioresour. Bioprod.* **2020**, *5*, 163–179. [CrossRef]
6. Culaba, A.B.; Mayol, A.P.; San Juan, J.L.G.; Vinoya, C.L.; Concepcion, R.S.; Bandala, A.A.; Vicerra, R.R.P.; Ubando, A.T.; Chen, W.H.; Chang, J.S. Smart sustainable biorefineries for lignocellulosic biomass. *Bioresour. Technol.* **2022**, *344*, 126215. [CrossRef]
7. Ubando, A.T.; Felix, C.B.; Chen, W.H. Biorefineries in circular bioeconomy: A comprehensive review. *Bioresour. Technol.* **2020**, *299*, 122585. [CrossRef]
8. Arpia, A.A.; Chen, W.H.; Lam, S.S.; Rousset, P.; de Luna, M.D.G. Sustainable biofuel and bioenergy production from biomass waste residues using microwave-assisted heating: A comprehensive review. *Chem. Eng. J.* **2021**, *403*, 126233. [CrossRef]
9. Kumar, R.; Tabatabaei, M.; Karimi, K.; Horváth, I.S. Recent updates on lignocellulosic biomass derived ethanol—A review. *Biofuel Res. J.* **2016**, *3*, 347–356. [CrossRef]
10. Tursi, A. A review on biomass: Importance, chemistry, classification, and conversion. *Biofuel Res. J.* **2019**, *6*, 962–979. [CrossRef]
11. Chuetor, S.; Barakat, A.; Rouau, X.; Ruiz, T. Analysis of ground rice straw with a hydro-textural approach. *Powder Technol.* **2017**, *310*, 74–79. [CrossRef]
12. Chuetor, S.; Ruiz, T.; Barakat, A.; Laosiripojana, N.; Champreda, V.; Sriariyanun, M. Evaluation of rice straw biopowder from alkaline-mechanical pretreatment by hydro-textural approach. *Bioresour. Technol.* **2021**, *323*, 124619. [CrossRef]

13. Areepak, C.; Jiradechakorn, T.; Chuetor, S.; Phalakornkule, C.; Sriariyanun, M.; Raita, M.; Champreda, V.; Laosiripojana, N. Improvement of lignocellulosic pretreatment efficiency by combined chemo–Mechanical pretreatment for energy consumption reduction and biofuel production. *Renew. Energy* **2021**, *182*, 1094–1102. [CrossRef]
14. Xu, Y.H.; Li, M.F. Hydrothermal liquefaction of lignocellulose for value-added products: Mechanism, parameter and production application. *Bioresour. Technol.* **2021**, *342*, 126035. [CrossRef]
15. Yue, P.; Hu, Y.; Tian, R.; Bian, J.; Peng, F. Hydrothermal pretreatment for the production of oligosaccharides: A review. *Bioresour. Technol.* **2022**, *343*, 126075. [CrossRef]
16. Li, Y.; Huang, X.; Lv, J.; Wang, F.; Jiang, S.; Wang, G. Enzymolysis-treated wood-derived hierarchical porous carbon for fluorescence-functionalized phase change materials. *Compos. Part B Eng.* **2022**, *234*, 109735. [CrossRef]
17. Kumar, A.; Biswas, B.; Kaur, R.; Krishna, B.B.; Bhaskar, T. Hydrothermal oxidative valorisation of lignin into functional chemicals: A review. *Bioresour. Technol.* **2021**, *342*, 126016. [CrossRef]
18. Steinbach, D.; Kruse, A.; Sauer, J. Pretreatment technologies of lignocellulosic biomass in water in view of furfural and 5-hydroxymethylfurfural production- A review. *Biomass Convers. Biorefin.* **2017**, *7*, 247–274. [CrossRef]
19. Ruiz, H.A.; Conrad, M.; Sun, S.-N.; Sanchez, A.; Rocha, G.J.M.; Romaní, A.; Castro, E.; Torres, A.; Rodríguez-Jasso, R.M.; Andrade, L.P.; et al. Engineering aspects of hydrothermal pretreatment: From batch to continuous operation, scale-up and pilot reactor under biorefinery concept. *Bioresour. Technol.* **2020**, *299*, 122685. [CrossRef]
20. Meng, X.; Yoo, C.G.; Pu, Y.; Ragauskas, A.J. Opportunities and challenges for flow-through hydrothermal pretreatment in advanced biorefineries. *Bioresour. Technol.* **2022**, *343*, 126061. [CrossRef]
21. Li, D.; Sun, Y.; Li, R.; Ao, T.; Liu, X.; Luo, Y. Selective conversion of corncob hemicellulose to xylose via hydrothermal treatment with $Fe_2(SO_4)_3$ and NaCl. *Biomass Convers. Biorefinery* **2021**, *2*, 1–10. [CrossRef]
22. Amamou, S.; Sambusiti, C.; Monlau, F.; Dubreucq, E.; Barakat, A. Mechano-Enzymatic Deconstruction with a New Enzymatic Cocktail to Enhance Enzymatic Hydrolysis and Bioethanol Fermentation of Two Macroalgae Species. *Molecules* **2018**, *23*, 174. [CrossRef]
23. Wang, G.; Qi, S.; Xia, Y.; Parvez, A.M.; Si, C.; Ni, Y. Mild One-Pot Lignocellulose Fractionation Based on Acid-Catalyzed Biphasic Water/Phenol System to Enhance Components' Processability. *ACS Sustain. Chem. Eng.* **2020**, *8*, 2772–2782. [CrossRef]
24. Vaezi, M.; Pandey, V.; Kumar, A.; Bhattacharyya, S. Lignocellulosic biomass particle shape and size distribution analysis using digital image processing for pipeline hydro-transportation. *Biosyst. Eng.* **2013**, *114*, 97–112. [CrossRef]
25. Juneja, A.; Kumar, D.; Singh, V.K.; Yadvika; Singh, V. Chemical Free Two-Step Hydrothermal Pretreatment to Improve Sugar Yields from Energy Cane. *Energies* **2020**, *13*, 5805. [CrossRef]
26. Wu, R.; Liu, W.; Li, L.; Ren, Q.; Jiang, C.; Hou, Q. Combination of hydrothermal and chemi-mechanical pretreatments to enhance enzymatic hydrolysis of poplar branches and insights on cellulase adsorption. *Bioresour. Technol.* **2021**, *342*, 126024. [CrossRef] [PubMed]
27. Sluiter, A.; Hames, B.; Ruiz, R.; Scarlata, C.; Sluiter, J.; Templeton, D.; Crocker, D. Determination of structural carbohydrates and lignin in biomass. *Lab. Anal. Proced.* **2008**, *1617*, 1–16.
28. Imman, S.; Arnthong, J.; Burapatana, V.; Champreda, V.; Laosiripojana, N. Effects of acid and alkali promoters on compressed liquid hot water pretreatment of rice straw. *Bioresour. Technol.* **2014**, *171*, 29–36. [CrossRef]
29. Imman, S.; Arnthong, J.; Burapatana, V.; Champreda, V.; Laosiripojana, N. Fractionation of rice straw by a single-step solvothermal process: Effects of solvents, acid promoters, and microwave treatment. *Renew. Energy* **2015**, *83*, 663–673. [CrossRef]
30. Phitsuwan, P.; Sakka, K.; Ratanakhanokchai, K. Improvement of lignocellulosic biomass in planta: A review of feedstocks, biomass recalcitrance, and strategic manipulation of ideal plants designed for ethanol production and processability. *Biomass Bioenergy* **2013**, *58*, 390–405. [CrossRef]
31. Heiss-Blanquet, S.; Zheng, D.; Ferreira, N.L.; Lapierre, C.; Baumberger, S. Effect of pretreatment and enzymatic hydrolysis of wheat straw on cell wall composition, hydrophobicity and cellulase adsorption. *Bioresour. Technol.* **2011**, *102*, 5938–5946. [CrossRef] [PubMed]
32. Monteiro, C.R.M.; Ávila, P.F.; Pereira, M.A.F.; Pereira, G.N.; Bordignon, S.E.; Zanella, E.; Stambuk, B.U.; de Oliveira, D.; Goldbeck, R.; Poletto, P. Hydrothermal treatment on depolymerization of hemicellulose of mango seed shell for the production of xylooligosaccharides. *Carbohydr. Polym.* **2021**, *253*, 117274. [CrossRef]
33. Sun, D.; Sun, S.C.; Wang, B.; Sun, S.F.; Shi, Q.; Zheng, L.; Wang, S.F.; Liu, S.J.; Li, M.F.; Cao, X.F.; et al. Effect of various pretreatments on improving cellulose enzymatic digestibility of tobacco stalk and the structural features of co-produced hemicelluloses. *Bioresour. Technol.* **2020**, *297*, 122471. [CrossRef] [PubMed]
34. Dimitrellos, G.; Lyberatos, G.; Antonopoulou, G. Does acid addition improve liquid hot water pretreatment of lignocellulosic biomass towards biohydrogen and biogas production? *Sustainability* **2020**, *12*, 8935. [CrossRef]
35. Bokhari, S.M.Q.; Chi, K.; Catchmark, J.M. Structural and physico-chemical characterization of industrial hemp hurd: Impacts of chemical pretreatments and mechanical refining. *Ind. Crops Prod.* **2021**, *171*, 113818. [CrossRef]
36. Jiang, J.; Wang, J.; Zhang, X.; Wolcott, M. Microstructure change in wood cell wall fracture from mechanical pretreatment and its influence on enzymatic hydrolysis. *Ind. Crops Prod.* **2017**, *97*, 498–508. [CrossRef]

polymers MDPI

Review

Progress in the Valorization of Fruit and Vegetable Wastes: Active Packaging, Biocomposites, By-Products, and Innovative Technologies Used for Bioactive Compound Extraction

Mohd Salahuddin Mohd Basri [1,2,3,*], Nor Nadiah Abdul Karim Shah [1,2], Alifdalino Sulaiman [1,2], Intan Syafinaz Mohamed Amin Tawakkal [1,2], Mohd Zuhair Mohd Nor [1,2], Siti Hajar Ariffin [1,2], Nur Hamizah Abdul Ghani [1] and Faiqa Shazeaa Mohd Salleh [1]

1 Department of Process and Food Engineering, Faculty of Engineering, University Putra Malaysia, Serdang 43400, Selangor, Malaysia; nadiahkarim@upm.edu.my (N.N.A.K.S.); alifdalino@upm.edu.my (A.S.); intanamin@upm.edu.my (I.S.M.A.T.); zuhair@upm.edu.my (M.Z.M.N.); hajarariffin@upm.edu.my (S.H.A.); nurhamizah@upm.edu.my (N.H.A.G.); faiqazea@upm.edu.my (F.S.M.S.)
2 Laboratory of Halal Science Research, Halal Products Research Institute, Universiti Putra Malaysia, Serdang 43400, Selangor, Malaysia
3 Laboratory of Biopolymer and Derivatives, Institute of Tropical Forestry and Forest Products (INTROP), Universiti Putra Malaysia, Serdang 43400, Selangor, Malaysia
* Correspondence: salahuddin@upm.edu.my

Citation: Mohd Basri, M.S.; Abdul Karim Shah, N.N.; Sulaiman, A.; Mohamed Amin Tawakkal, I.S.; Mohd Nor, M.Z.; Ariffin, S.H.; Abdul Ghani, N.H.; Mohd Salleh, F.S. Progress in the Valorization of Fruit and Vegetable Wastes: Active Packaging, Biocomposites, By-Products, and Innovative Technologies Used for Bioactive Compound Extraction. *Polymers* **2021**, *13*, 3503. https://doi.org/10.3390/polym13203503

Academic Editors: Domenico Acierno and Antonella Patti

Received: 15 September 2021
Accepted: 8 October 2021
Published: 12 October 2021

Abstract: According to the Food Wastage Footprint and Climate Change Report, about 15% of all fruits and 25% of all vegetables are wasted at the base of the food production chain. The significant losses and wastes in the fresh and processing industries is becoming a serious environmental issue, mainly due to the microbial degradation impacts. There has been a recent surge in research and innovation related to food, packaging, and pharmaceutical applications to address these problems. The underutilized wastes (seed, skin, rind, and pomace) potentially present good sources of valuable bioactive compounds, including functional nutrients, amylopectin, phytochemicals, vitamins, enzymes, dietary fibers, and oils. Fruit and vegetable wastes (FVW) are rich in nutrients and extra nutritional compounds that contribute to the development of animal feed, bioactive ingredients, and ethanol production. In the development of active packaging films, pectin and other biopolymers are commonly used. In addition, the most recent research studies dealing with FVW have enhanced the physical, mechanical, antioxidant, and antimicrobial properties of packaging and biocomposite systems. Innovative technologies that can be used for sensitive bioactive compound extraction and fortification will be crucial in valorizing FVW completely; thus, this article aims to report the progress made in terms of the valorization of FVW and to emphasize the applications of FVW in active packaging and biocomposites, their by-products, and the innovative technologies (both thermal and non-thermal) that can be used for bioactive compounds extraction.

Keywords: fruit waste; vegetable waste; waste valorization; bioactive compound; active packaging; biocomposites; by-product; extraction; thermal processing; non-thermal processing

1. Introduction

Food waste (FW) is already acknowledged as a major global issue that threatens the long-term viability of the food supply chain [1]. FW in the European Union is estimated to be at 89 million tonnes per year. This is expected to rise by 40% in the next four years [2]. According to the Food and Agriculture Organization of the United Nations [3] and the International Food Policy Research Institute [4] estimated one-third of food produced globally, or 1.3 billion tonnes, is thrown away each year. Over half is generated at the final consumption stage in food services (e.g., restaurants, school canteens) and households [5,6]. The Sustainable Development Goals (SDGs) recognize the importance of this issue. By 2030, SDG Target 12.3 calls for the reduction by half of the per-capita global FW at retail and consumer levels, diminishing food losses in production and supply chains, including post-harvest losses [7]. The call was agreed to by all 193 UN member states, which comes as no surprise.

Every year, more than 1748 million tonnes of fruit and vegetable waste (FVW) are produced worldwide [8]. Due to rapid economic growth, FW generation in Asia is predicted to increase by 278 to 416 million tonnes [9], resulting in an increase in world anthropogenic (carbon) emissions [10]. Malaysia is expected to produce 6.7 million tonnes of FW per year by 2020 [11]. In the EU, households account for over half of all FW [12]. Around 12 percent of the total EU food production is wasted in the United Kingdom [13], while the United States recorded the highest FW rates per capita at 278 kg [14]. FVW is a major component of FW, particularly in industrialized countries [15,16]. Figure 1 shows the FW management practices and the novel or emerging valorization approaches [17].

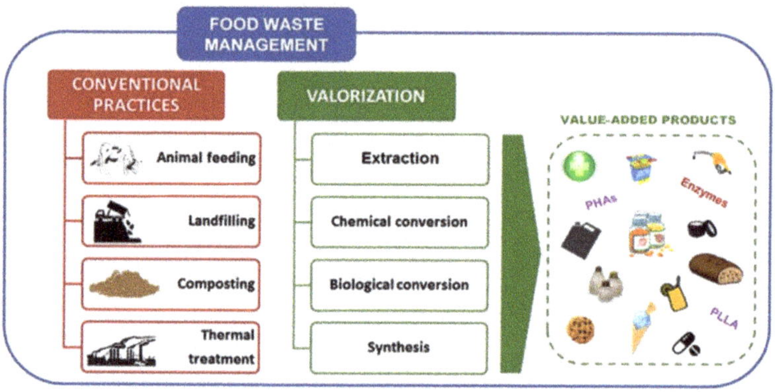

Figure 1. Food waste management practices and the novel or emerging valorization methods. Reproduced with permission from Ref. [17]. Copyright 2021 Elsevier.

According to Fabi et al. [18], median losses are estimated to be higher than 10 percent in Africa and Latin America for fruits and vegetables. In comparison, they range between 4 and 7 percent in Europe and North America. Figure 2 depicts the amounts of fruits produced globally, including 124.73 million metric tonnes (MMT) of citrus, 114.08 MMT of bananas, 84.63 MMT of apples, 74.49 MMT of grapes, 45.22 MMT of mangoes, mangosteens, and guavas, and 25.43 MMT of pineapples. Besides the production of potatoes, which was recorded at a considerable 3820.00 MMT, other vegetables are also shown in Figure 2, including tomatoes (171.00 MMT), cabbages and other brassicas (71.77 MMT), carrots and turnips (38.83 MMT), cauliflowers, broccolis (24.17 MMT), and peas (17.42 MMT) [19]. Due to losses in sales and transportation costs, such wastage, including subsequent generation of FVW, raises market operation costs, resulting in higher inflation.

The dairy, meat, fishery, and seafood processing sectors are the primary contributors of trash from animal wastes. Numerous types of residues can be identified from vegetable wastes, including grains, roots and tubers, oil crops and pulses, and fruits and vegetables, depending on the source [20]. Typically, FVW is characterized by a high water content and a high concentration of biodegradable organic substances (e.g., carbohydrates, lipids, and organic acids) [21]. As a result of these varied organic wastes, typical solid waste management systems (such as landfills and incineration) may cause serious environmental impacts, such as the discharge of leachate and the production of greenhouse gases [22].

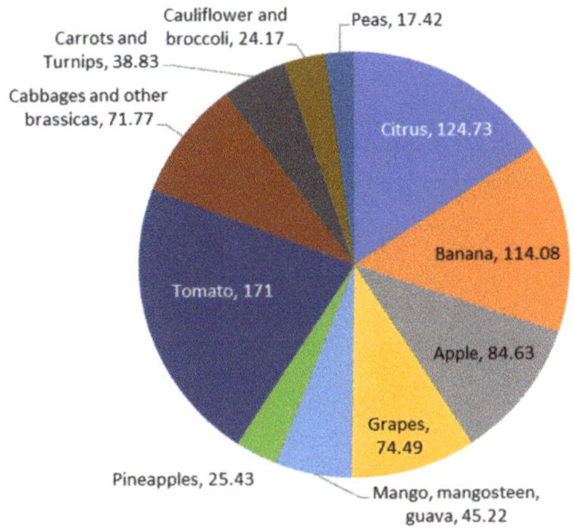

Figure 2. A significant amount of fruits and vegetables (in MMT) are produced globally.

FVW is generated in large quantities in open markets, but little information is available on the actual volume of waste generated. Since significant amounts of waste are produced by the fruit and vegetable value chains, biorefinery concepts have been developed to valorize these wastes [23]. Waste from the FVW can be reused or recycled, which is more environmentally beneficial than disposal through open dumpsites or incineration [24]. Wholesale marketplaces, supermarkets, and agricultural activities are the traditional sources of FVW. Fruits and vegetables produce FVW at every stage of the chain, including production, transportation, storage, distribution, and consumption [25]. The valorization of FVW and potential products are shown in Figure 3.

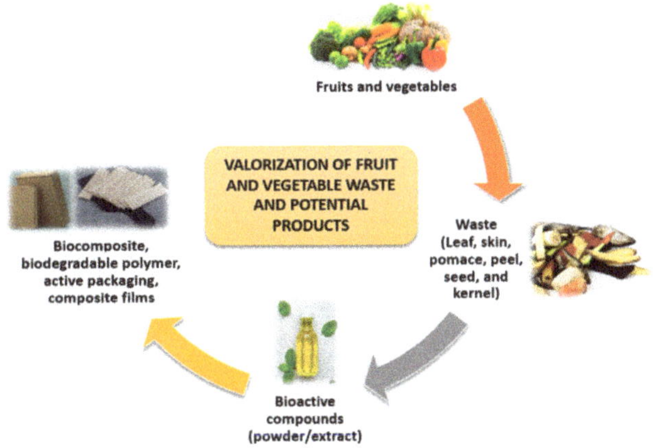

Figure 3. Valorization of FVW and potential products.

There are several reviews on the applications of fruit and vegetable wastes. A review conducted by Bayram et al. [26] focused on the potential applications of fruit- and vegetable-based by-products such as biopolymers, biocomposites, active or intelligent packaging, and edible films and coatings. The authors discussed the advantages, disadvantages, and applications of these by-products as well. Kadzińska et al. [27] reviewed the role of specific chemical compounds found in fruits and vegetables, particularly in designing the physicochemical and functional properties of edible packaging materials. The advantages and disadvantages of using fruit and vegetables as components in matrix-forming solutions and their potential applications, future trends, and issues to consider when commercializing these products were discussed within the context of sustainable development.

In the fruit and vegetable processing industry, by-products such as peels, seeds, and shells are produced in large quantities [28,29]. These by-products contain high concentration of bioactive components such as antioxidants (polyphenols and dietary fibers), pigments and flavor compounds, proteins, essential oils, enzymes, and dietary fibers [30]. Coman et al. [31] looked at the bioactive potential of fruit and vegetable by-products and how they could be used in the food industry (functional foods) or the health sector (nutraceuticals). Several applications of food vegetable waste incorporated into the meat and its derivatives were reviewed by Calderón-Oliver and López-Hernández [32]. The peels and seeds of fruits, such as grapes, pomegranates, avocados, and citrus, are the most commonly used vegetable by-products because they help inhibit oxidation (lipid and protein) and the growth of pathogenic and deteriorating bacteria in the food supply. Adding these by-products to meat products can sometimes improve the quality of the product while also extending its shelf life. In food processing of waste and biomass by-products, pectin is one of the most common constituents; therefore, improving pectin extraction and recovery is critical to completely valorize these significant feedstock resources [29]. As a result, it is crucial to investigate the composting process of FW, particularly FVW, within this framework.

Given the potential for the valorization of such organic waste, research on the proper use of waste materials obtained from horticulture commodities may generate sustainable development initiatives to minimize environmental problems [19]. Significant products can be produced from these organic wastes, including active packaging, biopolymers, biocomposites, and other by-products. This review also explores the innovative technology (thermal and non-thermal) for bioactive compounds extraction based on the natural resources from fruit and vegetable losses and waste.

2. Active Packaging and pH Indicator Film

2.1. Active Packaging

The integration of additives into the packaging system to maintain or enhance the quality of food products and shelf life is referred to as active packaging [33]. Active packaging concepts include moisture absorbers, gas scavengers, carbon dioxide emitters, antioxidant, and antimicrobial-releasing and-containing systems. For example, the addition of antimicrobial additives into the packaging materials can minimize the risk of food spoilage and contamination from microorganisms [34]. Several additives have been successfully incorporated into packaging materials, including organic acids and their salts, bacteriocins, enzymes, chelators, and a range of plant extracts [35–37]. Due to concern about the health hazards of synthetic additives, researchers have been implementing various natural plant extracts to the biopolymers as active components [38]. In addition, active ingredients from FVW have added another level to active packaging systems as a solution for reducing FW. Several studies have focused on utilizing these wastes to make active films and investigated their properties, as summarized in Table 1.

For example, Luchese et al. [39] added blueberry pomace into cassava starch films. The aromatic compounds in blueberry pomace improved the light barrier properties, indicating the films' ability to preserve food against UV lights. At the same time, the films were structurally stable when immersed in water for more than 24 h. The authors suggested the feasibility of the films for packaging aqueous food products.

In another study, fiber and ethanolic extracts from blueberry juice from processing waste were used to make active films from gelatin capsules wastes [40]. The study found that the films with fiber showed reduced tensile strength and increased water vapor permeability with improved light barrier activity against UV light, which effectively reduced the lipid oxidation of sunflower oil. Both films showed significant decreases in light transmission and stability in antioxidant activity over 28 days.

Table 1. Utilization of fruit and vegetable wastes in food packaging systems.

Film Type	Polymers	Fruit/Vegetable Waste	Applications	Findings	Ref.
Active film	Hydroxypropyl high-amylose starch	Pomegranate peel (PGP)	-	PGP inhibited the growth of *S. aureus* and *Salmonella* bacteria.	[34]
Active film	Cassava starch	Blueberry pomace (BP)	-	BP improved the barrier properties and released active compounds into acidic medium.	[39]
Active film	Gelatin capsule waste	Fiber and ethanolic extract from blueberry juice processing waste	Sunflower oil	Improved light barrier properties; reduced the lipid oxidation of sunflower oil and stability in antioxidant activity.	[40]
Active film	Tapioca starch thymol	Jackfruit skin and straw as filler	Cherry tomato	Jackfruit skin and straw resulted in higher tensile strength with lower water solubility and water vapor permeability.	[41]
Active packaging	Poly (butylene adipate-co-terephthalate) (PBAT) Cinnamon essential oil	Cellulose nanofibers (CNF)	Strawberry	High thermal stability, decreased water vapor permeability.	[37]
Active packaging	Chitosan	Apricot kernel essential oil (AKEO)	Bread slices	AKEO improved water resistance, water vapor barrier, and mechanical properties. Inhibited *E. coli*, *B. subtilis* and fungal growth.	[42]
Biodegradable film	Tapioca starch	Banana pseudostems (BP)	-	BP reduced the mechanical and optical properties, improved the barrier properties.	[43]
Active film	Mango kernel fat (MKF), phenolic extract from mango kernel (MKPE), mango kernel starch (MKS),	Mango kernel	-	Films exhibited antioxidant activity, UV-absorbing capacity, and good barrier properties.	[44]
Biodegradable film	Pectin, sodium carboxymethyl cellulose (CMC−Na) Thyme essential oil	Okara soluble dietary fiber	-	Higher mechanical, barrier, optical, and antioxidant properties. Antimicrobial activity against *E. coli* and *S. aureus* was not significant.	[45]
Active edible film	Basil seed gum *Zataria multiflora* essential oil	Basil seed gum (BSG)	-	Efficient antimicrobial activity against *E. coli* and *B. cereus*.	[46]

Ali et al. [34] utilized pomegranate peel as a filler and an antimicrobial agent in developing films with hydroxypropyl high-amylose starch. The results showed that the tensile properties of the films improved and the pomegranate peel inhibited the growth of both Gram-positive (*S. aureus*) and Gram-negative (*Salmonella*) bacteria. On the other hand, research done by dos Santos Caetano et al. [47] produced biodegradable films based on minimally processed pumpkin residue extract (PRE) (0 to 6%) incorporated with cassava starch, glycerol, and oregano essential oil (OEO) (0 to 2%). The addition of pumpkin residue is crucial, since it provides opacity to the films, but it was not effective in improving the antioxidant and antimicrobial properties.

Other researchers also used pomegranate peel as an antibacterial additive with sodium caseinate powder [48]. The peel resulted in a decrease in tensile strength and increased water vapor permeability (WVP). The films showed profound growth inhibition for Gram-positive (*S. aureus*) rather than Gram-negative (*E. coli*) bacteria. Shukor et al. [41] prepared tapioca-starch-based films incorporating thymol, jackfruit skin, and straw. Improvement in mechanical and barrier properties was observed for the films. The incorporation of skin and thymol retarded bacterial and fungal growth due to the antimicrobial activity of the films.

2.2. pH Indicator Films

Researchers are currently attracted to intelligent packaging systems, whereby active ingredients such as dyes and pigments are added into the film as pH indicators to trace and monitor food freshness throughout the storage period. This is related to the interactions between the food and its environment and the packaging material [49]. Table 2 shows some of the fruit and vegetables used as pH indicators in packaging films.

Table 2. Utilization of fruit and vegetable wastes as pH indicator in packaging films.

Polymers	Fruit/Vegetable Waste	Applications	Findings	Ref.
Cassava starch	Blueberry residue (BR) powder, two particle sizes	Distilled water, sucrose, sodium chloride, soybean protein, milk protein, whole milk powder, orange juice, corn oil, chicken meat	BR films displayed changes in color responding to the pH of tested samples.	[50]
Corn starch	Blueberry juice processing waste	-	The blueberry residue film showed visual color differences in different pH ranges.	[51]
Sweet potato starch	Sweet potato peel	Chicken flesh	SPP reduced the mechanical and barrier properties, did not influence the color response of the films.	[52]

The colorimetric pH indicator films display apparent color changes with alterations of the food pH due to food deterioration and extrinsic environmental changes [53]; thus, from these color changes, the consumers receive authentic information regarding the food's quality and its edibility, which is not achievable from the expiry date written on the package. As a pH indicator dye replacing synthetic pigments, natural dyes are now used mainly in biodegradable packaging materials [54,55]. Interestingly, the natural dyes and pigments from waste and by-products from fruits and vegetables have also been implemented as pH sensing dyes into packaging films to ensure food safety [56].

A study was conducted by developing colorimetric indicator film from mulberry based on gelatin and polyvinyl alcohol (PVA), whereby anthocyanin extract from the residue of mulberry processing was incorporated. This study showed that the mechanical properties were improved and visible color changes were shown when monitoring fish spoilage [57]. In addition, an earlier study produced films from cassava starch and blueberry residue that were rich in anthocyanin. Insertion of blueberry residue produced less compact films with high oxygen permeability. The films exhibited visual color changes in the pH range of 2 to 12 [56]. Luchese et al. [50] developed biodegradable pH indicator films based on cassava starch, utilizing the blueberry residue obtained after juice processing. The researchers deliberated on two different particle sizes for the blueberry residue powder for film preparation. They found that the films with smaller particles were more uniform and homogenous in appearance and the color change with pH was more intense than films with large particles.

In addition, the tensile strength of the films decreased whereas elongation increased; simultaneously, the water vapor permeability of the films increased due to the presence of the particles and their heterogeneity. In another study, blueberry agro-industrial waste, a co-product from juice processing, was successfully used to develop pH indicator films based on corn starch [51]. The films changed their colors in response to different pH, and the color changes were visually perceptible to the human eye.

Black chokeberry (*Aronia melanocarpa*) pomace extract (a residue material after juice pressing) was chosen as a pH sensing dye of chitosan films. The addition of pomace extract reduced the solubility and swelling of chitosan, and these indicator films maintained integrity in acidic pH environments [58]. Sohany et al. [52] utilized sweet potato peel powder (SPP) as a filler to develop sweet potato starch (SPS)-based pH indicator films incorporating purple sweet potato anthocyanin. The films with peel exhibited reduced tensile strength with higher swelling and WVP values. Visually, the films were dark maroon and changed their color in response to various pH buffers, as shown in Figure 4. The films successfully indicated chicken freshness by changing their color with changes in pH of the deteriorated chicken.

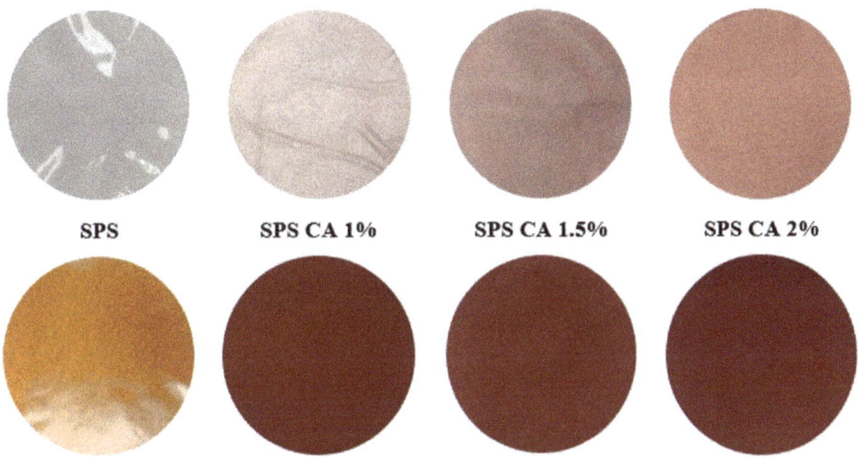

Figure 4. Visual appearance of SPS and SPS/SPP films at CA of 0, 1, 1.5, and 2%. Reproduced with permission from Ref. [52]. Copyright 2021 Taylor & Francis.

3. Biocomposites

Several biocomposites products are produced from agro-industrial wastes whereby the fibs, proteins, carbohydrates, organic acids, and oils are extracted from the wastes, followed by fermentation and enzymatic processing [59]. Biocomposites produced from natural fillers/fibers and biodegradable plastics are examples of biodegradable materials. Starch can be obtained through the extraction process [60], whereas polylactic acid (PLA), polybutylene succinate (PBS), and polyethylene (PE) are produced by fermenting, chemical processing, and polymerizing their monomers [61,62]. The polyhydroxyalkanoate (PHAs) are synthesized by bacterial fermentation [63]. The carboxymethyl cellulose (CMC) can be produced via etherification [64]. Starches are partially modified to make biopolymer; PLA is widely used in food packaging; PHAs used for water-resistant packaging and injection molding [65].

Further, the peel/skin, seeds, and pomace of fruits and vegetables contain fibers used as filler in the film matrix to improve the biopolymer properties [39,52]. Pectin is another compound extracted from wastes and used in the polymeric matrix [35]. Also, peel and seeds contain more phenolic compounds than their fleshy parts [26]. For example, mango peel has a high level of phenolic content compared to its flesh [66]. The presence of phytochemicals and bioactive compounds provide preservative effects such as antimicrobial activity with anti-inflammatory and antioxidant attributes [67]. The blending of additives to polymers is found to improve the mechanical and barrier properties. Depending on the presence of additives, the polymers also function as pH indicators (sensor) or antimicrobial films [34,39]. Moreover, abundant organic wastes are available in skin and pulp, including citrus fruits such as orange, grapefruit, pineapple, mandarin/tangerine, lemon, and lime; seed waste from mango, grape, and pumpkin; skin from potato, sweet potato, jackfruit, pomegranate, and banana [65]. The utilization of these wastes into packing materials offers an essential alternative to conventional plastic packaging and contributes to a sustainable environment.

Despite substantial research on natural fiber composites, little is known about incorporating FW into biocomposites. Most biocomposites from FW research have focused on the biomass such as olive, pineapple, and banana. The mechanical and thermal properties of composites made from these biomasses and from other fruits are thoroughly investigated. Fiber treatment, type of polymer matrix, amount of fiber, compatibilizer, and process techniques have significant impact on the properties of biocomposites. Table 3 summarizes the residues from fruits and vegetable waste used in the production of biocomposites.

Table 3. Residues derived from fruit and vegetable waste.

Residue	Fruit/Vegetable	Matrix	Properties	Ref.
Cellulose nanofibrils (CCNF)	Carrot pomace	PLA	Mechanical, hydrophilic, thermal, and antibacterial	[36]
Nanofiber	Durian skin	PLA	Tensile strength	[68]
Fiber	Banana peel	PP	Sound insulation	[69]
Fiber	Durian skin	PLA	Production energy	[70]
Fiber	Durian skin	PLA	Tensile strength, modulus of elasticity, and enzymatic degradation	[71]
Fiber	Sea mango peel	PP	Flexural strength, flexural modulus, and thermal	[72]
Pomace extract	Chokeberry, blackcurrant, apple, and raspberry	Rapeseed meal, microcrystalline cellulose	Flexural strength and water contact angle	[73]
Extract	Coconut shell	Polyvinyl alcohol (PVA) and corn starch	Antioxidant activity and thermal	[74]
Extract	Date fruit	Gelatin	Moisture content and water solubility	[75]
Powder	Cocopith	Wood powder, tapioca	Thermal conductivity	[76]
Powder	Date and tamarind seed	PLA	Tensile, flexural, and impact strength	[77]
Powder	Grape and acerola	Cassava starch	Antioxidant, physicochemical, and mechanical	[78]
Powder and fiber	Jackfruit skin	PLA	Tensile strength and tensile modulus	[79]
Powder	Sour cherry shell	PE	Elastic modulus, tensile strength, moisture absorption, and water vapor transmission rate	[80]
Husks	Chestnut	Starch	Elastic modulus and tensile strength	[81]

3.1. Lignocellulosic Fiber

Fibers extracted from agricultural residues such as fruit and vegetable waste, woodland residues, and farming deposits are rich in cellulose, hemicellulose, lignin and are termed lignocellulose. These lignocellulosic fibers are obtained from biosources such as bast, foliage, fruit, kernel, timber, farmed excess, lawn, etc. Natural fibers possess comparable or even better mechanical properties like glass or aramid fibers [82]. Nanocomposite films based on lemon waste, 3% cellulose nanofiber (CNF), and 3% savory essential oil (SEO) are fabricated and are shown to enhance the barrier and mechanical properties. Film from lemon waste showed antibacterial properties against five foodborne pathogens [83].

Szymańska-Chargot et al. [36] evaluated the mechanical, hydrophilic, thermal, and antibacterial properties of nanocomposite made of PLA and nanocellulose. The nanocellulose is a carrot CCNF isolated from carrot pomace modified with silver nanoparticles. The nanocellulose modified with metal nanoparticles at a concentration of 0.25 and 2 mM was prepared earlier before combining with PLA. Composite containing CCNF with 2 mM of AgNPs showed the most significant improvement in mechanical properties. The degradation temperature was lower for PLA with CCNF/AgNPs, and this addition also increased hydrophilicity. The addition also improved transmission rates of oxygen, nitrogen, and carbon dioxide. It also acquired antibacterial function against *Escherichia coli* and *Bacillus cereus*, suggesting the lack of migration of nanoparticles from the composite.

Mohd Nordin et al. [68] studied the effect of freeze-dried durian skin nanofiber on the physical properties of PLA biocomposites. Durian skin nanofiber (DSNF) was developed using a freeze-drying (FD) process from durian skin fiber (DSF), and cinnamon essential oil was added as a plasticizer for PLA biocomposites. The tensile strength of these composites showed significant changes in the presence of DSF and DSNF in PLA.

Fibers from bananas have the potential to be incorporated into sound insulation composites. Singh and Mukhopadhyay [69] studied the effect of hybridization on sound insulation of coir-banana-PE hybrid biocomposites. These were prepared as shown in Figure 5 with chopped and randomly oriented coir and banana fibers. PP was used as a matrix, and a compression molding technique for composite fabrication. Hybrid and nonhybrid composites from coir and banana fibers were prepared at total fiber loading of 5, 10, 15, 20, and 25% by volume. The ratio of both fibers in hybrid biocomposites was maintained at 1:1. It was found that an increase in fiber loading considerably improved sound insulation up to a certain limit. The difference in transmission loss at the minimum and maximum fiber loadings for nonhybrid was higher for the finer banana fibers.

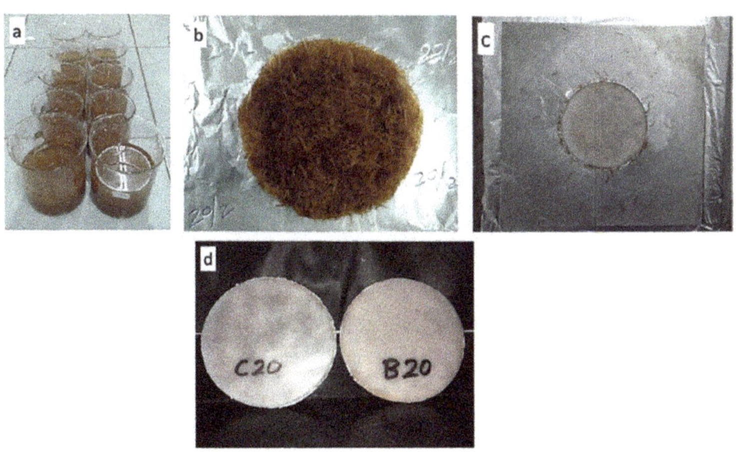

Figure 5. The process includes (**a**) fiber treatment followed by layering of fibers to form (**b**) fibrous bed, composite sample preparation through (**c**) compression molding, and (**d**) final composite samples. Reproduced with permission from Ref. [69]. Copyright 2021 Taylor & Francis.

Anuar et al. [70] developed durian skin fiber (DSF)-reinforced PLA biocomposites with the addition of epoxidized palm oil (EPO). The amount of energy required for the production of the biocomposites was studied. The results showed that the PLA/DSF/EPO biocomposites had lower negative impacts as compared to the PLA/DSF biocomposites because the EPO improved the workability and processability of the biocomposites. They concluded that the plasticized PLA/DSF biocomposites could be potential biodegradable food packaging materials, as they have acceptable properties and produce no waste.

Gisan et al. [71] investigated the tensile, water absorption, and biodegradation properties of PLA/durian husk fiber (DHF) biofilms. The PLA/DHF biofilms with different DHF contents (0, 5, 10, 15, and 20 wt.%) were prepared via a simple solvent casting method. The results revealed that the tensile strength and modulus of elasticity of the biofilms increased with increasing DHF content from 5 wt.% to 10 wt.%. The tensile strength and modulus of elasticity of the PLA/DHF biofilms decreased compared to the neat PLA film due to the plasticized effect in the biofilms; however, the enzymatic degradation with α-amylase and the water absorption properties of the PLA/DHF biofilms increased with the DHF content.

Sea mango (SM) fiber was used as a filler in PP polymer biocomposites. Ong et al. [72] investigated the flexural and thermal properties of 5 to 25 weight percentages of SM added into the PP. The results showed an improvement in the flexural strength and stiffness when the SM content increased. The thermal stability and degree of crystallinity results were positive when a compatibilizer (such as PP-g-MA) was incorporated into the biocomposites.

3.2. Extract

Several natural extracts have been used as active additives to develop antioxidant-enriched films for food packaging applications. Natural antioxidants of plant extracts (PE) derived from various non-edible portions of fruit and vegetable by-products, such as peels, stones, and seed extracts, often contain a high amount of phenolic substances [84] and have been used as active ingredients in the manufacture of active films.

The valorization of fruit pomace (chokeberry, blackcurrant, apple, and raspberry pomace) in biocomposites was achieved by Żelaziński [73]. The mechanical and physicochemical characteristics were studied. The results showed that adding 30% chokeberry, apple, raspberry, and currant pomace substantially contributed to the improvements in flexural strength (between 11.1 and 12.3 MPa) and the increase of the water contact angle of the surface by 40%.

Tanwar et al. [74] investigated the characteristics of PVA starch incorporated with coconut shell extract and sepiolite clay as an antioxidant film for active food packaging applications. An active antioxidant film was fabricated using polyvinyl alcohol (PVA) and corn starch (ST) and incorporated with 3, 5, 10, or 20% (*v*/*v*) coconut shell extract (CSE) and sepiolite clay (SP) for the first time. It was found that the addition of CSE to films enhanced their antioxidant activity properties by up to 80%. Further, increasing the amount of CSE resulted in color changes in the active films and improved their thermal properties.

Rangaraj et al. [75] investigated the effects of date fruit syrup waste extract (DSWE) on the physical properties of gelatin films. The results showed that the loading of DSWE did not affect the thickness of the material. The moisture content and water solubility, on the other hand, increased with an increase in DSWE from 5 to 25 wt.%. PE/sour cherry shell powder biocomposite was investigated as a potential food packaging by Farhadi and Javanmard [80]. The addition of 2.5% sour cherry shell increased the elastic modulus, tensile strength, and mechanical properties of the composite. The increased sour cherry shell powder loading from 2.5 to 7.5 wt.% increased the moisture absorption and water vapor transmission.

Grapefruit and pomelo are commonly consumed fruits with higher levels of essential oils in the peels than other fruits [85]. Previous studies have reported that 10% extract of *Citrus paradisi* (grapefruit) peel obtained by microwave-assisted extraction (MAE) incorporated with multilayer low-density polyethylene (LDPE)/polyethylene terephthalate (PET) showed high antioxidant levels and acts as a free radical scavenger [86]. Pumpkins seeds and peels are the waste types generated from the processing industry, with high potential for utilization as biodegradable films. Defatted pumpkin seeds (DPS) and pumpkins peels (PP) with glycerol and lecithin are produced in the process. The films exhibited the highest tensile strength values (1401 ± 5.4 kPa) and good elongation (9.74 ± 0.46%). The properties of the films were improved using ultrasound treatment. Generally, waste from the pumpkin processing industry is successfully developed as biodegradable films [87].

3.3. Powder and Husk

The incorporation of pomegranate (PMG), papaya (PPY), and jackfruit (JF) peel into gelatin/PE bilayer films led to significant increases in thickness, opacity, and moisture content ($p < 0.05$) but reduced film solubility in water. Films incorporated with pomegranate (PMG) exhibited high antimicrobial and antioxidant properties [88].

Mardijanti et al. [76] examined the material characteristics of cocopith and evaluated the potential of a mycelium-based biocomposite as an insulator. Dry cocopith, which is the residue from the coconut coir milling industry, was mixed with wood powder, pollard (bran), lime, tapioca, and Ganoderma mushroom seeds and then put into baglogs. The solid baglogs were then removed from the molds, dried, and compacted to the desired size using a hot press. The potential as an insulator was validated via a thermal conductivity test at temperatures of 13 to 40 °C. The test showed a thermal conductivity value range of 0.0887241 to 0.002964 W/mK. A value ranges of 0.01 to 1.00 W/mK is recommended for thermal conductivity insulators.

A study on the enhancement of the mechanical behavior of a PLA matrix using tamarind and date seed micro fillers was conducted by Nagarjun et al. [77]. The composites were manufactured using the compression molding technique. The tensile results showed that the seed filler reinforcement significantly improved the tensile strength of the PLA matrix. The maximum tensile strength values achieved with TI/PLA and PD/PLA were 72.42 MPa and 61.39 MPa, respectively. The particulate reinforcements of both tamarind and date almost doubled the flexural and impact strengths of the PLA matrix. Moreover, the date seed powder-incorporated composites showed a 34.68% improvement in microhardness. The uniform dispersion of the filler was evident in TI/PLA and PD/PLA with 2 wt.% filler, which contributed to their better tensile strength. Conversely, increasing the filler content to 4 wt.% resulted in agglomeration of the fillers and subsequently contributed to the low mechanical strength of the composites.

Reinaldo et al. [78] investigated the effects of grape and acerola residues on the antioxidant, physicochemical, and mechanical properties of cassava starch biocomposites. Various concentrations of grape skin (Gr) and acerola (Ac) residues (0.1, 1.0, 5.0, and 10.0 wt.%) were prepared using extrusion and injection molding processes. The large size distribution of cassava starch may favor the plasticization stage to obtain TPS compared to the different types of starch obtained from other plant sources. The results showed that the addition of grape skins and acerola residues to the cassava thermoplastic starch resulted in better antioxidant characteristics. The addition of grape residue in TPS resulted in decreased in elongation at the break compared with pure TPS, which was more significant with higher concentrations of grape residue (5.0 and 10.0 wt.% of Gr). The similar mechanical behavior was recorded by Gutiérrez et al. [89] and Deng and Zhao [90].

Marzuki et al. [79] studied the effects of jackfruit skin powder (JSP) and fiber bleaching treatment in PLA composites incorporating thymol. The insertion of 30 wt.% jackfruits fibers gave the best tensile performance. The elongation at the break decreased with increased fiber loading, regardless of treatment, but no significant changes from 10 to 30 wt.% loading of powder were observed. All JSP or bleached jackfruit skin fiber (BJSP) composites showed higher tensile modulus than pure PLA, and the results were in agreement with Suradi et al. [91]. The SEM micrographs in Figure 6 show the fiber surface differences between unbleached JSP and BJSP, respectively. Following bleaching treatment, a rougher fiber surface is shown, indicating effective removal of the non-cellulosic components' potential for good mechanical fiber locking with the matrix.

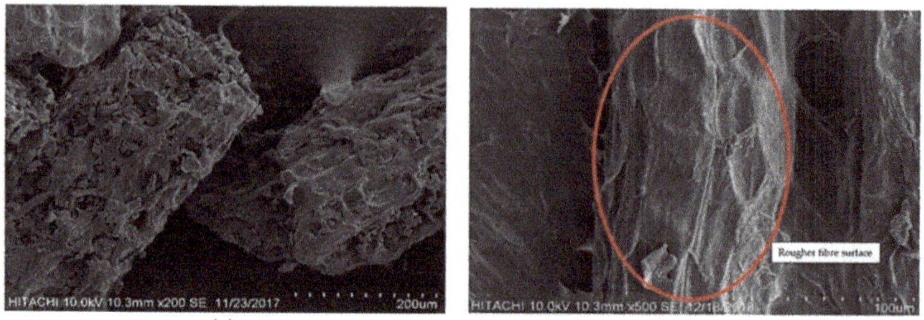

(a) (b)

Figure 6. Scanning electron microscopy (SEM) micrographs of (**a**) unbleached jackfruit skin powder (JSP) and (**b**) bleached jackfruit skin powder (BJSP). Reproduced with permission from Ref. [79]. Copyright 2021 Elsevier.

Torres et al. [81] evaluated the influence of chestnut husk content on the mechanical properties of novel starch/chestnut husk biocomposites. This was developed by incorporating 2.5, 5, and 7.5 wt.% chestnut husks via an extrusion molding procedure. The results showed that the pure starch samples had an average elastic modulus of 3.30 MPa, while starch samples with 7.5 wt.% chestnut husks displayed an average elastic modulus of 4.85 MPa. The ultimate tensile strength value was independent of the chestnut husk content, while the elongation at the break point decreased as the filler content increased. These results were in agreement with those of previous reports on starch-based biocomposites reinforced with cellulosic fillers such as cotton, hemp, and winceyette fibers [92–94].

Othman et al. [43] developed tapioca-starch-based biodegradable film incorporating banana pseudostems (waste) powder for the starch-based films. The mechanical and optical properties of the films were reduced but the barrier activity improved. The optimum percentage composition of 40 wt.% pseudostem powder can be used for incorporation into starch-based films as food packaging.

Cassava starch-based films have been investigated by Leites et al. [95] to determine the effect of waste from the production of orange juice on the properties of the films. The orange waste was added in two different forms, which are powder and aqueous extract (by soaking the powder in water followed by filtration). When comparing the moisture content, water solubility, and thickness of the extract to that of the powder orange residue, it was discovered that the extract had higher values.

3.4. Isolation of Fiber from FVW

For the extraction of cellulose from carrot waste, polysaccharides were extracted in acidic and alkaline environments before being treated with sodium hypochlorite solution to remove lignin and other lignin-containing compounds. With this treatment method, carrot cellulose was obtained as a 4% concentration in water suspension after the treatment. The nanocellulose was created by homogenizing cellulose with ultrasonically homogenized cellulose. Because the sonication system included a temperature probe, an ice bath was used to keep the samples from becoming too hot. The amplitude of the ultrasonic homogenizer was maintained at 90% of its maximum. Finally, carrot CCNF samples containing 0.1 wt.% carrots were obtained [36].

The extraction of fiber from durian skin required only a few simple procedures. The durian skin was cut into smaller pieces and thoroughly washed with tap water to remove any dust or adhering particles before being prepared. Next, the skin was dried at 70 °C for 24 h. The dried skin was ground to obtain fibers ranging in length from 100 to 150 μm. Approximately 300 g of raw durian skin fibers was used in this process. The alkali treatment of DSF was carried out with the help of sodium hydroxide (NaOH). [70].

The chemical treatment was used to isolate the orange peel. Eight grams was kept at room temperature under mechanical stirring for 16 h in a 5% KOH solution. Following the alkaline treatment, the insoluble residue was bleached with $NaClO_2$ solution for one hour at 70 °C, pH 5.0, which was adjusted with 10% acetic acid. The residue was neutralized, washed, and centrifuged at 6000 rpm for 20 min at 25 °C. The second alkaline treatment was repeated with a concentration of 5% KOH. The insoluble residue was subjected to acid hydrolysis for an hour at 80 °C with 1% H_2SO_4 for one hour. Following centrifugation and washing of the final residue, the diluted suspension was stored at 4 °C in a sealed container. The cellulose suspension was dried by lyophilization and stored at 4 °C, where it was designated as NFC [96].

The cellulose fiber from banana peels was obtained by removing the hemicellulose and lignin. The banana peel was treated with 10% (w/v) natural lye, which was made by soaking wood charcoal ash in water for 48 h before being applied. The extracted cellulose fiber was then bleached twice with 10% (v/v) hydrogen peroxide at 90 °C for 10 min each time. As previously described, the white cellulose fiber was retreated with a lye solution. Finally, cellulose was obtained [97].

4. By-Products

Fruits and vegetables are crucial for human nutrition, providing significant amounts of daily vitamin, mineral, and fiber needs. The fruit and vegetable processing industries generate significant amounts of waste in the form of liquid and solid, which contain several reusable high-value substances with significant economic potentials. Fruit and vegetable by-products accumulated from industrial operations. such as bagasse, peels, trimmings, stems, shells, bran, and seeds, make up more than half of all fresh fruit and have nutritional and functional contents that are sometimes higher than the finished product [98]. These by-products can be applied in food and other non-food applications, including medical, pharmaceutical, energy, and chemical production.

Typically, the fruit processing activities generate solid wastes (peel, skin, pulp, pomace, seeds, bunch, stems, and shells) and liquid wastes (crude extracts and wash water). With proper management, these wastes can be utilized for the production of by-products. Efforts have been made to explore the potential of these by-products in many applications, including in food and non-food industries.

Based on recent literature, most of the progress on utilizing fruit waste has been towards food and polymer applications, particularly in the production of flour, fiber, and pectin as fillers. Khoozani et al. [99] used banana pulps and peels to produce flour by varying the drying process using oven drying, a spouted bed drier, ultrasound, a pulsed vacuum oven, a microwave, spray drying, and lyophilization. Additionally, flour made from ripe "Prata" banana (*Musa* spp.) peels was used in the development of edible coatings, as shown in Figure 7 [100].

Figure 7. Development of edible coatings from banana peel. Reproduced with permission from Ref. [100]. Copyright 2021 Elsevier.

Reißner et al. [101] processed the pomace of different berries (blackcurrant, redcurrant, chokeberry, rowanberry, and gooseberry) into flour and fruit powder. de Andrade et al. [102] used orange, passion fruit, and watermelon to produce flour. The prebiotic potential of the fruit-based by-product flour obtained from the solid waste from fruit processing was evaluated after undergoing in vitro gastrointestinal digestion process.

In addition to flour, fiber is another food application for fruit waste by-products. Sharma et al. [103] processed apple pomace to produce fiber to prepare fiber-enriched products. The dried pomace was packed in gunny and PE bags and stored at low (0–4 °C) and ambient temperatures (13–26 °C) after washing, blanching, and drying in a polytunnel drier (45 ± 8 °C). The apple fiber can be used in product formulations with good cholesterol-lowering characteristics and for the establishment of polymer composites such as reinforcements with epoxy resin to form hybrid composites [104].

Begum and Deka [105] used banana bract to extract dietary fiber (DF) using an ultrasound-assisted extraction method combined with alkaline extraction. This antioxidant DF is vital because it is linked to various health benefits, including preventing and treating chronic and degenerative diseases. Additionally, Mir et al. [106] produced gluten-free crackers with high fiber from brown rice flour and apple pomace. Apple pomace flour blends were made by combining brown rice flour with 0, 3, 6, and 9% apple pomace.

Another common use of fruit waste is that of pectic substances for use in other applications. Khamsucharit et al. [107] extracted pectin from the peels of five banana varieties through a conventional hot acid extraction method with a citric acid solution. Another study on pectin extract from banana peel was conducted by Maran et al. [108] using an ultrasound-assisted, citric-acid-mediated extraction method, optimized through the response surface methodology (RSM) approach. Moorthy et al. [109] used jackfruit peels to produce pectin using an ultrasound-assisted extraction method, a similar approach to that used by Hosseini et al. [110] and Guandalini et al. [111], who respectively extracted pectin from orange and mango peels. Chaiwarit et al. [112] found that pectin extracted from mango peel (Figure 8) can be regarded as a potential biopolymer for an edible film for food packaging.

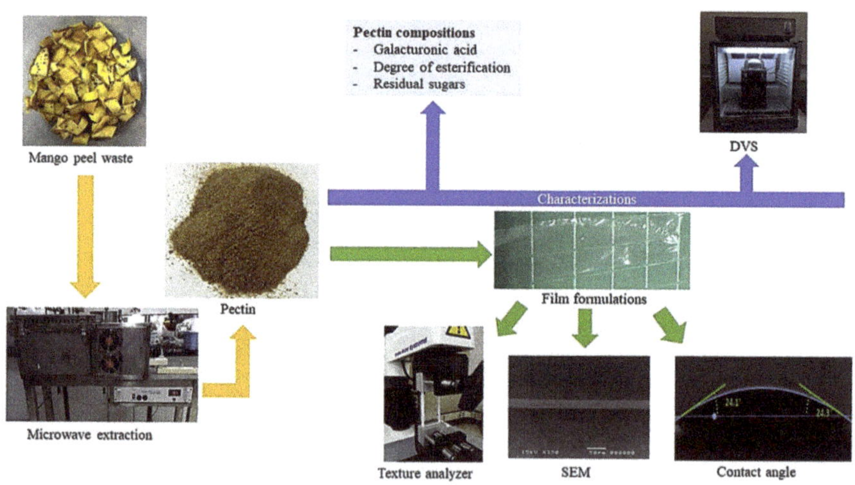

Figure 8. Fabrication of thin polymer films from mango peel waste. Reproduced with permission from Ref. [112]. Copyright 2021 Elsevier.

Fruit waste has also been used for the production of non-food by-products. One of the most common applications of fruit waste is for the production of bioadsorbents. Laysandra et al. [113] used durian skin to produce adsorbent by mixing it with acid-activated bentonite (AAB) and natural surfactant (rarasaponin). They found that the rarasaponin/bentonite-activated biochar from the durian shell composite (RBAB) was effective for use as a new composite adsorbent for the removal of crystal violet and Cr (VI) from aqueous solutions. Meanwhile, fruit waste is also utilized to make films. Mango peel was used in the production of fish-gelatin-based films [114]. The films were prepared via the solution casting method with three different concentrations of the mango peel extract (1 to 5%). The films produced were then tested for their physical, barrier, mechanical, and antioxidant qualities. Based on their findings, the addition of mango peel extract in the formulation produced a thicker, denser, and continuous structure with outstanding free radical scavenging activity (Figure 9).

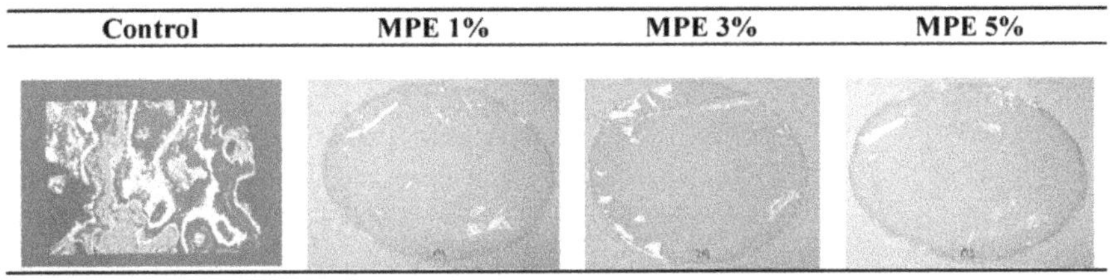

Figure 9. Gelatin films with various concentrations of mango peel extracts. Reproduced with permission from Ref. [114]. Copyright 2021 Elsevier.

Fruit waste has also been used in the production of edible coatings to improve their properties. Blackberry pomace has reported been used for its the antioxidants and antimicrobials sources in the production of edible coatings for foods and consisting of carboxymethylcellulose, bacterial cellulose fibrils, and pectin sweets, with high lipid contents [115]. Meanwhile, grape seed extract has been included in pectin–pullulan edible film production for the storage of peanuts [116]. This coating successfully extended the shelf life of the stored peanuts by delaying rancidity.

Similar to fruit wastes, flour is the common by-product produced from vegetable waste. de Andrade et al. [102] used solid products from eight types of vegetables, including carrot, courgette, cucumber, lettuce, mint, rocket, spinach, and taro, to produce flour, then tested the modulatory effects on the gut microbiota composition. Amofa-Diatuo et al. [117] produced flour from cauliflower stems and leaves as a source of isothiocyanates (ITC) in an apple juice beverage. Cauliflower waste was also utilized in producing fermentable sugar. Majumdar et al. [118] pre-treated cauliflower stalks and leaves with dilute phosphoric acid prior to enzymatic hydrolysis to better release fermentable sugars. Vegetable waste is also a source of fiber and pectin. Iwassa et al. [119] produced and characterized fiber concentrates from asparagus by-products. They concluded that the fiber concentrates had high potential for use in formulating functional food products. Kazemi et al. [120] used an ultrasonic extraction method to extract pectin from eggplant peels. Eggplant peels provide substantial extraction yields and are considered to have high potential in pectin production

Waste from vegetables was also used for the extraction of essential oils. Chiboub et al. [121] extracted essential oils from the tops of carrots. The result showed that the essential oil produced was a good source of natural antimicrobials or aromatic agents. A similar finding was reported by Caliceti et al. [122] for peptide extract from cauliflower leaves. In contrast, the tops and roots of carrots were reported to be used to produce biodegradable composite films [123]. To improve the composite properties, they mixed and optimized the formulation with hydroxypropyl methylcellulose (HPMC) and high-pressure microfluidized cellulose fibers. Garlic peel extract was reported to be added in the formulated gelatin film as a source of antioxidant and antibacterial agents [124]. The gelatin film was proven to maintain the qualities of the rainbow trout fillets during refrigerated storage. Additionally, vegetable waste was also used for the production of the highly antioxidant edible coatings. Asparagus waste extract was incorporated into a hydroxyethyl cellulose–sodium alginate edible coating to significantly extend the postharvest life and retain the quality of strawberry fruit [125].

5. Innovative Technologies Used for Bioactive Compound Extraction

5.1. Thermal Extraction

Thermal extraction includes decoction, reflux, Soxhlet, hydro, and steam distillation techniques [126]. Incorporating bioactive compounds such as flavonoids, phenolic acids, and polysaccharides into polymer materials improves both the antioxidant and antibacterial activity [127]. Temperature increases the solubility and diffusivity in solid–liquid extraction, resulting in high yields of extracted compounds. Thermal degradation of enzymes and desired bioactive compounds occurs during prolonged extraction at high temperatures [128]. Regarding the extraction of FVW and FVB for industrial applications, various methods are used, such as fermentation and separation (lactic acid extraction from potato peels), hot dilute acid treatment and alcohol precipitation (pectin extraction from citrus peels), and steam distillation and alkali treatment (D-limonene extraction from citrus peels) [129]. These methods are appropriate for batch and large-scale processing. Meanwhile, emerging thermal processes, such as microwave-assisted extraction (MAE), pressurized liquid extraction (PLE), and subcritical water extraction (SWE), offer additional benefits and have the potential to replace conventional methods due to being efficient, economical, and environmentally friendly [17].

5.1.1. Conventional Heat Extraction

One of the oldest extraction methods is conventional heating extraction (CHE). CHE involves direct contact with a thermal source either by conduction or convection as the heating mechanism. The CHE thermal source usually is an used oven, hot plate, or water bath, while acidic solutions are used as the agents. Recent research on the FVW and FVB extraction methods showed that CHE was not favored as compared to the new emerging extraction technologies, such as MAE [130–132] and ultrasonic-assisted extraction (UAE) [133–135]. Although the CHE method produced more bioactive compounds than the MAE method, such as pectin extracted from black carrot pomace [132] and lime peel [136], the CHE method has several drawbacks, including high energy requirements and long processing time, making it inefficient in industrial applications; however, the combination of CHE and other techniques has shown tremendous potential in FVW and FVB processing. High PLE [137] and UAE [133] are two technologies that can be integrated with CHE, which can increase the extracted yield by over 50%. Table 4 shows the optimal yields of bioactive compounds from FVW that have been achieved in recent years.

Table 4. Progress studies of FVW using conventional heating extraction.

Fruit and Vegetables Waste	Bioactive Extraction	Process Condition	Agent	Optimum Yield (%)	Ref.
Red and white dragon fruit peel, passion fruit peel	Pectin	43—107 min, 60–80 °C (Dragon Fruit), 60–120 °C (Passion Fruit)	Citric acid	15.12% (red dragon fruit), 14.11% (white dragon fruit), 13.18% (passion fruit)	[130]
Grape (white and red) skin	TPC	60–90 min, 40–70 °C	Ethanol	1.74–2.12 gGAE/L	[131]
Black carrot pomace	Pectin	90 min, 110 °C	Acidic solution	0.22 kg pectin/kg pomace	[132]
Wheat bran	Phenolics	3, 6, and 24 h, 50–90 °C	99% glycerol, citric acid, Folin–Ciacalteu	4.57–16.11 mgFAE/gdm	[133]
Grapefruit peel	Pectin	90 min, 80 °C	HCl	MW 385.55 kDa	[135]
Lime peel	Pectin	60 min, 95 °C	HCl or citric acid	16.12–23.52%	[136]

5.1.2. Soxhlet Extraction

Another traditional thermal extraction method is Soxhlet extraction (SE), which involves high temperatures, long extraction times, and a large amount of solvent, meaning this method triggers environmental concerns. Similar to conventional heating extraction, the SE method also results in highly extracted bioactive compounds from FVW and FVB, including papaya [138], grape [139], and soapberry seed oil [140]. This classical extraction method has been compared to MAE [141–143] and UAE [144,145]. Table 5 shows the findings from the latest studies on SE methods using FVW. Throughout the experiments, Fernandez-Pastor et al. [142] showed that olive peel took 30–90 min to yield around 19 to 23% of bioactive compounds as compared to other studies that required longer times for the extraction technique (4 to 8 h).

Table 5. Progress studies of FVW in the application of Soxhlet extraction.

Fruit and Vegetables Waste	Bioactive Extraction	Process Condition	Agent	Optimum Yield (%)	Ref.
Olive Oil (Alperujo) solid waste	TPC	4 h, 70 °C	n-Hexane	0.75–3.76 g/kg raw alperujo	[139]
Indian Soapberry seed	Oil	6 h, 80 °C	n-Hexane	40.63%	[140]
Mango peel, Soursop peel, Grape peel, Grape seed	AOA, TPC, TFC	8 h, 40 °C	60% Ethanol; 1:25 solid/liquid ratio	52.28% (Mango peel), 50.63% (Soursop peel), 64.65% (Grape peel), 18.45% (Grape seed)	[141]
Olive skin/peel	Triterpene acids	unmilled/milled, 30–90 min, 65–70 °C	Ethyl Acetate/Methanol (1:40, 1:20, 1:10 sample/solvent ratio)	unmilled, ratio 1:40 g/mL: 22.24%	[142]
Bitter Gourd peel	AOA, TPC, TFC	6 h, 40–50 °C	Methanol	26.48%	[145]

Sukatta et al. [146] studied the bioactivity levels and characterization of rambutan peel extract (RPE) and the feasibility of RPE as a bioactive compound for antimicrobial and antioxidant applications in whey protein isolate (WPI)/cellulose nanocrystal film. The RPE was extracted for 16 h in a Soxhlet extractor with 95% ethanol. Using the HPLC chromatogram, the main components of RPE were corilagin, ellagic acid, geraniin, and gallic acid. The main bioactive components were classified as phenolic compounds, which exhibited antioxidant and antimicrobial properties.

As an alternative to existing synthetic packaging films, Jridi et al. [147] developed grey triggerfish skin gelatin films containing phenolic extracts from blood orange (Citrus sinensis) peel. The dried orange peel extract (DOPE) was obtained by extracting the dried orange peels via Soxhlet extraction using 300 mL of ethanol for 6 h. Similarly, fresh orange peel extract (FOPE) was obtained by extracting the fresh orange peels. It was found that the extraction yield of the DOPE was 31.2% (*w*/*w*), which was significantly higher than that of FOPE (25.3%).

5.1.3. Microwave-Assisted Extraction

Microwave-assisted extraction (MAE) is a powerful alternative method for extracting bioactive compounds from FW. It is mainly used for extraction because it reduces solvent usage, energy consumption, extraction times, heating rates and increases extraction efficiency and selectivity, resulting in quality target products [148]. Many studies were conducted using MAE methods on FVW and FVB, including longan seed [149], tomato peel waste [150], apple pomace [151], banana peel [152], pitaya fruit peel [130,153], passion fruit peel [130], lemon peel, mandarin peel, and kiwi peel [154]; thus, MAE is mostly used to extract pectin from fruit waste. This approach has been used to extract high-quality pectin with biomasses such as mango peels [155], citrus mandarin peels [156], fig skin [157], orange peel, apple pomace, mango peel, carrot pulp [158], pumpkin biomass [148], and banana peels [159]. The study conducted by Zin et al. [160] showed that the highest microwave power used on the fruit waste (sour cherry pomace) was 700 W. Other biochemical compounds can also be obtained using this technique, such as antioxidants from black carrot pomace [161], pitaya peel [153], and mango seed kernels [162]; flavonoids from Jocote pomace [163]; anthocyanins from grape juice waste [164] and sour cherry pomace [165]; and essential oil from lemon peels waste [166]. Casas et al. [167] also showed the potential of cocoa butter produced from mango kernel butter by extracting discarded seed kernels. Table 6 shows the summary conditions used for optimal bioactive compound extraction in recent studies on FVW.

For optimum yield extraction of the targeted compound, focused microwave-assisted extraction is preferred to conventional or household microwave ovens, as the parameters, namely the pressure and temperature, can be monitored [163]. A previous study showed that the combination of MAE with other methods could help produce high-yield compounds. Sequential ultrasound-microwave assisted extraction (UMAE) of fig skin extract resulted in higher pectin yield ~14.0%, with citric acid used as the solvents. Brönsted acidic–ionic liquid-based ultrasound-microwave synergistic extraction (BUME) from pomelo peels achieved the highest pectin yield of 328.64 ± 4.19 mg/g with optimum conditions involving 10 mM $[HO_3S(CH_2)_4mim]HSO_4$ solvent, 15 min of extraction time, 360 W of microwave irradiation power, and 27 mL/g liquid–solid ratio compared as compared to MAE (210.39 ± 5.82 mg/g).

Utama-Ang et al. [168] studied MAE of dried ginger and developed a rice-based edible film containing ginger extract. The optimal MAE conditions were determined to be 400 W microwave power and one minute extraction time. At high temperatures and microwave power, 6-gingerol dehydrates water (H_2O) from its structure, resulting in the formation of 6-shogaol4. Microwave power accelerated the retro-aldol reaction of 6-gingerol, and it is suggested that zingerone constituents be generated with aldehyde to deliver the products. The optimized extract showed good results in terms of the levels of total phenolic compounds (198.2 ± 0.7 mg GAE/g); antioxidant activity as measured by DPPH (91.4 ± 0.6% inhibition), ABTS (106.4 ± 3.1 mgTE/g), and FRAP (304.6 ± 5.5 mgTE/g); and bioactive compounds, including 6-gingerol (71.5 ± 3.6 mg/g), 6-shogaol (12.5 ± 1.0 mg/g), paradol (23.1 ± 1.1 mg/g), and zingerone (5.0 ± 0.3 mg/g).

In contrast, industrial potato peel by-products allowed greater antioxidant extraction yields than in combinations with ultrasound treatment [169]. Radiofrequency-assisted extraction (RFAE) with a frequency range of 1 to 300 MHz is another method used in dielectric heating as compared to microwave-assisted extraction (300 to 3000 MHz) in electromagnetic field-based thermal processes. An analysis discovered that the optimum conditions for RFAE were similar to MAE for pectin extraction from apple pomace. Still, the physicochemical properties (DE, GA, color values, and thermal stability) of apple pomace showed better RFAE pectin values [151].

Table 6. Progress studies of FVW in the application of microwave-assisted extraction.

Fruit and Vegetables Waste	Bioactive Extraction	Process Condition	Agent	Optimum Yield (%)	Ref.
Red and White Dragon fruit peels, Passion Fruit peel	Pectin	10–12 min, 75 °C, 153–218 W	Methanol (pH: 2.9–3.0)	17.01 ± 0.32% (Red Dragon Fruit), 13.22 ± 1.42% (White Dragon Fruit), 18.73 ± 0.06 (Passion Fruit)	[130]
Black Carrot pomace	Phenolic, Antioxidants, Anthocyanins	5 min, 110 °C, 20% output power of 900 W	Hot Acidic Water (pH: 2.5)	phenolic content: 1692 ± 79.4 mg GAE/l (0.17 kg pectin/kg pomace); antioxidant: 60 ± 9.6 MTE/mL; anthocyanins: 456.8 ± 38.2 mg/L	[132]
Longan seeds	Pectin	3.5 min, 700 W	50% Ethanol	64.95 + 20.56 mgGAE/gdw	[149]
Lemon, Mandarin and Kiwi peels	Pectin	1–3 min, 60–75 °C, 360–600 W	HCl, Nitric Acid (HCl)	17.97% (kiwi peels), 7.47% (mandarin peels), 7.31% (lemon peels)	[154]
Apple pomace; Orange peel; Mango peel; Carrot pulp	Pectin	10–180 min, 90 °C, 50–200 W	Water	Orange peel, 60 min, 200 W: 12.9 ± 1.0%; Mango peel, 120 min, 200 W: 14.7 ± 0.6%; Apple pomace, 120 min, 200 W: 14.7 ± 0.1%; Carrot pulp, 60 min, 200 W: 6.3 ± 0.7%	[158]
Black Carrot pomace	Phenolic, Flavonoid, Anthocyanins, AOA	9.8 min, 348.07 W	20% Ethanol	polyphenolic content: 264.9 ± 10.025 mg GAE/100 mL; flavonoid: 1662.22 ± 47.3 mgQE/L; AOA: 13.14 ± 1.05 MTE/mL; anthocyanins: 753.40 ± 31.6 mg/L; color density; 68.63 ± 5.40	[161]

Table 6. *Cont.*

Fruit and Vegetables Waste	Bioactive Extraction	Process Condition	Agent	Optimum Yield (%)	Ref.
Jocote (*Spondias purpurea* L.) pomace	Pectin, Flavonoid	20 min, 68 °C, 100 W	80% Ethanol	phenol: 0.897 g GAE/g (3.42%), flavonoid: 1.271 g QE/g,	[163]
Mango kernel	Crude butter	3.5 min, 160 W	Water	48.85%	[167]
Lemon peel	Essential oil, Pigment	50 min, 20 °C/min, 500 W	80% Methanol	Essential oil: 2 wt.%, Pigment: 6 wt.%	[166]
Watermelon rind	Pectin	12 min, 279.3 W	Acetic Acid	3.93–5.77% (DE: 56.86–85.76%)	[170]

Table 7. Progress studies of FVW involving pressurized liquid extraction.

Fruit and Vegetables Waste	Bioactive Extraction	Process Condition	Agent	Optimum Yield (%)	Ref.
Grape pomace: skin, seed	Polyphenols, Antioxidants	5 min with 250 s nitrogen purge, 100–160 °C, ~10 atm	20–60% Ethanol	Polyphenols content, 160 °C, 60% ethanol, Skin:1.98 ± 0.06 mgGAE/gdw; Seeds: 12.54 8 ± 0.02 mgGAE/gdw; Antioxidant by DPPH, 100 °C, 20% Ethanol, skin: 121.91 8 ± 0.08 mg/mL, Seeds: 39.63 8 ± 0.01 mg/mL; Antioxidant by ORAC, 160 °C, 60% Ethanol, Skin: 36.33 8 ± 0.06 MTE/gdw, Seeds: 137.65 ± 0.11 MTE/gdw	[172]
Pomegranate peel	TPC, Punicalagin content, antimicrobial activity	200 °C	77% Ethanol	Polyphenols content: 164.3 ± 10.7 mgGAE/gdw; Punicalagin content: 17 ± 3.6 mg/gdw	[174]
Olive pomace	TPC, AOA, TFC	65–18 °C, supercritical carbon dioxide ($scCO_2$)	8–92% Ethanol	TPC, 160.7 °C, 75% Ethanol: 280.37 mgGAE/gDE; AOA, 125 °C, 50% Ethanol: 6.88 MTE/gDE; TFC, 89.3 °C, 25% Ethanol: 15.82 mgRE/gDE	[175]
Mulberry pulp	TPC, Anthocyanins	10 min, 75.5 °C, 200 atm, purge time 90 s	47.2% Methanol (pH3.01)	TPC: 2186.09 ug/g, AOA: 164.53 ug/g	[176]
Beetroot waste: residues, leaves and stems	TPC, AOA	40 °C, 7.5–12.5 MPa, 3 mL/min	70–100% Ethanol	Leaves- TPC, 40 °C, 10 MPa, 100% Ethanol: 252 ± 2 mgGAE/g; AOA by ABTS, 40 °C, 12.5 MPa, 100% Ethanol: 823 ± 48 MTE/g; Stems- TPC, 40 °C, 10 MPa, 100% Ethanol: 14 ± 2 mgGAE/g; AOA by DPPH, 40 °C, 10 MPa, 100% Ethanol: 515 ± 89 G/mL (Leaves > Stems)	[177]

Cejudo-Bastante et al. [178] investigated the extraction process used in developing bioactive jute-fiber-based food packaging using pressurized liquid extraction (PLE) and enhanced solvent extraction (ESE) techniques. The extraction yield and antioxidant capacity levels of the red grape pomace extract (RGPE) obtained using ESE and PLE were compared under varying pressure (10 and 20 MPa), temperatures (55–70 °C), and co-solvent (C_2H_5OH

or $C_2H_5OH:H_2O$) conditions. They discovered that the PLE technique produced the most bioactive extract with 20 MPa, 55 °C, and one hour residence time using $C_2H_5OH:H_2O$ (1:1 *v*/*v*), providing antibacterial capacity against *Escherichia coli*, *Staphylococcus aureus*, and *Pseudomonas aeruginosa*.

5.1.4. Subcritical Water Extraction

Compared to conventional extraction procedures, sub-critical water extraction (SWE) is a green extraction technology that yields superior quality extraction products and is cost-effective with a short extraction or treatment time [179–182]. SWE is also known as pressurized hot water or superheated water extraction, as it uses water at temperatures between 100 °C and 374 °C (critical temperature) and at pressures of up to 22.1 MPa (greater than vapor saturation) to keep the water molecules in a liquid state throughout the process [183]. Water is a polar solvent with a dielectric constant (ε) of 79.9 and a density of 1000 kg/m^3 [184,185]. When water is heated to higher temperatures, its hydrogen bonds break down, resulting in a drop in its dielectric constant (ε), demonstrating water's ability to act as a material reaction medium. Water has a density of 79.9% at ambient temperature and atmospheric pressure. Water may be lowered to 27–32.5% while remaining in liquid form by increasing the temperature to 250 °C and increasing the pressure to 5 MPa. Water has a similar density to methanol (32.5%) and ethanol (27%) at ambient temperature [186,187]. The latter enables water to interact with polar compounds, lowering the binding force and allowing substances to dissolve in water at greater temperatures and pressures.

The ionic constant of water (Kw) increases with increases in the reaction temperature according to Pourali et al. [188], and is nearly three times greater than at room temperature. Water's reactivity increases as the concentrations of H^+ and OH^- in the aqueous medium increase, causing it to act as an acid or base catalyst that is appropriate for hydrolysis reactions. Organic waste can, thus, be hydrolyzed and the necessary components contained inside can be removed using sub-critical water treatment [189]. Sub-critical water treatment is an environmentally favorable procedure because no chemical solvent is required. As a result, less effluent is created. Table 8 shows several examples of the application of SWE on FVW.

Table 8. Progress studies of FVW involving sub-critical water extraction.

Fruit and Vegetables Waste	Bioactive Extraction	Process Condition	Agent	Optimum Yield (%)	Ref.
Grape pomace	Phenolic compounds	50–190 °C	Water	29 g/100 g extracts	[190]
Kiwifruit peel	TPC, TFC, AOA	5–30 min (20 min) 120–160 °C (160 °C)	Aqueous mixture	TPC: 51.24 mg GAE/gdw, TFC: 22.49 mgCE/gde AOA by ABTS: 269.4 mM TE/gdw	[191]
Citrus (*C. unshiu*) peel	TFC	145–175 °C 15 min	Water	TFC: 59,490 g/gdb	[192]
Tamarind seed	Xyloglucan component, TPC, AOA	100–200 °C (175 °C) 5.03–13.55 min	Distilled Water	Xyloglucan component: 62.28%, TPC: 14.65–42.00 gGAE/g, AOA: 1.93–3.20 MTE/g	[193]
Dates seed	TPC, AOA, TFC, Dietary fiber	120–180 °C (144 °C) 10–30 min (18.4 min)	Aqueous mixture	TPC: 9.97 mgGAE/g, TFC: 3.52 mgQE/g, AOA 1.67 mgTE/g, Dietary fibers: 29 g/mg	[194]

The extraction mechanism of SWE begins with solute desorption under elevated pressure and temperature, followed by the diffusion of extracted chemicals into the solvent. Finally, the extracted solutions are eluted from the extraction cell and transferred to a collection container [195,196].

Several process parameters influence the extraction efficiency of SWE, including the reaction temperature, pressure, reaction duration, solid-to-water ratio, sample particle size, pH, solute properties, and surfactant addition [197]; however, the reaction temperature, reaction duration, and solid-to-water ratio have the greatest influence on the SWE process. Because the viscosity and surface tension of the extraction solvent diminish with increasing temperature, Thani et al. [198] discovered that increasing the treatment temperature enhances the mass transfer rate and solubility of bioactive chemicals; however, if the temperature is raised above a certain point, the selected chemicals may degrade. As a result, the closely associated process temperature and duration should be optimized for each unique situation and are highly reliant on the desired product's qualities.

Ho et al. [199] investigated the influences of *P. palatiferum* freeze-dried powder (PFP) using SWE on the antioxidant activities and physical properties of gelatin–sodium alginate (GSA)-based films. *P. palatiferum* (Nees) Radlk. leaves were extracted with subcritical water, which increased the total phenolic content (TPC) and antioxidant activity as the PFP concentrations increased. The increase in antioxidant activity occurred in parallel with TPC. The antioxidant activity was attributed to the phenolic compounds [200] and gelatin–sodium alginate [201]. In addition to TPC, *P. palatiferum* leaves also contained a variety of other compounds, including protein, saponin, total sugar, and phytosterol [202], all of which can contribute to the antioxidant activity of GSA-based films.

Mohd Thani et al. [203] conducted an exhaustive review on SWE of sugar from FW. Monosaccharides and oligosaccharides are important carbohydrate molecules that can be hydrolyzed from FW. Sugar extraction from bakery waste, for example, is an effective way of valorizing this type of FW [198]. The leftover croissants had the most fructose and glucose at 4.74 and 3.76 mg/g substrate, respectively [198]; however, the sugar yield increased to a maximum value following SWE and then steadily decreased over time, whereas the yield of the degradation products increased over time [204,205].

Additionally, SWE can be successfully employed to extract antioxidant-rich extracts from yarrow by-products. The obtained extracts are rich in total phenols and flavonoids and have high antioxidant activity [206]. Oilseed cake extracts derived using subcritical water show a significant amount of promise for application in the fortification of various food goods and cosmetics. Depending on the type of oilseed, specific components such as the flavor amino acids aspartic acid, glutamic acid, and alanine can be extracted. This biowaste's favorable chemical composition and high nutritional content provide it with high utilization potential [207]. The most significant yield of phenolic compounds (4855 mg/100 g dry weight) was obtained using subcritical water extraction from pomegranate seed remnants [208]. Similar results were obtained for phenolics in white wine grape pomace [209] and peach palm by-products [210].

5.2. Non-Thermal Extraction

Each non-thermal extraction process has its own set of advantages and disadvantages, as summarized in Table 9. Several factors must be considered to obtain the best results when extracting phenolic compounds from FVW. The operation's success depends on the understanding of the nature of the target compounds, source materials, and waste matrices. Additionally, the process type and operating parameters used in the recovery process are critical determinants of the yield and quality of the recovered chemicals. The FVW matrix, the materials' physicochemical properties, and the type of compounds extracted may affect the chosen approach. Conventional extraction processes with low yields require longer extraction times, large amounts of energy, and significant capital investment. Due to the difficulties associated with achieving high-purity target compounds, conventional technologies are regarded as inadequate compared to the emerging non-conventional approaches, such as ultrasound-assisted, HPP, and PEF.

Table 9. Comparison of innovative extraction methods.

Extraction Method	Working Principles	Parameters	Advantages	Disadvantages
High Pressure Processing (HPP)	• Works by increasing pressure up to 600 MPa. • Isotactic pressure cause rupture to the cell membrane and increase mass transfer from the inside cell to the outside environment and vice versa [211]. • Mild temperature improves further the mass transfer.	• Pressure • Temperature (mild) • Solid/liquid ratio • Processing time	• Green technology. • Good with thermo-sensitive bioactive components	• Reduces pressure-sensitive bioactive components [212]. • Need to be hurdled with other technologies to improve the extraction process.
Ultrasound-assisted extraction (UAE)	• UAE produces acoustic waves in the solvent that lead to cavitation bubbles. The developed cavitation bubbles burst at the surface of the plant sample matrix, disrupt the plant cell wall, and help in the release of the phenolic bioactive compound into the solvent [213]. • Suitable for phenolic compounds, lipids, chlorophyll, carotenoids [214]	• Concentrations of solvent • Solvent-to-sample ratio • Extraction time • Temperature • Frequency • Power • Particle size • Liquid height • Duty cycle	• Analytically simpler (reduce unit operations), more efficient, lower extraction temperature, and faster processing time [215]. • Reduction in energy and power usage. • Higher yield and less chemical solvent usage [214]. • Conducted under room temperature or heat and at atmospheric pressure. • Water, aqueous and non-aqueous solvents. • Quick return of investment	• Deterioration of phenolic compounds due to the generation of hydroxyl radicals with the formation of cavitation bubbles. • Lack of uniformity in UAE energy distribution and potential change in the constitutive molecules [216]. • Proper optimization in US frequency, the nominal power of the device, propagation of cycles, input power, system geometry is required for maximum yield [214].
Supercritical fluid extraction (SFE)	• A process based on the use of solvents above or near their critical temperature and pressure to recover extracts from solid matrices. • Uses supercritical fluids (CO_2 and H_2O) for the extraction of phenols from the plant matrices. • Suitable for volatile compounds [217].	• Temperature • Pressure • Time • Solvent-to-sample ratio • Sample particle size • Processing before extraction	• Faster, selective, and improved recovery of phenolics without using toxic organic solvents [215]. • Conducted under room temperature and at high pressure. • Automated system, no filtration required, recycle and reuse of the supercritical fluid, polarity of CO_2 can be retuned and extraction of thermolabile compounds possible at low temperature.	• Increased time due to solutes, lower diffusion rate from the solid matrix into the supercritical fluid. • Initial cost of the equipment is high. • Greatly impacted by the property of the fluid used [215]. • Elevated pressure required, risk of volatile compounds losses, many parameters to optimize.
Pulsed electric field (PEF)	• Causes electroporation, which involves the formation of localized pores in the cell membranes, increasing the extraction yield [218]. • Best for phytosterols and various polyphenols [219].	• Electric field strength • Treatment time • Pulse shape • Pulse width • Pulse frequency • Pulse polarity • Temperature • Treatment flow mode	• Shorter extraction time (μs), consumes less energy, environmentally sustainable [215]. • Conducted under room temperature or heat and at atmospheric pressure. • Water, aqueous and non-aqueous solvents. • Can be applied in continuous mode up to 10000 kg/h [19].	• Increases in electric field strength and treatment time could lead to increased energy consumption. • High equipment cost [19].

When combined with a properly chosen extraction method, media, and optimized parameters, these technologies have been shown to increase the yield of specific chemicals from FVW while minimizing carbon footprints [220]. Although selective for lipophilic and volatile compounds such as fats and oils, certain methods such as SFE utilize CO_2; thus, co-solvents are indicated to boost extract purity. As such, green extraction media should be thoroughly evaluated prior to extraction, especially when the desired chemicals are food-grade. In order to turn the FVW problem into a solution for recovering the valuable qualities of bioactive substances that are now being wasted, more research is required.

As such, this sub-section will focus on the innovative technologies used to extract bioactive compounds from FVW and will be limited to notable articles published in the previous decade.

5.2.1. High-Pressure Processing

High-pressure processing (HPP) is commonly used to produce commercially available commodities such as minimally processed fruit juices, guacamole, jellies, dips, salsas, meat and poultry, seafood, and ready-to-eat meals [221,222]. Pasteurization at high pressures ranging from 400 to 600 MPa and temperatures ranging from 20 to 70 °C is a common industrial process. On the other hand, high-pressure sterilization at pressures greater than 600 MPa and temperatures ranging from 90 to 120 °C is more commonly used to eliminate resistant food enzymes, bacteria, and spoilage spores [223,224]. In addition, this technology is being researched for various applications, such as reducing allergenicity in meals, inactivating fruit and vegetable enzymes, and valorizing food matrices. High pressure benefits both fresh produce and FW by-products [225,226]. High-pressure extraction and infusion technologies, for example, may show promise for agricultural and FW valorization.

High-pressure treatment is a green method for extracting bioactive compounds from agricultural commodities. Flavonoids, polyphenols, ginsenosides, anthocyanins, lycopene, caffeine, salidroside, corilagin, and momordicosides are bioactive compounds that have varying polarity levels [226,227]. The steps in specific bioactive compound extraction methods are breaking down plant cell walls to free intracellular molecules, isolating the bioactive compounds from auxiliary components, and purifying them [228]. High-pressure extraction improves the bioactive and heat-sensitive chemical extraction processes while requiring less time and energy [226].

Plant tissues, cellular membranes, and organelles are disrupted, allowing the solvent to enter the cell and dissolve the bioactive compounds [211]. The mass transfer rate is directly proportional to the applied pressure because solubility increases with pressure [229]. As a result, high-pressure treatments improve extraction rates and the availability of bioactive molecules, particularly from difficult-to-release matrices. Pressure has been shown to reduce intracellular pH [230], which aids in the extraction of acylated anthocyanins because they are more stable at low pH [231]. High-pressure treatments also reduce the dielectric constant of water and solvents, which aids in releasing phenolic compounds and the most stable anthocyanins in acylated form, which is less polar [231].

5.2.2. Supercritical Fluid Extraction (SFE)

Supercritical fluid extraction (SFE) is another non-conventional extraction method that obeys the principle of the Green Extraction of Natural Products (second principle) [232]. This is due to the use of supercritical fluids such as CO_2, which can reduce the solvent consumption and amount of waste [233]. SFE operates at temperatures and pressures above the critical points of the solvents used, whereby gas and liquid exist as separate phases. These fluids exhibit the properties of the liquid (density and salvation power) and gas (viscosity, diffusion, and surface tension), facilitating higher extraction yield within a short time [234].

Compounds that are soluble in CO_2, such as oil, fatty acids, carotenoids, and tocopherols, have benefited from the low critical temperature (31 °C) and pressure (7.4 MPa) of CO_2 [232]; however, due to CO_2's low polarity, co-solvents (methanol, ethanol, dichloromethane, acetone, ethylene glycol, water) are sometimes required to extract polar molecules from water-rich FVW [235]. These co-solvents are used to change the polarity of CO_2, improve its solvating power, and increase the extraction efficiency by reducing interactions between analytes and plant cell matrices [216].

The sample particle size, temperature, pressure, time, co-solvents, solvent-to-solid ratio, and processing before extraction are important SFE operational factors [216]. Pressure and temperature are essential elements in SFE, and adjusting them is critical to achieve optimal yield and economic performance. The performance of SFE is strongly temperature-dependent, as increasing temperature reduces the solvent density, lowering the yield, while increasing the solute vapor pressure increases the yield; however, temperatures that are too high can damage fragile molecules such as carotenoids, altering their structure and bioactivity [236]. Pressure, on the other hand, causes the fluid density to increase as the pressure increases so that the extraction yield improves. In general, high pressure, temperature, and flow rate maximize polyphenol extraction in FVW.

Despite its many benefits, SFE has some drawbacks, including limited solvent diffusibility into the matrix, prolonged extraction times, high pressure requirements, costly infrastructure, inconsistency, and lack of repeatability during continuous processes [234]. Furthermore, even with identical chemicals, SFE process conditions can differ amongst plant matrices. Pre-treatments such as lyophilization, micronization, maceration, and decoction often impact the final extraction yields and compositions; thus, before using SFE for bioactive extraction, it is imperative that the appropriate operating conditions and pre-treatment are thoroughly investigated.

5.2.3. Pulsed Electric Field (PEF)

PEF has been gaining traction for FW recycling and by-products due to its ability to extract valuable ingredients [237]. It is able to decrease energy costs, improve the extraction yield, lessen the degradation of heat-sensitive substances, and purify the extraction process with no environmental impacts [238]. In the PEF process, which occurs at ambient temperatures (20–25 °C), the sample is positioned in the middle of two or more electrodes before being exposed to high-voltage electric field pulses for short processing times with repeated frequency. This results in high electric field strengths (EFS) in batch mode of 100–300 V/cm and in continuous mode of 20–80 kV/cm [239].

PEF works by breaking down the structure of plant cell membranes with a high electric field. Due to their dipole nature, the electric charge separates the molecules of plant cell membranes. Because charged molecules repel each other, the pores on the weak sides of the membranes expand, causing permeability [240]. This electroporation or electropermeabilization element allows targeted chemical release from plant matrices [235]. For delicate plant tissues (e.g., pericarp or mesocarp of few fruits), a voltage of 0.1 to 10 kV/cm is sufficient, although for robust materials (e.g., seeds), a voltage of 10 to 20 kV/cm is required [241]. Another benefit of a low EFS (500–1000 V/cm) is the ability of the system to keep the temperature low [242]; thus, PEF reduces heat-labile chemical degradation [243].

The efficiency of PEF-assisted extraction is dependent on the PEF system configuration and extraction parameters. The intensity of the electric fields applied to the processed material is related to the electrode gap, the delivered voltage, the electrode geometry, and their placement in the reactor. Additionally, the extraction yield can be improved by considering the pulse width, number of pulses, treatment time, and total specific energy (kJ/kg). The physicochemical aspects of the treated matrix (size, shape, electric conductivity, cell structure, and membrane characteristics) and nature and cell location of the targeted molecules being extracted can also influence the extraction yield [218].

The EFS affects the physical properties of the targeted molecules, such as their diffusivity, surface tension, viscosity, and solubility [244]. The electric fields must be dispersed consistently across the treatment chamber. There are many waveforms that can be used to deliver electric field energy. In PEF extraction, high-energy exponential square wave pulses are typically used. Due to strong energy transfer in the plant cell matrix, boosting EFS also increases the chemical extraction. The treatment temperature also impacts the PEF extraction process [245], which is commonly conducted at room temperature; however, a high voltage electric field or inefficient delivery pump may increase the overall energy and sample temperature. Higher temperatures may reduce the solvent viscosity, affecting extraction. The treatment time (pulse numbers and width) and solvent selection are equally critical in assessing the PEF performance [128]. An increase in solvent conductivity allows for faster electroporation of the cell membrane. It also aids in mass transfer and increases the extraction rate due to its high solubility in the solvent [246].

Further applications of PEF extraction should be explored; however, the high investment cost is the major hindrance to this technology being widely employed in industry. Nevertheless, its principal benefits outweigh conventional extraction methods; namely, improved extraction yields with minimal thermal degradation while reducing the extraction time, temperature, and solvent usage, subsequently lowering the energy consumption and environmental effects.

5.2.4. Ultrasound-Assisted Extraction (UAE)

Ultrasound-assisted extraction (UAE) is an established extraction method that has been successfully employed to extract polyphenols, carotenoids, volatiles, and polysaccharides from various FVW [117,247–251]. This approach reduces the extraction time (saving energy) and solvent usage while increasing the bioactive component yield from FVW. It is also one of the most common green extraction methods because it is fast and straightforward. UAE utilizes mechanical waves from 20 to 100 kHz [252]. These waves are composed of compression and rarefaction cycles that can travel through any media, displacing and dislodging the treated FVW cell matrix. The cavitation bubbles implode forcefully at the end of the rarefaction cycle, releasing tremendous amounts of energy at temperatures up to 5000 K and pressures up to 50 MPa [253]. The collapsing cavitation bubbles will cause microjets, fragmentation due to rapid interparticle collision, localized erosion, pore creation, shear forces, enhanced absorption, and enhanced swelling index values in the treated plant cells. Reduced particle size, higher surface area, and high mass transfer rates in the border layer of the solid matrix contribute to the solubilization of bioactive components [254]; thus, by improving the mass transfer between the plant cells and solvent, UAE can improve extraction. Despite the considerable energy produced by collapsed cavitation bubbles, the timeframe for these processes is too short to affect the overall system; hence, UAE is the ideal method to extract heat-sensitive compounds [255].

Combining two or more operating factors (frequency, power, duty cycle, temperature, solvent type, and extraction time) creates synergistic effects [256]; thus, determining the extraction kinetics is critical to optimizing extraction times and lowering energy use. In experiments involving sonoporation, capillarity, and detexturization in plant cells, frequency has been shown to have a significant impact on the bioactive chemical yield and characteristics [257]. Low-frequency, high-intensity ultrasound produces strong shear and mechanical forces that are desirable in the extraction process, whereas high-frequency, low-power density ultrasound produces a large number of reactive radicals [247,249,258].

Low frequency is preferred due to large cavitation effects that diminish with ultrasound frequency. Extraction with more than 20 kHz energy affects the physicochemical properties of phytochemicals, causing chemical deterioration and free radical production [259]. Using response surface methods, González-Centeno et al. [248] determined that 40 kHz was most successful in extracting phenolics from grape pomace. The yields were high at both low and high frequencies but low at intermediate frequencies for all examined responses.

The power levels used for UAE bioactives from FVW vary depending on the component to be extracted and the plant matrix chosen for extraction [254]; however, UAE power is inversely proportional to the cavitation bubbles generated within the solvent or solid media. This relationship is significant because as cavitation bubbles collapse, they increase the contact area between the solid and solvent, the shear forces causing turbulence behavior, and ultimately cell wall rupture and solvent penetration. Although increasing the power increases the extraction yield, it should only be increased to a point where the cavitation effects do not diminish. The influence of power on yield also depends on other extraction parameters such as the temperature and solvent extraction time. Achat et al. [260] discovered that the UAE power (60 W) and solvent temperature (olive oil, 16 °C) had significant impacts on oleuropein extraction, with TPC extraction increasing by 53% (414 mg oleuropein eq./100 g). The maximum TPC extraction yield (30.7 mg GAE/g) was achieved by Martínez-Patiño et al. [261] using high amplitude percentages (70%) and extended ultrasonication periods (15 min); however, elevated temperatures (>75 °C) were reached at the end of the experiment, perhaps restricting the extraction yield.

The combined impacts of the sample particle size, solvent-to-solid ratio, pH, temperature, and extraction time on the yields of targeted bioactive chemicals should not be underestimated. More surface area means a higher yield, especially when combined with a higher reaction temperature and the correct solvent-to-solid ratio. The bigger concentration difference improves the solute diffusivity and solubility in the solvent, enhancing extraction [254]. Pectin recovery is high when the pH is low. Insoluble pectin is hydrolyzed into soluble pectin, and the molecular weight of the pectin is reduced, boosting dissolution into the surrounding medium and recovery [262]. The effect of increased sonication time on yield is similar to the effects from increasing the power and temperature. Longer exposure times and higher power input speed up the disintegration of dissolved pectin, producing simpler monosaccharides. According to Wang et al. [262] and Xu et al. [263], increasing the extraction time and decreasing the power intensity (lower energy expenditure) did not improve the pectin extraction yield in grapefruit peels.

It is essential to screen the optimal operating parameters that may help boost the extraction of the targeted chemicals and may further optimize the settings used in prior studies. Furthermore, secondary contamination from the UAE probe should be considered when extracting bioactive chemicals from FVW.

6. Conclusions and Future Perspectives

Fruit and vegetable wastes (FVW) from the food bioprocessing industry result in environmental pollution; however, these wastes can be valuable sources of polymer materials. This review focuses on recent advances in biocomposites, active packaging, and by-products and the innovative technologies used for bioactive compound extraction. The mechanical, thermal, antibacterial, and physicochemical properties of FVW-based biocomposites have recently shown improvement. Common matrices used for biocomposites are PLA and PP. Additionally, pectin is an extracted compound used in the polymeric matrix, and the films produced are both active and biodegradable. Additionally, blueberry, sweet potato, and black chokeberry dyes and pigments are added to the films as pH indicators to trace and monitor food freshness throughout storage. To extract bioactive compounds from FVW, various techniques are used, including thermal extraction. Traditional methods (conventional and Soxhlet heating extractions) and new emerging methods (pressurized liquid extraction, subcritical water extraction, and microwave-assisted extraction) have been reviewed in this paper. As a result of the reduced use of organic solvents in the extraction procedures, these well-developed and established methods have been recognized as green technology methods.

While the use of FVW has the potential to improve polymer properties, it is critical to maintain the low-cost benefits of using FVW while also maintaining the mechanical and thermal properties. Surface treatments with biopolymers or fibers improve these properties but are more expensive. The high temperatures involved in such processes may reduce the quality of the resulting bioactive compounds. The value of FVW for use in biocomposites is expected to grow for both industry and research applications due to global waste concerns. Additionally, the vast majority of FVW-based research has been conducted at the bench scale. The next level of valorization of FVW should be scaling up this process to the industrial level. Overall, the growth of FVW-based polymer materials has been rapid, and their applications in active packaging, biocomposites, by-products, and recent technologies for the extraction of bioactive compounds mean they appear to have a promising future in the coming years.

Author Contributions: Conceptualization, M.S.M.B. and N.N.A.K.S.; validation, A.S. and M.Z.M.N.; writing—original draft preparation, M.S.M.B., N.N.A.K.S., A.S., I.S.M.A.T., M.Z.M.N., S.H.A., N.H.A.G. and F.S.M.S.; supervision, N.N.A.K.S. and M.S.M.B.; project administration, M.S.M.B. All authors have read and agreed to the published version of the manuscript.

Funding: This research received no external funding.

Institutional Review Board Statement: Not applicable.

Informed Consent Statement: Not applicable.

Data Availability Statement: The data presented in this study are available on request from the corresponding author.

Acknowledgments: The authors gratefully acknowledge the technical and financial support from the Universiti Putra Malaysia (UPM).

Conflicts of Interest: The authors declare no conflict of interest.

References

1. Tonini, D.; Albizzati, P.F.; Astrup, T.F. Environmental impacts of food waste: Learnings and challenges from a case study on UK. *Waste Manag.* **2018**, *76*, 744–766. [CrossRef]
2. Stenmarck, Â.; Jensen, C.; Quested, T.; Moates, G.; Buksti, M.; Cseh, B.; Juul, S.; Parry, A.; Politano, A.; Redlingshofer, B. *Estimates of European Food Waste Levels*; IVL Swedish Environmental Research Institute: Stockholm, Sweden, 2016.
3. FAO. *The State of Food and Agriculture 2019: Moving forward on Food Loss and Waste Reduction*; FAO: Rome, Italy, 2019; pp. 2–13.
4. International Food Policy Research Institute. *2019 Global Food Policy Report*; International Food Policy Research Institute: Washington, DC, USA, 2019.
5. Fiore, M.; Pellegrini, G.; Sala, P.L.; Conte, A.; Liu, B. Attitude toward food waste reduction: The case of Italian consumers. *Int. J. Glob. Small Bus.* **2017**, *9*, 185–201. [CrossRef]
6. Mc Carthy, U.; Uysal, I.; Badia-Melis, R.; Mercier, S.; O'Donnell, C.; Ktenioudaki, A. Global food security–Issues, challenges and technological solutions. *Trends Food Sci. Technol.* **2018**, *77*, 11–20. [CrossRef]
7. FAO. *Tracking Progress on Food and Agriculture-Related SDG Indicators 2020—A Report on the Indicators under FAO Custodianship*; Food and Agriculture Organization of the United Nations: Rome, Italy, 2020.
8. Edwiges, T.; Frare, L.; Mayer, B.; Lins, L.; Triolo, J.M.; Flotats, X.; de Mendonça Costa, M.S.S. Influence of chemical composition on biochemical methane potential of fruit and vegetable waste. *Waste Manag.* **2018**, *71*, 618–625. [CrossRef]
9. Negri, C.; Ricci, M.; Zilio, M.; D'Imporzano, G.; Qiao, W.; Dong, R.; Adani, F. Anaerobic digestion of food waste for bio-energy production in China and Southeast Asia: A review. *Renew. Sustain. Energy Rev.* **2020**, *133*, 110138. [CrossRef]
10. Yang, J.; Cai, W.; Ma, M.; Li, L.; Liu, C.; Ma, X.; Li, L.; Chen, X. Driving forces of China's CO_2 emissions from energy consumption based on Kaya-LMDI methods. *Sci. Total. Environ.* **2020**, *711*, 134569. [CrossRef]
11. Bong, C.P.C.; Ho, W.S.; Hashim, H.; Lim, J.S.; Ho, C.S.; Tan, W.S.P.; Lee, C.T. Review on the renewable energy and solid waste management policies towards biogas development in Malaysia. *Renew. Sustain. Energy Rev.* **2017**, *70*, 988–998. [CrossRef]
12. De Laurentiis, V.; Corrado, S.; Sala, S. Quantifying household waste of fresh fruit and vegetables in the EU. *Waste Manag.* **2018**, *77*, 238–251. [CrossRef]
13. Barrera, E.L.; Hertel, T. Global food waste across the income spectrum: Implications for food prices, production and resource use. *Food Policy* **2021**, *98*, 101874. [CrossRef]
14. Brizi, A.; Biraglia, A. "Do I have enough food"? How need for cognitive closure and gender impact stockpiling and food waste during the COVID-19 pandemic: A cross-national study in India and the United States of America. *Personal. Individ. Differ.* **2021**, *168*, 110396. [CrossRef]
15. Li, Y.; Wang, L.-E.; Liu, G.; Cheng, S. Rural household food waste characteristics and driving factors in China. *Resour. Conserv. Recycl.* **2021**, *164*, 105209. [CrossRef]
16. McCarthy, B.; Kapetanaki, A.B.; Wang, P. Completing the food waste management loop: Is there market potential for value-added surplus products (VASP)? *J. Clean. Prod.* **2020**, *256*, 120435. [CrossRef]
17. Esparza, I.; Jiménez-Moreno, N.; Bimbela, F.; Ancín-Azpilicueta, C.; Gandía, L.M. Fruit and vegetable waste management: Conventional and emerging approaches. *J. Environ. Manag.* **2020**, *265*, 110510. [CrossRef] [PubMed]
18. Fabi, C.; Cachia, F.; Conforti, P.; English, A.; Moncayo, J.R. Improving data on food losses and waste: From theory to practice. *Food Policy* **2021**, *98*, 101934. [CrossRef]
19. Sagar, N.A.; Pareek, S.; Sharma, S.; Yahia, E.M.; Lobo, M.G. Fruit and vegetable waste: Bioactive compounds, their extraction, and possible utilization. *Compr. Rev. Food Sci. Food Saf.* **2018**, *17*, 512–531. [CrossRef]
20. Álvarez, C.; Mullen, A.M.; Pojić, M.; Hadnađev, T.D.; Papageorgiou, M. Classification and target compounds. In *Food Waste Recovery*; Elsevier: Amsterdam, The Netherlands, 2021; pp. 21–49.
21. Li, W.; Bhat, S.A.; Li, J.; Cui, G.; Wei, Y.; Yamada, T.; Li, F. Effect of excess activated sludge on vermicomposting of fruit and vegetable waste by using novel vermireactor. *Bioresour. Technol.* **2020**, *302*, 122816. [CrossRef]
22. Chen, D.M.-C.; Bodirsky, B.L.; Krueger, T.; Mishra, A.; Popp, A. The world's growing municipal solid waste: Trends and impacts. *Environ. Res. Lett.* **2020**, *15*, 074021. [CrossRef]
23. Liakou, V.; Pateraki, C.; Palaiogeorgou, A.-M.; Kopsahelis, N.; de Castro, A.M.; Freire, D.M.G.; Nychas, G.-J.E.; Papanikolaou, S.; Koutinas, A. Valorisation of fruit and vegetable waste from open markets for the production of 2, 3-butanediol. *Food Bioprod. Process.* **2018**, *108*, 27–36. [CrossRef]
24. Plazzotta, S.; Manzocco, L.; Nicoli, M.C. Fruit and vegetable waste management and the challenge of fresh-cut salad. *Trends Food Sci. Technol.* **2017**, *63*, 51–59. [CrossRef]
25. Chatterjee, B.; Mazumder, D. New approach of characterizing fruit and vegetable waste (FVW) to ascertain its biological stabilization via two-stage anaerobic digestion (AD). *Biomass Bioenergy* **2020**, *139*, 105594. [CrossRef]

26. Bayram, B.; Ozkan, G.; Kostka, T.; Capanoglu, E.; Esatbeyoglu, T. Valorization and Application of Fruit and Vegetable Wastes and By-Products for Food Packaging Materials. *Molecules* **2021**, *26*, 4031. [CrossRef]
27. Kadzińska, J.; Janowicz, M.; Kalisz, S.; Bryś, J.; Lenart, A. An overview of fruit and vegetable edible packaging materials. *Packag. Technol. Sci.* **2019**, *32*, 483–495. [CrossRef]
28. Osorio, L.L.D.R.; Flórez-López, E.; Grande-Tovar, C.D. The Potential of Selected Agri-Food Loss and Waste to Contribute to a Circular Economy: Applications in the Food, Cosmetic and Pharmaceutical Industries. *Molecules* **2021**, *26*, 515. [CrossRef]
29. Figueroa, K.H.N.; García, N.V.M.; Vega, R.C. Cocoa By-Products. In *Food Wastes and By-Products: Nutraceutical and Health Potential*; John Wiley & Sons: Hoboken, NJ, USA, 2020; pp. 373–411.
30. Trigo, J.P.; Alexandre, E.M.; Saraiva, J.A.; Pintado, M.E. High value-added compounds from fruit and vegetable by-products–Characterization, bioactivities, and application in the development of novel food products. *Crit. Rev. Food Sci. Nutr.* **2020**, *60*, 1388–1416. [CrossRef]
31. Coman, V.; Teleky, B.-E.; Mitrea, L.; Martău, G.A.; Szabo, K.; Călinoiu, L.-F.; Vodnar, D.C. Bioactive potential of fruit and vegetable wastes. In *Advances in Food and Nutrition Research*; Elsevier: Amsterdam, The Netherlands, 2020; Volume 91, pp. 157–225.
32. Calderón-Oliver, M.; López-Hernández, L.H. Food vegetable and fruit waste used in meat products. *Food Rev. Int.* **2020**, 1–27. [CrossRef]
33. Lee, S.Y.; Lee, S.J.; Choi, D.S.; Hur, S.J. Current topics in active and intelligent food packaging for preservation of fresh foods. *J. Sci. Food Agric.* **2015**, *95*, 2799–2810. [CrossRef]
34. Ali, A.; Chen, Y.; Liu, H.; Yu, L.; Baloch, Z.; Khalid, S.; Zhu, J.; Chen, L. Starch-based antimicrobial films functionalized by pomegranate peel. *Int. J. Biol. Macromol.* **2019**, *129*, 1120–1126. [CrossRef]
35. Dash, K.K.; Ali, N.A.; Das, D.; Mohanta, D. Thorough evaluation of sweet potato starch and lemon-waste pectin based-edible films with nano-titania inclusions for food packaging applications. *Int. J. Biol. Macromol.* **2019**, *139*, 449–458. [CrossRef] [PubMed]
36. Szymańska-Chargot, M.; Chylińska, M.; Pieczywek, P.M.; Walkiewicz, A.; Pertile, G.; Frąc, M.; Cieślak, K.J.; Zdunek, A. Evaluation of nanocomposite made of polylactic acid and nanocellulose from carrot pomace modified with silver nanoparticles. *Polymers* **2020**, *12*, 812. [CrossRef] [PubMed]
37. Montero, Y.; Souza, A.G.; Oliveira, É.R.; dos Santos Rosa, D. Nanocellulose functionalized with cinnamon essential oil: A potential application in active biodegradable packaging for strawberry. *Sustain. Mater. Technol.* **2021**, *29*, e00289.
38. Lozano-Navarro, J.I.; Díaz-Zavala, N.P.; Velasco-Santos, C.; Melo-Banda, J.A.; Páramo-García, U.; Paraguay-Delgado, F.; García-Alamilla, R.; Martínez-Hernández, A.L.; Zapién-Castillo, S. Chitosan-starch films with natural extracts: Physical, chemical, morphological and thermal properties. *Materials* **2018**, *11*, 120. [CrossRef] [PubMed]
39. Luchese, C.L.; Garrido, T.; Spada, J.C.; Tessaro, I.C.; de la Caba, K. Development and characterization of cassava starch films incorporated with blueberry pomace. *Int. J. Biol. Macromol.* **2018**, *106*, 834–839. [CrossRef]
40. De Moraes Crizel, T.; Costa, T.M.H.; de Oliveira Rios, A.; Flôres, S.H. Valorization of food-grade industrial waste in the obtaining active biodegradable films for packaging. *Ind. Crop. Prod.* **2016**, *87*, 218–228. [CrossRef]
41. Shukor, U.; Nordin, N.; Tawakkal, I.; Talib, R.; Othman, S. Utilization of jackfruit peel waste for the production of biodegradable and active antimicrobial packaging films. In *Biopolymers and Biocomposites from Agro-Waste for Packaging Applications*; Elsevier: Amsterdam, The Netherlands, 2021; pp. 171–192.
42. Priyadarshi, R.; Kumar, B.; Deeba, F.; Kulshreshtha, A.; Negi, Y.S. Chitosan films incorporated with Apricot (*Prunus armeniaca*) kernel essential oil as active food packaging material. *Food Hydrocoll.* **2018**, *85*, 158–166. [CrossRef]
43. Othman, S.H.; Tarmiti, N.A.N.; Shapi'i, R.A.; Zahiruddin, S.M.M.; Tawakkal, I.S.M.A.; Basha, R.K. Starch/banana pseudostem biocomposite films for potential food packaging applications. *BioResources* **2020**, *15*, 3984–3998. [CrossRef]
44. Gaikwad, K.K.; Lee, J.Y.; Lee, Y.S. Development of polyvinyl alcohol and apple pomace bio-composite film with antioxidant properties for active food packaging application. *J. Food Sci. Technol.* **2016**, *53*, 1608–1619. [CrossRef]
45. Lin, D.; Zheng, Y.; Wang, X.; Huang, Y.; Ni, L.; Chen, X.; Wu, Z.; Huang, C.; Yi, Q.; Li, J. Study on physicochemical properties, antioxidant and antimicrobial activity of okara soluble dietary fiber/sodium carboxymethyl cellulose/thyme essential oil active edible composite films incorporated with pectin. *Int. J. Biol. Macromol.* **2020**, *165*, 1241–1249. [CrossRef]
46. Gahruie, H.H.; Ziaee, E.; Eskandari, M.H.; Hosseini, S.M.H. Characterization of basil seed gum-based edible films incorporated with Zataria multiflora essential oil nanoemulsion. *Carbohydr. Polym.* **2017**, *166*, 93–103. [CrossRef] [PubMed]
47. Dos Santos Caetano, K.; Lopes, N.A.; Costa, T.M.H.; Brandelli, A.; Rodrigues, E.; Flôres, S.H.; Cladera-Olivera, F. Characterization of active biodegradable films based on cassava starch and natural compounds. *Food Packag. Shelf Life* **2018**, *16*, 138–147. [CrossRef]
48. Emam-Djomeh, Z.; Moghaddam, A.; Yasini Ardakani, S.A. Antimicrobial activity of pomegranate (*Punica granatum* L.) peel extract, physical, mechanical, barrier and antimicrobial properties of pomegranate peel extract-incorporated sodium caseinate film and application in packaging for ground beef. *Packag. Technol. Sci.* **2015**, *28*, 869–881. [CrossRef]
49. Madhusudan, P.; Chellukuri, N.; Shivakumar, N. Smart packaging of food for the 21st century–A review with futuristic trends, their feasibility and economics. *Mater. Today Proc.* **2018**, *5*, 21018–21022. [CrossRef]
50. Luchese, C.L.; Abdalla, V.F.; Spada, J.C.; Tessaro, I.C. Evaluation of blueberry residue incorporated cassava starch film as pH indicator in different simulants and foodstuffs. *Food Hydrocoll.* **2018**, *82*, 209–218. [CrossRef]
51. Luchese, C.L.; Sperotto, N.; Spada, J.C.; Tessaro, I.C. Effect of blueberry agro-industrial waste addition to corn starch-based films for the production of a pH-indicator film. *Int. J. Biol. Macromol.* **2017**, *104*, 11–18. [CrossRef]
52. Sohany, M.; Tawakkal, I.S.M.A.; Ariffin, S.H.; Shah, N.N.A.K.; Yusof, Y.A. Characterization of Anthocyanin Associated Purple Sweet Potato Starch and Peel-Based pH Indicator Films. *Foods* **2021**, *10*, 2005. [CrossRef]

53. Chen, S.; Wu, M.; Lu, P.; Gao, L.; Yan, S.; Wang, S. Development of pH indicator and antimicrobial cellulose nanofibre packaging film based on purple sweet potato anthocyanin and oregano essential oil. *Int. J. Biol. Macromol.* **2020**, *149*, 271–280. [CrossRef] [PubMed]
54. Chen, H.-Z.; Zhang, M.; Bhandari, B.; Yang, C.-H. Novel pH-sensitive films containing curcumin and anthocyanins to monitor fish freshness. *Food Hydrocoll.* **2020**, *100*, 105438. [CrossRef]
55. Ezati, P.; Tajik, H.; Moradi, M.; Molaei, R. Intelligent pH-sensitive indicator based on starch-cellulose and alizarin dye to track freshness of rainbow trout fillet. *Int. J. Biol. Macromol.* **2019**, *132*, 157–165. [CrossRef]
56. Andretta, R.; Luchese, C.L.; Tessaro, I.C.; Spada, J.C. Development and characterization of pH-indicator films based on cassava starch and blueberry residue by thermocompression. *Food Hydrocoll.* **2019**, *93*, 317–324. [CrossRef]
57. Zeng, P.; Chen, X.; Qin, Y.-R.; Zhang, Y.-H.; Wang, X.-P.; Wang, J.-Y.; Ning, Z.-X.; Ruan, Q.-J.; Zhang, Y.-S. Preparation and characterization of a novel colorimetric indicator film based on gelatin/polyvinyl alcohol incorporating mulberry anthocyanin extracts for monitoring fish freshness. *Food Res. Int.* **2019**, *126*, 108604. [CrossRef]
58. Halász, K.; Csóka, L. Black chokeberry (*Aronia melanocarpa*) pomace extract immobilized in chitosan for colorimetric pH indicator film application. *Food Packag. Shelf Life* **2018**, *16*, 185–193. [CrossRef]
59. Ventorino, V.; Robertiello, A.; Viscardi, S.; Ambrosanio, A.; Faraco, V.; Pepe, O. Bio-based chemical production from Arundo donax feedstock fermentation using Cosenzaea myxofaciens BPM1. *BioResources* **2016**, *11*, 6566–6581. [CrossRef]
60. Kringel, D.H.; Dias, A.R.G.; Zavareze, E.d.R.; Gandra, E.A. Fruit wastes as promising sources of starch: Extraction, properties, and applications. *Starch-Stärke* **2020**, *72*, 1900200. [CrossRef]
61. Hassan, M.A.; Yee, L.-N.; Yee, P.L.; Ariffin, H.; Raha, A.R.; Shirai, Y.; Sudesh, K. Sustainable production of polyhydroxyalkanoates from renewable oil-palm biomass. *Biomass Bioenergy* **2013**, *50*, 1–9. [CrossRef]
62. Sulaiman, A.; Othman, N.; Baharuddin, A.S.; Mokhtar, M.N.; Tabatabaei, M. Enhancing the halal food industry by utilizing food wastes to produce value-added bioproducts. *Procedia-Soc. Behav. Sci.* **2014**, *121*, 35–43. [CrossRef]
63. Reis, M.; Albuquerque, M.; Villano, M.; Majone, M. *Mixed Culture Processes for Polyhydroxyalkanoate Production from Agro-Industrial Surplus/Wastes as Feedstocks*; Elsevier: Amsterdam, The Netherlands, 2011.
64. Chumee, J.; Khemmakama, P. Carboxymethyl cellulose from pineapple peel: Useful green bioplastic. *Adv. Mater. Res.* **2014**, *979*, 366–369. [CrossRef]
65. Nayak, A.; Bhushan, B. An overview of the recent trends on the waste valorization techniques for food wastes. *J. Environ. Manag.* **2019**, *233*, 352–370. [CrossRef]
66. Kim, H.; Moon, J.Y.; Kim, H.; Lee, D.-S.; Cho, M.; Choi, H.-K.; Kim, Y.S.; Mosaddik, A.; Cho, S.K. Antioxidant and antiproliferative activities of mango (*Mangifera indica* L.) flesh and peel. *Food Chem.* **2010**, *121*, 429–436. [CrossRef]
67. Wang, S.; Marcone, M.F.; Barbut, S.; Lim, L.-T. Fortification of dietary biopolymers-based packaging material with bioactive plant extracts. *Food Res. Int.* **2012**, *49*, 80–91. [CrossRef]
68. Mohd Nordin, N.; Anuar, H.; Buys, Y.F.; Ali, F.; Thomas, S.; Mohd Nasir, N.A. Effect of freeze-dried durian skin nanofiber on the physical properties of poly(lactic acid) biocomposites. *Polym. Compos.* **2021**, *42*, 842–848. [CrossRef]
69. Singh, V.K.; Mukhopadhyay, S. Studies on the effect of hybridization on sound insulation of coir-banana-polypropylene hybrid biocomposites. *J. Nat. Fibers* **2020**, 1–10. [CrossRef]
70. Anuar, H.; Abd Rashid, S.M.S.; Nordin, N.M.; Ali, F.; Buys, Y.F.; Thomas, S.; Nasir, N.A.M.; Asri, S.E.A.M. Potential of fabrication of durian skin fiber biocomposites for food packaging application through the electricity impact analysis. *IIUM Eng. J.* **2021**, *22*, 294–305. [CrossRef]
71. Gisan, K.; Chan, M.; Koay, S. Solvent-cast Biofilm from Poly(lactic) Acid and Durian Husk Fiber: Tensile, Water Absorption, and Biodegradation Behaviors. *J. Nat. Fibers* **2020**, 1–12. [CrossRef]
72. Ong, H.; Kam, K.; Islam, A.; Bautista-Patacsil, L.; Villagracia, A. Flexural and thermal properties of biocomposites from *Cerbera manghas* L. and polypropylene. In Proceedings of the 1st International Conference on Biomass Utilization and Sustainable Energy, Virtual Conference, Malaysia, 15–16 December 2020; p. 012050.
73. Żelaziński, T. Properties of biocomposites from rapeseed meal, fruit pomace and microcrystalline cellulose made by press pressing: Mechanical and physicochemical characteristics. *Materials* **2021**, *14*, 890. [CrossRef]
74. Tanwar, R.; Gupta, V.; Kumar, P.; Kumar, A.; Singh, S.; Gaikwad, K.K. Development and characterization of PVA-starch incorporated with coconut shell extract and sepiolite clay as an antioxidant film for active food packaging applications. *Int. J. Biol. Macromol.* **2021**, *185*, 451–461. [CrossRef]
75. Rangaraj, V.M.; Rambabu, K.; Banat, F.; Mittal, V. Effect of date fruit waste extract as an antioxidant additive on the properties of active gelatin films. *Food Chem.* **2021**, *355*, 129631. [CrossRef]
76. Mardijanti, D.S.; Megantara, E.N.; Bahtiar, A.; Sunardi, S. Turning the Cocopith Waste into Myceliated Biocomposite to Make an Insulator. *Int. J. Biomater.* **2021**, *2021*, 6630657. [CrossRef]
77. Nagarjun, J.; Kanchana, J.; RajeshKumar, G.; Manimaran, S.; Krishnaprakash, M. Enhancement of Mechanical Behavior of PLA Matrix Using Tamarind and Date Seed Micro Fillers. *J. Nat. Fibers* **2021**, 1–13. [CrossRef]
78. Reinaldo, J.S.; Milfont, C.H.; Gomes, F.P.; Mattos, A.L.; Medeiros, F.G.; Lopes, P.F.; Matsui, K.N.; Ito, E.N. Influence of grape and acerola residues on the antioxidant, physicochemical and mechanical properties of cassava starch biocomposites. *Polym. Test.* **2021**, *93*, 107015. [CrossRef]

79. Marzuki, M.N.A.; Tawakkal, I.S.M.A.; Basri, M.S.M.; Othman, S.H.; Kamarudin, S.H.; Lee, C.H.; Khalina, A. The effect of jackfruit skin powder and fiber bleaching treatment in pla composites with incorporation of thymol. *Polymers* **2020**, *12*, 2622. [CrossRef]
80. Farhadi, S.; Javanmard, M. Mechanical and physical properties of polyethylene/sour cherry shell powder bio-composite as potential food packaging. *Food Sci. Nutr.* **2021**, *9*, 3071–3077. [CrossRef]
81. Torres, F.G.; Mayorga, J.P.; Vilca, C.; Arroyo, J.; Castro, P.; Rodriguez, L. Preparation and characterization of a novel starch–chestnut husk biocomposite. *SN Appl. Sci.* **2019**, *1*, 1–7. [CrossRef]
82. Peças, P.; Carvalho, H.; Salman, H.; Leite, M. Natural fibre composites and their applications: A review. *J. Compos. Sci.* **2018**, *2*, 66. [CrossRef]
83. Soofi, M.; Alizadeh, A.; Hamishehkar, H.; Almasi, H.; Roufegarinejad, L. Preparation of nanobiocomposite film based on lemon waste containing cellulose nanofiber and savory essential oil: A new biodegradable active packaging system. *Int. J. Biol. Macromol.* **2021**, *169*, 352–361. [CrossRef]
84. Rasid, N.; Nazmi, N.; Isa, M.; Sarbon, N. Rheological, functional and antioxidant properties of films forming solution and active gelatin films incorporated with *Centella asiatica* (L.) urban extract. *Food Packag. Shelf Life* **2018**, *18*, 115–124. [CrossRef]
85. Viuda-Martos, M.; Ruiz-Navajas, Y.; Fernández-López, J.; Pérez-Álvarez, J. Antifungal activity of lemon (*Citrus lemon* L.), mandarin (*Citrus reticulata* L.), grapefruit (*Citrus paradisi* L.) and orange (*Citrus sinensis* L.) essential oils. *Food Control* **2008**, *19*, 1130–1138. [CrossRef]
86. Kaanin-Boudraa, G.; Brahmi, F.; Wrona, M.; Nerín, C.; Moudache, M.; Mouhoubi, K.; Madani, K.; Boulekbache-Makhlouf, L. Response surface methodology and UPLC-QTOF-MSE analysis of phenolic compounds from grapefruit (Citrus × paradisi) by-products as novel ingredients for new antioxidant packaging. *LWT* **2021**, *151*, 112158. [CrossRef]
87. Lalnunthari, C.; Devi, L.M.; Amami, E.; Badwaik, L.S. Valorisation of pumpkin seeds and peels into biodegradable packaging films. *Food Bioprod. Process.* **2019**, *118*, 58–66. [CrossRef]
88. Hanani, Z.N.; Husna, A.A.; Syahida, S.N.; Khaizura, M.N.; Jamilah, B. Effect of different fruit peels on the functional properties of gelatin/polyethylene bilayer films for active packaging. *Food Packag. Shelf Life* **2018**, *18*, 201–211. [CrossRef]
89. Gutiérrez, T.J.; Herniou-Julien, C.; Álvarez, K.; Alvarez, V.A. Structural properties and in vitro digestibility of edible and pH-sensitive films made from guinea arrowroot starch and wastes from wine manufacture. *Carbohydr. Polym.* **2018**, *184*, 135–143. [CrossRef]
90. Deng, Q.; Zhao, Y. Physicochemical, nutritional, and antimicrobial properties of wine grape (cv. Merlot) pomace extract-based films. *J. Food Sci.* **2011**, *76*, E309–E317. [CrossRef]
91. Suradi, S.; Yunus, R.; Beg, M.; Rivai, M.; Yusof, Z. Oil palm bio-fiber reinforced thermoplastic composites-effects of matrix modification on mechanical and thermal properties. *J. Appl. Sci.* **2010**, *10*, 3271–3276. [CrossRef]
92. Prachayawarakorn, J.; Sangnitidej, P.; Boonpasith, P. Properties of thermoplastic rice starch composites reinforced by cotton fiber or low-density polyethylene. *Carbohydr. Polym.* **2010**, *81*, 425–433. [CrossRef]
93. Cheng, W. Preparation and properties of lignocellulosic fiber/$CaCO_3$/thermoplastic starch composites. *Carbohydr. Polym.* **2019**, *211*, 204–208. [CrossRef] [PubMed]
94. Ochi, S. Fabrication of Manila Hemp Fiber Reinforced Cross Ply Biodegradable Composites and Their Tensile Properties. *Open J. Compos. Mater.* **2018**, *8*, 75. [CrossRef]
95. Leites, L.C.; Frick, P.J.M.; Cristina, T.I. Influence of the incorporation form of waste from the production of orange juice in the properties of cassava starch-based films. *Food Hydrocoll.* **2021**, *117*, 106730. [CrossRef]
96. Tavker, N.; Yadav, V.K.; Yadav, K.K.; Cabral-Pinto, M.; Alam, J.; Shukla, A.K.; Ali, F.A.A.; Alhoshan, M. Removal of cadmium and chromium by mixture of silver nanoparticles and nano-fibrillated cellulose isolated from waste peels of citrus sinensis. *Polymers* **2021**, *13*, 234. [CrossRef]
97. Phirom-on, K.; Apiraksakorn, J. Development of cellulose-based prebiotic fiber from banana peel by enzymatic hydrolysis. *Food Biosci.* **2021**, *41*, 101083. [CrossRef]
98. Ayala-Zavala, J.; Vega-Vega, V.; Rosas-Domínguez, C.; Palafox-Carlos, H.; Villa-Rodriguez, J.; Siddiqui, M.W.; Dávila-Aviña, J.; González-Aguilar, G. Agro-industrial potential of exotic fruit byproducts as a source of food additives. *Food Res. Int.* **2011**, *44*, 1866–1874. [CrossRef]
99. Khoozani, A.A.; Birch, J.; Bekhit, A.E.-D.A. Production, application and health effects of banana pulp and peel flour in the food industry. *J. Food Sci. Technol.* **2019**, *56*, 548–559. [CrossRef]
100. De Faria Arquelau, P.B.; Silva, V.D.M.; Garcia, M.A.V.T.; de Araújo, R.L.B.; Fante, C.A. Characterization of edible coatings based on ripe "Prata" banana peel flour. *Food Hydrocoll.* **2019**, *89*, 570–578. [CrossRef]
101. Reißner, A.M.; Al-Hamimi, S.; Quiles, A.; Schmidt, C.; Struck, S.; Hernando, I.; Turner, C.; Rohm, H. Composition and physicochemical properties of dried berry pomace. *J. Sci. Food Agric.* **2019**, *99*, 1284–1293. [CrossRef] [PubMed]
102. De Andrade, R.M.S.; Silva, S.; Costa, C.M.d.S.F.; Veiga, M.; Costa, E.; Ferreira, M.S.L.; de Andrade Gonçalves, E.C.B.; Pintado, M.E. Potential prebiotic effect of fruit and vegetable byproducts flour using in vitro gastrointestinal digestion. *Food Res. Int.* **2020**, *137*, 109354. [CrossRef] [PubMed]
103. Sharma, P.; Gupta, A.; Issar, K. Effect of packaging and storage on dried apple pomace and fiber extracted from pomace. *J. Food Process. Preserv.* **2017**, *41*, e12913. [CrossRef]
104. Mohankumararadhya, H.; Wadappi, P.; Chandrashekar, A.; Naik, Y. Studies on bio waste product particle reinforced polymer composites. In *AIP Conference Proceedings*; AIP Publishing LLC.: Melville, NY, USA, 2020; p. 030047.

105. Begum, Y.A.; Deka, S.C. Effect of processing on structural, thermal, and physicochemical properties of dietary fiber of culinary banana bracts. *J. Food Process. Preserv.* **2019**, *43*, e14256. [CrossRef]
106. Mir, S.A.; Bosco, S.J.D.; Shah, M.A.; Santhalakshmy, S.; Mir, M.M. Effect of apple pomace on quality characteristics of brown rice based cracker. *J. Saudi Soc. Agric. Sci.* **2017**, *16*, 25–32. [CrossRef]
107. Khamsucharit, P.; Laohaphatanalert, K.; Gavinlertvatana, P.; Sriroth, K.; Sangseethong, K. Characterization of pectin extracted from banana peels of different varieties. *Food Sci. Biotechnol.* **2018**, *27*, 623–629. [CrossRef]
108. Maran, J.P.; Priya, B.; Al-Dhabi, N.A.; Ponmurugan, K.; Moorthy, I.G.; Sivarajasekar, N. Ultrasound assisted citric acid mediated pectin extraction from industrial waste of Musa balbisiana. *Ultrason. Sonochem.* **2017**, *35*, 204–209. [CrossRef]
109. Moorthy, I.G.; Maran, J.P.; Ilakya, S.; Anitha, S.; Sabarima, S.P.; Priya, B. Ultrasound assisted extraction of pectin from waste Artocarpus heterophyllus fruit peel. *Ultrason. Sonochem.* **2017**, *34*, 525–530. [CrossRef]
110. Hosseini, S.S.; Khodaiyan, F.; Kazemi, M.; Najari, Z. Optimization and characterization of pectin extracted from sour orange peel by ultrasound assisted method. *Int. J. Biol. Macromol.* **2019**, *125*, 621–629. [CrossRef]
111. Guandalini, B.B.V.; Rodrigues, N.P.; Marczak, L.D.F. Sequential extraction of phenolics and pectin from mango peel assisted by ultrasound. *Food Res. Int.* **2019**, *119*, 455–461. [CrossRef]
112. Chaiwarit, T.; Masavang, S.; Mahe, J.; Sommano, S.; Ruksiriwanich, W.; Brachais, C.-H.; Chambin, O.; Jantrawut, P. Mango (cv. Nam Dokmai) peel as a source of pectin and its potential use as a film-forming polymer. *Food Hydrocoll.* **2020**, *102*, 105611. [CrossRef]
113. Laysandra, L.; Santosa, F.H.; Austen, V.; Soetaredjo, F.E.; Foe, K.; Putro, J.N.; Ju, Y.-H.; Ismadji, S. Rarasaponin-bentonite-activated biochar from durian shells composite for removal of crystal violet and Cr (VI) from aqueous solution. *Environ. Sci. Pollut. Res.* **2018**, *25*, 30680–30695. [CrossRef]
114. Adilah, A.N.; Jamilah, B.; Noranizan, M.; Hanani, Z.N. Utilization of mango peel extracts on the biodegradable films for active packaging. *Food Packag. Shelf Life* **2018**, *16*, 1–7. [CrossRef]
115. Isopencu, G.O.; Stoica-Guzun, A.; Busuioc, C.; Stroescu, M.; Deleanu, I.M. Development of antioxidant and antimicrobial edible coatings incorporating bacterial cellulose, pectin, and blackberry pomace. *Carbohydr. Polym. Technol. Appl.* **2021**, *2*, 100057.
116. Priyadarshi, R.; Riahi, Z.; Rhim, J.-W. Antioxidant pectin/pullulan edible coating incorporated with Vitis vinifera grape seed extract for extending the shelf life of peanuts. *Postharvest Biol. Technol.* **2022**, *183*, 111740. [CrossRef]
117. Amofa-Diatuo, T.; Anang, D.M.; Barba, F.J.; Tiwari, B.K. Development of new apple beverages rich in isothiocyanates by using extracts obtained from ultrasound-treated cauliflower by-products: Evaluation of physical properties and consumer acceptance. *J. Food Compos. Anal.* **2017**, *61*, 73–81. [CrossRef]
118. Majumdar, S.; Naha, A.; Bhattacharyya, D.; Bhowal, J. Effective delignification and decrystallization of cauliflower wastes by using dilute phosphoric acid for efficient enzymatic digestibility to produce fermentable sugars. *Biomass Bioenergy* **2019**, *125*, 169–179. [CrossRef]
119. Iwassa, I.J.; Piai, J.F.; Bolanho, B.C. Fiber concentrates from asparagus by-products: Microstructure, composition, functional and antioxidant properties. *Ciência Agrotecnol.* **2019**, *43*, e007319. [CrossRef]
120. Kazemi, M.; Khodaiyan, F.; Hosseini, S.S. Eggplant peel as a high potential source of high methylated pectin: Ultrasonic extraction optimization and characterization. *LWT* **2019**, *105*, 182–189. [CrossRef]
121. Chiboub, W.; Sassi, A.B.; Amina, C.M.h.; Souilem, F.; El Ayeb, A.; Djlassi, B.; Ascrizzi, R.; Flamini, G.; Harzallah-Skhiri, F. Valorization of the green waste from two varieties of fennel and carrot cultivated in tunisia by identification of the phytochemical profile and evaluation of the antimicrobial activities of their essentials oils. *Chem. Biodivers.* **2019**, *16*, e1800546. [CrossRef]
122. Caliceti, C.; Capriotti, A.; Calabria, D.; Bonvicini, F.; Zenezini Chiozzi, R.; Montone, C.; Piovesana, S.; Zangheri, M.; Mirasoli, M.; Simoni, P. Peptides from cauliflower by-products, obtained by an efficient, ecosustainable, and semi-industrial method, exert protective effects on endothelial function. *Oxidative Med. Cell. Longev.* **2019**, *2019*, 1046504. [CrossRef]
123. Otoni, C.G.; Lodi, B.D.; Lorevice, M.V.; Leitão, R.C.; Ferreira, M.D.; de Moura, M.R.; Mattoso, L.H. Optimized and scaled-up production of cellulose-reinforced biodegradable composite films made up of carrot processing waste. *Ind. Crop. Prod.* **2018**, *121*, 66–72. [CrossRef]
124. Uçak, I. Physicochemical and antimicrobial effects of gelatin-based edible films incorporated with garlic peel extract on the rainbow trout fillets. *Prog. Nutr.* **2019**, *21*, 232–240.
125. Liu, C.; Jin, T.; Liu, W.; Hao, W.; Yan, L.; Zheng, L. Effects of hydroxyethyl cellulose and sodium alginate edible coating containing asparagus waste extract on postharvest quality of strawberry fruit. *LWT* **2021**, *148*, 111770. [CrossRef]
126. Zhang, Q.-W.; Lin, L.-G.; Ye, W.-C. Techniques for extraction and isolation of natural products: A comprehensive review. *Chin. Med.* **2018**, *13*, 1–26. [CrossRef] [PubMed]
127. Rakariyatham, K.; Zhou, D.; Rakariyatham, N.; Shahidi, F. Sapindaceae (Dimocarpus longan and Nephelium lappaceum) seed and peel by-products: Potential sources for phenolic compounds and use as functional ingredients in food and health applications. *J. Funct. Foods* **2020**, *67*, 103846. [CrossRef]
128. Zia, S.; Khan, M.R.; Shabbir, M.A.; Aslam Maan, A.; Khan, M.K.I.; Nadeem, M.; Khalil, A.A.; Din, A.; Aadil, R.M. An inclusive overview of advanced thermal and nonthermal extraction techniques for bioactive compounds in food and food-related matrices. *Food Rev. Int.* **2020**, 1–31. [CrossRef]
129. RedCorn, R.; Fatemi, S.; Engelberth, A.S. Comparing end-use potential for industrial food-waste sources. *Engineering* **2018**, *4*, 371–380. [CrossRef]

130. Dao, T.A.T.; Webb, H.K.; Malherbe, F. Optimization of pectin extraction from fruit peels by response surface method: Conventional versus microwave-assisted heating. *Food Hydrocoll.* **2021**, *113*, 106475.
131. Kwiatkowski, M.; Kravchuk, O.; Skouroumounis, G.K.; Taylor, D.K. Response surface parallel optimization of extraction of total phenolics from separate white and red grape skin mixtures with microwave-assisted and conventional thermal methods. *J. Clean. Prod.* **2020**, *251*, 119563. [CrossRef]
132. Misra, N.; Yadav, S.K. Extraction of pectin from black carrot pomace using intermittent microwave, ultrasound and conventional heating: Kinetics, characterization and process economics. *Food Hydrocoll.* **2020**, *102*, 105592.
133. Cherif, M.M.; Grigorakis, S.; Halahlah, A.; Loupassaki, S.; Makris, D.P. High-efficiency extraction of phenolics from wheat waste biomass (bran) by combining deep eutectic solvent, ultrasound-assisted pretreatment and thermal treatment. *Environ. Process.* **2020**, *7*, 845–859. [CrossRef]
134. Marić, M.; Grassino, A.N.; Zhu, Z.; Barba, F.J.; Brnčić, M.; Brnčić, S.R. An overview of the traditional and innovative approaches for pectin extraction from plant food wastes and by-products: Ultrasound-, microwaves-, and enzyme-assisted extraction. *Trends Food Sci. Technol.* **2018**, *76*, 28–37. [CrossRef]
135. Wang, W.; Ma, X.; Jiang, P.; Hu, L.; Zhi, Z.; Chen, J.; Ding, T.; Ye, X.; Liu, D. Characterization of pectin from grapefruit peel: A comparison of ultrasound-assisted and conventional heating extractions. *Food Hydrocoll.* **2016**, *61*, 730–739. [CrossRef]
136. Rodsamran, P.; Sothornvit, R. Microwave heating extraction of pectin from lime peel: Characterization and properties compared with the conventional heating method. *Food Chem.* **2019**, *278*, 364–372. [CrossRef] [PubMed]
137. De Oliveira, C.F.; Gurak, P.D.; Cladera-Olivera, F.; Marczak, L.D.F.; Karwe, M. Combined effect of high-pressure and conventional heating on pectin extraction from passion fruit peel. *Food Bioprocess Technol.* **2016**, *9*, 1021–1030. [CrossRef]
138. Hossain, S.; Taher, S.; Khan, A.; Sultana, N.; Irfan, M.; Haq, B.; Razzak, S. Experimental study and modeling approach of response surface methodology coupled with crow search algorithm for optimizing the extraction conditions of papaya seed waste oil. *Arab. J. Sci. Eng.* **2020**, *45*, 7371–7383. [CrossRef]
139. De Menezes, M.L.; Johann, G.; Diório, A.; Pereira, N.C.; da Silva, E.A. Phenomenological determination of mass transfer parameters of oil extraction from grape biomass waste. *J. Clean. Prod.* **2018**, *176*, 130–139. [CrossRef]
140. Hu, B.; Xi, X.; Li, H.; Qin, Y.; Li, C.; Zhang, Z.; Liu, Y.; Zhang, Q.; Liu, A.; Liu, S. A comparison of extraction yield, quality and thermal properties from Sapindus mukorossi seed oil between microwave assisted extraction and Soxhlet extraction. *Ind. Crop. Prod.* **2021**, *161*, 113185. [CrossRef]
141. Flórez Montes, C.; Felipe Rojas-González, A.; Rodríguez-Barona, S. Evaluation of Extracts Obtained from Fruit Wastes Using Different Methods. *Ingeniería* **2021**, *26*, 1–11.
142. Fernandez-Pastor, I.; Fernandez-Hernandez, A.; Perez-Criado, S.; Rivas, F.; Martinez, A.; Garcia-Granados, A.; Parra, A. Microwave-assisted extraction versus Soxhlet extraction to determine triterpene acids in olive skins. *J. Sep. Sci.* **2017**, *40*, 1209–1217. [CrossRef]
143. Alias, N.H.; Abbas, Z. Microwave-assisted extraction of phenolic compound from pineapple skins: The optimum operating condition and comparision with soxhlet extraction. *Malays. J. Anal. Sci.* **2017**, *21*, 690–699.
144. Khandare, R.D.; Tomke, P.D.; Rathod, V.K. Kinetic modeling and process intensification of ultrasound-assisted extraction of d-limonene using citrus industry waste. *Chem. Eng. Process. -Process. Intensif.* **2021**, *159*, 108181. [CrossRef]
145. Garude, H.; Ade, V.; Gadhave, R. Comparative Study of Various Methods for Extraction of Antioxidants Compound from Bitter Gourd Peel. *Adv. Life Sci.* **2016**, *24*, 11189–11192.
146. Sukatta, U.; Rugthaworn, P.; Khanoonkon, N.; Anongjanya, P.; Kongsin, K.; Sukyai, P.; Harnkarnsujarit, N.; Sothornvit, R.; Chollakup, R. Rambutan (*Nephelium lappaceum*) peel extract: Antimicrobial and antioxidant activities and its application as a bioactive compound in whey protein isolate film. *Songklanakarin J. Sci. Technol.* **2021**, *43*, 37–44.
147. Jridi, M.; Boughriba, S.; Abdelhedi, O.; Nciri, H.; Nasri, R.; Kchaou, H.; Kaya, M.; Sebai, H.; Zouari, N.; Nasri, M. Investigation of physicochemical and antioxidant properties of gelatin edible film mixed with blood orange (Citrus sinensis) peel extract. *Food Packag. Shelf Life* **2019**, *21*, 100342. [CrossRef]
148. Košťálová, Z.; Aguedo, M.; Hromádková, Z. Microwave-assisted extraction of pectin from unutilized pumpkin biomass. *Chem. Eng. Process. Process. Intensif.* **2016**, *102*, 9–15. [CrossRef]
149. Narkprasom, K.; Tanongkankit, Y.; Saenscharoenrat, P.; Narkprasom, N. Optimization of Total Phenolic from Euphoria longana Lam. Seed by Microwave Assisted Extraction. *Burapha Sci. J.* **2019**, *24*, 48–63.
150. Bakić, M.T.; Pedisić, S.; Zorić, Z.; Dragović-Uzelac, V.; Grassino, A.N. Effect of microwave-assisted extraction on polyphenols recovery from tomato peel waste. *Acta Chim. Slov.* **2019**, *66*, 367–377. [CrossRef]
151. Zheng, J.; Li, H.; Wang, D.; Li, R.; Wang, S.; Ling, B. Radio frequency assisted extraction of pectin from apple pomace: Process optimization and comparison with microwave and conventional methods. *Food Hydrocoll.* **2021**, *121*, 107031. [CrossRef]
152. Rivadeneira, J.P.; Wu, T.; Ybanez, Q.; Dorado, A.A.; Migo, V.P.; Nayve, F.R.P.; Castillo-Israel, K.A.T. Microwave-assisted extraction of pectin from "Saba" banana peel waste: Optimization, characterization, and rheology study. *Int. J. Food Sci.* **2020**, *2020*, 8879425. [CrossRef]
153. Zain, N.; Nazeri, M. Antioxidant and mineral content of pitaya peel extract obtained using microwave assisted extraction (MAE). *Aust. J. Basic Appl. Sci.* **2016**, *10*, 63–68.
154. Karbuz, P.; Tugrul, N. Microwave and ultrasound assisted extraction of pectin from various fruits peel. *J. Food Sci. Technol.* **2021**, *58*, 641–650. [CrossRef]

155. Wongkaew, M.; Sommano, S.R.; Tangpao, T.; Rachtanapun, P.; Jantanasakulwong, K. Mango peel pectin by microwave-assisted extraction and its use as fat replacement in dried Chinese sausage. *Foods* **2020**, *9*, 450. [CrossRef]
156. Hayat, K.; Hussain, S.; Abbas, S.; Farooq, U.; Ding, B.; Xia, S.; Jia, C.; Zhang, X.; Xia, W. Optimized microwave-assisted extraction of phenolic acids from citrus mandarin peels and evaluation of antioxidant activity in vitro. *Sep. Purif. Technol.* **2009**, *70*, 63–70. [CrossRef]
157. Gharibzahedi, S.M.T.; Smith, B.; Guo, Y. Ultrasound-microwave assisted extraction of pectin from fig (*Ficus carica* L.) skin: Optimization, characterization and bioactivity. *Carbohydr. Polym.* **2019**, *222*, 114992. [CrossRef] [PubMed]
158. Mao, Y.; Robinson, J.; Binner, E. Understanding heat and mass transfer processes during microwave-assisted and conventional solvent extraction. *Chem. Eng. Sci.* **2021**, *233*, 116418. [CrossRef]
159. Swamy, G.J.; Muthukumarappan, K. Optimization of continuous and intermittent microwave extraction of pectin from banana peels. *Food Chem.* **2017**, *220*, 108–114. [CrossRef] [PubMed]
160. Zin, M.M.; Anucha, C.B.; Bánvölgyi, S. Recovery of phytochemicals via electromagnetic irradiation (microwave-assisted-extraction): Betalain and phenolic compounds in perspective. *Foods* **2020**, *9*, 918. [CrossRef]
161. Kumar, M.; Dahuja, A.; Sachdev, A.; Kaur, C.; Varghese, E.; Saha, S.; Sairam, K. Valorisation of black carrot pomace: Microwave assisted extraction of bioactive phytoceuticals and antioxidant activity using Box–Behnken design. *J. Food Sci. Technol.* **2019**, *56*, 995–1007. [CrossRef]
162. Torres-León, C.; Rojas, R.; Serna-Cock, L.; Belmares-Cerda, R.; Aguilar, C.N. Extraction of antioxidants from mango seed kernel: Optimization assisted by microwave. *Food Bioprod. Process.* **2017**, *105*, 188–196. [CrossRef]
163. Reis, L.C.; Carneiro, L.M.; Branco, C.R.; Branco, A. Comparison of conventional microwave and focused microwave-assisted extraction to enhance the efficiency of the extraction of antioxidant flavonols from jocote pomace (*Spondias purpurea* L.). *Plant Foods Hum. Nutr.* **2015**, *70*, 160–169. [CrossRef]
164. Varadharajan, V.; Shanmugam, S.; Ramaswamy, A. Model generation and process optimization of microwave-assisted aqueous extraction of anthocyanins from grape juice waste. *J. Food Process. Eng.* **2017**, *40*, e12486. [CrossRef]
165. Popovic, B.M.; Micic, N.; Potkonjak, A.; Blagojevic, B.; Pavlovic, K.; Milanov, D.; Juric, T. Novel extraction of polyphenols from sour cherry pomace using natural deep eutectic solvents–Ultrafast microwave-assisted NADES preparation and extraction. *Food Chem.* **2022**, *366*, 130562. [CrossRef]
166. Martínez-Abad, A.; Ramos, M.; Hamzaoui, M.; Kohnen, S.; Jiménez, A.; Garrigós, M.C. Optimisation of sequential microwave-assisted extraction of essential oil and pigment from lemon peels waste. *Foods* **2020**, *9*, 1493. [CrossRef] [PubMed]
167. Casas, E.V.; Comedia, V.J.G.; Gilbuena, A.G.; Yaptenco, K.F. Optimizing Microwave-assisted Crude Butter Extraction from Carabao Mango (Mangifera indica) Kernels. *Sci. Diliman* **2015**, *27*, 41–75.
168. Utama-Ang, N.; Sida, S.; Wanachantararak, P.; Kawee-Ai, A. Development of edible Thai rice film fortified with ginger extract using microwave-assisted extraction for oral antimicrobial properties. *Sci. Rep.* **2021**, *11*, 14870. [CrossRef] [PubMed]
169. Sepelevs, I.; Zagorska, J.; Galoburda, R. A food-grade antioxidant production using industrial potato peel by–products. *Agron. Res.* **2020**, *18*, 2.
170. Sari, A.; Ishartani, D.; Dewanty, P. Effects of microwave power and irradiation time on pectin extraction from watermelon rinds (*Citrullus lanatus*) with acetic acid using microwave assisted extraction method. In *IOP Conference Series: Earth and Environmental Science*; IOP Publishing: Bristol, UK, 2018; p. 012085.
171. Herrero, M.; del Pilar Sánchez-Camargo, A.; Cifuentes, A.; Ibáñez, E. Plants, seaweeds, microalgae and food by-products as natural sources of functional ingredients obtained using pressurized liquid extraction and supercritical fluid extraction. *TrAC Trends Anal. Chem.* **2015**, *71*, 26–38. [CrossRef]
172. Allcca-Alca, E.E.; León-Calvo, N.C.; Luque-Vilca, O.M.; Martínez-Cifuentes, M.; Pérez-Correa, J.R.; Mariotti-Celis, M.S.; Huamán-Castilla, N.L. Hot Pressurized Liquid Extraction of Polyphenols from the Skin and Seeds of *Vitis vinifera* L. cv. Negra Criolla Pomace a Peruvian Native Pisco Industry Waste. *Agronomy* **2021**, *11*, 866. [CrossRef]
173. Viganó, J.; Brumer, I.Z.; de Campos Braga, P.A.; da Silva, J.K.; Júnior, M.R.M.; Reyes, F.G.R.; Martínez, J. Pressurized liquids extraction as an alternative process to readily obtain bioactive compounds from passion fruit rinds. *Food Bioprod. Process.* **2016**, *100*, 382–390. [CrossRef]
174. García, P.; Fredes, C.; Cea, I.; Lozano-Sánchez, J.; Leyva-Jiménez, F.J.; Robert, P.; Vergara, C.; Jimenez, P. Recovery of Bioactive Compounds from Pomegranate (*Punica granatum* L.) Peel Using Pressurized Liquid Extraction. *Foods* **2021**, *10*, 203. [CrossRef] [PubMed]
175. Katsinas, N.; Bento da Silva, A.; Enríquez-de-Salamanca, A.; Fernández, N.; Bronze, M.R.; Rodríguez-Rojo, S. Pressurized Liquid Extraction Optimization from Supercritical Defatted Olive Pomace: A Green and Selective Phenolic Extraction Process. *ACS Sustain. Chem. Eng.* **2021**, *9*, 5590–5602. [CrossRef]
176. Espada-Bellido, E.; Ferreiro-González, M.; Barbero, G.F.; Carrera, C.; Palma, M.; Barroso, C.G. Alternative extraction method of bioactive compounds from mulberry (*Morus nigra* L.) pulp using pressurized-liquid extraction. *Food Anal. Methods* **2018**, *11*, 2384–2395. [CrossRef]
177. Lasta, H.F.B.; Lentz, L.; Rodrigues, L.G.G.; Mezzomo, N.; Vitali, L.; Ferreira, S.R.S. Pressurized liquid extraction applied for the recovery of phenolic compounds from beetroot waste. *Biocatal. Agric. Biotechnol.* **2019**, *21*, 101353. [CrossRef]

178. Cejudo-Bastante, C.; Arjona-Mudarra, P.; Fernández-Ponce, M.T.; Casas, L.; Mantell, C.; Martínez de la Ossa, E.J.; Pereyra, C. Application of a Natural Antioxidant from Grape Pomace Extract in the Development of Bioactive Jute Fibers for Food Packaging. *Antioxidants* **2021**, *10*, 216. [CrossRef] [PubMed]
179. Abdelmoez, W.; Abdelfatah, R.; Tayeb, A.; Yoshida, H. Extraction of cottonseed oil using subcritical water technology. *AIChE J.* **2011**, *57*, 2353–2359. [CrossRef]
180. Tunchaiyaphum, S.; Eshtiaghi, M.; Yoswathana, N. Extraction of bioactive compounds from mango peels using green technology. *Int. J. Chem. Eng. Appl.* **2013**, *4*, 194. [CrossRef]
181. Ravber, M.; Knez, Ž.; Škerget, M. Simultaneous extraction of oil-and water-soluble phase from sunflower seeds with subcritical water. *Food Chem.* **2015**, *166*, 316–323. [CrossRef] [PubMed]
182. Zakaria, S.M.; Kamal, S.M.M. Subcritical water extraction of bioactive compounds from plants and algae: Applications in pharmaceutical and food ingredients. *Food Eng. Rev.* **2016**, *8*, 23–34. [CrossRef]
183. King, J.W.; Grabiel, R.D. Isolation of Polyphenolic Compounds from Fruits or Vegetables Utilizing Sub-Critical Water Extraction. U.S. Patent 7,208,181, 24 April 2007.
184. Lu, J.; Feng, X.; Han, Y.; Xue, C. Optimization of subcritical fluid extraction of carotenoids and chlorophyll a from Laminaria japonica Aresch by response surface methodology. *J. Sci. Food Agric.* **2014**, *94*, 139–145. [CrossRef] [PubMed]
185. Cheigh, C.-I.; Yoo, S.-Y.; Ko, M.-J.; Chang, P.-S.; Chung, M.-S. Extraction characteristics of subcritical water depending on the number of hydroxyl group in flavonols. *Food Chem.* **2015**, *168*, 21–26. [CrossRef] [PubMed]
186. Amashukeli, X.; Pelletier, C.C.; Kirby, J.P.; Grunthaner, F.J. Subcritical water extraction of amino acids from Atacama Desert soils. *J. Geophys. Res. Biogeosci.* **2007**, *112*, G04S16. [CrossRef]
187. Singh, P.P.; Saldaña, M.D. Subcritical water extraction of phenolic compounds from potato peel. *Food Res. Int.* **2011**, *44*, 2452–2458. [CrossRef]
188. Pourali, O.; Asghari, F.S.; Yoshida, H. Simultaneous rice bran oil stabilization and extraction using sub-critical water medium. *J. Food Eng.* **2009**, *95*, 510–516. [CrossRef]
189. Marcet, I.; Álvarez, C.; Paredes, B.; Díaz, M. The use of sub-critical water hydrolysis for the recovery of peptides and free amino acids from food processing wastes. Review of sources and main parameters. *Waste Manag.* **2016**, *49*, 364–371. [CrossRef]
190. Pedras, B.M.; Regalin, G.; Sá-Nogueira, I.; Simões, P.; Paiva, A.; Barreiros, S. Fractionation of red wine grape pomace by subcritical water extraction/hydrolysis. *J. Supercrit. Fluids* **2020**, *160*, 104793. [CrossRef]
191. Guthrie, F.; Wang, Y.; Neeve, N.; Quek, S.Y.; Mohammadi, K.; Baroutian, S. Recovery of phenolic antioxidants from green kiwifruit peel using subcritical water extraction. *Food Bioprod. Process.* **2020**, *122*, 136–144. [CrossRef]
192. Kim, D.-S.; Lim, S.-B. Semi-continuous subcritical water extraction of flavonoids from citrus unshiu peel: Their antioxidant and enzyme inhibitory activities. *Antioxidants* **2020**, *9*, 360. [CrossRef]
193. Limsangouan, N.; Milasing, N.; Thongngam, M.; Khuwijitjaru, P.; Jittanit, W. Physical and chemical properties, antioxidant capacity, and total phenolic content of xyloglucan component in tamarind (*Tamarindus indica*) seed extracted using subcritical water. *J. Food Process. Preserv.* **2019**, *43*, e14146. [CrossRef]
194. Li, B.; Akram, M.; Al-Zuhair, S.; Elnajjar, E.; Munir, M.T. Subcritical water extraction of phenolics, antioxidants and dietary fibres from waste date pits. *J. Environ. Chem. Eng.* **2020**, *8*, 104490. [CrossRef]
195. Teo, C.C.; Tan, S.N.; Yong, J.W.H.; Hew, C.S.; Ong, E.S. Pressurized hot water extraction (PHWE). *J. Chromatogr. A* **2010**, *1217*, 2484–2494. [CrossRef]
196. Kronholm, J.; Hartonen, K.; Riekkola, M.-L. Analytical extractions with water at elevated temperatures and pressures. *TrAC Trends Anal. Chem.* **2007**, *26*, 396–412. [CrossRef]
197. Ndlela, S.; De Moura, J.; Olson, N.; Johnson, L. Aqueous extraction of oil and protein from soybeans with subcritical water. *J. Am. Oil Chem. Soc.* **2012**, *89*, 1145–1153. [CrossRef]
198. Thani, N.M.; Kamal, S.M.M.; Taip, F.S.; Sulaiman, A.; Omar, R. Effect of sub-critical water hydrolysis on sugar recovery from bakery leftovers. *Food Bioprod. Process.* **2019**, *117*, 105–112. [CrossRef]
199. Ho, T.C.; Kim, M.H.; Cho, Y.-J.; Park, J.-S.; Nam, S.Y.; Chun, B.-S. Gelatin-sodium alginate based films with Pseuderanthemum palatiferum (Nees) Radlk. freeze-dried powder obtained by subcritical water extraction. *Food Packag. Shelf Life* **2020**, *24*, 100469. [CrossRef]
200. Adilah, Z.M.; Jamilah, B.; Hanani, Z.N. Functional and antioxidant properties of protein-based films incorporated with mango kernel extract for active packaging. *Food Hydrocoll.* **2018**, *74*, 207–218. [CrossRef]
201. Kchaou, H.; Benbettaïeb, N.; Jridi, M.; Abdelhedi, O.; Karbowiak, T.; Brachais, C.-H.; Léonard, M.-L.; Debeaufort, F.; Nasri, M. Enhancement of structural, functional and antioxidant properties of fish gelatin films using Maillard reactions. *Food Hydrocoll.* **2018**, *83*, 326–339. [CrossRef]
202. Ho, T.C.; Chun, B.S. Extraction of bioactive compounds from Pseuderanthemum palatiferum (Nees) Radlk. using subcritical water and conventional solvents: A Comparison Study. *J. Food Sci.* **2019**, *84*, 1201–1207. [CrossRef] [PubMed]
203. Mohd Thani, N.; Mustapa Kamal, S.M.; Sulaiman, A.; Taip, F.S.; Omar, R.; Izhar, S. Sugar Recovery from Food Waste via Sub-critical Water Treatment. *Food Rev. Int.* **2020**, *36*, 241–257. [CrossRef]
204. Khuwijitjaru, P.; Watsanit, K.; Adachi, S. Carbohydrate content and composition of product from subcritical water treatment of coconut meal. *J. Ind. Eng. Chem.* **2012**, *18*, 225–229. [CrossRef]
205. Cantero, D.A.; Bermejo, M.D.; Cocero, M.J. Reaction engineering for process intensification of supercritical water biomass refining. *J. Supercrit. Fluids* **2015**, *96*, 21–35. [CrossRef]

206. Vladić, J.; Jakovljević, M.; Molnar, M.; Vidović, S.; Tomić, M.; Drinić, Z.; Jokić, S. Valorization of Yarrow (*Achillea millefolium* L.) By-Product through Application of Subcritical Water Extraction. *Molecules* **2020**, *25*, 1878. [CrossRef] [PubMed]
207. Švarc-Gajić, J.; Morais, S.; Delerue-Matos, C.; Vieira, E.F.; Spigno, G. Valorization Potential of Oilseed Cakes by Subcritical Water Extraction. *Appl. Sci.* **2020**, *10*, 8815. [CrossRef]
208. He, L.; Zhang, X.; Xu, H.; Xu, C.; Yuan, F.; Knez, Ž.; Novak, Z.; Gao, Y. Subcritical water extraction of phenolic compounds from pomegranate (*Punica granatum* L.) seed residues and investigation into their antioxidant activities with HPLC–ABTS+ assay. *Food Bioprod. Process.* **2012**, *90*, 215–223. [CrossRef]
209. Pedras, B.; Salema-Oom, M.; Sa-Nogueira, I.; Simoes, P.; Paiva, A.; Barreiros, S. Valorization of white wine grape pomace through application of subcritical water: Analysis of extraction, hydrolysis, and biological activity of the extracts obtained. *J. Supercrit. Fluids* **2017**, *128*, 138–144. [CrossRef]
210. Giombelli, C.; Iwassa, I.J.; da Silva, C.; Barros, B.C.B. Valorization of peach palm by-product through subcritical water extraction of soluble sugars and phenolic compounds. *J. Supercrit. Fluids* **2020**, *165*, 104985. [CrossRef]
211. Yan, L.-G.; Xi, J. Micro-mechanism analysis of ultrahigh pressure extraction from green tea leaves by numerical simulation. *Sep. Purif. Technol.* **2017**, *180*, 51–57. [CrossRef]
212. Briones-Labarca, V.; Giovagnoli-Vicu, C.; Figueroa-Alvarez, P.; Quispe-Fuentes, I.; Pérez-Won, M. Extraction of β-carotene, vitamin C and antioxidant compounds from Physalis peruviana (Cape Gooseberry) assisted by high hydrostatic pressure. *Food Nutr. Sci.* **2013**, *4*, 35284.
213. Chemat, F.; Rombaut, N.; Meullemiestre, A.; Turk, M.; Perino, S.; Fabiano-Tixier, A.-S.; Abert-Vian, M. Review of green food processing techniques. Preservation, transformation, and extraction. *Innov. Food Sci. Emerg. Technol.* **2017**, *41*, 357–377. [CrossRef]
214. Azmir, J.; Zaidul, I.S.M.; Rahman, M.; Sharif, K.; Mohamed, A.; Sahena, F.; Jahurul, M.; Ghafoor, K.; Norulaini, N.; Omar, A. Techniques for extraction of bioactive compounds from plant materials: A review. *J. Food Eng.* **2013**, *117*, 426–436. [CrossRef]
215. Sharma, M.; Usmani, Z.; Gupta, V.K.; Bhat, R. Valorization of fruits and vegetable wastes and by-products to produce natural pigments. *Crit. Rev. Biotechnol.* **2021**, *41*, 535–563. [CrossRef]
216. Panzella, L.; Moccia, F.; Nasti, R.; Marzorati, S.; Verotta, L.; Napolitano, A. Bioactive phenolic compounds from agri-food wastes: An update on green and sustainable extraction methodologies. *Front. Nutr.* **2020**, *7*, 60. [CrossRef] [PubMed]
217. Mendiola, J.; Jaime, L.; Santoyo, S.; Reglero, G.; Cifuentes, A.; Ibañez, E.; Señoráns, F. Screening of functional compounds in supercritical fluid extracts from Spirulina platensis. *Food Chem.* **2007**, *102*, 1357–1367. [CrossRef]
218. Puértolas, E.; Barba, F.J. Electrotechnologies applied to valorization of by-products from food industry: Main findings, energy and economic cost of their industrialization. *Food Bioprod. Process.* **2016**, *100*, 172–184. [CrossRef]
219. Puértolas, E.; Luengo, E.; Álvarez, I.; Raso, J. Improving mass transfer to soften tissues by pulsed electric fields: Fundamentals and applications. *Annu. Rev. Food Sci. Technol.* **2012**, *3*, 263–282. [CrossRef]
220. Galanakis, C.M. Recovery of high added-value components from food wastes: Conventional, emerging technologies and commercialized applications. *Trends Food Sci. Technol.* **2012**, *26*, 68–87. [CrossRef]
221. Houska, M.; Pravda, P. Examples of Commercial Fruit and Vegetable Juices and Smoothies Cold Pasteurized by High Pressure. In *High Pressure Processing of Fruit and Vegetable Products*; CRC Press: Boca Raton, FL, USA, 2017; pp. 147–154.
222. Silva, F.V. Heat assisted HPP for the inactivation of bacteria, moulds and yeasts spores in foods: Log reductions and mathematical models. *Trends Food Sci. Technol.* **2019**, *88*, 143–156.
223. Silva, F.V.M.; Sulaiman, A. *Polyphenoloxidase in Fruit and Vegetables: Inactivation by Thermal and Non-Thermal Processes*; Elsevier: Amsterdam, The Netherlands, 2019.
224. Da Silva, F.V.M. High-Pressure Processing Effect on Microorganisms in Fruit and Vegetable Products. In *High Pressure Processing of Fruit and Vegetable Products*; CRC Press: Boca Raton, FL, USA, 2017; pp. 3–38.
225. Zhu, Z.; Gavahian, M.; Barba, F.J.; Roselló-Soto, E.; Kovačević, D.B.; Putnik, P.; Denoya, G.I. Valorization of waste and by-products from food industries through the use of innovative technologies. In *Agri-Food Industry Strategies for Healthy Diets and Sustainability*; Elsevier: Amsterdam, The Netherlands, 2020; pp. 249–266.
226. El-Shamy, S.; Farag, M.A. Novel trends in extraction and optimization methods of bioactives recovery from pomegranate fruit biowastes: Valorization purposes for industrial applications. *Food Chem.* **2021**, *365*, 130465. [CrossRef]
227. Jun, X. High-pressure processing as emergent technology for the extraction of bioactive ingredients from plant materials. *Crit. Rev. Food Sci. Nutr.* **2013**, *53*, 837–852. [CrossRef] [PubMed]
228. Scepankova, H.; Martins, M.; Estevinho, L.; Delgadillo, I.; Saraiva, J.A. Enhancement of bioactivity of natural extracts by non-thermal high hydrostatic pressure extraction. *Plant Foods Hum. Nutr.* **2018**, *73*, 253–267. [CrossRef] [PubMed]
229. Richard, J. *High Pressure Phase Behaviour of Multicomponent Fluid Mixtures*; Elsevier: Amsterdam, The Netherlands, 1992.
230. Samaranayake, C.P.; Sastry, S.K. In-situ pH measurement of selected liquid foods under high pressure. *Innov. Food Sci. Emerg. Technol.* **2013**, *17*, 22–26. [CrossRef]
231. Corrales, M.; Toepfl, S.; Butz, P.; Knorr, D.; Tauscher, B. Extraction of anthocyanins from grape by-products assisted by ultrasonics, high hydrostatic pressure or pulsed electric fields: A comparison. *Innov. Food Sci. Emerg. Technol.* **2008**, *9*, 85–91. [CrossRef]
232. Carciochi, R.A.; D'Alessandro, L.G.; Vauchel, P.; Rodriguez, M.M.; Nolasco, S.M.; Dimitrov, K. Valorization of agrifood by-products by extracting valuable bioactive compounds using green processes. In *Ingredients Extraction by Physicochemical Methods in Food*; Elsevier: Amsterdam, The Netherlands, 2017; pp. 191–228.

233. Armenta, S.; Garrigues, S.; Esteve-Turrillas, F.A.; de la Guardia, M. Green extraction techniques in green analytical chemistry. *TrAC Trends Anal. Chem.* **2019**, *116*, 248–253. [CrossRef]
234. Ameer, K.; Shahbaz, H.M.; Kwon, J.H. Green extraction methods for polyphenols from plant matrices and their byproducts: A review. *Compr. Rev. Food Sci. Food Saf.* **2017**, *16*, 295–315. [CrossRef]
235. Wijngaard, H.; Hossain, M.B.; Rai, D.K.; Brunton, N. Techniques to extract bioactive compounds from food by-products of plant origin. *Food Res. Int.* **2012**, *46*, 505–513. [CrossRef]
236. Baldino, L.; Scognamiglio, M.; Reverchon, E. Supercritical fluid technologies applied to the extraction of compounds of industrial interest from *Cannabis sativa* L. and to their pharmaceutical formulations: A review. *J. Supercrit. Fluids* **2020**, *165*, 104960. [CrossRef]
237. Oroian, M.; Escriche, I. Antioxidants: Characterization, natural sources, extraction and analysis. *Food Res. Int.* **2015**, *74*, 10–36. [CrossRef]
238. Evrendilek, G.A. Change regime of aroma active compounds in response to pulsed electric field treatment time, sour cherry juice apricot and peach nectars, and physical and sensory properties. *Innov. Food Sci. Emerg. Technol.* **2016**, *33*, 195–205. [CrossRef]
239. Nowosad, K.; Sujka, M.; Pankiewicz, U.; Kowalski, R. The application of PEF technology in food processing and human nutrition. *J. Food Sci. Technol.* **2021**, *58*, 397–411. [CrossRef]
240. Vaessen, E.; Timmermans, R.; Tempelaars, M.; Schutyser, M.; den Besten, H. Reversibility of membrane permeabilization upon pulsed electric field treatment in lactobacillus plantarum WCFS1. *Sci. Rep.* **2019**, *9*, 19990. [CrossRef] [PubMed]
241. Sarkis, J.R.; Boussetta, N.; Tessaro, I.C.; Marczak, L.D.F.; Vorobiev, E. Application of pulsed electric fields and high voltage electrical discharges for oil extraction from sesame seeds. *J. Food Eng.* **2015**, *153*, 20–27. [CrossRef]
242. Fincan, M.; Dejmek, P. In Situ visualization of the effect of a pulsed electric field on plant tissue. *J. Food Eng.* **2002**, *55*, 223–230. [CrossRef]
243. Ade-Omowaye, B.; Taiwo, K.; Eshtiaghi, N.; Angersbach, A.; Knorr, D. Comparative evaluation of the effects of pulsed electric field and freezing on cell membrane permeabilisation and mass transfer during dehydration of red bell peppers. *Innov. Food Sci. Emerg. Technol.* **2003**, *4*, 177–188. [CrossRef]
244. Aadil, R.M.; Zeng, X.-A.; Sun, D.-W.; Wang, M.-S.; Liu, Z.-W.; Zhang, Z.-H. Combined effects of sonication and pulsed electric field on selected quality parameters of grapefruit juice. *LWT-Food Sci. Technol.* **2015**, *62*, 890–893. [CrossRef]
245. Zhou, Y.; Zhao, X.; Huang, H. Effects of pulsed electric fields on anthocyanin extraction yield of blueberry processing by-products. *J. food Process. Preserv.* **2015**, *39*, 1898–1904. [CrossRef]
246. Pintać, D.; Majkić, T.; Torović, L.; Orčić, D.; Beara, I.; Simin, N.; Mimica–Dukić, N.; Lesjak, M. Solvent selection for efficient extraction of bioactive compounds from grape pomace. *Ind. Crop. Prod.* **2018**, *111*, 379–390. [CrossRef]
247. Ahmad-Qasem, M.H.; Cánovas, J.; Barrajón-Catalán, E.; Micol, V.; Cárcel, J.A.; García-Pérez, J.V. Kinetic and compositional study of phenolic extraction from olive leaves (var. Serrana) by using power ultrasound. *Innov. Food Sci. Emerg. Technol.* **2013**, *17*, 120–129. [CrossRef]
248. González-Centeno, M.R.; Knoerzer, K.; Sabarez, H.; Simal, S.; Rosselló, C.; Femenia, A. Effect of acoustic frequency and power density on the aqueous ultrasonic-assisted extraction of grape pomace (*Vitis vinifera* L.)—A response surface approach. *Ultrason. Sonochem.* **2014**, *21*, 2176–2184. [CrossRef]
249. Drosou, C.; Kyriakopoulou, K.; Bimpilas, A.; Tsimogiannis, D.; Krokida, M. A comparative study on different extraction techniques to recover red grape pomace polyphenols from vinification byproducts. *Ind. Crop. Prod.* **2015**, *75*, 141–149. [CrossRef]
250. Milella, R.A.; Basile, T.; Alba, V.; Gasparro, M.; Giannandrea, M.A.; Debiase, G.; Genghi, R.; Antonacci, D. Optimized ultrasonic-assisted extraction of phenolic antioxidants from grape (*Vitis vinifera* L.) skin using response surface methodology. *J. Food Sci. Technol.* **2019**, *56*, 4417–4428. [CrossRef] [PubMed]
251. Sengar, A.S.; Rawson, A.; Muthiah, M.; Kalakandan, S.K. Comparison of different ultrasound assisted extraction techniques for pectin from tomato processing waste. *Ultrason. Sonochem.* **2020**, *61*, 104812. [CrossRef] [PubMed]
252. Medina-Meza, I.G.; Barbosa-Cánovas, G.V. Assisted extraction of bioactive compounds from plum and grape peels by ultrasonics and pulsed electric fields. *J. Food Eng.* **2015**, *166*, 268–275. [CrossRef]
253. Barba, F.J.; Zhu, Z.; Koubaa, M.; Sant'Ana, A.S.; Orlien, V. Green alternative methods for the extraction of antioxidant bioactive compounds from winery wastes and by-products: A review. *Trends Food Sci. Technol.* **2016**, *49*, 96–109. [CrossRef]
254. Kumar, K.; Srivastav, S.; Sharanagat, V.S. Ultrasound assisted extraction (UAE) of bioactive compounds from fruit and vegetable processing by-products: A review. *Ultrason. Sonochem.* **2021**, *70*, 105325. [CrossRef]
255. Roselló-Soto, E.; Barba, F.J.; Parniakov, O.; Galanakis, C.M.; Lebovka, N.; Grimi, N.; Vorobiev, E. High voltage electrical discharges, pulsed electric field, and ultrasound assisted extraction of protein and phenolic compounds from olive kernel. *Food Bioprocess Technol.* **2015**, *8*, 885–894. [CrossRef]
256. Carciochi, R.A.; Manrique, G.D.; Dimitrov, K. Optimization of antioxidant phenolic compounds extraction from quinoa (*Chenopodium quinoa*) seeds. *J. Food Sci. Technol.* **2015**, *52*, 4396–4404. [CrossRef]
257. Rodríguez De Luna, S.L.; Ramírez-Garza, R.; Serna Saldívar, S.O. Environmentally friendly methods for flavonoid extraction from plant material: Impact of their operating conditions on yield and antioxidant properties. *Sci. World J.* **2020**, *2020*, 6792069. [CrossRef]
258. Gonzales, G.B.; Smagghe, G.; Raes, K.; Van Camp, J. Combined alkaline hydrolysis and ultrasound-assisted extraction for the release of nonextractable phenolics from cauliflower (*Brassica oleracea* var. botrytis) waste. *J. Agric. Food Chem.* **2014**, *62*, 3371–3376. [CrossRef]
259. Leong, T.; Ashokkumar, M.; Kentish, S. The fundamentals of power ultrasound—A review. *Acoust. Aust.* **2011**, *39*, 2.

260. Achat, S.; Tomao, V.; Madani, K.; Chibane, M.; Elmaataoui, M.; Dangles, O.; Chemat, F. Direct enrichment of olive oil in oleuropein by ultrasound-assisted maceration at laboratory and pilot plant scale. *Ultrason. Sonochem.* **2012**, *19*, 777–786. [CrossRef] [PubMed]
261. Martínez-Patiño, J.C.; Gullón, B.; Romero, I.; Ruiz, E.; Brnčić, M.; Žlabur, J.Š.; Castro, E. Optimization of ultrasound-assisted extraction of biomass from olive trees using response surface methodology. *Ultrason. Sonochem.* **2019**, *51*, 487–495. [CrossRef] [PubMed]
262. Wang, W.; Ma, X.; Xu, Y.; Cao, Y.; Jiang, Z.; Ding, T.; Ye, X.; Liu, D. Ultrasound-assisted heating extraction of pectin from grapefruit peel: Optimization and comparison with the conventional method. *Food Chem.* **2015**, *178*, 106–114. [CrossRef]
263. Xu, Y.; Zhang, L.; Bailina, Y.; Ge, Z.; Ding, T.; Ye, X.; Liu, D. Effects of ultrasound and/or heating on the extraction of pectin from grapefruit peel. *J. Food Eng.* **2014**, *126*, 72–81. [CrossRef]

polymers

MDPI

Article

Supramolecular Structure and Mechanical Performance of κ-Carrageenan–Gelatin Gel

Anastasiya O. Makarova [1], Svetlana R. Derkach [2], Aidar I. Kadyirov [3], Sufia A. Ziganshina [4], Mariia A. Kazantseva [1,5], Olga S. Zueva [6], Aidar T. Gubaidullin [7,8] and Yuriy F. Zuev [1,8,*]

1 Kazan Institute of Biochemistry and Biophysics, FRC Kazan Scientific Center of RAS, Lobachevsky St. 2/31, 420111 Kazan, Russia
2 Department of Chemistry, Murmansk State Technical University, Sportivnaya Str. 13, 183010 Murmansk, Russia
3 Institute of Power Engineering and Advanced Technologies, FRC Kazan Scientific Center of RAS, Lobachevsky St. 2/31, 420111 Kazan, Russia
4 Zavoisky Physical-Technical Institute, FRC Kazan Scientific Center of RAS, Sibirsky Tract 10/7, 420029 Kazan, Russia
5 HSE Tikhonov Moscow Institute of Electronics and Mathematics, Tallinskaya St. 34, 123458 Moscow, Russia
6 Department of Physics, Kazan State Power Engineering University, Krasnoselskaya St. 51, 420066 Kazan, Russia
7 Arbuzov Institute of Organic and Physical Chemistry, FRC Kazan Scientific Center of RAS, Arbuzov St. 8, 420088 Kazan, Russia
8 A. Butlerov Chemical Institute, Kazan Federal University, Kremlevskaya St. 18, 420008 Kazan, Russia
* Correspondence: yufzuev@mail.ru

Citation: Makarova, A.O.; Derkach, S.R.; Kadyirov, A.I.; Ziganshina, S.A.; Kazantseva, M.A.; Zueva, O.S.; Gubaidullin, A.T.; Zuev, Y.F. Supramolecular Structure and Mechanical Performance of κ-Carrageenan–Gelatin Gel. *Polymers* **2022**, *14*, 4347. https://doi.org/10.3390/polym14204347

Academic Editors: Domenico Acierno and Antonella Patti

Received: 1 September 2022
Accepted: 11 October 2022
Published: 15 October 2022

Publisher's Note: MDPI stays neutral with regard to jurisdictional claims in published maps and institutional affiliations.

Abstract: In this work, by means of complex physicochemical methods the structural features of a composite κ-carrageenan–gelatin system were studied in comparison with initial protein gel. The correlation between the morphology of hydrogels and their mechanical properties was demonstrated through the example of changes in their rheological characteristics. The experiments carried out with PXRD, SAXS, AFM and rheology approaches gave new information on the structure and mechanical performance of κ-carrageenan–gelatin hydrogel. The combination of PXRD, SAXS and AFM results showed that the morphological structures of individual components were not observed in the composite protein–polysaccharide hydrogels. The results of the mechanical testing of initial gelatin and engineered κ-carrageenan–gelatin gel showed the substantially denser parking of polymer chains in the composite system due to a significant increase in intermolecular protein–polysaccharide contacts. Close results were indirectly followed from the SAXS estimations—the driving force for the formation of the common supramolecular structural arrangement of proteins and polysaccharides was the increase in the density of network of macromolecular chains entanglements; therefore, an increase in the energy costs was necessary to change the conformational rearrangements of the studied system. This increase in the macromolecular arrangement led to the growth of the supramolecular associate size and the growth of interchain physical bonds. This led to an increase in the composite gel plasticity, whereas the enlargement of scattering particles made the novel gel system not only more rigid, but also more fragile.

Keywords: κ-carrageenan–gelatin hydrogel; supramolecular structure; mechanical performance

1. Introduction

Biomolecular hydrogels are attracting great attention as "smart" materials for different biotechnology and biomedicine applications [1–7].

Hydrogels form three-dimensional polymer networks capable of accumulating and holding up to 99% of water in their volume space [8,9]. Technologically more attractive are the "physical" hydrogels with the polymer network maintained by the mechanical weaving of polymer molecules and the combination of intermolecular interactions, such as ionic bridges, hydrogen bonding and hydrophobic forces [10]. Intriguingly, that supramolecular polymer network is preserved both in the gel and sol states, although with definite structural differences [11]. Because of biocompatibility, structural similarities to the native extracellular matrix and excellent permeability for nutrients and metabolites, hydrogels based on natural polysaccharides and proteins have gained considerable attention [12]. In particular, hydrogels are widely applied in tissue engineering, dressing and wound healing, where among others their mechanical properties are of great importance [13]. The development of hydrogel systems in tissue and organ regeneration, such as bones, cartilage and intervertebral discs presents a significant challenge. Engineered hydrogels have to meet certain requirements in biomedicine including definite fluidity under moderate pressure for use in injections, quick coagulation at the target site and maintenance of sufficient integrity and mechanical strength [14,15]. Thus, the mechanical performance is one of the most important properties of engineered implants.

Among natural biopolymers, biodegradable gelatin and κ-carrageenan are the universal candidates for the food and pharmacy industries and tissue engineering. Gelatin is the peptide product of partially hydrolyzed animal protein collagen [16]. Gelatin forms thermoreversible hydrogels through the coil-to-helix conformational transition [17]. κ-Carrageenan is a linear galactan, built of repeating disaccharide units of 1,4-linked α-D-galactose and 1,3-linked -β-D-galactose with a variable proportion of sulfate groups at different positions [18]. The hydrogel formed from κ-carrageenan occurs via the entanglement of polymer chains with the formation of ordered structures in the junction domains. The gelation of κ-carrageenan is temperature-dependent, with gelation at cooling and melting upon heating.

In the scientific literature there are many different examples of the engineering of mechanical properties of hydrogels to retain their favorable biomedical characteristics. It may be the mechanical reinforcing of the hydrogel structure by fibers, for instance by collagen in molecular form or by collagen fibrils or fiber bundles [15]. Another popular approach to modulate the initial mechanical properties of the biopolymer systems is the initiation of the additional chemical and physical cross-linking [19]. Because the subject of the present study is physical hydrogels, we shall refer here only to the possibilities of physical cross-linking in the mechanical performance of hydrogels, for example, such as the artificial increase in hydrogen bonding by the physical incorporation of H-bond donors of a polymer nature [20]. One other modifying approach is the use of synthetic nanosized additivities, such as carbon nanotubes which operate as structural and mechanical modifiers of the biopolymer network [11,21]. The combination of polysaccharides and proteins is one of the additional engineering approaches that influences the gel structure, its jelling conditions and the alteration of physical–chemical properties [22]. In gelatin/carrageenan hydrogels, the synergistic effects have been determined [23], for example, the increase in mechanical strength [24], flexibility and porosity and the water retention capacity [25].

In this work, the novel information on the modulation of gelatin structure and mechanical performance in the gel state were studied. The goal of the present study was to investigate the influence of κ-carrageenan on gelatin hydrogel structure and mechanical properties below the jelling temperature. The structure was analyzed with a complex of experimental approaches including small-angle X-ray scattering (SAXS), powder X-ray diffraction (XRD) and atomic force (AFM) microscopy. The interconnection between structural changes in gelatin hydrogel and the gel mechanical properties under the admixture of κ-carrageenan were demonstrated by rheology experiments.

In the present study, we compared the structure and mechanical behavior of the initial gelatin system with the data obtained earlier for the mixed κ-carrageenan–gelatin gel [11]. For this purpose, in this work we used a part of the graphical data and numerical characteristics of the mixed system obtained and given.

2. Materials and Methods

2.1. Materials

The sample of gelatin (gel strength 300, Type A, produced by Sigma-Aldrich, St. Louis, MO, USA) from porcine skin with M_η = 100 kDa was used as the protein component of hydrogel. κ-Carrageenan (Sigma-Aldrich, St. Louis, MO, USA) with $M\eta$ = 430 kDa was used in experiments as the polysaccharide. All measurements were carried out in ultrapure water purified using the "Arium mini" water purification system (Sartorius, Gottingen, Germany).

2.2. Preparation of Solutions and Gels

Aqueous solutions of κ-carrageenan and gelatin were prepared separately. Initially, the biopolymers were swollen in distilled water at 20 °C for 15 h. Solutions of κ-carrageenan and gelatin were prepared separately and dissolved at 70 and 50 °C, respectively. Aqueous mixtures of κ-carrageenan–gelatin complexes were prepared by stirring the initial solutions using an ultrasound bath for 60 min at 50 °C to obtain an aqueous mixture with the desired ratio of κ-carrageenan–gelatin Z = 0.8 (*w/w*). The pH values of the mixture of gelatin with κ-carrageenan were in the range of 5.3–5.5. At the given pH and biopolymer ratio, there was no macrophase separation in the system [26]. To study κ-carrageenan–gelatin hydrogel in the gel state (the jelling temperature of the studied composition is about 25 °C [27]), the temperature was decreased from 40 to 14 °C. Before all experiments in the gel state the samples were stored at 14 °C for 1 h.

2.3. X-ray Powder Diffraction

The PXRD diffractograms of initial gelatin powder and gelatin gel were determined in the reflection mode using a MiniFlex 600 diffractometer (Kazan Federal University, Kazan, Russia) equipped with a D/teX Ultra detector (Rigaku, Akishima-shi, Japan) (CuK_α, λ = 1.54178 Å, Ni-filter). The PXRD experiments with κ-carrageenan–gelatin gel were performed with the automatic Bruker D8 Advance diffractometer with the Vantec linear PSD (Bruker Corporation, Billerica, MA, USA) (λ $CuK_{\alpha 1}$ 1.5406 Å). The scattering data in all experiments were collected in the reflection mode with a flat-plate samples, which were placed on the standard silicon plate (Bruker Corporation, Billerica, MA, USA) with zero diffraction and were kept spinning (15 rpm) throughout the data collection. The diffraction patterns were recorded in the 2θ range between 3° and 90° in 0.008° steps with a step time of 0.1–4.0 s. Data processing was performed with software packages EVA Version 11 (Bruker AXS: Karlsruhe, Germany) and TOPAS Version 3 (Bruker AXS: Karlsruhe, Germany) [28,29].

2.4. Small-Angle X-ray Scattering

The SAXS experiments were fulfilled with the Nanostar diffractometer (Bruker AXS, Billerica, MA, USA) (CuK_α, λ = 1.5418 Å), coupled with the Gobbel mirrors optics and the HiStar 2D area detector (Bruker AXS, Billerica, MA, USA). The parameters of the diffractometer and the conditions and methods for performing experiments were similar to those used by us earlier and are described in detail in our publication [11]. The 2D scattering data processing was carried out with the SAXS program [30] and the calculations of structural parameters and simulation were performed with the SASView [31] and PRIMUS [32] programs. The studied hydrogels are described by the Gauss –Lorentz gel model [33], which well describes the scattering from physical networks, which is the characteristic of the gelatin and κ-carrageenan–gelatin gels. The SAXS response was modeled as the sum of exponential decays at low *s* values and the Lorentzian at higher values of *s* [34,35].

2.5. Atomic Force Microscopy (AFM)

The surface morphology of the samples was detected by atomic force microscope Titanium (NT-MDT, Zelenograd, Russia). Measurements were carried out under open air in the semi-contact mode. Silicon cantilevers NSG-10 (NT-MDT, Zelenograd, Russia) with a force constant of 3.1–37.6 Nm^{-1} and a resonant frequency of 140–390 kHz were used. The software Nova PX (NT-MDT, Zelenograd, Russia) was used to operate the microscope. The 2 μL volume of hydrogel was placed on a freshly cleaved mica surface and dried under ambient conditions. The AFM images were obtained at room temperature and processed and analyzed with the Image Analysis program (NT-MDT, Moscow, Russia).

2.6. Rheological Measurements

Rheological properties of the studied systems were evaluated using an MCR102 (Anton Paar, Graz, Austria) rotational rheometer with a "plate–plate" measuring system (two 50 mm diameter plates with a 0.5 mm gap between them). The temperature control of samples was fulfilled with the lower heating system and active casing both using the Peltier elements P-PTD200 (Anton Paar, Graz, Austria). The variation of given temperature was within ± 0.1 °C. Measurements were carried out in the following deformation modes: periodic oscillations at constant temperature (14 °C) with different amplitudes, ω, at a constant frequency, $\omega = 6.28\ s^{-1}$, or varying frequency, γ, at constant amplitude, $\gamma = 1\%$, the range was 0.9–139% and ω was 0.0671–23.8 s^{-1}. To exclude the initial gelation time from the final results, all samples of gels were thermally stabilized during 60 min at 14.0 °C under a weak dynamic oscillation.

3. Results

3.1. PXRD Overview of the Hydrogel Phase State

The powder X-ray diffraction method (PXRD) was used to compare the phase states of the initial gelatin powder product and gelatin gel under its natural drying with the structure of the κ-carrageenan–gelatin hydrogel. Original gelatin is the nanostructured sample (Figure 1A, black curve). A wide peak on the gelatin diffraction pattern at $2\theta = 20°$ can be defined as an "amorphous halo" and corresponds to the amorphous nature of protein [25,36]. The gelatin gel state was characterized by the same type of diffraction pattern as the κ-carrageenan-gelatin hydrogel (Figure 1B), depicting two broadened amorphous halos with maximums near the 30 and 40°, which characterized the average interatomic distances. To determine the initial structure of components in the powder phase, we studied the PXRD patterns under the drying of prepared solutions. During the natural drying of the samples placed on the surface of the silicon plate, these peaks degenerated into one small, broadened peak in the interval of diffraction angles of 10–15 degrees. Its low intensity was the result of a small amount of scattering substance after the drying. It is clear that under gelatin drying, it is structured like the original powdered phase. At the same time, the κ-carrageenan–gelatin gel had the less structured state, closer to an amorphous one. It is worth noting that no crystallization of initial crystallizing components in the κ-carrageenan–gelatin hydrogel was observed, which indirectly points out the homogenization of the system without the separation of the initial components into individual associates [37].

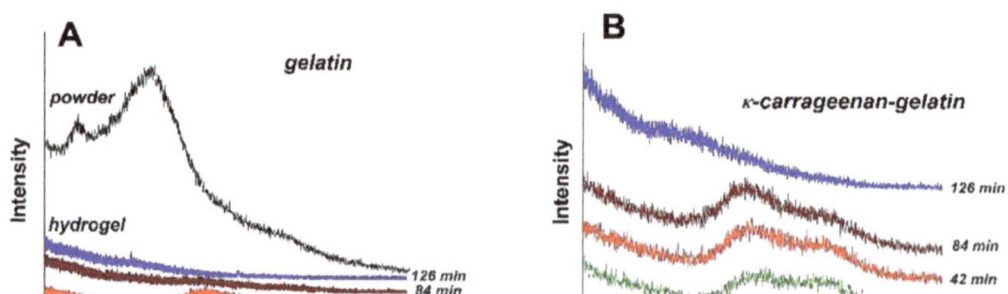

Figure 1. Diffraction patterns from original dry gelatin powder (black curve) and the gel state of gelatin (upon drying) (**A**), time evolution of X-ray diffraction for κ-carrageenan–gelatin hydrogel upon drying (**B**).

Thus, the integration of protein with polysaccharide into a composite hydrogel suppressed their individual structure and the formation of integral supramolecular structure. The phase homogeneity of the composite hydrogel, confirmed by the PXRD method cannot exclude the aggregation and segregation processes at all size scales, which, in turn, can be estimated in a hierarchical order by SAXS and AFM experiments. The influence of such segregation on the mechanical characteristics of κ-carrageenan–gelatin gel was evaluated by studying the rheology of this system.

3.2. SAXS Structural Characterization of Hydrogel Sol and Gel States

The integration of two-dimensional experimental patterns of small-angle scattering resulted in the one-dimensional SAXS curves with the shape which is typical for systems of non-interacting particles (Figure 2). The obtained high scattering intensity is evidence of the structural microheterogeneity of the studied gels—the presence of randomly oriented scattering particles (domains of increased density), with their size corresponding to the range of SAXS techniques (1–100 nm) [38]. To analyze the morphology of the studied gels, a number of structural characteristics were calculated from the obtained SAXS experimental data.

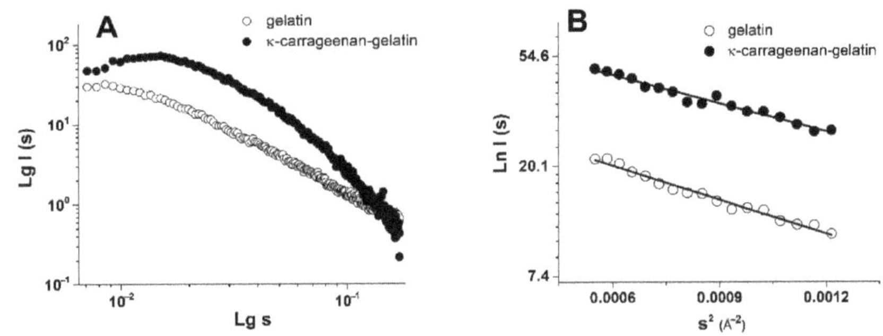

Figure 2. Experimental SAXS data for gelatin and κ-carrageenan–gelatin gel (**A**) and linear approximation of the Guinier plot (**B**) after background substruction.

The combined κ-carrageenan–gelatin hydrogel was characterized by more intense scattering (Figure 2A), which is associated with the participation of both the protein and polysaccharide in the common supramolecular structure. The SAXS data gave the information on the supramolecular organization of the gel-forming protein and the protein–polysaccharide systems, namely, on the size of static inhomogeneities in the studied gel structures. To analyze the morphology of studied systems, a number of parameters and dimensional characteristics were calculated on the basis of the experimental SAXS data. One of the important characteristics obtained from the SAXS experiment was the particle gyration radius R_g [38]. This parameter is the rms distance of all scattering sources from the center of particle. The R_g value can be used to estimate the size of macromolecular or supramolecular aggregates. The particle gyration radius R_g was determined in two ways. The first one was based on the use of Guinier approximation, which is valid under the condition (sR_g) < 1.3 (Figure 2B, Table 1). The second way was to estimate R_g from the distance distribution function *P(r)* (Figure 3). Thus, assuming the spherical shape of particles, it was possible to calculate the effective average particle radius R_{sph} from R_g (Table 1) using the equation $R_{sph} = \sqrt{(5/3)}\, R_g$ [38].

Quite indicative for the determination of the studied supramolecular structure and the morphology is the character of the Kratky plots (Figure 4), which are also widely used to characterize the structure of hydrogels [39]. The shape of the depicted curves is a qualitative estimation of the folding or compactness of macromolecules. The bell-shaped Kratky plots (Figure 4) characterize the scattering pattern from a system of densely folded globules. Obviously, the obtained Kratky plots are also evidence of the joint participation of individual components in formation of common supramolecular structure of combined gel. With all this going on, the curve shape of the $P(r)$ functions provides information about the particle shape and gives an alternative way to determine R_g. In this case, the radius of gyration R_g, determined using the $P(r)$ curve turned out to be a much more stable estimate for the presence of admixture of polydisperse particles than the R_g value from the data presented as the Guinier plot.

Table 1. Structural parameters for hydrogels at 14 °C.

Sample	*Guinier Analysis*					*P(r) Analysis*	
	R_g, Å	R_{sph}, Å	R_c, Å	r_c, Å	L, Å	R_g, Å	D_{max}, Å
gelatin	47.5	61.3	–	–	–	42.5	149
κ-carrageenan-gelatin	56	72.3	21.2	27.4	179.5	63.4	222.7

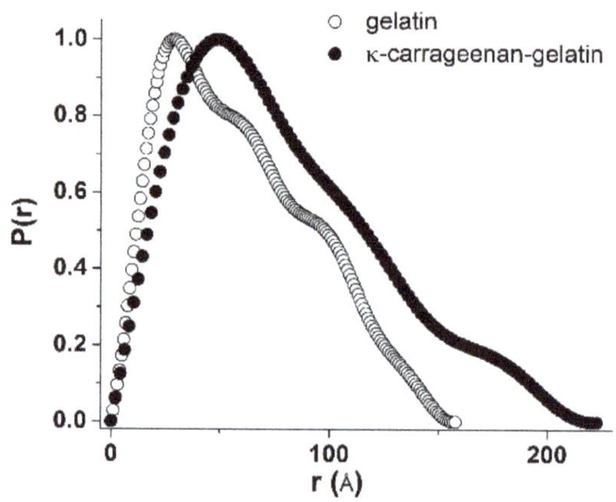

Figure 3. Normalized distance distribution functions *P(r)* derived from SAXS data for gelatin (open symbols) and κ-carrageenan–gelatin (filled symbols) gels.

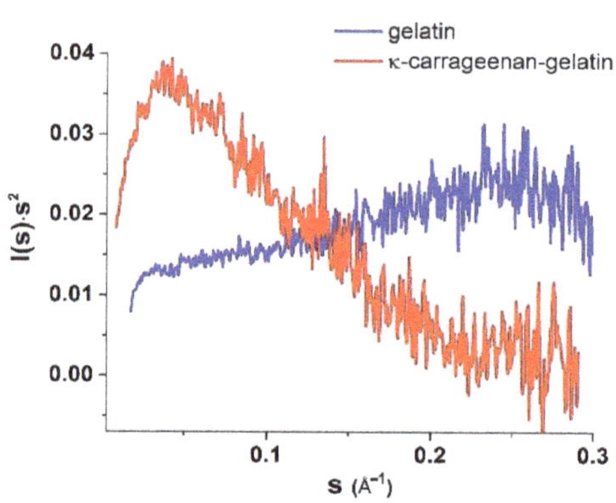

Figure 4. Kratky plots for gelatin and κ-carrageenan–gelatin gels.

In our previous fundamental work on the SAXS study of κ-carrageenan–gelatin hydrogels, which was referred to above, we substantiated two options for estimating the shape and size of scattering particles. Based on the obtained X-ray scattering data, we supposed two options for calculating the morphology of these hydrogels. The first one is the use of the spherical symmetry model of formed aggregates, when the radius of such spherical particles is quite unambiguously calculated from the obtained R_g values of particles. However, the comparison of obtained R_g values with the calculated maximum distances

D_{max} in particles (Table 1) characterizes the shape of particles as noticeably deviating from spherical, i.e., an elongated (cylindrical) shape (Figure 5).

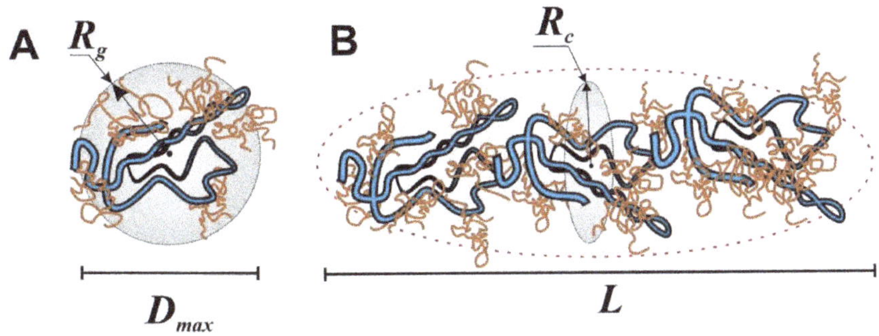

Figure 5. Structural schemes of spherical (**A**) and cylindrical (**B**) models indication of determined parameters.

Additional confirmation of this assumption is presented in our other article [17]. On the base of FTIR spectroscopy and molecular docking it was proposed that in the κ-carrageenan–gelatin hydrogels the most energetically favorable configuration of junctions was the interaction of the gelatin triple helix segments and the κ-carrageenan double helix. Both components of the composite hydrogel and their junctions formed the rod-shaped and rigid local structures, resulting in an anisotropic structural network, which can be hypothetically imagined as elongated particles between nodes. Such a highly anisotropic system can be depicted as the ensemble of cylindrical particles with an average length L and cross section r_c, which can be obtained from the experimental value of the gyration dimension of the cross-section R_c using the modified Guinier plots ($\ln(s \cdot I(s))$ vs. s^2) [38]. In the special case of anisometric particles, the two-dimensional analogue of gyration radius R_g is called the cross-section gyration radius R_c [38]. These parameters are also presented in Table 1. Note that for a gel sample based on pure gelatin, it was not possible to identify sections on low-angle curves that would be suitable for calculating the cross-section gyration radius R_c. Moreover, this is true, since pure gelatin gel is characterized by the formation of particles that are close to spherical, and only the addition of κ-carrageenan leads to the formation of elongated cylindrical particles in the mixed composition.

The analysis of the SAXS analysis shows that the admixture of κ-carrageenan to gelatin resulted in the formation of an integral supramolecular structure, where the determinative role was played by more sizable and flexible κ-carrageenan molecules. The initial gelatin gel has an amorphous nature, which can only be analyzed within spherical models. The formed integral structure of a combined polysaccharide–protein gel was characterized by elongated (cylindrical) scattering particles (domains with increased density), giving an increase in characteristic sizes independently of the used model.

3.3. AFM Study of Gelatin and κ-Carrageenan–Gelatin Gels

The AFM experiments [40] gave structural visualization of polysaccharide–protein hydrogels at the nanoscale. Figure 6 (A, B) depicts the AFM images of the surface of gelatin and κ-carrageenan–gelatin xerogel films. The xerogel was obtained by the drying of hydrogels, which was necessary for AFM experiment. Upon the drying of the hydrogels one can see (Figure 6) the formation of a biopolymer network or the biopolymer scaffold [41]. As can be seen from the presented data, gelatin has a smooth surface with a roughness value of R_q = 0.36 nm. The admixture of κ-carrageenan to gelatin led to a change in the supramolecular structure, which is represented by an inhomogeneous rough surface consisting of densely packed spheroidal particles, and the root-mean-square value of the surface roughness of κ-carrageenan–gelatin R_q = 10.7 nm. Thus, the AFM results also confirmed the formation of a common supramolecular structure in the protein–polysaccharide system.

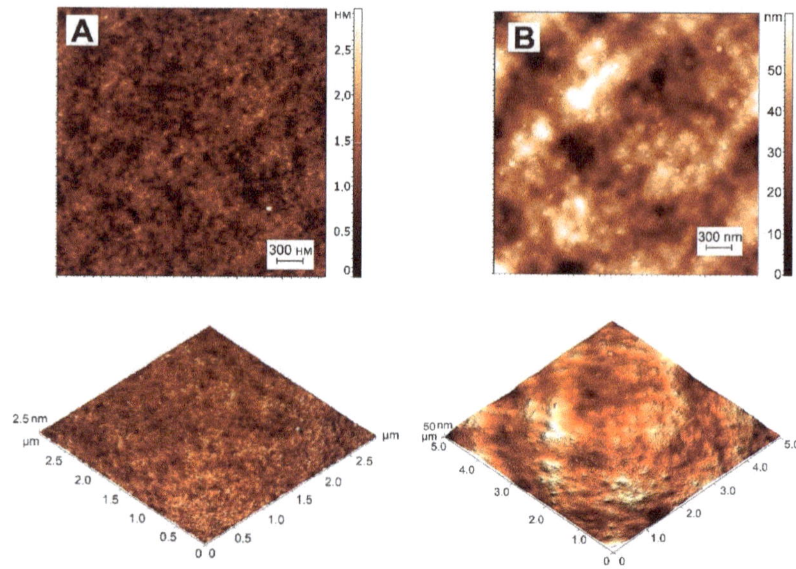

Figure 6. AFM pictures of xerogel (dried hydrogel) films and corresponding 3D surface of gelatin (**A**) and κ-carrageenan–gelatin (**B**) systems.

3.4. *Rheological Characterization of Gelatin and κ-Carrageenan–Gelatin Gels*

The rheological study of gelatin and κ-carrageenan–gelatin gels was used to look at the kinetics of structure formation in these systems (Figure 7). The experiments consisted of measuring the isothermal evolution of dynamic modulus of elasticity G′ at a constant frequency $\omega = 0.2\ s^{-1}$ and $\gamma = 0.2\%$ until the stability of the data (elastic modulus) was achieved, which means the sample gelation.

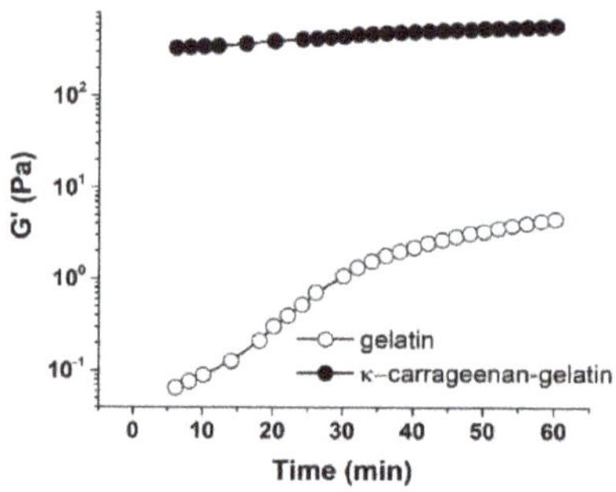

Figure 7. Time dependence of elastic modulus (G′) of gelatin (–) and κ-carrageenan–gelatin (,) hydrogels at 14 °C and $\omega = 0.2\ s^{-1}$, deformation amplitude $\gamma = 0.2\%$.

The experiments were performed at a temperature of 14 °C, i.e., under the melting point of gels [27]. It will be shown below that the elasticity modulus at this temperature was practically independent of frequency, so that we were dealing with a highly elastic state of studied materials. Furthermore. the measured amplitude dependence of elasticity modulus was absent (see below), so that the corresponding modulus values refer to the region of linear viscoelasticity, and the measurement results refer to the structure of systems not destroyed by mechanical action. Therefore, the measured modulus values can be considered as analogous to the equilibrium values of the Ge modulus. Then, according to Flory's theory [42], a semi-quantitative estimation of average degree of polymerization between neighboring grid nodes Z_c is expressed as

$$Z_c = \frac{cRT}{G_e M_n}$$

where c is the polymer concentration with a weight-average molecular mass M_n, R is the universal gas constant and T is absolute temperature. Then, as follows from this formula, with an increase in the elastic modulus the length of chain segment between the mesh nodes decreases, i.e., the number of nodes increases. In the case of a mesh, knots should be understood not as chemical bonds (as in rubbers), but as secondary bonds of various types and, possibly just intermolecular entanglements (entanglements). We obtained the estimated values of Z_c 5.96 and 0.4 for gelatin and κ-carrageenan–gelatin gels, correspondingly. The introduction of polysaccharide to protein led to a decrease in Z, i.e., to increase in the grid nodes number. As a consequence, these knots prevented the usual chain folding effect observed for high-molecular compounds—(in our case, the helicalization of the gelatin chain also decreased, which was confirmed by IR spectroscopy [17]. This led to the apparent chain length becoming longer in the composite κ-carrageenan–gelatin gel. This directly correlated with an increase in the chain parameters determined by the SAXS method.

Figure 7 shows the measured time-dependence of the elastic modulus G′ for the gelatin and κ-carrageenan–gelatin systems. The kinetics of the gelation process were estimated from the time-dependence of the elastic modulus G′ (Figure 7) [43]. The gelation of gelatin occurred gradually, since there was a distinct time-dependence of the elastic modulus G′ on time (Figure 7), similar to how it is described in [44]. In contrast, the elasticity model in the κ-carrageenan–gelatin system almost instantly reached equilibrium values, which then changed very little with time, i.e., in this case, a fairly rapid formation of the structural gel network took place.

The difference between the two compared systems was clearly manifested when the elastic G′(ω) and loss G″(ω) moduli were measured in a wide range of strain amplitudes (Figure 8). It follows from the data obtained that at small amplitudes the elastic modulus G′ confidently revealed a linear region of the mechanical behavior of gels [45], i.e., there was a fairly wide area of deformations in which the studied gels behaved like solid media. At the same time, this behavior is typical both for the κ-carrageenan –gelatin system and pure gelatin without polysaccharide additivity.

Consideration of the amplitude-dependence of modulus components made it possible to determine the limits of linear viscoelasticity, i.e., the range of strain values in which the structure of test sample was not destroyed under the action of deformation [46]. Data presented in Figure 8 allowed the estimation of strain amplitude, γ^*, (and stress, σ^*, as the product of the elastic modulus value and corresponding strain) corresponding to the linearity boundary. At a threshold value of amplitude exceeding a certain critical value γ^*, there was a sharp drop in the elastic modulus G′ and an increase in the loss modulus G″. This is a classic picture of destruction of gel structure, leading to transition of hydrogel from a solid to a fluid state [47,48], i.e., this point corresponded to the yield strength of the material, which can be considered as a viscous–plastic medium [49].

As one can see from Figure 8, the gelatin gel retained the linearity of the mechanical properties up to the deformation amplitude $\gamma^* \approx 370\%$, while for the κ-carrageenan –gelatin gel, the transition to a fluid state had occurred already at $\gamma \approx 20\%$. Thus, the introduction of κ-carrageenan to gelatin not only promotes an increase in elastic G′ and loss G″ moduli, but also leads to a sharp decrease in the limit of linear viscoelasticity region γ^*. This means that the interaction of gelatin with κ-carrageenan provides the appearance of not only a more rigid but also a more fragile structural network in the gel. These results correlate with our observations that indicate that a change in the rheological properties of κ-carrageenan –gelatin gels occurs when the polysaccharide/protein mass ratio exceeds Z = 0.1, which is due to changes in the interactions in the structural network of gel.

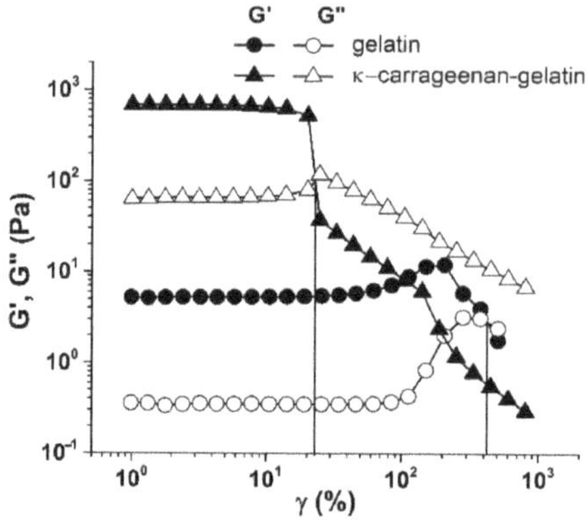

Figure 8. Amplitude dependences of elastic (G′) and loss (G″) moduli for gelatin (gel) („ –) and κ-carrageenan–gelatin (7, 8) at T = 14°C and $\omega = 6.28\ s^{-1}$.

Figure 9 depicts the frequency-dependence of the elastic G′ and loss G″ moduli and complex viscosity (η*) for the systems under study. A weak dependence of G′ and G″ on frequency was found for both systems, which, as was already mentioned above, corresponds to the pronounced solid-like behavior of gels as viscous–plastic media at low stresses. This is also associated with the higher values of the elastic modulus compared to the loss one (G′ > G″) in the entire frequency range. This behavior of studied gels is typical for soft materials with a gel-like structure, which are the viscous–plastic media with a relatively low limit [50,51]. The elastic modulus G′ of κ-carrageenan–gelatin exceeded the elastic modulus G′ of gelatin, which indicates an increase in the gelling properties of gelatin when it is complexed by a polysaccharide [52–54].

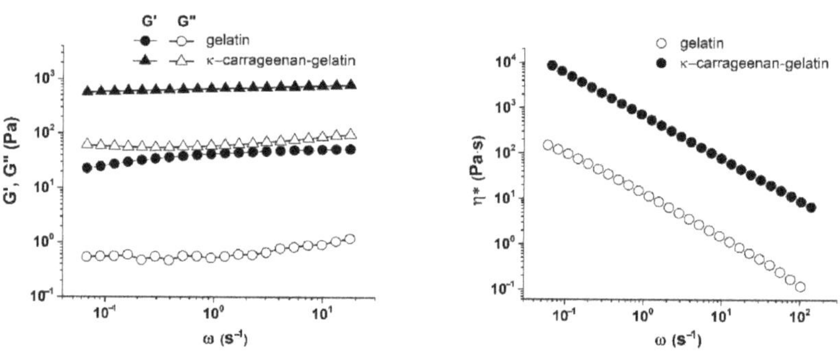

Figure 9. Frequency dependences of A: elastic (G′) and loss (G″) moduli and B: complex viscosity (η*) of gelatin and κ-carrageenan-gelatin gels at T = 14 °C and γ = 1%.

It can be seen from the obtained results that as the frequency increased, the complex viscosity decreased. Such dependences correspond to a situation when the systems under study exhibit non-Newtonian behavior with an increase in the shear flow velocity. The complex viscosity of hydrogel based on gelatin was an order of magnitude lower than that of hydrogel based on κ-carrageenan and gelatin. According to the considerations expressed in [55], the increase in complex viscosity indicates an increase in the density of the network of macromolecular chains entanglements; therefore, an increase in the energy costs was necessary to change conformational rearrangements within the system, which corresponds to the non-Newtonian behavior of the medium [56]. This conclusion regarding the increase in the density of network links (decrease in average distance between nodes) corresponds to the results presented above in Figure 7.

4. Discussion

Previously, we proposed a mechanism of interaction and the structure of the κ-carrageenan–gelatin polyelectrolyte complex [57]. The methods used in this work were aimed at obtaining information about the magnitude of static inhomogeneities in the systems under study. As the main source of quantitative and numerical information on a particular subject of the study, we used the philosophy and techniques of the SAXS method, in accordance with the scheme and dimensional definitions presented in Figure 10.

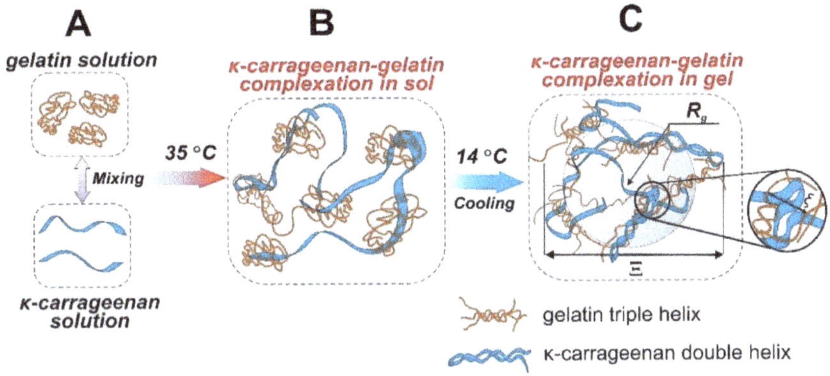

Figure 10. Structural schemes of κ-carrageenan and gelatin initial conformation (**A**), structural details of κ-carrageenan and gelatin complexation (**B**) and supramolecular polysaccharide–protein hydrogel (**C**).

The inhomogeneities of colloidal size (1–100 nm) in gelatin and κ-carrageenan gels were studied using the SAXS method. As such, density inhomogeneities with their specific dimensions we considered as the structural units of the κ-carrageenan–gelatin system shown in Figures 9 and 10. To our knowledge, the main structural element of this system is a polysaccharide–protein polyelectrolyte complex (PEC). Previously, one of our co-authors [45] showed that at the used weight ratio of κ-carrageenan–gelatin (0.8) there are no free protein or polysaccharide molecules, but almost all of them are bound in paired polyelectrolyte complexes. Their size and averaged spatial arrangement can be characterized by the parameters shown in Figure 10B. They are the PEC radius of gyration (R_g), the PEC maximum dimension (D_{max}), Ξ is the characteristic mean size of the static heterogeneities in the system and ξ is the correlation length of intramolecular interactions between the fluctuating chains of polymer, i.e., some zones of intramolecular hardening [33]. The main distinction of the system in the gel state consists in the presence of connectors (or crosslinking) (Figure 10C), combining individual PECs into a spatial 3D network via the complexation of gelatin triple helixes with one chain or a double helix of κ-carrageenan molecules.

The digital estimation of the polysaccharide–protein PEC spatial structure is presented in Table 1. One can see that admixture of polysaccharide to protein resulted in an increase in the apparent size of the scattering particles. Another deduction from the obtained SAXS results is the strong probability of the elongation of scattering of the structural elements of the newly formed hybrid κ-carrageenan –gelatin hydrogel. The results of the mechanical testing of the initial gelatin and engineered κ-carrageenan –gelatin gel showed the substantially denser parking of polymer chains in the composite system due to the huge number of new close intermolecular contacts between the polysaccharide and protein, such as the hydrophobic interaction, electrostatic ion pairing and hydrogen bonding [40]. Obviously, the close result indirectly followed from the SAXS estimations—the driving force for the formation the common supramolecular structural arrangement of protein and polysaccharide was the increase in the density of network of macromolecular chains entanglements; therefore, an increase in the energy costs necessary to change the conformational rearrangements within the system. This increase in the macromolecular arrangement led to increase in supramolecular associate size and the growth of interchain physical bonds. From the one side, the increase in network density led to an increase in the composite gel plasticity, and from another one the enlargement of scattering particles, making the novel gel system not only more rigid, but also more fragile.

5. Conclusions

Gelatin is one of the important hydrocolloids that is actively used in modern food and pharmacy technologies. One of the widely distributed methods for altering the mechanical properties of gelatin is the admixture of polysaccharide to gelatin systems. In this study we studied the supramolecular structure and mechanical performance of κ-carrageenan–gelatin gel in comparison with the initial gelatin one.

The presented experiments offer an overview of the structural and mechanical properties of the nanocomposite κ-carrageenan –gelatin hydrogel. The obtained results indicated that the combination of SAXS, powder X-ray diffractometry, AFM microscopy and rheology completely characterized the structure and properties of a protein–polysaccharide hydrogel. The phase homogeneity of the studied systems was confirmed by the PXRD approach. Using the SAXS method, it was shown that morphological structures based on individual components do not form in protein –polysaccharide hydrogels. The rheological study of gelatin and κ-carrageenan–gelatin gels was used to look at the kinetics of structure formation in these systems. The results of the mechanical testing of the initial gelatin and the engineered κ-carrageenan –gelatin gel confirmed the substantially denser parking of polymer chains in the composite system, obviously due to a significant increase in the intermolecular protein–polysaccharide contacts. A close result indirectly followed from the SAXS estimations—the driving force for the formation of the common supramolecular structural arrangement of the protein and polysaccharide was the increase in the density of the network of macromolecular chains entanglements; therefore, an increase in the energy costs was necessary to change the conformational rearrangements of the system. This increase in the macromolecular arrangement led to an increase in the supramolecular associate size and the growth of interchain physical bonds. This led to an increase in the composite gel plasticity, whereas the enlargement of scattering particles made the novel gel system not only more rigid, but also more fragile.

The obtained information on the structural and mechanical performance of hydrogels will open up the possibility of developing new materials with improved properties for targeted applications in biomedicine, biotechnology, as well as in any other relevant areas.

Author Contributions: Conceptualization, Y.F.Z., A.T.G. and S.R.D.; methodology, A.T.G., A.O.M., A.I.K., S.A.Z. and S.R.D.; investigation, A.T.G., A.O.M., A.I.K. and S.A.Z.; formal analysis, A.T.G., S.R.D., A.O.M., O.S.Z. and Y.F.Z.; writing—original draft preparation, Y.F.Z., A.O.M., A.T.G., S.R.D. and M.A.K.; writing—review and editing, Y.F.Z., A.T.G. and S.R.D.; visualization, A.T.G., A.O.M., A.I.K., S.A.Z., S.R.D. and M.A.K.; supervision and project administration Y.F.Z. and A.T.G. All authors have read and agreed to the published version of the manuscript.

Funding: This research received no external funding.

Institutional Review Board Statement: Not applicable.

Informed Consent Statement: Not applicable.

Data Availability Statement: The data in this study are available on reasonable request from the corresponding author.

Acknowledgments: A.T.G.: A.O.M., A.I.K., S.A.Z., M.A.K. and Y.F.Z. thanks for financial support from the government assignment for FRC Kazan Scientific Center of RAS. The contribution of Y.F.Z. was partly supported by Kazan Federal University Strategic Academic Leadership Program ("PRIORITY-2030"). O.S.Z. gives thanks for the support of Kazan State Power Engineering University Strategic Academic Leadership Program ("PRIORITY-2030"). The X-ray measurements have been carried out using the equipment of Distributed Spectral-Analytical Center of Shared Facilities for Study of Structure, Composition and Properties of Substances and Materials of FRC Kazan Scientific Center of RAS for the opportunity to fulfill the SAXS and rheology experiments.

Conflicts of Interest: The authors declare no conflict of interest.

References

1. Makshakova, O.N.; Zuev, Y.F. Interaction-Induced Structural Transformations in Polysaccharide and Protein-Polysaccharide Gels as Functional Basis for Novel Soft-Matter: A Case of Carrageenans. *Gels* **2022**, *8*, 287. [CrossRef] [PubMed]
2. Rial-Hermida, M.I.; Rey-Rico, A.; Blanco-Fernandez, B.; Carballo-Pedrares, N.; Byrne, E.M.; Mano, J.F. Recent Progress on Polysaccharide-Based Hydrogels for Controlled Delivery of Therapeutic Biomolecules. *ACS Biomater. Sci. Eng.* **2021**, *7*, 4102–4127. [CrossRef] [PubMed]
3. Bogdanova, L.R.; Zelenikhin, P.V.; Makarova, A.O.; Zueva, O.S.; Salnikov, V.V.; Zuev, Y.F.; Ilinskaya, O.N. Alginate-Based Hydrogel as Delivery System for Therapeutic Bacterial RNase. *Polymers* **2022**, *14*, 2461. [CrossRef]
4. Qi, X.; Su, T.; Zhang, M.; Tong, X.; Pan, W.; Zeng, Q.; Zhou, Z.; Shen, L.; He, X.; Shen, J. Macroporous Hydrogel Scaffolds with Tunable Physicochemical Properties for Tissue Engineering Constructed Using Renewable Polysaccharides. *ACS Appl. Mater. Interfaces* **2020**, *12*, 13256–13264. [CrossRef] [PubMed]
5. Ozel, B.; Aydin, O.; Grunin, L.; Oztop, M.H. Physico-Chemical Changes of Composite Whey Protein Hydrogels in Simulated Gastric Fluid Conditions. *J. Agric. Food Chem.* **2018**, *66*, 9542–9555. [CrossRef] [PubMed]
6. Farhoudi, N.; Leu, H.-Y.; Laurentius, L.B.; Magda, J.J.; Solzbacher, F.; Reiche, C.F. Smart Hydrogel Micromechanical Resonators with Ultrasound Readout for Biomedical Sensing. *ACS Sens.* **2020**, *5*, 1882–1889. [CrossRef] [PubMed]
7. Bovone, G.; Dudaryeva, O.Y.; Marco-Dufort, B.; Tibbitt, M.W. Engineering Hydrogel Adhesion for Biomedical Applications via Chemical Design of the Junction. *ACS Biomater. Sci. Eng.* **2021**, *7*, 4048–4076. [CrossRef] [PubMed]
8. Peppas, N.A.; Hilt, J.Z.; Khademhosseini, A.; Langer, R. Hydrogels in Biology and Medicine: From Molecular Principles to Bionanotechnology. *Adv. Mater.* **2006**, *18*, 1345–1360. [CrossRef]
9. Qian, C.; Higashigaki, T.; Asoh, T.-A.; Uyama, H. Anisotropic Conductive Hydrogels with High Water Content. *ACS Appl. Mater. Interfaces* **2020**, *12*, 27518–27525. [CrossRef] [PubMed]
10. Yahia, L. History and Applications of Hydrogels. *J. Biomed. Sci.* **2015**, *4*, 1–23. [CrossRef]
11. Gubaidullin, A.T.; Makarova, A.O.; Derkach, S.R.; Voron'ko, N.G.; Kadyirov, A.I.; Ziganshina, S.A.; Salnikov, V.V.; Zueva, O.S.; Zuev, Y.F. Modulation of Molecular Structure and Mechanical Properties of κ-Carrageenan-Gelatin Hydrogel with Multi-Walled Carbon Nanotubes. *Polymers* **2022**, *14*, 2346. [CrossRef]
12. Kanduč, M.; Kim, W.K.; Roa, R.; Dzubiella, J. How the Shape and Chemistry of Molecular Penetrants Control Responsive Hydrogel Permeability. *ACS Nano* **2021**, *15*, 614–624. [CrossRef]
13. Singh, S.R. Advanced Hydrogels for Biomedical Applications. *BJSTR* **2018**, *5*, 1–5. [CrossRef]
14. Guvendiren, M.; Burdick, J.A. Stiffening Hydrogels to Probe Short- and Long-Term Cellular Responses to Dynamic Mechanics. *Nat. Commun.* **2012**, *3*, 792. [CrossRef]
15. Beckett, L.E.; Lewis, J.T.; Tonge, T.K.; Korley, L.T.J. Enhancement of the Mechanical Properties of Hydrogels with Continuous Fibrous Reinforcement. *ACS Biomater. Sci. Eng.* **2020**, *6*, 5453–5473. [CrossRef] [PubMed]
16. Alipal, J.; Mohd Pu'ad, N.A.S.; Lee, T.C.; Nayan, N.H.M.; Sahari, N.; Basri, H.; Idris, M.I.; Abdullah, H.Z. A Review of Gelatin: Properties, Sources, Process, Applications, and Commercialisation. *Mater. Today Proc.* **2021**, *42*, 240–250. [CrossRef]
17. Derkach, S.R.; Voron'ko, N.G.; Kuchina, Y.A.; Kolotova, D.S.; Gordeeva, A.M.; Faizullin, D.A.; Gusev, Y.A.; Zuev, Y.F.; Makshakova, O.N. Molecular Structure and Properties of κ-Carrageenan-Gelatin Gels. *Carbohydr. Polym.* **2018**, *197*, 66–74. [CrossRef] [PubMed]
18. Yermak, I.M.; Khotimchenko, Y.S. Chemical Properties, Biological Activities and Applications of Carrageenans from Red Algae. *Recent Adv. Mar. Technol.* **2003**, *9*, 207–255.
19. De France, K.J.; Chan, K.J.W.; Cranston, E.D.; Hoare, T. Enhanced Mechanical Properties in Cellulose Nanocrystal–Poly(Oligoethylene Glycol Methacrylate) Injectable Nanocomposite Hydrogels through Control of Physical and Chemical Cross-Linking. *Biomacromolecules* **2016**, *17*, 649–660. [CrossRef]
20. Yang, Y.; Zhao, X.; Yu, J.; Chen, X.; Chen, X.; Cui, C.; Zhang, J.; Zhang, Q.; Zhang, Y.; Wang, S.; et al. H-Bonding Supramolecular Hydrogels with Promising Mechanical Strength and Shape Memory Properties for Postoperative Antiadhesion Application. *ACS Appl. Mater. Interfaces* **2020**, *12*, 34161–34169. [CrossRef] [PubMed]
21. Yu, H.; Zhao, H.; Huang, C.; Du, Y. Mechanically and Electrically Enhanced CNT–Collagen Hydrogels As Potential Scaffolds for Engineered Cardiac Constructs. *ACS Biomater. Sci. Eng.* **2017**, *3*, 3017–3021. [CrossRef] [PubMed]
22. Cai, W.; Huang, W.; Chen, L. Soluble Pea Protein Aggregates Form Strong Gels in the Presence of κ-Carrageenan. *ACS Food Sci. Technol.* **2021**, *1*, 1605–1614. [CrossRef]

23. Tytgat, L.; Vagenende, M.; Declercq, H.; Martins, J.C.; Thienpont, H.; Ottevaere, H.; Dubruel, P.; Van Vlierberghe, S. Synergistic Effect of κ-Carrageenan and Gelatin Blends towards Adipose Tissue Engineering. *Carbohydr. Polym.* **2018**, *189*, 1–9. [CrossRef] [PubMed]
24. Wen, C.; Lu, L.; Li, X. Enzymatic and Ionic Crosslinked Gelatin/K-Carrageenan IPN Hydrogels as Potential Biomaterials. *J. Appl. Polym. Sci.* **2014**, *131*, 21. [CrossRef]
25. Padhi, J.R.; Nayak, D.; Nanda, A.; Rauta, P.R.; Ashe, S.; Nayak, B. Development of Highly Biocompatible Gelatin & I-Carrageenan Based Composite Hydrogels: In Depth Physiochemical Analysis for Biomedical Applications. *Carbohydr. Polym.* **2016**, *153*, 292–301. [CrossRef] [PubMed]
26. Cao, Y.; Wang, L.; Zhang, K.; Fang, Y.; Nishinari, K.; Phillips, G.O. Mapping the Complex Phase Behaviors of Aqueous Mixtures of κ-Carrageenan and Type B Gelatin. *J. Phys. Chem. B* **2015**, *119*, 9982–9992. [CrossRef] [PubMed]
27. Shimada, R.; Kumeno, K.; Akabane, H.; Nakahama, N. Gelation and Melting of a Mixed Carrageenan-Gelatin Gel. *J. Home Econ. Jpn.* **1993**, *44*, 999–1005. [CrossRef]
28. *DIFFRAC Plus Evaluation Package EVA, Version 11, User's Manual*; Bruker AXS: Karlsruhe, Germany, 2005.
29. *TOPAS V3: General Profile and Structure Analysis Software for Powder Diffraction Data, Technical Reference*; Bruker AXS: Karlsruhe, Germany, 2005.
30. *Small Angle X-Ray Scattering*, Version 4.0. Software Reference Manual, M86-E00005-0600; Bruker AXS Inc.: Madison, WI, USA, 2000.
31. UMD, UTK, NIST, ORNL, ISIS, ESS, ILL. 2009. Available online: https://zenodo.org/record/825675#.Y0o3lUxBxPY (accessed on 30 August 2022).
32. Konarev, P.V.; Volkov, V.V.; Sokolova, A.V.; Koch, M.H.J.; Svergun, D.I. PRIMUS: A Windows PC-Based System for Small-Angle Scattering Data Analysis. *J. Appl. Crystallogr.* **2003**, *36*, 1277–1282. [CrossRef]
33. Evmenenko, G.; Theunissen, E.; Mortensen, K.; Reynaers, H. SANS Study of Surfactant Ordering in κ-Carrageenan/Cetylpyridinium Chloride Complexes. *Polymer* **2001**, *42*, 2907–2913. [CrossRef]
34. Shibayama, M.; Tanaka, T.; Han, C.C. Small Angle Neutron Scattering Study on Poly(N-isopropyl Acrylamide) Gels near Their Volume-phase Transition Temperature. *J. Chem. Phys.* **1992**, *97*, 6829–6841. [CrossRef]
35. Yeh, F.; Sokolov, E.L.; Walter, T.; Chu, B. Structure Studies of Poly(Diallyldimethylammonium Chloride- *Co* -Acrylamide) Gels/Sodium Dodecyl Sulfate Complex. *Langmuir* **1998**, *14*, 4350–4358. [CrossRef]
36. Zhuang, C.; Tao, F.; Cui, Y. Anti-Degradation Gelatin Films Crosslinked by Active Ester Based on Cellulose. *RSC Adv.* **2015**, *5*, 52183–52193. [CrossRef]
37. Yang, F.; Li, G.; He, Y.-G.; Ren, F.-X.; Wang, G. Synthesis, Characterization, and Applied Properties of Carboxymethyl Cellulose and Polyacrylamide Graft Copolymer. *Carbohydr. Polym.* **2009**, *78*, 95–99. [CrossRef]
38. Guinier, A.; Fournet, G. *Small-Angle Scattering of X-Rays*; Wiley: New York, NY, USA, 1955.
39. Wei, Y.; Hore, M.J.A. Characterizing Polymer Structure with Small-Angle Neutron Scattering: A Tutorial. *J. Appl. Phys.* **2021**, *129*, 171101. [CrossRef]
40. Liu, S.; Wang, Y. A Review of the Application of Atomic Force Microscopy (AFM) in Food Science and Technology. In *Advances in Food and Nutrition Research*; Elsevier: Amsterdam, The Netherlands, 2011; Volume 62, pp. 201–240. ISBN 978-0-12-385989-1.
41. Yang, Z.; Yang, H.; Yang, H. Effects of Sucrose Addition on the Rheology and Microstructure of κ-Carrageenan Gel. *Food Hydrocoll.* **2018**, *75*, 164–173. [CrossRef]
42. Ferry, J.D.; Rice, S.A. Viscoelastic Properties of Polymers. *Phys. Today* **1962**, *15*, 76–78. [CrossRef]
43. Rial, R.; Soltero, J.F.A.; Verdes, P.V.; Liu, Z.; Ruso, J.M. Mechanical Properties of Composite Hydrogels for Tissue Engineering. *CTMC* **2018**, *18*, 1214–1223. [CrossRef]
44. Zhu, W.; Chu, C.; Kuddannaya, S.; Yuan, Y.; Walczak, P.; Singh, A.; Song, X.; Bulte, J.W.M. In Vivo Imaging of Composite Hydrogel Scaffold Degradation Using CEST MRI and Two-Color NIR Imaging. *Adv. Funct. Mater.* **2019**, *29*, 1903753. [CrossRef]
45. Derkach, S.R.; Ilyin, S.O.; Maklakova, A.A.; Kulichikhin, V.G.; Malkin, A.Y. The Rheology of Gelatin Hydrogels Modified by κ-Carrageenan. *LWT Food Sci. Technol.* **2015**, *63*, 612–619. [CrossRef]
46. He, Q.; Yu, W.; Wu, Y.; Zhou, C. Shear Induced Phase Inversion of Dilute Smectic Liquid Crystal/Polymer Blends. *Soft Matter* **2012**, *8*, 2992. [CrossRef]
47. Pan, A.; Roy, S.G.; Haldar, U.; Mahapatra, R.D.; Harper, G.R.; Low, W.L.; De, P.; Hardy, J.G. Uptake and Release of Species from Carbohydrate Containing Organogels and Hydrogels. *Gels* **2019**, *5*, 43. [CrossRef]
48. Yang, K.; Liu, Z.; Wang, J.; Yu, W. Stress Bifurcation in Large Amplitude Oscillatory Shear of Yield Stress Fluids. *J. Rheol.* **2018**, *62*, 89–106. [CrossRef]
49. Malkin, A.; Kulichikhin, V.; Ilyin, S. A Modern Look on Yield Stress Fluids. *Rheol. Acta* **2017**, *56*, 177–188. [CrossRef]
50. Chae, B.S.; Lee, Y.S.; Jhon, M.S. The Scaling Behavior of a Highly Aggregated Colloidal Suspension Microstructure and Its Change in Shear Flow. *Colloid Polym. Sci.* **2004**, *282*, 236–242. [CrossRef]
51. Nishinari, K. Rheological and DSC Study of Sol-Gel Transition in Aqueous Dispersions of Industrially Important Polymers and Colloids. *Colloid. Polym. Sci.* **1997**, *275*, 1093–1107. [CrossRef]
52. Derkach, S.R.; Kuchina, Y.A.; Kolotova, D.S.; Voron'ko, N.G. Polyelectrolyte Polysaccharide–Gelatin Complexes: Rheology and Structure. *Polymers* **2020**, *12*, 266. [CrossRef]

53. Goudoulas, T.B.; Germann, N. Phase Transition Kinetics and Rheology of Gelatin-Alginate Mixtures. *Food Hydrocoll.* **2017**, *66*, 49–60. [CrossRef]
54. Picard, J.; Giraudier, S.; Larreta-Garde, V. Controlled Remodeling of a Protein-Polysaccharide Mixed Gel: Examples of Gelatin-Hyaluronic Acid Mixtures. *Soft Matter* **2009**, *5*, 4198. [CrossRef]
55. Dobrynin, A.V. Solutions of Charged Polymers. In *Polymer Science: A Comprehensive Reference*; Elsevier: Amsterdam, The Netherlands, 2012; pp. 81–132. ISBN 978-0-08-087862-1.
56. Ali, I.; Shah, L.A. Rheological Investigation of the Viscoelastic Thixotropic Behavior of Synthesized Polyethylene Glycol-Modified Polyacrylamide Hydrogels Using Different Accelerators. *Polym. Bull.* **2021**, *78*, 1275–1291. [CrossRef]
57. Makshakova, O.N.; Faizullin, D.A.; Zuev, Y.F. Interplay between Secondary Structure and Ion Binding upon Thermoreversible Gelation of κ-Carrageenan. *Carbohydr. Polym.* **2020**, *227*, 115342. [CrossRef]

polymers

MDPI

Review

Small Diameter Cell-Free Tissue-Engineered Vascular Grafts: Biomaterials and Manufacture Techniques to Reach Suitable Mechanical Properties

María A. Rodríguez-Soto [1,*], Camilo A. Polanía-Sandoval [2,3], Andrés M. Aragón-Rivera [1], Daniel Buitrago [1], María Ayala-Velásquez [1], Alejandro Velandia-Sánchez [2,3], Gabriela Peralta Peluffo [1], Juan C. Cruz [1], Carolina Muñoz Camargo [1], Jaime Camacho-Mackenzie [2], Juan Guillermo Barrera-Carvajal [2] and Juan Carlos Briceño [1,4,*]

1 Department of Biomedical Engineering, Universidad de los Andes, Bogotá 111711, Colombia
2 Vascular and Endovascular Surgery Research Group, Fundación Cardio Infantil—Instituto de Cardiología, Bogotá 111711, Colombia
3 School of Medicine and Health Sciences, Universidad del Rosario, Bogotá 111711, Colombia
4 Department of Research, Fundación Cardio Infantil Instituto de Cardiología, Bogotá 111711, Colombia
* Correspondence: ma.rodriguezs1@uniandes.edu.co (M.A.R.-S.); jbriceno@uniandes.edu.co (J.C.B.); Tel.: +57-(1)-3394949 (ext. 1789) (M.A.R.-S.)

Abstract: Vascular grafts (VGs) are medical devices intended to replace the function of a blood vessel. Available VGs in the market present low patency rates for small diameter applications setting the VG failure. This event arises from the inadequate response of the cells interacting with the biomaterial in the context of operative conditions generating chronic inflammation and a lack of regenerative signals where stenosis or aneurysms can occur. Tissue Engineered Vascular grafts (TEVGs) aim to induce the regeneration of the native vessel to overcome these limitations. Besides the biochemical stimuli, the biomaterial and the particular micro and macrostructure of the graft will determine the specific behavior under pulsatile pressure. The TEVG must support blood flow withstanding the exerted pressure, allowing the proper compliance required for the biomechanical stimulation needed for regeneration. Although the international standards outline the specific requirements to evaluate vascular grafts, the challenge remains in choosing the proper biomaterial and manufacturing TEVGs with good quality features to perform satisfactorily. In this review, we aim to recognize the best strategies to reach suitable mechanical properties in cell-free TEVGs according to the reported success of different approaches in clinical trials and pre-clinical trials.

Keywords: tissue engineered vascular grafts; biomaterials; biodegradable; biomechanical stimulation; mechanical properties

Citation: Rodríguez-Soto, M.A.; Polanía-Sandoval, C.A.; Aragón-Rivera, A.M.; Buitrago, D.; Ayala-Velásquez, M.; Velandia-Sánchez, A.; Peralta Peluffo, G.; Cruz, J.C.; Muñoz Camargo, C.; Camacho-Mackenzie, J.; et al. Small Diameter Cell-Free Tissue-Engineered Vascular Grafts: Biomaterials and Manufacture Techniques to Reach Suitable Mechanical Properties. *Polymers* **2022**, *14*, 3440. https://doi.org/10.3390/polym14173440

Academic Editors: Domenico Acierno and Antonella Patti

Received: 1 June 2022
Accepted: 6 July 2022
Published: 23 August 2022

Publisher's Note: MDPI stays neutral with regard to jurisdictional claims in published maps and institutional affiliations.

1. Introduction

Cardiovascular diseases (CVDs) are a group of conditions affecting the heart and blood vessels. According to the World Health Organization, CVDs are currently the leading cause of death worldwide, representing 32% of all global deaths [1]. Accordingly, almost 18 million people died from a CVD in 2019. Among CVDs, vascular diseases compromising the blood vessel structure take hold of more than 8.5 million people worldwide. For instance, coronary artery disease (CAD) is the most common type of CVD, affecting 6.7% worldwide. CAD alters the mechanical properties of the coronary artery supplying the heart muscle, resulting in a decrease in the blood flow [2]. Moreover, other conditions also affect the mechanical properties of the blood vessels, such as the Peripheral Artery Disease (PAD), which compromises the blood flow in the limbs, and aortic aneurysms, identified by a balloon-like bulge in the aorta [3].

Currently, surgical approaches for vascular reconstruction focus on using autologous or synthetic vascular grafts (VGs) to recover the blood flow. Autologous grafts such as the great saphenous vein (GSV) and Internal Mamary Artery (IMA) can maintain long-term patency due to their regenerative properties (86% up to five years). However, GSV has limited availability and represents additional risks due to possible post-surgical complications with morbidity rates up to 50% [4]. Furthermore, synthetic VGs in the market are made from Polytetrafluoroethylene (PTFE—known as Teflon®) and Polyethylene terephthalate (PET—known as Dacron®). Although their performance is suitable in straight blood vessels and large diameters, they are inefficient in replacing small diameter vessels (>6 mm diameter) and under non-unidirectional or non-fully developed flows, as seen in their decreased patency rates below 32% after two years [5]. These low patency rates have been correlated to a foreign body response promoting thrombogenesis and stenosis, compromising the blood flow. With 3 out of 100,000 adults requiring peripheral vascular segment repairs through small diameter VGs [6], new approaches to overcome the current challenges include the use of bio-based VGs, such as allogeneic cryopreserved blood vessels (Cryograft®—CryoLife) [7], xenografts from bovine blood vessels (Artegraft®—LeMaitre) [8], bovine pericardium [9], or xenografts from sheep extracellular matrix (ECM) (Omniflow® II—LeMaitre) [10]. Nevertheless, they still do not show a clear advantage compared to commercial synthetic non-biodegradable VGs [11].

These bio-based VGs are part of Tissue-engineered vascular grafts (TEVGs); these scaffolds are intended to guide the tissue regeneration of the vascular wall and are considered as a strategy to get closer to the biological response of native blood vessels. Under physiological conditions, native arteries have a complex, multilayered structure with different ECM compositions and microstructures to contribute to the required compliance precisely responding to pulsatile pressure. Components in each layer will determine the total vessel elastic, viscous, and inertial properties that provide the critical biomechanical signals required to regulate cell adhesion, growth, and differentiation [12].

Although there is promising research on TEVGs, several barriers are still to be overcome. In this sense, the rational design of TEVGs must consider the self-defeating cellular response to biomaterial surfaces and structures under the hemodynamic operative conditions leading to graft patency loss. Accordingly, the inflammatory process after implantation of a TEVG is required for the vascular wall regeneration and depends on the ECM composition and mechanical properties regarding the presence of bioactive molecules and their micro/macrostructure [13]. For instance, one of the most reported causes for patency loss in TEVGs arises due to differences between the mechanical properties of the TEVGs and the native vessels affecting the flow pattern and the biomechanical stimulation. Furthermore, the inability of the cells to infuse into the scaffold from the perivascular tissues can also promote failure [14,15].

The failure of TEVGs depends on the alteration of the complex interactions between the biomaterials, cells, and hemodynamic conditions in the blood flow. Namely, the most reported failure of TEVGs is related to the development of intimal hyperplasia, in which smooth muscle cells over-proliferate, and also atherogenesis, in which foreign body responses promote graft calcification [16]. However, other failure causes include aneurysms, thrombogenesis, and bacterial infection, affecting the flow pattern and the biomechanical stimulation. Consequently, for a TEVG to provide continuity between the native tissue and the required biomechanical signals to induce regeneration, the rational design must consider the suitable mechanical properties of biodegradable structures and biomaterials similar to the native arteries in which the TEVG will be anastomosed, and to the current gold standards (i.e., GSV and IMA) [17].

Herein, we aim to recognize the best strategies to reach suitable mechanical properties in TEVGs according to the success of different approaches in clinical and pre-clinical trials. Furthermore, we aim to identify the latest trends in cell-free TEVGs development regarding the manufacturing methods and biomaterials in the context of ideal mechanical properties.

2. Database Information Extraction and Data Analysis

An integrative review of the literature on tissue-engineered vascular grafts was conducted and reported according to the elements described in the PRISMA guidelines. To this end, an electronic search was performed in the following databases: MEDLINE, SCOPUS, Web of Knowledge, and ClinicalTrials.gov. The indexing terms for the search strategy were in MeSH terminology. These terms included "Tissue engineering", "Vascular graft", "Biodegradable" and "Regenerative" from the last five years (2017–2022). A total of 4592 records were identified, with 2476 removed before the screening, as they were considered duplicated records or ineligible by automation tools.

For the selection process, an initial title review was performed; if the title indicated that the study might be relevant, abstracts were reviewed, or they were otherwise excluded. Finally, identified eligible studies were read on a full-text basis. Inclusion and exclusion criteria were applied to the list of articles obtained; those that did not meet these criteria were eliminated, and the documents corresponding to the remaining articles were reviewed. Thereof, five registries in Clinicaltrials.gov were included for TEVGs, 43 records were included for TEVGs on pre-clinical models (Figure S1), and 46 studies found in databases were included for TEVGs on in vitro testing (Figure S2)

The inclusion criteria for the screening and choice were covered by TEVGs on clinical trials registered on clinicaltrials.gov, with pre-clinical testing on in vivo models and data related to in vitro studies found in different databases, reporting at least two mechanical properties according to the requirements of the ISO 7198:2016 and from the last five years; otherwise, manuscripts were excluded. Additional exclusion criteria considered TEVGs with pre-seeded cells, articles in languages different from English or Spanish, or if the retrieved information was incomplete, did not correspond to the scope of the review, or was unavailable for retrieval for any reason. A total of 2160 records were screened, 1745 were removed, and classification was performed considering three categories: TEVGs on clinical trials, TEVGs on pre-clinical models, and TEVGs with in vitro data.

Subsequently, data collection and extraction were performed regarding the mechanical properties reported in the TEVGs analyzed from the clinical trials, pre-clinical models, or in vitro models and the relationship with patency and regeneration potential. Reviewers had independent access and the data were fed as the literature review process was carried out. The data were classified according to the study type and organized according to title, authors, DOI or CT registry number, and year of publication, including manufacturing techniques, materials, and any functionalization used for developing the TEVG. Detailed information about the study, animal model, and physicochemical and biological characterization was incorporated, considering the regenerative potential through the endothelialization success and inflammatory response.

As the manuscripts reporting mechanical properties were included, we compared available biodegradable TEVGs to native blood vessels (GSV and IMA) as the least defining suitable mechanical properties to be fulfilled to the TEVGs. Then, all summarized data were analyzed using descriptive analysis, reporting central tendency measures for quantitative variables and frequencies for qualitative variables. Finally, all the information was tabulated. This article's relevant data are in tables or graphics presented in this review.

3. Physiology of Blood Vessels

The vascular system is comprised of blood vessels that can be classified into arteries, capillaries, and veins, with differences according to their size and physiological function. Blood rich in oxygen and nutrients is transported from the heart to different tissues in the body by a network of capillaries, branching from the arteries. The capillaries then transport the blood containing carbon dioxide and other metabolic wastes from the tissues to the venules and veins to remove and re-oxygenate the blood in the lungs.

Besides their physiological differences, blood vessels have a complex vascular wall in terms of composition and structure, which gives them a wide variety of mechanical properties according to the different hemodynamic conditions to which the blood vessel can be exposed. However, these properties are potentially affected by pathologies that change the vascular wall, such as atherosclerosis or aneurysms, representing the world's leading causes of death [12]. Currently, Coronary Artery Disease (CAD) and Peripheral Arterial Disease (PAD) represent the leading cause of mortality worldwide, with an estimated annual incidence increase of 23.3 million by 2030 [18]. Angioplasty, stent implantation, and surgical bypass grafting are the current treatment options. For the latter, autologous veins, or arteries such as the GSV or IMA, represent the gold standard for this and are preferred due to their long-term patency up to 80% at five years [18].

3.1. Micro and Macrostructure of Blood Vessels

At the macrostructural level, the vascular wall of blood vessels presents a multilayered structure composed of 70% water and 30% collagen, elastin, proteoglycans, and vascular cells [12]. The vascular wall is distinguishable in three layers: tunica intima, media, and adventitia (Figure 1). The innermost layer–the tunica intima–is constituted by endothelial cells (ECs) as a monolayer on the blood vessel's lumen and maintains homeostasis through a carefully regulated interaction with the other types of cells. Following this layer, and separated by an inner membrane composed of elastin, there is the tunica media composed mainly of smooth muscle cells (SMCs), elastin, and collagen fibers. Finally, the tunica adventitia is a layer of fibroblasts and other components of the extracellular matrix (EMC)–mainly collagen fibers [19].

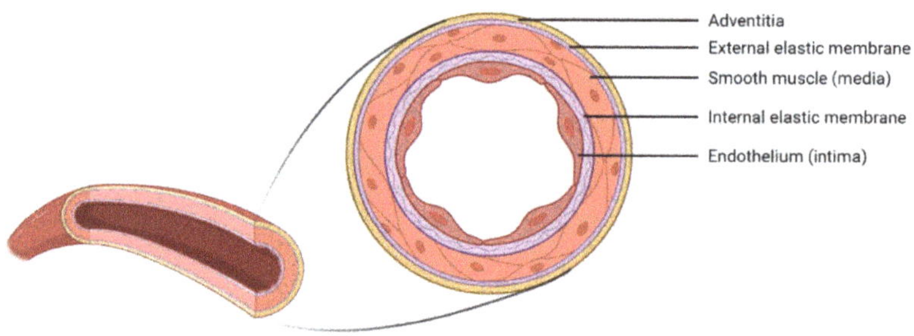

Figure 1. Structure and layers of blood vessels. Graph created with BioRender.com (accessed on 29 May 2022).

Nevertheless, the wall thickness and its components may vary according to the type of blood vessel. Due to the aim of this study, we will focus on small arteries with diameters ranging between 1–6 mm and a wall thickness between 125–800 μm [12]. Figure 2 illustrates the composition of this kind of small-sized blood vessels.

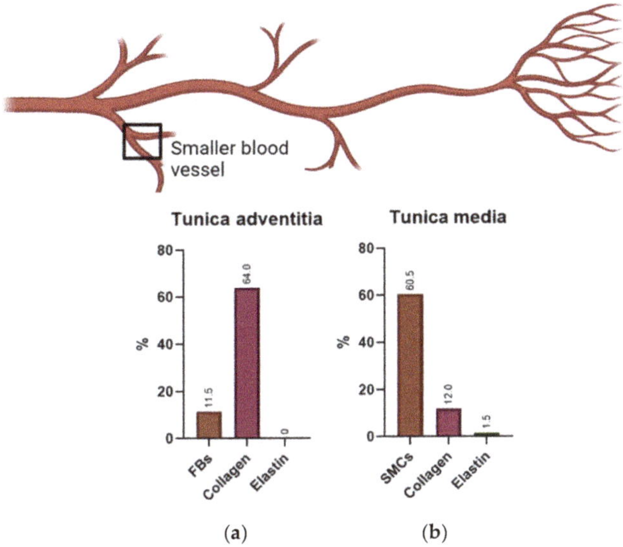

Figure 2. Component percentages on small-sized blood vessels for (**a**) tunica adventitia and (**b**) tunica media. Data from small-sized vessels retrieved from D.B. Camasão et al. [12]. Fibroblasts (FBs) and smooth muscle cells (SMCs). Image created with BioRender.com and graph constructed with GraphPad Prism (accessed on 29 May 2022).

3.2. Mechanical Properties of Blood Vessels Dsed as Gold Standards on Vascular Grafts

The mechanical analyses of blood vessels are of utmost importance to understand their properties under different hemodynamic conditions and to guide the rational design and production of TEVGs capable of withstanding physiological stresses and pressures. The blood vessel wall presents a complex structure and composition, allowing responses to different strengths and pressures caused by blood flow. Due to the anisotropic behavior of blood vessels, they support a significant load in the circumferential direction due to the collagen components in this direction, while allowing compliance given the elastin fibers. Therefore, at high pressures in the blood vessels, the elastin fibers will allow expansion, and collagen fibers will be stiffer to permit the change in diameter but prevent damage or rupture when the pressure increases. Moreover, the compliance of the blood vessels is a property that allows us to measure the storage of blood that the artery can support and release to the vascular network in order to reach regions of lower pressure in a pulsatile flow through its stretching due to the elastin fibers [12].

Currently, for the substitution of small diameter vessels (<6 mm), autografts from GSV and IMA are used due to their excellent compliance and compatibility with the native vessels, remaining as the gold standard [20]. For this study, GSV will be considered as a standard due to its great usefulness as a graft, being a blood vessel that presents a significant length and is easily accessible to the surgeon [21]. However, the selection of blood vessels as grafts varies according to the circumstances outlined by the American College of Cardiology and the American Heart Association. Table 1 summarizes the mechanical properties reported for the GSV and IMA.

As shown in Table 1 it is possible to identify that the GSV presents better mechanical properties such as burst pressure, suture retention strength, and longitudinal tensile strength compared to IMA. However, other properties such as dynamic compliance, internal diameter, wall thickness, and circumferential tensile strength are lower due to the wall composition in veins presenting some differences compared to the arteries. For example, a lower wall thickness provides a difference in the structure of the veins because of the thinner media layer, lower amount of elastin, and relatively high collagen content [12]. For better comparison, details on the mechanical properties of IMA and GSV are shown in a radar chart in Figure 3.

Table 1. Mechanical properties in blood vessels: internal mammary artery (IMA) and great saphenous vein (GSV).

Test Performed	IMA	GSV	Reference
Internal diameter (mm)	3.50	3	[22,23]
Wall Thickness (μM)	710	518	[24,25]
Circumferential tensile Strength (KPa)	4100	2405	[12,19,26]
Longitudinal tensile Strength (KPa)	4300	9760	[12,19,26]
Burst pressure (KPa)	266	371.96	[12,26]
Suture Retention Strength (g)	138	327	[22,27]
Dynamic Compliance (%/100 mmHg)	5.22	4.40	[12,28]

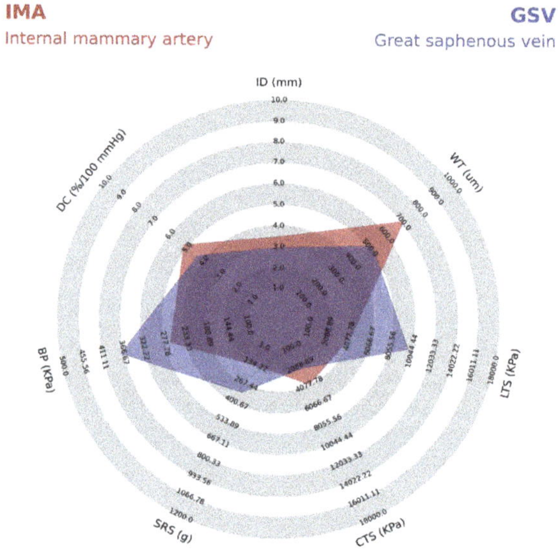

Figure 3. Comparison of the mechanical properties of the internal mammary artery (IMA) and great saphenous vein (GSV) as vascular grafts. Identified mechanical properties include internal diameter (ID), wall thickness (WT), longitudinal tensile strength (LTS), circumferential tensile strength (CTS), suture retention strength (SRS), burst pressure (BP), and dynamic compliance (DC). Chart generated from the Python library developed by StatsBomb/Anmol Durgapal (accessed on 29 May 2022) [29].

4. Polytetrafluoroethylene (PTFE) Has the Highest Share on VGs Market

The global vascular graft market is increasing due to the prevalence of cardiovascular diseases related to low levels of physical activity and a sedentary lifestyle. It is expected to reach 3.3 billion by 2026 at a CAGR of 6.4%. Polytetrafluoroethylene (PTFE or Teflon) based vascular grafts are currently the most used. PTFE as a raw material reached 710.3 million USD in 2018, and it is expected to present the fastest CARG over the forecast period due to the graft advantages related to low delamination, minimal blood loss, and excellent mechanical properties [30].

PTFE vascular grafts are the nonwoven type. They were developed in 1972, with their first use as a venous prosthesis in a swine model, where they have since been used for more than 30 years. As PTFE-based VG has been the most successful type for small diameter applications; we have considered it a reference for superior mechanical properties.

4.1. Micro and Macrostructure of PTFE VGs

PTFE is a thermoplastic and crystalline fluorocarbon polymer like polyethylene; fluoride atoms have substituted hydrogen atoms. These Fluorinated carbons have a high affinity for avoiding reactions with other molecules, for which the PTFE is inert, biocompatible, and avoids thrombogenesis. Implantable PTFE devices were commonly manufactured from a single film of PTFE and subsequently stretched to produce a microporous structure from the non-porous film; the degree of porosity depends on the stretch. PTFE VGs nowadays are produced by extrusion and sintering in a porous tube with fibrils and nodules of controllable sizes. Therefore, the final PTFE VG comprises a fibrous structure defined by interconnected interspaced nodes [31]. Since patency loss results from a lack of tissue regeneration, the internodal distance of PTFE grafts can be modified to enhance tissue ingrowth and endothelization. However, higher porosity reduces the overall circumferential tensile strength and suture retention strength, negatively affecting the handling [32].

4.2. Mechanical Properties of PTFE VGs

Despite the wide use of PTFE-based vascular grafts, PTFE is a non-biodegradable biomaterial with low compliance and moderate stiffness. Due to this factor, PTFE VGs have been evolving through different modifications to improve the handling properties and mechanical properties. To improve mechanical properties, especially for lower limb bypass, reconstructions above or below the knee, extra-anatomic procedures, and vascular access grafts, PTFE VGs include internal or external support in the form of rings or spirals offering kink and compression resistance [33]. On the other hand, they have also been coated to improve blood compatibility, avoiding thrombus formation and related patency loss; these coated PTFE VGs are often used for limb reconstruction and vascular access, or for early cannulation vascular access [34,35]. Even multilayered PTFE grafts include an anti-thrombogenic coating and an elastomeric layer intended to mechanically seal the needle entry hole after its removal [32]; PTFE grafts have also been modified to improve longitudinal stretching to adapt to different anatomies. Table 2 summarizes the mean mechanical properties in average commercial PTFE vascular grafts for Arteriovenous Fistula with a 6 mm internal diameter.

Table 2. Mechanical properties in average commercial PTFE vascular grafts for Arteriovenous Fistula with 6 mm internal diameter.

Test Performed	Average Value	Reference
Relaxed internal diameter (mm)	6 [2]	[36]
Pressurized Internal Diameter (mm)	6 [2]	[36]
Wall Thickness (mm)	0.39 [2]	[36]
Porosity (%)	50	[37]
Void area (μM) [1]	20–500	[37]
Water permeability (mL·cm^2/min)	4320	[37]
Circumferential tensile Strength (KPa)	16,000 [2]–20,590 [3]	[37]
Longitudinal tensile Strength (KPa)	15,630 [2]–41,480 [3]	[37,38]
Burst Strength (kPa)	361	[37,38]
Suture Retention Strength (g)	480	[37]
Dynamic Compliance (%/mmHg)	2.1	[37]
Strength After Puncture (KPa)/ # Punctures	19,220/0 9230/8 8730/16 6370/24	[27]

[1] Minimum value and maximum value to maintain tissue integration. [2] Non-stretchable PTFE VGs. [3] Stretchable PTFE VGs.

As shown in Table 2, PTFE grafts have a high isotropic modulus, whereas the native arteries usually have a lower modulus and are anisotropic. The stiffer feature of the PTFE VGs has been shown to cause a high impedance within the pulsatile flow as the propagation velocity of pressure and flow waves increase. Therefore, in the anastomosis, the reflected waves from pressure and flow transform into different phases that increase the risk of turbulence and unstable flow regions [37]. These forces create a high wall shear stress, ultimately leading to the gene expression related to the development of intimal hyperplasia and atherogenesis. This idea is supported by further compliance reduction of 50% to 86% after 12 weeks of implantation related to foreign body response and fibrous capsule formation [39]. As shown in Figure 4, when comparing the mechanical properties of PTFE vascular grafts with the most successful vascular graft from the great saphenous vein (GSV), it is possible to identify that the compliance is almost two times greater in the GSV. In this sense, Salacinski et al. performed a linear regression analysis comparing the compliance of host arteries, saphenous vein, umbilical veins, bovine xenograft, PET, and PTFE-based grafts with their patency rates at three years of clinical trials. The main results are that the graft patency decreases as the compliance mismatch increases due to the low or absent increase in compliance under the changes in blood pressure due to the lack of viscoelastic properties. Details on the mechanical properties of the PTFE synthetic VG compared with those of the Great Saphenous Vein are shown in Figure 4.

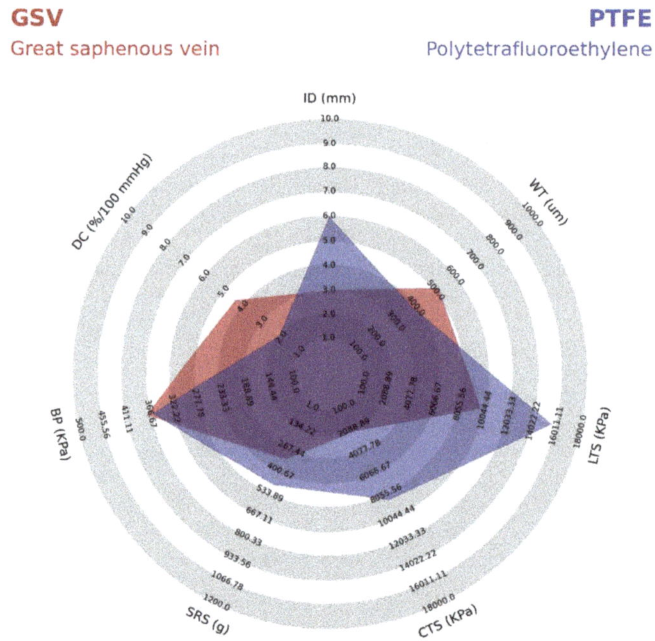

Figure 4. Mechanical properties comparison of the great saphenous vein, as a gold standard, and PTFE. Mechanical properties identified include Internal diameter (ID), wall thickness (WT), longitudinal tensile strength (LTS), circumferential tensile strength (CTS), suture retention strength (SRS), burst pressure (BP), and dynamic compliance (DC). Chart generated from the Python library developed by StatsBomb/Anmol Durgapal (accessed on 29 May 2022) [29].

5. The Potential of TEVGs on Clinical Success

Even though plenty of development has been reached over the last 50 years in the field of synthetic materials for the design of vascular grafts and favorable results have been obtained for aortic and wide arteries replacement, there have not been any satisfactory results in small caliber grafts due to thrombus formations and poor patency rates [40]. Nevertheless, substantial efforts have been undertaken to overcome these limitations. The objective of developing TEVGs is to create a vascular graft that integrates with the native tissue and behaves like a native vascular vessel, providing the right biochemical and biomechanical stimuli required for the growth and self-regeneration [41].

As stated above, patency loss is a severe complication of small diameter grafts. In this sense, PTFE VGs have been reported to present patency rates of 75% at one year, which will decrease to 44% at five years. This behavior is non-comparable to the saphenous vein with a reported patency rate of 91% at one year, decreasing to 76% at five years, superior to PTFE in all stages. On the other hand, some types of TEVGs show reasonable patency rates. One example is Artegraft®, with a mean patency rate of 73% over 18 months. Another example can be the Humacyte graft, with a patency rate of 83% at three months and 60% after six months [42]. Therefore, regenerating the vascular wall through TEVGs represents an opportunity to maintain long-term patency.

5.1. Failure Causes of TEVG's Related to the Mechanical Properties

The leading cause of failure of VG and TEVGs is patency loss where blood flow becomes compromised and can occur through different mechanisms. The most common mechanism is related to thrombogenesis when the surface of the biomaterial lacks the required hemocompatibility [15]. Furthermore, this is the most common cause of failure on VGs and TEVGs used as arteriovenous fistulas, given that clots are formed after repeated puncture, and the thrombus might spread if there is not an anticoagulant surface or treatment. In these cases, secondary patency in a VG is achieved by removing the clot with a catheter, and the lumen of the graft can be restored, allowing blood perfusion. However, this procedure is not always successful if the main thrombogenic conditions are still present, in which case the condition may evolve towards a fibrotic thrombus that will make it impossible to perform a new procedure to recover patency [14].

However, the most common cause of failure related to the mechanical properties of the VG or TEVG includes intimal hyperplasia, in which the overproliferation of smooth muscle cells thickens the tunica intima in the blood vessel, causing the contraction of the construct and the loss of patency [16]. On the other hand, the formation of fibrotic tissue surrounding the graft–in this case, the microstructure of the VG or TEVG–does not allow cell infiltration, and a dense layer of collagen is formed surrounding the graft. In this case, stem cells differentiate toward myofibroblasts, causing the contraction of the VG or TEVG, and stenosis is developed [43]. However, another critical failure cause is the formation of aneurysms in which the chemical or biological degradation of the graft and/or structural defects may lead to the dilatation or rupture of the graft [44].

Unstable Flow Conditions and Intimal Hyperplasia Development in TEVGs

Intimal hyperplasia has been reported to present in 10–30% of failure causes of VGs. This high failure rate has been correlated with the compliance mismatch between the VG and the native vessel. In this sense, it has also been reported that the patency loss is directly proportional to the compliance mismatch [45]. The low compliance not only depends on the biomaterial origin but also depends on the microstructure. For instance, it has been reported that there is a strong correlation between low porosity and increasing TEVG wall thickness with a low compliance [46].

The flow stability is key to maintaining the graft patency. The flow regime, velocity profile, and cyclical deformation caused by the pulsatile flow create determinant wall shear stresses (WSS), which are mechanical signals in the cells interacting and repopulating the TEVG [46]. From this perspective, low compliances have also been strongly correlated with low WSS due to the low-compliant grafts presenting a significant difference in diameter under the pulsatile pressure [45]. Furthermore, while the VG maintains a constant diameter, the artery will dilate and contract. When the artery dilates, the VG maintains its low diameter, and the blood flow profile produces a sizeable corresponding effect. This profile is characterized by zones of blood recirculation negatively affecting the velocity profile, inducing turbulent flows, and altering the pressure wave, decreasing the WSS [45]. Figure 5 summarizes the effect of low, medium, and high compliance on the blood flow pattern.

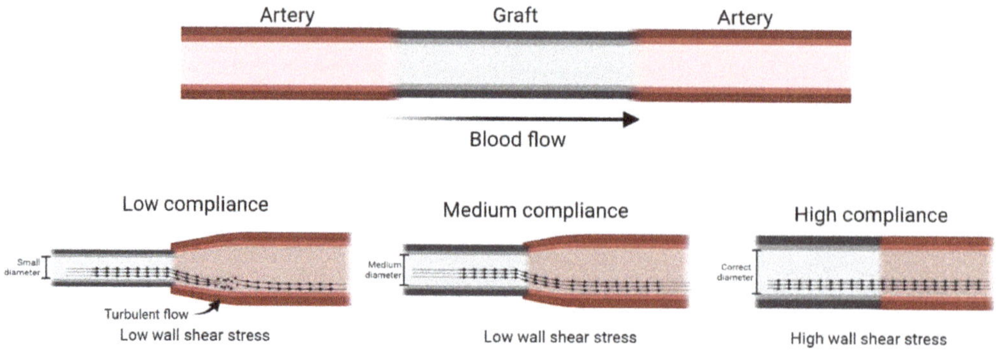

Figure 5. Effect of compliance on flow patterns and wall shear stress at the distal anastomosis in a VGs. Data retrieved from Post A, et al. [45] Image created with BioRender.com (accessed on 29 May 2022).

Regarding the compliance mismatch effect on WSS, normal WSS reported on small diameter vessels—such as the coronary artery—is 0.68 N/m^2 ranging between 0.3 to 1.24 N/m^2 [47]. It has been reported that non-compliant VGs present low shear stress near 0.03 N/m^2, medium compliance VGs display similar shear stresses with an average of 1.04 N/m^2, and high compliance grafts present higher WSS with an average of 3.6 N/m^2 [45]. Based on this correlation, it could be expected that low-compliant TEVGs will induce intimal hyperplasia.

Accordingly, TEVGs for different applications must consider the compliance values and the physiological shear stress in which the TEVG will be anastomosed. For instance, GSV has an average compliance value of 4.4%/100 mmHg, and it has been reported that implanted GSV grafts for coronary artery bypass present a range of 1.22 N/m^2 to 1.73 N/m^2, whereas GSVs grafts below 0.71 N/m^2 are predictors for the VG failure [48]. Recent reports have also shown that the left internal mammary artery (IMA), with compliance of 5.22%/100 mmHg, used as a VG for a coronary bypass, maintained its patency for one year and presented high WSS values near 4.43 N/m^2, contrasted with those occluded with a WSS 2.56 N/m^2 [49]. Although the differences between the compliance of the GSV and IMA are not significantly different, the WSS produced as VGs for CABG are very different due to differences between the layer's composition in the media and adventitia layers.

Although GSV has a different structure from arteries, the vein adapts to the arterial environment due to the different hemodynamic conditions and increased oxygen tension [50]. The most significant change is the increase in the vein diameter between 20% and 85% for arteriovenous fistulas or lower extremity grafts, respectively. The vein wall also augments the wall thickness due to the increased pressure from the proliferation of smooth muscle cells and adventitial fibroblasts derived from bone marrow progenitor cells. These changes have been associated with the normalization of shear stress, occurring during the first months where the initial shear stress will reach up to 9.6 N/m^2, leading to a patency rate between 75–90% at one year and 50% at 15 years [51].

5.2. Biomechanical Stimulation for Physiological Regenerative Responses; Physiological Wall Shear Stresses

When developing TEVGs, it is essential to induce the regeneration of the vascular wall within the anastomosis; not only will the bioactive chemical properties of the graft induce this process, but biomechanical signals are required. Aside from the microstructure allowing the cell infiltration and proliferation, the wall shear stress will generate different responses in the cells interacting with the biomaterial in the hemodynamic context.

Therefore, physiological WSS will lead to the regeneration of the vascular wall, whereas the low WSS induced by the lack of compliance will have a detrimental effect on the gene expression of immune cells, smooth muscle cells, and endothelial cells. For instance, for regeneration to occur, the inflammatory response must be modulated towards the differentiation of Macrophages to the M2 type, recognized by the cytokine secretion to induce extracellular matrix deposition. Low WSS has been correlated to the maintenance of the M1 phenotype leading to chronic inflammatory responses.

On the other hand, smooth muscle cells must begin their differentiation towards a contractile phenotype rather than a synthetic phenotype. The synthetic phenotype related to low WSS has been associated with lower blood pressure responses affecting the construct's overall compliance. Finally, endothelial cells should differentiate towards a functional phenotype able to release nitric oxide (NO) as a regulator of the vascular wall tone. Low WSS has been shown to induce senescence on endothelial cells without proliferation capacity and limited release of NO. Table 3 summarizes the effect of baseline and lows WSS on the endothelial cells, smooth muscle cells, and macrophage gene expression.

Table 3. Effect pf baseline and low WSS over endothelial cells, smooth muscle cells, and macrophages gene expression.

Wall Shear Stress (N/m^2)		Endothelial Cells Genes		Smooth Muscle Cells Genes		Macrophage Genes	
		Down-Regulated	**Up-Regulated**	**Down-Regulated**	**Up-Regulated**	**Down-Regulated**	**Up-Regulated**
Physiological Limits	1.5–2.4	E-Selectin TXNIP PKC JNK TNFa	KLF-2 ERK P38 ERK CD31 eNOS vWF	P21 BMP4 SMAD	Ciclyn D1 AKT IL-11	Leukocyte	M2 phenotype CD206 IL-10 TGF-β1
Low limits	0.1–1	eNOS NOX4 NOTCH1	VCAM-1 ICAM-1 EDN-1 MCP-1 PDGF	a-SMA SM22 SMTN CNN	Prolifferative MMP2 TGF-β1 PDGF	M2 phenotype	M1 Phenotype NF-κB IL-1 MCP-1 Selectin MMP2-9

Data obtained and complemented from Rodriguez-Soto, et al. [14].

According to the data registered in Table 3, it can be observed that the low WSS generates the down-regulation in the expression of eNOS (endothelial nitric oxide synthase), the enzyme responsible for the NO release and the maintenance of the overall homeostasis in the vascular wall [52]. In addition, it generates the down-regulation of NOTCH1 (Neurogenic locus notch homolog protein 1), which is a mechanical sensor that maintains the junctional integrity of endothelium [53], and NOX 4 (NADPH oxidase 4), which produces H_2O_2 as a signaling molecule for endothelial cell proliferation [54]. Without the expression of these genes, it is likely improbable that the endothelial lining can be regenerated over the TEVG surface.

On the other hand, low WSS on endothelial cells also increases the expression of MCP-1 (Monocyte Chemoattractant Protein-1), which is a chemoattractant for proinflammatory monocytes. At the same time, the expression of VCAM-1 (Vascular cell adhesion protein-1), ICAM-1 (Intercellular Adhesion Molecule-1), and EDN-1 (Endothelin-1) increases, which are adhesive molecules for monocytes, and PDGF (platelet-derived growth factor) also increases, promoting thrombus formation.

Regarding smooth muscle cells, a low WSS decreases the expression of a-SMA (smooth muscle actin), SM22 (Transgelin), SMTN (Smothelin), and CNN (calponin); all are genes related to the contractile function of smooth muscle cells required to respond to contractile and dilating signals from endothelial cells [16]. Furthermore, low WSS induces the up-regulation of proliferative genes, responsible for intimal hyperplasia, TGF-β1 (transforming growth factor-beta) inducing inflammation, and the MMP2 (matrix metalloproteinase-2) that not only degrades the vascular wall but also increase the migration and proliferation of the smooth muscle cells with the synthetic phenotype [51,55].

Finally, macrophage modulation due to low WSS has been shown to reduce the expression of M2 phenotype-related genes such as CD206 (macrophage mannose receptor 1), IL-10 (Interleukin 10), which blocks the NF-κB (nuclear factor kappa-light-chain-enhancer of activated B cells) proinflammatory pathway, and TGF-B1. Furthermore, genes related to increased inflammation are reported to be activated, such as the signaling pathway related to NF-κB activation, MCP-1, and Selectin as chemotactic for new inflammatory cells, as well as present an increase in MMP9, related to the degradation of the scaffold [56,57].

6. Biomechanical Design Requirement for Vascular Grafts

Due to the significant differences between vascular grafts and native arteries in their mechanical properties and behavior under implantation conditions, several requirements have been studied to develop mechanical characterizations of these new grafts. The requirements to evaluate vascular grafts are outlined in International Standards ANSI/AAMI VP20: 1994 (American National Standard for cardiovascular implants and vascular prostheses) [58], ISO 7198:2016 (Cardiovascular implants and extracorporeal systems—Vascular prostheses—Tubular vascular grafts and vascular patches 94, and ASTM F3225-17 (Standard Guide for Characterization and Assessment of Vascular Graft Tissue Engineered Medical Products (TEMPs) [59].

ANSI/AAMI VP20: 1994, ISO 7198:2016 and ASTM F3225-17

Test guidelines for international standards are defined for all vascular grafts: synthetic, biological, and coated [59]. According to ASTM F3225-17, mechanical testing must be performed in an environment that emulates the vascular grafting conditions of use, most commonly in buffered saline solutions at 37 °C. Otherwise, non-physiological test conditions must be justified. In addition, parameters such as the time between tissue collection and testing and sample storage may modify the mechanical properties [59]. A minimum of three samples from at least three manufactured lots is necessary to have an appropriate variability of the characteristics in the samples studied.

Due to the anisotropic properties associated with vascular grafts, strength tests should be performed, including more than one axis. Therefore, longitudinal, and circumferential tensile strength tests are performed to determine whether the axial and radial yield and/or breakpoint are reached, respectively. Stress vs. strain plots must be made to include and analyze the data. In addition, suture retention strength tests should be developed to determine the force required to pull a suture from the vascular graft to simulate clinical techniques (straight-across, oblique, and longitudinal procedures). The suture is usually placed 2 mm from the edge and tested in various directions to determine the strength that the prosthesis withstands before mechanical failure.

Moreover, a burst strength test is performed to determine the pressure rate change until sample bursting occurs. This method allows to report the diameter of the sample when it is pressurized directly with fluid or gas. A repeated puncture test with a dialysis needle (16G) should be performed to measure the strength that supports the prosthesis through a force test such as pressurized burst strength or circumferential tensile strength [59].

It is necessary to perform tests to determine the thickness of the wall and the relaxed and pressurized internal diameter to observe changes in diameter under different hemodynamic conditions. Furthermore, the vascular graft discontinuity should be reported when the lumen diameter decreases during kinking and the radius of curvature that impedes normal flow through the graft. Another way to evaluate conditions that approach the preclinical environment is by measuring diameter change simulated under cardiac cycle conditions to determine the radial dynamic compliance. In addition, the rate of water leakage through the prosthesis wall should be characterized to avoid leakage at the time of implantation [59].

7. Trends on TEVGs Design, Biomaterials, and Manufacture Techniques in the Context of Desired Mechanical Properties

The ideal graft must have comparable mechanical properties to those of the native vessels that will be anastomosed. For instance, it has been shown that the gold standard for the repair of peripheral vessels is the GSV, whereas in other applications such as in Coronary Artery Bypass Graft Surgery (CABG), the gold standard corresponds to the IMA. This is implemented to reduce the level of mechanical decoupling and levels of failure that can lead to stenosis or aneurysms and, therefore, long-term permeability loss. The most important factors involve mechanical strength and compliance concerning the viscoelasticity of the scaffold. Likewise, the grafts must have sufficient mechanical strength and compliance to withstand the changes in blood pressure. At the same time, they should adjust correctly to the adjacent vessels when completing the suture procedure, enabling the correct velocity profiles and continuity in the pressure waves. In the following sections, the current trends in TEVGs design, biomaterials, and manufacturing techniques are reviewed according to the TEVGs that have been reached, including clinical trials, pre-clinical studies on animal models, and techniques that have been tested on in vitro conditions during the last five years.

7.1. TEVGs That Have Reached Clinical Trials

After analyzing the current trends in TEVGs, five clinical trials were identified regarding TEGVs for human use in different contexts. The main mechanical properties of each vascular graft compared to GSV and IMA grafts are summarized in Table S1. The mean values of the mechanical properties of the TEVGs on clinical trial are summarized and compared in Table 4 and Figure 6.

The most recent TEVG that has entered to clinical trial phase is a Polyhedral oligomeric silsesquioxane poly (carbonate-urea) urethane (POSS-PCU) small-diameter vascular graft, fabricated through extrusion and phase inversion method using sodium bicarbonate as a porogen. This TEVG is currently being tested as arteriovenous fistulas for hemodialysis; the study began in 2021 and is expected to reach completion in April of 2025. This TEVG has a small diameter (<5 mm) with a mean wall thickness of 0.94 mm that is placed in adult female patients undergoing hemodialysis and in need of new arteriovenous access but without any viable access for dialysis. This study has an inclusion criteria of patients on oral contraceptives or with an intrauterine device, aiming to demonstrate patency in patients with increased risk of thrombosis. It is expected to measure patency rates at 18 months by Doppler ultrasonography (d-US) and compare it to the patency of the PTFE grafts. They also will measure the occurrence of any Serious Adverse Event (SAE) related to the implantation of the POSS-PCU TEVG. For the initial characterization of the POSS-PCU vascular graft, they used Delfino's strain energy potential to capture the viscoelastic properties of the materials. Later, they conducted different experimental and computational experiments to show shear stress within the graft wall in the context of increased uniaxial strength. They concluded that besides the overall resistance of the graft, the stiffness and long-term viscoelastic properties could be improved with better manufacturing techniques; detailed mechanical properties of this TEVG can be found in the Supplementary Data [60]. The POSS-PCU TEVG offers similar longitudinal tensile strength compared to GSV and IMA grafts (3210 vs. 2405 and 4300 KPa), however, its suture resistance strength is markedly higher (4460 vs. 3200 and 1350 KPa, respectively). Furthermore, it presents reduced dynamic compliance compared to the allografts (1.59 vs. 4.4 and 5.22%/100 mmHg respectively) may explain its increased stiffness and poor vascular physiology resemblance (Figure 6).

A Human Acellular Vessel (HAV) from Humacyte is a graft also developed for patients with need of hemodialysis access but who are not candidates for a fistula. This TEVG is created in vitro through the culture of human smooth muscle cells and fibroblast cultured over a biodegradable polymer. Further decellularization process creates a structure that retains the extracellular matrix protein and provides corresponding mechanical properties. Animal models were performed in baboons as arteriovenous grafts with 80% patency at six months. This vascular graft has been included in different clinical trials. One of the most relevant, made in 2016 as a phase 2 trial in 40 elected patients, demonstrated primary patency rates of 63% (95% CI 47–72) at six months and 28% (95% CI 17–40) at 12 months. Conversely, secondary assisted patency rates at six and 12 months were 97% (95% CI 85–98) and 89% (95% CI 74–93), respectively. Secondary patency rates in extensive multicenter cohort studies with PTFE were 55–66% at one year, making HAV a suitable option. In the group of patients, they also measured serious adverse events (SAEs) that occurred 155 times in 33 patients, leading to increased reintervention rates. SAEs mainly included patency loss, infection, and one case of steal syndrome. HAV provides an exciting conclusion, as they have better secondary patency rates than PTFE-based VGs, leading to a well-tolerated TEVG with no signs of aneurysm formation or degradation with low immunogenicity. For which it has been included in different clinical trials from 2015 to 2020. Researchers found that Artegraft®has primary, primary-assisted, and secondary patency rates of immunogenicity. However, HAV has the lower reported tensile strengths (circumferential and longitudinal) along all TEVGs in clinical trial phases (1400 and 1200 KPa) and low dynamic compliance compared to human allografts (1.5 vs. 4.4 and 5.22%/100 mmHg), leading to increased stiffness and a poor vascular physiology resemblance; complete data can be found in the Supplementary Material.

Likewise, the TRUE vascular graft is also a decellularized arteriovenous TEVG designed by neonatal human dermal fibroblasts seeded in a bovine fibrin gel. A baboon pre-clinical model was performed on 10 subjects to determine its safety in terms of occurrence of adverse events and to find patency rates. They found patency rates at 3 and 6 months of 83% (5 of 6) and 60% (3 of 5), respectively, with evidence of recellularization of smooth muscle cells and endothelium formation. This TEVG is currently being tested as an arteriovenous graft for hemodialysis access on 10 participants beginning in May of 2021 and it is expected to finish in June of 2022. The mechanical properties of this kind of graft can be seen in the Supplementary Data. However, circumferential tensile strength and suture resistance strength were like the IMA (3800 and 1950 vs. 4100 and 1350 KPa, respectively), making it a newly suitable option for further study.

Another kind of TEVG for hemodialysis accesses is Artegraft®, a decellularized bovine carotid artery used for arteriovenous fistula generation. These TEVG have been included in different clinical trials from 2015 to 2020. Researchers found that Artegraft® has primary, primary-assisted, and secondary patency rates of 73.3%, 67%, and 89%, respectively. SAE was present in one immunocompromised patient that presented a resistant infection, leading to the early removal of Artegaft® at two months. A relevant finding is that anastomotic venous stenosis occurred in some grafts and was the most common indication of graft removal. Artegraft® demonstrated similar tensile strength properties compared to PTFE VGs. However, its low dynamic compliance [1,5] can explain the occurrence of stenotic venous anastomosis in most grafts (Figure 6).

BioIntegral Surgical No-React ® are bovine pericardial xenografts used as a strategy for vascular graft infections due to their regenerative properties; these series of clinical trials began in 2019. A prospective study of six patients with infected aortoiliac segments treated with BioIntegral Surgical No-React ® demonstrated that four out of the six patients are still alive with complete patency demonstrated by d-US. Unfortunately, two died from acute myocardial infarction, and the other due to sepsis secondary to the vascular infection. Although this graft has a remarkably increased longitudinal tensile strength compared to GSVs and IMA (10,000 vs. 2405 and 4300 KPa), it conferred to most patients (4 out of 6) a proper control of the infection with reasonable long-term patency rates.

According to the reported data, a mean value was extracted from the mechanical properties of the current vascular grafts undergoing on clinical trials and were compared to the mechanical properties of the GSV, as shown in Table 4 and Figure 6; individual data are shown on a bar graph in Figure 6.

Table 4. Mechanical properties in blood vessels: grafts in clinical trials (CT grafts) and great saphenous vein (GSV).

Test Performed	CT Grafts	GSV	Reference
Internal Diameter (mm)	5	3	[22,23,61–64]
Wall Thickness (μM)	690	518	[24,25,61–65]
Circumferential Tensile Strength (KPa)	2380	2405	[12,19,26,62–65]
Longitudinal Tensile Strength (KPa)	4230	9760	[61–63,65]
Burst Pressure (KPa)	405.17	371.96	[12,26,62,64]
Suture Retention Strength (g)	306	327	[22,27,61,62,64]
Dynamic Compliance (%/100 mmHg)	1.53	4.40	[12,28,61–63]

(A)

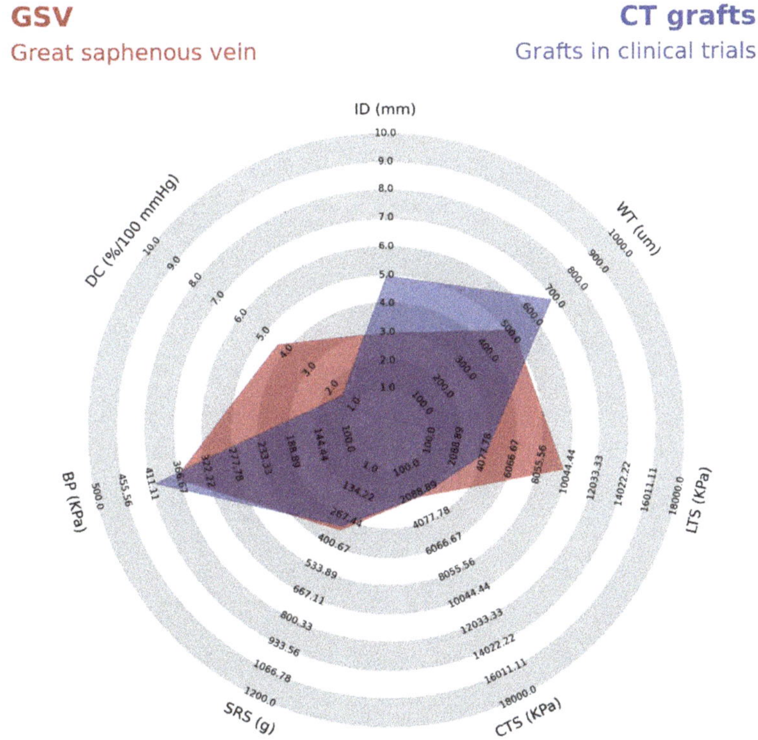

Figure 6. *Cont.*

(B)

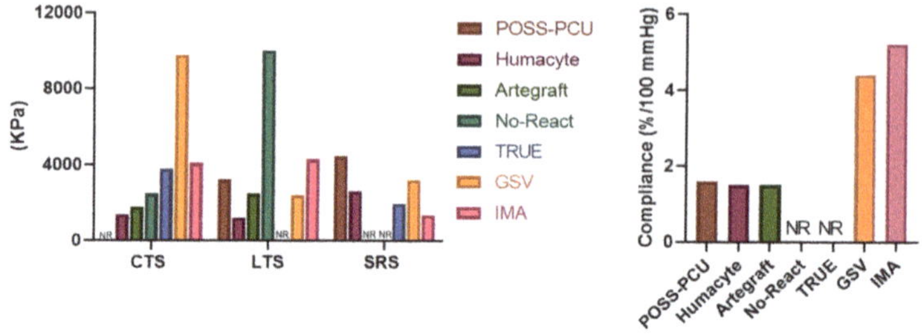

Figure 6. (**A**) Mechanical properties comparison of great saphenous vein, as a gold standard, and grafts in clinical trials. Mechanical properties identified include internal diameter (ID), wall thickness (WT), longitudinal tensile strength (LTS), circumferential tensile strength (CTS), suture retention strength (SRS), burst pressure (BP), and dynamic compliance (DC). Chart generated from the Python library developed by StatsBomb/Anmol Durgapal (accessed on 29 May 2022) [29]. (**B**) Mechanical properties comparison of great saphenous vein, as a gold standard, internal mammary artery, and each graft made in clinical trials. Circumferential tensile strength (CTS), longitudinal tensile strength (LTS), suture retention strength (SRS), internal mammary artery (IMA), and great saphenous vein (GSV). Data are according to literature reports of a representative data set. Additional information regarding descriptive analysis can be found in Table S1. Graph generated with GraphPad Prism (accessed on 29 May 2022).

7.2. TEVGs That Have Reached Pre-Clinical Animal Models

After the analysis of the recruited data, a total of 43 studies fulfilled inclusion criteria for TEVGs that have reached pre-clinical animal models [66–107]. It was found that 46.43% (n = 26; 46.46%) of the studies reported the use of Poly ε-caprolactone (PCL), mostly manufactured through electrospinning (n = 23, 88.46%) and more frequently tested on murine animal models (rats). PCL has been shown to be biocompatible, and the constructs exhibit slow degradation rates of about 1–2 years [67,68,70,72,74,76,78–82,89,90,92,95–97, 99,102–105,107]. This has been correlated with the ability to maintain the TEVGs stability under hemodynamic operative conditions and less acidic breakdown products compared to other polyesters because it can be easily degraded by lipases and macrophages, presenting a low inflammatory profile and the potential for loadbearing [29]. However, although it has been proven that PCL is a relatively easy material to work with, PCL-Based TEVGs usually present low compliance compared to native vessels (GSV). From this perspective, the raw PCL has a compliance value close to 2%/mmHg, and PCL-Based TEVGs present a value close to 3%/mmHg; both values are lower than those of the GSV or IMA, closer to 5%/mmHg.

To overcome these limitations, various authors have chosen to use a variation of PCL such as poly (L-lactide-co-ε-caprolactone) or PLCL with a softer and more elastic nature. Compared to PCL, PLCL-based TEVGs present higher compliance values close to 8%/mmHg [108]. It has also been reported that PLCL has excellent mechanical properties such as high plasticity and higher degradation rates than PCL. Nevertheless, the lactide group inclusion generates low biocompatibility, poor hydrophilicity, and presents acidic degradation.

Polyurethane-Based TEVGs (PUs) are the second most reported TEVGs ($n = 6$; 10.71%). PUs has been considered as a good base biomaterial due to its relatively high tensile and flexural strength [37,77,81,91,93,100,104]. Furthermore, it has been shown that PU has relatively high compliance compared to other materials. PU-Based TEVGs present an average compliance value of 6.5%/mmHg [37], being closer to the value found in native arteries. However, although none of the consulted manuscripts had a specific value for the circumferential tensile strength of the material, it has been reported that repeated puncture of PU-Based TEVGs might lead to aneurism generation, meaning that the material's circumferential strength might be compromised, and it could thus not be comparable to the native arteries. Therefore, for its use in TEVGs, different alternatives should be included to improve the circumferential strength of the material while maintaining the required elasticity for compliance.

The third most common biomaterial for TEVGs fabrication corresponds to decellularized arteries ($n = 4$; 7.14%), which can be correlated to the advantages of maintaining the structure and mechanical properties of native tissues. Moreover, these biomaterials have a relatively low immune response from the patient, cause minor damage to other bioactive components, and usually do not use chemical reagents to reduce adverse reactions in the body [109]. Regarding the compliance of the material, it has been reported that the compliance of this type of graft is close to 9.7%/mmHg, which is a high value compared to previously described materials. However, it has also been reported that most decellularized grafts involve high costs, and they can also require two or more surgeries.

All these data are summarized in Table 5 and Figure 7, comparing the mechanical properties the average TEVG entered to pre-clinical models with GSV, this is a PCL-Based TEVG and fabricated with Electrospinning. Supplementary Data are summarized in Tables S4 and S10.

Table 5. Mechanical properties in blood vessels: Poly(ε-Caprolactone) tested in vivo (PCL in vivo) and great saphenous vein (GSV).

Test Performed	PCL In Vivo	GSV	Reference
Internal Diameter (mm)	2.17	3	[22,23,67,68,70,72,74,76,78–83,88–90,95–99,102–105]
Wall Thickness (µM)	400	518	[24,25,67,68,70,72,74,76,78–83,88–90,95–99,102–105]
Circumferential Tensile Strength (KPa)	6440	2405	[12,19,26,67,70,74,76,78,88–90,92,96,98]
Longitudinal Tensile Strength (KPa)	8400	9760	[11,19,26,67,68,72,74,76,78,79,81–83,95,97,102–104]
Burst Pressure (KPa)	348.22	371.96	[12,26,67,74,76,79,81,89,90,97,99,102,104,105,107]
Suture Retention Strength (g)	124	327	[22,27,67,74,76,95,96,99,104,107]
Dynamic Compliance (%/100 mmHg)	2.63	4.40	[12,28,74,76,78,81,104]

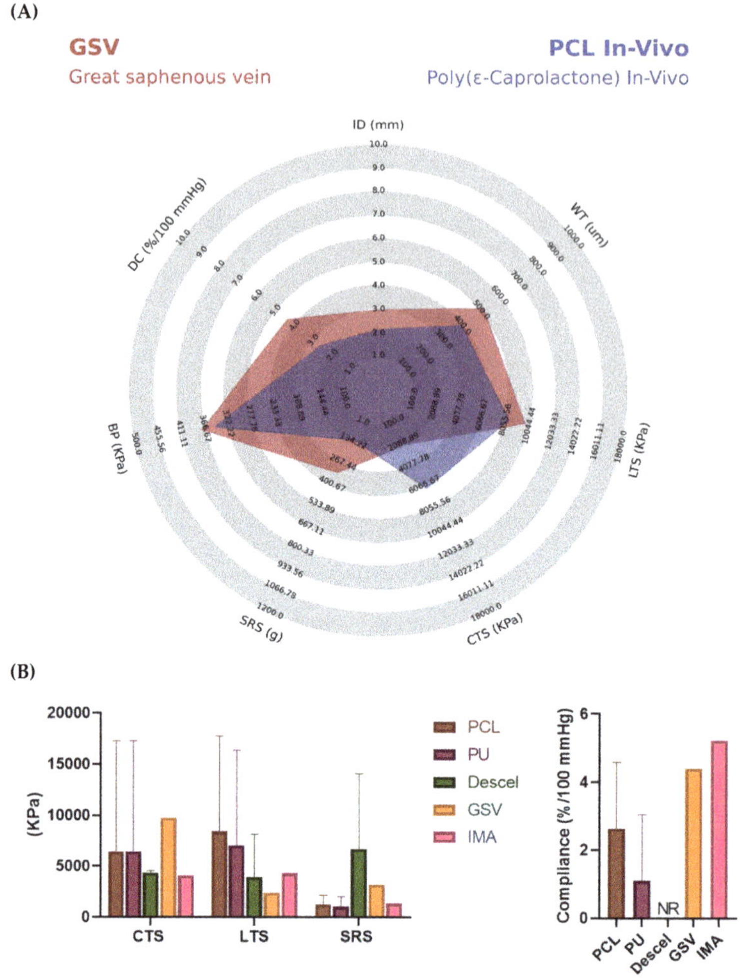

Figure 7. (**A**) Mechanical properties comparison of great saphenous vein, as a gold standard, and Poly(ε-Caprolactone) tested in vivo. Mechanical properties identified include internal diameter (ID), wall thickness (WT), longitudinal tensile strength (LTS), circumferential tensile strength (CTS), suture retention strength (SRS), burst pressure (BP), and dynamic compliance (DC). Chart generated from the Python library developed by StatsBomb/Anmol Durgapal (accessed on 29 May 2022). (**B**) Mechanical properties comparison of great saphenous vein, as a gold standard, internal mammary artery and each graft made in clinical trials. Circumferential tensile strength (CTS), longitudinal tensile strength (LTS), suture retention strength (SRS), Decellularized (Descel), internal mammary artery (IMA), and great saphenous vein (GSV). Data are shown as mean ± standard deviation, according to the number of studies reporting the data. Additional information regarding descriptive analysis can be found on Table S2. Graph generated with GraphPad Prism (accessed on 29 May 2022).

7.3. Current Strategies of TEVGs on In Vitro Testing

To succeed with international standards, various authors have overseen the innovation of new techniques, materials, and practices for developing vascular grafts. A total of 46 articles fulfilled the inclusion criteria regarding TEVGs undergoing in vitro testing. Among the most used materials in in vitro processes, the PCL, TPU (Thermoplastic Polyurethane), and

PLGA (poly(lactic-co-glycolic acid)) are the most used, together with natural biomaterials such as collagen and gelatin [109–156].

Based on these biomaterials, there has been a significant trend of various manufacturing methods, in which the electrospinning method is one of the most used. This methodology is based on the formation of nanofibers through the application of electricity to the material. This process is usually used due to its versatility in adjusting different parameters such as porosity, size, thickness, and density of the graft. About 37% of the articles mentioned used the electrospinning method in manufacturing [114,116,118,120–123,125,127,128,132–135,137,139,141–144,146,147,150,151,154,155]. On the other hand, the second most used process is the Solvent Casting method, with 13.04% of articles using this process [117,136,148,149,156].

However, different authors carry out mechanical and cell viability tests to prove the possible behavior of the grafts in a patient. Due to the mechanical tests, a trend has been found regarding the tests carried out, of which the longitudinal tensile strength tests are most often reported. This measurement helps to determine the vascular graft resistance against longitudinal tensile forces. Along with the revised manuscripts, an average of 94,300 KPa (n = 37) was evidenced, thus indicating that most of the grafts resist high longitudinal forces [113–156]. Nonetheless, circumferential tensile strength is not often reported, same as compliance, both being highly important data required to establish suitable mechanical properties for TEVGs applications. On the other hand, the second most reported measurement is the Burst Strength since it is essential to determine the pressure that can resist substantial flows. The authors recorded a mean value of 2161.14 mmHg (n = 17) [118,120,122,123,126,128,142,143,146,147,151,152,154–157].

Likewise, some articles have also included other relevant tests to comply with international standards. Another of the most important tests is circumferential traction resistance, with an average result of 9210 KPa (n = 13), suture retention resistance with an average result of 7.34 N (n = 11), and finally dynamic compliance with a mean result of 327.28 mmHg (n = 6). This can be evidenced in Figures 7 and 8, where it is also possible to observe the morphology of the synthetic vascular grafts compared to the great saphenous vein. Regarding the radar chart, it is possible to analyze the role of PCL in the mechanical properties provided to the grafts, where it stands out over TPU and PLGA. These materials show different factors such as longitudinal tensile strength, circumferential tensile strength, suture retention strength, and dynamic compliance, thus making it a material with the necessary characteristics of durability and hardness for a device that is going to be subjected to such a changing environment. It was previously mentioned that the ideal graft should have mechanical characteristics similar to a vein. In this way, it can be argued that the PCL meets these characteristics, even surpassing the behavior of the superior saphenous vein in said properties. For these reasons, there has been a trend among authors to use this material by combining it with one that can help improve properties such as burst pressure, and thus generate an ideal replacement material [120,123–125,127,132,133,135,139,140,142,145,147,149,154–157].

Another critical aspect of in vitro tests is cell analysis. Cell viability, cytotoxicity, and cell adhesion, among others, are necessary tests to comply with standards. Thus, the researchers have developed several tests to reach a physiological environment and understand what the possible behavior of the graft will be in an authentic context.

This review has found interesting data regarding the biomaterials used for TEVGs that have been reported in in vitro testing (Figure 8). The three most common materials in manufacturing TEGVs (Table S27) and tested in vitro were PCL (37.5%), TPU (8.93%), and PLGA (7.14%). PCL-based TEGVs tend to be more like the internal mammary artery in circumferential (3594 ± 2486 vs. 4100 KPa), longitudinal (4107 ± 2376 vs. 4300 KPa) tensile strengths, and dynamic compliance (5.42 ± 2.65 vs. 5.22%/100 mmHg), respectively. However, its suture resistance strength (4520 ± 4130 KPa) is much higher than that of saphenous vein grafts and the internal mammary artery (3200 and 1350 KPa), respectively, leading to decreased adaptability of the TEVG to suture tension. In addition, 12 out of

22 TEGVs in vitro studies reported the presence of endothelization of the vascular graft lumen. The principal manufacturing technique was electrospinning (16 out of 22) and most products were monolayered (11 out of 22) [118,127,132,137,139–141,145,147,154–157].

On the other hand, TPU-based TEGVs showed the highest mechanical properties regarding longitudinal tensile strength (15,750 ± 4546 KPa), suture resistance strength (7860 ± 1600 KPa) compared to saphenous vein grafts (2405 and 3200 KPa), and internal mammary artery (4300 and 1300 KPa), respectively. These results show that TPU is a highly rigid material that does not resemble vascular physiology. However, when used as part of a mixture of other materials, TPU can provide the rigidity necessary for more flexible materials rather than being used alone. Regardless of there being no information regarding dynamic compliance, three out of five TPU-based TEGVs had endothelization. All TEGVs were manufactured by electrospinning [118,137,141,154].

Finally, PLGA-based TEGVs had intermediate mechanical properties. Its circumferential tensile strength (8350 KPa) and suture resistance strength (1950 ± 1670 KPa) was higher than the internal mammary artery properties (4100 and 1350 KPa), but lower than the saphenous vein graft (9760 and 3200 KPa), respectively. Its longitudinal tensile strength (5008 ± 3468 KPa) was higher than the saphenous vein graft (2450 KPa) and the internal mammary artery (4300 KPa), but its dynamic compliance (3.41%/100 mmHg) was lower than both vessels (4.4 and 5.22%/100 mmHg). In addition, endothelization was found in two out of four PLGA-based TEGVs and the principal manufacturing technique was electrospinning. These results conclude that PLGA is a rigid material that, as with TPU, can give rigidity to a TEVG, but more flexible materials need to be used to offer compliance to the vascular graft [118,119,137,141,154].

All these data are summarized in Table 6 and Figure 8, comparing the mechanical properties average most common TEVGs from in vitro studies (PCL, TPU, and PLGA) with the GSV. Complete data is summarized in Tables S27 and S32.

Table 6. Mechanical properties in blood vessels: PCL, TPU, and PLGA tested in vitro (PCL in vivo) and great saphenous vein (GSV).

Test Performed	PCL In Vitro	TPU In Vitro	PLGA In Vitro	GSV	Reference
Internal Diameter (mm)	4.59	3.83	4	3	[22,23,113,115,117,120,122–125,132,133,139–142,145,154,157]
Wall Thickness (μM)	580	540	330	518	[24,25,113,115,117,118,120,123,124,133,140,141,143,145,147,151]
Circumferential Tensile Strength (KPa)	13,360	2570	8350	2405	[12,19,26,113,115,120,122,125,141,143,145,151]
Longitudinal Tensile Strength (KPa)	15,170	11,320	5010	9760	[12,26,115,117,118,122,124,127,132,135,137,139–143,145,147,151,154]
Burst Pressure (KPa)	264.51	451.29	312.84	371.96	[12,26,118,120,122,123,142,143,147,151,154,157]
Suture Retention Strength (g)	1034	803	1050	327	[22,27,115,117,118,123,143,151]
Dynamic Compliance (%/100 mmHg)	5.42	NR	3.41	4.40	[12,28,117,120,124]

(A)

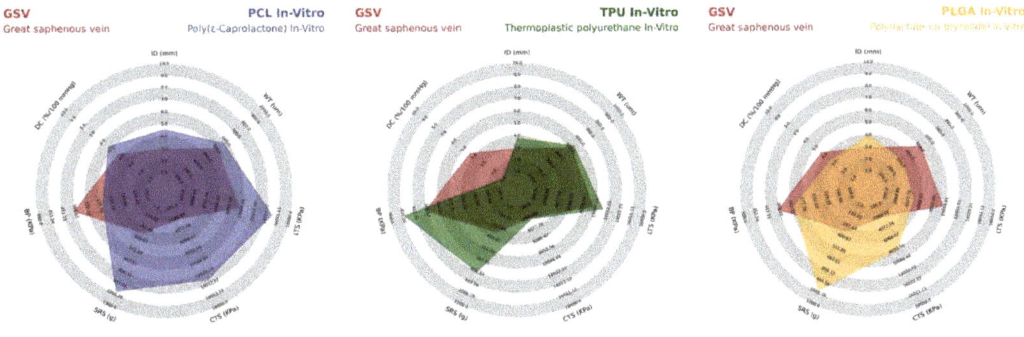

(a) **(b)** **(c)**

(B)

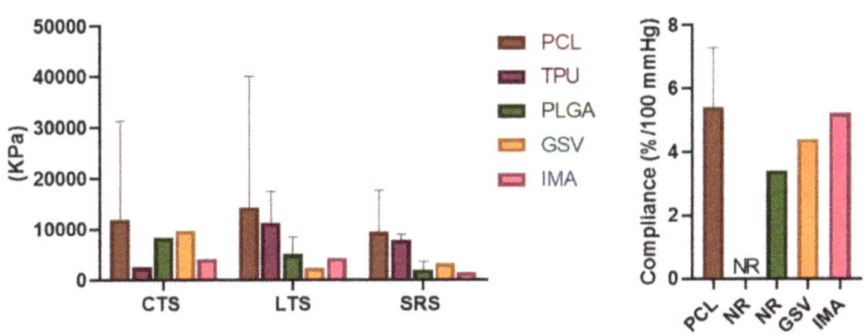

Figure 8. (**A**) Mechanical properties comparison of great saphenous vein, as a gold standard, (**a**) PCL, (**b**) TPU, and (**c**) PLGA tested in vitro. Mechanical properties identified include internal diameter (ID), wall thickness (WT), longitudinal tensile strength (LTS), circumferential tensile strength (CTS), suture retention strength (SRS), burst pressure (BP), and dynamic compliance (DC). Chart generated from the Python library developed by StatsBomb/Anmol Durgapal (accessed on 29 May 2022). (**B**) Mechanical properties comparison of great saphenous vein, as a gold standard, internal mammary artery and PCL, TPU, and PLGA subgroups of in vitro studies. Circumferential tensile strength (CTS), longitudinal tensile strength (LTS), suture retention strength (SRS), internal mammary artery (IMA), and great saphenous vein (GSV). Data are shown as mean ± standard deviation, according to the number of studies reporting the data. Additional information regarding descriptive analysis can be found on Table S3. Graph generated with GraphPad Prism (accessed on 29 May 2022).

7.4. Promising Biomaterials and Clinical Practice Correlation on TEVGs Applications

Despite the advances made in TEVGs development, the pursuit of the optimal mechanical properties that approximate those of native arteries is still ongoing. Multiple factors are essential to achieve suitable surgical and mechanical properties, including the proper selection of biomaterials, manufacturing techniques, and surface functionalizations. For instance, new biomaterials have been developed to increase patency rates and avoid complications such as pulmonary thromboembolism [158]. In these sections, we are only talking about newer approaches that have been proposed for use on TEVGs development.

The article published by Bai et al. showed structured scaffolds with biodegradable polyester-polydepsipeptid and silk fibroin (PCL-PIBMD/SF). The manufacturing process was on a sandwich-like composite delivering plasmid complexes aiming to induce endothelialization and was manufactured through layer-by-layer electrospinning and electrospraying techniques. The scaffold was tested with human umbilical vein endothelial cell cultures (HUVECs). The average diameter of nanofibers decreased from 573.8 nm to 285.1 nm, with SF content increasing from 0 to 90%. The mechanical properties found were suitable for vascular scaffolds when the weight ratio of SF was 10%, as the found tensile strength and elongation were 7050.34 kPa and 210 ± 21%. A porosity of 27.5 ± 7.4% with a weight ratio of 90/10 was reported. The HUVECs coverage ratio on day three was 66 ± 3%. This scaffold showed mechanical properties that suggest promising application for TEVGs and also demonstrated the promotion of cell proliferation, adhesion, spreading, and migration [159].

Adding to the research, the study published by Xie et al. designed a tissue-engineered vascular scaffold with portulaca flavonoid (PTF) via electrospinning and was integrated with PCL. It was tested on a culture of human vascular smooth muscle cells (HVSMCs). The scaffold exhibited biomimetic net-like fiber structures, an elastic modulus of 2–20 MPa, the ultimate tensile stress of 2000 kPa, and a fracture strain of 60% in the transverse direction. In addition, it showed that compared to the PCL scaffold, the integration of bioactive PTF had better hydrophilicity and degradability. In addition, inhibition of abnormal intimal hyperplasia was observed [160].

Moreover, one of the most critical factors in understanding the clinical use of each native artery/vein is the clinical practice guidelines that continue to support using native arteries or veins. However, we have observed that native arteries/veins are used for the clinical condition and the patient's requirements, whereas even for the same pathology, different vessels with very different mechanical properties are used. In pathologies involving small vessels, the evidence supports the individualized and specific use of different types of native vessels. The clearest example is the variability between myocardial revascularization per se and infrainguinal revascularization.

The ACC/AHA/SCAI Coronary Revascularization Guidelines published in 2021 recommend using the radial artery in isolated Coronary Artery Bypass Grafting (CABG) and the left mammary artery in multiple bypasses over the saphenous vein. The evidence shows that using the left mammary artery prolongs the survival of the patient who only requires isolated CABG in a significantly stenosed non–Left Anterior Descending (LAD) vessel. In addition, the case of the radial artery has shown that in the medium and long term, it has higher patency rates and better clinical results at ten years of undergoing CABG to bypass the LAD [161].

On the other hand, the Society for Vascular Surgery practice guidelines for the atherosclerotic occlusive disease of the lower extremities recommends infrainguinal bypass using the saphenous vein [162].

Depending on the patient's need and underlying pathology, this variability in mechanical property requirements also applies to vascular grafts. Thus, evaluating the different mechanical properties in each clinical indication is necessary since it is possible to hypothesize that not only a single small vascular graft with specific mechanical properties will be required.

7.5. Approaches/Techniques for Fabrication of Small Diameter TEVGs

Currently, there is a great diversity of manufacturing techniques for small diameter TEVGs. Several biomaterials have the plasticity required for application on different manufacturing techniques. Most studies reported the use of electrospinning as the primary method of fabrication (n = 25; 58.14%); this is because it offers the ability to fine-tune mechanical properties during the fabrication processes while also offering precise control over the composition, dimension, and alignment of the fiber of the material [66,69,70,72,78,80–84,88,93,95–97,99–101,103,104,107,110]. Furthermore, it can combine synthetic and natural materials, meeting specific needs like high mechanical durability in terms of high burst strength and compliance. Finally, this method allows the incorporation of natural polymers that promote the proliferation of different cell types in the matrix of the graft's wall.

Although electrospinning as the primary manufacturing technique is one of the most common for the development of grafts, the fiber diameters should be considered. The reasoning is that if the diameter is too small, the TEVG will have low porosity, meaning that cell infiltration will be limited. On the other hand, if the diameter of the fiber is too big, the biomechanical graft properties could be compromised due to an increase in the porosity, leading to blood leakages. It has been reported that thicker-fiber grafts (5–6 μm) tend to polarize into the immunomodulatory and tissue remodeling (M2) phenotype, while thinner-fiber grafts (2–3 μm) express a pro-inflammatory (M1) phenotype [111]. In this case, the average fiber diameter reported for the PCL TEVGs data was 2.4, indicating that this diameter should be improved to enhance regenerative outcomes.

Another relatively common technique typically used for the manufacture of TEVGs is Freeze tawing (n = 3; 6.98%) [67,76,86,163–165]; some of the advantages of this method are that the structure of the material is maintained, moisture is removed at low temperatures, the stability of the material is increased during storage, and the fast transition of the moisturized material to be dehydrated minimized several degradation reactions [112]. Regarding the use of this method in the manufacture of TEVGs, results have reported that the porosity of the graft is high enough to polarize into the immunomodulatory and tissue remodeling (M2) phenotype, meaning that there should be an improvement in regenerative outcome. Nevertheless, this technique has not been widely used now that some mechanical properties seem to be affected. For example, the circumferential tensile strength seems to be lowered compared to other methods such as electrospinning.

8. Biomechanical Properties of Surfaces and Vascular Infections on TEVGs

Another topic related to the physicochemical properties of TEVGs and its mechanical properties are the surfaces of the designed scaffolds. Depending on these properties, biomechanical stimuli–the same as the effect of WSS–might have an impact on cell adhesion and regeneration, or even might promote bacterial colonization. For instance, some micropatterns as well as stiffer surfaces promote cell adhesion and migration, and these data have been widely reported.

Nevertheless, one of the least studied topics on TEVGs is how to prevent vascular infections, which is one of the most detrimental complications. Due to the increase in antimicrobial resistance and infection of different types of implants, multiple strategies have been studied to prevent bacterial colonization and biofilm formation. One of the most efficient strategies is a colonization-resistant surface by coating the surface with the inclusion of superhydrophobic, uncharged, or highly hydrophilic molecules that do not allow bacterial adhesion. For example, polyethylene glycol and zwitterionic polymers have been shown to inhibit bacterial adhesion because they generate a stearic repulsion and movement phenomenon. Hydrophilic molecules such as heparin have also been used to inhibit bacterial adhesion. Superhydrophobic structures with specific topographies have shown efficiency in decreasing adhesion [166–168].

Although less studied, microparticles and nanoparticles with antimicrobial peptides are also used for functionalization, with a wide range of inhibitory effects against bacteria, fungi, parasites, and viruses. Other alternatives include coatings with silver, aluminum, cobalt, zinc, and copper, given that the ions destabilize the bacterial membrane. The same principle applies to Cationic polymers such as chitosan and polyethyleneimine (PEI). Other strategies include local mechanisms of controlled release of antibiotics such as gentamicin, amoxicillin, and vancomycin, among others. For this purpose, multiple transport media such as alginate nanoparticles, hydrogels loaded with these molecules, and even mixing biomaterials with these antibiotics have been employed [169–173].

A 2003 meta-analysis found that autologous vein repairs at any site have a better prognosis and better long-term outcomes regarding graft patency than PTFE grafts in peripheral vessels. In addition, evidence regarding the use of other materials such as bovine pericardium in abdominal aortic repairs, and rifampicin-embedded or silver-embedded prostheses has come to be considered. However, reinfection rates are often higher than

10%. So far, the only regenerative vascular graft for application in aortic aneurysm repair and large-diameter peripheral vessel repair is Omniflow II from the commercial house LeMaitré. Omniflow II is a biosynthetic vascular graft composed of a polyester mesh as an endoskeleton covered by cross-linked ovine collagen. Its simple fabrication is based on the immune response to foreign body and fibrous tissue formation on the polyester mesh implanted in the animal after a few months.

Collagen-based biosynthetic vascular grafts have demonstrated graft patency rates approaching 70–80% at one year, with favorable rates (57–64%) in below-knee reconstructions. Infection rates are meager, being lower than prosthetic grafts due to better healing patterns linked to their collagen structure, better integration with host tissue, and evidence of capillary growth or micro vascularization of the graft. In the diabetic population–who are at higher risk of infection–low incidences of reinfection have been found, ranging from 2.8 to 4.8%–much lower than in other types of materials. In this regard, the evidence associated with collagen biosynthetic grafts tends to be very promising. However, their use has only been described in case reports and very small cohorts, especially in repairs associated with the abdominal aorta, which is why more evidence is needed. For this reason, a regenerative vascular graft that reduces reoperation rates and ensures laminar flow through the anastomoses will guarantee long-term graft patency and limb salvage [173–175].

Various biodegradable biomaterials with bacteriostatic/bactericidal activity have been developed. For this purpose, electrospinning–an additive manufacturing technique for the controlled deposition of nanofibers–has been one of the most widely used techniques. Using antibacterial agents such as tetracycline, chlorhexidine, triclosan, and even eugenol loaded on fibers obtained by electrospinning biodegradable polymers as a wound dressing in tissue engineering has been reported. In 2018, Zhenguang, L et al. designed membranes by electrospinning polycaprolactone (PCL) and gelatin incorporating eugenol and adhesive peptides for endothelial cells as a method for fabricating a regenerative vascular graft with bactericidal properties. Its inhibitory antimicrobial activity against S.aureus was 78% and against E.coli was 74%, which was attributed to the ability to destabilize the cell membrane [135,176–178].

As for bacteriostatic/bactericidal vascular grafts, none are currently on the market with this indication, and the literature reports few approaches. However, one of the most common is the fabrication of vascular grafts with hydrogels or polymers, including chitosan. Due to their bacteriostatic properties, they have helped stop the microbial activity. In addition, some of these TEVGs include other molecules such as silver and even heparin, which, although their general purpose is not to contribute to antimicrobial activity, do have this function.

9. Conclusions

This literature review aimed to identify the different trends in the use of materials to develop readily available TEVGs without pre-seeded cells. We show that PCL is currently one of the most used polymers in both in vivo and in vitro studies. However, lack of compliance with PCL-based TEVG has been identified as one of the main limitations of its long-term implementation in clinical trials. Other polymers such as TPU, PU, and PLGA show relevant properties to offer the required circumferential and longitudinal tensile strength for TEVGs applications. However, their use should be carefully implemented, including other materials to offer the required flexibility required for improved compliance and to provide the biomechanical properties suitable for TEVGS applications.

On the other hand, electrospinning is currently the most used manufacturing method, offering standardization and industrialization of the TEVGs. Furthermore, it is easy to use for various polymers while providing proper microstructure and macrostructure. Nonetheless, more studies are still required in manufacturing to reach an ideal fiber thickness and porosity for the grafts since cell migration can be affected while maintaining the long-term graft mechanical properties.

Strategies such as decellularized blood vessels offer a suitable alternative for TEVGs since their mechanical properties are very similar to the native tissues and are successful in the regeneration process. However, standardization and industrialization are more complex and will require multiple steps to be commercialized.

Personalized medicine marks the future of health sciences, and the use of vascular grafts from tissue engineering allows us not only to create alternatives for each of the vascular pathologies but also to offer quality and durable elements that meet each patient's requirements concerning their vascular physiology and rheology.

Finally, the biomaterials and manufacture methods should ensure the compliance of the anastomosed artery or blood vessel, considering that the increase in the compliance mismatch is directly correlated to the long-term patency loss due to the low wall shear stresses generated by the flow pattern disturbances along the blood vessel and the vascular graft. Therefore, the main objective of a TEVG to provide the required biomechanical signals for regeneration is to reach proper compliance.

Supplementary Materials: The following supporting information can be downloaded at: https://www.mdpi.com/article/10.3390/polym14173440/s1, Figure S1: PRISMA algorithm for pre-clinical trials; Figure S2: PRISMA algorithm for in vitro trials; Tables S1–S3: Summary tables of descriptive analysis; Tables S4–S26: Tables from descriptive analysis of pre-clinical trials; Tables S27–S42: Tables from descriptive analysis of in vitro tested TEGVs.

Author Contributions: Conceptualization, M.A.R.-S., C.M.C., J.C.C., J.C.-M., J.G.B.-C. and J.C.B.; Formal analysis, M.A.R.-S., C.A.P.-S., A.M.A.-R., D.B., M.A.-V., A.V.-S. and G.P.P.; Funding acquisition, C.M.C., J.C.C., J.C.-M., J.G.B.-C. and J.C.B.; Investigation, M.A.R.-S., C.A.P.-S., A.M.A.-R., D.B., M.A.-V., A.V.-S. and G.P.P.; Methodology, M.A.R.-S. and C.A.P.-S.; Project administration, J.C.-M., J.G.B.-C. and J.C.B.; Resources, C.M.C., J.C.C., J.C.-M., J.G.B.-C. and J.C.B.; Supervision, C.M.C., J.C.C., J.C.-M., J.G.B.-C. and J.C.B.; Validation C.M.C., J.C.C., J.C.-M., J.G.B.-C. and J.C.B.; Visualization, M.A.R.-S., C.A.P.-S., A.M.A.-R., D.B., M.A.-V., A.V.-S. and G.P.P.; Writing—original draft, M.A.R.-S., C.A.P.-S., A.M.A.-R., D.B., M.A.-V., A.V.-S. and G.P.P.; Writing—review & editing, M.A.R.-S., C.A.P.-S., A.M.A.-R., D.B., M.A.-V., A.V.-S., G.P.P. and J.C.C. All authors have read and agreed to the published version of the manuscript.

Funding: This research was funded by MINCIENCIAS, grant number RC # 635-2021 Code 1544101252100 with the project "Multigraft-Bac: Injerto vascular bacteriostático/bactericida y regenerativo para la reconstrucción de extremidades y el tratamiento de infección de aneurismas micóticos".

Data Availability Statement: Not applicable.

Acknowledgments: The authors would like to thank Departments of Biomedical Engineering at Universidad de los Andes. The authors acknowledge Fundación Cardioinfantil Instituto de Cardiología for providing financial and technical support.

Conflicts of Interest: The authors declare no conflict of interest.

References

1. World Health Organization. Cardiovascular Diseases (CVDs). Available online: https://www.who.int/news-room/fact-sheets/detail/cardiovascular-diseases-(cvds) (accessed on 22 May 2022).
2. Ralapanawa, U.; Sivakanesan, R. Epidemiology and the Magnitude of Coronary Artery Disease and Acute Coronary Syndrome: A Narrative Review. *J. Epidemiol. Glob. Health* **2021**, *11*, 169–177. [CrossRef] [PubMed]
3. Shu, J.; Santulli, G. Update on peripheral artery disease: Epidemiology and evidence-based facts. *Atherosclerosis* **2018**, *275*, 379–381. [CrossRef] [PubMed]
4. Royse, A.; Pamment, W.; Pawanis, Z.; Clarke-Errey, S.; Eccleston, D.; Ajani, A.; Wilson, W.; Canty, D.; Royse, C. Patency of conduits in patients who received internal mammary artery, radial artery and saphenous vein grafts. *BMC Cardiovasc. Disord.* **2020**, *20*, 148. [CrossRef] [PubMed]
5. Skovrind, I.; Harvald, E.B.; Belling, H.J.; Jørgensen, C.D.; Lindholt, J.S.; Andersen, D.C. Concise Review: Patency of Small-Diameter Tissue-Engineered Vascular Grafts: A Meta-Analysis of Preclinical Trials. *STEM CELLS Transl. Med.* **2019**, *8*, 671–680. [CrossRef] [PubMed]
6. MedSuite. Peripheral Vascular Devices Market Analysis, Size, Trends. iData Research: Intellenge behind the Data. Available online: https://idataresearch.com/product/peripheral-vascular-devices-market-united-states/ (accessed on 24 May 2022).

7. CryoGraft®. Descending Thoracic Aorta. Cryolife: Life Restoring Technology. Available online: https://www.cryolife.com/products/cardiac-allografts/cryograft/ (accessed on 22 May 2022).
8. Artegraft®. Collagen Vascular Graft*. LeMaitre. Available online: https://www.lemaitre.com/products/artegraft-collagen-vascular-graft (accessed on 22 May 2022).
9. Aroz, S.G.; Spaggiari, M.; Jeon, H.; Oberholzer, J.; Benedetti, E.; Tzvetanov, I. The use of bovine pericardial patch for vascular reconstruction in infected fields for transplant recipients. *J. Vasc. Surg. Cases, Innov. Tech.* **2017**, *3*, 47–49. [CrossRef]
10. Omniflow®II Vascular Prosthesis*. LeMaitre. Available online: https://www.lemaitre.com/products/omniflow-ii-vascular-prosthesis (accessed on 22 May 2022).
11. Gupta, P.; Mandal, B.B. Tissue-Engineered Vascular Grafts: Emerging Trends and Technologies. *Adv. Funct. Mater.* **2021**, *31*, 2100027. [CrossRef]
12. Camasão, D.B.; Mantovani, D. The mechanical characterization of blood vessels and their substitutes in the continuous quest for physiological-relevant performances. A critical review. *Mater. Today Bio* **2021**, *10*, 100106. [CrossRef]
13. Zhang, F.; King, M.W. Immunomodulation Strategies for the Successful Regeneration of a Tissue-Engineered Vascular Graft. *Adv. Health Mater.* **2022**, *11*, 2200045. [CrossRef]
14. Rodriguez-Soto, M.A.; Riveros, A.; Vargas, N.S.; Garcia-Brand, A.J.; Camargo, C.M.; Cruz, J.C.; Sandoval, N.; Briceño, J.C. Failure Analysis of TEVG's II: Late Failure and Entering the Regeneration Pathway. *Cells* **2022**, *11*, 939. [CrossRef]
15. Rodriguez-Soto, M.A.; Vargas, N.S.; Riveros, A.; Camargo, C.M.; Cruz, J.C.; Sandoval, N.; Briceño, J.C. Failure Analysis of TEVG's I: Overcoming the Initial Stages of Blood Material Interaction and Stabilization of the Immune Response. *Cells* **2021**, *10*, 3140. [CrossRef]
16. Jeong, Y.; Yao, Y.; Yim, E.K.F. Current understanding of intimal hyperplasia and effect of compliance in synthetic small diameter vascular grafts. *Biomater. Sci.* **2020**, *8*, 4383–4395. [CrossRef] [PubMed]
17. Maleckis, K.; Kamenskiy, A.; Lichter, E.Z.; Oberley-Deegan, R.; Dzenis, Y.; MacTaggart, J. Mechanically tuned vascular graft demonstrates rapid endothelialization and integration into the porcine iliac artery wall. *Acta Biomater.* **2021**, *125*, 126–137. [CrossRef] [PubMed]
18. Cao, X.; Maharjan, S.; Ashfaq, R.; Shin, J.; Zhang, Y.S. Bioprinting of Small-Diameter Blood Vessels. *Engineering* **2020**, *7*, 832–844. [CrossRef]
19. Öztemur, J. The Role of Biopolymer Selection in the Design of Electrospun Small Caliber Vascular Grafts to Replace the Native Arterial Structure. Chapter 9. Book: Theory and Research in Engineering. p. 27. Available online: https://www.researchgate.net/publication/351007141_THE_ROLE_OF_BIOPOLYMER_SELECTION_IN_THE_DESIGN_OF_ELECTROSPUN_SMALL_CALIBER_VASCULAR_GRAFTS_TO_REPLACE_THE_NATIVE_ARTERIAL_STRUCTURE (accessed on 24 May 2022).
20. Carrabba, M.; Madeddu, P. Current Strategies for the Manufacture of Small Size Tissue Engineering Vascular Grafts. *Front. Bioeng. Biotechnol.* **2018**, *6*, 41. [CrossRef] [PubMed]
21. Altshuler, P.; Nahirniak, P.; Welle, N.; Saphenous Vein Grafts. National Library of Medicine (NIH). 2022. Available online: https://www.ncbi.nlm.nih.gov/books/NBK537035/ (accessed on 23 May 2022).
22. Konig, G.; McAllister, T.N.; Dusserre, N.; Garrido, S.A.; Iyican, C.; Marini, A.; Fiorillo, A.; Avila, H.; Wystrychowski, W.; Zagalski, K.; et al. Mechanical properties of completely autologous human tissue engineered blood vessels compared to human saphenous vein and mammary artery. *Biomaterials* **2009**, *30*, 1542–1550. [CrossRef]
23. Ishii, Y.; Gossage, J.A.; Dourado, R.; Sabharwal, T.; Burnand, K.G. Minimum Internal Diameter of the Greater Saphenous Vein Is an Important Determinant of Successful Femorodistal Bypass Grafting that Is Independent of the Quality of the Runoff. *Vascular* **2004**, *12*, 225–233. [CrossRef]
24. van Andel, C.J.; Pistecky, P.V.; Borst, C. Mechanical properties of porcine and human arteries: Implications for coronary anastomotic connectors. *Ann. Thorac. Surg.* **2003**, *76*, 58–64. [CrossRef]
25. Hamedani, B.A.; Navidbakhsh, M.; Tafti, H.A. Comparison between mechanical properties of human saphenous vein and umbilical vein. *Biomed. Eng. Online* **2012**, *11*, 59. [CrossRef]
26. Stekelenburg, M.; Rutten, M.C.; Snoeckx, L.H.; Baaijens, F.P. Dynamic Straining Combined with Fibrin Gel Cell Seeding Improves Strength of Tissue-Engineered Small-Diameter Vascular Grafts. *Tissue Eng. Part A* **2009**, *15*, 1081–1089. [CrossRef]
27. Ohst, J.J.D.; Johnson, D.O.J. Development of Novel, Bioresorbable, Small-Diameter Electrospun Vascular Grafts. *J. Tissue Sci. Eng.* **2015**, *6*, 1000151. [CrossRef]
28. Montini-Ballarin, F.; Calvo, D.; Caracciolo, P.C.; Rojo, F.; Frontini, P.M.; Abraham, G.A.; Guinea, G.V. Mechanical behavior of bilayered small-diameter nanofibrous structures as biomimetic vascular grafts. *J. Mech. Behav. Biomed. Mater.* **2016**, *60*, 220–233. [CrossRef] [PubMed]
29. Soccerplots 1.0.0. PyPI. Available online: https://pypi.org/project/soccerplots/ (accessed on 29 May 2022).
30. Vascular Grafts Market Share, Size, Trends Industry Analysis Report by Product (Hemodialysis Access, Peripheral, Endovascular Stent), by Application (Vascular Occlusion, Cardiac Aneurysms, Kidney Failure), by Raw Material (PTFE, Polyester, Polyurethane, Biological Materials), by Region, and Segment Forecasts. 2014—2025. Polaris Marjet Research. 2019. Available online: https://www.polarismarketresearch.com/industry-analysis/vascular-graft-market (accessed on 24 May 2022).
31. Lentz, D.; PTFE Vascular Graft and Method of Manufacture. Google Patents. Available online: https://patents.google.com/patent/US6719783B2/en (accessed on 27 May 2022).

32. Lentz, D.; Self-sealing PTFE Vascular Graft and Manufacturing Methods. Google Patents. Available online: https://patents.google.com/patent/US20040193242A1/en (accessed on 27 May 2022).
33. GORE-TEX®Stretch Vascular Graft for Vascular Access. Gore. Available online: https://www.goremedical.com/products/vgavaccess (accessed on 27 May 2022).
34. Uhl, C.; Grosch, C.; Hock, C.; Töpel, I.; Steinbauer, M. Comparison of Long-term Outcomes of Heparin Bonded Polytetrafluoroethylene and Autologous Vein Below Knee Femoropopliteal Bypasses in Patients with Critical Limb Ischaemia. *Eur. J. Vasc. Endovasc. Surg.* **2017**, *54*, 203–211. [CrossRef] [PubMed]
35. Samson, R.H.; Morales, R.; Showalter, D.P.; Lepore, M.R.; Nair, D.G. Heparin-bonded expanded polytetrafluoroethylene femoropopliteal bypass grafts outperform expanded polytetrafluoroethylene grafts without heparin in a long-term comparison. *J. Vasc. Surg.* **2016**, *64*, 638–647. [CrossRef] [PubMed]
36. Jørgensen, C.; Paaske, W. Physical and mechanical properties of ePTFE stretch vascular grafts determined by time-resolved scanning acoustic microscopy. *Eur. J. Vasc. Endovasc. Surg.* **1998**, *15*, 416–422. [CrossRef]
37. Kim, L.; Scott, R.A.P.; Ashton, H.A.; Thompson, S.G. For the Multicentre Aneurysm Screening Study Group A Sustained Mortality Benefit from Screening for Abdominal Aortic Aneurysm. *Ann. Intern. Med.* **2007**, *146*, 699–706. [CrossRef]
38. Famaey, N.; Verhoeven, J.; Jacobs, S.; Pettinari, M.; Meyns, B. In situ Evolution of the Mechanical Properties of Stretchable and Non-Stretchable ePTFE Vascular Grafts and Adjacent Native Vessels. *Int. J. Artif. Organs* **2014**, *37*, 900–910. [CrossRef]
39. Salacinski, H.J.; Goldner, S.; Giudiceandrea, A.; Hamilton, G.; Seifalian, A.M.; Edwards, A.; Carson, R.J. The Mechanical Behavior of Vascular Grafts: A Review. *J. Biomater. Appl.* **2001**, *15*, 241–278. [CrossRef]
40. Lovett, M.; Eng, G.; Kluge, A.J.; Cannizzaro, C.; Vunjak-Novakovic, G.; Kaplan, D.L. Tubular silk scaffolds for small diameter vascular grafts. *Organogenesis* **2010**, *6*, 217–224. [CrossRef]
41. Leal, B.B.J.; Wakabayashi, N.; Oyama, K.; Kamiya, H.; Braghirolli, D.I.; Pranke, P. Vascular Tissue Engineering: Polymers and Methodologies for Small Caliber Vascular Grafts. *Front. Cardiovasc. Med.* **2021**, *7*, 592361. [CrossRef]
42. Fayon, A.; Menu, P.; El Omar, R. Cellularized small-caliber tissue-engineered vascular grafts: Looking for the ultimate gold standard. *npj Regen. Med.* **2021**, *6*, 46. [CrossRef]
43. Mehta, R.I.; Mukherjee, A.K.; Patterson, T.D.; Fishbein, M.C. Pathology of explanted polytetrafluoroethylene vascular grafts. *Cardiovasc. Pathol.* **2011**, *20*, 213–221. [CrossRef]
44. Hamilton, G.; Megerman, J.; L'Italien, G.J.; Warnock, D.F.; Schmitz-Rixen, T.; Brewster, D.C.; Abbott, W.M. Prediction of aneurysm formation in vascular grafts of biologic origin. *J. Vasc. Surg.* **1988**, *7*, 400–408. [CrossRef]
45. Post, A.; Diaz-Rodriguez, P.; Balouch, B.; Paulsen, S.; Wu, S.; Miller, J.; Hahn, M.; Cosgriff-Hernandez, E. Elucidating the role of graft compliance mismatch on intimal hyperplasia using an ex vivo organ culture model. *Acta Biomater.* **2019**, *89*, 84–94. [CrossRef] [PubMed]
46. He, F.; Hua, L.; Gao, L.-J. A Computational Model for Biomechanical Effects of Arterial Compliance Mismatch. *Appl. Bionics Biomech.* **2015**, *2015*, 213236. [CrossRef] [PubMed]
47. Reneman, R.S.; Hoeks, A.P.G. Wall shear stress as measured in vivo: Consequences for the design of the arterial system. *Med Biol. Eng. Comput.* **2008**, *46*, 499–507. [CrossRef] [PubMed]
48. Khan, M.O.; Tran, J.S.; Zhu, H.; Boyd, J.; Packard, R.R.S.; Karlsberg, R.P.; Kahn, A.M.; Marsden, A.L. Low Wall Shear Stress Is Associated with Saphenous Vein Graft Stenosis in Patients with Coronary Artery Bypass Grafting. *J. Cardiovasc. Transl. Res.* **2021**, *14*, 770–781. [CrossRef] [PubMed]
49. Zhu, L.; Pan, Z.; Li, Z.; Chang, Y.; Zhu, Y.; Yan, F.; Tu, S.; Yang, W. Can the Wall Shear Stress Values of Left Internal Mammary Artery Grafts during the Perioperative Period Reflect the One-Year Patency? *Thorac. Cardiovasc. Surg.* **2020**, *68*, 723–729. [CrossRef]
50. Owens, C.D. Adaptive changes in autogenous vein grafts for arterial reconstruction: Clinical implications. *J. Vasc. Surg.* **2010**, *51*, 736–746. [CrossRef]
51. Lu, D.; Kassab, G.S. Role of shear stress and stretch in vascular mechanobiology. *J. R. Soc. Interface* **2011**, *8*, 1379–1385. [CrossRef]
52. Heiss, C.; Rodriguez-Mateos, A.; Kelm, M. Central Role of eNOS in the Maintenance of Endothelial Homeostasis. *Antioxidants Redox Signal.* **2015**, *22*, 1230–1242. [CrossRef]
53. Mack, J.J.; Iruela-Arispe, M.L. NOTCH regulation of the endothelial cell phenotype. *Curr. Opin. Hematol.* **2018**, *25*, 212–218. [CrossRef] [PubMed]
54. Hakami, N.Y.; Ranjan, A.K.; Hardikar, A.A.; Dusting, G.J.; Peshavariya, H.M. Role of NADPH oxidase-4 in human endothelial progenitor cells. *Front. Physiol.* **2017**, *8*, 150. [CrossRef] [PubMed]
55. Seo, K.W.; Lee, S.J.; Kim, Y.H.; Bae, J.U.; Park, S.Y.; Bae, S.S.; Kim, C.D. Mechanical Stretch Increases MMP-2 Production in Vascular Smooth Muscle Cells via Activation of PDGFR-β/Akt Signaling Pathway. *PLoS ONE* **2013**, *8*, e70437. [CrossRef] [PubMed]
56. Seneviratne, A.N.; Cole, J.; Goddard, M.E.; Park, I.; Mohri, Z.; Sansom, S.; Udalova, I.; Krams, R.; Monaco, C. Low shear stress induces M1 macrophage polarization in murine thin-cap atherosclerotic plaques. *J. Mol. Cell. Cardiol.* **2015**, *89*, 168–172. [CrossRef] [PubMed]
57. Seneviratne, A.N.; Cole, J.E.; Goddard, M.E.; Mohri, Z.; Krams, R.; Monaco, C. Macrophage polarisation in shear stress modulated atherosclerotic plaque vulnerability. *Atherosclerosis* **2012**, *225*, e2. [CrossRef]
58. *ANSI/AAMI/ISO 7198:1998/2001 (R2010)*; Cardiovascular Implants—Tubular Vascular Prostheses. American National Standards Institute: Arlington, MA, USA, 2010.

59. *ISO 7198:2016*; Cardiovascular Implants and Extracorporeal Systems—Vascular Prostheses—Tubular Vascular Grafts and Vascular Patches. ISO: Geneva, Switzerland, 2016.
60. Ahmed, M.; Ghanbari, H.; Cousins, B.G.; Hamilton, G.; Seifalian, A.M. Small calibre polyhedral oligomeric silsesquioxane nanocomposite cardiovascular grafts: Influence of porosity on the structure, haemocompatibility and mechanical properties. *Acta Biomater.* **2011**, *7*, 3857–3867. [CrossRef] [PubMed]
61. A Clinical Investigation to Assess the Safety and Performance of POSS-PCU Small Diameter Grafts as Conduits in Arteriovenous (AV) Access. *Natl. Libr. Med. US* 2021, NCT02301312. Available online: https://www.clinicaltrials.gov/ct2/show/NCT02301312 (accessed on 25 June 2022).
62. A Pilot Study for Evaluation of the Safety and Efficacy of Humacyte's Human Acellular Vascular Graft as an Above-Knee Femoro-Popliteal Bypass Graft in Patients With Peripheral Arterial Disease. *Natl. Libr. Med. US* 2013, NCT01872208. Available online: https://clinicaltrials.gov/ct2/show/NCT01872208 (accessed on 25 June 2022).
63. A Prospective, Randomized Study of the Artegraft vs Propaten Vascular Grafts for Renal Failure Patients Requiring Dialysis Access. A Prospective, Randomized Study of the Artegraft vs. Propaten Vascular Grafts for Renal Failure Patients Requiring Dialysis Access. *Natl. Libr. Med. US* 2014, NCT02099344. Available online: https://clinicaltrials.gov/ct2/show/NCT02099344 (accessed on 25 June 2022).
64. Preliminary Safety of The TRUE Vascular Graft for Hemodialysis Access (TRUE HD I Study). *Natl. Libr. Med. US* 2021, NCT04905511. Available online: https://clinicaltrials.gov/ct2/show/NCT04905511 (accessed on 25 June 2022).
65. Vascular No-React Graft Against Infection. *Natl. Libr. Med. US* 2017, NCT04031001. Available online: https://clinicaltrials.gov/ct2/show/NCT04031001 (accessed on 25 June 2022).
66. Lorentz, K.L.; Gupta, P.; Shehabeldin, M.S.; Cunnane, E.M.; Ramaswamy, A.K.; Verdelis, K.; DiLeo, M.V.; Little, S.R.; Weinbaum, J.S.; Sfeir, C.S.; et al. CCL2 loaded microparticles promote acute patency in silk-based vascular grafts implanted in rat aortae. *Acta Biomater.* **2021**, *135*, 126–138. [CrossRef]
67. Zhang, C.; Xie, Q.; Cha, R.; Ding, L.; Jia, L.; Mou, L.; Cheng, S.; Wang, N.; Li, Z.; Sun, Y.; et al. Anticoagulant Hydrogel Tubes with Poly(ε-Caprolactone) Sheaths for Small-Diameter Vascular Grafts. *Adv. Health Mater.* **2021**, *10*, 2100839. [CrossRef]
68. Antonova, L.; Krivkina, E.; Sevostianova, V.; Mironov, A.; Rezvova, M.; Shabaev, A.; Tkachenko, V.; Krutitskiy, S.; Khanova, M.; Sergeeva, T.; et al. Tissue-Engineered Carotid Artery Interposition Grafts Demonstrate High Primary Patency and Promote Vascular Tissue Regeneration in the Ovine Model. *Polymers* **2021**, *13*, 2637. [CrossRef]
69. Wang, Y.; Wu, T.; Zhang, J.; Feng, Z.; Yin, M.; Mo, X. A bilayer vascular scaffold with spatially controlled release of growth factors to enhance in situ rapid endothelialization and smooth muscle regeneration. *Mater. Des.* **2021**, *204*, 109649. [CrossRef]
70. Wu, Y.-L.; Szafron, J.M.; Blum, K.M.; Zbinden, J.C.; Khosravi, M.R.; Best, C.A.; Reinhardt, J.W.; Zheng, Q.; Yi, T.; Shinoka, T.; et al. Electrospun Tissue-Engineered Arterial Graft Thickness Affects Long-Term Composition and Mechanics. *Tissue Eng. Part A* **2020**, *27*, 593–603. [CrossRef] [PubMed]
71. Kimicata, M.; Allbritton-King, J.D.; Navarro, J.; Santoro, M.; Inoue, T.; Hibino, N.; Fisher, J.P. Assessment of decellularized pericardial extracellular matrix and poly(propylene fumarate) biohybrid for small-diameter vascular graft applications. *Acta Biomater.* **2020**, *110*, 68–81. [CrossRef] [PubMed]
72. Shi, J.; Chen, S.; Wang, L.; Zhang, X.; Gao, J.; Jiang, L.; Tang, D.; Zhang, L.; Midgley, A.; Kong, D.; et al. Rapid endothelialization and controlled smooth muscle regeneration by electrospun heparin-loaded polycaprolactone/gelatin hybrid vascular grafts. *J. Biomed. Mater. Res. Part B: Appl. Biomater.* **2019**, *107*, 2040–2049. [CrossRef] [PubMed]
73. Nakayama, Y.; Furukoshi, M.; Terazawa, T.; Iwai, R. Development of long in vivo tissue-engineered "Biotube" vascular grafts. *Biomaterials* **2018**, *185*, 232–239. [CrossRef] [PubMed]
74. Szafron, J.M.; Khosravi, R.; Reinhardt, J.; Best, C.A.; Bersi, M.R.; Yi, T.; Breuer, C.K.; Humphrey, J.D. Immuno-driven and Mechano-mediated Neotissue Formation in Tissue Engineered Vascular Grafts. *Ann. Biomed. Eng.* **2018**, *46*, 1938–1950. [CrossRef]
75. Zhu, M.; Wu, Y.; Li, W.; Dong, X.; Chang, H.; Wang, K.; Wu, P.; Zhang, J.; Fan, G.; Wang, L.; et al. Biodegradable and elastomeric vascular grafts enable vascular remodeling. *Biomaterials* **2018**, *183*, 306–318. [CrossRef]
76. Lee, K.-W.; Gade, P.S.; Dong, L.; Zhang, Z.; Aral, A.M.; Gao, J.; Ding, X.; Stowell, C.E.; Nisar, M.U.; Kim, K.; et al. A biodegradable synthetic graft for small arteries matches the performance of autologous vein in rat carotid arteries. *Biomaterials* **2018**, *181*, 67–80. [CrossRef]
77. Wang, W.; Nie, W.; Zhou, X.; Feng, W.; Chen, L.; Zhang, Q.; You, Z.; Shi, Q.; Peng, C.; He, C. Fabrication of heterogeneous porous bilayered nanofibrous vascular grafts by two-step phase separation technique. *Acta Biomater.* **2018**, *79*, 168–181. [CrossRef]
78. Wu, T.; Zhang, J.; Wang, Y.; Sun, B.; Yin, M.; Bowlin, G.L.; Mo, X. Design and Fabrication of a Biomimetic Vascular Scaffold Promoting in Situ Endothelialization and Tunica Media Regeneration. *ACS Appl. Bio Mater.* **2018**, *1*, 833–844. [CrossRef]
79. Huang, R.; Gao, X.; Wang, J.; Chen, H.; Tong, C.; Tan, Y.; Tan, Z. Triple-Layer Vascular Grafts Fabricated by Combined E-Jet 3D Printing and Electrospinning. *Ann. Biomed. Eng.* **2018**, *46*, 1254–1266. [CrossRef]
80. Duijvelshoff, R.; Van Engeland, N.C.A.; Gabriels, K.M.R.; Söntjens, S.H.M.; Smits, A.I.P.M.; Dankers, P.Y.W.; Bouten, C.V.C. Host Response and Neo-Tissue Development during Resorption of a Fast Degrading Supramolecular Electrospun Arterial Scaffold. *Bioengineering* **2018**, *5*, 61. [CrossRef] [PubMed]
81. Jirofti, N.; Mohebbi-Kalhori, D.; Samimi, A.; Hadjizadeh, A.; Kazemzadeh, G.H. Small-diameter vascular graft using co-electrospun composite PCL/PU nanofibers. *Biomed. Mater.* **2018**, *13*, 055014. [CrossRef] [PubMed]

82. Shafiq, M.; Zhang, Q.; Zhi, D.; Wang, K.; Kong, D.; Kim, D.-H.; Kim, S.H. In Situ Blood Vessel Regeneration Using SP (Substance P) and SDF (Stromal Cell–Derived Factor)-1α Peptide Eluting Vascular Grafts. *Arter. Thromb. Vasc. Biol.* **2018**, *38*, e117–e134. [CrossRef] [PubMed]
83. Gao, J.; Jiang, L.; Liang, Q.; Shi, J.; Hou, D.; Tang, D.; Chen, S.; Kong, D.; Wang, S. The grafts modified by heparinization and catalytic nitric oxide generation used for vascular implantation in rats. *Regen. Biomater.* **2018**, *5*, 105–114. [CrossRef]
84. Wu, T.; Zhang, J.; Wang, Y.; Li, D.; Sun, B.; El-Hamshary, H.; Yin, M.; Mo, X. Fabrication and preliminary study of a biomimetic tri-layer tubular graft based on fibers and fiber yarns for vascular tissue engineering. *Mater. Sci. Eng. C* **2017**, *82*, 121–129. [CrossRef]
85. Jiang, B.; Suen, R.; Wang, J.-J.; Zhang, Z.J.; Wertheim, J.A.; Ameer, G.A. Vascular scaffolds with enhanced antioxidant activity inhibit graft calcification. *Biomaterials* **2017**, *144*, 166–175. [CrossRef]
86. Aussel, A.; Thébaud, N.B.; Bérard, X.; Brizzi, V.; Delmond, S.; Bareille, R.; Siadous, R.; James, C.; Ripoche, J.; Durand, M.; et al. Chitosan-based hydrogels for developing a small-diameter vascular graft: In vitro and in vivo evaluation. *Biomed. Mater.* **2017**, *12*, 065003. [CrossRef]
87. Kristofik, N.J.; Qin, L.; Calabro, N.E.; Dimitrievska, S.; Li, G.; Tellides, G.; Niklason, L.E.; Kyriakides, T.R. Improving in vivo outcomes of decellularized vascular grafts via incorporation of a novel extracellular matrix. *Biomaterials* **2017**, *141*, 63–73. [CrossRef]
88. Wang, Z.; Wu, Y.; Wang, J.; Zhang, C.; Yan, H.; Zhu, M.; Wang, K.; Li, C.; Xu, Q.; Kong, D. Effect of Resveratrol on Modulation of Endothelial Cells and Macrophages for Rapid Vascular Regeneration from Electrospun Poly(ε-caprolactone) Scaffolds. *ACS Appl. Mater. Interfaces* **2017**, *9*, 19541–19551. [CrossRef]
89. Emechebe, A.G.; Obiweluozor, O.F.; Jeong, I.S.; Park, J.-K.; Park, C.H.; Kim, C.S. Merging 3D printing with electrospun biodegradable small-caliber vascular grafts immobilized with VEGF. *Nanomed. Nanotechnol. Biol. Med.* **2020**, *30*, 102306. [CrossRef]
90. Wang, H.; Xing, M.; Deng, W.; Qian, M.; Wang, F.; Wang, K.; Midgley, A.C.; Zhao, Q. Anti-Sca-1 antibody-functionalized vascular grafts improve vascular regeneration via selective capture of endogenous vascular stem/progenitor cells. *Bioact. Mater.* **2022**, *16*, 433–450. [CrossRef] [PubMed]
91. Inoue, T.; Kanda, K.; Yamanami, M.; Kami, D.; Gojo, S.; Yaku, H. Modifications of the mechanical properties of in vivo tissue-engineered vascular grafts by chemical treatments for a short duration. *PLoS ONE* **2021**, *16*, e0248346. [CrossRef] [PubMed]
92. Antonova, L.; Kutikhin, A.; Sevostianova, V.; Velikanova, E.; Matveeva, V.; Glushkova, T.; Mironov, A.; Krivkina, E.; Shabaev, A.; Senokosova, E.; et al. bFGF and SDF-1α Improve In Vivo Performance of VEGF-Incorporating Small-Diameter Vascular Grafts. *Pharmaceuticals* **2021**, *14*, 302. [CrossRef] [PubMed]
93. Guo, H.-F.; Dai, W.-W.; Qian, D.-H.; Qin, Z.-X.; Lei, Y.; Hou, X.-Y.; Wen, C. A simply prepared small-diameter artificial blood vessel that promotes in situ endothelialization. *Acta Biomater.* **2017**, *54*, 107–116. [CrossRef]
94. Negishi, J.; Hashimoto, Y.; Yamashita, A.; Zhang, Y.; Kimura, T.; Kishida, A.; Funamoto, S. Evaluation of small-diameter vascular grafts reconstructed from decellularized aorta sheets: Vascular grafts from decellularized aorta sheets. *J. Biomed. Mater. Res. Part A* **2017**, *105*, 1293–1298. [CrossRef]
95. Stowell, C.E.; Li, X.; Matsunaga, M.H.; Cockreham, C.B.; Kelly, K.M.; Cheetham, J.; Tzeng, E.; Wang, Y. Resorbable vascular grafts show rapid cellularization and degradation in the ovine carotid. *J. Tissue Eng. Regen. Med.* **2020**, *14*, 1673–1684. [CrossRef]
96. Matsuzaki, Y.; Iwaki, R.; Reinhardt, J.W.; Chang, Y.-C.; Miyamoto, S.; Kelly, J.; Zbinden, J.; Blum, K.; Mirhaidari, G.; Ulziibayar, A.; et al. The effect of pore diameter on neo-tissue formation in electrospun biodegradable tissue-engineered arterial grafts in a large animal model. *Acta Biomater.* **2020**, *115*, 176–184. [CrossRef]
97. Wen, M.; Zhi, D.; Wang, L.; Cui, C.; Huang, Z.; Zhao, Y.; Wang, K.; Kong, D.; Yuan, X. Local Delivery of Dual MicroRNAs in Trilayered Electrospun Grafts for Vascular Regeneration. *ACS Appl. Mater. Interfaces* **2020**, *12*, 6863–6875. [CrossRef]
98. Fukunishi, T.; Ong, C.S.; Yesantharao, P.; Best, C.A.; Yi, T.; Zhang, H.; Mattson, G.; Boktor, J.; Nelson, K.; Shinoka, T.; et al. Different degradation rates of nanofiber vascular grafts in small and large animal models. *J. Tissue Eng. Regen. Med.* **2019**, *14*, 203–214. [CrossRef]
99. Shojaee, M.; Sameti, M.; Vuppuluri, K.; Ziff, M.; Carriero, A.; Bashur, C.A. Design and characterization of a porous pouch to prevent peritoneal adhesions during in vivo vascular graft maturation. *J. Mech. Behav. Biomed. Mater.* **2020**, *102*, 103461. [CrossRef]
100. Yang, L.; Li, X.; Wu, Y.; Du, P.; Sun, L.; Yu, Z.; Song, S.; Yin, J.; Ma, X.; Jing, C.; et al. Preparation of PU/Fibrin Vascular Scaffold with Good Biomechanical Properties and Evaluation of Its Performance in vitro and in vivo. *Int. J. Nanomed.* **2020**, *15*, 8697–8715. [CrossRef] [PubMed]
101. Gostev, A.A.; Chernonosova, V.S.; Murashov, I.S.; Sergeevichev, D.S.; Korobeinikov, A.A.; Karaskov, A.M.; Karpenko, A.A.; Laktionov, P.P. Electrospun polyurethane-based vascular grafts: Physicochemical properties and functioning in vivo. *Biomed. Mater.* **2020**, *15*, 015010. [CrossRef] [PubMed]
102. Li, W.; Wu, P.; Zhang, Y.; Midgley, A.C.; Yuan, X.; Wu, Y.; Wang, L.; Wang, Z.; Zhu, M.F.; Kong, D. Bilayered Polymeric Micro- and Nanofiber Vascular Grafts as Abdominal Aorta Replacements: Long-Term in Vivo Studies in a Rat Model. *ACS Appl. Bio Mater.* **2019**, *2*, 4493–4502. [CrossRef] [PubMed]
103. Yang, Y.; Lei, D.; Zou, H.; Huang, S.; Yang, Q.; Li, S.; Qing, F.-L.; Ye, X.; You, Z.; Zhao, Q. Hybrid electrospun rapamycin-loaded small-diameter decellularized vascular grafts effectively inhibit intimal hyperplasia. *Acta Biomater.* **2019**, *97*, 321–332. [CrossRef]

104. Johnson, R.; Ding, Y.; Nagiah, N.; Monnet, E.; Tan, W. Coaxially-structured fibres with tailored material properties for vascular graft implant. *Mater. Sci. Eng. C* **2019**, *97*, 1–11. [CrossRef] [PubMed]
105. Yang, S.; Zheng, X.; Qian, M.; Wang, H.; Wang, F.; Wei, Y.; Midgley, A.C.; He, J.; Tian, H.; Zhao, Q. Nitrate-Functionalized poly(ε-Caprolactone) Small-Diameter Vascular Grafts Enhance Vascular Regeneration via Sustained Release of Nitric Oxide. *Front. Bioeng. Biotechnol.* **2021**, *9*, 770121. [CrossRef]
106. Wang, C.; Li, Z.; Zhang, L.; Sun, W.; Zhou, J. Long-term results of triple-layered small diameter vascular grafts in sheep carotid arteries. *Med. Eng. Phys.* **2020**, *85*, 1–6. [CrossRef]
107. Hui, X.; Geng, X.; Jia, L.; Xu, Z.; Ye, L.; Gu, Y.; Zhang, A.-Y.; Feng, Z.-G. Preparation and in vivo evaluation of surface heparinized small diameter tissue engineered vascular scaffolds of poly(ε-caprolactone) embedded with collagen suture. *J. Biomater. Appl.* **2020**, *34*, 812–826. [CrossRef]
108. Yang, X.; Wang, Y.; Zhou, Y.; Chen, J.; Wan, Q. The Application of Polycaprolactone in Three-Dimensional Printing Scaffolds for Bone Tissue Engineering. *Polymers* **2021**, *13*, 2754. [CrossRef]
109. Liao, J.; Xu, B.; Zhang, R.; Fan, Y.; Xie, H.; Li, X. Applications of decellularized materials in tissue engineering: Advantages, drawbacks and current improvements, and future perspectives. *J. Mater. Chem. B* **2020**, *8*, 10023–10049. [CrossRef]
110. Wan, Y.; Yang, S.; Peng, M.; Gama, M.; Yang, Z.; Deng, X.; Zhou, J.; Ouyang, C.; Luo, H. Controllable synthesis of biomimetic nano/submicro-fibrous tubes for potential small-diameter vascular grafts. *J. Mater. Chem. B* **2020**, *8*, 5694–5706. [CrossRef] [PubMed]
111. Wang, Z.; Cui, Y.; Wang, J.; Yang, X.; Wu, Y.; Wang, K.; Gao, X.; Li, D.; Li, Y.; Zheng, X.-L.; et al. The effect of thick fibers and large pores of electrospun poly(ε-caprolactone) vascular grafts on macrophage polarization and arterial regeneration. *Biomaterials* **2014**, *35*, 5700–5710. [CrossRef] [PubMed]
112. Boss, E.A.; Filho, R.M.; de Toledo, E.C.V. Freeze drying process: Real time model and optimization. *Chem. Eng. Process. Process Intensif.* **2004**, *43*, 1475–1485. [CrossRef]
113. Li, C.; Wang, F.; Chen, P.; Zhang, Z.; Guidoin, R.; Wang, L. Preventing collapsing of vascular scaffolds: The mechanical behavior of PLA/PCL composite structure prostheses during in vitro degradation. *J. Mech. Behav. Biomed. Mater.* **2017**, *75*, 455–462. [CrossRef]
114. Lee, S.J.; Yoo, J.J.; Lim, G.J.; Atala, A.; Stitzel, J. In vitro evaluation of electrospun nanofiber scaffolds for vascular graft application. *J. Biomed. Mater. Res. Part A* **2007**, *83A*, 999–1008. [CrossRef]
115. Li, C.; Wang, F.; Douglas, G.; Zhang, Z.; Guidoin, R.; Wang, L. Comprehensive mechanical characterization of PLA fabric combined with PCL to form a composite structure vascular graft. *J. Mech. Behav. Biomed. Mater.* **2017**, *69*, 39–49. [CrossRef]
116. Zhang, Y.; Li, X.S.; Guex, A.G.; Liu, S.S.; Müller, E.; Malini, R.I.; Zhao, H.J.; Rottmar, M.; Maniura, K.; Rossi, R.; et al. A compliant and biomimetic three-layered vascular graft for small blood vessels. *Biofabrication* **2017**, *9*, 025010. [CrossRef]
117. Wang, N.; Zheng, W.; Cheng, S.; Zhang, W.; Liu, S.; Jiang, X. In Vitro Evaluation of Essential Mechanical Properties and Cell Behaviors of a Novel Polylactic-co-Glycolic Acid (PLGA)-Based Tubular Scaffold for Small-Diameter Vascular Tissue Engineering. *Polymers* **2017**, *9*, 318. [CrossRef]
118. Maleki, S.; Shamloo, A.; Kalantarnia, F. Tubular TPU/SF nanofibers covered with chitosan-based hydrogels as small-diameter vascular grafts with enhanced mechanical properties. *Sci. Rep.* **2022**, *12*, 6179. [CrossRef]
119. Du, J.; Hu, X.; Su, Y.; Wei, T.; Jiao, Z.; Liu, T.; Wang, H.; Nie, Y.; Li, X.; Song, K. Gelatin/sodium alginate hydrogel-coated decellularized porcine coronary artery to construct bilayer tissue engineered blood vessels. *Int. J. Biol. Macromol.* **2022**, *209*, 2070–2083. [CrossRef]
120. Pimenta, F.A.; Carbonari, R.C.; Malmonge, S.M. Nanofibrous tubular scaffolds for tissue engineering of small-diameter vascular grafts—Development using SBS fabrication technique and mechanical performance. *Res. Biomed. Eng.* **2022**, 1–15. [CrossRef]
121. Zakeri, Z.; Salehi, R.; Mahkam, M.; Rahbarghazi, R.; Abbasi, F.; Rezaei, M. Electrospun POSS integrated poly(carbonate-urea)urethane provides appropriate surface and mechanical properties for the fabrication of small-diameter vascular grafts. *J. Biomater. Sci. Polym. Ed.* **2022**, 1–20. [CrossRef] [PubMed]
122. Zhang, X.; Bai, S.; Zang, L.; Chen, X.; Yuan, X. Preparation of Poly(ε-caprolactone)/Poly(ester amide) Electrospun Membranes for Vascular Repair. *Chem. Res. Chin. Univ.* **2022**, 1–7. [CrossRef]
123. Radakovic, D.; Reboredo, J.; Helm, M.; Weigel, T.; Schürlein, S.; Kupczyk, E.; Leyh, R.G.; Walles, H.; Hansmann, J. A multilayered electrospun graft as vascular access for hemodialysis. *PLoS ONE* **2017**, *12*, e0185916. [CrossRef] [PubMed]
124. Cao, Y.; Jiang, J.; Jiang, Y.; Li, Z.; Hou, J.; Li, Q. Biodegradable highly porous interconnected poly(ε-caprolactone)/poly(L-lactide-co-ε-caprolactone) scaffolds by supercritical foaming for small-diameter vascular tissue engineering. *Polym. Adv. Technol.* **2021**, *33*, 440–451. [CrossRef]
125. Montoya, Y.; Cardenas, J.; Bustamante, J.; Valencia, R. Effect of sequential electrospinning and co-electrospinning on morphological and fluid mechanical wall properties of polycaprolactone and bovine gelatin scaffolds, for potential use in small diameter vascular grafts. *Biomater. Res.* **2021**, *25*, 38. [CrossRef] [PubMed]
126. Helms, F.; Haverich, A.; Böer, U.; Wilhelmi, M. Transluminal compression increases mechanical stability, stiffness and endothelialization capacity of fibrin-based bioartificial blood vessels. *J. Mech. Behav. Biomed. Mater.* **2021**, *124*, 104835. [CrossRef] [PubMed]
127. Sun, B.; Hou, L.; Sun, B.; Han, Y.; Zou, Y.; Huang, J.; Zhang, Y.; Feng, C.; Dou, X.; Xu, F. Use of Electrospun Phenylalanine/Poly-ε-Caprolactone Chiral Hybrid Scaffolds to Promote Endothelial Remodeling. *Front. Bioeng. Biotechnol.* **2021**, *9*, 773635. [CrossRef]

128. Yang, T.; Xu, H.; Zhao, C.; Tang, D.; Mu, F.; Lu, H.; Rao, Z.; Wang, S. Construction and evaluation of co-electrospun poly (butylene succinate)/gelatin materials as potential vascular grafts. *Chin. J. Chem. Eng.* **2021**, *39*, 297–305. [CrossRef]
129. Frazão, L.P.; Fernandes, A.M.; Oliveira, C.; Martins, A.; Silva, T.H.; de Castro, J.V.; Nogueira-Silva, C.; Neves, N.M. New Vascular Graft Using the Decellularized Human Chorion Membrane. *ACS Biomater. Sci. Eng.* **2021**, *7*, 3423–3433. [CrossRef]
130. Asadpour, S.; Yeganeh, H.; Ai, J.; Ghanbari, H. A novel polyurethane modified with biomacromolecules for small-diameter vascular graft applications. *J. Mater. Sci.* **2018**, *53*, 9913–9927. [CrossRef]
131. Silva, J.M.; Rodrigues, L.C.; Silva, S.S.; Reis, R.L.; Duarte, A.R.C. Engineered tubular structures based on chitosan for tissue engineering applications. *J. Biomater. Appl.* **2018**, *32*, 841–852. [CrossRef] [PubMed]
132. Kharazi, A.Z.; Atari, M.; Vatankhah, E.; Javanmard, S.H. A nanofibrous bilayered scaffold for tissue engineering of small-diameter blood vessels. *Polym. Adv. Technol.* **2018**, *29*, 3151–3158. [CrossRef]
133. Horakova, J.; Mikes, P.; Saman, A.; Jencova, V.; Klapstova, A.; Svarcova, T.; Ackermann, M.; Novotny, V.; Suchy, T.; Lukas, D. The effect of ethylene oxide sterilization on electrospun vascular grafts made from biodegradable polyesters. *Mater. Sci. Eng. C* **2018**, *92*, 132–142. [CrossRef]
134. Braghirolli, D.; Caberlon, B.; Gamba, D.; Petry, J.; Dias, M.; Pranke, P. Poly(trimethylene carbonate-co-L-lactide) electrospun scaffolds for use as vascular grafts. *Braz. J. Med. Biol. Res.* **2019**, *52*, e8318. [CrossRef]
135. Wang, Y.; He, C.; Feng, Y.; Yang, Y.; Wei, Z.; Zhao, W.; Zhao, C. A chitosan modified asymmetric small-diameter vascular graft with anti-thrombotic and anti-bacterial functions for vascular tissue engineering. *J. Mater. Chem. B* **2020**, *8*, 568–577. [CrossRef]
136. Generali, M.; Casanova, E.A.; Kehl, D.; Wanner, D.; Hoerstrup, S.P.; Cinelli, P.; Weber, B. Autologous endothelialized small-caliber vascular grafts engineered from blood-derived induced pluripotent stem cells. *Acta Biomater.* **2019**, *97*, 333–343. [CrossRef]
137. Shokraei, N.; Asadpour, S.; Shokraei, S.; Sabet, M.N.; Faridi-Majidi, R.; Ghanbari, H. Development of electrically conductive hybrid nanofibers based on CNT-polyurethane nanocomposite for cardiac tissue engineering. *Microsc. Res. Tech.* **2019**, *82*, 1316–1325. [CrossRef]
138. Nguyen, T.-U.; Shojaee, M.; Bashur, C.; Kishore, V. Electrochemical fabrication of a biomimetic elastin-containing bi-layered scaffold for vascular tissue engineering. *Biofabrication* **2018**, *11*, 015007. [CrossRef]
139. Rychter, M.; Milanowski, B.; Grześkowiak, B.F.; Jarek, M.; Kempiński, M.; Coy, E.L.; Borysiak, S.; Baranowska-Korczyc, A.; Lulek, J. Cilostazol-loaded electrospun three-dimensional systems for potential cardiovascular application: Effect of fibers hydrophilization on drug release, and cytocompatibility. *J. Colloid Interface Sci.* **2019**, *536*, 310–327. [CrossRef]
140. Zhang, X.; Shi, J.; Chen, S.; Dong, Y.; Zhang, L.; Midgley, A.C.; Kong, D.; Wang, S. Polycaprolactone/gelatin degradable vascular grafts simulating endothelium functions modified by nitric oxide generation. *Regen. Med.* **2019**, *14*, 1089–1105. [CrossRef] [PubMed]
141. Yan, S.; Napiwocki, B.; Xu, Y.; Zhang, J.; Zhang, X.; Wang, X.; Crone, W.C.; Li, Q.; Turng, L.-S. Wavy small-diameter vascular graft made of eggshell membrane and thermoplastic polyurethane. *Mater. Sci. Eng. C* **2019**, *107*, 110311. [CrossRef] [PubMed]
142. Li, X.; Huang, L.; Li, L.; Tang, Y.; Liu, Q.; Xie, H.; Tian, J.; Zhou, S.; Tang, G. Biomimetic dual-oriented/bilayered electrospun scaffold for vascular tissue engineering. *J. Biomater. Sci. Polym. Ed.* **2019**, *31*, 439–455. [CrossRef] [PubMed]
143. Hu, Q.; Wu, C.; Zhang, H. Preparation and Optimization of a Biomimetic Triple-Layered Vascular Scaffold Based on Coaxial Electrospinning. *Appl. Biochem. Biotechnol.* **2020**, *190*, 1106–1123. [CrossRef] [PubMed]
144. Wu, P.; Wang, L.; Li, W.; Zhang, Y.; Wu, Y.; Zhi, D.; Wang, H.; Wang, L.; Kong, D.; Zhu, M. Construction of vascular graft with circumferentially oriented microchannels for improving artery regeneration. *Biomaterials* **2020**, *242*, 119922. [CrossRef] [PubMed]
145. Oliveira, S.; Felizardo, T.; Amorim, S.; Mithieux, S.M.; Pires, R.; Reis, R.L.; Martins, A.; Weiss, A.; Neves, N.M. Tubular fibrous scaffold functionalized with tropoelastin as a small-diameter vascular graft. *Biomacromolecules* **2020**, *21*, 3582–3595. [CrossRef]
146. Hodge, J.; Quint, C. Tissue engineered vessel from a biodegradable electrospun scaffold stimulated with mechanical stretch. *Biomed. Mater.* **2020**, *15*, 055006. [CrossRef]
147. Woods, I.; Black, A.; Molloy, E.J.; Jockenhoevel, S.; Flanagan, T.C. Fabrication of blood-derived elastogenic vascular grafts using electrospun fibrinogen and polycaprolactone composite scaffolds for paediatric applications. *J. Tissue Eng. Regen. Med.* **2020**, *14*, 1281–1295. [CrossRef]
148. Ryan, A.J.; Ryan, E.J.; Cameron, A.R.; O'Brien, F.J. Hierarchical biofabrication of biomimetic collagen-elastin vascular grafts with controllable properties via lyophilisation. *Acta Biomater.* **2020**, *112*, 52–61. [CrossRef]
149. Le, A.N.-M.; Tran, N.M.-P.; Phan, T.B.; Tran, P.A.; Tran, L.D.; Nguyen, T.H. Poloxamer additive as luminal surface modification to modulate wettability and bioactivities of small-diameter polyurethane/polycaprolactone electrospun hollow tube for vascular prosthesis applications. *Mater. Today Commun.* **2021**, *26*, 101771. [CrossRef]
150. Zhang, Y.; Liu, Y.; Jiang, Z.; Wang, J.; Xu, Z.; Meng, K.; Zhao, H. Poly(glyceryl sebacate)/silk fibroin small-diameter artificial blood vessels with good elasticity and compliance. *Smart Mater. Med.* **2021**, *2*, 74–86. [CrossRef]
151. Hu, Q.; Shen, Z.; Zhang, H.; Liu, S.; Feng, R.; Feng, J.; Ramalingam, M. Designed and fabrication of triple-layered vascular scaffold with microchannels. *J. Biomater. Sci. Polym. Ed.* **2021**, *32*, 714–734. [CrossRef] [PubMed]
152. Zhang, F.; Bambharoliya, T.; Xie, Y.; Liu, L.; Celik, H.; Wang, L.; Akkus, O.; King, M.W. A hybrid vascular graft harnessing the superior mechanical properties of synthetic fibers and the biological performance of collagen filaments. *Mater. Sci. Eng. C* **2021**, *118*, 111418. [CrossRef] [PubMed]
153. Kuźmińska, A.; Kwarta, D.; Ciach, T.; Butruk-Raszeja, B. Cylindrical Polyurethane Scaffold Fabricated Using the Phase Inversion Method: Influence of Process Parameters on Scaffolds' Morphology and Mechanical Properties. *Materials* **2021**, *14*, 2977. [CrossRef]

154. Chen, X.; Yao, Y.; Liu, S.; Hu, Q. An integrated strategy for designing and fabricating triple-layer vascular graft with oriented microgrooves to promote endothelialization. *J. Biomater. Appl.* **2021**, *36*, 297–310. [CrossRef]
155. Malik, S.; Sundarrajan, S.; Hussain, T.; Nazir, A.; Ramakrishna, S. Fabrication of Highly Oriented Cylindrical Polyacrylonitrile, Poly(lactide-*co*-glycolide), Polycaprolactone and Poly(vinyl acetate) Nanofibers for Vascular Graft Applications. *Polymers* **2021**, *13*, 2075. [CrossRef]
156. Zumbardo-Bacelis, G.A.; Meza-Villegas, L.A.; Pérez-Aranda, C.A.; Vargas-Coronado, R.; Castillo-Cruz, O.; Montaño-Machado, V.; Mantovani, D.; Cauich-Rodríguez, J.V. On arginine-based polyurethane-blends specific to vascular prostheses. *J. Appl. Polym. Sci.* **2021**, *138*, 51247. [CrossRef]
157. Wang, D.; Xu, Y.; Li, Q.; Turng, L.-S. Artificial small-diameter blood vessels: Materials, fabrication, surface modification, mechanical properties, and bioactive functionalities. *J. Mater. Chem. B* **2020**, *8*, 1801–1822. [CrossRef]
158. Stoiber, M.; Grasl, C.; Frieberger, K.; Moscato, F.; Bergmeister, H.; Schima, H. Impact of the testing protocol on the mechanical characterization of small diameter electrospun vascular grafts. *J. Mech. Behav. Biomed. Mater.* **2020**, *104*, 103652. [CrossRef]
159. Bai, L.; Li, Q.; Duo, X.; Hao, X.; Zhang, W.; Shi, C.; Guo, J.; Ren, X.; Feng, Y. Electrospun PCL-PIBMD/SF blend scaffolds with plasmid complexes for endothelial cell proliferation. *RSC Adv.* **2017**, *7*, 39452–39464. [CrossRef]
160. Xie, C.; Guo, T.; Wang, W.; Li, G.; Cai, Z.; Chen, S.; Wang, X.; Liu, Z.; Wang, Z. Scaffold engineering with flavone-modified biomimetic architecture for vascular tissue engineering applications. *Tissue Eng. Regen. Med.* **2022**. [CrossRef] [PubMed]
161. Lawton, J.S.; Tamis-Holland, J.E.; Bangalore, S.; Bates, E.R.; Beckie, T.M.; Bischoff, J.M.; Bittl, J.A.; Cohen, M.G.; DiMaio, J.M.; Don, C.W.; et al. 2021 ACC/AHA/SCAI Guideline for Coronary Artery Revascularization. *J. Am. Coll. Cardiol.* **2022**, *79*, e21–e129. [CrossRef] [PubMed]
162. Conte, M.S.; Pomposelli, F.B.; Clair, D.G.; Geraghty, P.J.; McKinsey, J.F.; Mills, J.L.; Moneta, G.L.; Murad, M.H.; Powell, R.J.; Reed, A.B.; et al. Society for Vascular Surgery practice guidelines for atherosclerotic occlusive disease of the lower extremities: Management of asymptomatic disease and claudication. *J. Vasc. Surg.* **2015**, *61*, 2S–41S. [CrossRef] [PubMed]
163. Durán-Rey, D.; Crisóstomo, V.; Sánchez-Margallo, J.A.; Sánchez-Margallo, F.M. Systematic Review of Tissue-Engineered Vascular Grafts. *Front. Bioeng. Biotechnol.* **2021**, *9*, 771400. [CrossRef]
164. Weekes, A.; Bartnikowski, N.; Pinto, N.; Jenkins, J.; Meinert, C.; Klein, T.J. Biofabrication of small diameter tissue-engineered vascular grafts. *Acta Biomater.* **2022**, *138*, 92–111. [CrossRef]
165. Jafarihaghighi, F.; Ardjmand, M.; Mirzadeh, A.; Hassani, M.S.; Parizi, S.S. Current challenges and future trends in manufacturing small diameter artificial vascular grafts in bioreactors. *Cell Tissue Bank.* **2020**, *21*, 377–403. [CrossRef]
166. Ahmed, W.; Zhai, Z.; Gao, C. Adaptive antibacterial biomaterial surfaces and their applications. *Mater. Today Bio.* **2019**, *2*, 100017. [CrossRef]
167. Nalawade, T.M.; Sogi, S.H.P.; Bhat, K. Bactericidal activity of propylene glycol, glycerine, polyethylene glycol 400, and polyethylene glycol 1000 against selected microorganisms. *J. Int. Soc. Prev. Community Dent.* **2015**, *5*, 114–119. [CrossRef]
168. Mahlapuu, M.; Håkansson, J.; Ringstad, L.; Björn, C. Antimicrobial Peptides: An Emerging Category of Therapeutic Agents. *Front. Cell. Infect. Microbiol.* **2016**, *6*, 194. [CrossRef]
169. Yilmaz Atay, H. Antibacterial Activity of Chitosan-Based Systems. In *Functional Chitosan*; Jana, S., Ed.; Springer: Berlin, Germany, 2020; pp. 457–489.
170. Gibney, K.A.; Sovadinova, I.; Lopez, A.I.; Urban, M.; Ridgway, Z.; Caputo, G.A.; Kuroda, K. Poly(ethylene imine)s as Antimicrobial Agents with Selective Activity. *Macromol. Biosci.* **2012**, *12*, 1279–1289. [CrossRef]
171. Andercou, O.; Marian, D.; Olteanu, G.; Stancu, B.; Cucuruz, B.; Noppeney, T. Complex treatment of vascular prostheses infections. *Medicine* **2018**, *97*, e11350. [CrossRef] [PubMed]
172. Pietsch, F.; O'Neill, A.; Ivask, A.; Jenssen, H.; Inkinen, J.; Kahru, A.; Ahonen, M.; Schreiber, F. Selection of resistance by antimicrobial coatings in the healthcare setting. *J. Hosp. Infect.* **2020**, *106*, 115–125. [CrossRef] [PubMed]
173. Chakfé, N.; Diener, H.; Lejay, A.; Assadian, O.; Berard, X.; Caillon, J.; Fourneau, I.; Glaudemans, A.W.; Koncar, I.; Lindholt, J.; et al. Editor's Choice—European Society for Vascular Surgery (ESVS) 2020 Clinical Practice Guidelines on the Management of Vascular Graft and Endograft Infections. *Eur. J. Vasc. Endovasc. Surg.* **2020**, *59*, 339–384. [CrossRef] [PubMed]
174. Ishii, Y.; Sakamoto, S.-I.; Kronengold, R.T.; Virmani, R.; Rivera, E.A.; Goldman, S.M.; Prechtel, E.J.; Hill, J.G.; Damiano, R.J. A novel bioengineered small-caliber vascular graft incorporating heparin and sirolimus: Excellent 6-month patency. *J. Thorac. Cardiovasc. Surg.* **2008**, *135*, 1237–1246. [CrossRef]
175. Zhang, J.; Huang, H.; Ju, R.; Chen, K.; Li, S.; Wang, W.; Yan, Y. In vivo biocompatibility and hemocompatibility of a polytetrafluoroethylene small diameter vascular graft modified with sulfonated silk fibroin. *Am. J. Surg.* **2017**, *213*, 87–93. [CrossRef] [PubMed]
176. Woods, I.; Flanagan, T.C. Electrospinning of biomimetic scaffolds for tissue-engineered vascular grafts: Threading the path. *Expert Rev. Cardiovasc. Ther.* **2014**, *12*, 815–832. [CrossRef]
177. Li, Z.; Zhou, P.; Zhou, F.; Zhao, Y.; Ren, L.; Yuan, X. Antimicrobial eugenol-loaded electrospun membranes of poly(ε-caprolactone)/gelatin incorporated with REDV for vascular graft applications. *Colloids Surfaces B Biointerfaces* **2018**, *162*, 335–344. [CrossRef]
178. Fujita, M.; Kinoshita, M.; Ishihara, M.; Kanatani, Y.; Morimoto, Y.; Simizu, M.; Ishizuka, T.; Saito, Y.; Yura, H.; Matsui, T.; et al. Inhibition of vascular prosthetic graft infection using a photocrosslinkable chitosan hydrogel. *J. Surg. Res.* **2004**, *121*, 135–140. [CrossRef]

polymers

MDPI

Article

Impact of Drying Regimes and Different Coating Layers on Carboxymethyl Cellulose Cross-Linked with Citric Acid on Cotton Thread Fibers for Wound Dressing Modification

Mohamad Khalid Khairunnisa-Atiqah [1], Kushairi Mohd Salleh [1,*], A. H. Ainul Hafiza [1,2], Nyak Syazwani Nyak Mazlan [1], Marhaini Mostapha [3] and Sarani Zakaria [1,*]

1 Bioresources and Biorefinery Laboratory, Faculty of Science and Technology, Universiti Kebangsaan Malaysia, Bangi 43600, Selangor, Malaysia; kat.atiq@gmail.com (M.K.K.-A.); ainulhafiza@uitm.edu.my (A.H.A.H.); nyaksyazwani@gmail.com (N.S.N.M.)
2 Centre of Foundation Studies, Universiti Teknologi MARA, Cawangan Selangor, Kampus Dengkil, Dengkil 43800, Selangor, Malaysia
3 Centre for Biofuel and Biochemical Research, Institute of Self-Sustainable Building, Universiti Teknologi Petronas, Seri Iskandar 32610, Perak, Malaysia; marhainimostapha@gmail.com
* Correspondence: authors: kushairisalleh@ukm.edu.my (K.M.S.); szakaria@ukm.edu.my (S.Z.)

Abstract: The oldest preservation techniques used are drying techniques, which are employed to remove moisture and prevent microorganisms' growths, prolonging a material's shelf life. This study evaluates the effects of drying methods on carboxymethyl cellulose (CMC) + citric acid (CA) coating layers on cotton threads. For this reason, cotton threads were washed and then coated with different layers of CMC cross-linked with CA, followed by drying using an oven (OD), infrared (IR), and a combination of oven + IR (OIR) drying methods at 65 °C. Our investigations revealed that CMC + CA yields a pliable biopolymer. The differences in drying regimes and coating layers of CMC + CA have a significant effect on the coated cotton thread strength and absorption capability. The study concluded that the IR drying regime is more effective to dry a single-layered cotton thread with a single layer of CMC + CA coating to enhance desirable properties for wound dressing modification.

Keywords: carboxymethyl cellulose; drying effects; wound dressing; coating layers; fiber-based material; mechanical properties

Citation: Khairunnisa-Atiqah, M.K.; Salleh, K.M.; Ainul Hafiza, A.H.; Nyak Mazlan, N.S.; Mostapha, M.; Zakaria, S. Impact of Drying Regimes and Different Coating Layers on Carboxymethyl Cellulose Cross-Linked with Citric Acid on Cotton Thread Fibers for Wound Dressing Modification. *Polymers* **2022**, *14*, 1217. https://doi.org/10.3390/polym14061217

Academic Editors: Domenico Acierno and Antonella Patti

Received: 20 December 2021
Accepted: 14 March 2022
Published: 17 March 2022

1. Introduction

Wound healing is a complex, systemic, and regulated process to maintain human body functions. In wound healing, the major problem would be wound infections, which impair the healing process. As an act of prevention of this problem, antibiotic medications are prescribed to patients. However, this is ineffective in long-term treatments, especially for chronic wound patients, as this will cause multiresistant bacteria emergence. Therefore, in modern-day wound care, the functionality of wound dressings must be enhanced and addressed to properly manage and care for the wound as it may severely threaten an individual's quality of life and health.

Cotton fiber is the most recognized natural fiber, which has been widely used in the textile industry. In the biomedical field, cotton fiber has been modified to have antimicrobial properties [1], water repellency [2], and other enhancements of mechanical properties to work efficiently as a wound dressing. Cotton threads have a high specific area, are adjustable in shape, and have high absorption capability, which is good to absorb exudate, reduce blood loss, and keep the wound area free from other debris [3,4]. A minor setback is when a cotton thread wound dressing dries up, it tends to stick to the wound area during the removal process [5]. The removal of the dressing might cause discomfort to patients, causing secondary damage to newly developed tissue, and requires a tedious cleaning process. Therefore, the cotton thread should be modified to have properties that permit balanced and maintaining moisture at the wound site to prevent the dressing from adhering to the wound, reducing pain and discomfort, thus accelerating wound healing.

Carboxymethyl cellulose (CMC) is used in the biomedical and pharmacology fields as an enzyme immobilizer, absorbent, r wound healing, and drug delivery. CMC is biocompatible with other biomaterials. Hence, CMC-based biomaterials with antibacterial properties for wound healing and tissue engineering could be fabricated [6]. CMC-coated wound dressing is more flexible, absorbs exudate, retains moisture to boost angiogenesis and autolytic debridement [7], is nontoxic to humans, and is water soluble [8]. Salt CMC derivatives, such as sodium CMC (Na-CMC) and calcium CMC (Ca-CMC), are commonly used. Na-CMC has an excellent compatibility with human skin, which shows a high film-forming capability and is proven to effectively eliminate microbial growths from wound beds through microbe adhesion and promote wound healing [9]. It is due to Na-CMC substituted carboxylated group (-COO) in the backbone along its cellulose chain. Besides, Na-CMC has been used to fabricate cellulose-based coating through chemical cross-linking as it can form a cross-linked network through ionic bonding, hydrogen bonding, or polymer–polymer interaction [10]. With these advantages, CMC-based biomaterials are widely used in wound dressing applications [7] to prevent wound infections. As coating agent, CMC improve thread quality and mechanical strength [11]. However, CMC on its own has poor mechanical properties. Therefore, citric acid (CA) was used as a crosslinker in this study. CA is present in citrus fruit, such as lemons and oranges. This natural organic acid consists of three carboxylic groups. CA is used in food processing and is now known in the biomedical field for its excellent antimicrobial and antioxidant properties. CA is also beneficial in the pharmaceutical application as it is used in various applications, such as cross-linking, interaction between molecules, and coating agents [12].

Oven drying (OD) is a conventional technique used with a standard temperature–time combination [13]. It is considered convenient since it can accommodate a large number of samples and has relatively rapid and precise temperature control to reach the desired temperature. However, the drawback is that temperature variations might occur due to the samples' size, weight, and position in the oven, which affect the total removal of moisture from samples, the risk of losing volatile substances or compounds, and sample decomposition during the long drying process [14–16]. In addition, sample decomposition is undesirable in wound dressing materials as it may have an immense degradable effect on temperature-sensitive medicinal substances. A more time-saving method, such as an infrared (IR) dryer, is viewed as an alternative to overcome this problem. IR drying provides rapid heating and moisture removal. It serves many advantages, such as high energy efficiency, faster heat transfer rate, and maintaining high-quality products [17,18]. During the drying process, the IR wavelength radiation emitted from the heat source passes through the wet sample and increases the temperature internally without heating the surrounding air [15]. The IR penetration causes water molecules to vibrate, which leads to heating [19]. The heat provided by IR interacts with the samples' internal structure and facilitates heating from the inner to the outer layer through the radiation and convection thermal phenomena, causing a decrease in moisture content through the evaporation process [19,20]. IR has been shown to provide more advantages in the drying method because it provides uniform heating, has a higher heat transfer rate, reduces processing time and energy, and improves the material quality [15,20]. Table 1 shows a comparison between IR drying and other drying techniques based on the cost-effectiveness, drying time, product quality, and advantages and disadvantages of the drying techniques. Therefore, a distinct comparison between oven and IR drying and a combination of both with regard to the quality of cotton thread wound dressing will be measured.

Table 1. Comparison between drying techniques.

Drying Techniques	Cost-Effective	Drying Time	Product Quality	Mechanism	Advantages	Disadvantages	Ref.
Oven/convective	Low	Long. Depends on hot air temperature and air velocity	Depends on the parameter of drying	Moisture exchange between sample and the hot air flow through drying chamber	Long shelf life, simple and easy operation	Exposure to oxidation, crust formation of sample surface due to high temperature	[20,21]
Infrared	Low	Short. Increases with air velocity and decreases with IR intensity	High	Sample is exposed to electromagnetic radiation where heat is transferred from heat source to product surface	Environmentally friendly, provides heat homogeneity, short heating time and low energy consumption, has high accuracy, and increases manufacturing efficiency	Exposure to high heat leads to burns; the penetration depth is limited. Prolonged exposure causes tissues rupture. Infrared is not sensitive to reflective properties.	[18,20,22]
Microwave	Low	Reduces drying time	High	Volumetric heating occurs when electromagnetic waves pass through sample, leading to molecule oscillation that generates thermal energy to remove water	Volumetric heating spread through sample reduces drying time	Leads to damaged product because of improper heat control and mass transfer	[21]
Freezing	High	Slow	Low	Two steps: (1) Freezing water molecules in sample (2) Frozen solid samples are heated to induce moisture sublimation	Prevents oxidation, minimal shrinkage and soluble solid shift, volatile compound retention, maintaining porous structure	High facility cost. Could cause major loss of aromatic components	[22–24]

Our study aims to evaluate the effect of drying methods to CMC cross-linked with CA on cotton thread fibers' mechanical strength and its wettability properties for wound dressing enhancement. To achieve this aim, we investigate how different coating layers of CMC cross-linked with CA onto cotton thread fibers are influenced by different drying regimes. Our experimental approach evaluated whether drying regimes and two different applied coating layers can potentially affect the cotton thread absorption capability, surface morphology, and mechanical properties. The coated cotton thread moisture content and water absorption were determined. The CMC-coated cotton thread surface was visualized using an optical microscope (OM), and the mechanical properties, such as tensile strength, were assessed. Our results stipulate that a single-layer coat of CMC cross-linked with CA coated cotton thread undergoing IR drying shows better property enhancement of the cotton thread fibers than the oven or oven + IR drying method.

2. Materials and Methods

2.1. Raw Materials

Mercerized cotton thread (3-ply with 100% cotton) was purchased from Coats Cotton (Hungary, Europe). Solid powder sodium carboxymethyl cellulose (Na-CMC) (CAT. NO: 419273) with medium viscosity and a degree of substitution of 0.65–0.90 and citric acid (anhydrous, Mw: 192.12) (CAS. NO: 77-92-9) were purchased from Sigma-Aldrich (St. Louis, MO, USA).

2.2. Preparation of CMC and Citric Acid Coating Solution

About 2% (w/v) of CMC as a coating material was dissolved in distilled water (dH_2O) by stirring with a magnetic stirrer, forming a homogeneous aqueous solution. Then 4M CA was dissolved in dH_2O. The 2M and 3M CA were prepared by diluting 4M CA.

2.3. Design of Coating Processes

Cotton thread fibers were washed before coating with 2% detergent solution and 80% ethanol and were oven-dried at 60 °C for 15 min. The coating process was based on previously published results [22] with modifications. The 2% CMC was prepared and allocated in 25 mL syringes with a 20-gauge needle size. Cotton thread fibers were soaked in the syringes with CMC solution for 18 h.

After the threads were soaked, the threads were pulled out from the syringe and were immediately submerged into beakers containing heated 2M, 3M, and 4M CA. The threads were left immersed for 1 h at 90 °C with constant stirring to promote CA and CMC interaction [23,24]. After the immersion, the excess CA was removed using filtered paper, and the thread groups were separately dried using OD and IR drying techniques at 65 °C for 30 min. In addition to that, OD drying was conducted in a convection oven (Memmert UFTS, Germany). For IR drying, the distance between the IR lamp and the sample in the drying machine was fixed at 7 cm (Desktop Infrared Drying Machine (GW-200H), Hoystar®, Guandong, China). During the sequential drying regime (OIR), drying was first conducted in the oven at 65 °C for 15 min, followed by IR drying for another 15 min, similarly at 65 °C, making a total drying time of 30 min.

The coating process was altered in six samples, which were divided into three sets. The first three sample groups were set as control, where no coating of CMC and CA was be performed. The second set consisted of the sample groups where one layer of CMC was performed. The third set is where two CMC layers were applied under the same conditions, with respective drying regimes after each coat. Once all samples were dried, the dry coat weight was measured as a control sample by weighing. Table 2 lists the coating parameters used to design the coating process.

Table 2. Design of coating trial with a defined number of layers and drying regimes. Trials with one layer of CMC are indicated by OD-1, IR-1, and OIR-1, while trials with two CMC layers are indicated by OD-2, IR-2, and OIR-2. CT-A, CT-B, and CT-C are the uncoated samples.

Samples	Coating Layer	Drying Method
CT-A		Oven
CT-B	0	IR
CT-C		Oven + IR
OD-1		Oven
IR-1	1	IR
OIR-1		Oven + IR
OD-2		Oven
IR-2	2	IR
OIR-2		Oven + IR

2.4. Determination of Physical Characteristic of Uncoated and Coated Cotton Threads

Basic physical properties, such as basis weight, thickness, and moisture content, were assessed using uncoated and CMC-coated samples. The basis weight of uncoated and coated samples was determined using an analytical balance (A&D Compact Analytical Balance, HR-250AZ), and the sample thickness was measured with an ABS Digital Thickness Gauge (Code: 547-301) (Mitutoyo, Kanagawa, Japan). After samples were dried according to the regimes, samples' moisture content was evaluated using a moisture analyzer machine (A&D Moisture Analyzer, MX-50).

2.5. Evaluation of Mechanical Properties

Tensile strength (TS), modulus, and percent elongation at break were measured using the Servo Control System Desktop Tensile Strength Tester AI-3000N (Qingdao, China). The test was performed based on the ASTM D3822 standard (Standard Test Method for Tensile Properties of Single Textile Fibers). Before running the tensile test, all samples were conditioned overnight in a drying cabinet at room temperature and 59.4% relative humidity (Weifo, Taiwan). The distance between the two clamps was set to 50 mm, and the strain rate speed was constant for all samples at 10 mm/min with a 10 kg load cell. The samples were cut at a length of 70 mm. Both coated and uncoated thread samples were secured with masking tape at both ends and inserted between the clamps to avoid fiber slipping during testing. The average fibers' strength and percent elongation were calculated based on five measurements. The tensile strength was obtained by dividing the average tensile load and average fiber diameter.

2.6. Determination of Water Absorption and Moisture Content

2.6.1. Water Absorption

Water absorption was evaluated by immersing samples into distilled water. Each thread sample was immersed for 0.5, 1, 3, 5, 7, and 30 min. The samples were weighed before and after immersion, and then the water absorption was calculated using Equation (1).

$$\mathrm{W}\ (\%) = \left[\frac{m_a}{m_o}\right] \times 100 \tag{1}$$

where W is the water absorption of the samples in (%), m_a is the sample mass (mg) after water absorption, and m_o is the initial mass of the samples before water absorption [25].

2.6.2. Moisture Content

Moisture content was determined by recording the sample mass at similar time intervals, and moisture content was calculated as shown in Equation (2).

$$MC\ (\%) = \frac{Mcws - Mcs}{Mcs - Mc} \times 100 \quad (2)$$

where *Mcws* is the container plus wet sample mass (mg), *Mcs* is the mass of the container plus dried sample (mg), and *Mc* is the container mass (mg) [26].

2.7. Surface Analysis

2.7.1. Surface Morphology

The surface topography of coated threads was observed under a Dino-Lite Optical Microscope and analyzed with the DinoCapture V2 program. The samples were observed under 440× magnification.

2.7.2. Attenuated Total Reflection–Fourier-Transform Infrared Spectroscopy (ATR–FTIR)

FTIR was carried out using Alpha Platinum-ATR (Bruker, Massachusetts, United States). FTIR spectra of the samples were recorded at 32 scans with a resolution of 4 cm^{-1} and a wavenumber range between 400 and 4000 cm^{-1}.

2.8. Antibacterial Analysis

Fabricated cotton threads with different concentrations of CA and different coating layers were investigated based on a previous published study [27], which was conducted using agar plate diffusion method (GB/T 20944.1-2007). Gram-negative (*Escherichia coli*) and Gram-positive (*Staphylococcus aureus*) bacteria were diluted into a suspension (1×10^6 CFU/mL) and spread on the sterile agar plate. The fabricated cotton threads were placed transversely onto the agar surface to ensure contact, and the plates were incubated at 37 °C for 24 h. The zone of inhibition was evaluated by the inhibition zone diameter values of the samples and calculated based on the formula shown in Equation (3).

$$H\ (mm) = \frac{D - d}{d} \quad (3)$$

where H refers to the zone of inhibition (mm), *D* refers to the total diameter of the sample and inhibition zone (mm), and *d* refers to the diameter of the cotton thread (mm).

2.9. Statistical Analysis

The selected data were given as mean values with standard deviations. The number of replicates was constant, where $n = 3$ replicates for each observation. Analysis of the data obtained from the experiments was performed using the ANOVA function in Microsoft Excel with a confidence level of $p < 0.05$.

3. Results and Discussion

3.1. ATR–FTIR Characterization of Uncoated CT and CT Coated with CMC Cross-Linked with CA (CT/CMC + CA)

The presence of CMC and CMC + CA coating on CT was confirmed by FTIR analysis. The ATR–FTIR measured in this study is within the mid-IR spectrum of 400–4000 cm^{-1}. Figure 1 shows the spectrum of the pristine CT and CT coated with CMC cross-linked with 2M CA samples at different coating layers with different drying regimes. CA spectra showed characteristic peaks at 3290 cm^{-1} corresponding to O–H stretching for H_2O and 1745 and 1698 cm^{-1}, which matched with C=O stretching of carboxylic acid [28]. As regards CMC, FTIR peaks were observed at 3356, 1587, and 1051 cm^{-1} corresponding to O–H stretching, carboxylate C=O stretching, and C–O–C stretching, respectively [29]. Based on Figure 1, there is no significant difference between coating layers at $p > 0.05$, which proves that coating layers have no adverse effect on the chemical functionalities of the samples. Additionally, similarities among the spectrum prove that CA was homogeneously cross-linked with CMC. However, as regards different drying regimes, there are slightly different transmittance intensities to the CMC + CA spectra.

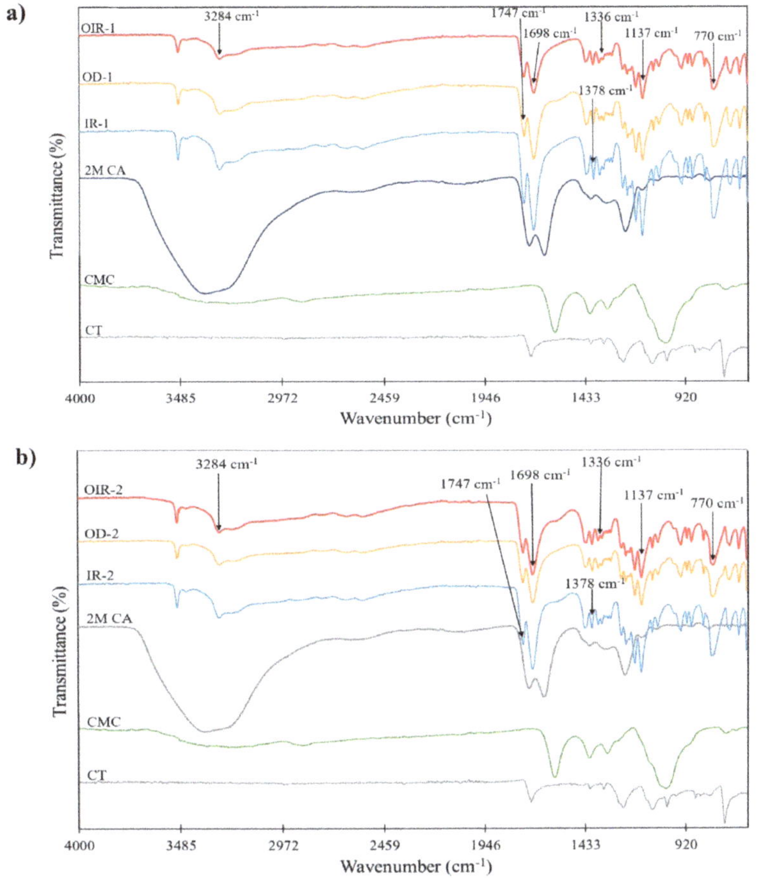

Figure 1. FTIR spectra of (**a**) single-coated and (**b**) double-coated CMC cross-linked with 2M CA in different drying regimes.

After different concentrations of CA were added to the CT/CMC samples, peaks can be observed at 3284, 1747, and 1698 cm^{-1}, and the cross-linking mechanism involves the attachment of carboxylate groups of CA to the hydroxyl group of CMC, which could be attributed to the esterification between citric acid and CMC, demonstrating chemical linkage formation among them [30]. In addition, the peaks at 1747 and 1698 cm^{-1} became more intense, with the increase in CA concentrations, as shown in Figure 2a,b, stipulating a higher cross-linking reaction. The ATR–FTIR results suggested that there was an occurrence of cross-linking interaction between CT/CMC and CA. Figure 2a,b demonstrates a similar observation at 3284 cm^{-1}, where peaks at lower concentration showed less intense peaks due to the esterification reaction during cross-linking [31]. Thus, the results obtained are in tune with those of previous studies [28,32].

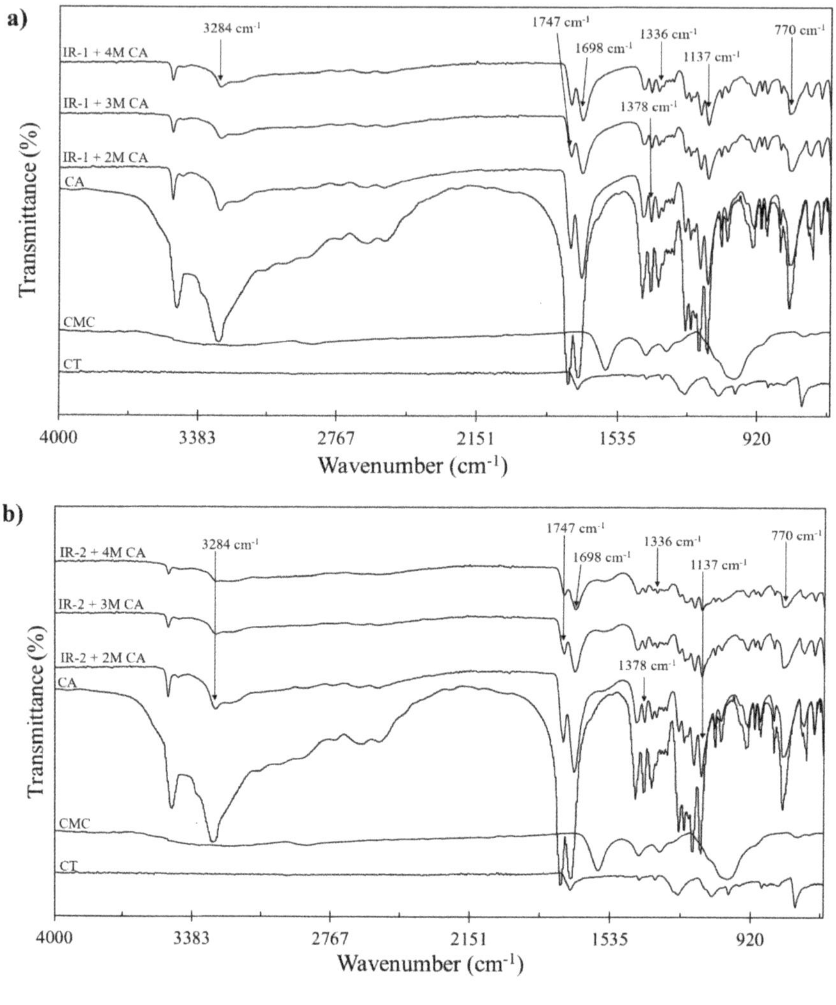

Figure 2. FTIR spectrum comparison between (**a**) single-coated CMC cross-linked with 2M–4M CA, (**b**) double-coated CMC cross-linked with 2M–4M CA.

The slight differences in transmittance intensity among different drying regimes are due to the transmittance value among the drying regimes. IR has the highest transmittance values, followed by OD and OIR drying regimes. A high transmittance value indicates a large population of specific functional groups that emits vibrational energies, which corresponds to the reflected light [33]. The large population of specific bonds suggests that IR is an effective drying regime since CMC + CA is more concentrated due to due to the efficient moisture removal. As the moisture is efficiently removed from the coated CT, macromolecular compaction occurs, and the population of specific bonds per mm^2 increases. The concentrated amount of CMC + CA coating in IR-dried samples causes them to emit a high transmission value to a specific functional group population bond. More apparent differences are shown in Figure 1 of OD and OIR samples, regardless of layers. Even though the amounts of CMC and CA used were controlled, distinct transmittance values among different drying regimes indicated different efficacies of moisture removal. The efficiency of moisture removal will be highlighted in Section 3.2.

In Figure 3, four enlargements of Figure 1, which exhibit the FTIR spectra of dried samples by different drying regimes on a specific range of wavelengths, are shown. The stretching bands of the functional groups of the IR-dried samples were similar to those of the OD and OIR samples. According to Figure 2, the OH group appeared within the broad adsorption peak in all drying methods. A slight decrease in the wavenumber was observed by comparing the CT/CMC + CA samples and uncoated CT samples. The carbonyl peaks of CMC + CA emerged at 1747 and 1698 cm^{-1} in the spectral region between 1750 and 1600 cm^{-1} for all samples, both in single and double coats. There was a shift towards a smaller wavenumber within this wavenumber, and a similar trend was observed at peaks of 1378 and 1137 cm^{-1}. When the peaks shift to a smaller wavenumber, this indicates that the molecule within this wavenumber has increased in its mass. The vibration frequency is inversely proportional to the mass of the vibrating molecule. Therefore, the heavier the molecule, the lower the vibration frequency, thus the smaller the wavenumbers [34]. At a wavenumber of 770 cm^{-1}, no changes were observed between the drying regimes. This result helps confirm that different drying regimes did not affect the sample compositions.

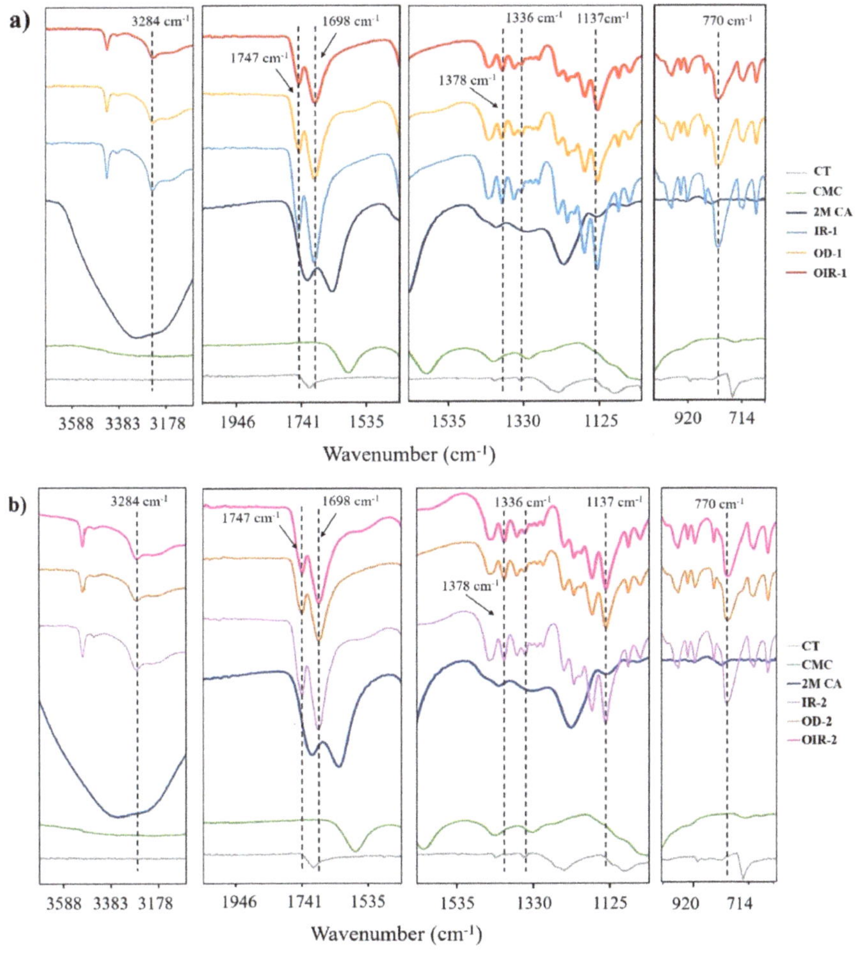

Figure 3. Enlargement of the FTIR spectra of (**a**) single-coated and (**b**) double-coated CMC cross-linked with 2M CA in different drying regimes.

3.2. Physical Properties of Uncoated CT and Coated CT/CMC + CA

3.2.1. Basis Weight and Thickness

As seen in Table 3, the basis weight and thickness of coated CT are directly proportional to the coating layer and concentration of solids in the coating formulation. Coated samples gained the basis weight and thickness as expected compared with uncoated samples, with double-coated samples accumulating the highest weight gain and thickness. CT coated with CMC increased its weight and thickness but not as much as CT coated with CMC + CA. The substantial increase in CMC + CA samples' weight proved that cross-linked CMC + CA has better coating efficiency due to the accumulation of CA attached to the -OH group of CMC after the esterification process. It was also found that the CT/CMC weight and average thickness increased significantly with CA concentration. This is because the higher the CA concentration, the higher the probability of CA to cross-link with CMC. At a higher cross-linking degree, CMC + CA becomes more viscous, allowing more intermolecular bonds to form between CMC and CA [35], thus increasing their weight and average thickness. For medical purposes, such as skin grafting, it is proposed that the dressing film should be thinner than 400 μm to allow vascularization for nutrient delivery to the epidermis [36]. Therefore, the coating thickness of the samples was fabricated within the desirable range except for the samples OIR-1 and OIR-2 at 4M CA.

Table 3. Physical measurement of uncoated and coated cotton thread samples.

Samples	Coating Layers	CA Concentration (M)	Drying Regime	Basis Weight (UCS) (mg)	Basis Weight (CS) (mg)	Average Thickness (μm)
CT-A	0	n.a.	Oven	3.4 ± 0.2	-	112.5 ± 1.1
CT-B		n.a.	IR	3.3 ± 0.2	-	112.8 ± 1.0
CT-C		n.a.	Oven + IR	3.2 ± 0.3	-	106.5 ± 1.3
OD-1	1	CMC	Oven	3.2 ± 0.2	3.5 ± 0.1	127.8 ± 1.1
		2		3.3 ± 0.1	4.7 ± 0.3	183.5 ± 1.2
		3		3.2 ± 0.2	5.1 ± 0.3	249.0 ± 1.0
		4		3.4 ± 0.1	6.0 ± 0.2	277.0 ± 1.1
IR-1	1	CMC	IR	3.1 ± 0.2	3.6 ± 0.3	125.3 ± 1.2
		2		3.2 ± 0.1	4.9 ± 0.6	178.5 ± 1.0
		3		3.0 ± 0.3	5.9 ± 0.6	182.3 ± 1.0
		4		3.3 ± 0.1	6.7 ± 0.4	226.0 ± 1.1
OIR-1	1	CMC	Oven + IR	3.3 ± 0.1	3.5 ± 0.3	247.8 ± 1.2
		2		3.0 ± 0.2	3.9 ± 0.3	336.0 ± 1.2
		3		3.2 ± 0.1	4.3 ± 0.1	362.5 ± 1.3
		4		3.0 ± 0.3	4.8 ± 0.4	444.8 ± 1.3
OD-2	2	CMC	Oven	3.4 ± 0.1	6.3 ± 0.2	136.5 ± 1.1
		2		3.4 ± 0.1	10.0 ± 0.6	256.1 ± 1.3
		3		3.5 ± 0.1	11.7 ± 0.1	267.8 ± 1.4
		4		3.4 ± 0.1	14.4 ± 0.1	309.0 ± 1.8
IR-2	2	CMC	IR	3.2 ± 0.1	6.5 ± 0.2	129.5 ± 1.3
		2		3.1 ± 0.2	10.8 ± 0.1	241.7 ± 1.5
		3		3.3 ± 0.1	12.1 ± 0.2	283.8 ± 1.1
		4		3.3 ± 0.1	18.1 ± 0.2	286.8 ± 1.1
OIR-2	2	CMC	Oven + IR	3.1 ± 0.1	6.4 ± 0.2	285.8 ± 1.3
		2		3.4 ± 0.2	9.3 ± 0.1	374.0 ± 1.5
		3		3.0 ± 0.1	10.0 ± 0.1	381.0 ± 1.1
		4		3.5 ± 0.2	14.4 ± 0.2	446.5 ± 1.3

For the curing process via several drying regimes, IR samples showed the smallest thickness, followed by OD and OIR samples in single- and double-coated CT. A higher sample thickness on OIR samples is due to the unequal heating energy imposed in the transitioning process between the samples from OD and IR in the sequential drying regimes of OIR. Meanwhile, samples dried via IR and OD provided more uniform heating, allowing moisture from the samples to be removed efficiently and causing a substantial difference reduction in sample size [15]. Nevertheless, based on sample thickness, IR is the most effective, followed by OD and, finally, OIR. OIR is not as effective as expected, particularly as a solitary drying regime. It is due to the uneven energy used to dry the samples, and the samples are prone to absorb moisture when transitioning from one drying regime to another. Therefore, the ineffectiveness of OIR will have to prolong the drying time or increase the heating energy for the sample to be dried effectively.

3.2.2. Moisture Content

Moisture content was determined to evaluate the amount of moisture in the conditioned uncoated and coated CT. After the drying process, the extent of destruction to the lamellar structure of the macromolecule network will differ among different drying regimes. Hence, the ability to absorb the surrounding moisture will determine the severity of drying regimes to the CMC-CA polymer networks. The ability to retain moisture is crucial for the CMC + CA coating on CT to function as a functional wound dressing. According to the Gibbs–Donnan effect theory, in regard to the transportation across cell membranes, if the Gibbs–Donnan equilibrium between the exterior surroundings and the wound bed is achieved, the increase in intracellular ions would cause cells to swell due to the osmotic influx of water. Low ionic strength can lead to a high distribution of ion mobility, establishing the Gibbs–Donnan equilibrium between the external environment and wound bed, and thus providing a stable structure for the coated CT to retain moisture and/or for drug delivery systems for wound healing [37].

Based on Table 4, the moisture content of CT/CMC + CA gradually decreases with the increase in the coating layer in all drying regimes. Increases in coating layers resulted in an increase in ionic strength, which reduced the number of mobile ions within the CMC + CA coating matrix, thus reducing the CMC + CA coating structure and limiting the moisture content of CT/CMC + CA [38]. The result is similar to the findings of other biomaterials, such as membrane [39] and hydrogel films [29,40], that are cross-linked with CA. Meanwhile, IR-1 shows the highest moisture content among the single-coated samples, followed by OD-1 and OIR-1. The same trend can be observed for the double-coated samples. The increase in moisture content at a low concentration of 2M CA in all drying regimes shows that the CMC + CA coated on CT can retain moisture. This is caused by the hygroscopic nature of CT/CMC + CA with hydroxyl and polar groups available and the interfacial area between CT and the CMC + CA matrix [41]. On the other hand, at a higher CA concentration (>2M) and double-coated CT/CMC + CA samples, a low moisture content was obtained, and it correlates with the cross-linking effect, which reduces the number of free hydrophilic groups in the coating, thus limiting molecular mobility, and binds the polymer together tightly [40].

Table 4. Moisture content (%) of CMC cross-linked with different concentrations of CA-coated cotton thread samples.

Samples	Coating Layer	(CA) (M)	Drying Regime	Moisture Content (%)	
				Dry	Wet
CT-A	0	n.a	Oven	3.08 ± 0.04	92.3 ± 0.02
CT-B		n.a	IR	3.05 ± 0.03	100 ± 0.02
CT-C		n.a	Oven + IR	3.11 ± 0.03	84.6 ± 0.03
OD-1	1	CMC	Oven	3.15 ± 0.04	88.9 ± 0.06
		2		3.43 ± 0.06	95.0 ± 0.03
		3		3.29 ± 0.06	45.8 ± 0.04
		4		3.17 ± 0.03	19.0 ± 0.05
IR-1		CMC	IR	3.14 ± 0.02	63.2 ± 0.03
		2		3.39 ± 0.03	120.9 ± 0.04
		3		3.28 ± 0.02	70.9 ± 0.02
		4		3.15 ± 0.04	34.1 ± 0.03
OIR-1		CMC	Oven + IR	3.33 ± 0.06	33.3 ± 0.03
		2		3.45 ± 0.04	90.3 ± 0.02
		3		3.38 ± 0.03	44.6 ± 0.03
		4		3.33 ± 0.04	18.0 ± 0.06
OD-2	2	CMC	Oven	3.12 ± 0.02	19.5 ± 0.02
		2		3.38 ± 0.03	36.1 ± 0.05
		3		3.22 ± 0.04	32.1 ± 0.06
		4		3.11 ± 0.03	12.5 ± 0.03
IR-2		CMC	IR	3.09 ± 0.06	70.6 ± 0.03
		2		3.31 ± 0.04	71.9 ± 0.04
		3		3.21 ± 0.03	50.0 ± 0.05
		4		3.11 ± 0.06	34.0 ± 0.03
OIR-2		CMC	Oven + IR	3.31 ± 0.04	20.0 ± 0.06
		2		3.43 ± 0.03	24.4 ± 0.02
		3		3.37 ± 0.06	17.2 ± 0.06
		4		3.31 ± 0.03	7.1 ± 0.05

The findings are consistent with the basis weight of coated samples portrayed in Table 2. A higher moisture content by IR showed that IR is more effective in removing the moisture than OD and OIR. At a similar temperature and time, unlike OD, IR minimized the capillary force driven by the surface tension of water, hindering the destruction of its lamellar structure. For OIR, the lowest moisture content absorbed by them is due to the sequential drying regime unable to effectively remove moisture during the drying process, resulting in high capillary force and the destruction of OIR samples' lamellar structure, rendering them unable to retain moisture.

3.3. *Surface Morphology of Uncoated CT and Coated CT/CMC + CA*

The surface morphology of uncoated and coated samples is shown in Figures 4 and 5, respectively. In Figure 4, the uncoated samples (CT-A, CT-B, and CT-C) showed no significant difference in terms of morphology and embodied plenty of empty spaces among the cotton fibers, making them porous. This will allow CMC + CA solutions to penetrate easily during the coating process.

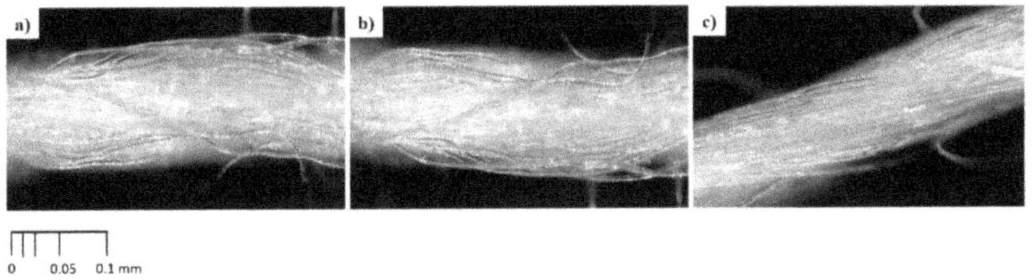

Figure 4. Uncoated cotton thread surface of (**a**) CT-A, (**b**) CT-B, and (**c**) CT-C.

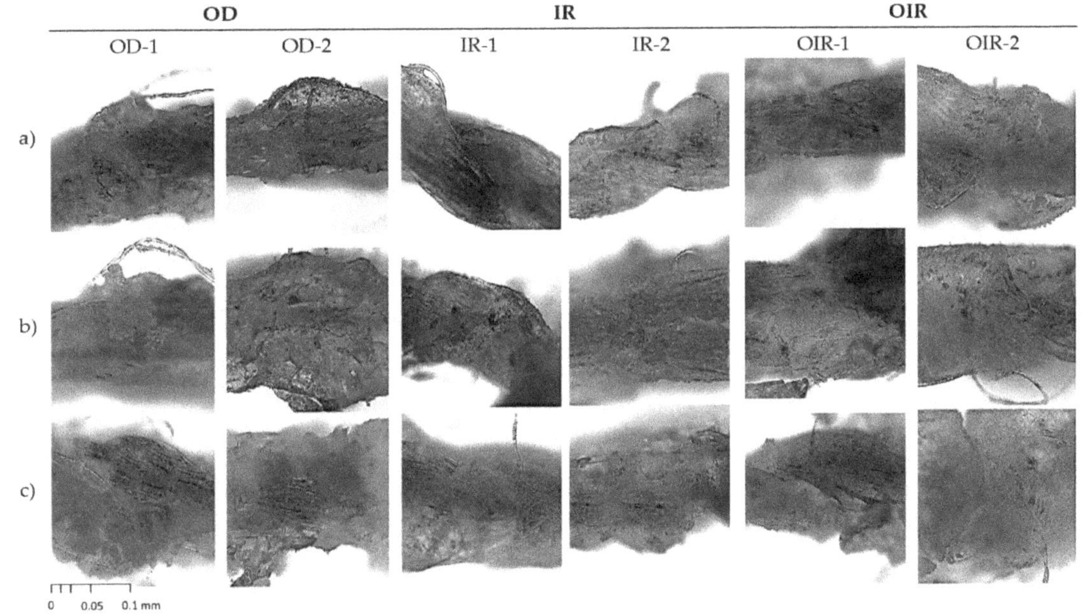

Figure 5. Observation under an optical microscope of a single coat (OD-1, IR-1, and OIR-1) and double coat (OD-2, IR-2, and OIR-2) cross-linked with (**a**) 2M CA, (**b**) 3M CA, and (**c**) 4M CA under different drying regimes with a scale at 0.1 mm.

Figure 5 shows CT morphology after coating (single and double coat) and the respective drying process. In comparison with Figure 4, Figure 5 shows that the empty spaces between the cotton thread fibers of single- and double-coated samples are filled with coated materials. In Figure 5, all samples show that the CMC-CA were homogeneously and evenly coated on the CT surfaces. The cross-linking between CT/CMC + CA involves the interaction of covalent esters composited as a coating layer that enhances the physicochemical properties of the CT. This is a crucial feature for CT/CMC + CA to achieve a stable structure. The stable structure permits CT/CMC + CA moisture retention enhancement, as shown in Section 3.2.2, and further improvement of CT/CMC + CA mechanical properties and absorption capability. Therefore, CT/CMC + CA can be utilized as a functional wound dressing.

As shown in Figure 5, the cotton threads coated with CMC cross-linked with CA presented textured surfaces, particularly when a high concentration of CA (4M CA) was used, which affected their physical properties. CA-cross-linked CT/CMC thread had more protrusions than uncoated cotton threads, especially at a high CA concentration, indicating increased surface roughness. On the other hand, double-coated samples (OD-2, IR-2, and OIR-2) showed excessive citric acid crystal formations due to the high CA amount present in the coating, promoting a coarser surface on cotton thread. The enhancement of surface roughness in a medical application provides a better interaction zone between the dressing with the cells and tissue surface, allowing dermal fibroblast adhesion [42], which enhances cell adhesion and cell proliferation [43]. This aids in strong protein adhesion to the wound dressing surfaces, implying that a rough surface can stimulate cell signaling by imitating similar cues as the extracellular matrix network [44–46].

IR-1 and IR-2 samples visually showed a smoother coating topography than OD and OIR drying samples, which exhibited rougher surface coverage. This was due to IR drying involving a more uniform heating mechanism where the temperature of the samples was increased internally [15]. When the samples were heated internally via the IR drying process, disturbance to the polymer interactions scarcely occurred, facilitating a uniform solidification of coating materials interpreted by a smoother topography as depicted in Figure 5. Meanwhile, for OD and OIR samples in Figure 5, nonuniform topography is triggered by the sample's placement during the drying process. Samples' positioning promotes temperature variations in the oven environment, which subsequently cause propagation of polymer interactions during the solidification process of coating materials [15,16]. Besides, a sequential drying process of OIR is inefficient due to the change of heating energy between the two drying regimes. In addition, the surrounding factor during the transition of OD to IR drying is exacerbated by moisture absorption and the sudden temperature drop and, thereupon, disturbs the solidification process of coating materials. In essence, due to the uneven heating energy, OD- and OIR-dried samples exhibit a heterogeneous structure and texture. With this in mind, different drying regimes gives different morphology depending on the drying mechanism efficiency [47].

3.4. Mechanical Properties of Uncoated CT and Coated CT/CMC + CA

The tensile test was performed to investigate the effects of coating layers and drying regimes on the mechanical performances of CT/CMC + CA. The samples' tensile strength and elongation percentage (%) are presented in Figures 6 and 7.

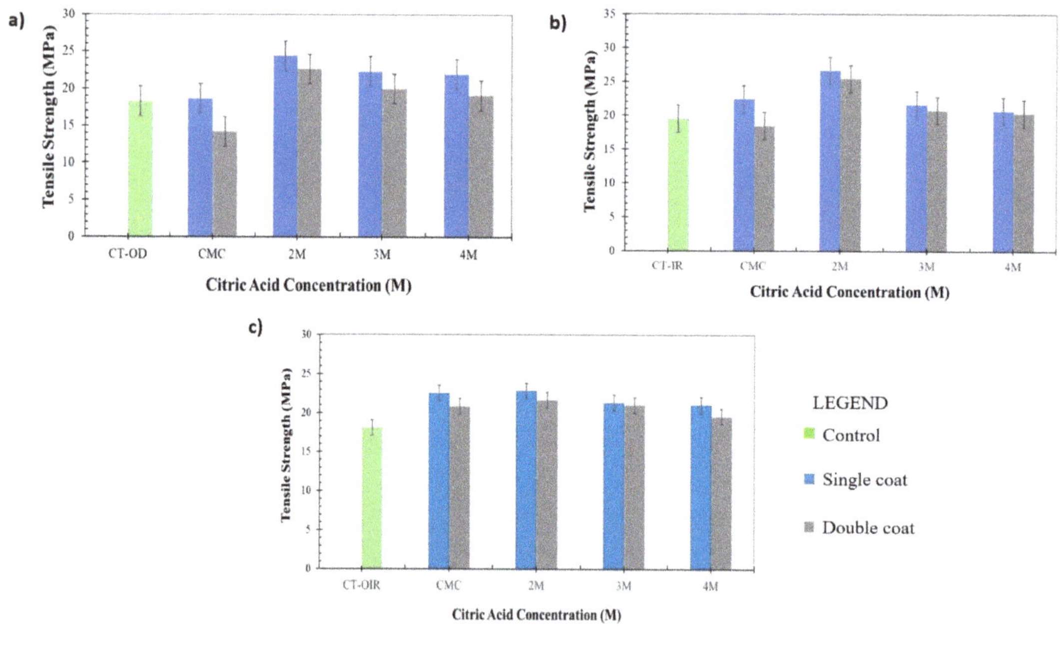

Figure 6. Tensile strength of CT/CMC cross-linked with different CA concentrations dried by (**a**) OD, (**b**) IR, and (**c**) OIR.

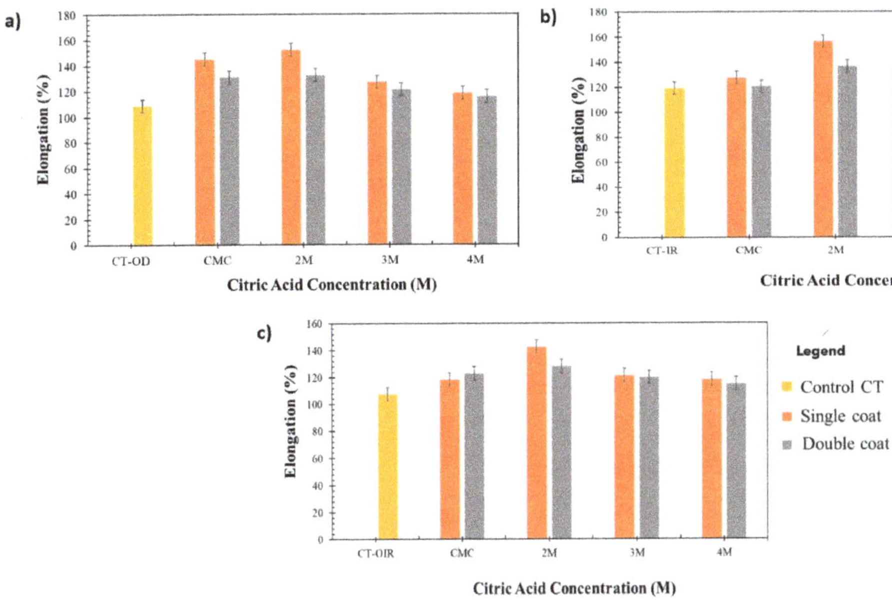

Figure 7. Elongation percentage of CT/CMC samples cross-linked with different CA concentrations dried by (**a**) OD, (**b**) IR, and (**c**) OIR.

Figures 6 and 7 show that the tensile strength and elongation percentage of CT/CMC + CA are higher than those of uncoated samples. It is well understood that as tensile strength increases, the elongation percentage decreases and vice versa [48]. However, CA, as the crosslinker in the coating system, increases tensile strength and elongation due to a higher potential of interaction between material and matrix polymer, resulting in increased of intra- and intermolecular cross-linking between polymers [49]. The CT fiber twisting orientation also plays a role in the increase in tensile strength and elongations percentage [50] as the twisting provides extra length, allowing CT to stretch more when a load is applied. The extension increased CT lateral pressure and compaction and, as a result, CT became denser and more coherent to strain. CT has a slightly loose twisting orientation that allows CMC + CA to penetrate and coat individual fiber surfaces, which further improves CT load bearing capacity by intercepting higher loads, thus leading to the improvement of elongation and avoiding early breakage [51].

The interactions contributed to a more coherent structure, improving the tensile strength and elongation percentage of CT/CMC + CA. When comparing single- and double-coated samples, single-coated CT had higher tensile strength than double-coated samples. Generally, the tensile strength of composites increases with composite content until a maximum or optimum value is achieved. Once the maximum value is reached, the tensile strength will decrease. A higher tensile strength for single-coated CT for all drying regimes is due to the penetration of the coating solution between the fibers. CMC + CA acts as a space filler, which also contributes to the load distribution, increasing the tensile strength and elongation percentage [52]. However, the reversed scenario was observed for double-coated CT/CMC + CA samples. This is due to the high amount of CMC + CA content in the double-coating layer that reduces the matrix mobility, making it stiffer, hence indicating that the amount of CMC + CA in double-coated CT might have exceeded the optimum composite content of CT/CMC + CA strength [53].

The presence of CA caused significant differences ($p < 0.05$) in the tensile strength. In all drying regimes, regardless of coating layers, the tensile strength and elongation percentage are the highest at 2M CA and continuously decrease afterward. At higher concentrations of CA, namely, 3M and 4M, clumping is prone to occur due to the quick cross-linking process between polymers. When cross-linking occurs at a higher rate, it causes a higher number of CA that might have cross-linked with CMC in the matrix, causing CMC + CA agglomeration. With that being said, a higher concentration of CA leads to clumping [49,54], thus acting as stress concentrators, causing CT to be stiff and resulting in lower mechanical performances [55]. Based on the results obtained, the mixture performed best at a lower concentration of CA (2M) to improve the tensile properties of CT because of the intramolecular and intermolecular interaction forces, which enhanced the coated CT strength due to the formation of a stronger network force in the matrices. In addition, at higher CA concentrations (>2M) and coating layers, high CA content results in the formation of CMC + CA crystals (see Figure 6). This causes the CT/CMC + CA structure to be brittle [56]. Brittleness degrades the mechanical performances of CT by disrupting the elasticity and intermolecular force network in the CMC + CA matrices. This explains the low tensile properties of the above-stated determinant. CA concentrations of more than 2M were shown to be unsatisfactory when conjoined with the double coating process.

Apart from the effect of the coating layer, an apparent distinction between different drying regimes can also be seen in Figure 6. Overall, IR-dried single- and double-coated CTs showed the highest increment in tensile strength and elongation percentage, followed by OD and OIR. The tensile strength of single-coat samples OD and IR improved by 33% and 36%, respectively. In contrast, for double-coated samples, the tensile strength increased by 24% and 30%, respectively, compared with the uncoated samples. As regards OIR, it showed the lowest tensile strength improvement at 26% for single coat and 18% for double coat. The tensile strength between the single- and double-coated samples had a significant difference of $p < 0.05$ through all drying regimes. The significant differences were attributed to a uniform heating pattern of OD and IR drying in the single-coated samples, which allowed moisture to evaporate efficiently [15] compared with the double-coated samples. However, for OIR, despite the significant differences, the regime did not effectively remove moisture from the coated samples, both single and double, as shown in Table 4, under the dry moisture content percentage. A rapid removal of water induced by IR is more focused than the OIR technique, where the differences in heating energy during OIR transitioning affect samples' tensile properties. In OIR, environmental factors, such as moisture absorption that ensued with a temperature drop during the samples' transitioning process, should also be considered. Due to this matter, incomplete drying can cause the formation of a void of a channel structure in the coating matrix, which also leads to the presence of moisture during OIR sample fabrication, which disrupts the solidification of CMC + CA due to air entrapment in the coating structure [57]. A heated CT via OD prior samples transferred to IR creates osmotic differences between samples and the environment. Thus, moisture will be absorbed by the CT, which leads to a temperature drop. Consequently, OIR CT/CMC + CA samples have irregularities in their chemical and physical interactions due to the fluctuating sample temperature, which were later signified by the OIR sample morphological structure and tensile strength.

3.5. Effects of Drying and Coating Layers on Water Absorption

The average water absorption of CT/CMC + CA samples versus time is plotted in Figure 8. It can be observed that the water absorption of CT/CMC + CA samples increased with an increasing time and equilibrium after 5 min of immersion for single-coated samples and 3 min of immersion for double-coated samples in all drying regimes. Due to the relatively higher CMC + CA content in double-coated samples than single-coated samples, their water absorption capability is reduced, causing it to reach its maximum capacity faster than single-coated samples. When CMC + CA double-coats the CT, the solid content increases due to the cross-linking interactions, creating more networks between CMC and CA, thus reducing the mobility of the ionized ionic group in the CMC + CA matrix due to a stiffer and denser structure [31,56]. On the other hand, in a single CMC + CA coating, the carboxyl groups (ionic group) in CMC are known to be highly hydrophilic. When ionized (COO-), it increases the electrostatic repulsion between intramolecular and intermolecular interaction forces to open the polymer matrix, increasing the water absorption properties to the materials [58].

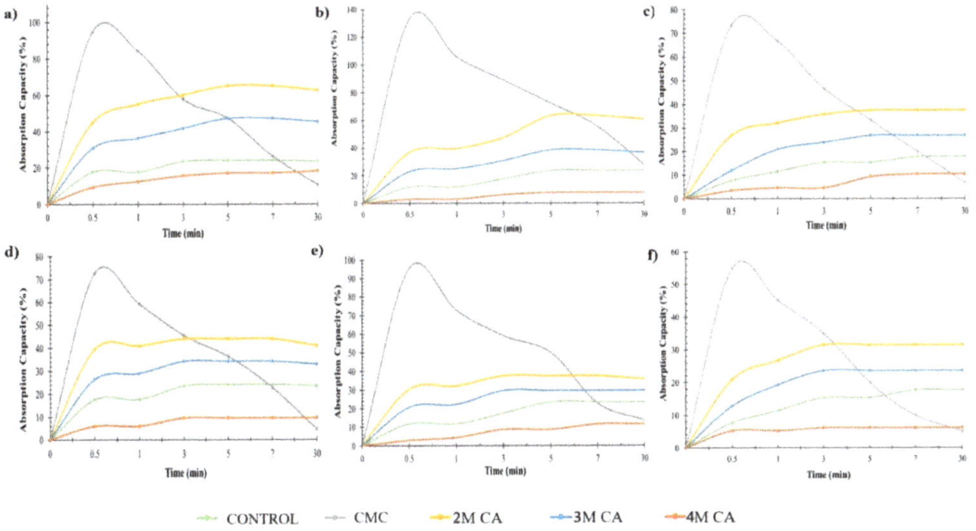

Figure 8. Water absorption behavior of CT/CMC + CA based on different drying methods. (**a**) OD-1, (**b**) IR-1, (**c**) OIR-1, (**d**) OD-2, (**e**) IR-2, and (**f**) OIR-2.

According to Figure 8, an obvious pattern is seen between different concentrations of CA. A higher CA causes a continual reduction in water absorption, with the most conspicuous at 4M CA. The CT coated with CMC + CA at 4M CA water absorption is even lower than the CT itself. Even though CMC and CA are naturally hydrophilic, excessive cross-linking events between polymers can induce the hydrophobic character. The induction is triggered by physical changes, where the cross-linking effect of CA results in the coat being hydrophobic with a high-density structure [28]. A high CA concentration can increase the solid content of the CT/CMC + CA matrix, which causes the cross-linking interactions to create a stronger bond between CMC and CA, resulting in a compact structure and higher density [34,58]. Besides, the carboxyl groups (ionic group) in CMC are known to be highly hydrophilic. When ionized (COO-), it increases the electrostatic repulsion between intramolecular and intermolecular interaction forces to open the polymer matrix, increasing the water absorption properties to the materials [58]. As proven in the surface morphology analysis, the different drying methods affect the reinforcement of CMC + CA into the CT structure. For this reason, water absorption is also altered, and the hydrophilicity of CMC + CA is influenced by different drying methods. However, other factors, such as coating layers, contribute to the result and must be considered. In this study, a single layer of CMC + CA was sufficient to coat the CT and penetrate the CT fiber. However, by adding double layers, the coating structure became stiffer and denser. This resulted in moderate changes in the CT/CMC + CA properties, such as tensile strength, moisture absorption, and water absorption, when a single layer was applied, and a significant impact when double layers were added, regardless of the drying regimes used.

3.6. Antibacterial Activity of Coated CT

From a medical perspective, *Escherichia coli* (*E. coli*) and *Staphylococcus aureus* (*S. aureus*) are the most common representatives for Gram-negative and Gram-positive bacteria, respectively, that cause infections in the human body. Gram-negative bacteria (*E.coli*) tend to be more resistant to antimicrobial agents that Gram-positive bacteria (*S. aureus*) because of the presence of the additional protection afforded by the outer membrane (peptidoglycan layer). However, in this study, *E. coli* growth was inhibited due to the presence of CA, which weakened the outer membrane of Gram-negative bacteria [59]. The result is supported by that of a previous study, where CA exhibited more inhibition to *E. coli* compared with *S. aureus* [60]. *S. aureus* was more resistant to the acidic environment since *S. aureus* is better at tolerating and adaptive to stress [61].

As shown in Figure 9, the zone of inhibition increases with the increase in coating layer and CA concentration for OD and OIR, which is attributed to the increase in reactive oxygen species (ROS). ROS is responsible for the inhibition and causes interruption of the bacterial cell wall synthesis process, growth biosynthesis inhibition, DNA transcription process interference, and metabolic pathway chain reaction disruption occurring in the bacteria cell [62]. However, IR samples showed lower CA concentrations able to inhibit *E.coli* growth. At 2M CA, both IR-1 and IR-2 samples showed the highest inhibition for *E.coli*. It is due to CA, which acts as a permeabilization agent able to inhibit *E. coli* growth by causing cell aggregation [59] and bacterial toxicity by blocking the permeability of the outer membrane [63,64]. Both bacterial species are resistant to CMC except for the IR-2 samples, while the IR-1 CMC sample only showed inhibition to *S. aureus* bacteria. The effectiveness of IR drying might also be the reason why a lower CA concentration is sufficient to inhibit bacterial growths. This finding might be due to the effective drying of CMC through IR drying, which permits its antimicrobial property to be preserved. It is achieved due to the uniform heating provided by the IR drying technique, as mentioned in Section 3.3. These findings indicate that CMC with CA as a cross-linker coating dried by IR improved the antibacterial property of cotton threads.

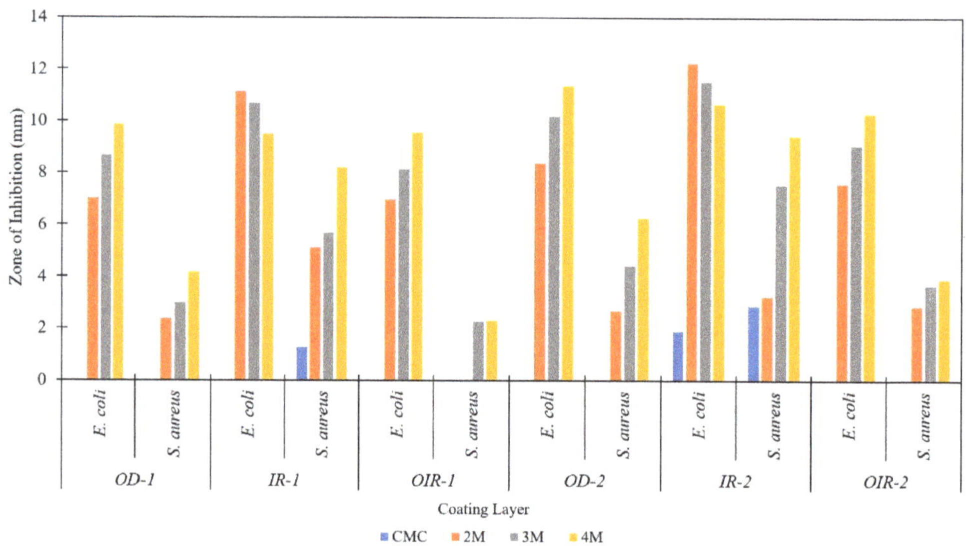

Figure 9. Antibacterial activity of fabricated cotton thread with different coating layers of CT/CMC cross-linked with different CA concentrations.

4. Conclusions

Based on these findings, the mechanical properties of coated samples are significantly affected by the drying method used at $p < 0.05$. However, the IR drying method is the best-method approach. A single coat of CT/CMC + CA showed the best properties among all samples, with single-coated samples dried via IR drying, exhibiting a less brittle and less dense structure with a slightly smoother surface as well as enhanced water absorption and tensile strength in contrast to double-coated samples. The ATR–FTIR analysis spectra corresponding to C=O from COOH of CA indicated that cross-linking between CMC and CA interaction occurred between CT/CMC-coated samples. Single-coated CMC cross-linked with 2M CA cotton thread dried with IR proved to be a successful candidate for the fabrication of cotton thread with beneficial properties for wound dressings and biomedical applications.

Author Contributions: Conceptualization, M.K.K.-A., N.S.N.M. and A.H.A.H.; methodology, M.K.K.-A.; software, S.Z.; validation, M.K.K.-A., A.H.A.H. and M.M.; formal analysis, M.K.K.-A., N.S.N.M. and A.H.A.H.; investigation, M.K.K.-A., N.S.N.M. and A.H.A.H.; data curation, M.K.K.-A.; writing—original draft preparation, M.K.K.-A.; writing—review and editing, M.K.K.-A. and K.M.S.; supervision, K.M.S. and M.M.; project administration, S.Z. All authors have read and agreed to the published version of the manuscript.

Funding: This research was funded by TRGS/1/2019/UKM/02/1/2 of the Ministry of Higher Education Malaysia and 600-RMC/MYRA 5/3/LESTARI (041/2020) of Universiti Teknologi MARA.

Informed Consent Statement: Not applicable.

Acknowledgments: We thank the Ministry of Higher Education Malaysia for research grant no. TRGS/1/2019/UKM/02/1/2 and Universiti Teknologi MARA for research grant no. 600-RMC/MYRA 5/3/LESTARI (041/2020), which funded this project, and Universiti Kebangsaan Malaysia (UKM) for providing the facilities for conducting this study. We also thank the Institute of Biology System, UKM, for providing the facility for conducting the antibacterial analysis.

Conflicts of Interest: The authors declare no conflict of interest.

References

1. Gadkari, R.R.; Ali, S.W.; Joshi, M.; Rajendran, S.; Das, A.; Alagirusamy, R. Leveraging Antibacterial Efficacy of Silver Loaded Chitosan Nanoparticles on Layer-by-Layer Self-Assembled Coated Cotton Fabric. *Int. J. Biol. Macromol.* **2020**, *162*, 548–560. [CrossRef]
2. El-Naggar, M.E.; Khattab, T.A.; Abdelrahman, M.S.; Aldalbahi, A.; Hatshan, M.R. Development of antimicrobial, UV blocked and photocatalytic self-cleanable cotton fibers decorated with silver nanoparticles using silver carbamate and plasma activation. *Cellulose* **2021**, *28*, 1105–1121. [CrossRef]
3. Li, Y.D.; Li, W.Y.; Chai, H.H.; Fang, C.; Kang, Y.J.; Li, C.M.; Yu, L. Chitosan functionalization to prolong stable hydrophilicity of cotton thread for thread-based analytical device application. *Cellulose* **2018**, *25*, 4831–4840. [CrossRef]
4. Liu, Z.; Zhou, W.; Wang, C.; Hu, W.; Chen, Z. Cotton thread modified with ionic liquid copolymerized polymer for online in-tube solid-phase microextraction and HPLC analysis of nonsteroidal anti-inflammatory drugs. *J. Sep. Sci.* **2020**, *43*, 2827–2833. [CrossRef] [PubMed]
5. Montaser, A.S.; Rehan, M. Designing strategy for coating cotton gauze fabrics and its application in wound healing. *Carbohydr. Polym.* **2020**, *244*, 116479. [CrossRef]
6. Zainal, S.H.; Mohd, N.H.; Suhaili, N.; Anuar, F.H.; Lazim, A.M.; Othaman, R. Preparation of cellulose-based hydrogel: A review. *J. Mater. Res. Technol.* **2021**, *10*, 935–952. [CrossRef]
7. Vinklárková, L.; Masteiková, R.; Foltýnová, G. Film wound dressing with local anesthetic based on insoluble carboxymethycellulose matrix. *J. Appl. Biomed.* **2017**, *15*, 313–320. [CrossRef]
8. Basu, P.; Narendrakumar, U.; Arunachalam, R.; Devi, S.; Manjubala, I. Characterization and Evaluation of Carboxymethyl Cellulose-Based Films for Healing of Full-Thickness Wounds in Normal and Diabetic Rats. *ACS Omega* **2018**, *3*, 12622–12632. [CrossRef] [PubMed]
9. Wong, T.W.; Ramli, N.A. Carboxymethylcellulose film for bacterial wound infection control and healing. *Carbohydr. Polym.* **2014**, *112*, 1–35. [CrossRef] [PubMed]
10. Salleh, K.M.; Zakaria, S.; Sajab, M.S.; Gan, S.; Kaco, H. Superabsorbent hydrogel from oil palm empty fruit bunch cellulose and sodium carboxymethylcellulose. *Int. J. Biol. Macromol.* **2019**, *131*, 50–59. [CrossRef] [PubMed]
11. Salleh, K.M.; Armir, N.A.Z.; Mazlan, N.S.N.; Wang, C.; Zakaria, S. *Cellulose and Its Derivatives in Textiles: Primitive Application to Current Trend*; Semantic Scholar: Seattle, WA, USA, 2021; ISBN 9780128214831.
12. Nangare, S.; Vispute, Y.; Tade, R.; Dugam, S.; Patil, P. Pharmaceutical Applications of Citric Acid. *Future J. Pharm.* **2021**, *7*, 1–23. [CrossRef]
13. Nirmaan, A.M.C.; Rohitha Prasantha, B.D.; Peiris, B.L. Comparison of Microwave Drying and Oven-Drying Techniques for Moisture Determination of Three Paddy (*Oryza sativa* L.) Varieties. *Chem. Biol. Technol. Agric.* **2020**, *7*, 1–7. [CrossRef]
14. Park, Y.W. Chapter 3 Moisture and Water Activity. In *Handbook of Processed Meats and Poultry Analysis*; Nollet, L., Toldra, F., Eds.; CRC Press: Boca Raton, FL, USA, 2008; pp. 35–67.

15. Rihana, H.; Lubna; Shravya, K.; Jhansi, A.; Tasleem, P.; Vellanki, B. Comparative Study of Tray Dryer, Infrared Dryer and Hot Air Oven for Drying of Chickpeas. *Int. J. Food Sci. Nutr.* **2019**, *4*, 90–92.
16. Mazandarani, Z.; Mirsaeidghazi, N.; Kaviani, M.; Shariati, M.A. Drying of Agriculture Products Using Hot Air Oven and Microwave Method. *Indian J. Res. Pharm. Biotechnol.* **2014**, *5674*, 1522–1523.
17. Lyu, J.; Chen, Q.; Bi, J.; Zeng, M.; Wu, X. Drying Characteristics and Quality of Kiwifruit Slices with/without Osmotic Dehydration under Short-and Medium-Wave Infrared Radiation Drying. *Int. J. Food Eng.* **2017**, *13*, 1–15. [CrossRef]
18. Wu, B.; Pan, Z.; Xu, B.; Bai, J.; El-Mashad, H.M.; Wang, B.; Zhou, C.; Ma, H. Drying Performance and Product Quality of Sliced Carrots by Infrared Blanching Followed by Different Drying Methods. *Int. J. Food Eng.* **2018**, *14*, 1–12. [CrossRef]
19. Aboud, S.A.; Altemimi, A.B.; Al-Hilphy, A.R.S.; Yi-Chen, L.; Cacciola, F. A Comprehensive Review on Infrared Heating Applications in Food Processing. *Molecules* **2019**, *24*, 4125. [CrossRef] [PubMed]
20. Ruiz Celma, A.; Rojas, S.; Lopez-Rodríguez, F. Mathematical Modelling of Thin-Layer Infrared Drying of Wet Olive Husk. *Chem. Eng. Process. Process Intensif.* **2008**, *47*, 1810–1818. [CrossRef]
21. Calín-Sánchez, Á.; Lipan, L.; Cano-Lamadrid, M.; Kharaghani, A.; Masztalerz, K.; Carbonell-Barrachina, Á.A.; Figiel, A. Comparison of Traditional and Novel Drying Techniques and Its Effect on Quality of Fruits, Vegetables and Aromatic Herbs. *Foods* **2020**, *9*, 1261. [CrossRef] [PubMed]
22. Zheng, W.J.; Gao, J.; Wei, Z.; Zhou, J.; Chen, Y.M. Facile Fabrication of Self-Healing Carboxymethyl Cellulose Hydrogels. *Eur. Polym. J.* **2015**, *72*, 514–522. [CrossRef]
23. Uranga, J.; Nguyen, B.T.; Si, T.T.; Guerrero, P.; De la Caba, K. The Effect of Cross-Linking with Citric Acid on the Properties of Agar/Fish Gelatin Films. *Polymers* **2020**, *12*, 291. [CrossRef]
24. Capanema, N.S.V.; Mansur, A.A.P.; Mansur, H.S.; de Jesus, A.C.; Carvalho, S.M.; Chagas, P.; de Oliveira, L.C. Eco-Friendly and Biocompatible Cross-Linked Carboxymethylcellulose Hydrogels as Adsorbents for the Removal of Organic Dye Pollutants for Environmental Applications. *Environ. Technol.* **2017**, *39*, 2856–2872. [CrossRef] [PubMed]
25. Syafiuddin, A.; Fulazzaky, M.A.; Salmiati, S.; Roestamy, M.; Fulazzaky, M.; Sumeru, K.; Yusop, Z. Sticky Silver Nanoparticles and Surface Coatings of Different Textile Fabrics Stabilised by Muntingia Calabura Leaf Extract. *SN Appl. Sci.* **2020**, *2*, 1–10. [CrossRef]
26. Nadeem, H.; Naseri, M.; Shanmugam, K.; Dehghani, M.; Browne, C.; Miri, S.; Garnier, G.; Batchelor, W. An Energy Efficient Production of High Moisture Barrier Nanocellulose/Carboxymethyl Cellulose Films via Spray-Deposition Technique. *Carbohydr. Polym.* **2020**, *250*, 116911. [CrossRef] [PubMed]
27. Fu, S.; Zhang, P. Surface Coating Modified Polyglycolide (PGA) Braided Threads as Potential Thread-Embedding Materials. *Fibers Polym.* **2020**, *21*, 2401–2406. [CrossRef]
28. Wu, H.; Lei, Y.; Lu, J.; Zhu, R.; Xiao, D.; Jiao, C.; Xia, R.; Zhang, Z.; Shen, G.; Liu, Y.; et al. Effect of Citric Acid Induced Crosslinking on the Structure and Properties of Potato Starch/Chitosan Composite Films. *Food Hydrocoll.* **2019**, *97*, 105208. [CrossRef]
29. Ghorpade, V.S.; Yadav, A.V.; Dias, R.J.; Mali, K.K.; Pargaonkar, S.S.; Shinde, P.V.; Dhane, N.S. Citric Acid Crosslinked Carboxymethylcellulose-Poly(Ethylene Glycol) Hydrogel Films for Delivery of Poorly Soluble Drugs. *Int. J. Biol. Macromol.* **2018**, *118*, 783–791. [CrossRef] [PubMed]
30. Seligra, P.G.; Medina Jaramillo, C.; Famá, L.; Goyanes, S. Biodegradable and Non-Retrogradable Eco-Films Based on Starch-Glycerol with Citric Acid as Crosslinking Agent. *Carbohydr. Polym.* **2016**, *138*, 66–74. [CrossRef] [PubMed]
31. Sun, L.; Sun, J.; Chen, L.; Niu, P.; Yang, X.; Guo, Y. Preparation and Characterization of Chitosan Film Incorporated with Thinned Young Apple Polyphenols as an Active Packaging Material. *Carbohydr. Polym.* **2017**, *163*, 81–91. [CrossRef]
32. Kanafi, N.M.; Rahman, N.A.; Rosdi, N.H. Citric Acid Cross-Linking of Highly Porous Carboxymethyl Cellulose/Poly(Ethylene Oxide) Composite Hydrogel Films for Controlled Release Applications. *Mater. Today Proc.* **2019**, *7*, 721–731. [CrossRef]
33. Faghihzadeh, F.; Anaya, N.M.; Schifman, L.A. Fourier Transform Infrared Spectroscopy to Assess Molecular-Level Changes in Microorganisms Exposed to Nanoparticles. *Nanotechnol. Environ. Eng.* **2016**, *1*, 1–16. [CrossRef]
34. Coates, J. Interpretation of Infrared Spectra, A Practical Approach. *Encycl. Anal. Chem* **2006**, 10815–10837. [CrossRef]
35. Hu, Z.; Xiang, Z.; Song, T.; Lu, F. Effects of Crosslinking Degree on the Coating Properties of Arabinoxylan. *BioResources* **2019**, *14*, 70–86. [CrossRef]
36. Kimna, C.; Tamburaci, S.; Tihminlioglu, F. Novel Zein-Based Multilayer Wound Dressing Membranes with Controlled Release of Gentamicin. *J. Biomed. Mater. Res.—Part B Appl. Biomater.* **2018**, *107*, 2057–2070. [CrossRef] [PubMed]
37. Nguyen, M.K.; Kurtz, I. Quantitative Interrelationship between Gibbs-Donnan Equilibrium, Osmolality of Body Fluid Compartments, and Plasma Water Sodium Concentration. *J. Appl. Physiol.* **2006**, *100*, 1293–1300. [CrossRef]
38. Salleh, K.M.; Zakaria, S.; Gan, S.; Baharin, K.W.; Ibrahim, N.A.; Zamzamin, R. Interconnected Macropores Cryogel with Nano-Thin Crosslinked Network Regenerated Cellulose. *Int. J. Biol. Macromol.* **2020**, *148*, 11–19. [CrossRef]
39. Zhuang, L.; Zhi, X.; Du, B.; Yuan, S. Preparation of Elastic and Antibacterial Chitosan-Citric Membranes with High Oxygen Barrier Ability by in Situ Cross-Linking. *ACS Omega* **2020**, *5*, 1086–1097. [CrossRef] [PubMed]
40. Dharmalingam, K.; Anandalakshmi, R. Fabrication, Characterization and Drug Loading Efficiency of Citric Acid Crosslinked NaCMC-HPMC Hydrogel Films for Wound Healing Drug Delivery Applications. *Int. J. Biol. Macromol.* **2019**, *134*, 815–829. [CrossRef]
41. Kamble, Z.; Behera, B.K. Mechanical Properties and Water Absorption Characteristics of Composites Reinforced with Cotton Fibres Recovered from Textile Waste. *J. Eng. Fibers Fabr.* **2020**, *15*, 1–8. [CrossRef]

42. Hassan, M.A.; Tamer, T.M.; Valachová, K.; Omer, A.M.; El-Shafeey, M.; Mohy Eldin, M.S.; Šoltés, L. Antioxidant and Antibacterial Polyelectrolyte Wound Dressing Based on Chitosan/Hyaluronan/Phosphatidylcholine Dihydroquercetin. *Int. J. Biol. Macromol.* **2021**, *166*, 18–31. [CrossRef] [PubMed]
43. Tamer, T.M.; Valachová, K.; Hassan, M.A.; Omer, A.M.; El-Shafeey, M.; Mohy Eldin, M.S.; Šoltés, L. Chitosan/Hyaluronan/Edaravone Membranes for Anti-Inflammatory Wound Dressing: In Vitro and in Vivo Evaluation Studies. *Mater. Sci. Eng. C* **2018**, *90*, 227–235. [CrossRef] [PubMed]
44. Hassan, A.A.; Radwan, H.A.; Abdelaal, S.A.; Al-radadi, N.S.; Ahmed, M.K.; Shoueir, K.; Hady, M.A. Polycaprolactone Based Electrospun Matrices Loaded with Ag/Hydroxyapatite as Wound Dressings: Morphology, Cell Adhesion and Antibacterial Activity. *Int. J. Pharm.* **2021**, *596*, 1–30. [CrossRef]
45. Huang, X.; Fu, Q.; Deng, Y.; Wang, F.; Xia, B.; Chen, Z.; Chen, G. Surface Roughness of Silk Fibroin/Alginate Microspheres for Rapid Hemostasis in Vitro and in Vivo. *Carbohydr. Polym.* **2021**, *253*, 1–41. [CrossRef] [PubMed]
46. Rosqvist, E.; Niemelä, E.; Venu, A.P.; Kummala, R.; Ihalainen, P.; Toivakka, M.; Eriksson, J.; Peltonen, J. Human Dermal Fibroblast Proliferation Controlled by Surface Roughness of Two-Component Nanostructured Latex Polymer Coatings. *Colloids Surf. B Biointerfaces* **2019**, *174*, 136–144. [CrossRef] [PubMed]
47. Mazlan, N.S.N.; Zakaria, S.; Hu, C.C.; Gan, S. Effect of Coagulant and Drying Methods on Regenerated Cellulose Membrane. *AIP Conf. Proc.* **2019**, *2111*, 030007. [CrossRef]
48. Ozsoy, I.; Demirkol, A.; Mimaroglu, A.; Unal, H.; Demir, Z. The Influence of Micro- And Nano-Filler Content on the Mechanical Properties of Epoxy Composites. *Stroj. Vestn./J. Mech. Eng.* **2015**, *61*, 601–609. [CrossRef]
49. Liang, J.; Wang, R.; Chen, R. The Impact of Cross-Linking Mode on the Physical and Antimicrobial Properties of a Chitosan/Bacterial Cellulose Composite. *Polymers* **2019**, *11*, 491. [CrossRef] [PubMed]
50. Mathangadeera, R.W.; Hequet, E.F.; Kelly, B.; Dever, J.K.; Kelly, C.M. Importance of Cotton Fiber Elongation in Fiber Processing. *Ind. Crops Prod.* **2020**, *147*, 112217. [CrossRef]
51. Dalfi, H.K.; Tausif, M.; Yousaf, Z. Effect of Twist Level on the Mechanical Performance of S-Glass Yarns and Non-Crimp Cross-Ply Composites. *J. Ind. Text.* **2021**, 1528083720987206. [CrossRef]
52. Tania, I.S.; Ali, M.; Islam, Z. Solaiman Development of Antimicrobial Activity and Mechanical Performances of Cotton Fabric Treated with Silver Nano Particles (AgNPs). *AIP Conf. Proc.* **2019**, *2121*, 150003. [CrossRef]
53. Liang, J.Z.; Wang, F. Tensile Properties of Polyformaldehyde Blends and Nanocomposites. *J. Polym. Eng.* **2015**, *35*, 417–422. [CrossRef]
54. Pereira, P.H.F.; Waldron, K.W.; Wilson, D.R.; Cunha, A.P.; Brito, E.S.d.; Rodrigues, T.H.S.; Rosa, M.F.; Azeredo, H.M.C. Wheat Straw Hemicelluloses Added with Cellulose Nanocrystals and Citric Acid. Effect on Film Physical Properties. *Carbohydr. Polym.* **2017**, *164*, 317–324. [CrossRef]
55. Sivaranjana, P.; Nagarajan, E.R.; Rajini, N.; Jawaid, M.; Rajulu, A.V. Cellulose Nanocomposite Films with in Situ Generated Silver Nanoparticles Using Cassia Alata Leaf Extract as a Reducing Agent. *Int. J. Biol. Macromol.* **2017**, *99*, 223–232. [CrossRef]
56. Kopacic, S.; Walzl, A.; Hirn, U.; Zankel, A.; Kniely, R.; Leitner, E.; Bauer, W. Application of Industrially Produced Chitosan in the Surface Treatment of Fibre-Based Material: Effect of Drying Method and Number of Coating Layers on Mechanical and Barrier Properties. *Polymers* **2018**, *10*, 1232. [CrossRef] [PubMed]
57. Ismail, A.S.; Jawaid, M.; Naveen, J. Void Content, Tensile, Vibration and Acoustic Properties of Kenaf/Bamboo Fiber Reinforced Epoxy Hybrid Composites. *Materials* **2019**, *12*, 2094. [CrossRef] [PubMed]
58. Salleh, K.M.; Zakaria, S.; Sajab, M.S.; Gan, S.; Chia, C.H.; Jaafar, S.N.S.; Amran, U.A. Chemically Crosslinked Hydrogel and Its Driving Force towards Superabsorbent Behaviour. *Int. J. Biol. Macromol.* **2018**, *118*, 1422–1430. [CrossRef] [PubMed]
59. Plotka, M.; Kapusta, M.; Dorawa, S.; Kaczorowska, A.K.; Kaczorowski, T. Ts2631 Endolysin from the Extremophilic Thermus Scotoductus Bacteriophage VB_Tsc2631 as an Antimicrobial Agent against Gram-Negative Multidrug-Resistant Bacteria. *Viruses* **2019**, *11*, 657. [CrossRef] [PubMed]
60. Eliuz, E.A.E. Antimicrobial Activity of Citric Acid against Escherichia Coli, Staphylococcus Aureus and Candida Albicans as a Sanitizer Agent. *Eurasian J. For. Sci.* **2020**, *8*, 295–301. [CrossRef]
61. Al-Rousan, W.M.; Olaimat, A.N.; Osaili, T.M.; Al-Nabulsi, A.A.; Ajo, R.Y.; Holley, R.A. Use of Acetic and Citric Acids to Inhibit Escherichia Coli O157:H7, Salmonella Typhimurium and Staphylococcus Aureus in Tabbouleh Salad. *Food Microbiol.* **2018**, *73*, 61–66. [CrossRef] [PubMed]
62. Khan, S.; Shah, Z.H.; Riaz, S.; Ahmad, N.; Islam, S.; Raza, M.A.; Naseem, S. Antimicrobial Activity of Citric Acid Functionalized Iron Oxide Nanoparticles –Superparamagnetic Effect. *Ceram. Int.* **2020**, *46*, 10942–10951. [CrossRef]
63. Epand, R.M.; Walker, C.; Epand, R.F.; Magarvey, N.A. Molecular Mechanisms of Membrane Targeting Antibiotics. *Biochim. Biophys. Acta—Biomembr.* **2016**, *1858*, 980–987. [CrossRef] [PubMed]
64. Park, K.M.; Yoon, S.; Choi, T.; Kim, H.J.; Park, K.J.; Koo, M. The Bactericidal Effect of a Combination of Food-Grade Compounds and Their Application as Alternative Antibacterial Agents for Food Contact Surfaces. *Foods* **2020**, *9*, 59. [CrossRef] [PubMed]

polymers

MDPI

Article

Gellan Gum Hydrogels Filled Edible Oil Microemulsion for Biomedical Materials: Phase Diagram, Mechanical Behavior, and In Vivo Studies

Muhammad Zulhelmi Muktar [1], Muhammad Ameerul Amin Bakar [2], Khairul Anuar Mat Amin [1,*], Laili Che Rose [1], Wan Iryani Wan Ismail [1], Mohd Hasmizam Razali [1], Saiful Izwan Abd Razak [3] and Marc in het Panhuis [4,*]

1 Faculty of Science and Marine Environment, Universiti Malaysia Terengganu, Kuala Nerus 21030, Terengganu, Malaysia; gsk2761@student.umt.edu.my (M.Z.M.); laili@umt.edu.my (L.C.R.); waniryani@umt.edu.my (W.I.W.I.); mdhasmizam@umt.edu.my (M.H.R.)

2 Pharmaniaga LifeScience Sdn. Bhd., Lot 7, Jalan PPU 3, Taman Perindustrian Puchong Utama, Puchong 47100, Selangor, Malaysia; ameeramin110@gmail.com

3 BioInspired Device and Tissue Engineering Research Group, Faculty of Engineering, School of Biomedical Engineering and Health Sciences, Universiti Teknologi Malaysia, Skudai 81300, Johor, Malaysia; saifulizwan@utm.my

4 Faculty Science, Medicine and Health, School of Chemistry and Molecular Bioscience, University of Wollongong, Wollongong, NSW 2500, Australia

* Correspondence: kerol@umt.edu.my (K.A.M.A.); panhuis@uow.edu.au (M.i.h.P.)

Citation: Muktar, M.Z.; Bakar, M.A.A.; Amin, K.A.M.; Che Rose, L.; Wan Ismail, W.I.; Razali, M.H.; Abd Razak, S.I.; in het Panhuis, M. Gellan Gum Hydrogels Filled Edible Oil Microemulsion for Biomedical Materials: Phase Diagram, Mechanical Behavior, and In Vivo Studies. *Polymers* **2021**, *13*, 3281. https://doi.org/10.3390/polym13193281

Academic Editor: Domenico Acierno

Received: 24 August 2021
Accepted: 15 September 2021
Published: 26 September 2021

Abstract: The demand for wound care products, especially advanced and active wound care products is huge. In this study, gellan gum (GG) and virgin coconut oil (VCO) were utilized to develop microemulsion-based hydrogel for wound dressing materials. A ternary phase diagram was constructed to obtain an optimized ratio of VCO, water, and surfactant to produce VCO microemulsion. The VCO microemulsion was incorporated into gellan gum (GG) hydrogel (GVCO) and their chemical interaction, mechanical performance, physical properties, and thermal behavior were examined. The stress-at-break (σ) and Young's modulus (YM) of GVCO hydrogel films were increased along with thermal behavior with the inclusion of VCO microemulsion. The swelling degree of GVCO hydrogel decreased as the VCO microemulsion increased and the water vapor transmission rate of GVCO hydrogels was comparable to commercial dressing in the range of 332–391 g m^{-2} d^{-1}. The qualitative antibacterial activities do not show any inhibition against Gram-negative (*Escherichia coli* and *Klebsiella pneumoniae*) and Gram-positive (*Staphylococcus aureus* and *Bacillus subtilis*) bacteria. In vivo studies on Sprague–Dawley rats show the wound contraction of GVCO hydrogel is best (95 $\pm$ 2%) after the 14th day compared to a commercial dressing of Smith and Nephew Opsite post-op waterproof dressing, and this result is supported by the ultrasound images of wound skin and histological evaluation of the wound. The findings suggest that GVCO hydrogel has the potential to be developed as a biomedical material.

Keywords: gellan gum; virgin coconut oil; hydrogels; biomaterials; wound dressing

1. Introduction

There are an overwhelming number of wound dressings available in the market. The high demand for wound dressing is due to the increasing number of wound cases recorded. It is reported that the treatment costs for chronic wounds are substantial and are estimated to account for approximately 1–3% of the total healthcare expenditure in developed countries [1,2]. For example, Wales was estimated to have a chronic wound prevalence of 6% in the year 2012–2013, corresponding to 5.5% of National Health Service expenditure [1], and in the United Kingdom as a whole, the cost associated with wound management was estimated to be GBP 4.5 and GBP 5.1 billion in 2012 [3]. In the United

States, it has been reported that over 6.5 million patients with wounds cost the health care system USD 25 billion annually [4]. The point of using wound dressing is to cover the wound from infection and provide appropriate conditions to enhance the healing process of the wounds. However, there is a concern about the usage of hazardous chemicals in the development of wound dressing such as various inorganic nanoparticles. Inappropriate chemicals in wound dressing may lead to other diseases such as skin cancer [5]. The usage of biopolymer and natural products for a better and safer wound dressing is needed. Hence, gellan gum produced by the Gram-negative bacterium of Pseudomonas elodea is chosen due to its biocompatibility and unique properties [6,7].

Gellan gum (GG) is nontoxic, biocompatible, and biodegradable, and the resulting hydrogel is transparent and stable [8]. GG hydrogels are three-dimensional crosslinked polymer networks, a result of transformation from a disordered single coiled structure at high temperature (80 °C) to a double helix and bonded by internal hydrogen bonding between D-glucoronate and D-glucose C residues upon cooling between 30 to 50 °C [9]. In a swollen state these are soft and elastic, resembling the living tissue and exhibiting excellent biocompatibility. The unique properties of hydrogels lead to the wide use of these biomaterials in different fields, including pharmaceutical and biomedical [10,11]. Gellan gum has been proven by the United States Food and Drug Administration (US FDA) and the European Union (EU) to be safely utilized in the food industry. Gellan gum has been studied as matrices to repair and regenerate a wide variety of tissues and organs [12]. The material has also been used as scaffold materials for the application of tissue engineering [6], in the development of wound dressing materials [13], as a vehicle for drug delivery [14], and in eye drops [15]. Gellan gum also has shown good compatibility against various live cells such as mouse fibroblast (L929 cell line) [13], mouse fibroblast cell (3T3) [16], human fetal osteoblast (HFOBs 1.19) [17], human skin fibroblast (CRL2522) [6], and human nasal cartilage [18].

Virgin coconut oil (VCO) is essentially colorless, free from rancidity, and endowed with natural antioxidant and vitamin E that prevents the peroxidation reaction. VCO mainly consists of medium-chain triglycerides (MCT) and differs from animal fats that consist of long-chain saturated fatty acids, which is the one main risk factor for cardiac compilation. Based on studies, VCO has been reported to have the potential in promoting the healing process [19]. The oil has been applied in treating wounds in young rats and healed faster by decreasing time for complete epithelization, and results in a significant increase of collagen production, which indicates higher collagen crosslinking. VCO also showed a significant effect in reducing inflammation in acute and chronic inflammation on ethyl phenylpropionate-induced ear edema in rats [20]. A few studies have been reporting the use of biopolymers with essential oil to produce a dressing material [21]. Gellan gum hydrogel films with lavender/tea tree oil showed 98% wound contraction in rats after ten days of treatment and histological images displayed completely healed epidermis [21]. Another study used poly(lactic acid) (PLA) polymer and babassu oil and reported that this material provides a good option for use as wound dressings—films showed a recommended value of the water vapor transmission rate (WVTR), maintained a humid environment above the wound, had good cytotoxicity on normal human keratinocytes (HaCaT), stimulated the keratinocytes migration, and inhibited Pseudomonas aeruginosa growth [22].

Based on the demands to produce a more efficient and safer wound dressing, this study optimizes the ratios of VCO, water, and surfactant by developing a ternary phase diagram and producing a VCO microemulsion. The optimum concentration of VCO microemulsion is selected for incorporation into GG solution and characterized for their chemical interaction, mechanical performance, physical properties, and thermal behavior. Furthermore, the qualitative in vitro antibacterial activities were examined against two Gram-negative (*Escherichia coli* and *Klebsiella pneumoniae*) and two Gram-positive (*Staphylococcus aureus* and *Bacillus subtilis*) bacteria. The in vivo studies were carried out to study the healing properties of the samples on Sprague–Dawley rats, observing the ultrasound images of wound skin and the histological evaluation after the 14th day of post-wound.

2. Materials and Chemicals

2.1. Materials

Low-acyl gellan gum (GG, batch no: 5C1574A) was obtained from CP Kelco, Atlanta, GA, USA. Glycerin (product number—G2289), anhydrous calcium chloride, $CaCl_2$ (product number—C5670), Tween 80 (product number—P1754), and Triton X-100 (product number—T9284) were obtained from Sigma Aldrich, St. Louis, MO, USA. Virgin coconut oil (product number—VCO0216) was obtained from Phyto Biznet Sdn Bhd, Skudai, Johor, Malaysia. All materials were used as received without further purification.

2.2. Construction of Phase Diagrams

Phase diagrams were constructed by mixing two of the components (VCO and water) and titrated using surfactant (Tween 80—Method I or TritonX-100—Method II) as a third component. The surfactant, i.e., Tween 80, was added into the VCO containing distilled water at different ratios (Table 1) and was vortexed (Vortex 3, Eppendorf, Germany) for 3 min. The mixtures were then centrifuged (Minispin Eppendorf, Germany) at 400 rpm for 10 min and later placed in a water bath at room temperature (26 ± 2 °C) for 24 h to record the stability of the mixtures produced. The same procedure was carried out with TritonX-100 (Method II) surfactant at the specific ratios (Table 1).

Table 1. Ratios of VCO, water, and surfactants (Tween 80—Method I and TritonX-100—Method II) to develop a ternary phase diagram and produce a stable VCO microemulsion.

	Method I				Method II			
Ratio VCO: Water	**VCO (%)**	**Water (%)**	**Tween 80 (%)**	**Total (%)**	**VCO (%)**	**Water (%)**	**TritonX-10 (%)**	**Total (%)**
90:10	27.45	3.06	69.49	100	12.54	1.39	86.07	100
80:20	35.52	8.88	55.59	100	13.41	3.35	83.24	100
70:30	23.98	10.29	65.73	100	11.46	4.91	83.63	100
60:40	21.38	14.25	64.37	100	12.23	8.15	79.62	100
50:50	16.43	16.43	67.14	100	9.66	9.66	80.68	100
40:60	17.27	25.91	56.84	100	7.82	11.74	80.44	100
30:70	9.15	21.35	69.50	100	6.57	15.34	78.09	100
20:80	6.96	27.86	65.18	100	3.88	15.55	80.57	100
10:90	4.03	36.23	59.74	100	2.29	16.80	80.91	100

2.3. Preparation of GVCO Hydrogel

The gellan gum (GG) solutions were prepared by dissolving 2% (w/v) gellan gum in deionized water (18 MΩ) with 50% (w/w) glycerin under continuous stirring at 500 rpm for 2 h at 80 °C. VCO microemulsions were prepared by mixing the VCO: Water with Tween 80 surfactant at a specific percentage, as shown in Table 1. For example, to produce the VCO microemulsion of VCO60, 21.38% of VCO was mixed with 14.25% of water with an addition of 64.37% of Tween 80 (Table 1). The same process was carried out for VCO70 and VCO80. To produce GVCO60 hydrogel, 5% (v/v) of the VCO60 microemulsions were added into the GG solution and stirred for 20 min at 80 °C. The methods were repeated for VCO70 and VCO80 microemulsions. The mixtures were then poured into a Petri dish and allowed to cool at room temperature for 24 h before use for characterization. The GG solution with VCO60 was then known as GVCO60 hydrogels, and the same naming was applied for VCO70 and VCO80 hydrogels.

2.4. Characterisation of the Sample

2.4.1. Ultraviolet–Visible Spectroscopy

Ultraviolet-visible (UV–Vis) absorption and transmission spectra of solutions and hydrogels were performed using a spectrophotometer (Varian, Cary 50 UV–Vis NIR) with data interval = 5 nm, scan speed = 24,000 nm/min, and wavelength range 500–800 nm. The UV–Vis transmittance was conducted by attaching the hydrogel to the cuvette surface.

2.4.2. ATR–FTIR Spectroscopy

ATR–FTIR spectra were collected using a Perkin Elmer Spectrum 100 FTIR spectrophotometer with a PIKE Miracle ATR accessory (single-bounce beam path, 45° incident angle, 16 scans, 4 cm^{-1} resolution). An advanced ATR correction was applied to all spectra in the region from 4000 to 600 cm^{-1}.

2.4.3. Mechanical Properties

Mechanical properties of hydrogel films were performed using an Instron Universal Testing Machine (model 3366) with a load capacity and cross-speed according to ASTM standard method D882 (ASTM, 2001). Each sample was cut to 2.0 × 2.0 cm^2 for stress–strain measurements. The thicknesses of hydrogel films were measured using a handheld micrometer (Mitutoyo Corporation, Mitutoyo, Japan). Stress-at-break (σ), strain-at-break (γ), and Young's modulus (E) were recorded and a minimum of three independent measurements were obtained per sample.

2.4.4. Swelling Properties

The swelling properties were determined according to the ASTM Standard Test Methods for One-Dimensional Swell (D4546-08). The swelling was measured by weighing dried films (W_{dry}) before immersion into 50 mL phosphate buffer solutions of pH 7.2 at 37 ± 0.5 °C in a water bath. The hydrogels were removed after 24 h, gently wiped with a tissue to expel the surface water, and weighed (W_{wet}). Swelling degree (%) was determined according to Equation (1):

$$\text{Swelling degree (\%)} = (W_{wet} - W_{dry})/W_{dry} \times 100 \quad (1)$$

where W_{dry} and W_{wet} are the initial weight and final weight, respectively. A minimum of three independent measurements was obtained per sample.

2.4.5. Water Vapor Transmission Rate (WVTR)

The water vapor transmission rate (WVTR) was measured following a modified ASTM International standard method ASTM E96-95. Each hydrogel was fixed on the circular opening of a permeation bottle with an effective transfer area (A) of 1.33 cm^2. The permeation bottle was placed in the temperature–humidity chamber at 37 °C and 50 ± 5% relative humidity. The equilibrium moisture penetration was determined by weighing the bottles at 0 and 24 h. The WVTR was calculated according to Equation (2):

$$\text{WVTR} = (m/\Delta T)/A. \quad (2)$$

where $m/\Delta T$ is the amount of water gain per unit time of transfer and A is the area exposed to water transfer (m^2).

2.4.6. Thermogravimetric Analysis

Thermogravimetric analyses were performed on a Pyris 6, Perkin-Elmer-TGA6. Hydrogel samples were analyzed in platinum pans at a heating rate of 10 °C/min to 275 °C in an atmosphere of N_2 atmosphere at a flow rate of 50 mL/min. The sample used was approximately 10 mg.

2.4.7. Scanning Electron Microscopy (SEM)

The cross-section morphology of the hydrogels was acquired using a JOEL JSM 6360 LA electron microscope. Scanning electron microscopy (SEM) images of cross-sections were obtained by freeze-dried technique. Samples were freeze-dried in liquid nitrogen (−160 °C) and fractured at −150 °C at the middle of hydrogels. It was then coated with Auto Fine Coats (JFC-1600) before microscopic observation.

2.5. Antibacterial Study

2.5.1. Culture Conditions

Gram-positive (*Staphylococcus aureus* and *Bacillus subtilis*) and Gram-negative (*Escherichia coli* and *Klebsiella Pneumoniae*) microbes were used for antibacterial assays. The standard growth medium Mueller–Hinton (MH, Difco, Sparks, MD, USA) agar was used for the growth of both bacterial types and prepared by sterilization using an autoclave (15 min, 120 °C). These two species of microbes were evenly spread on the solid MH agar and incubated aerobically at 37 °C for 24 h. The bacterial suspension was measured by using Spectrophotometer BiomerieuxDensicheck Plus at 600 nm to obtain optical density at 0.5.

2.5.2. Qualitative Study

The Gram-positive and Gram-negative bacteria suspensions were evenly spread on the solid MH agar in the sterile Petri plates. Using a sterile cotton swab, both bacteria were swabbed over the surface of the agar plates and dried in a laminar flow air chamber. The hydrogel sample was gently pressed on the agar with bacteria and incubated at 37 °C for 24 h in triplicates. The presence of any clear zone around the disc on the MH agar was recorded as an indication of inhibition against the bacteria.

2.6. In Vivo Studies

2.6.1. In Vivo Wound Healing Experiments

Animals

In this study, a total number of 20 six-week-old female Sprague–Dawley rats with a range of body weight from 200–250 g were used. They were randomly divided into three experimental groups of five rats each. The animals were acclimatized to the laboratory conditions for one week before the onset of the experiment. All rats were individually caged with a 12-h light/dark cycle, given adequate commercial pellets and water ad libitum throughout the study. All animal experiments were carried out under protocols approved by the Animal Ethics Committee (AEC)—UMT/JKEPHMK/2020/48, Universiti Malaysia Terengganu.

Establishment of Wound Skin

Rats were anaesthetized using an intraperitoneal (i.p.) injection of ketamine (90 mg/kg) and xylazine (10 mg/kg). The dorsal skin was prepared by removing the hair with a razor blade and the surgical area was disinfected with 70% ethanol. Since the shaving procedure produced marked edema of the skin, the prepared rats were left for twenty-four hours before the wound was inflicted. After the rats were anesthetized with a combination of ketamine and xylazine via i.p., a full-thickness wound was created by using an 8-mm sterile skin biopsy punch. Each rat received two full-thickness wounds at their back dorsal.

Treatment of Hydrogel

All wounds in the treatment groups were dressed with GG and GVCO80 hydrogels, and followed by Opsite post-op waterproof film dressing (Smith and Nephew, Hull, England) as the secondary dressing. The dressings were then held in place with gauze to give mechanical protection to the dressing. The changing of dressing was done every 3 days to minimize the infection to the wound site. The Opsite film dressing acted as a positive control for comparison to the other treatments. The treated wound with GG dressing was considered a negative control. The rat's wound was photographed using a 13.1-megapixel Sony camera for evaluation of wound closure with the actual measurement.

Macroscopic Observation of Wound

The wound measurement of the size taken at the time of biopsy was used to calculate the percent of wound contraction using equation:

$$\%\text{ wound contraction } = \frac{W_0 - W_t}{W_0} \times 100 \quad (3)$$

where W_0 is the original wound area and W_t is the wound area on the selected day after the biopsy. The measurements of wound size were taken on days 2, 4, 7, 11, and 14 consecutively throughout the study. The wound area was measured by placing the 1 mm^2 graph over the wound pictures. The squares were counted, and the area was recorded. The wound area was accessed by the same blinded observer.

2.6.2. Ultrasound Imaging

The wound area also was analyzed by using real-time high-resolution 20 MHz ultrasound imaging equipment (Dermalab Combo, Cortex, Denmark) skin analyzer to produce images representing the cross-section of the wound skin. A standard echographic gel was applied and used as a medium between the probe and wound skin surface. The images produced were recorded.

2.6.3. Histological Examination

Rats were euthanized on day 14 and skin samples that contained the wound area were taken for histological study. The skin samples were fixed with 10% buffered formalin for 24 h. The samples were embedded in paraffin and cut into 6 mm-thick sections for the middle part of the wounds. The sections were subsequently stained with hematoxylin and eosin (H & E) staining procedure. The H & E slides were visualized using a light microscope at 20x magnification.

Statistical Analysis

All data are presented as the mean ± standard deviation (SD). The data were processed by two-way ANOVA using statistical software analysis SPSS (version 20). The p-value < 0.05 was considered statistically significant.

3. Results and Discussion

3.1. Formulation of Stabilizing VCO Microemulsion

In general, water and oil were separated due to their high interfacial tension. Water is immiscible with virgin coconut oil (VCO) due to its high interfacial tension, which is typical for oils with water. The interfacial tension was reduced by introducing surfactant, which allowed the microemulsion polymerization process to take place. Nonionic surfactants, Triton X-100 and Tween 80, were used in preparing a stabilized VCO microemulsion through a ternary phase diagram. The optimum ratio from the ternary phase diagram was chosen in preparing the gellan gum–virgin coconut oil hydrogels (GVCO). The microemulsion has a characteristic of a clear solution, stable at room temperature and forming a one layer solution. The pseudo ternary phase diagrams consisting of virgin coconut oil microemulsion, distilled water, and surfactant with different hydrophilic–lipophilic balance (HLB) values were constructed using the surfactant titration method. Nine ratios of VCO: Water were constructed, i.e., 90:10, 80:20, 70:30, 60:40, 50:50, 40:60, 30:70, 20:80, and 10:90 (w/w). Two types of nonionic surfactants were used, which were Tween 80 and Triton X-100 (Table 1). The mixture was observed using visual inspection after each addition of surfactant to the VCO and water. The samples were identified as microemulsions when they appeared transparent, and the liquid easily flowed. The results were plotted on a triangular graph as a ternary phase diagram, as shown in Figure 1.

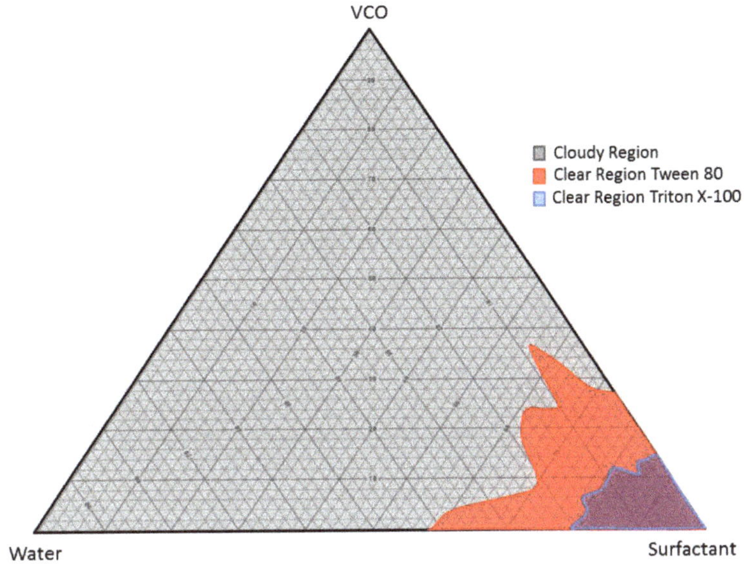

Figure 1. The ternary phase diagram of VCO microemulsion by using Tween 80 and TritonX-100 surfactants.

The ternary phase diagram shows that different amounts of VCO, water, and Tween 80 can produce a stable oil-in-water microemulsion. The ratio of VCO: Water required a different amount of Tween 80 in the range from 55.59% to 69.50% to achieve a stable phase of microemulsions (Table 1). The ratios of 80:20 needed 55.59% of Tween 80 compared to the ratio of 70:30 and 60:40, which only needed 65.73% and 64.37%, respectively (Table 1). The volume of Triton X-100 required to produce a stable formulation in VCO: Water was higher than in Tween 80. The Triton X-100 amount ranged from 78.09% to 86.07% to produce stable microemulsions (Table 1). The amount of Triton X-100 was 83.24%, 83.63%, and 79.62% for ratios 80:20, 70:30, and 60:40, respectively. The formulation of VCO:Water: Tween 80 microemulsion was adopted due to the large area of microemulsion stability in the ternary phase diagram and lower amount of surfactant needed to form a stable mixture, compared to Triton X-100. Moreover, Tweens were reported to have been used widely in food, cosmetics, and pharmaceutical application due to minimal toxicity and low cost [23,24].

3.2. Physical Properties of the Hydrogels

The addition of VCO microemulsion to the gellan gum does not change the physical appearance of transparent free-standing GG hydrogel (Figure 2). The GVCO hydrogels at every concentration were transparent but becoming cloudy due to the increased amount of VCO microemulsion and supported by the transmittance result (Figure 2e). The transmittance for pure GG hydrogels was ≈ 99% compared to GVCO60 ≈ 85%, GVCO70 ≈ 70%, and GVCO80 ≈ 65% at λ = 700 nm. The transparent behavior of the GVCO hydrogel gives an advantage for the materials to be used as wound dressing material, and the healing process could be monitored with ease.

3.3. Fourier Transform Infrared Spectroscopy (ATR–FTIR)

ATR–FTIR spectra of the GVCO hydrogels confirm the presence of characteristic peaks of GG and VCO (Figure 3). The peaks at 2941 and 2895 cm^{-1} are related to the saturated alkyl and the carbonyl groups of the fatty acids in VCO, respectively [25]. Peaks appearing at 1758, 1470, and 1419 cm^{-1} were from the stretching of carboxylic (C=O), bending of methylene (CH_2), and bending of methyl (CH_3) of VCO. The 1182 and 1142 cm^{-1} peaks resulted from the stretching of ester C-O [26].

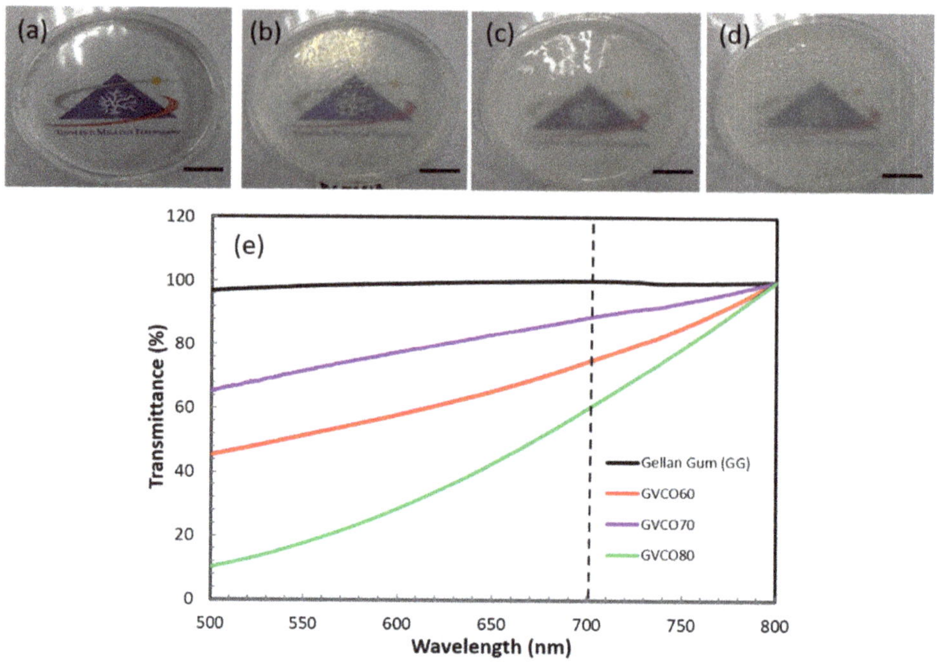

Figure 2. The physical appearance of (**a**) GG hydrogel, (**b**) GVCO60 (**c**) GVCO70, (**d**) GVCO80 hydrogels, and (**e**) transmittance of GG and GG–VCO hydrogels.

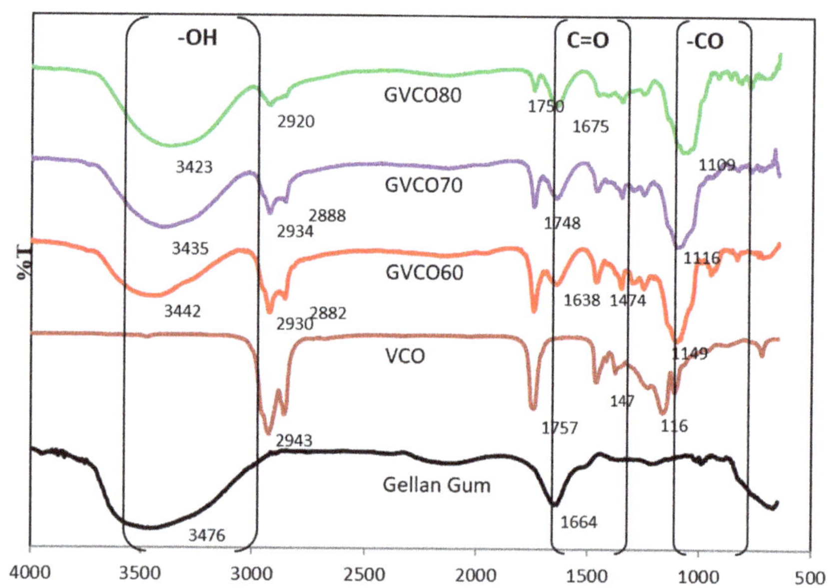

Figure 3. ATR spectra of pure GG hydrogels, VCO, and GVCO hydrogels at different concentrations.

GG has a broad band at 3476 cm^{-1}, which is a typical region for stretching of the O-H group (3000–3500 cm^{-1}) [27]. A new distinguishing peak observed at 1638–1675 cm^{-1} in GVCO hydrogels correlates to the shifting of -C-H (-CH_2) bending scissoring of VCO and C=C due to the appearance of phenolic content and stretching of that aromatic ring [28]. The stretching alkyl signal in VCO microemulsion at 2943 cm^{-1} shifted to 2930–2920 cm^{-1} in GVCO hydrogels confirming the hydrogen bonding interaction in GVCO blends [28].

A significant change of enlargement peak of GVCO hydrogel can also be observed around 1149–1109 cm^{-1}, due to an increased stretching vibration of esters group of gellan gum and VCO. The ATR–FTIR spectra of GVCO show all prominent peaks of VCO with significant changes in variation and shifting of transmittance intensities, indicating that GG and VCO microemulsion interact to produce a stable network hydrogel.

3.4. Mechanical Performances

Mechanical performance is crucial to evaluate and understand the strength of the materials. Table 2 shows the tensile properties of GVCO hydrogels at the different ratios on the stress-at-break (σ), strain-at-break (γ), and Young's modulus (YM). Free-standing gellan gum (GG) hydrogel was brittle and almost impossible for use in biomedical applications such as a wound dressing material. To overcome this problem, VCO microemulsions were incorporated into GG as a plasticizer and to improve the flexibility of the materials. The incorporation of VCO microemulsion into GG caused the hydrogels to become durable, flexible, and easy to handle due to an increase in stress-at-break (σ) and Young's modulus (YM), but decreased slightly for strain-at-break (γ) values (Figure 4a). The σ and YM values of GG hydrogel were 4 $\pm$ 0.2 and 85 $\pm$ 7 kPa, respectively, and the addition of VCO microemulsion (GVCO80) increased both values to 11 $\pm$ 0.3 and 259 $\pm$ 7 kPa, respectively (Table 2). Although the γ value of the GVCO60 decreased to 7.9 $\pm$ 0.2% from 10.7 $\pm$ 0.4% (GG hydrogel), the addition of higher content of VCO microemulsion resulted in the γ value increasing to 9.3 $\pm$ 0.4% compared to GVCO60 hydrogel films. Thus, the plasticizer effect of the VCO microemulsion in GCVO hydrogel films was proven.

Table 2. Summary of the stress-at-break (σ), strain-at-break (γ), Young's modulus (YM), swelling degree, and water vapor transmission rate (WVTR) of gellan gum (GG) and GVCO hydrogels at different ratios. Control for WVTR is a test without hydrogel film.

	Stress (σ) (kPa)	**Strain (γ) (%)**	**Modulus (YM) (kPa)**	**Swelling Degree (%)**	**WVTR (g m^{-2} d^{-1})**
Control	-	-	-	-	1547 $\pm$ 32
GG	4 $\pm$ 0.2	10.7 $\pm$ 0.4	85 $\pm$ 7	3 $\pm$ 0.6	964 $\pm$ 47
GVCO60	6 $\pm$ 0.1	7.9 $\pm$ 0.2	134 $\pm$ 6	12 $\pm$ 4.0	391 $\pm$ 13
GVCO70	7 $\pm$ 0.4	8.5 $\pm$ 0.2	175 $\pm$ 8	9 $\pm$ 1.5	344 $\pm$ 23
GVCO80	11 $\pm$ 0.3	9.3 $\pm$ 0.4	259 $\pm$ 7	6 $\pm$ 1.5	332 $\pm$ 11

The mechanism of GG hydrogel formation is closely related to the conformational transition from the coil to helix structures [13]. Normally, GG hydrogels are produced by physical crosslinking methods induced by temperature variation or by the presence of divalent cations [29]. Normal hydrogels are usually spontaneously formed by weak secondary forces such as hydrogen bonding, van der Waals interactions, and ionic bonding [30]. In an aqueous solution at high temperatures ($\approx$60 °C), the gellan gum chain is in a disordered single-coil state. Upon cooling from 70 to 30 °C, the gellan solution promotes the formation of double helices stabilized by internal hydrogen bonding [31]. This facilitates the tight packaging of gellan gum chains, resulting in fragile hydrogels. However, the addition of VCO microemulsion that contains lauric acid (C12 $\approx$ 78%) and a hydroxyl group is responsible for promoting the formation of hydrogen bonds between gellan gum and VCO microemulsion [28]. This bonding replaces the hydrogen bonds between gellan gum chains and thus decreases the intermolecular bonds along polymer chains and improves strain-at-break [6] of the hydrogels regarding the content of VCO microemulsion. The bonding is expected to increase the stress-at-break values and give more strength to the hydrogels. Another factor that was contributing to improving flexibility of the GVCO hydrogels could be due the polymerization process. The packed and solid structure of GG and GVCO hydrogels was observed in the cross-section of hydrogels. This shows that VCO microemulsion and GG were homogenous, and no separation within the mixture occurred.

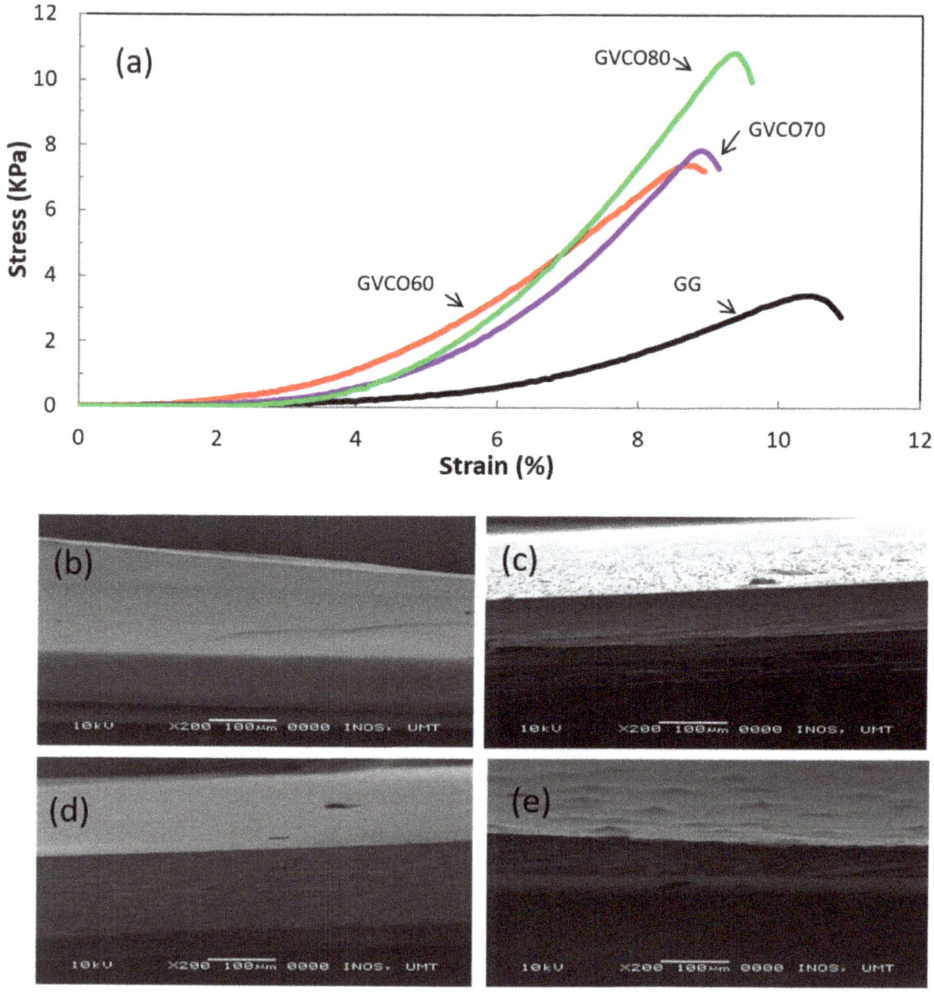

Figure 4. (**a**) Typical stress–strain curves of GG and GVCO hydrogels, (**b**–**e**) scanning electron microscopy images of the cross-sectional area of (**b**) pure gellan gum (GG) hydrogels, (**c**) GVCO60, (**d**) GVCO70, and (**e**) GVCO80 hydrogels.

3.5. Swelling Ratio, and Water Vapor Transmission Rate (WVTR)

The swelling degree of GVCO hydrogels is summarized in Table 2. The standard buffer solution with pH $\approx$ 7.2 was used to mimic human body fluid. At pH $\approx$ 7.2, the number of ionic groups is highest in the solution. The GVCO60 absorbed the maximum swelling rates $\approx 12 \pm 4$, which is an increment of four-fold more than GG hydrogel. The increase of swelling is expected due to the reaction of –OH groups of VCO with the –OH and –COOH of the polymeric chains [32]. These functional groups can relieve the entanglement of polymeric chains, thus weakening the hydrogen bonding among the hydrophilic groups. This decreases the degree of physical crosslinking and therefore improves the water uptake of the hydrogels. However, the swelling degree for GVCO70 and GVCO80 hydrogels decreased slightly upon adding a higher concentration of VCO microemulsion, 9 $\pm$ 1.5% and 6 $\pm$ 1.5%, respectively. The other reason could be due to the hydrophobicity property of VCO itself; the greater the amount of VCO, the greater water resistance of the hydrogels expected.

The water vapor transmission rate (WVTR) of control and GVCO hydrogels are in the range of 332–1547 g m^{-2} d^{-1} (Table 2). The value decreased upon the addition of VCO microemulsion that can be expected due to tight packaging of VCO and GG blends, which then interrupted the water vapor transmission rate. The tight packaging of GG and VCO microemulsion resulted in reducing the surface tension and increasing the stability of water retention in GVCO hydrogels. However, these values remain within the range of WVTR values (90–2893 g m^{-2} d^{-1}), as reported for eight commercially available synthetic wound dressings [33].

3.6. Thermal Analysis

Thermogravimetric analysis (TGA) was conducted to study the thermal stability of GG and GVCO hydrogels. The thermogram and derivative thermogram of GG and GVCO hydrogels are presented in Figure 5. The degradation below $T \leq 100$ °C, which occurred for GG hydrogels, is common due to evaporation of moisture in the hydrogels [6]. The degradation of GG and GVCO hydrogels shows a similar trend, in which the degradation occurred at the onset temperature $\approx$ 54 °C, and the offset temperature at $\approx$ 160 °C (Table 3). However, there are differences in the peak of GVCO hydrogels at $\approx$ 130 °C. The GVCO80 hydrogels (containing the highest concentration of VCO) show three small peaks at $\approx$ 130 °C and are assumed due to a few degradation processes of polymer materials. The GVCO60 and GVCO70 hydrogels have two peaks (at $\approx$ 130 °C), and a broader peak at the center (Figure 5). The appearance of these small peaks at $\approx$ 130 °C shows that the addition of VCO exhibits a better outcome in increasing the thermal stability of hydrogels due to a certain degree of interaction between GG and VCO. The weight loss of the GVCO hydrogels is further supported the thermal behavior of the hydrogels, in which the weight loss of GVCO60, GVCO70, and GVCO80 decreased to 86%, 89%, and 88%, respectively, compared to GG hydrogel at 95% (Table 3). Decreased weight loss reflects the increases in the thermal behavior of a material [34]. Similar phenomena have been observed for other oil-based materials [35,36].

Table 3. Thermal gravimetric properties of gellan gum (GG) and GVCO hydrogels.

Sample	T_o (°C)	Temperature Offset (°C)	Weight Loss (%)
GG	54	157	95
GVCO60	48	165	86
GVCO70	53	155	89
GVCO80	50	157	88

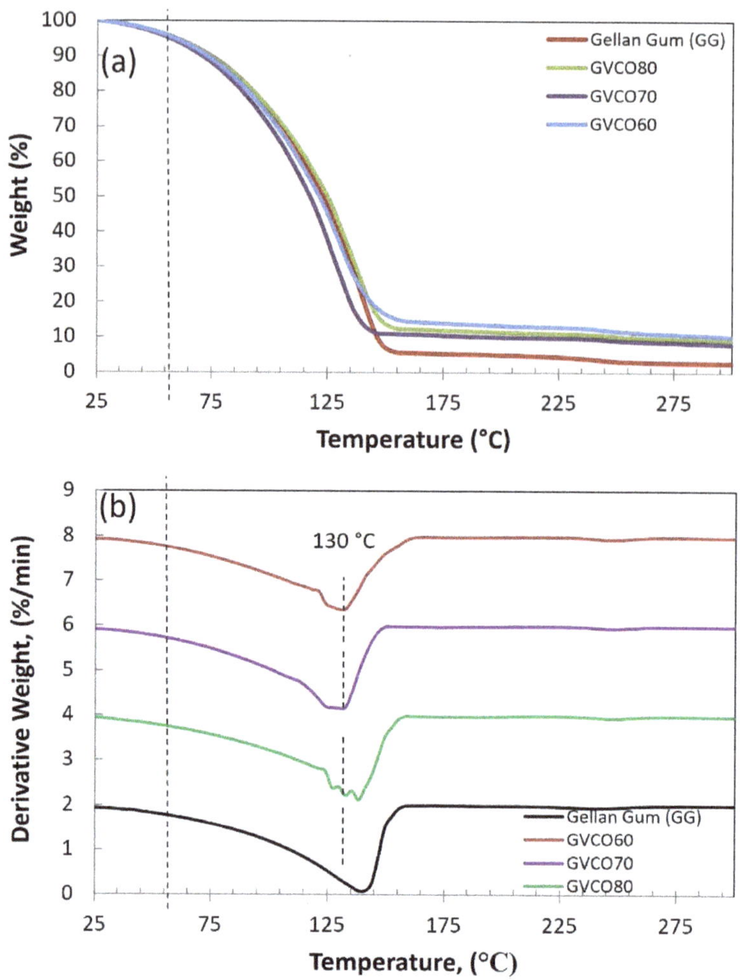

Figure 5. (**a**) Thermogravimetric thermograms and (**b**) derivative thermograms of GG and GVCO hydrogels at different concentrations.

3.7. Antibacterial Studies

GG is well known for its behavior to promote cell growth. For that reason, the result confirming that the control (GG hydrogel) shows no antibacterial activities examined through a qualitative method (inhibition zones) after incubating for 24 h against Gram-negative (*Escherichia coli* and *Klebsiella pneumoniae*) and Gram-positive (*Staphylococcus aureus* and *Bacillus subtilis*) bacteria (Figure 6a–d). The addition of VCO microemulsion (GVCO60, GVCO70, and GVCO80) in gellan gum hydrogels, however, results in no clear zone of inhibition around the samples against all bacteria (Figure 6). It has been elaborated that the VCO does not possess antibacterial activity on its own, but rather is induced by its free fatty acids, particularly lauric acid (C12), and to a small extent by capric acid (C10) and caprylic acid (C8) [37]. In other words, the VCO must be metabolized to release those components and exert its antimicrobial effects [38]. Another mechanism proposed for the antibacterial effect of VCO is the lipolyze process with lipase and water to form monoglyceride, for which the structure consists of glyceride molecules attached to either sn-1 or sn-2 position of the glycerol [39]. The derivative of monoglycerides, known as monolaurin or monoester of the lauric acid is the most effective in inhibiting the microorganism by disrupting the cell membrane and the cytoplasm of cells [40].

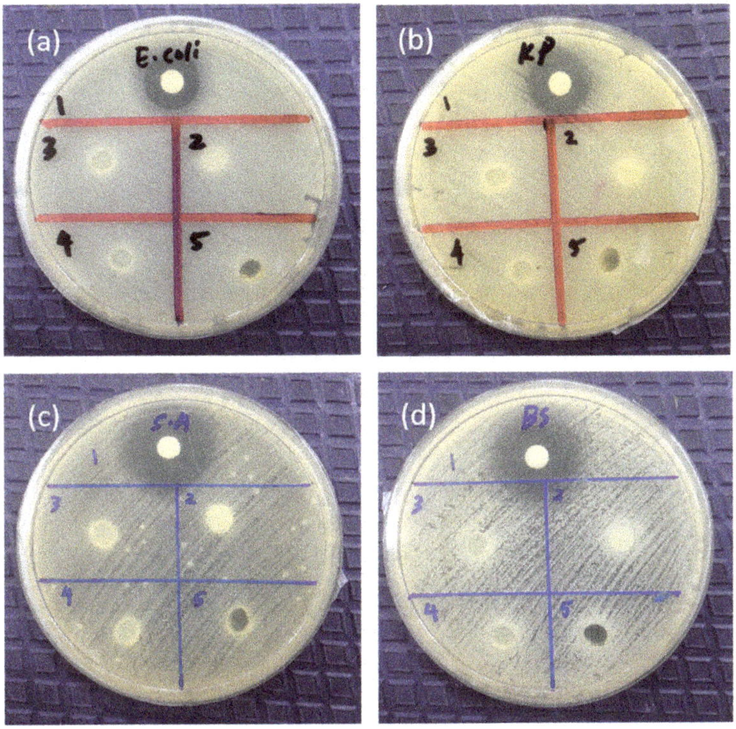

Figure 6. (**a**) *Escherichia coli* (*E. coli*), (**b**) *Klebsiella pneumoniae* (KP), (**c**) *Staphylococcus aureus* (S.A), and (**d**) *Bacillus subtilis* (BS) shown in the figure. Note: The number refers to samples: (1) penicillin (positive control), (2) GVCO80, (3) GVCO70, (4) GVCO60, and (5) gellan gum (negative control).

In this particular study, the GVCO hydrogels do not show antibacterial properties against Gram-negative (*Escherichia coli* and *Klebsiella pneumoniae*) and Gram-positive (*Staphylococcus aureus* and *Bacillus subtilis*) bacteria, maybe due to lesser contact of VCO with the agar, which reduces the chances of the latter to diffuse into the agar. As discussed earlier, the antibacterial activities of VCO are triggered by the monolaurin of lactic acid, which depends on the concentration of VCO. Higher content of VCO may contain a higher content of monoglyceride to transform to monolaurin derivatives and thus exert the antimicrobial effects. However, our result is contrary to a few studies reported previously [36,41]. Although these studies examined the film on the agar spread with Gram-negative and Gram-positive bacteria, which has the lesser contact of essential oil to the agar, but their observation shows promising antibacterial activities. Lee and coworkers reported the antibacterial activities of hydroxypropyl methylcellulose/ oregano essential oil nanoemulsion (HPMC/ORNE) composite films against Gram-positive (*S. aureus*, *B. cereus*, and *L. monocytogenes*) and Gram-negative bacteria (*E. coli*, *S. typhimurium*, *P. aeruginosa*, and *V. parahaemolyticus*) [41]. Their results show that the diameters of the inhibition zones increased significantly ($p < 0.05$) with increasing oregano essential oil concentration, with films containing over 5.0% (v/v) ORNE showing antibacterial activity against all tested strains, particularly against *S. typhimurium*. The authors claimed the antibacterial activity was attributed to the concentration of its most active compound, namely carvacrol, and other phenolics such as thymol and p-cymene, which can act synergistically, and later disrupting the outer membrane of bacteria, resulting in increased permeability and loss of ions, ATP, and other cytoplasm contents, thus inhibiting bacterial growth [41]. Another study reported by Fangfang and coworkers also shows the antibacterial activity of virgin coconut oil (VCO) incorporated into potato starch-based biodegradable (PSBB) films [36]. According to the authors, the antibacterial effect of films with different VCO concentrations is due to metabolization of the lauric acid in VCO and produce mononucleotides, which have antibacterial properties against *E. coli* and *S. aureus* [36]. Further study should be carried out to understand the possible interaction of VCO with bacteria on the agar at higher concentrations.

3.8. In Vivo Studies

For in vivo studies, one sample of GVCO hydrogel was chosen, i.e., GVCO80 hydrogel. This sample was chosen due to the optimum properties shown by the latter compared to GVCO60 and GVCO70 hydrogels. The GVCO80 hydrogels show a transparent appearance, optimum mechanical properties, and thermal behavior, together with acceptable swelling ratio and WVTR values. Table 4 shows the wound contraction of the GG and GVCO80 hydrogels compared to the commercial product, Opsite dressing. For the first 7 days, the Opsite dressing shows a significant acceleration of healing at 49 ± 11%, followed by GVCO80 hydrogel at 46 ± 7%. The healing process is further enhanced on day 11 when the GVCO80 surpassed the wound closure, compared Opsite dressing, and achieved 95 ± 2% on day 14. The Opsite dressing completed the closure at 93 ± 4 % on day 14. The GG hydrogel shows the lowest percentage of wound closure, 91 ± 4%, compared to other treatments.

Table 4. The percentage of wound contraction of experiment groups on days 2, 4, 7, 11, and 14; the means of two replicates where Opsite acts as a control; * ($p < 0.05$ compare with the control group).

Day/Treatment	2nd	4th	7th	11th	14th
	Percentage (%) of Wound Closure (Mean ± SD)				
GG	9 ± 5	25 ± 7	35 ± 2	81 ± 6	91 ± 4
Opsite	25 ± 3 *	31 ± 3	49 ± 11	88 ± 5	93 ± 4
GVCO80	11 ± 3	28 ± 2	46 ± 7	89 ± 3	95 ± 2

In this study, no skin irritation is observed for all treatments. This indicated that the GG is a safe material and a good candidate in biomedical applications [29]. The wound healing process was monitored on days 2, 4, 7, 11, and 14 by capturing images of each animal (Figure 7a). The results are in agreement with the wound contraction (Table 4) in which the GVCO80 hydrogel accelerated the wound closure and the wound gradually disappeared through time. The ultrasound images of the thickness growing on the wound skin are shown in Figure 7b. It shows that the skin formation of a wound treated with GVCO80 hydrogel exhibited the optimum recovery compared to other samples on days 2, 4, 7, 11, and 14. The intensity of white/yellowish/green color indicates the good formation of epidermis, dermis, and subcutis of GVCO80 hydrogel and followed by Opsite. The dermis layer was characterized by varying intensities (different colors) present on the wound while the subcutis layer referred to the low-intensity areas due to the homogenous composition. Subcutis areas are described as the black areas, which are referred to the homogeny structure, such as fat, water, and blood. The GG hydrogel as control showed less intensity among the others. The epidermal regeneration was observed in all experimental groups after the 14th day of treatment.

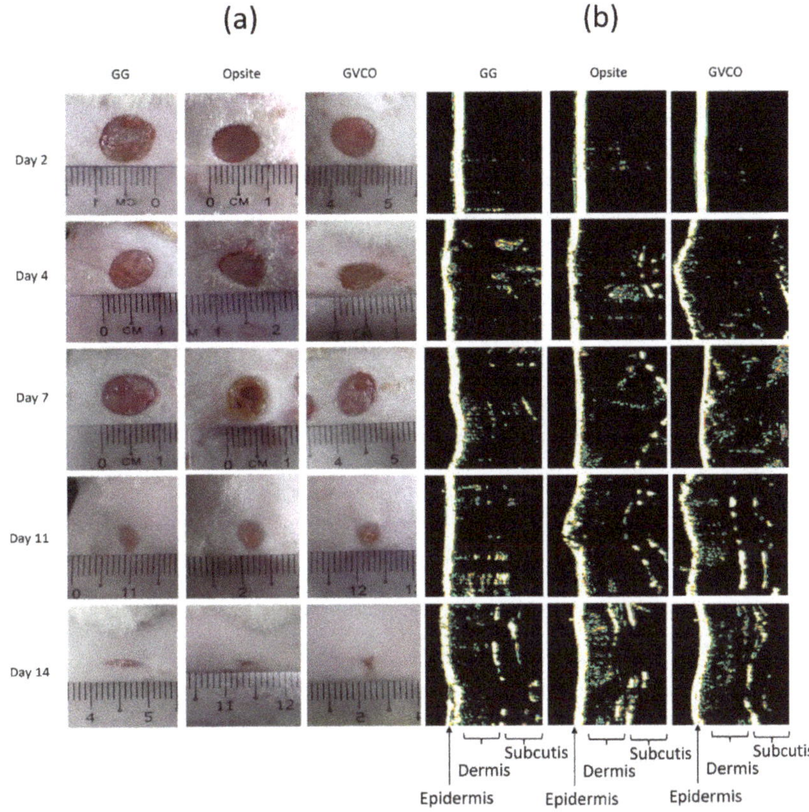

Figure 7. (**a**) Wound healing studies and (**b**) typical ultrasound images of wound skin on gellan gum (GG), Opsite, and GVCO80 hydrogels on days 2, 4, 7, 11, and 14.

Hematoxylin and eosin (H & E) staining was performed to evaluate the quality of the wound tissue. Histological evaluation results show that the GVCO80 hydrogel resulted in better re-epithelization as compared to other samples (Figure 8). For Opsite and GVCO80 hydrogel, the formation of epithelial growth was observed and the number of inflammatory cells reduced. Meanwhile, in the control group (GG hydrogel), the new epithelium was noted to regenerate and a little scab was spotted and the necrotic tissue was found under defect. The skin treated with GVCO80 hydrogel and Opsite presented a better result than the control.

The moist environment characteristic of hydrogel is most suited for re-epithelization and enhancing wound healing mechanism on the skin [42]. For a long time, VCO has been a well-known and powerful substance for treating wounds, mainly due to its anti-bacterial, anti-inflammatory, and anti-oxidant properties [43]. An in vivo study conducted by Soliman and coworkers evaluated the effects of topical application of VCO on wound healing in diabetes-induced Sprague–Dawley rats [44]. They found that the wound closure rate in the VCO group was higher on all days compared to diabetic nontreated rats, and VCO was found to be better than silver sulfadiazine cream in the healing of diabetic wounds via promoting re-epithelialization and collagen synthesis, as well as increasing WCR and total protein content. A few other studies reported the effectiveness of pure VCO (liquid form) in promoting the healing process [45–47]. The wound healing potency of fermented virgin coconut oil was also verified against human umbilical vein endothelial (HUVEC), fibroblast (CCD-18), and retinal ganglion (RGC-5) cells, as well as a wound excision model in Sprague–Dawley rats [48]. Their finding shows that the expression of phospho-VEGFR2 (vascular endothelial growth factor receptor 2) in HUVECs was detected by Western blot; rats in the VCO group had significantly smaller wound size, higher wound healing percentage, and shorter wound closure time when compared with a control group. Their study also confirmed that a high angiogenic and wound healing potency of VCO contributed to the regulation of the VEGF signing pathway [48]. These past studies show that the VCO significantly affected the healing of the wounds and show promising results to be applied in biomedical applications. With this noted, our study shows that the inclusion of VCO into a biopolymer does not stop the effectiveness of the oil to promote and enhance the healing process. Similar results are obtained, thus indicating this GVCO hydrogel as a promising candidate to be used as dressing materials.

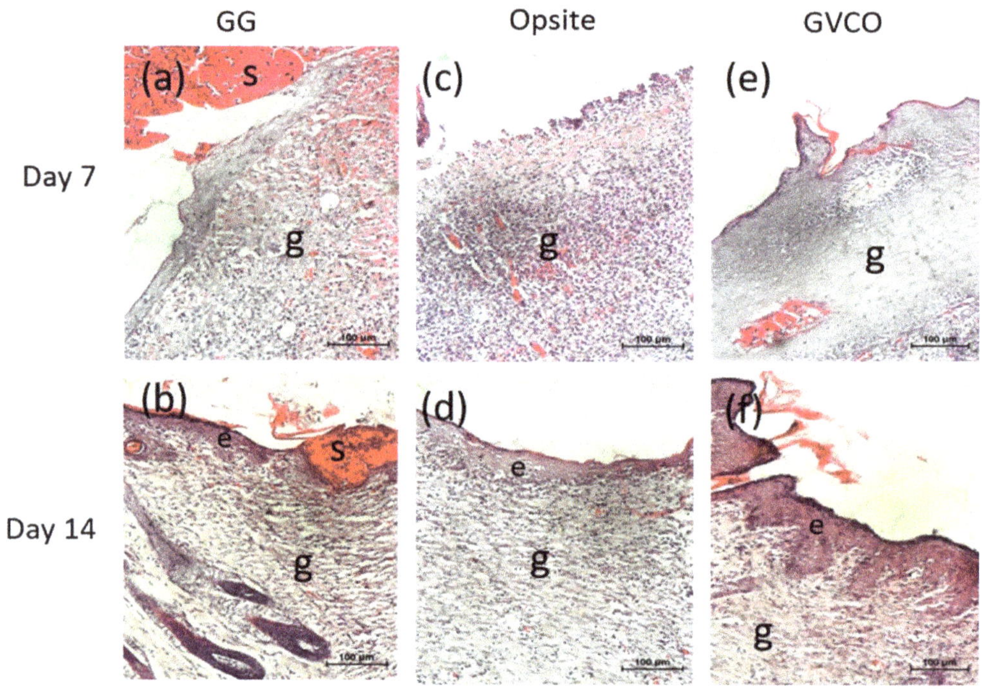

Figure 8. Representative images of the histological evaluation section on day 7 and 14 post-wound, stained with H & E (**a**,**b**) GG, (**c**,**d**) Opsite, and (**e**,**f**) GVCO80 hydrogel. Note: e = epidermis, g = granulation, s= scab; 20x magnification. The bar on the micrograph represents 100 μm.

4. Conclusions

This study successfully prepared gellan gum (GG) hydrogel incorporated with virgin coconut oil (VCO) microemulsion with the addition of surfactants, i.e., Tween 80 or TritonX-100. A ternary phase diagram was constructed to obtain an optimized ratio of VCO, water, and Tween 80, which was chosen due to optimum concentration to produce a stable VCO microemulsion. The VCO microemulsion was incorporated into gellan gum (GG) hydrogel (GVCO). Its chemical interaction, mechanical performance, physical properties, and thermal behavior were examined. The stress-at-break (σ) and Young's modulus (YM) of GVCO hydrogel films increased, together with thermal behavior, following the inclusion of VCO microemulsion. The swelling value of GVCO hydrogel decreased as the VCO microemulsion increased and the water vapor transmission rate of GVCO hydrogels was comparable to commercial dressing in the range of 332–391 g m^{-2} d^{-1}. Meanwhile, the in vitro qualitative antibacterial study of GVCO hydrogels against Gram-negative

(*Escherichia coli* and *Klebsiella pneumoniae*) and Gram-positive (*Staphylococcus aureus* and *Bacillus subtilis*) bacteria showed that VCO possesses a weak antibacterial effect. In vivo studies on Sprague–Dawley rats show the wound contraction of GVCO80 hydrogel is the best (95 $\pm$ 2%) after the 14th day compared to Smith & Nephew Opsite post-op waterproof dressing at 93 $\pm$ 4%, and supported by the ultrasound images of wound skin and histological evaluation on the wound. This study demonstrated that the GVCO hydrogels had the potential to be used as wound dressing materials.

Author Contributions: Conceptualization, K.A.M.A. and M.H.R.; methodology, K.A.M.A., L.C.R., W.I.W.I., and S.I.A.R.; formal analysis, M.Z.M. and K.A.M.A.; investigation, M.Z.M. and M.A.A.B.; resources, M.Z.M. and K.A.M.A.; data curation, M.Z.M., M.A.A.B. and K.A.M.A.; writing—original draft preparation, M.Z.M., M.A.A.B., and K.A.M.A.; writing—review and editing, K.A.M.A. and M.i.h.P.; visualization, M.Z.M. and K.A.M.A.; supervision, K.A.M.A.; project administration, K.A.M.A. and M.H.R.; and funding acquisition, K.A.M.A. All authors have read and agreed to the published version of the manuscript.

Funding: This research received a funding from the Ministry of Higher Education Malaysia (FRGS/1/2016/STG07/UMT/02/1).

Institutional Review Board Statement: The study was conducted according to the guidelines of the Declaration of Helsinki and approved by the Research Ethics Approval of Universiti Malaysia Terengganu (UMT/JKEPHMK/2020/48) on 3 November 2020.

Informed Consent Statement: Not applicable.

Data Availability Statement: The data presented in this study are available on request from the corresponding author.

Acknowledgments: Special thanks to the Ministry of Higher Education Malaysia for research grants (FRGS/1/2016/STG07/UMT/02/1). The authors are also grateful to the Faculty of Science and Marine Environment, Centre Laboratory, and University Malaysia Terengganu for providing facilities.

Conflicts of Interest: The authors declare no conflict of interest.

References

1. Phillips, C.J.; Humphreys, I.; Fletcher, J.; Harding, K.; Chamberlain, G.; Macey, S. Estimating the costs associated with the management of patients with chronic wounds using linked routine data. *Int. Wound J.* **2016**, *13*, 1193–1197. [CrossRef] [PubMed]
2. Frykberg, R.G.; Banks, J. Management of diabetic foot ulcers: A review. *Fed. Pract.* **2016**, *33*, 16. [PubMed]
3. Moher, D.; Shamseer, L.; Clarke, M.; Ghersi, D.; Liberati, A.; Petticrew, M.; Shekelle, P.; Stewart, L.A.; PRISMA-P Group. Preferred reporting items for systematic review and meta-analysis protocols (PRISMA-P) 2015 statement. *Syst. Rev. Pharm.* **2015**, *4*, 1. [CrossRef] [PubMed]
4. Sen, C.K.; Gordillo, G.M.; Roy, S.; Kirsner, R.; Lambert, L.; Hunt, T.K.; Gottrup, F.; Gurtner, G.C.; Longaker, M.T. Human skin wounds: A major and snowballing threat to public health and the economy. *Wound Rep. Reg.* **2009**, *17*, 763–771. [CrossRef]
5. Augustine, R.; Hasan, A. Emerging applications of biocompatible phytosynthesized metal/metal oxide nanoparticles in healthcare. *J. Drug Deliv. Sci. Technol.* **2020**, *56*, 101516. [CrossRef]
6. Mohd, S.S.; Abdullah, M.A.A.; Mat Amin, K.A. Gellan gum/clay hydrogels for tissue engineering application: Mechanical, thermal behavior, cell viability, and antibacterial properties. *J. Bioact. Compat. Polym.* **2016**, *31*, 648–666. [CrossRef]
7. Mohammadinejad, R.; Kumar, A.; Ranjbar-Mohammadi, M.; Ashrafizadeh, M.; Han, S.S.; Khang, G.; Roveimiab, Z. Recent Advances in Natural Gum-Based Biomaterials for Tissue Engineering and Regenerative Medicine: A Review. *Polymers* **2020**, *12*, 176. [CrossRef]
8. Sebria, N.J.M.; Amin, K.A.M. Gellan Gum/Ibuprofen Hydrogel for Dressing Application: Mechanical Properties, Release Activity and Biocompatibility Studies. *Int. J. Appl. Chem.* **2016**, *12*, 483–498.
9. Picone, C.S.F.; Cunha, R.L. Influence of pH on formation and properties of gellan gels. *Carbohydr. Polym.* **2011**, *84*, 662–668. [CrossRef]
10. Karvinen, J.; Koivisto, J.T.; Jönkkäri, I.; Kellomäki, M. The production of injectable hydrazone crosslinked gellan gum-hyaluronan-hydrogels with tunable mechanical and physical properties. *J. Mech. Behav. Biomed. Mater.* **2017**, *71*, 383–391. [CrossRef] [PubMed]
11. Choi, J.; Lee, J.; Shin, M.E.; Been, S.; Lee, D.H.; Khang, G. Eggshell Membrane/Gellan Gum Composite Hydrogels with Increased Degradability, Biocompatibility, and Anti-Swelling Properties for Effective Regeneration of Retinal Pigment Epithelium. *Polymers* **2020**, *12*, 2941. [CrossRef] [PubMed]
12. Yi, H.G.; Lee, H.; Cho, D.W. 3D printing of organs-on-chips. *Bioengineering* **2017**, *4*, 10. [CrossRef] [PubMed]

13. Amin, K.A.M.; in het Panhuis, M. Polyelectrolyte complex materials from chitosan and gellan gum. *Carbohydr. Polym.* **2011**, *86*, 352–358. [CrossRef]
14. Oliveira, C.V.M.; Stringhetti, F.C.B.; Evangelista, R.C.; Daflon, G.M.P. Development and characterization of cross-linked gellan gum and retrograded starch blend hydrogels for drug delivery applications. *J. Mech. Behav. Biomed. Mater.* **2017**, *65*, 317–333. [CrossRef] [PubMed]
15. Fernández-Ferreiro, A.; Barcia, M.G.; Gil-Martínez, M.; Vieites-Prado, A.; Lema, L.; Argibay, B.; Méndez, J.B.; Lamas, M.J.; Otero-Espinar, F.J. In vitro and in vivo ocular safety and eye surface permanence determination by direct and Magnetic Resonance Imaging of ion-sensitive hydrogels based on gellan gum and kappa-carrageenan. *Eur. J. Pharm. Biopharm.* **2015**, *94*, 342–351. [CrossRef] [PubMed]
16. Razali, M.H.; Ismail, N.A.; Amin, K.A.M. Titanium Dioxide Nanotubes Incorporated Gellan Gum Bio-Nanocomposite Film for Wound Healing: Effect of TiO_2 Nanotubes Concentration. *Int. J. Biol. Macromol.* **2020**, *153*, 1117–1135. [CrossRef] [PubMed]
17. Wang, C.; Gong, Y.; Lin, Y.; Shen, J.B.; Wang, D.A. A novel gellan gel-based microcarrier for anchorage-dependent cell delivery. *Acta Biomater.* **2008**, *4*, 1226–1234. [CrossRef]
18. Vilela, C.A.; Correia, C.; Morais, A.D.S.; Santos, T.C.; Gertrudes, A.C.; Moreira, E.S.; Frias, A.M.; Learmonth, D.A.; Oliveira, P.; Oliveira, J.M.; et al. In vitro and in vivo performance of methacrylated gellan gum hydrogel formulations for cartilage repair. *J. Biomed. Mater. Res. Part A* **2018**, *106*, 1987–1996. [CrossRef]
19. Nevin, K.G.; Rajamohan, T. Effect of Topical Application of Virgin Coconut Oil on Skin Components and Antioxidant Status during Dermal Wound Healing in Young Rats. *Skin Pharmacol. Physiol.* **2010**, *23*, 290–297. [CrossRef]
20. Intahphuak, S.; Khonsung, P.; Panthong, A. Anti-inflammatory, analgesic, and antipyretic activities of virgin coconut oil. *Pharm. Biol.* **2010**, *48*, 151–157. [CrossRef]
21. Mahmood, H.; Khan, I.U.; Asif, M.; Khan, R.U.; Asghar, S.; Khalid, L.; Khalid, S.H.; Irfan, M.; Rehman, F.; Shahzad, Y.; et al. In vitro and in vivo evaluation of gellan gum hydrogel films: Assessing the co impact of therapeutic oils and ofloxacin on wound healing. *Int. J. Biol. Macromol.* **2021**, *166*, 483–495. [CrossRef] [PubMed]
22. Fernandes, D.M.; Barbosa, W.S.; Rangel, W.S.P.; Valle, I.M.M.; Matos, A.P.D.S.; Melgaço, F.G.; Dias, M.L.; Júnior, E.R.; Silva, L.C.P.D.; Abreu, L.C.L.D.; et al. Polymeric membrane based on polyactic acid and babassu oil for wound healing. *Mater. Today Commun.* **2021**, *26*, 102173. [CrossRef]
23. Kaur, G.; Mehta, S.K. Developments of Polysorbate (Tween) based microemulsions: Preclinical drug delivery, toxicity and antimicrobial applications. *Int. J. Pharm.* **2017**, *529*, 134–160. [CrossRef] [PubMed]
24. Yin, Y.; Gu, J.; Wang, X.; Zhang, Y.J.; Zheng, W.; Chen, R.; Wang, X.C. Effects of rhamnolipid and Tween-80 on cellulase activities and metabolic functions of the bacterial community during chicken manure composting. *Bioresour. Technol.* **2019**, *288*, 121507. [CrossRef] [PubMed]
25. Amit; Jamwal, R.; Kumari, S.; Dhaulaniya, A.S.; Balanet, B.; Kelly, S.; Cannavanal, A.; Singh, D.K. Utilizing ATR-FTIR spectroscopy combined with multivariate chemometric modelling for the swift detection of mustard oil adulteration in virgin coconut oil. *Vib. Spectrosc.* **2020**, *109*, 103066. [CrossRef]
26. Rohman, A.; Cheman, Y.B.I.N.; Ali, M.D.E. The authentication of virgin coconut oil from grape seed oil and soybean oil using ftir spectroscopy and chemometrics. *Int. J. Appl. Pharm.* **2019**, *11*, 259–263. [CrossRef]
27. Ismail, N.A.; Amin, K.A.M.; Razali, M.H. Novel gellan gum incorporated TiO_2 nanotubes film for skin tissue engineering. *Mater. Lett.* **2018**, *228*, 116–120. [CrossRef]
28. Ismail, N.A.; Mohamad, S.F.; Ibrahim, M.A.; Amin, K.A.M. Evaluation of gellan gum film containing virgin coconut oil for transparent dressing materials. *Adv. Biomater.* **2014**, *2014*, 351248. [CrossRef]
29. Amin, K.A.M.; Gilmore, K.J.; Walker, M.J.; Wilson, M.R.; Panhuis, M.I.H. Polyelectrolyte complex materials consisting of antibacterial and cell-supporting layers. *Macromol. Biosci.* **2012**, *12*, 374–382. [CrossRef]
30. Coutinho, D.F.; Sant, S.V.; Shin, H.; Oliveira, J.T.; Gomes, M.E.; Neves, N.M.; Khademhosseini, A.; Reis, R.L. Modified Gellan Gum hydrogels with tunable physical and mechanical properties. *Biomaterials* **2010**, *31*, 7494–7502. [CrossRef]
31. Souza, F.S.D.; Ferreira, I.L.D.M.; Costa, M.A.D.S.; Costa, M.P.M.; Silva, G.M.D. Effect of pH variation and crosslinker absence on the gelling mechanism of high acyl gellan: Morphological, thermal and mechanical approaches. *Carbohydr. Polym.* **2021**, *251*, 117002. [CrossRef] [PubMed]
32. Nitta, Y. Gelation and gel properties of gellan gum and xyloglucan. *J. Biol. Macromol.* **2005**, *5*, 47–52.
33. Wu, P.; Fisher, A.C.; Foo, P.P.; Queen, D.; Gaylor, J.D.S. In vitro assessment of water vapour transmission of synthetic wound dressings. *Biomaterials* **1995**, *16*, 171–175. [CrossRef]
34. Chong, E.W.N.; Jafarzadeh, S.; Paridah, M.T.; Gopakumar, D.A.; Tajarudin, H.A.; Thomas, S.; Abdul Khalil, H.P.S. Enhancement in the Physico-Mechanical Functions of Seaweed Biopolymer Film via Embedding Fillers for Plasticulture Application—A Comparison with Conventional Biodegradable Mulch Film. *Polymers* **2019**, *11*, 210.
35. Edhirej, A.; Sapuan, S.M.; Jawaid, M.; Zahari, N.I. Preparation and characterization of cassava bagasse reinforced thermoplastic cassava starch. *Fibers Polym.* **2017**, *18*, 162–171. [CrossRef]
36. Fangfang, Z.; Xinpeng, B.; Wei, G.; Wang, G.D.; Shi, Z.Z.; Jun, G. Effects of virgin coconut oil on the physicochemical, morphological and antibacterial properties of potato starch-based biodegradable films. *Int. J. Food Sci. Technol.* **2020**, *55*, 192–200. [CrossRef]

37. Hussain, M.S.; Al-Alaq, F.T.; Al-Khafaji, N.S.; Al-Dahmoshi, H.O.M. Antibacterial Effect of Virgin and Refined Coconut Oils on Pathogenic Bacteria: A Review. *Indian J. Forensic Med. Toxicol.* **2020**, *14*, 6042–6048.
38. Nasir, N.A.M.M.; Abllah, Z.; Jalaludin, A.A.; Shahdan, I.A.; Abd Manan, W.N.H.W. Virgin Coconut Oil and Its Antimicrobial Properties against Pathogenic Microorganisms: A Review. In Proceedings of the International Dental Conference of Sumatera Utara 2017 (IDCSU 2017); Atlantis Press: Berlin/Heidelberg, Germany, 2018; pp. 192–199.
39. Nair, M.K.M.; Vasudevan, P.; Hoagland, T.; Venkitanarayanan, K. Inactivation of *Escherichia coli* O157: H7 and Listeria monocytogenes in milk by caprylic acid and monocaprylin. *Food Microbiol.* **2004**, *21*, 611–616. [CrossRef]
40. Shilling, M.; Matt, L.; Rubin, E.; Visitacion, M.P.; Haller, N.A.; Grey, S.F.; Woolverton, C.J. Antimicrobial effects of virgin coconut oil and its medium-chain fatty acids on Clostridium difficile. *J. Med. Food* **2013**, *16*, 1079–1085. [CrossRef]
41. Lee, J.Y.; Garcia, C.V.; Shin, G.H.; Kima, J.T. Antibacterial and antioxidant properties of hydroxypropyl methylcellulose-based active composite films incorporating oregano essential oil nanoemulsions. *Lwt* **2019**, *106*, 164–171. [CrossRef]
42. Li, J.; Yu, F.; Chen, G.; Liu, J.; Li, X.L.; Cheng, B.; Mo, X.M.; Chen, C.; Pan, J.F. Moist-retaining, self-recoverable, bioadhesive, and transparent in situ forming hydrogels to accelerate wound healing. *ACS Appl. Mater. Interfaces* **2020**, *12*, 2023–2038. [CrossRef] [PubMed]
43. Gani, A.; Benjakul, S. Effect of antioxidants in combination of VCO nanoemulsion on gel properties and storage stability of refrigerated sardine surimi gel. *Int. J. Food Sci. Technol.* **2020**, *55*, 2451–2461. [CrossRef]
44. Soliman, A.M.; Lin, T.S.; Ghafar, N.A.; Das, S. Virgin Coconut Oil and Diabetic Wound Healing: Histopathological and Biochemical Analysis. *Eur. J. Anat.* **2018**, *22*, 135–144.
45. Andriana, N.; Lister, I.N.E.; Fachrial, E.; Ginting, C.N.; Lie, S. Effectiveness Test of Wound Healing Based Virgin Coconut Oil toward Commercial Products on Rabbits. In Proceedings of the 2020 3rd International Conference on Mechanical, Electronics, Computer, and Industrial Technology (MECnIT), Medan, Indonesia, 25–26 June 2020; pp. 104–107.
46. Yuniati, R.; Subchan, P.; Riawan, W.; Khrisna, M.B.; Restiwijaya, M.; Dyan, N.S.; Nur, M. Topical Ozonated Virgin Coconut Oil Improves Wound Healing and Increases HSP90α, VEGF-A, EGF, BFGF and CD34 in Diabetic Ulcer Mouse Model of Wound Healing. *F1000Research* **2020**, *9*, 580. [CrossRef]
47. Rajagukguk, H.; Syukur, S.; Ibrahim, S.; Syafrizayanti, S. Beneficial Effect of Application of Virgin Coconut Oil (VCO) Product from Padang West Sumatra, Indonesia on Palatoplasty Wound Healing. *Am. Sci. Res. J. Eng. Technol. Sci. (ASRJETS)* **2017**, *34*, 231–236.
48. Ibrahim, A.H.; Li, H.; Al-Rawi, S.S.; Majid, A.S.A.; Al-Habib, O.A.; Xia, X.B.; Majid, A.M.A.; Ji, D. Angiogenic and wound healing potency of fermented virgin coconut oil: In vitro and in vivo studies. *Am. J. Transl. Res.* **2017**, *9*, 4936.

polymers

MDPI

Article

The Effect of Salinized Nano ZrO_2 Particles on the Microstructure, Hardness, and Wear Behavior of Acrylic Denture Tooth Nanocomposite

Kawkb M. El-Tamimi [1], Dalia A. Bayoumi [2], Mohamed M. Z. Ahmed [3,4,*], Ibrahim Albaijan [3] and Mohammed E. El-Sayed [1]

1 Department of Removable Prosthodontics, Faculty of Dentistry, Suez Canal University, Ismailia 41522, Egypt; kowkb_eltamimi@dent.suez.edu.eg (K.M.E.-T.); mohamed_ezzat@dent.suez.edu.eg (M.E.E.-S.)
2 Department of Dental Biomaterials, Faculty of Dentistry, Suez Canal University, Ismailia 41522, Egypt; dalia_bayoumi@dent.suez.edu.eg
3 Mechanical Engineering Department, College of Engineering at Al Kharj, Prince Sattam Bin Abdulaziz University, Al Kharj 16273, Saudi Arabia; i.albaijan@psau.edu.sa
4 Department of Metallurgical and Materials Engineering, Faculty of Petroleum and Mining Engineering, Suez University, Suez 43512, Egypt
* Correspondence: moh.ahmed@psau.edu.sa

Abstract: The wear of acrylic denture teeth is a serious problem that can change the vertical dimensions of dentures. This study evaluates the effect of adding salinized nano ZrO_2 particles on the microstructure, hardness, and wear resistance of acrylic denture teeth. Heat polymerizing polymethyl methacrylate resin was mixed with salinized ZrO_2 at concentrations of 5 wt.% and 10 wt.%. Acrylic resin specimens without filler addition were used as a control group. SEM/EDS analyses were performed and the Vickers' hardness was evaluated. Two-body wear testing was performed using a chewing simulator with a human enamel antagonist. After subjecting the samples to 37,500 cycles, both height loss and weight loss were used to evaluate the wear behavior. The microstructural investigation of the reinforced-denture teeth indicates sound nanocomposite preparation using the applied regime without porosity or macro defects. The addition of zirconium oxide nanofillers to PMMA at both 5% and 10% increased the microhardness, with values of up to 49.7 HV. The wear mechanism in the acrylic base material without nanoparticle addition was found to be fatigue wear; a high density of microcracks were found. The addition of 5 wt.% ZrO_2 improved the wear resistance. Increasing the nanoparticles to 10 wt.% ZrO_2 further improved the wear resistance, with no microcracks found.

Keywords: nanocomposite; acrylic denture teeth; nano ZrO_2 particles; wear behavior; microstructure; microhardness; polymerization; polymethyl methacrylate resin

Citation: El-Tamimi, K.M.; Bayoumi, D.A.; Ahmed, M.M.Z.; Albaijan, I.; El-Sayed, M.E. The Effect of Salinized Nano ZrO_2 Particles on the Microstructure, Hardness, and Wear Behavior of Acrylic Denture Tooth Nanocomposite. *Polymers* **2022**, *14*, 302. https://doi.org/10.3390/polym14020302

Academic Editors: Domenico Acierno and Antonella Patti

Received: 20 November 2021
Accepted: 10 January 2022
Published: 12 January 2022

Publisher's Note: MDPI stays neutral with regard to jurisdictional claims in published maps and institutional affiliations.

1. Introduction

Denture teeth are currently made of either methacrylate-based resins (acrylic resin) or porcelain, but acrylic teeth have nearly eliminated porcelain teeth from the market [1] due to a number of advantages, including their chemical bonding to the denture base [2,3], lower susceptibility to fracture [4,5] and decreased clicking [6,7]. Acrylic resin tooth wear is a serious complication during denture service and can change the vertical dimension of dentures. This process harms the denture and exerts an impact on facial aesthetics and the function of the masticatory muscles, resulting in less efficient mastication. This can result in temporomandibular disorders, digestive disturbances and decreased patient comfort [6–9]. Efforts were made to enhance the wear resistance of the acrylic resin denture teeth, such as the formation of cross-linked polymer teeth (interpenetrated polymer network) [8,9]. Another possible solution is to add nanofillers to enhance mechanical properties. Despite this, these materials are softer than composite resin or porcelain teeth [8–10]. Nano-fillers, such as metal oxides, carbon, and glass fibers have also been used to improve the mechanical behavior of the acrylic resin denture base materials [11–16]. Recently, zirconium oxide nanoparticles (ZrO_2 NPs) have been recognized for their high biocompatibility. However, because they are white in color, they are thought to be less likely to change aesthetics than other metal oxide nanoparticles [17]. ZrO_2 NPs are not only biocompatible but are also resistant to wear and corrosion. In addition to these characteristics, ZrO_2 NPs offer high toughness and mechanical strength [18–20]. ZrO_2 NPs are frequently used to mechanically reinforce polymers [12]. Saline coupling agents are applied to the surface of ZrO_2 NPs to decrease the risk of aggregation and enhance compatibility with the polymer matrix [21]. The stresses caused by the dispersion of ZrO_2 NPs are transferred from the weak Polymethylmethacrylate (PMMA) matrix to the strong nanoparticles [20]. Ayad et al. [18] reported a minor increase in the surface hardness and the impact strength of zirconium dioxide acrylic composite when compared to the unreinforced resin (control specimen). Another study reported a reduction in impact strength as well as surface hardness when 10 wt.% ZrO_2 and 20 wt.% ZrO_2 were added [22]. Alternatively, a study found that increasing the amount of modified (ZrO_2) added to nano-ZrO_2/PMMA composite increased the hardness by up to 10%, and increased the hardness of groups 5% and 10% (water storage groups) compared with the zero and 20% ($p < 0.01$). The highest hardness value was observed in group 5% (TC− 36.8, TC+ 35.2) [23,24]. Furthermore, after incorporating salinized zirconia NPs in acrylic resin, there was a significant increase in hardness and a minor improvement in surface roughness, as well as a decrease in porosity [25,26]. However, investigations into the addition of salinized nano-ZrO_2 powder to acrylic resin teeth are lacking. The overall aim of this study is to develop nanocomposite denture teeth with high wear resistance. To this end, the effect of incorporating salinized nano-ZrO_2 particles into acrylic resin on the acrylic denture tooth microstructure, hardness, and wear resistance was investigated.

2. Materials and Methods

2.1. ZrO_2 Nanoparticles Surface Treatment

To enhance the ZrO_2 nanoparticle's wettability with the resin matrix through the formation of a reactive group on their surface, the salinization process was carried out in the laboratory of Inorganic Chemistry, Chemistry Department, Faculty of Science, Suez Canal University. The ZrO_2 nanoparticle powder of 99.9% purity, with 9 ± 2 m^2/g surface area and an average particle size of 40 ± 3 nm, (Nanogate, Cairo, Egypt) was saline-treated using the saline coupling agent 3-trimethoxysilyl propylmethacrylate (TMSPM) (Sigma-Aldrich, Berlin, Germany) [27]. This procedure was accomplished by mixing 30 g of ZrO_2 nanoparticles with 0.3 g of TMSPM dissolved in 100 mL of acetone for 1 h with the help of a magnetic stirrer at 300 rpm (Hot plate with a magnetic stirrer, MSH-A 30A, Seoul, Korea) for 60 min. Subsequently, the solvent was removed using a rotary evaporator under vacuum for 30 min at 60 °C and 150 rpm, followed by heating for 2 h at 120 °C. Next, the treated powder was allowed to cool to 28 °C [28–30].

2.2. PMMA Powder of Artificial Teeth/ZrO_2 Nanocomposite Preparation

The salinized nano-Zirconium oxide particles and PMMA powder of artificial teeth (Pigeon Dental Co Shanghai, Beijing, China) (Figure 1a) were pre-weighed using an electronic balance (Sartorius, Done Biopharmaceutical and Laboratories, Berlin, Germany) with an accuracy of 0.0001 gm so that the Nano-filler (Figure 1b) concentrations were 5% and 10% by weight. Pre-weighed Nano-filler ZrO_2 powder with two concentrations (5% and 10%) was added individually to the heat polymerized PMMA powder of artificial teeth and thoroughly mixed using an electric mixer to produce a uniform blend and ensure that the homogeneous mix of PMMA/ZrO_2, was mixed independently [18,20,29,31,32].

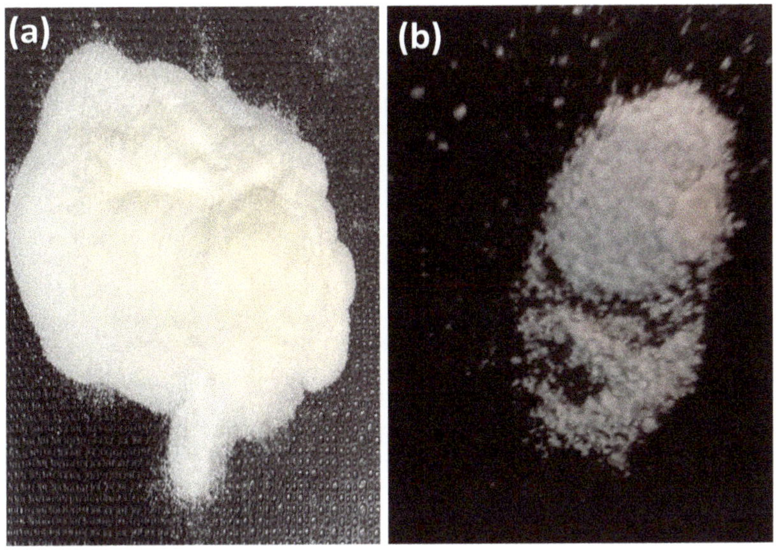

Figure 1. Macro images for the raw material powder used in this work (**a**) acrylic resin powder of artificial teeth and (**b**) ZrO_2 powder.

Twenty-four cylindrical samples 4 mm in diameter × 8 mm in height, were prepared from acrylic resin of teeth material. The control group (GI) was made from the conventional PMMA without zirconium oxide nano-particles. Two other groups GII, GIII, featured concentrations of 5 and 10 wt.% of zirconium oxide reinforced acrylic resin tooth-forming material. These samples were used to evaluate the wear behavior by using the chewing simulator and also to evaluate the hardness. Wax cylinders, of 8 mm × 100 mm shown in Figure 2, (Cavex, Haarlem, The Netherlands) were used for wax pattern fabrication that invested in a metal flask with dental stone (Zeta Mufle, Nevilicure, Italy).

Figure 2. An image for the metal flask with the wax specimens invested and a magnified region of the wax specimen.

After setting the stone, the wax was melted by placing the flasks in a wax elimination machine for 10 min. After removing the softened wax and all wax traces, a separating medium (Acrostone(A), Anglo-Egyptian Company. Hegaz, Cairo, Egypt, Batch No.505/04) was applied on the warm stone mold surfaces. When the flasks cooled to room temperature, heat-polymerized resin (pure or reinforced with nano-ZrO_2) was mixed and packed into the mold spaces at the dough stage. The flask parts were brought together under pressure using a hydraulic bench press for 5 min and then allowed to rest for 30 min. To polymerize the resin, the flasks were placed in a laboratory curing bath for 90 min at 74 °C, and then 30 min at 100 °C. After curing, the flasks were left for slow bench cooling till 28 °C before opening. Acrylic samples were retrieved from the stone, finished, and polished. Initially, the excess resin was trimmed using progressively finer grits of silicon carbide papers (grits 120–500) in wet conditions, followed by polishing using a rag wheel and pumice on a dental lathe and felt disc on a polishing machine, and the samples were cut to the required dimensions (4 × 8 mm). A digital micrometer (Mitutoyo, Digimatic Caliper 25SB, Mitutoyo Corporation, Tokyo, Japan) with a precision of 1/1000 was used to measure specimen dimensions [27]. An electronic balance (Sartorius, Biopharmaceutical and Laboratories, Berlin, Germany) with an accuracy of 0.001 gm was used to measure samples weight before and after the wear test.

2.3. Micro-Hardness Test Procedure

Micro-hardness of all the acrylic samples was investigated using Digital Display Vickers Micro-hardness Tester (Model HVS-50, Laizhou Huayin Testing Instrument Co., Ltd. Beijing, China) using a Vickers diamond indenter with a 20× objective lens. A load of 0.98 N was applied to the surface of the specimens for a 10 s dwell time. Three indentations, which were equally placed over a circle and not closer than 0.5 mm to the adjacent indentations, were made on the surface of each sample. The diagonal length of the indentations was measured by a built-in scaled microscope and Vickers micro-hardness values were obtained [30]. The micro-hardness mean value was then calculated for each sample. The micro-hardness was obtained using Equation (1).

$$HV = 1.854\ P/d^2 \tag{1}$$

where HV is Vickers hardness in Kgf/mm^2, P is the load in Kgf, and d is the length of the diagonals in mm.

2.4. Wear Testing with Natural Teeth Surface Antagonist

Carie-free human premolar teeth without fractures or worn cusps that were recently extracted for orthodontic treatment plans were chosen for this study and stored in 0.1% thymol solution (approved by the accredited ethics committee of the Faculty of Dentistry Suez Canal University, no297/2020). The buccal cusps with buccal surface were cut using no. 943 Miniflex diamond disc (Brasseler, Lemgo, Germany). They were then inserted into a self-cured acrylic resin mold (Acrostone Co., Cairo, Egypt) inside copper specimen holder with diameter15 mm. The insertion of the natural tooth surfaces was guided by a dental surveyor to ensure proper alignment of the tooth surface to the long axis of the specimen holder [33,34]. Figure 3 shows the assembly of the natural teeth within the self-cured acrylic resin mold.

Figure 3. Image of assembly of the natural teeth within the self-cured acrylic resin mold.

As a preparation for the wear test, the buccal surface of each tooth specimen was wet abraded and finished with grit abrasive paper (1000, 2500, and 4000), to a total depth of 0.5 mm. As a result, a flat area of about 2–3 mm was obtained for loading during the wear test [34,35]. Next, the two-body wear test was performed using the newly developed four stations multimodal Dual-axis ROBOTA chewing simulator integrated with thermo-cyclic protocol operated on servo-motor (Model ACH-09075DC-T, AD-TECH Technology Co., LTD., Berlin, Germany). The device allows the simulation of vertical and horizontal movements simultaneously in the thermodynamic condition in a range of 5 °C to 55 °C with a dwell time of 10 s. The ROBOTA chewing simulator features four chambers, each consisting of an upper part as a sample holder and a lower plastic antagonist holder Figure 4a. Samples were mounted into the metal receptacle present in the chewing simulator upper part Figure 4b. All the samples were tested under standard conditions in which the buccal surface of premolar teeth was used in the lower plastic holder of the device acting as antagonist Figure 4c. The acrylic samples were positioned on the upper samples holder in point contact with the teeth surfaces Figure 4d. A weight of 5 kg, which is comparable to 49 N of chewing force, was exerted. The samples were subjected to 37,500 cycles to simulate three months of clinical function. After 37,500 cycles, the samples were removed from the holder, cleaned with running water, and followed by cleaning in an ultrasonic cleaner for 2 min to remove any abraded particles from the surface before measurements.

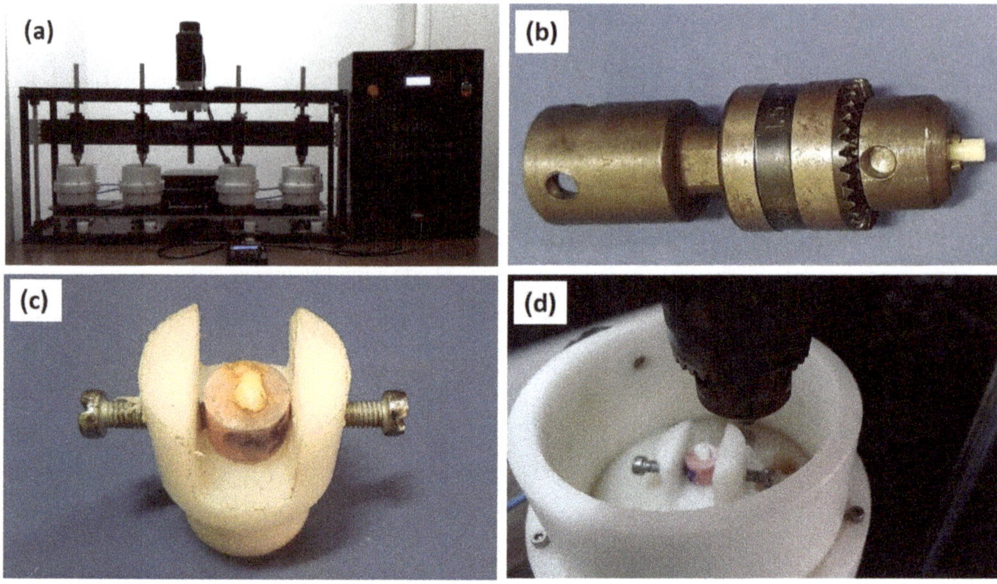

Figure 4. Images of the wear testing stages: (**a**) chewing simulator setup, (**b**) fixing the acrylic teeth sample in the holder, (**c**) plastic holder of the device acting as the antagonist, (**d**) positioning of the upper samples holder in point contact with the teeth surfaces.

Both height loss and weight loss were used to evaluate the wear behavior of the pure PMMA- and ZrO_2-reinforced tooth samples. Height loss was measured using the digital micrometer (Mitutoyo, Digimatic Caliper 25SB, Mitutoyo Corporation, Tokyo, Japan) with a precision of 1/1000. The difference between the readings before and after the wear simulation gave the amount of vertical sample height loss. Furthermore, weight loss was measured by weighing samples in the electronic analytical balance (Sartorius, Biopharmaceutical and Laboratories, Berlin, Germany) with an accuracy of 0.001 gm to show the difference in weight before and after the wear test. As this electronic balance featured a fully automated calibration technology and a micro weighing scale, values of all the mounted discs and antagonist samples were accurately measured. Each mounted sample was cleaned and dried with tissue paper before weighing. All data were calculated, tabulated, and statistically analyzed using SPSS for Windows, version 22.0 (Statistical Package for Social Science, Armonk, NY, USA: IBM Corp) at significant levels 0.05 (p-Value ≤ 0.5). A normality test was done to check the normal distribution of the sample, and all groups and subgroups showed normal distributions.

(A) Descriptive data: Descriptive statistics were calculated in the form of Mean $\pm$ Standard deviation (SD), median, range (Max-Min).
(B) ANOVA—test: One-way ANOVA (Analysis of variance) was used to compare between the three groups under study. Tukey Post hoc test was performed for the evaluation of statistical significance among the groups. p-value ≤ 0.05 is considered to be statistically significant.

2.5. *Microstuctur Investigation*

The microstructure of the acrylic nanocomposites was investigated using optical microscopy (OM) and scanning electron microscopy (SEM) equipped with EDS analysis. For the microstructural analysis, the samples were prepared using grinding with conventional silicon carbide emery papers of different grit sizes and then mechanically polished using alumina suspension of 0.05 μm [36,37]. Furthermore, the samples were investigated with SEM after the wear testing to study the worn surface and determine the wear mechanisms.

3. Results

3.1. *Acrylic/ZrO_2 Nanocomposites Microstructure*

Figure 5 shows an SEM micrograph of the raw material powder of the ZrO_2 nanoparticles that were used as reinforcing materials for the acrylic matrix. It can be observed that the powder is of nano-scale, with an almost homogenous particle size that is below 50 nm, except for some agglomerated particles of submicron or micron size. Figure 6 shows the SEM macro and micrographs of the base acrylic material (Figure 6a,b), 5 wt.% ZrO_2 nanocomposites (Figure 6c,d) and 10 wt.% ZrO_2 nanocomposites (Figure 6e,f). At the macro level, the samples showed no porosity defects, which indicates the effectiveness of the production procedure at producing sound acrylic nanocomposites at different percentages. On the other hand, the presence of macro-size particles, which increased in line with the increase in the ZrO_2 weight percentage, can be observed. The sources of these macro-sized particles were the ZrO_2 particles, as some macro-sized particles were noted within the SEM micrograph of the raw powder. In addition, at the micro-level, no defects were observed either, and nanoparticles can only be observed in the sample of 10 wt.% ZrO_2 (Figure 6f). Figure 7 shows the optical microstructure at different magnifications (Figure 7a,b) and the SEM microstructure at different positions (Figure 7c,d) for the acrylic with 10 wt.% ZrO sample. The micrographs show a macro-sized particle, with some particle measurements indicated. The EDS analysis of the micro-sized particle is presented in Figure 8a and the EDS of the nano-composite is shown in Figure 8b. The EDS analysis confirmed that the large size particles were mainly ZrO_2 particles and also confirmed the presence of ZrO_2 nanoparticles in the acrylic matrix, as indicated in Figure 8b. In addition, the peak carbon can be seen in both EDS charts. This may have been due to the polymeric material and may have resulted from interference from the instrument [38,39].

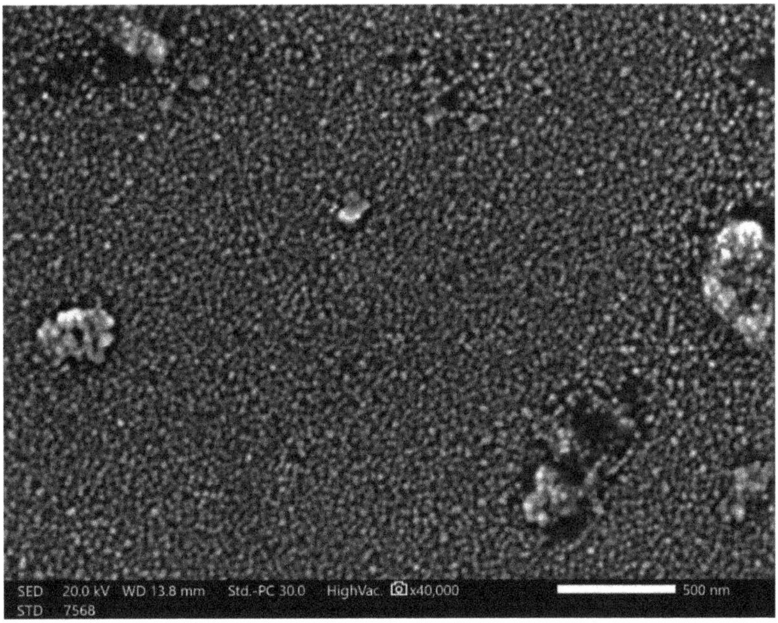

Figure 5. SEM micrograph for the raw material powder of the ZrO_2 nanoparticles used as a reinforcing material for the acrylic matrix.

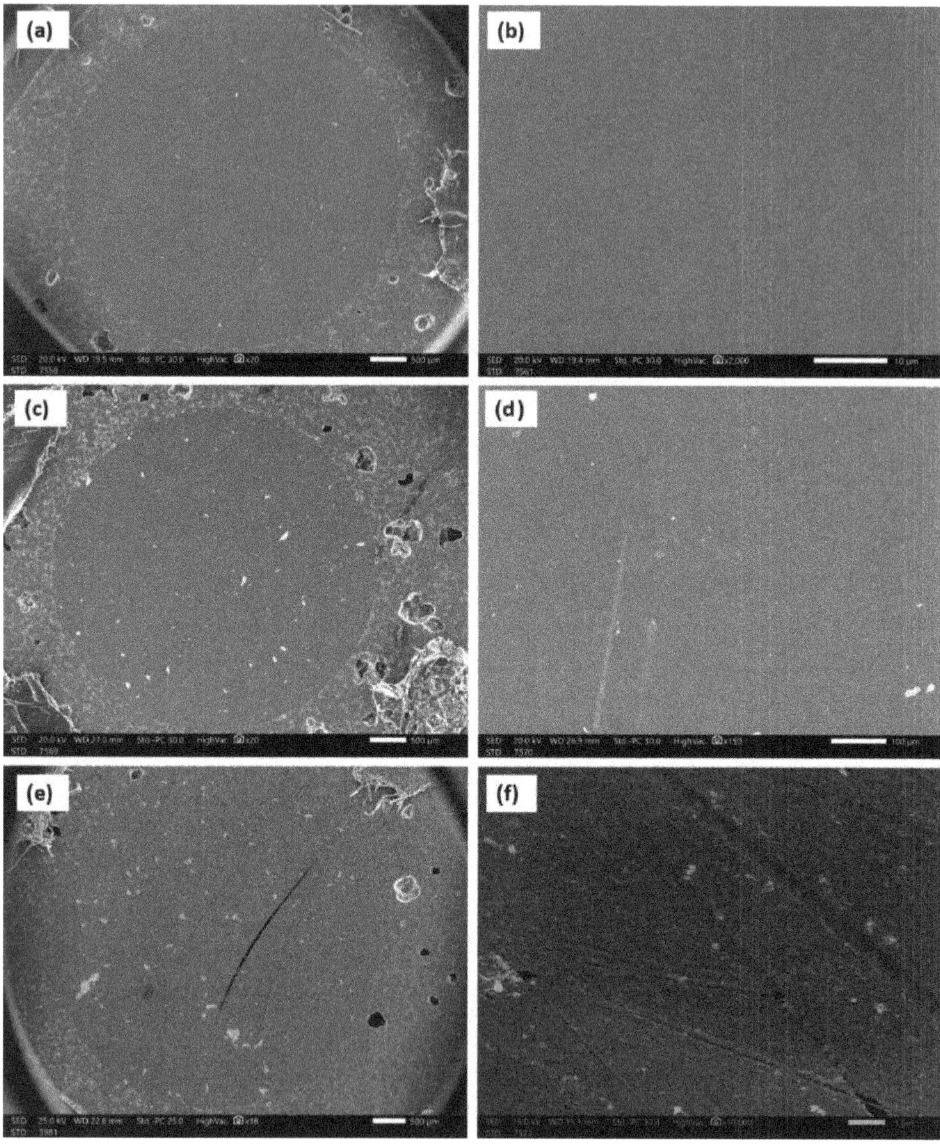

Figure 6. SEM macro and micrographs of the base acrylic material(**a**,**b**), (**c**,**d**) 5 wt.% ZrO_2 nano-composites (**c**,**d**) and 10 wt.% ZrO_2 nano-composites (**e**,**f**).

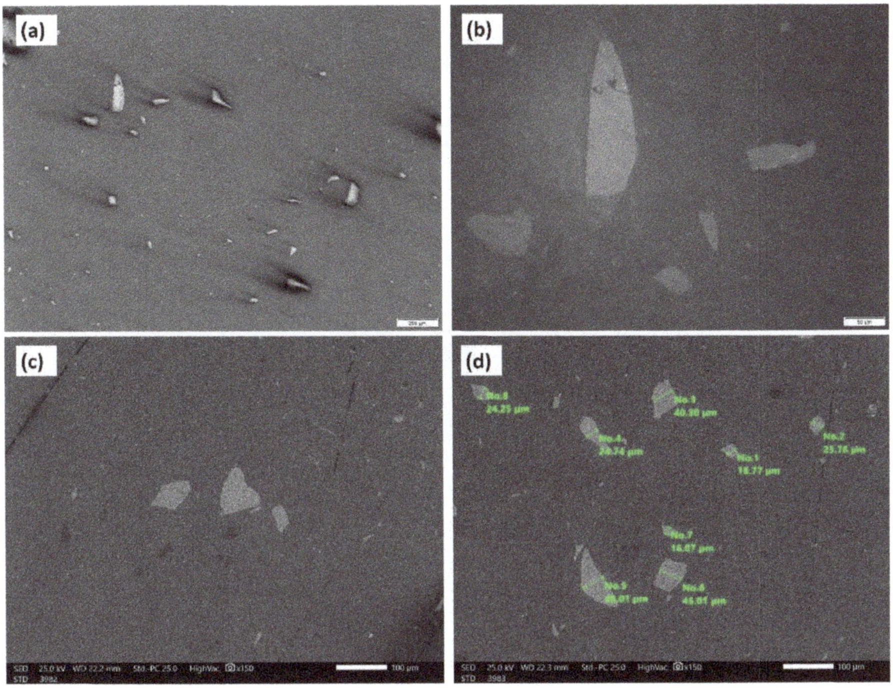

Figure 7. Optical microstructure at different magnifications (**a**,**b**) and SEM microstructure at different positions (**c**,**d**) for the acrylic with 10 wt.% ZrO sample.

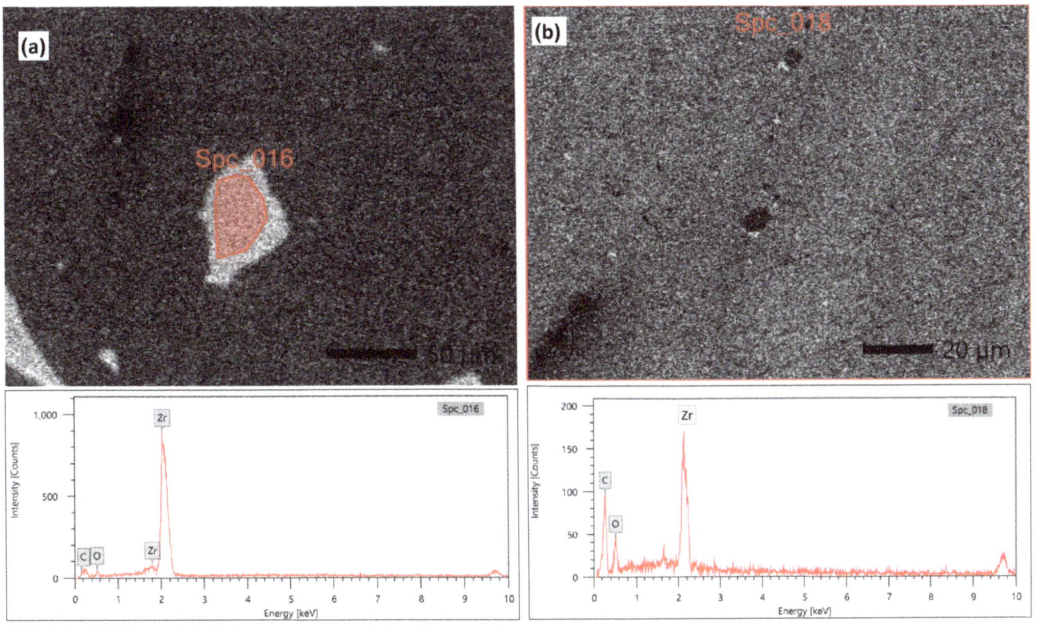

Figure 8. The EDS analysis of the micro-sized particle is presented in (**a**) and the EDS of the nanocomposite (**b**).

3.2. Micro-Hardness Analysis

The mean microhardness values and their statistical data were obtained by analysis of variance (one-way ANOVA) at *p*-value < 0.05, for the base acrylic material and the ZrO_2-reinforced groups. Furthermore, the data are illustrated in Figure 9 as a bar chart of microhardness against the wt.% of the ZrO_2 nanoparticles. The microhardness of the base acrylic material is 45.59 HV that increased to 48.59 HV after adding 5 wt.% ZrO_2 nanoparticles. A further increase in microhardness, up to 49.68 HV, was produced by increasing the concentration of the ZrO_2 nanoparticles to 10 wt.%.

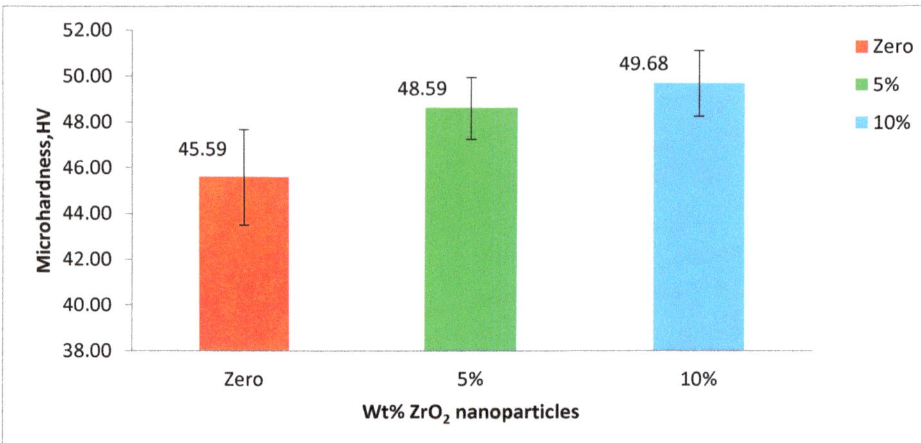

Figure 9. Mean hardness (VHN) of the tested groups of PMMA.

3.3. Wear Behavior

The mean height loss values and their statistical data were obtained by analysis of variance (one-way ANOVA) at *p*-value < 0.05 for the base acrylic material and the ZrO_2-reinforced groups. The data are illustrated in Figure 10 as a bar chart of the mean height loss against the wt.% of the ZrO_2 nanoparticles. Furthermore, the mean weight loss values and their statistical data were obtained by analysis of variance (one-way ANOVA) at *p*-value < 0.05 for the base acrylic material and the ZrO_2-reinforced groups. The data are illustrated in Figure 11 as a bar chart of the mean weight loss against the wt.% of ZrO_2 nanoparticles. The results of the wear tests of the height loss and weight loss both indicate a significant improvement in the wear resistance due to the addition of the ZrO_2 nanoparticles and that the wear resistance increased by increasing the ZrO_2 nanoparticle concentration.

The current investigation showed a significant enhancement of the wear resistance by adding ZrO_2 nanoparticles. Figures 12–14 show the SEM micrographs at different magnifications, (a) 150×, (b) 2000×, (c) 5000×, and (d) 20,000× (Figure 12) and 10,000× (Figures 13 and 14), for the wear surface of the different groups after conducting the wear test. The wear surface of the acrylic base material without ZrO_2 nanoparticles (Figure 12) showed severe wear features with deep wear tracks that can be observed on the low magnifications micrograph (a) 150×. Furthermore, a high density of microcracks (indicated by red arrows (d) 20,000×) can be observed on the highest magnification micrograph. The incorporation of the ZrO_2 nanoparticles changed the wear features of the wear surface, as can be observed in Figure 13 for the 5 wt.% ZrO_2 and in Figure 14 for the 10 wt.% ZrO_2. Less severe wear features were observed as the smoothness of the surface increased, in line with the increasing ZrO_2 content. In addition, no microcracks were observed, especially at 10 wt.% ZrO_2.

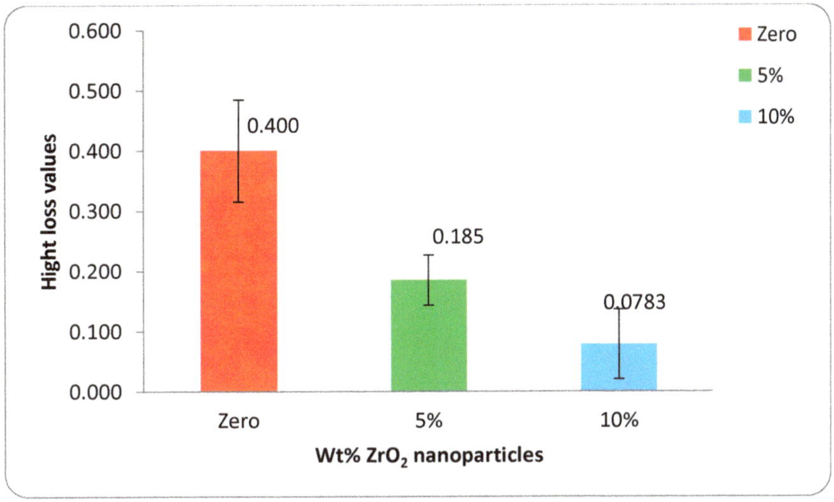

Figure 10. Mean height loss of the tested groups after wear testing.

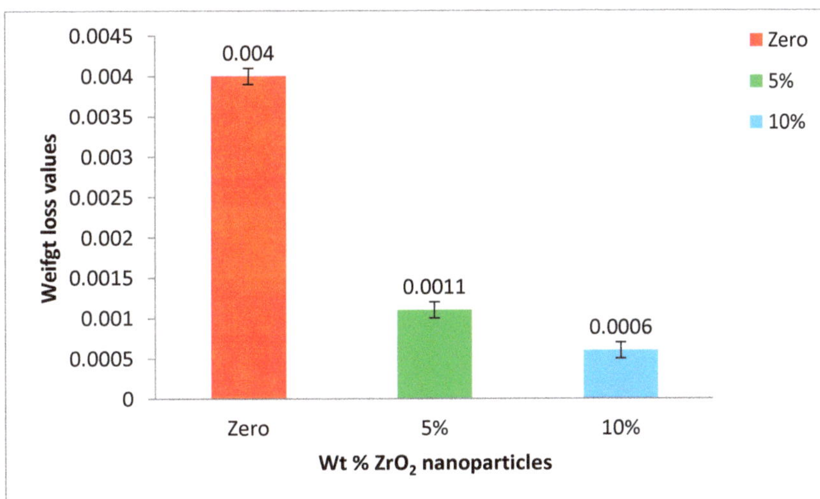

Figure 11. Mean weight loss of the tested groups after wear testing.

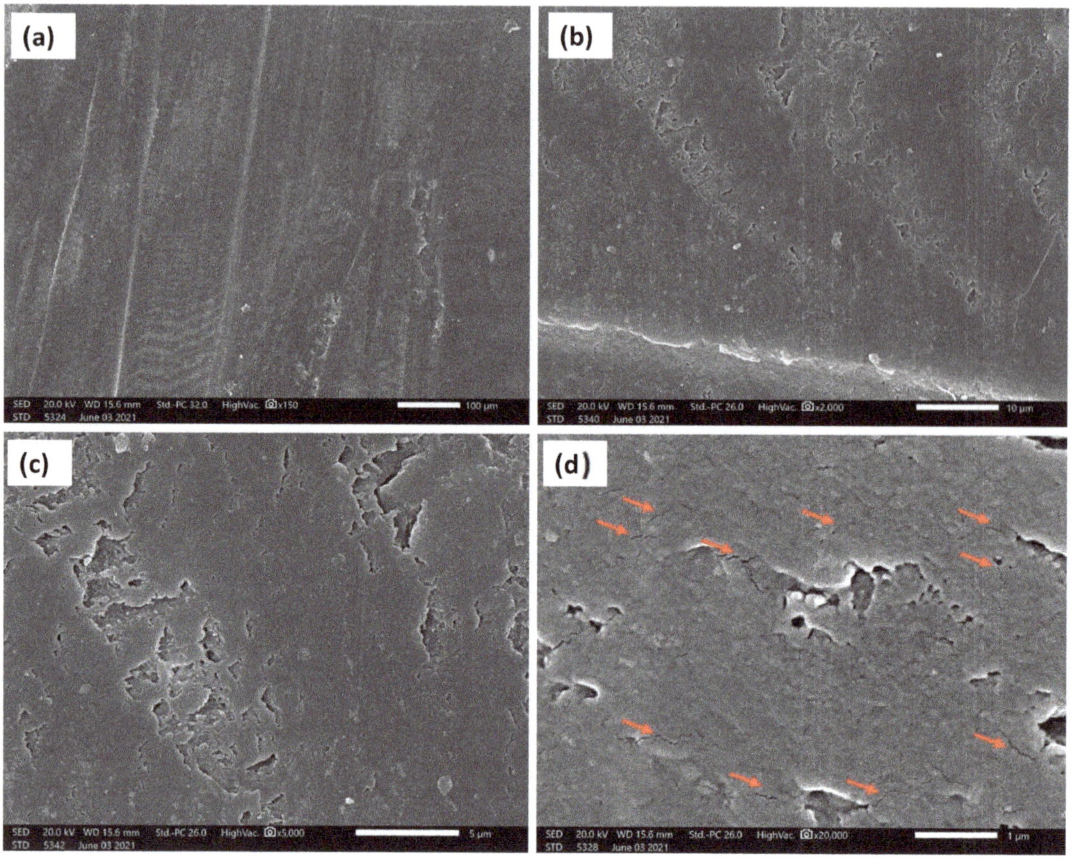

Figure 12. SEM micrographs at different magnifications and positions of the base acrylic material after wear test. Red arrows indicate the microcracks. (**a**) 150×, (**b**)2000×, (**c**) 5000× and (**d**) 20,000×.

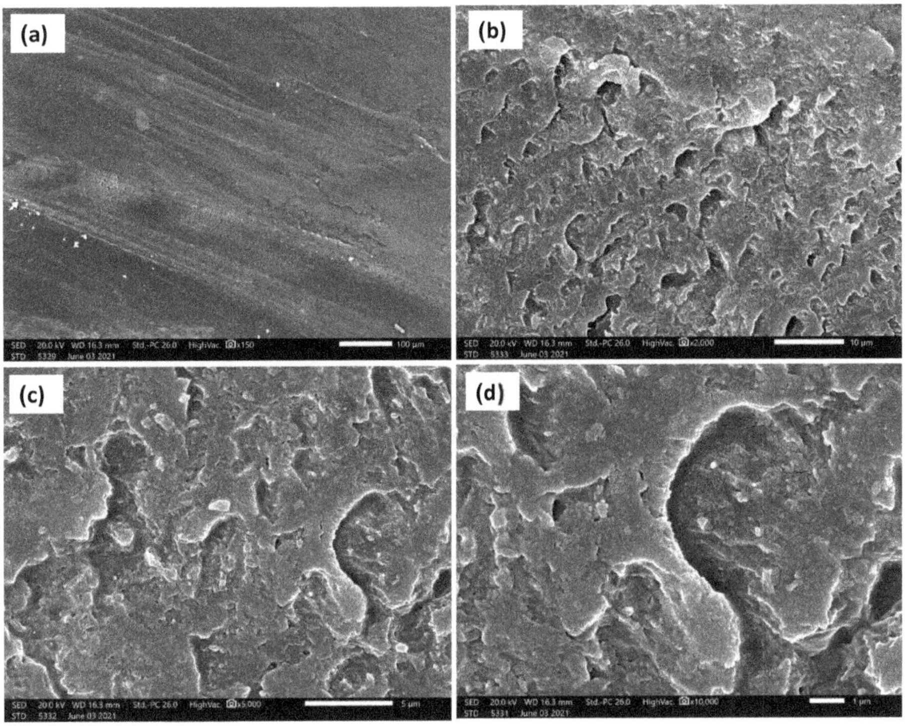

Figure 13. SEM micrographs at different magnifications and positions of the 5 wt.% ZrO_2 acrylic nanocomposite material after wear test. (**a**) 150×, (**b**)2000×, (**c**) 5000× and (**d**) 10,000×.

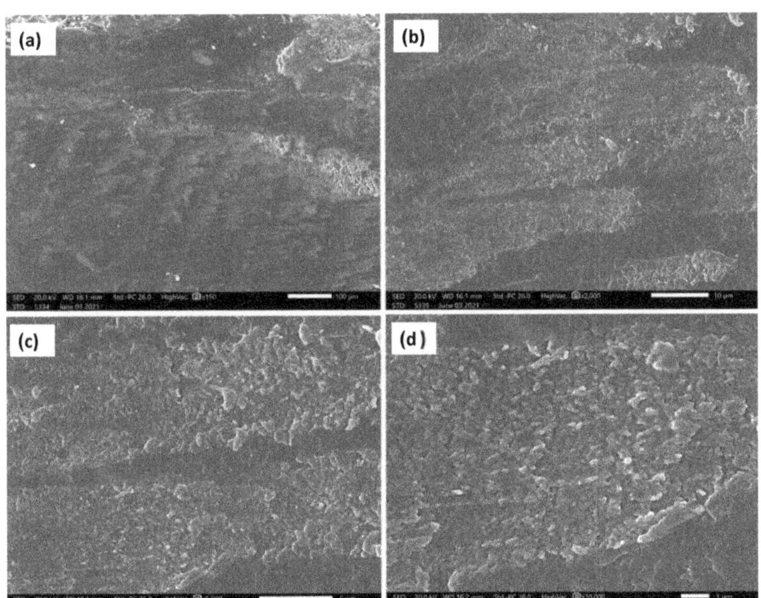

Figure 14. SEM micrographs at different magnifications and positions of the 10wt.% ZrO_2 acrylic nanocomposite material after wear test. (**a**) 150×, (**b**)2000×, (**c**) 5000× and (**d**)10,000×.

4. Discussion

Developing new materials with improved properties, especially high wear resistance, is an important focus of research. Thus, this work focuses on the improvement of wear resistance of acrylic denture teeth through the use of nanoparticle filler material. Different weight percentages of ZrO_2 nanofillers (5% and 10%) were used with the acrylic matrix. The addition of zirconia filler significantly increased the Vicker's hardness of the tested materials. This may have been due to the characteristics of the ZrO_2 particles, as well as high interfacial shear strength between the nanofiller and resin matrix, which in turn increased the resistance of the material for penetration [40]. Similar results have been reported in previous studies [9,12]. The increase in microhardness by about 6–8% relative to the base material is similar to that reported in previous studies. Hameed and Abd-Elrahaman [25] investigated the effect of ZrO_2 nanoparticles on the properties of acrylic denture base and reported an increase of about 5% in microhardness when increasing ZrO_2 nanoparticles up to 7%.

Wear resistance is an important requirement of prosthetic materials, such as denture teeth, which do not provide a durable functional and esthetic restoration unless they exhibit high wear resistance [41]. Intraoral wear is a complex process related to both the composition of denture teeth and the patient's oral condition. The ability to replicate and control these factors intraorally is challenging. Therefore, in vitro two-body wear testing is a valid model to measure antagonist wear without involving an intermediate medium that may influence the results [42]. The height and weight loss of the samples were evaluated. In dentistry, height loss is a more clinically relevant parameter, because the vertical distance between the maxilla and mandible is stabilized by the occlusal contact points. However, weight loss provides more accurate measurements, as indicated by lower variation. The wear resistance was significantly improved after the addition of zirconia nanofillers to the polymer matrix at both 5% and 10% concentrations. This result agrees with previous studies, in which concentrations of saline-treated alumina 0.1% [41] and silver nanoparticles 0.2–8% [43] considerably improved the resin wear resistance of PMMA. In terms of the effect of surface roughness on wear resistance, Benli et al. [44] investigated the relationship between the wear rate and surface roughness of occlusal splint materials made of contemporary and high-performance polymers and found that it is directly proportional. Furthermore, Nayyer et al. [45] reported that surface roughness is affected by clinical adjustments such as polishing, because polishing leads to smooth surfaces that undergo less wear and provides the advantage of extended longevity of the restoration [46,47]. It should be mentioned here that in this study, the samples were examined for wear behavior without polishing; thus it was expected that the polishing would significantly improve the wear resistance of the developed nanocomposites.

This improvement can be attributed to the fact that the ZrO_2 particles exhibited greater hardness than the surrounding resin matrix and, therefore, were not as easily abraded by the antagonist [40]. The SEM of the acrylic base material without ZrO_2 particles (Figure 12) showed high-density microcracks, which may have been the main mechanism behind the deterioration of the surface and the high reduction in both the height and weight observed in the measurements above. This might indicate that the type of wear was fatigue wear due to the formation of cracks under the subsurface. It has been reported that fatigue wear occurs when microcracks form below the surface of a material and eventually coalesce to free a small fragment of the material. Materials with a lower elastic modulus, such as unfilled resins, are susceptible to fatigue wear [42]. Furthermore, increasing the concentration of ZrO_2 from 5% to 10% produced more improvement in the wear resistance, which can be attributed to the change in the wear mechanism from fatigue wear to abrasive wear type. This was confirmed by the elimination of microcracks at high ZrO_2 particle concentrations, as can be observed in Figure 14. The results suggest that wear resistance may be influenced not only by the presence or absence of fillers, but also by the type, amount and size of the filler particles, as well as by how the filler particles are integrated into the matrix. However, the results of this study can provide guidance because the applied in vitro test arrangement cannot fully replicate the intraoral conditions. Although the use of natural enamel as an antagonist is considered the gold standard, disadvantages do exist, including variability in the form, fluoride content, and amount of aprismatic enamel; therefore, standardizing each stylus is not easy [48]. Further in vivo studies are needed to investigate the effect of adding nano ZrO_2 with different concentrations on wear resistance and other mechanical properties intraorally in acrylic resin denture teeth.

5. Conclusions

This study investigated the effect of using ZrO_2 nanoparticles as a filler material in acrylic denture teeth, aiming to enhance the wear resistance. Based on the obtained results, the following conclusions can be drawn:

(1) The microstructural investigation of reinforced denture teeth indicates sound material preparation using the applied regime without porosity or macro defects. No segregation was observed for the nanoparticles; only submicron-sized ZrO_2 particles were found.
(2) The addition of zirconium oxide nanofillers to PMMA at both 5% and 10% increased the microhardness, with values up to 49.7 HV found after the addition of 10 wt.% ZrO_2.
(3) The wear mechanism in the acrylic base material without nanoparticle addition was found to be fatigue wear; a high density of microcracks was found.
(4) The addition of 5 wt.% ZrO_2 significantly reduced weight loss and height loss relative to the base material, which indicates improved wear resistance.
(5) Increasing the nanoparticles to 10 wt.% ZrO_2 further improved the wear resistance, with no microcracks found.

Author Contributions: Conceptualization, K.M.E.-T., D.A.B. and M.E.E.-S.; methodology, K.M.E.-T., D.A.B. and M.E.E.-S.; software, K.M.E.-T. and D.A.B.; validation, K.M.E.-T., D.A.B. and M.E.E.-S.; formal analysis, K.M.E.-T., D.A.B. and M.E.E.-S.; investigation K.M.E.-T., M.M.Z.A., D.A.B. and M.E.E.-S.; resources, K.M.E.-T., D.A.B. and M.E.E.-S.; data curation, K.M.E.-T., D.A.B. and M.E.E.-S.; writing—original draft preparation, K.M.E.-T., M.M.Z.A., D.A.B. and M.E.E.-S.; writing—review and editing, K.M.E.-T., M.M.Z.A. and D.A.B.; visualization K.M.E.-T., M.M.Z.A., D.A.B. and M.E.E.-S.; supervision, K.M.E.-T., M.M.Z.A., I.A. and D.A.B.; project administration, K.M.E.-T., M.M.Z.A. and D.A.B.; funding acquisition, M.M.Z.A. and I.A. All authors have read and agreed to the published version of the manuscript.

Funding: This research received no external funding.

Institutional Review Board Statement: Not applicable.

Informed Consent Statement: Not applicable.

Data Availability Statement: This will be made available upon request through the corresponding author.

Acknowledgments: Sabry A. El-Korashy at Suez Canal University is greatly acknowledged for his support in carrying out the salinization process in the laboratory of Inorganic Chemistry, Chemistry Department, Faculty of Science. Furthermore, Mohamed M. Elsayed Seleman at Suez University is greatly acknowledged for his support in conducting the microstructural investigations.

Conflicts of Interest: The authors declare no conflict of interest.

References

1. Stober, T.; Lutz, T.; Gilde, H.; Rammelsberg, P. Wear of resin denture teeth by two-body contact. *Dent. Mater.* **2006**, 22, 243–249. [CrossRef]
2. Kawano, F.; Ohguri, T.; Ichikawa, T.; Mizuno, I.; Hasegawa, A. Shock absorbability and hardness of commercially available denture teeth. *Int. J. Prosthodont.* **2002**, 15, 243–247. [PubMed]
3. Beall, J.R. Wear of acrylic resin teeth (progress report). *J. Am. Dent. Assoc.* **1943**, 30, 252–256. [CrossRef]
4. Adams, L.P.; Jooste, C.H.; Thomas, C.J.; Harris, A.M.P. Biostereometric quantification of clinical denture tooth wear. *J. Oral Rehabil.* **1996**, 23, 667–674. [CrossRef]

5. Satoh, Y.; Nagai, E.; Maejima, K.; Azaki, M. Wear of denture teeth by use of metal plates part 2: Abrasive wear of posterior teeth. *J. Nihon Univ.* **1992**, *34*, 16–27. [CrossRef]
6. Akin, H.; Tugut, F.; Guney, U.; Akar, T. Shear bond strength of denture teeth to two chemically different denture base resins after various surface treatments. *J. Prosthodont.* **2014**, *23*, 152–156. [CrossRef]
7. Mello, P.C.; Coppedê, A.R.; Macedo, A.P.; Da, M.; De Mattos, G.C.; Cristina, R.; Rodrigues, S.; Ribeiro, R.F. Abrasion wear resistance of different artificial teeth opposed to metal and composite antagonists. *J. Appl. Oral Sci.* **2009**, *17*, 451–457. [CrossRef]
8. Kamonwanon, P.; Yodmongkol, S.; Chantarachindawong, R.; Thaweeboon, S.; Thaweeboon, B.; Srikhirin, T. Wear resistance of a modified polymethyl methacrylate artificial tooth compared to five commercially available artificial tooth materials. *J. Prosthet. Dent.* **2015**, *114*, 286–292. [CrossRef] [PubMed]
9. Stober, T.; Henninger, M.; Schmitter, M.; Pritsch, M.; Rammelsberg, P. Three-body wear of resin denture teeth with and without nanofillers. *J. Prosthet. Dent.* **2010**, *103*, 108–117. [CrossRef]
10. Suwannaroop, P.; Chaijareenont, P.; Koottathape, N.; Takahashi, H.; Arksornnukit, M. In vitro wear resistance, hardness and elastic modulus of artificial denture teeth. *Dent. Mater. J.* **2011**, *30*, 461–468. [CrossRef]
11. Gürbüz, Ö.A.; Ünalan, F.; Kursoglu, P. In vitro wear of denture teeth acrylic resin milled glass fiber composite. *Oral Health Dent. Manag. Black Sea Ctries.* **2005**, *4*, 46–51.
12. Gad, M.M.; Rahoma, A.; Al-Thobity, A.M.; ArRejaie, A.S. Influence of incorporation of ZrO_2 nanoparticles on the repair strength of polymethyl methacrylate denture bases. *Int. J. Nanomed.* **2016**, *11*, 5633–5643. [CrossRef] [PubMed]
13. Ghaffari, T.; Barzegar, A.; Rad, F.H.; Moslehifard, E.; Author, C. Effect of nanoclay on thermal conductivity and flexural strength of polymethyl methacrylate acrylic resin. *J. Dent. Shiraz. Univ. Med. Sci.* **2016**, *17*, 121–127.
14. Labella, R.; Braden, M.; Deb, S. Novel hydroxyapatite-based dental composites. *Biomaterials* **1994**, *15*, 1197–1200. [CrossRef]
15. Ghahremani, L.; Shirkavand, S.; Akbari, F.; Sabzikari, N. Tensile strength and impact strength of color modified acrylic resin reinforced with titanium dioxide nanoparticles. *J. Clin. Exp. Dent.* **2017**, *9*, e661–e665. [CrossRef]
16. Mc Nally, L.; O'Sullivan, D.J.; Jagger, D.C. An in vitro investigation of the effect of the addition of untreated and surface treated silica on the transverse and impact strength of poly(methyl methacrylate) acrylic resin. *Biomed. Mater. Eng.* **2006**, *16*, 93–100.
17. Gad, M.M.; Al-Thobity, A.M.; Rahoma, A.; Abualsaud, R.; Al-Harbi, F.A.; Akhtar, S. Reinforcement of PMMA denture base material with a mixture of ZrO_2 nanoparticles and glass fibers. *Int. J. Dent.* **2019**, *2019*, 2489393. [CrossRef] [PubMed]
18. Ayad, N.M.; Dawi, M.F.; Fatah, A.A. Effect of reinforcement of high-impact acrylic resin with micro-Zirconia on some physical and mechanical properties. *Rev. Clínica Pesqui. Odontológica* **2008**, *4*, 145–151.
19. Gad, M.; Arrejaie, A.S.; Abdel-Halim, M.S.; Rahoma, A. The reinforcement effect of nano-zirconia on the transverse strength of repaired acrylic denture base. *Int. J. Dent.* **2016**, *2016*, 7094056. [CrossRef]
20. Gad, M.M.; Abualsaud, R.; Rahoma, A.; Al-Thobity, A.M.; Al-Abidi, K.S.; Akhtar, S. Effect of zirconium oxide nanoparticles addition on the optical and tensile properties of polymethyl methacrylate denture base material. *Int. J. Nanomed.* **2018**, *13*, 283–292. [CrossRef]
21. Tang, E.; Cheng, G.; Pang, X.; Ma, X.; Xing, F. Synthesis of nano-ZnO/poly(methyl methacrylate) composite microsphere through emulsion polymerization and its UV-shielding property. *Colloid Polym. Sci.* **2006**, *284*, 422–428. [CrossRef]
22. Asopa, V.; Suresh, S.; Khandelwal, M.; Sharma, V.; Asopa, S.S.; Kaira, L.S. A comparative evaluation of properties of zirconia reinforced high impact acrylic resin with that of high impact acrylic resin. *Saudi J. Dent. Res.* **2015**, *6*, 146–151. [CrossRef]
23. Safi, I.N.; Hassanen, K.A.; Ali, N.A. Assessment of zirconium oxide nano-fillers incorporation and silanation on impact, tensile strength and color alteration of heat polymerized acrylic resin. *J. Baghdad Coll. Dent.* **2012**, *24*, 36–42.
24. Ergun, G.; Sahin, Z.; Ataol, A. The effects of adding various ratios of zirconium oxide nanoparticles to poly (methyl methacrylate) on physical and mechanical properties. *J. Oral Sci.* **2018**, *60*, 304–315. [CrossRef]
25. Ali Sabri, B.; Satgunam, M.; Abreeza, N.M.; Abed, A.N. A review on enhancements of PMMA denture base material with different nano-fillers. *Cogent Eng.* **2021**, *8*, 1875968. [CrossRef]
26. Mohammed, D.; Mudhaffar, M. Effect of modified zirconium oxide nano-fillers addition on some properties of heat cure acrylic denture base material. *J. Baghdad Coll. Dent.* **2012**, *24*, 1–7.
27. Zidan, S.; Silikas, N.; Alhotan, A.; Haider, J.; Yates, J. Investigating the mechanical properties of ZrO_2-impregnated PMMA nanocomposite for denture-based applications. *Materials* **2019**, *12*, 1344. [CrossRef] [PubMed]
28. Eltamimi, K.M.; Bayoumi, D. Microtesile bond strength of zirconia Nano-fillers reinforced acrylic resin to denture teeth after thermocylcling. *Egypt. Dent. J.* **2021**, *67*, 767–774. [CrossRef]
29. Gad, M.M.; Fouda, S.M.; ArRejaie, A.S.; Al-Thobity, A.M. Comparative effect of different polymerization techniques on the flexural and surface properties of acrylic denture bases. *J. Prosthodont.* **2019**, *28*, 458–465. [CrossRef]
30. Yuzugullu, B.; Acar, O.; Cetinsahin, C.; Celik, C. Effect of different denture cleansers on surface roughness and microhardness of artificial denture teeth. *J. Adv. Prosthodont.* **2016**, *8*, 333–338. [CrossRef]
31. Zhang, X.Y.; Zhang, X.J.; Huang, Z.L.; Zhu, B.S.; Rong-Rong, R.R.C. Hybrid effects of zirconia nanoparticles with aluminum borate whiskers on mechanical properties of denture base resin PMMA. *Dent. Mater. J.* **2014**, *33*, 141–146. [CrossRef] [PubMed]
32. Pero, A.C.; Scavassin, P.M.; Leite, A.R.; Marin, D.O.; Paleari, A.G.; Compagnoni, M.A. Effect of immersion cleansers on the bond strength between a denture base resin and acrylic resin teeth. *Int. J. Adhes. Adhes.* **2013**, *44*, 180–183. [CrossRef]
33. Fisher, R.M.; Moore, B.K.; Swartz, M.L.; Dykema, R.W. The effects of enamel wear on the metal-porcelain interface. *J. Prosthet. Dent.* **1983**, *50*, 627–631. [CrossRef]

34. Ghazal, M.; Yang, B.; Ludwig, K.; Kern, M. Two-body wear of resin and ceramic denture teeth in comparison to human enamel. *Dent. Mater.* **2008**, *24*, 502–507. [CrossRef] [PubMed]
35. Ghazal, M.; Hedderich, J.; Kern, M. Wear of feldspathic ceramic, nano-filled composite resin and acrylic resinartificial teeth when opposed todifferent antagonists. *Eur. J. Oral Sci.* **2008**, *116*, 585–592. [CrossRef]
36. Ahmed, M.M.Z.; Elnaml, A.; Shazly, M.; El-Sayed Seleman, M.M. The effect of top surface lubrication on the friction stir welding of polycarbonate sheets. *Int. Polym. Process.* **2021**, *36*, 94–102. [CrossRef]
37. Shazly, M.; Ahmed, M.M.Z.; El-Raey, M. Friction stir welding of polycarbonate sheets. In *Characterization of Minerals, Metals, and Materials*; TMS Annual Meeting; Carpenter, J.S., Bai, C., Hwan, -Y.J., Ikhmayies, S., Li, B., Monteiro, S.N., Peng, Z., Zhang, M., Eds.; Wiley: Hoboken, NJ, USA, 2014; ISBN 9781118887868.
38. Wang, S.; Gao, H.; Li, J.; Wang, Y.; Chen, C.; Yu, X.; Tang, S.; Zhao, X.; Sun, G.; Li, D. Comparative study of the photoluminescence performance and photocatalytic activity of $CeO_2/MgA_{l2}O_4$ composite materials with an n-n heterojunction prepared by one-step synthesis and two-step synthesis methods. *J. Phys. Chem. Solids* **2021**, *150*, 109891. [CrossRef]
39. Parish, C.M.; Brewer, L.N. Multivariate statistics applications in phase analysis of STEM-EDS spectrum images. *Ultramicroscopy* **2010**, *110*, 134–143. [CrossRef] [PubMed]
40. Ahmed, M.A.; Ebrahim, M.I. Effect of zirconium oxide nano-fillers addition on the flexural strength, fracture toughness, and hardness of heat-polymerized acrylic resin. *World J. Nano Sci. Eng.* **2014**, *4*, 50–57. [CrossRef]
41. Chaijareenont, P.; Takahashi, H.; Nishiyama, N.; Arksornnukit, M. Effect of different amounts of 3-methacryloxypropyltrimethoxysilane on the flexural properties and wear resistance of alumina reinforced PMMA. *Dent. Mater. J.* **2012**, *31*, 623–628. [CrossRef]
42. Esquivel, J.; Lawson, N.C.; Kee, E.; Bruggers, K.; Blatz, M.B. Wear of resin teeth opposing zirconia. *J. Prosthet. Dent.* **2020**, *124*, 488–493. [CrossRef]
43. Chladek, G.; Pakieła, K.; Pakieła, W.; Zmudzki, J.; Adamiak, M.; Krawczyk, C. Effect of antibacterial silver-releasing filler on the physicochemical properties of poly(methyl methacrylate) denture base material. *Materials* **2019**, *12*, 4146. [CrossRef] [PubMed]
44. Benli, M.; Eker, B.; Yusuf, G.; Bilge, K.; Rohlig, G.; Evlioğlu, G.; Huck, O. Surface roughness and wear behavior of occlusal splint materials made of contemporary and high-performance polymers. *Odontology* **2020**, *108*, 240–250. [CrossRef]
45. Nayyer, M.; Zahid, S.; Hassan, S.H.; Mian, S.A.; Mehmood, S.; Khan, H.A.; Kaleem, M.; Zafar, M.S.; Khan, A.S. Comparative abrasive wear resistance and surface analysis of dental resin-based materials. *Eur. J. Dent.* **2018**, *12*, 57–66. [CrossRef] [PubMed]
46. Matzinger, M.; Hahnel, S.; Preis, V.; Rosentritt, M. Polishing effects and wear performance of chairside CAD/CAM materials. *Clin. Oral Investig.* **2019**, *23*, 725–727. [CrossRef]
47. Preis, V.; Behr, M.; Handel, G.; Schneider-Feyrer, S.; Hahnel, S.; Martin, R. Wear performance of dental ceramics after grind- ing and polishing treatments. *J. Mech. Behav. Biomed. Mater.* **2012**, *10*, 13–22. [CrossRef]
48. Heintze, S.; Zellweger, G.; Cavalleri, A.; Ferracane, J. Influence of the antagonist material on the wear of different composites using two different wear simulation methods. *Dent. Mater.* **2006**, *22*, 166–175. [CrossRef] [PubMed]

polymers

MDPI

Article

Effect of Moisture Content on the Processing and Mechanical Properties of a Biodegradable Polyester

Vincenzo Titone [1,2,*], Antonio Correnti [3] and Francesco Paolo La Mantia [1,2,*]

1 Department of Engineering, University of Palermo, Viale delle Scienze, 90128 Palermo, Italy
2 INSTM Consortium for Materials Science and Technology, Via Giusti 9, 50125 Florence, Italy
3 Joeplast S.p.A., Zona Industriale S.S. 189, 92025 Casteltermini, Italy; info.antocor@gmail.com
* Correspondence: vincenzo.titone@unipa.it (V.T.); francescopaolo.lamantia@unipa.it (F.P.L.M.)

Abstract: This work is focused on the influence of moisture content on the processing and mechanical properties of a biodegradable polyester used for applications in injection molding. The pellets of the biodegradable polyester were exposed under different relative humidity conditions at a constant temperature before being compression molded. The compression-molded specimens were again placed under the above conditions before the mechanical testing. With all these samples, it is possible to determine the effect of moisture content on the processing and mechanical properties separately, as well as the combined effect of moisture content on the mechanical properties. The results obtained showed that the amount of absorbed water—both before processing and before mechanical testing—causes an increase in elongation at break and a slight reduction of the elastic modulus and tensile strength. These changes have been associated with possible hydrolytic degradation during the compression molding process and, in particular, with the plasticizing action of the moisture absorbed by the specimens.

Keywords: biodegradable polymers; mechanical properties; rheology; processing; moisture content

Citation: Titone, V.; Correnti, A.; La Mantia, F.P. Effect of Moisture Content on the Processing and Mechanical Properties of a Biodegradable Polyester. *Polymers* **2021**, *13*, 1616. https://doi.org/10.3390/polym13101616

Academic Editors: Domenico Acierno and Antonella Patti

Received: 27 April 2021
Accepted: 13 May 2021
Published: 17 May 2021

1. Introduction

The increasing use of plastics in agriculture and the growing amount of land and sea debris have led to the design and development of new polymer materials more "friendly" to the environment [1–3].

Generally, these biodegradable polymers are more expensive than traditional ones but are more environmentally friendly and more suitable for several specific applications, such as food packaging, agricultural mulching films, etc. [4–8]. In particular, biodegradable and compostable polymers are very appealing when used for some applications in agriculture such as films for mulching and plastic pots because these products can be left in the ground where they are transformed into water, CO_2, and biomass. This biomass is useful as a soil improver.

Most parts of the biodegradable polymer systems are made of polyesters; it is well known that the polyesters absorb humidity, and the presence of moisture gives rise to dangerous degradation phenomena. The degradation phenomena are mainly due to the hydrolysis of the macromolecular chains with a severe decrease of molecular weight [9–11]. In addition, the presence of moisture strongly modifies the mechanical properties [11–13].

Although the influence of temperature was widely studied in the literature [14–19], the influence of relative humidity on the processing and mechanical properties is still an open field of research.

Harris and Lee [11] investigated the hydrolytic degradation of PLA and a PLA/polycarbonate blend exposed at high temperature and humidity, finding a significant moisture absorption and hydrolysis, resulting in degradation of properties. Similarly, Muthuraj et al. [12] reported studies on the hydrolytic degradation of poly(butylene succinate), PBS, poly(butylene adipate-*co*-terephthalate), PBAT, and PBS/PBAT blend, finding that, as a result of chain scission because of the hydrolysis mechanism, mechanical performance was significantly affected after conditioning.

In our previous work [13], we demonstrated how the presence of moisture significantly affects the mechanical properties of biodegradable polyesters subjected to UV irradiation.

Of course, the application of fully amorphous biodegradable polymers is limited by the fact that the glass transition temperature (Tg) of the polymer is strongly influenced by the relative humidity, especially for hydrophilic polymers.

The aim of this study was to investigate the influence of moisture on the processing and mechanical properties of a biodegradable polyester used for injection molding. In particular, the effect of moisture has been investigated on both the pellets before processing and on the specimens before the mechanical testing. In particular, pellets of a biodegradable polyester were exposed at different relative humidity values and fixed temperatures before compression molding to investigate the effect of this treatment on the rheological behavior of this polymer. The compression-molded specimens were again treated under the same conditions before the mechanical testing in order to investigate the effect of the presence of different moisture contents both before and after processing on the mechanical properties.

The experimental results clearly indicate that the processing of humid samples led to a decrease in molecular weight and consequent decrease in viscosity. The mechanical properties are, of course, influenced by the presence of moisture absorbed both before and/or after the compression molding. In the conditions adopted in this work, the more important effect on the mechanical properties is the plasticizing effect of the moisture.

2. Materials and Methods

The material used in this study was a biodegradable polyester Mater-Bi TF01U supplied by Novamont (Novamont, Novara, Italy). It is a bioplastic based on an aliphatic polyester with a melting point of 72–75 °C and a glass transition temperature between −40 and −35 °C, used for injection molding.

The specimens for the rheological and mechanical characterization were prepared by compression molding in a Carver (Carver, Wabash, IN, USA) laboratory hydraulic press at the temperature of 180 °C under a mold pressure of 300 psi and for about 3 min. Before compression molding, the pellets were subjected to three different pretreatments in the environmental conditions reported in Table 1.

Table 1. Environmental conditions used for conditioning the pellets and the compression-molded specimens.

Condition 1	Condition 2	Condition 3
T = 38 °C RH = 0%	T = 38 °C RH = 50%	T = 38 °C RH = 90%

Condition 1 means, of course, a treatment in the dry state, while condition 3 is relative to treatment in an almost water-saturated condition. The temperature has been chosen to accelerate the sorption of the humidity.

Before testing the mechanical properties, the specimens were subjected to the same environmental conditions reported in Table 1.

Figure 1 depicts the processing undergone by the pellets and by the compression-molded specimens.

The conditioning of the pellets and of the compression-molded samples was carried out in a climate chamber KBF 115–Binder (Binder, Tuttlingen, Germany).

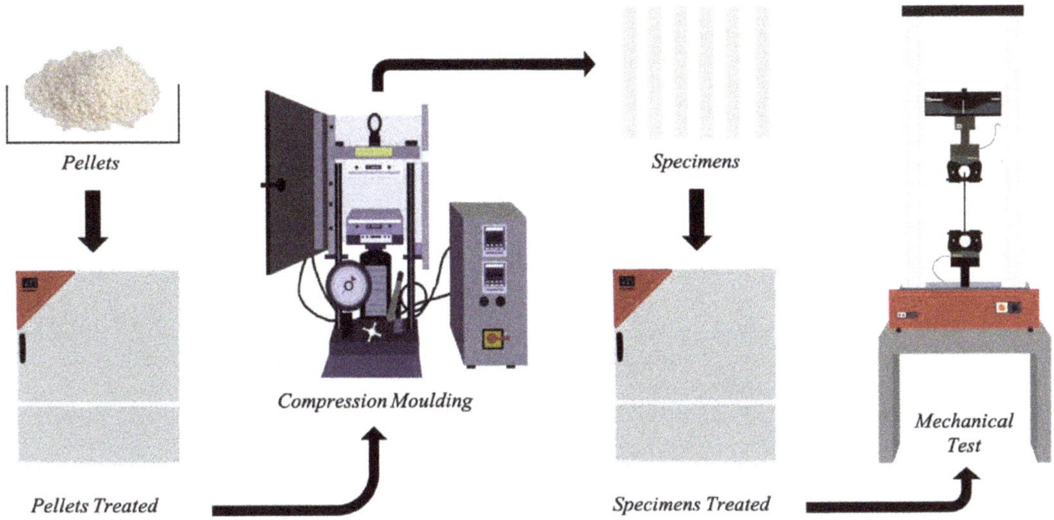

Figure 1. Scheme of the procedure adopted in this work.

Samples conditioned in nine different conditions (Table 2) were then investigated.

Table 2. Specimens for rheological and mechanical characterization.

	Specimens Code	
MC_{11}	MC_{21}	MC_{31}
MC_{12}	MC_{22}	MC_{32}
MC_{13}	MC_{23}	MC_{33}

The first subscript indicates under which conditions (Table 1) the pellets have been conditioned before processing and the second subscript indicates under which conditions the compression-molded specimens have been treated. For example, for the sample code MC_{11}, the pellets have been treated under condition 1 before compression molding, and the compression-molded specimens were treated again under condition 1 before testing the mechanical properties, while, for the sample MC_{23}, the pellets have been treated under condition 2 before compression molding and the compression-molded specimens were treated under condition 3 before testing the mechanical properties. With all these samples it is then possible to determine separately the effect of moisture content on the processing and mechanical properties, as well as the combined effect—processing and environmental exposure—on the mechanical properties of moisture content.

Moisture content was calculated according to ASTM D570-98 [20], using Equation (1).

$$MC,\ \% = \frac{(W_w - W_d)}{W_d} * 100 \quad (1)$$

where W_W and W_d are the weight of the sample after and before moisture exposure, respectively.

The rheological characterization was performed on disk-shaped samples, using an ARES G2 (TA Instruments, New Castle, DE, USA). The experiments were carried out in parallel plates with a gap of about 1.5 mm and a diameter of 25 mm. The shear viscosity values of the samples were measured at 180 °C and in a frequency range from 0.1 and 100 rad/s.

Tensile properties tests were carried out using an Instron (Instron, High Wycombe, UK) universal testing machine mod. 3365 equipped with a 1 kN load cell and 30 mm gauge length extensimeter. The tensile strength specimens were rectangular sheets according to ASTM D638-14 [21] (length: 90 mm, width: 10 mm, thickness: ≃0.4 mm). The mechanical tests were carried out on the conditioned specimens immediately after the end of the conditioning in order to avoid any significant change in the value of the moisture.

Elastic modulus, E, tensile strength, TS, and elongation at break, EB, were measured, and the reported data were determined as an average of 12 samples. The elastic modulus was measured at a deformation speed of 1 mm/min. When the deformation achieved 10%, the crosshead speed was increased to 100 mm/min until final breaking.

3. Results and Discussion

In Table 3, the values of moisture content at equilibrium for all three conditioning treatments are reported. As expected, the amount of moisture increases with increasing the relative humidity.

Table 3. Moisture content of the polymer conditioned under different relative humidity.

Relative Humidity, %	Moisture Content, %
0	0.00
50	0.35
90	0.42

Stress–strain curves obtained from the tensile tests of all the specimens are reported in Figures 2–4.

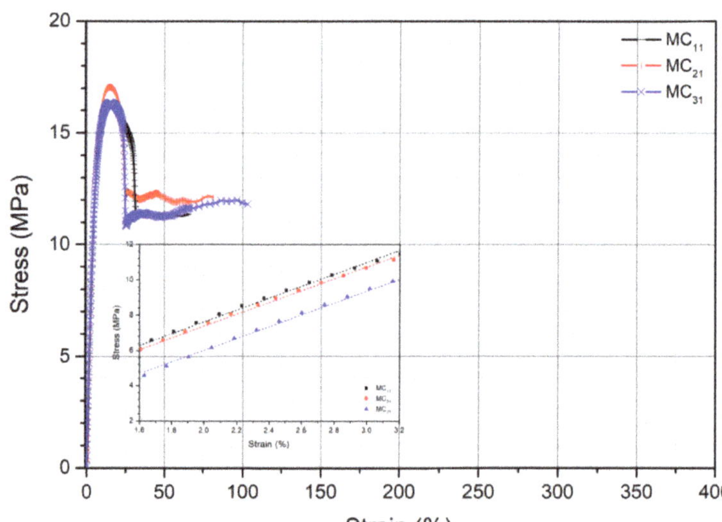

Figure 2. Stress–strain curve of samples coming from pellets conditioned under the three conditions and specimens for mechanical testing conditioned at RH = 0%.

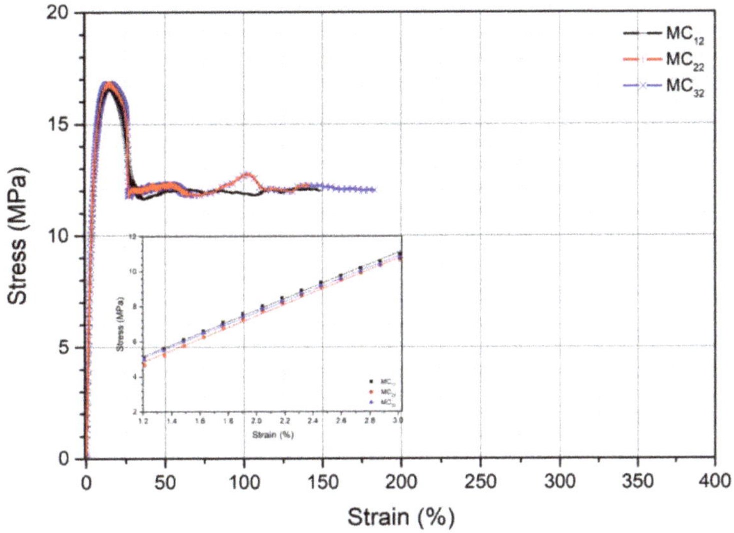

Figure 3. Stress–strain curve of samples coming from pellets conditioned under the three conditions and specimens for mechanical testing conditioned at RH = 50%.

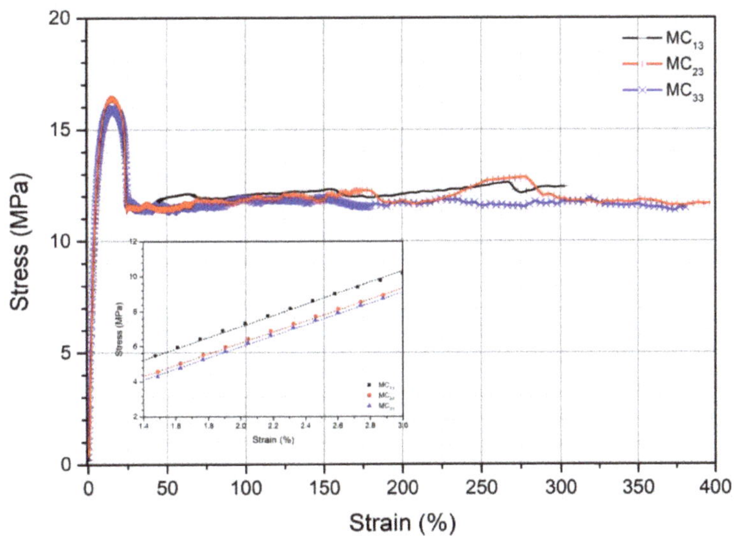

Figure 4. Stress–strain curve of samples coming from pellets conditioned under the three conditions and specimens for mechanical testing conditioned at RH = 90%.

All the curves show a ductile behavior, but the values of elastic modulus, E, tensile strength, TS, and elongation at break, EB are different according to the pretreatment, as evident in Table 4, in which the values of these mechanical properties for all the investigated samples are reported.

Table 4. Elastic modulus, E, tensile strength, TS, elongation at break, EB, of all the specimens.

Specimens Code	E, MPa	TS, MPa	EB, %
MC_{11}	336 ± 18	13.0 ± 0.8	65 ± 6
MC_{21}	334 ± 13	13.0 ± 0.7	82 ± 4
MC_{31}	333 ± 13	12.0 ± 0.6	103 ± 10
MC_{12}	329 ± 22	11.0 ± 0.8	147 ± 24
MC_{22}	328 ± 17	11.5 ± 0.6	143 ± 15
MC_{32}	323 ± 11	11.0 ± 0.8	182 ± 12
MC_{13}	318 ± 10	10.5 ± 0.9	301± 29
MC_{23}	314 ± 13	10.1 ± 1.1	395 ± 34
MC_{33}	313 ± 18	9.0 ± 0.3	381 ± 16

The elastic modulus is only slightly dependent on the pretreatment conditions both before the processing and before the mechanical testing. Tensile strength, on the contrary, depends significantly on the pretreatment before processing and before the mechanical testing. In particular, the tensile strength decreases with increasing the content of absorbed moisture in both the pretreatments. The elongation at break is strongly dependent on both the pretreatments, but, in this case, the elongation at break increases with increasing the amount of the absorbed moisture on both the pretreatments. In fact, the elongation at break is the mechanical property more dependent on the modifications of the molecular structure, on the modification of the morphology, and on the possible presence of water. The elongation at break increases, as expected, with increasing the amount of water in the specimen being the same as the pretreatment of the pellets before processing (from 65% for MC_{11} to 147% for MC_{12}, to 301% for MC_{13}), as well as with increasing the moisture absorbed before compression molding, being the same the pretreatment before testing (from 65% for MC_{11} to 82% for MC_{21}, to 105% for MC_{31}). The increase of the elongation at break is, however, larger when the moisture is absorbed before processing. The mechanical properties depend, therefore, on both the conditioning of the pellets before processing and on the conditioning of the specimens before mechanical testing. The effect of moisture content during the compression molding is to be attributed to the hydrolytic degradation with reduction of the molecular weight. The presence of moisture in the solid state gives rise to a plasticizing effect of the polymer.

In Figure 5, the flow curves of the dry specimens MC_{11}, MC_{21}, MC_{31}, and of the sample MC_{13} are reported in order to verify a possible degradation of the polymer during processing as a consequence of the preliminary treatment of the pellets in different humidity conditions. The sample MC_{13} is reported for comparison and in order to verify a possible hydrolytic degradation when the moisture is absorbed after the processing. For the sake of simplicity, the flow curves of all the other specimens are not reported. The flow curves show that the viscosity decreases with increasing moisture content absorbed by the pellets before the processing. On the contrary, no significant effect on the viscosity is observed for the specimen conditioned after processing. This means that no significant effect of the conditioning before the measurement is observed on the viscosity.

As is well known, the Newtonian viscosity is strongly dependent on the molecular weight (see Equation (2)).

$$\eta_0 = K\, Mw^{3.4} \tag{2}$$

and, then, the reduction of the Newtonian viscosity means a reduction of the molecular weight, which can be calculated using Equation (3) [22] as follows:

$$(\eta_0(MC_{ij})/\eta_0\,(MC_{11}))^{1/3.4} = Mw(MC_{11})/Mw(MC_{ij}) \tag{3}$$

where i represents the condition of the pretreatment of the pellet and j represents the condition of the pretreatment of the specimens. This means that the decrease of the molecular weight for MC_{21} and MC_{31} is about 6.2% and 9.3%, respectively. The decrease of the molecular weight can be attributed to the hydrolysis undergone by the macromolecular chains due to the presence of water. As for the sample MC_{13}, the flow curve is almost superimposable to that of the sample MC_{11}, and this means that the humidity absorbed on the molded sample MC_{13} does not change the molecular structure of the polymer.

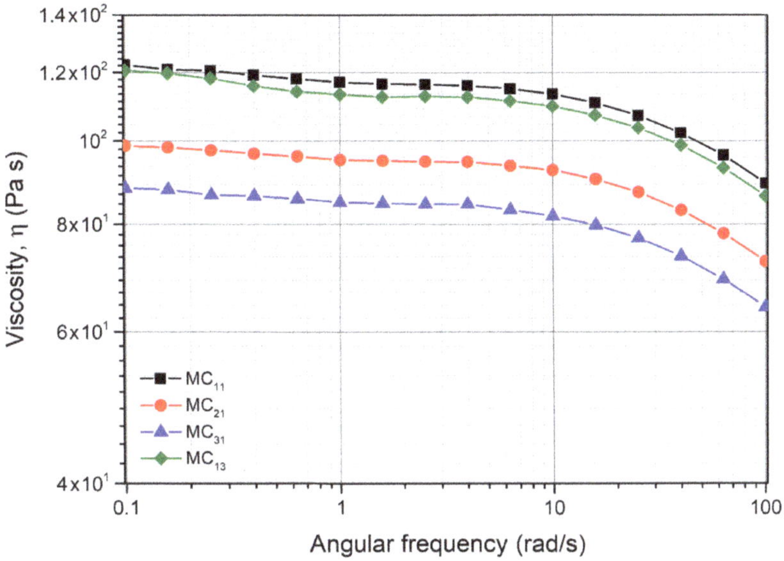

Figure 5. Flow curves of the samples from pellets conditioned under the three humidity conditions and specimen conditioned at RH = 0 and 90% after processing.

As already said, the increase of the elongation at break and the decrease of the rigidity observed for the humid specimens could be attributed to the presence of moisture in the specimens which acts as a plasticizer [23].

In order to verify better the effect of the pretreatment—before processing or after processing—the dimensionless values of the elongation at break were considered for all the samples. As previously mentioned, the elongation at break is the mechanical property more sensible to the change of molecular structure and morphology and to the presence of the water acting as a plasticizer.

In Figure 6, the dimensionless values of EB are plotted for all the samples. The dimensionless values have been calculated as the elongation at break for each sample divided that of the sample MC_{11}, i.e., the sample processed and characterized in a dry state.

It is possible to put in evidence three "blocks." The largest increase of the elongation at break is observed for the samples treated in condition 3, i.e., samples exposed to the humidity after processing and before the testing. The lowest values of the dimensionless elongation at break are shown by the specimen conditioned in dry conditions. This, of course, means that in the investigated pretreatment conditions, the plasticizing effect of the water is more important than the decrease of the molecular weight as a consequence of the hydrolytic degradation.

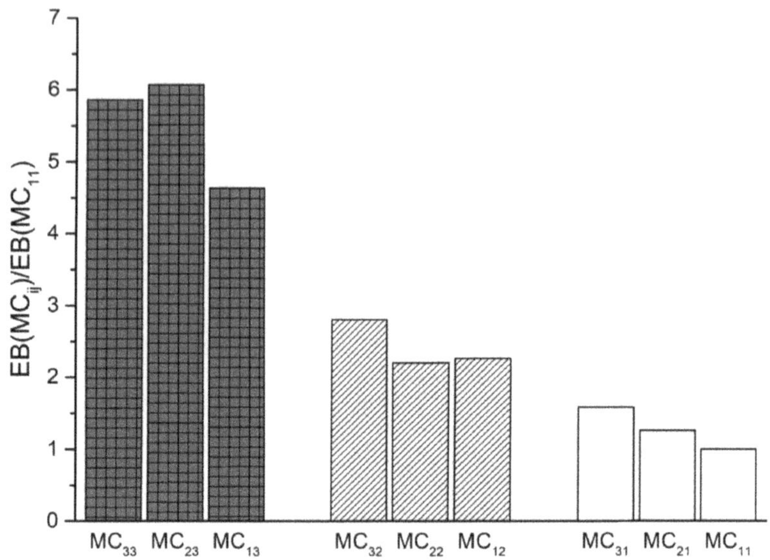

Figure 6. Histogram of the EB dimensionless values of all samples.

4. Conclusions

The presence of moisture in the biodegradable polyesters both before processing and during their lifetime can strongly modify the rheological and mechanical properties of the polymers. In this work, the effect of moisture during the processing and the effect of different level of moisture in molded samples has been investigated. During processing, the presence of moisture decreases the molecular weight of the polymer and then the viscosity of the melt. As expected, the reduction of the molecular weight is higher with increasing the level of moisture because of the hydrolysis cleavage of the macromolecular chains. The presence of moisture in solid state slightly reduces the rigidity of the polymer and remarkably increases the deformability. This behavior has been attributed to the plasticizing effect of moisture. As for the effect of moisture on the mechanical behavior and, in particular, on the elongation at break, the effect of the presence of moisture in the samples seems to be predominant over the effect of the hydrolytic degradation during processing.

Author Contributions: Conceptualization, F.P.L.M.; methodology, F.P.L.M.; validation, A.C. and V.T.; formal analysis, V.T. and A.C.; investigation, V.T. and A.C.; data curation, V.T., A.C., and F.P.L.M.; writing—original draft preparation, V.T., A.C., and F.P.L.M.; writing—review and editing, F.P.L.M. and V.T.; supervision, F.P.L.M. All authors have read and agreed to the published version of the manuscript.

Funding: This work has been financially supported by INSTM under the contract: INSTMPA004.

Institutional Review Board Statement: Not applicable.

Informed Consent Statement: Not applicable.

Data Availability Statement: The data presented in this study are available on request from the corresponding author.

Acknowledgments: Thanks are due to Novamont for supplying the polymer. Thanks are due to Ing. A. Rallis (Novamont) for valuable discussions.

Conflicts of Interest: The authors declare no conflict of interest.

References

1. Agarwal, S. Biodegradable Polymers: Present Opportunities and Challenges in Providing a Microplastic-Free Environment. *Macromol. Chem. Phys.* **2020**, *221*. [CrossRef]
2. Zhao, L.; Rong, L.; Zhao, L.; Yang, J.; Wang, L.; Sun, H. Plastics of the Future? The Impact of Biodegradable Polymers on the Environment. *Microplast. Terrestr. Environ.* **2020**, *95*, 423–445.
3. Ahmed, T.; Shahid, M.; Azeem, F.; Rasul, I.; Shah, A.A.; Noman, M.; Hameed, A.; Manzoor, N.; Manzoor, I.; Muhammad, S. Biodegradation of Plastics: Current Scenario and Future Prospects for Environmental Safety. *Environ. Sci. Pollut. Res. Int.* **2018**, *25*, 7287–7298. [CrossRef] [PubMed]
4. Mangaraj, S.; Yadav, A.; Bal, L.M.; Dash, S.K.; Mahanti, N.K. Application of Biodegradable Polymers in Food Packaging Industry: A Comprehensive Review. *J. Packag. Technol.* **2019**, *3*, 77–96. [CrossRef]
5. Touchaleaume, F.; Martin-Closa, L.; Angellier-Coussy, H.; Chevillard, A.; Cesar, G.; Gontard, N.; Gastaldi, E. Performance and Environmental Impact of Biodegradable Polymers as Agricultural Mulching Films. *Chemosphere* **2016**, *144*, 433–439. [CrossRef] [PubMed]
6. Zhong, Y.; Godwin, P.; Jin, Y.; Xiao, H. Biodegradable Polymers and Green-Based Antimicrobial packaging Material: A Mini-Review. *Adv. Ind. Eng. Polym.* **2020**, *3*, 27–35. [CrossRef]
7. Kapanen, A.; Schettini, E.; Vox, G.; Itävaara, M. Performance and Environmental Impact of Biodegradable Films in Agriculture: A Field Study on Protected Cultivation. *J. Polym. Environ.* **2008**, *16*, 109–122. [CrossRef]
8. Reichert, C.L.; Bugnicourt, E.; Coltelli, M.B.; Cinelli, P.; Lazzeri, A.; Canesi, I.; Braca, F.; Martínez, B.M.; Alonso, R.; Agostinis, L.; et al. Bio-Based Packaging: Materials, Modifications, Industrial Applications and Sustainability. *Polymers* **2020**, *12*, 1558. [CrossRef]
9. Bonartsev, A.P.; Boskhomodgiev, A.P.; Iordanskii, A.L.; Bonartseva, G.A.; Rebrov, A.V.; Makhina, T.K.; Myshkina, V.L.; Yakovlev, S.A.; Filatova, E.A.; Ivanov, E.A.; et al. Hydrolytic Degradation of Poly(3-hydroxybutyrate), Polylactide and their Derivatives: Kinetics, Crystallinity, and Surface Morphology. *Mol. Cryst. Liq. Cryst.* **2012**, *556*, 288–300. [CrossRef]
10. Gorrasi, G.; Pantani, R. Hydrolysis and biodegradation of Poly(lactic acid). *Adv. Polym. Sci.* **2017**, *117*, 1–33.
11. Harris, A.M.; Lee, E.C. Durability of Polylactide-Based Polymer Blends for Injection-Molded Applications. *J. Appl. Polym. Sci.* **2012**, *128*. [CrossRef]
12. Muthuraj, R.; Misra, M.; Mohanty, A.K. Hydrolytic Degradation of Biodegradable Polyesters Under Simulated Environmental Conditions. *J. Appl. Polym. Sci.* **2015**, *132*. [CrossRef]
13. Mistretta, M.C.; la Mantia, F.P.; Titone, V.; Botta, L.; Pedeferri, M.; Morreale, M. Effect of ultraviolet and moisture action on biodegradable polymers and their blend. *J. Appl. Bio. Funct. Mater.* **2020**, *18*. [CrossRef]
14. Mistretta, M.C.; la Mantia, F.P.; Titone, V.; Megna, B.; Botta, L.; Morreale, M. Durability of Biodegradable Polymers for the Conservation of Cultural Heritage. *Front. Mater.* **2019**, *6*. [CrossRef]
15. Rizzarelli, P.; Carroccio, S. Thermo-oxidative processes in biodegradable poly(butylene succinate). *Polym. Degrad. Stab.* **2009**, *94*, 1825–1838. [CrossRef]
16. Cuadri, A.A.; Martín-Alfonso, J.E. Thermal, thermo-oxidative and thermomechanical degradation of PLA: A comparative study based on rheological, chemical and thermal properties. *Polym. Degrad. Stab.* **2018**, *150*, 37–45. [CrossRef]
17. Tsuji, H.; Echizen, Y.; Nishimura, Y. Photodegradation of biodegradable polyesters: A comprehensive study on poly(L-lactide) and poly(3-caprolactone). *Polym. Degrad. Stab.* **2006**, *91*, 1128–1137. [CrossRef]
18. Carroccio, S.; Rizzarelli, P.; Scaltro, G.; Puglisi, C. Comparative investigation of photo- and thermal-oxidation processes in poly(butylene terephthalate). *Polymer* **2008**, *49*, 3371–3381. [CrossRef]
19. Titone, V.; la Mantia, F.P.; Mistretta, M.C. The Effect of Calcium Carbonate on the Photo-Oxidative Behavior of Poly(butylene adipate- co -terephthalate). *Macromol. Mater. Eng.* **2020**, *305*. [CrossRef]
20. ASTM D570-98. *Standard Test Method for Water Absorption of Plastics*; ASTM International: West Conshohocken, PA, USA, 2018.
21. ASTM D638-14. *Standard Test Method for Tensile Properties of Plastics*; ASTM International: West Conshohocken, PA, USA, 2014.
22. La Mantia, F.P.; Mistretta, M.C.; Titone, V. An Additive Model to Predict the Rheological and Mechanical Properties of Polypropylene Blends Made by Virgin and Reprocessed Components. *Recycling* **2021**, *6*, 2. [CrossRef]
23. Levine, H.; Slade, L. Water as a plasticizer: Physico-chemical aspects of low-moisture polymeric systems. In *Water Science Reviews*; Franks, F., Ed.; Cambridge University Press: Cambridge, UK, 1988; Volume 3, p. 79.

polymers MDPI

Article

Crazing Effect on the Bio-Based Conducting Polymer Film

Pei-Yi Wong [1], Akiyoshi Takeno [2], Shinya Takahashi [2], Sook-Wai Phang [3,*] and Azizah Baharum [1,4,*]

1 Department of Chemical Sciences, Faculty of Science and Technology, Universiti Kebangsaan Malaysia, Bangi 43600, Selangor, Malaysia; wvyerwong@gmail.com
2 Department of Chemistry and Biomolecular Science, Faculty of Engineering, Gifu University, Gifu 501-1193, Japan; takeno@gifu-u.ac.jp (A.T.); shinyat@gifu-u.ac.jp (S.T.)
3 Department of Physical Science, Faculty of Applied Sciences, Tunku Abdul Rahman University College, Jalan Genting Kelang, Kuala Lumpur 53300, Malaysia
4 Polymer Research Center, Faculty of Science and Technology, Universiti Kebangsaan Malaysia, Bangi 43600, Selangor, Malaysia
* Correspondence: pinkyphang@gmail.com (S.-W.P.); azeiss@ukm.edu.my (A.B.)

Abstract: The biodegradability problem of polymer waste is one of the fatal pollutFions to the environment. Enzymes play an essential role in increasing the biodegradability of polymers. In a previous study, antistatic polymer film based on poly(lactic acid) (PLA) as a matrix and polyaniline (PAni) as a conductive filler, was prepared. To solve the problem of polymer wastes pollution, a crazing technique was applied to the prepared polymer film (PLA/PAni) to enhance the action of enzymes in the biodegradation of polymer. This research studied the biodegradation test based on crazed and non-crazed PLA/PAni films by enzymes. The presence of crazes in PLA/PAni film was evaluated using an optical microscope and scanning electron microscopy (SEM). The optical microscope displayed the crazed in the lamellae form, while the SEM image revealed microcracks in the fibrils form. Meanwhile, the tensile strength of the crazed PLA/PAni film was recorded as 19.25 MPa, which is almost comparable to the original PLA/PAni film with a tensile strength of 20.02 MPa. However, the Young modulus decreased progressively from 1113 MPa for PLA/PAni to 651 MPa for crazed PLA/PAni film, while the tensile strain increased 150% after crazing. The significant decrement in the Young modulus and increment in the tensile strain was due to the craze propagation. The entanglement was reduced and the chain mobility along the polymer chain increased, thus leading to lower resistance to deformation of the polymer chain and becoming more flexible. The presence of crazes in PLA/PAni film showed a substantial change in weight loss with increasing the time of degradation. The weight loss of crazed PLA/PAni film increased to 42%, higher than that of non-crazed PLA/PAni film with only 31%. The nucleation of crazes increases the fragmentation and depolymerization of PLA/PAni film that induced microbial attack and led to higher weight loss. In conclusion, the presence of crazes in PLA/PAni film significantly improved enzymes' action, speeding up the polymer film's biodegradability.

Keywords: crazing; biodegradation; polyaniline; polylactic acid

Citation: Wong, P.-Y.; Takeno, A.; Takahashi, S.; Phang, S.-W.; Baharum, A. Crazing Effect on the Bio-Based Conducting Polymer Film. *Polymers* **2021**, *13*, 3425. https://doi.org/10.3390/polym13193425

Academic Editors: Domenico Acierno and Antonella Patti

Received: 16 August 2021
Accepted: 7 September 2021
Published: 6 October 2021

1. Introduction

Over the past, trillion tons of plastic packaging derived from fossil fuel were invented to fulfill the demands of society. Consequently, high tensile performance and low cost of plastic packaging contributed to a significant raise in municipal plastic wastes. In 2015, two-thirds out of 8.3 billion tons of plastic packaging accumulated and remained intact in the environment. The accumulation of plastic packaging threatens the health risks of human and marine life in the world [1]. Nevertheless, the presence of plastic packaging pollution also remarked in greenhouse gas emissions towards the lifecycle of the ecosystems. The environmental climate change issues due to fossil fuel-based plastic packaging should be a focus, and practical action is urged to be taken.

An awareness of the environmental problem, poly(lactic acid) (PLA), one of the bio-based polymers, was developed. The past study reported that PLA could be used to substitute fossil fuel-derived polymers, especially in packaging applications [2], which is due to the fact that PLA exhibited good mechanical properties, high biodegradability, and good biocompatibility. Thus, PLA played an essential role in green chemistry and reduced the carbon footprint effectively [3]. PLA degradation releases non-toxic substances, including water, gas, and biomass, as the end products [4]. Despite the degradation of PLA in such a direction, the challenges of the ingested PLA still existed, which is due to the unmanaged natural environment, which failed to provide a proper condition for the microbial degradation of PLA [5]. For instance, biodegradable PLA constricts to be degraded in seawater. Indeed, the low degradation efficiency of PLA due to unsuitable condition need to be overcome [6]. Meanwhile, PLA behaves as a non-conductive polymer that also restricts the application of PLA in electronic devices packaging.

The non-conductive challenges of PLA have sparked studies in targeted developing of conductive properties in PLA through incorporating conductive materials [7]. Polyaniline (PAni) is one of the conducting polymers widely investigated due to its good biocompatibility and conductivity. Based on the previous studies, researchers have demonstrated that the blending of PAni into the PLA with optimum proportion displayed excellent performance [8]. Thus, researchers hypothesize that the application of craze technology in PLA/PAni film would be highly desirable to promote degradation, conserving the mechanical properties and be used in electronic packaging applications.

Craze is commonly found in the ceramic industries, where it is applied to the clay pieces before being fired in a kiln for curing [9]. The craze is known as forming the microvoids or minor cracks of the materials [10]. The craze tends to transform the non-oriented glassy or semi-crystalized polymeric solid into fibrous states form in polymeric solid. Generally, the polymer crazing phenomenon is observed during the creep period when the load is applied [11]. Continuous propagation of the craze zone along the tip of the crack promotes the surface disclosure of the polymer. Large surface disclosure of the polymer allows microbial degradation and increases the polymer degradation efficiency [12]. Up to date, less quantitative data has been reported on the effect of crazing on the polymer by enzymatic degradation.

In the early work of this research, antistatic polymer film was developed by incorporating PAni into the PLA polymer film. The PLA/PAni film successfully inherited the antistatic properties compatible with the ESD standard to prevent the static charges trapped on the packaging surface to avoid static charges accidents. In this research, the craze technique was implemented to solve the biodegradation problem of PLA/PAni film. The crazed PLA/PAni film was evaluated using optical and scanning electron microscopes before and after the biodegradation test. The biodegradable property for the crazed PLA/PAni film was discussed. The effect of the crazes on the mechanical properties of PLA/PAni film was analyzed using a tensile machine. Subsequently, the results of the biodegradation rate of the crazed and non-crazed PLA/PAni films with different interval times were analyzed.

2. Characterization Techniques

2.1. Materials

The chemicals used in the synthesis of PAni, such as aniline monomer (Ani) (99%), ammonium persulfate (APS) oxidant (98%), and dioctyl sodium sulfosuccinate (AOT) dopant (96%), were purchased from Sigma-Aldrich, USA. Toluene (99.5%), used as a solvent to extract PAni precipitates, was purchased from Chemiz. Hydrochloric acid (HCl, 37%) dopant used to dissolve the APS was provided from R&M Chemicals. Tetrahydrofuran (THF) (99.8%) used as the medium to dissolve PLA was provided by R&M Chemicals. Glycerol (Gly) (99.5%) in the analytical grade was purchased from Friendemann Schmidt, Washington, USA. PLA resin with a melt flow index of 6.0 g/10 min at 210 °C and a specific gravity of 1.24, was purchased from Nature Works® PLA, 2003 D, USA. Sodium azide in analytical reagent grade (AR) was supplied by Systerm. The *Proteinase K* solution from *tritirachium* album ≥ 600 units/mL was supplied by Thermo Fisher Scientific (USA). In addition, the tris hydrochloride solution (pH 8.0) with a concentration of 1 M was supplied by Solarbio. Distilled water was obtained and purified by simple distillation. All the chemicals were used without further purification unless noted.

2.2. Preparation of Crazed PLA/PAni Film and Non-Crazed PLA/PAni Film

Two types of PLA/PAni film were prepared, including crazed PLA/PAni film and non-crazed PLA/PAni film. First, PAni was synthesized using Ani monomer, APS as oxidant, HCl as dopant, and AOT as surfactant through chemical oxidation at 0 °C for 24 h. The solution casting method was used to prepare the PLA/PAni film. Meanwhile, glycerol acts as the plasticizer in the PLA/PAni film to improve the interaction of PLA and PAni. In order to prepare the PLA/PAni, PLA in the amount of ~6.0 g was dissolved in 100 mL of THF solution with continuous stirring at 60 °C. Then, the glycerol with the amount of ~1.8 g was added to the dissolved PLA solution. After the glycerol was mixed well with the PLA solution, the synthesized PAni was added slowly into the mixture. The mixed solution was stirred at 60 °C for 24 h. The PLA/PAni film was produced by casting the mixtures onto a glass dish and dried at room temperature overnight. The PLA/PAni film was synthesized and optimized based on the crazing process applied to the prepared PLA/PAni film.

Next, the strip of PLA/PAni film was cut into a rectangular shape with a dimension of 70 mm (length) × 10 mm (width) and a thickness of 0.090 mm. Then, the PLA/PAni strip was clamped at the edges of the custom-made drawing device, as shown in Figure 1 [13]. The devices are associated with two film roll-up rollers, a bending blade, and stress control devices. The PLA/PAni strip was allocated in the position, as shown in Figure 1. Crazing stress applied on the strips was 6 MPa with 80° of bending angles. The PLA/PAni strip was in contact with the bending blade and created tension to form the crazes on the strip. The crazing process was conducted at room temperature with a processing rate of 20 mm/min. The presence of the crazes on the strip was identified by the annealing process. The crazed PLA/PAni strip was placed in an incubator at 60 °C for 30 min for the craze's healing process. The strip was then cold at room temperature before the length of the strip was measured. The strip was observed under an optical microscope before and after the crazing process to confirm the formation of crazes.

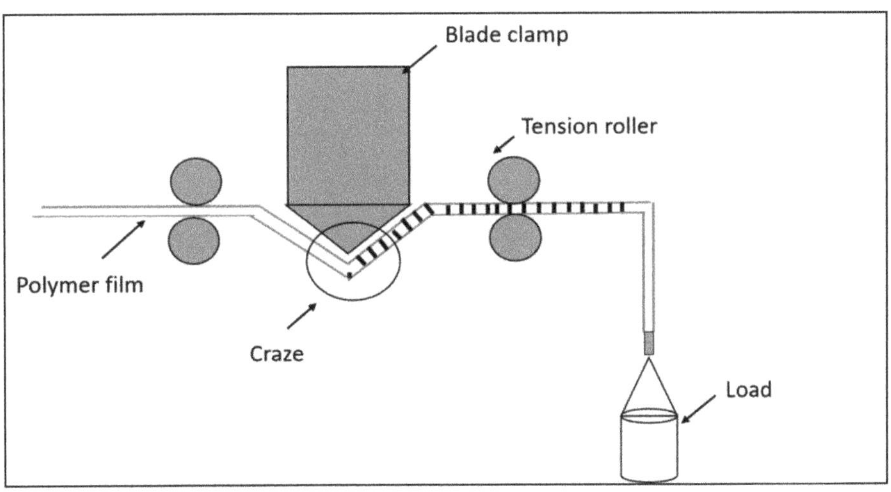

Figure 1. Instrument set-up used for the crazing process for the PLA/PAni film.

2.3. Characterization Techniques

The crazes formation on the PLA/PAni strip was observed using Nikon optical microscopes MM-400 equipped with a camera. The surface of the PLA/PAni strip before and after the crazing process was captured. The mechanical test for crazed PLA/PAni and non-crazed PLA/PAni was performed using the EZ-L Shimadzu Tensile Tester equipped with a load cell of 1 kN. The crazed PLA/PAni and non-crazed PLA/PAni were analyzed in the condition at 23 ± 2 °C and relative humidity (RH) of 50 ± 5%. The initial gauge separation and crosshead speed were set as 15 mm and 0.5 mm/s, respectively. The mechanical properties of the crazed PLA/PAni and non-crazed PLA/PAni were recorded accordingly. The morphology images of the crazed PLA/PAni and non-crazed PLA/PAni before and after the biodegradation test was observed using scanning electron microscopy, SEM (Hitachi HiTechnologies, model SEM-4800) with accelerating voltage of 1.0 kV. The crazed PLA/PAni and non-crazed PLA/PAni were coated with platinum using a sputtering coater before being observed under SEM at a magnification of 400× and 10,000×.

Furthermore, the enzymatic degradation test was conducted using Proteinase K to study the effect of the crazes towards the crazed PLA/PAni and non-crazed PLA/PAni. Enzymatic degradation was monitored based on the weight loss of PLA/PAni film at different time intervals. The initial weight of the crazed PLA/PAni strip and non-crazed PLA/PAni strip (dimension: 10 × 70 mm) was incubated at 37 ± 1 °C in the sampling tube. The crazed PLA/PAni and non-crazed PLA/PAni strips were placed in separate sampling tubes with different contents, as shown in Table 1. At regular time intervals during the biodegradation test, the strips were collected. They were washed gently with methanol followed by distilled water and dried in a glass vacuum dryer at room temperature for 3 days to obtain the dry mass of crazed PLA/PAni and non-crazed PLA/PAni after the biodegradation test.

Table 1. Content of each sample tube during the biodegradation test.

Sample	Content of Sampling Tube
Non-crazed PLA/PAni	0.5 mg of Proteinase K and 1.0 g of sodium azide in 10 mL Tris-HCl (pH 8.0)
Control sample	1.0 g Sodium azide in 10 mL Tris-HCl (pH 8.0)
Crazed PLA/PAni	0.5 mg Proteinase K and 1.0 g sodium azide in 10 mL Tris-HCl (pH 8.0)

3. Results and Discussion

The FTIR spectra for pristine PLA, pristine PAni, and PLA/PAni film is shown in Figure 2. Based on the FTIR spectra, the synthesized PAni revealed the characteristics of pristine PAni by giving the stretching of N−H (3215 cm^{-1}), quinoid and benzenoid ring (~1400–1500 cm^{-1}), C−N (1244 cm^{-1}), and C−H of para-disubstituted rings of PAni (810 cm^{-1}), respectively [14]. Meanwhile, the pristine PLA exhibited characteristics of C=O stretching (1746 cm^{-1}). Additionally, C−H stretching (~1300–1400 cm^{-1}) and C−H bending at 872 cm^{-1} indicate the characteristics of pristine PLA [15]. The PLA/PAni film was confirmed by the peaks observed at 1752 cm^{-1} (C=O), ~1400 cm^{-1} (quinoid and benzenoid ring), and ~800 cm^{-1} (C−H), respectively. All the absorption bands corresponding to the functional group of pristine PLA, pristine PAni, and PLA/PAni film were tabulated in Table 2.

The optical microscopy examination on non-crazed PLA/PAni film, crazed PLA/PAni film before annealing, and crazed PLA/PAni film after annealing allows the detection and labeling of the regions attributed to the crazing process as shown in Figure 3. Figure 3a showed the PLA/PAni film without undergoing the crazing process, while Figure 3b showed the crazed PLA/PAni before the annealing treatment. Figure 3a showed the normal homogenous distribution of PLA and PAni, which is due to the fact that PAni was added slowly to the PLA and glycerol solution mixture with constant stirring to get a homogenous solution. Thus, a homogenous and consistent micrograph of PLA/PAni (Figure 3a) indicated a good dispersion of PAni in PLA [16,17].

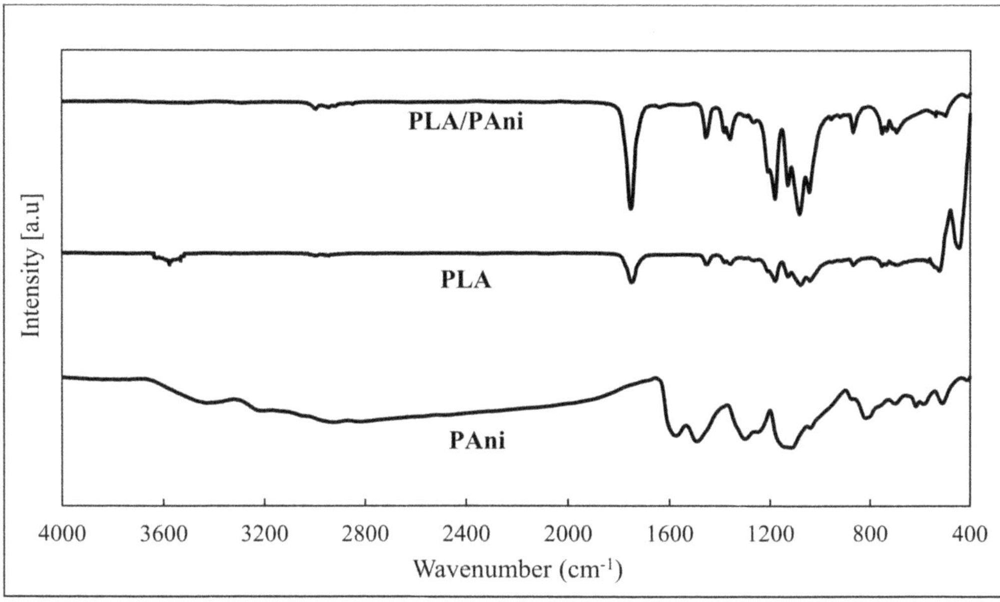

Figure 2. FTIR spectra of pristine PLA, pristine PAni, and PLA/PAni film.

Table 2. Functional group for pristine PLA, pristine PAni, and PLA/PAni film.

Functional Group	Wavenumber (cm^{-1})		
	PLA	**PAni**	**PLA/PAni**
N−H stretching	-	3215	-
C=O stretching	1746	-	1574–1576
C=C stretching of quinoid and benzenoid	-	1584, 1487	1457, 1354
C−N stretching	-	1244	1260–1301
C−H stretching	867	810	872

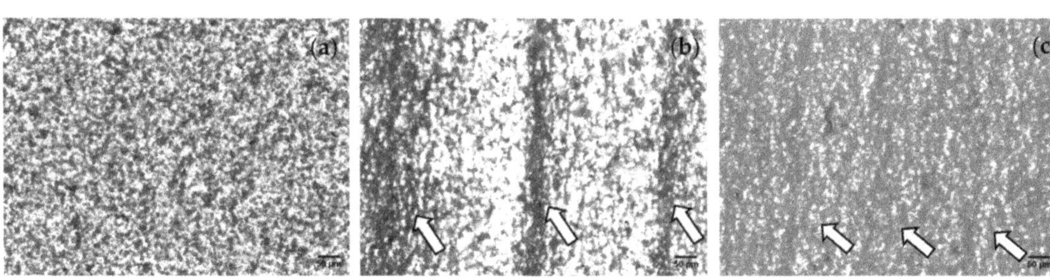

Figure 3. Microscope image of (**a**) non-crazed PLA/PAni film, (**b**) crazed PLA/PAni film, (**c**) crazed PLA/PAni film after the annealing process.

Figure 3b showed the lamellae lines formed in a perpendicular orientation on the crazed PLA/PAni. As indicated by the arrows in Figure 3b, the lamellae lines revealed the craze region of crazed PLA/PAni film [18]. This craze region indicated the spreading of internal stress and propagation of the craze zone along the tip of the craze. Hence, the fibrous and porous network strain formed an interval along with the PLA/PAni film and revealed in lamellae lines as shown in the optical image (Figure 3b). Thus, the crazes were successfully implemented on the PLA/PAni film after the crazing process. The fibrous and porous network strain that formed during the crazing process can be illustrated as in Figure 4.

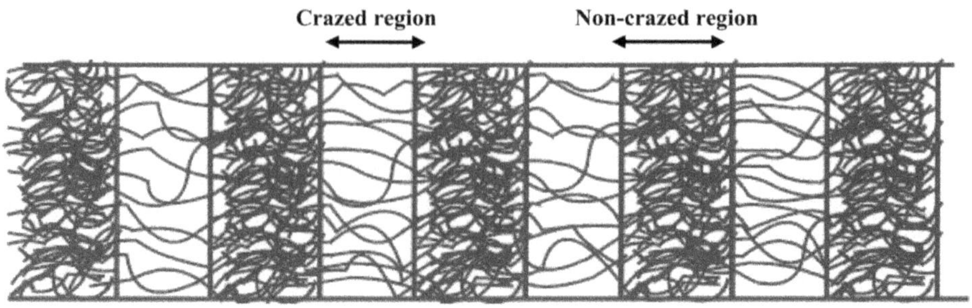

Figure 4. An illustration of fibrous and porous network strain that formed during the crazing process.

The annealing treatment was conducted to further confirm the presence of the crazes on the PLA/PAni film. The optical image of the crazed PLA/PAni film after the annealing treatment was observed in Figure 3c. After the annealing treatment, the craze width decreased significantly compared to the crazed PLA/PAni without the annealing treatment (Figure 3b). Generally, the internal stress created during the crazing process leads to the formation of a fibrous and porous network strain, which is reflected in the lamellae line, as observed in Figure 3b. However, the fibrous and porous network strain formed during the crazing process responded to the temperature and reduced the internal stress of the polymer film during the annealing process [19]. Finally, the changes in the temperature lead to the recovery of the porous network strain in crazed PLA/PAni film and significantly decrease the number of crazes as shown in Figure 3c. Hence, the formation of crazes on the PLA/PAni film was confirmed by the annealing process.

The mechanical properties of the crazed PLA/PAni and non-crazed PLA/PAni films are shown in Figure 5 in terms of the stress-strain curve. The curve pattern is generally almost similar but with a higher strain value for the crazed PLA/PAni film. Compared to the non-crazed film, the crazed film can reach an elongation of 150% more than the non-crazed PLA/PAni film. Generally, a higher strain rate indicates the longer time required for plastic deformation to occur [20]. The plastic deformation region of the crazed PLA/PAni film became more ductile and flexible than the non-crazed PLA/PAni film, which showed much brittle behavior. The phenomenon can be explained and supported by Figure 4, reflecting the ductile and more flexible behavior when the fibrous and porous network strain formed during the crazing process. The chain becomes easier to move and expand, hence giving more space to sustain stress. Thus, the crazed PLA/PAni film tends to have a higher elongation at break and a higher strain over time. The increasing strain for the crazed PLA/PAni film shows increases of the area under the stress-strain curve and shows a less stiffness behavior for the crazed PLA/PAni film [21]. The outcomes are in line with the result obtained in the Young modulus in this study.

From Figure 5, the non-crazed PLA/PAni film presents a higher Young's modulus with 1113 MPa than the crazed PLA/PAni film with 651 MPa of Young's modulus. It can be explained by the changes in the orientation of the polymer. Theoretically, the interaction between PLA and PAni created high entanglement between the polymer chain. However, the presence of crazes in the PLA/PAni film causes the mobility of polymer chains to increase. As a result, this decreases the polymer chain entanglement in the crazed PLA/PAni film and leads to lower resistance to deformation [22]. Hence, the crazed PLA/PAni exhibited less stiffness behavior and eventually decreased Young's modulus compared to the non-crazed PLA/PAni film.

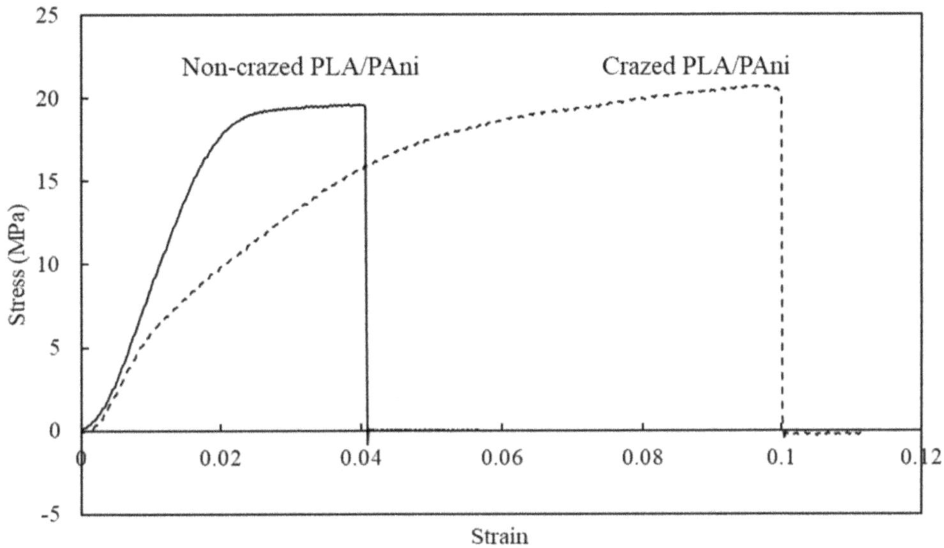

Figure 5. Stress–strain curve graph of non-crazed PLA/PAni film and crazed PLA/PAni film.

The tensile strength of crazed PLA/PAni film and non-crazed PLA/PAni film showed the resembled reading as 20.02 and 19.25 MPa, respectively. These phenomena can be explained by the structure of the polymer. Principally, the applied stress during the crazing process tends to change the orientation of the crazed polymer films rather than the backbone linkage along with the polymer [23]. Consequently, the backbone linkage for both PLA/PAni films with and without crazes remains unchanged. Thus, the amount of energy required to change the area of the film for both PLA/PAni films with and without crazes is the same. Therefore, both PLA/PAni films with and without crazes showed almost identical tensile strength reading.

Figure 6 shows the morphology images of non-crazed PLA/PAni film and crazed PLA/PAni film, respectively. By referring to Figure 6(ai), the PLA/PAni film without the crazing treatment showed lesser porosity and smooth surfaces. Meanwhile, the crazed PLA/PAni film in Figure 6(bi) can be observed to have a more porous structure than PLA/PAni film without the crazing treatment. This result showed the crazes-formation and reflected via the increases of porosity in the crazed PLA/PAni film morphology images. The porous structure found in the PLA/PAni film was formed by the fibrils separated by the nanosized pores as observed by the morphological images [24]. This finding is consistent with the generation of craze formation in polymer films under the action of electric discharge plasma done by Kurbanov et al. [25].

Meanwhile, by comparing Figure 6(aii,bii), it can be observed that there are diagonal cracks on the crazed PLA/PAni film. Theoretically, the crazes on the film can be identified by bright-field microscopy [26]. The diagonal cracks of the PLA/PAni film presented the craze zone, as outlined in Figure 6(bii). This eventually confirmed the craze-formation in the PLA/PAni film. The disruption of the lamellae, voiding, and fibril formation, as observed in the region of the diagonal crack, further proved the successful development of crazes in polymer film [27]. Additionally, the diagonal crack region is also recognized as a plastic deformation region where shear bands formed around the craze zone [28]. This also demonstrated that the crystalline behavior of PLA/PAni film was approaching to become an amorphous behavior [29]. As a result, the formation of crazes on the PLA/PAni film also desired to induce a higher biodegradation rate of the PLA/PAni film.

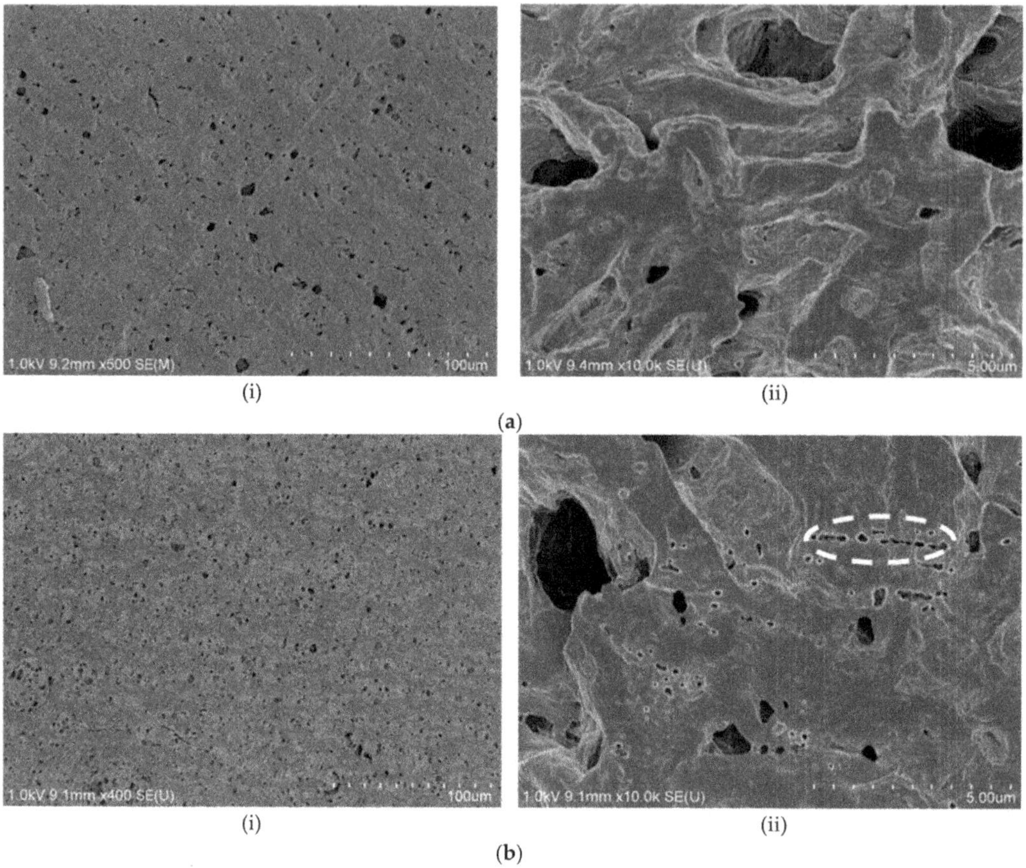

Figure 6. SEM morphology images of (**a**) non-crazed PLA/PAni film (i) 500× and (ii) 10,000×; (**b**) crazed PLA/PAni film (i) 400× and (ii) 10,000×.

This research studied the enzymatic degradation of PLA/PAni film with and without crazes within 21 days. Meanwhile, a reference film was conducted to investigate the hydrolysis of plasticizers in the PLA/PAni film. The enzymatic degradation of the films was analyzed by monitoring the weight loss of control PLA/PAni, non-crazed PLA/PAni, and crazed PLA/PAni films at different intervals of degradation time. Figure 7 showed the weight loss changes for the films under enzymatic degradation. In general, all the weight of the sample films reduced steadily by ~24% to 26% in the first 7 days. The decrease in weight of the samples for the first 7 days was due to the hydrolytic degradation of the glycerol plasticizer [30], which is attributed to the fact that glycerol played the plasticizer role in upsetting and restructuring the intermolecular polymer chain of PLA and PAni by hydrogen bonding [31]. Hydrolysis degradation of glycerol prior to happening during the degradation test as glycerol consists of hydrogen groups. Thus, the hydroxyl groups tend to have strong attraction with the hydrogen ion and form the hydrogen bonds within its structure [32]. Therefore, the weight loss in the prior days was due to the hydrolysis degradation of the glycerol in PLA/PAni film.

The weight loss of control PLA/PAni, non-crazed PLA/PAni, and crazed PLA/PAni films decrease to ~30% to 34% after 14 days of the enzymatic degradation process. At this stage, the control film reached a maximum weight loss of 30%. Thus, this indicates that the glycerol plasticizer in all the sample films had fully degraded. By comparison, the extra weight loss of 32% by the non-crazed PLA/PAni film and 34% by crazed the PLA/PAni samples showed after 14 days, which is due to the fact that the hydrolytic chain scission of ester bonds of PLA takes place after the degradation of glycerol [33]. Penetration of water into the PLA/PAni film hydrolyses at the ester group of PLA, which caused the long chains of PLA to convert into shorter chains and produce a high number of carboxyl end and hydroxyl end groups of PLA. This phenomenon resembled the explanation of Gupta and Kumar (2007) on the degradation of PLA [34].

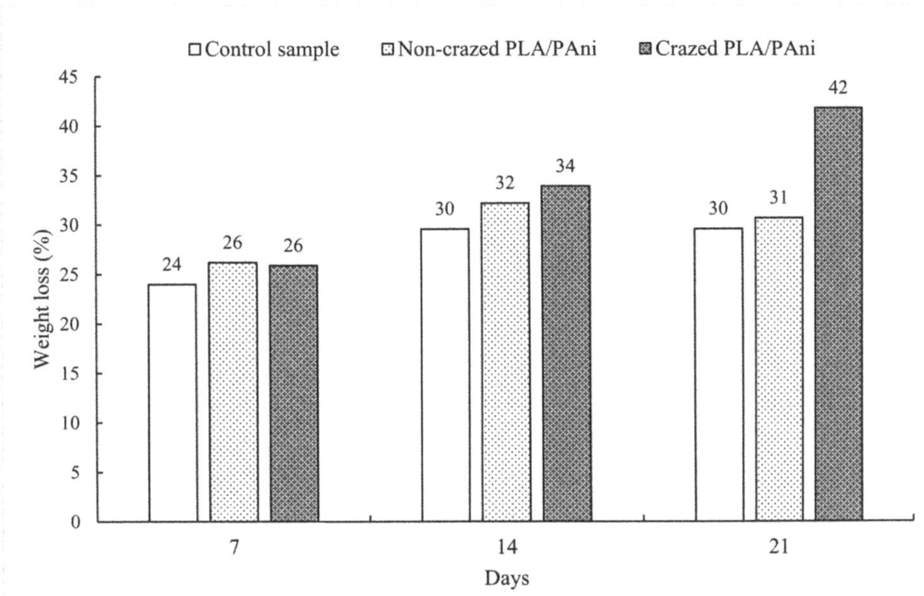

Figure 7. Biodegradation of non-crazed PLA/PAni and crazed PLA/PAni films at different time intervals.

Furthermore, the weight loss of crazed PLA/PAni film showed the highest percentages, which is 42% after 21 days of enzymatic degradation. It also found that the physical structure of the film collapsed entirely. Meanwhile, the weight loss of non-crazed PLA/PAni film showed almost linear, which is 31% after 21 days of enzymatic degradation. The highest weight reduction of 42% in crazed PLA/PAni film is correlated with the progressive deterioration of film due to the formation of crazes, which is due to the fact that crazes (in terms of pores) promote the diffusion of water in the PLA/PAni film. The high accessibility of water in PLA/PAni film caused random chain scission of the polymer chain to have occurred. Consequently, high fragmentation of polymer resulted in low molecules weight of the polymer chains [35]. Therefore, the formation of crazes accelerated the fragmentation and enhanced the depolymerization action of the polymer [36]. Hence, the crazed PLA/PAni film showed a significant decrease in the weight of the film after 21 days of degradation. This result is also in agreement with the research done by Mukhamed et al. (2020), which investigated the presence of crazes on the degradation properties of the PLA-based fibers.

At the same time, the highest weight loss of crazed PLA/PAni film also contributed to the physical erosion when exposed to the biological environment [37]. Generally, the degradation process happened on the surface of the samples as enzyme molecules are too large and hardly diffuse in the PLA/PAni film [38]. However, the limitation was overcome by the formation of crazes (in terms of pores) on the PLA/PAni film. The enzymes can efficiently adsorb on the surface, diffuse into the PLA/PAni film through the pores, and accelerate the degradation of the film [39]. Meanwhile, the enzyme used in this study is Proteinase K since it showed high efficiency in the degradation of PLA, especially in contact with the water [40]. Hence, the presence of crazes in the PLA/PAni film significantly enhanced the action of enzymes and improved the biodegradability of the polymer films, which could reduce the impact of polymer on environmental pollution. The diffusion action of the enzyme into the crazed PLA/PAni film is illustrated in Figure 8.

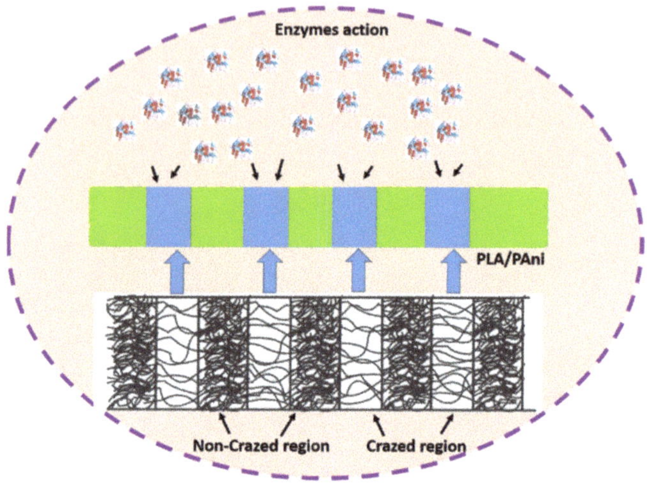

Figure 8. The diffusion of the enzyme in the crazed PLA/PAni film.

After the biodegradation test, the morphology images of crazed and non-crazed PLA/PAni films was observed and shown in Figure 9. It is well known that PLA is an aliphatic polyester, which is susceptible to enzymatic degradation. The increase in porosity of the crazed PLA/PAni film, as shown in Figure 9b, indicates the erosion degradation of enzymes towards the polymer film. Higher porosity in the polymer film leads to a higher rate of degradation, which is aligned with the result found in Figure 7. Higher porosity in crazed PLA/PAni film promotes greater immobilized enzyme loading within the polymer film [41]. Thus, the existence of crazes in polymer film supports the assessment of enzymes for the degradation action in the polymer film. As a result, high biodegradability of crazed PLA/PAni film was produced. This result is also in line with the selective enzymatic degradation of Poly (ε-caprolactone) done by Kulkarni et al. in 2008 [42].

Meanwhile, the morphology images of non-crazed PLA/PAni film after degradation also shows a compact and firm surface in Figure 9a, while the crazed PLA/PAni film shows inflate and expand surface in Figure 9b. Principally, the enzymatic degradation is related to the chemical structure and the hydrophilic/hydrophobic nature, as well as the degree of crystallinity of the polymer. The appearance of crazes in crazed PLA/PAni film tends to increase the film's amorphous nature. Consequently, the random hydrolytic chain scission of ester bonds in crazed PLA/PAni film allows higher diffusion of water into the amorphous region [43]. Thus, the high diffusion of water increases the degradation rate of the crazed PLA/PAni film [44]. Hence, the amorphous nature of crazed PLA/PAni film is more desired for the degradation of enzymes than the crystalline nature of non-crazed PLA/PAni film. Therefore, this also explained the highest weight loss of crazed PLA/PAni film compared to the non-crazed PLA/PAni film.

Overall, the crazed PLA/PAni film showed lamellae lines that confirmed the presence of crazes (optical microscope). At the same time, SEM morphology images reflected high porosity and diagonal cracks for the crazed PLA/PAni film, which responded to the presence of crazes in the film. On the other hand, the tensile strength is not affected due to the presence of crazes. Conversely, the flexibility of crazed PLA/PAni improved by 150%, giving a lower Young's modulus value. Meanwhile, biodegradation of the crazed PLA/PAni film was enhanced with the presence of crazes. The summarized characterization properties of non-crazed PLA/PAni and crazed PLA/PAni films were shown in Table 3.

Figure 9. SEM morphology images of (**a**) non-crazed PLA/PAni film and (**b**) crazed PLA/PAni film after the biodegradation test.

Table 3. The summarized characterization properties of non-crazed PLA/PAni and crazed PLA/PAni films.

Properties	Non-Crazed PLA/PAni	Crazed PLA/PAni	Remarks
Optical microscope	-	Lamellae presents	Confirmed crazes presented
SEM	↓ porosity and smooth	↑ porosity & diagonal crack	Confirmed crazes presented
Tensile strength	20.02 MPa (similar)	19.25 MPa (similar)	After crazed, high tensile strength remained
Young's modulus	1113 MPa	651 MPa	After crazed, modulus ↓, stiffness decreased
Tensile Strain	100%	250%	After crazed, 150% increment, better flexibility
Biodegradation	31%	42%	After crazed, % biodegradation ↑

4. Conclusions

Crazing was successfully developed in the PLA/PAni film to improve the biodegradability of the polymer film. The presence of crazes in the PLA/PAni film was confirmed by optical microscope and SEM. Crazes in the PLA/PAni film did not show significant tensile strength changes but showed progressive increases in tensile strain by 150% and decreases in Young's modulus. The non-crazed PLA/PAni film showed Young's modulus of 1113 MPa, while the crazed PLA/PAni film showed Young's modulus of 651 MPa. The differences that occurred in tensile strain and Young's modulus were due to the increment in chain mobility and decrement in chain entanglement of the polymer film. Meanwhile, the crazed PLA/PAni film improved the biodegradability of the polymer film. The significant weight loss of the crazed PLA/PAni film was observed after 21 days of biodegradation as 42%. This was due to the loss of crystalline nature of the polymer film in the presence of crazes, which induced the erosion degradation of enzymes. Thus, the appearance of crazes in the PLA/PAni film enhanced the biodegradability of the polymer film. The polymer film with high biodegradability based on PLA and PAni was formulated and this study could be useful in a packaging application to lower the polymeric pollution to the environment.

Author Contributions: Conceptualization, S.-W.P. and A.B.; Formal analysis, P.-Y.W., S.-W.P., A.T., S.T. and A.B.; Funding acquisition, A.B.; Investigation, P.-Y.W., S.-W.P., A.T., S.T and A.B.; Methodology, P.-Y.W., S.-W.P., A.T., S.T. and A.B.; Resources, A.T. and S.T.; Supervision, S.-W.P. and A.B.; Validation, P.-Y.W., S.-W.P. and A.B.; Visualization, P.-Y.W. and A.B.; Writing—original draft, P.-Y.W.; Writing—review & editing, S.-W.P., A.T., S.T. and A.B. All authors have read and agreed to the published version of the manuscript.

Funding: Research University Grant: GUP-2020-041 and Japan Student Services Organization: 2019.

Institutional Review Board Statement: Not applicable.

Informed Consent Statement: Not applicable.

Data Availability Statement: Not applicable.

Acknowledgments: The authors would like to acknowledge the Department of Chemistry, Faculty of Science and Technology, Universiti Kebangsaan Malaysia (UKM) and GUP-2020-041 grant for the funding support in this study. This work was performed, in part, at Gifu University (Japan). The scholarship support from Japan Student Services Organization (JASSO),2019 is gratefully acknowledged.

Conflicts of Interest: The authors declare no conflict of interest.

References

1. Bolman, B.; Boon, A.R.; Briere, C.; Group, E.; Prins, T. *Oceans Report Addressing SDG14 Issues with Factual Data and State of the Art Knowledge*; Spring: Deltares, The Netherlands, 2018; pp. 1–93.
2. Zhong, Y.; Godwin, P.; Jin, Y.; Xiao, H. Biodegradable Polymers and Green-Based Antimicrobial Packaging Materials: A Mini-Review. *Adv. Ind. Eng. Polym. Res.* **2020**, *3*, 27–35. [CrossRef]
3. Bertuoli, P.T.; Ordono, J.; Armelin, E.; Pérez-Amodio, S.; Baldissera, A.F.; Ferreira, C.A.; Puiggalí, J.; Engel, E.; Del Valle, L.J.; Alemán, C. Electrospun Conducting and Biocompatible Uniaxial and Core-Shell Fibers Having Poly(Lactic Acid), Poly(Ethylene Glycol), and Polyaniline for Cardiac Tissue Engineering. *ACS Omega* **2019**, *4*, 3660–3672. [CrossRef] [PubMed]
4. Harrison, J.P.; Boardman, C.; O'Callaghan, K.; Delort, A.M.; Song, J. Biodegradability Standards for Carrier Bags and Plastic Films in Aquatic Environments: A Critical Review. *R. Soc. Open Sci.* **2018**, *5*, 171792. [CrossRef]
5. Rocca-Smith, J.R.; Whyte, O.; Brachais, C.H.; Champion, D.; Piasente, F.; Marcuzzo, E.; Sensidoni, A.; Debeaufort, F.; Karbowiak, T. Beyond Biodegradability of Poly(Lactic Acid): Physical and Chemical Stability in Humid Environments. *ACS Sustain. Chem. Eng.* **2017**, *5*, 2751–2762. [CrossRef]
6. Rosli, N.A.; Ahmad, I.; Anuar, F.H.; Abdullah, I. The Contribution of Eco-Friendly Bio-Based Blends on Enhancing the Thermal Stability and Biodegradability of Poly(Lactic Acid). *J. Clean. Prod.* **2018**, *198*, 987–995. [CrossRef]
7. Baker, C.O.; Huang, X.; Nelson, W.; Kaner, R.B. Polyaniline Nanofibers: Broadening Applications for Conducting Polymers. *Chem. Soc. Rev.* **2017**, *46*, 1510–1525. [CrossRef] [PubMed]
8. Wong, P.Y.; Phang, S.W.; Baharum, A. Effects of Synthesised Polyaniline (PAni) Contents on the Antistatic Properties of PAni-Based Polylactic Acid (PLA) Films. *RSC Adv.* **2020**, *10*, 39693–39699. [CrossRef]
9. Hodges, H.W.M. The Formation of Crazing in Some Early Chinese Glazed Wares. *Stud. Conserv.* **1988**, *33*, 155–157. [CrossRef]
10. Yarysheva, A.Y.; Bagrov, D.V.; Bakirov, A.V.; Yarysheva, L.M.; Chvalun, S.N.; Volynskii, A.L. Effect of Initial Polypropylene Structure on Its Deformation via Crazing Mechanism in a Liquid Medium. *Eur. Polym. J.* **2018**, *100*, 233–240. [CrossRef]
11. Diani, J.; Gall, K. Finite Strain 3D Thermoviscoelastic Constitutive Model. *Society* **2006**, *46*, 1–10.
12. Barton-Pudlik, J.; Czaja, K.; Grzymek, M.; Lipok, J. Evaluation of Wood-Polyethylene Composites Biodegradability Caused by Filamentous Fungi. *Int. Biodeterior. Biodegrad.* **2017**, *118*, 10–18. [CrossRef]
13. Horiguchi, Y.; Takahashi, S.; Takeno, A. Growth of Craze Phase and Control of Void Diameter by Laplace-Pressure in Crazing Films. *Jpn. J. Appl. Phys.* **2019**, *58*, SAAD05. [CrossRef]
14. Rohani, R.; Izni Yusoff, I.; Adlyna Mey Efdi, F.; Mohd Junaidi, M.U. Polyaniline Composite Membranes Synthesis in Presence of Various Acid Dopants for Pressure Filtration Preparation of Polyaniline Composite Membranes Polyaniline Membranes Were Synthesized onto Microporous PVDF Supports in Presence of Various Acid Dopants. *UKM Eng. J.* **2017**, *29*, 1–12.
15. Wan Ishak, W.H.; Rosli, N.A.; Ahmad, I. Influence of Amorphous Cellulose on Mechanical, Thermal, and Hydrolytic Degradation of Poly(Lactic Acid) Biocomposites. *Sci. Rep.* **2020**, *10*, 1–13.
16. Shahdan, D.; Ahmad, S.; Chen, R.S.; Omar, A.; Zailan, F.D.; Abu Hassan, N.A. Mechanical Performance, Heat Transfer and Conduction of Ultrasonication Treated Polyaniline Bio-Based Blends. *Int. Commun. Heat Mass Transf.* **2020**, *117*, 104742. [CrossRef]
17. Zhang, H. Characterization of Electrospun Poly(Lactic Acid)/Polyaniline Blend Nanofibers. *Adv. Mater. Res.* **2011**, *332–334*, 317–320. [CrossRef]
18. Luo, W.; Liu, W. Incubation Time to Crazing in Stressed Poly(Methyl Methacrylate). *Polym. Test.* **2007**, *26*, 413–418. [CrossRef]
19. Hart, K.R.; Dunn, R.M.; Sietins, J.M.; Hofmeister Mock, C.M.; Mackay, M.E.; Wetzel, E.D. Increased Fracture Toughness of Additively Manufactured Amorphous Thermoplastics via Thermal Annealing. *Polymer* **2018**, *144*, 192–204. [CrossRef]

20. Deblieck, R.A.C.; Van Beek, D.J.M.; Remerie, K.; Ward, I.M. Failure Mechanisms in Polyolefines: The Role of Crazing, Shear Yielding and the Entanglement Network. *Polymer* **2011**, *52*, 2979–2990. [CrossRef]
21. Liu, X.; Luo, W.; Yin, B. Strain-Amplitude and Strain-Rate Dependent Craze Damage of Poly(Methyl Methacrylate). *Polym. Polym. Compos.* **2014**, *22*, 737–742.
22. Lim, H.; Hoag, S.W. Plasticizer Effects on Physical-Mechanical Properties of Solvent Cast Soluplus® Films. *AAPS PharmSciTech* **2013**, *14*, 903–910. [CrossRef]
23. Yarysheva, A.Y.; Arzhakova, O.V.; Yarysheva, L.M.; Volynskii, A.L. Biaxial Tensile Drawing of Poly(Ethylene Terephthalate) via Environmental Crazing as a Method for Creating a Porous Structure. *Polymer* **2018**, *158*, 243–253. [CrossRef]
24. Arzhakova, O.V.; Dolgova, A.A.; Yarysheva, L.M.; Volynskii, A.L.; Bakeev, N.F. Development of a Stable Open-Porous Structure in the Solvent-Crazed High-Density Polyethylene. *Inorg. Mater. Appl. Res.* **2011**, *2*, 493–498. [CrossRef]
25. Kurbanov, M.A.; Gol'dade, V.A.; Zotov, S.V.; Ramazanova, I.S.; Nuraliev, A.F.; Yakhyaev, F.F.; Yusifova, U.V.; Khudayarov, B.G. Generation of Craze-Formation Centers in Polymer Films under the Action of Electric Discharge Plasma. *Tech. Phys.* **2018**, *63*, 965–969. [CrossRef]
26. Newman, S.B.; Wolock, I. Optical Studies of Crazed Plastic Surfaces. *J. Res. Natl. Bur. Stand.* **1957**, *58*, 339. [CrossRef]
27. Kausch, H.H.; Gensler, R.; Grein, C.; Plummer, C.J.G.; Scaramuzzino, P. Crazing in Semicrystalline Thermoplastics. *J. Macromol. Sci. Phys.* **1999**, *38*, 37–41. [CrossRef]
28. Mohsenzadeh, M.S. Investigation of Damage Mechanisms of Polyethylene Film during Stable Crack Growth. *Mater. Res. Express* **2019**, *6*, 125343. [CrossRef]
29. Donth, E.; Michler, G.H. Discussion of Craze Formation and Growth in Amorphous Polymers in Terms of the Multiplicity of Glass Transition at Low Temperatures. *Colloid Polym. Sci.* **1989**, *267*, 557–567. [CrossRef]
30. Stoffel, F.; Weschenfelder, E.F.; Camassola, M.; Piemolini-Barreto, L.T.; Zeni, M. Influence of Plasticizers in Enzymatic Degradation and Water Resistance of Starch Foam Trays Obtained by Thermal Expansion. *J. Polym. Environ.* **2019**, *27*, 739–746. [CrossRef]
31. Tarique, J.; Sapuan, S.M.; Khalina, A. Effect of Glycerol Plasticizer Loading on the Physical, Mechanical, Thermal, and Barrier Properties of Arrowroot (Maranta Arundinacea) Starch Biopolymers. *Sci. Rep.* **2021**, *11*, 1–17. [CrossRef] [PubMed]
32. Cerqueira, M.A.; Souza, B.W.S.; Teixeira, J.A.; Vicente, A.A. Effect of Glycerol and Corn Oil on Physicochemical Properties of Polysaccharide Films—A Comparative Study. *Food Hydrocoll.* **2012**, *27*, 175–184. [CrossRef]
33. Klein, F.; Sommerfeld, A. Technical Applications. In *The Theory of the Top. Volume IV*; Birkhäuser: Basel, Switzerland, 2014; pp. 761–936. [CrossRef]
34. Gupta, A.P.; Kumar, V. EUROPEAN POLYMER New Emerging Trends in Synthetic Biodegradable Polymers—Polylactide: A Critique. *Eur. Polym. J.* **2007**, *43*, 4053–4074. [CrossRef]
35. Bagrov, D.V.; Yarysheva, A.Y.; Rukhlya, E.G.; Yarysheva, L.M.; Volynskii, A.L.; Bakeev, N.F. Atomic Force Microscopic Study of the Structure of High-Density Polyethylene Deformed in Liquid Medium by Crazing Mechanism. *J. Microsc.* **2014**, *253*, 151–160. [CrossRef] [PubMed]
36. Rosli, N.A.; Ahmad, I.; Anuar, F.H.; Abdullah, I. Effectiveness of Cellulosic Agave Angustifolia Fibres on the Performance of Compatibilised Poly(Lactic Acid)-Natural Rubber Blends. *Cellulose* **2019**, *26*, 3205–3218. [CrossRef]
37. Woodard, L.N.; Grunlan, M.A. Hydrolytic Degradation and Erosion of Polyester Biomaterials. *ACS Macro Lett.* **2018**, *7*, 976–982. [CrossRef] [PubMed]
38. Hegyesi, N.; Zhang, Y.; Kohári, A.; Polyák, P.; Sui, X.; Pukánszky, B. Enzymatic Degradation of PLA/Cellulose Nanocrystal Composites. *Ind. Crops Prod.* **2019**, *141*, 111799. [CrossRef]
39. Pathak, V.M.; Navneet. Review on the Current Status of Polymer Degradation: A Microbial Approach. *Bioresour. Bioprocess.* **2017**, *4*, 15. [CrossRef]
40. Huang, Q.; Hiyama, M.; Kabe, T.; Kimura, S.; Iwata, T. Enzymatic Self-Biodegradation of Poly(l-Lactic Acid) Films by Embedded Heat-Treated and Immobilized Proteinase K. *Biomacromolecules* **2020**, *21*, 3301–3307. [CrossRef]
41. Mohamad, N.R.; Marzuki, N.H.C.; Buang, N.A.; Huyop, F.; Wahab, R.A. An overview of technologies for immobilization of enzymes and surface analysis techniques for immobilized enzymes. *Biotechnol. Biotechnol. Equip.* **2015**, *29*, 205–220. [CrossRef]
42. Kulkarni, A.; Reiche, J.; Hartmann, J.; Kratz, K.; Lendlein, A. Selective Enzymatic Degradation of Poly(ε-Caprolactone) Containing Multiblock Copolymers. *Eur. J. Pharm. Biopharm.* **2008**, *68*, 46–56. [CrossRef]
43. Kim, K.; Elsawy, M.A.; Kim, K.; Park, J.; Deep, A. Hydrolytic Degradation of Polylactic Acid (PLA) and Its Composites Hydrolytic Degradation of Polylactic Acid (PLA) and Its Composites. *Renew. Sustain. Energy Rev.* **2017**, *79*, 1346–1352.
44. Benali, S.; Aouadi, S.; Dechief, A.L.; Murariu, M.; Dubois, P. Key Factors for Tuning Hydrolytic Degradation of Polylactide/Zinc Oxide Nanocomposites. *Nanocomposites* **2015**, *1*, 51–61. [CrossRef]

polymers

MDPI

Article

Lignin Inter-Diffusion Underlying Improved Mechanical Performance of Hot-Pressed Paper Webs

Amanda Mattsson [1,*], Tove Joelsson [1,2], Arttu Miettinen [3,4], Jukka A. Ketoja [1,4], Gunilla Pettersson [1] and Per Engstrand [1]

1 Department of Chemical Engineering, Mid Sweden University, SE-85170 Sundsvall, Sweden; tove.joelsson@more.se (T.J.); jukka.ketoja@vtt.fi (J.A.K.); gunilla.pettersson@miun.se (G.P.); per.engstrand@miun.se (P.E.)
2 MoRe Research Örnsköldsvik AB, Box 70, SE-89122 Örnsköldsvik, Sweden
3 Department of Physics, University of Jyvaskyla, P.O. Box 35, FI-40014 Jyvaskyla, Finland; arttu.i.miettinen@jyu.fi
4 VTT Technical Research Centre of Finland Ltd., P.O. Box 1000, FI-02044 Espoo, Finland
* Correspondence: amanda.mattsson@miun.se

Abstract: Broader use of bio-based fibres in packaging becomes possible when the mechanical properties of fibre materials exceed those of conventional paperboard. Hot-pressing provides an efficient method to improve both the wet and dry strength of lignin-containing paper webs. Here we study varied pressing conditions for webs formed with thermomechanical pulp (TMP). The results are compared against similar data for a wide range of other fibre types. In addition to standard strength and structural measurements, we characterise the induced structural changes with X-ray microtomography and scanning electron microscopy. The wet strength generally increases monotonously up to a very high pressing temperature of 270 °C. The stronger bonding of wet fibres can be explained by the inter-diffusion of lignin macromolecules with an activation energy around 26 kJ mol^{-1} after lignin softening. The associated exponential acceleration of diffusion with temperature dominates over other factors such as process dynamics or final material density in setting wet strength. The optimum pressing temperature for dry strength is generally lower, around 200 °C, beyond which hemicellulose degradation begins. By varying the solids content prior to hot-pressing for the TMP sheets, the highest wet strength is achieved for the completely dry web, while no strong correlation was observed for the dry strength.

Keywords: hot-pressing; paper web; fibre; lignin; diffusion; activation energy

Citation: Mattsson, A.; Joelsson, T.; Miettinen, A.; Ketoja, J.A.; Pettersson, G.; Engstrand, P. Lignin Inter-Diffusion Underlying Improved Mechanical Performance of Hot-Pressed Paper Webs. *Polymers* **2021**, *13*, 2485. https://doi.org/10.3390/polym13152485

Academic Editors: Domenico Acierno and Antonella Patti

Received: 3 July 2021
Accepted: 23 July 2021
Published: 28 July 2021

1. Introduction

Microplastic emissions are one of the world's greatest environmental threats. The amount of these emissions has been steadily increasing for many years and is expected to continue to do so [1]. Thus, material options that are both renewable and biodegradable have been extensively searched for. A particular challenge is to develop materials that have similar or better properties in humid or wet conditions as their oil-based counterparts. This should be the case not only for strength but also for dimensional stability and barrier properties, which are important, e.g., in packaging and construction applications [2].

Recent studies have shown that hot-pressing of lignin-rich paper webs could provide at least a partial solution to the above challenge. Clear improvements are observed for both wet and dry tensile strength (later also referred to as only wet and dry strength) compared to non-treated webs [3,4], enabling applications in several packaging areas. Joelsson et al. [5] showed that the tensile strength of paper based on chemithermomechanical pulp (CTMP) could be improved even by 100% when passing the paper through a hot nip (200 °C, 6 MPa) with a pressing time of 1.5 s and 70 s after hold. Moreover, by hot-pressing, the wet strength increased dramatically to a value of about 16 kNm/kg from the level of 2 kNm/kg found for non-heat treated paper. It could be seen that the amount of lignin was of great importance [5,6]. Thus, such a hot-pressing technology could provide innovative product solutions once both suitable raw materials and optimal process conditions are defined by a deeper understanding of underlying strengthening mechanisms. Also, other properties that are important for the final applications of packaging papers, such as water resistance, have shown promising results. Contact angle measurements showed increased values for the hot-pressed paper samples [5], which suggests a more hydrophobic surface [7]. Similar results and conclusions, that heat treatment of this material increases the hydrophobicity, have been observed in the area of thermal modification and welding of wood [8,9].

Joelsson et al. [5] postulate that softened lignin from fibres redistributes within the consolidated structure, enabling strong inter-fibre bonding even in a wet fibre network. In other words, lignin acts as a natural wet-strength additive. Similar heat-induced bonding was already found by Gupta et al. in 1962 [10]. They applied isolated lignin to paper samples and pressed them together at high temperature. The appearing inter-layer bonding increased the strength properties. The optimal bonding temperature depended on lignin type and differed for wet or dry paper. This was explained by the thermal transition of lignin and by the plasticizing effect of water, which reduces the glass transition temperature (Tg) by 70–165 °C depending on the type of lignin [11].

The importance of water for the viscoelastic properties of wood was reviewed in 1982 by Back and Salmén [12], who concluded that water-saturated native lignin has a softening temperature of about 115 °C. This could be further lowered by sulphonation. Joelsson et al. [13] have recently shown that the softening effect of sulfonation also occurs when a sulfite-enriched paper is hot-pressed. In this case, a lower temperature is required to maintain strength.

Lignin is often referred to as a by-product in industries such as the production of paper, ethanol, biomass, etc. [14]. However, the polymer is seen to have a huge technological potential, and related research has expanded in recent years. Nevertheless, there are currently only a few commercial products based on lignin. For example, there are pulping processes where the lignin is not totally removed, leading to so-called high yield pulps (HYP). Their yield can be as high as 95%, which can be compared to the yield of about 50% for chemical pulp with removed lignin [15]. Thus, high yield pulping is a preferred option from the viewpoint of efficient utilisation of wood raw material. The main reason for removing the lignin in chemical pulping is to achieve high brightness and strength, which are important properties, e.g., white packaging, copy paper and some heavily-coated brochure papers. On the other hand, HYP and particularly thermomechanical pulp (TMP) are mainly used for magazine paper, newspaper and book paper, for which high opacity and light scattering are more important than brightness or strength. However, the share of paper usage has declined rapidly during recent years, which leaves a considerable amount of space for new markets. At the same time, the process targets should be reconsidered based on the changed product requirements.

The aim of this work is to investigate mechanisms underlying the above improved mechanical properties obtained by hot-pressing. In particular, we would like to know how to control lignin redistribution in fibre networks without deteriorating other fibre polymers such as hemicelluloses. Moreover, the high temperature and moisture content of fibres may introduce also other structural changes that affect the mechanical properties of the hot-pressed web. These changes are characterized by X-ray tomography and scanning electron microscopy (SEM). The experimental results are interpreted with the help of theoretical ideas on polymer inter-diffusion. In addition to analysing the results carefully for webs containing TMP, we show that the same diffusion mechanisms explain the wet-strength improvement for a wide variety of other pulps despite their different lignin content.

2. Materials and Methods

2.1. Materials

The paper materials used in this study were based on different mechanical and chemical kraft pulps obtained from Swedish mills, together with some pulps produced in a laboratory. Mechanical pulps with a lignin content of 26–28% included the following types: TMP (Holmen AB Braviken mill, Norrköping, Sweden), CTMP (Rottneros AB Rottneros mill, Sunne, Sweden, and SCA AB Östrand mill, Timrå, Sweden), and high-temperature chemithermomechanical pulp (HTCTMP) produced at the test pilot refinery at Valmet AB, Sundsvall, Sweden. Chemical kraft pulps with a lignin content of 0–12% were unbleached kraft liner (SCA AB Obbola mill, Umeå, Sweden), bleached kraft liner (Metsäboard Husum mill, Örnsköldsvik, Sweden), bleached kraft (Södra Cell Värö mill, Varberg, Sweden), and unbleached kraft with different rest-lignin contents produced at the laboratory pilot of MoRe Research (Örnsköldsvik AB, Örnsköldsvik, Sweden). The mechanical pulps and the pilot-produced chemical kraft pulps were based on Norway spruce, and the rest of the chemical pulps were based on softwood (a mixture of spruce and pine). The lignin content was measured by the Klason method (T222).

All paper material except that containing TMP were prepared using a Rapid Köthen sheet former (Paper Testing Instruments, Pettenbach, Austria) according to ISO 5269-2:2004, resulting in uniform fibre orientation. The TMP paper was produced in an XPM Fourdrinier paper machine at the laboratory of MoRe Research (Örnsköldsvik AB, Örnsköldsvik, Sweden). The web width was 0.225 m, the machine speed was 1.4 m/min, and the fibre orientation ratio was 1.7 between machine direction (MD) and cross-machine direction (CD). The grammage of the paper materials was in the range of 100–150 g/m^2. In both of the above production methods, the structure forming step is followed by water removal with wet-pressing at relatively low temperatures, which significantly affects the density of the formed paper material. However, the largest changes in density take place during the final hot-pressing process.

2.2. Pressing Methods

Two different pressing methods were applied in the experiments (Figure 1). Firstly, test points pressed at temperatures equal to or lower than 200 °C were performed using an oil-heated cylinder press (Figure 1a). Moist sheets were fed into the press on a felted fabric with a rate of 1 m/min and a nip pressure of 6 MPa. The pressing time in the nip was 1500 ms (at a nip length of about 25 mm) and after-hold was 70 s. Secondly, test points hot-pressed at temperatures higher than 200 °C were performed using a test pilot press with an infrared-heated steel belt carrying the paper samples through a nip shown in Figure 1b. The speed was 3 m/min, corresponding to a pressing time of 40 ms (nip length was about 2 mm) and the after-hold was 23.5 s. The nip load was estimated to be 8 MPa, and the press load of the steel belt was 0.15 MPa. In both cases, nip lengths were measured with sensor films from Fujifilm Holdings Corporation (Tokyo, Japan), Prescale LW 2.5–10 MPa. Thin blotter papers on both sides of an actual paper sample were used in all tests to prevent sticking. The solids content of paper sheets was 50–60% before pressing at the cylinder press, and at the infrared-heated steel belt press test pilot solids content of TMP sheets was 50–100%.

2.3. Sheet Testing

Sheet testing was carried out after conditioning according to ISO 187. Grammage and density were determined according to ISO 536 and ISO 534 respectively. The standard sheet thickness was measured according to ISO 5270. Dry tensile strength was determined according to ISO 1924-3. Wet tensile strength was measured according to ISO 3781 after immersion in water for one hour.

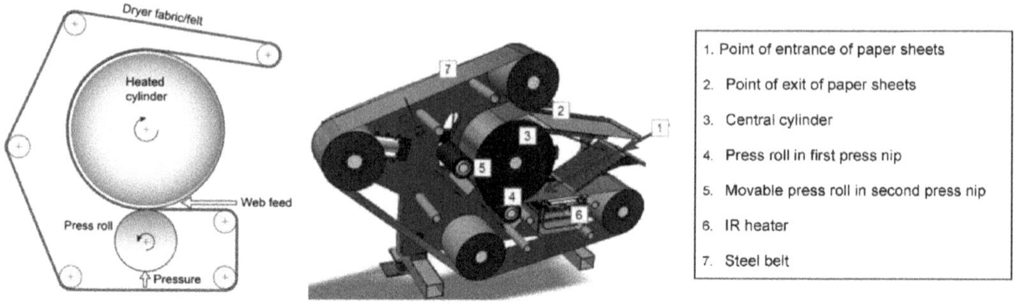

(a) (b)

Figure 1. (**a**) Felted, oil-heated cylinder press [16] operated at MoRe Research Örnsköldsvik AB, Örnsköldsvik, Sweden. Illustration: Mats Rundlöf, Capisco, Norrköping, Sweden. (**b**) Infrared-heated press based on a steel belt [17] produced by Ipco AB, Sandviken, Sweden. Both pictures are reproduced under the terms of the CC BY license.

2.4. Characterisation

2.4.1. Scanning Electron Microscope (SEM)

Image analyses using a high-resolution SEM (Tescan Maya3-2016, TESCAN Brno, s.r.o., Brno, Czechia) were performed on TMP sheets with different pressing conditions. The applied electron beam voltage was 3.00 kV and the beam intensity was 1.00. To obtain images of the structures at different scales, magnifications 500×, 2000× and 10,000× were used. These magnifications correspond to pixel sizes of 270 nm, 67 nm and 13 nm, respectively. The corresponding beam spot sizes were 26 nm, 26 nm and 4 nm, respectively. A secondary electron detector was used to capture the images. The working distance to the sample, which ranges from 5 to 7 mm, was adjusted for each image to achieve the best possible image quality.

The cross-sections were polished either using an argon ion milling system (Hitachi IM4000Plus, Hitachi High-Tech Co., Tokyo, Japan) or by freeze-drying the specimens at −110 °C and vacuum for 12 h followed by crushing to produce the transverse sections. Lastly, the samples were prepared by sputtering them with a 5–10 nm thin layer of iridium prior to imaging.

2.4.2. X-ray Microtomography

X-ray tomography images of the sheets pressed at different temperature levels were acquired using an X-ray microtomograph (CT) (Xradia MicroXCT-400, Xradia Inc., Concord, CA, USA). A sample approximately 1 mm wide was cut from the sheet with a surgical knife and glued to the top of a carbon fibre rod, which served as a sample holder. Images were acquired with 0.6 µm pixel size, corresponding to 1.5 µm spatial resolution (MTF10%), at 40 kV X-ray tube accelerating voltage and 4 W power. 1750 projection images per sample were acquired with an exposure time of 10 s per projection. The projections were reconstructed into a 3D volume image using the filtered backprojection algorithm. The volume images show an area of approximately 1.1 mm × 1.1 mm of the sheet.

The reconstructed images were denoised using bilateral filter (spatial sigma = 1.5 µm, radiometric sigma ≈ 7% of dynamic range) [18]. The filtered images showed a high contrast-to-noise ratio (typically ≈ 40) and could therefore be segmented using the simple Otsu thresholding method [19]. After the thresholding procedure, the remaining small image artefacts were removed by deleting all contiguous regions whose size was less than 100 voxels. This procedure results in a visually correct segmentation, as shown in Figure 2.

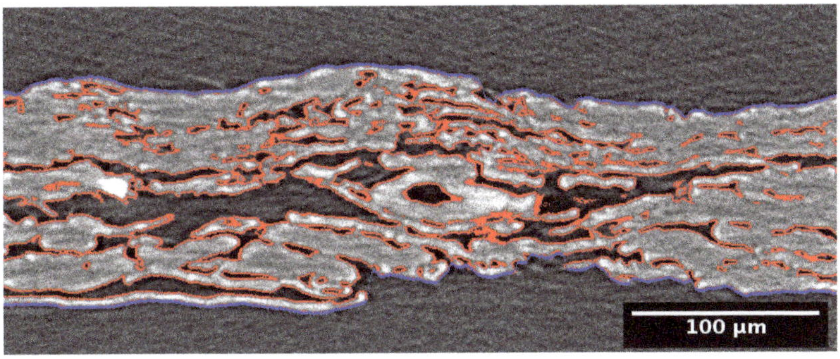

Figure 2. Visualisation of the X-ray microtomograph (CT) image of a sample pressed at 190 °C temperature (grayscale), edges of segmented regions (red) and surfaces of the sheet (blue).

The surfaces of the paper sheets were defined using the Carpet method [20] which works by lowering a surface following quenched noise Edwards-Wilkinson dynamics towards the segmented paper sheet. The bright pixels corresponding to the paper eventually slow down and stop the evolution of the surface. The paper surface is then defined by the position where the motion of the dynamic surface stops. An example of the surfaces is shown in Figure 2.

The total volumes of the sheet, pores, and fibres were determined by counting the number of pixels classified to each material phase. The pore size distribution was determined using the local thickness algorithm [21]. Image analysis was performed using the freely available software pi2 (https://www.github.com/arttumiettinen/pi2, accessed on 28 July 2021), and 3D visualisations were created using MeVisLab (MeVis Medical Solutions AG, Bremen, Germany).

3. Results

3.1. Porosity of the Fibre Networks from X-ray Microtomography

Hot-pressing narrows the pore-size distribution of a sheet significantly as can be seen in Figure 3. This effect is strongest at very high temperatures. Still, the mean pore size in all cases is several micrometres and thus clearly higher than the resolution of X-ray imaging. Therefore, it is reasonable to assume that the measurement of total pore volume gives reliable results.

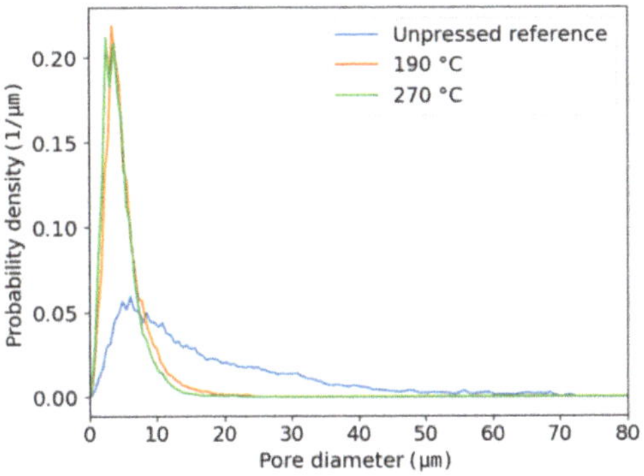

Figure 3. Pore-size distributions for the unpressed reference sheet and hot-pressed sheets with temperature of 190 °C (cylinder press) and 270 °C (steel belt press).

The average porosity, 0.74, is quite high for the unpressed reference sheet. In this case, lumens are still partly open, and the above value for porosity is similar as in earlier similar measurements [22]. During hot-pressing, density increases and porosity decreases significantly as lumina collapse and fibres soften (Figure 4). Moreover, a slight decrease in sheet porosity is also observed when temperature and pressure are increased, from 0.34 at 190 °C and 6 MPa (cylinder press) to 0.32 at 270 °C and 8 MPa (steel belt press), despite the much shorter pressing time in the latter case. Thus, plastic fibre deformations take place very rapidly at high temperatures when the polymer components of fibres soften.

(a) (b) (c)

Figure 4. 3D visualisations of the CT images of samples. (**a**) Unpressed reference, (**b**) pressed at 190 °C, and (**c**) pressed at 270 °C.

Table 1 shows the resulting sheet densities. The values obtained from 3D structural images are higher than those obtained with standard sheet density measurements mainly because of the surface roughness volume excluded when calculating the effective value from the CT data.

Table 1. Effective density of the TMP sheets obtained from the X-ray microtomography (CT) compared with the standard measurement.

Sample	Effective Sheet Density (CT)	Sheet Density (ISO 534)
Unpressed	367 kg/m^3	313 kg/m^3
Pressed 190 °C	955 kg/m^3	694 kg/m^3
Pressed 270 °C	1000 kg/m^3	734 kg/m^3

The effective fibre density ρ_f can be measured from sheet grammage G, area of sample A, and total volume of fibres V_f determined from the 3D images,

$$\rho_f = \frac{GA}{V_f} \quad (1)$$

Equation (1) gives the values 1440 kg/m^3 (unpressed reference), 1450 kg/m^3 (190 °C, cylinder press) and 1460 kg/m^3 (270 °C, steel belt press) for the density of the fibres. The wall density without lumen is about 1500 kg/m^3 for natural wood fibres [23]. The slightly lower values can be explained by a small total volume of pores whose size is below the imaging resolution. However, the main conclusion is that the hot-pressing does not induce any noticeable density change in the fibre walls, despite a large reduction in the network porosity and mean pore size.

3.2. Visual Observations on Pressing-Induced Changes in Fibres

Figure 5 shows SEM images of the TMP paper sheets pressed at different temperatures. The unpressed sample (Figure 5a) has a porous structure, with fibres having their characteristic oval shape. For the sheets pressed at higher temperatures, 190 °C (Figure 5b), and 270 °C (Figure 5c), fibres consolidate into ribbon-like structures with an almost perfectly closed lumen. The sample pressed at 190 °C (cylinder press) is treated for a much longer time, 1.5 s in the nip and 70 s after hold, compared to the sample at 270 °C (steel belt press), treated 40 ms in the nip and 23 sec in after hold. This difference appears as a more closed surface for the 190 °C sample, despite its slightly higher overall porosity (see Section 3.1).

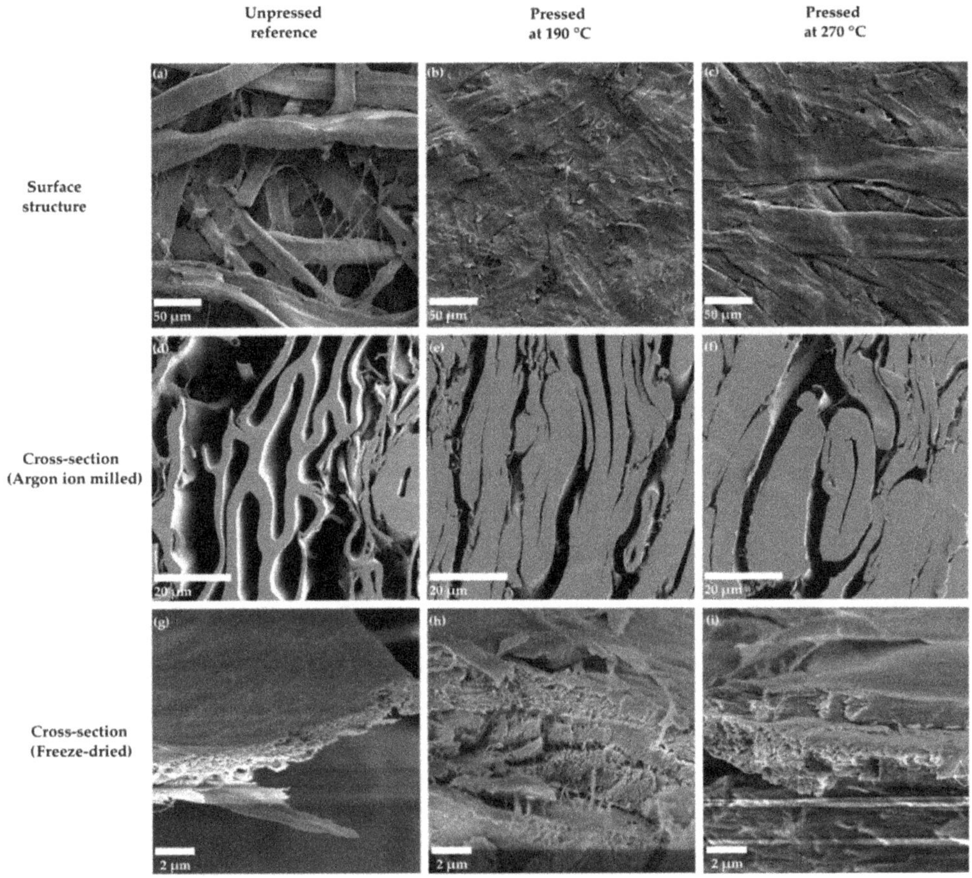

Figure 5. SEM images of the structures for three different pressing temperatures for the material, (**a**) unpressed reference, (**b**) pressed at 190 °C, and (**c**) pressed at 270 °C. SEM images of the cross-sections polished with an argon ion miller for three different pressing temperatures for the material, (**d**) unpressed reference, (**e**) pressed at 190 °C, and (**f**) pressed at 270 °C. SEM images of the freeze-dried cross-sections of the fibre wall for three different pressing temperatures for the material, (**g**) unpressed reference, (**h**) pressed at 190 °C, and (**i**) pressed at 270 °C. The working distance for the samples was in the range from 5 to 7 mm.

The porosity differences in different samples are best visible in SEM cross-sections of these structures, obtained after polishing the samples with an argon ion milling machine. In addition to inter-fibre pores, also fibre lumens stay partly open for the unpressed sheets (Figure 5d). On the other hand, the highest 270 °C temperature causes an almost complete disappearance of lumen space due to thermal softening (Figure 5f), whereas most of the collapsed lumens are still visible at the lower 190 °C temperature (Figure 5e).

In order to look closer at the nano-/microstructure inside the fibre wall, cross-sections were prepared also by freeze-drying the material prior to breaking the sheets. However, in these cross-sections (Figure 5g–i), it is not possible to observe any substantial differences in the porous structure when comparing the unpressed sample and the ones pressed at high temperatures. This suggests that lignin and other matrix polymers are not extracted from the fibre wall to the same extent as for some chemical treatments of wood fibres [24], where the extra microporosity is clearly visible. This observation is in alignment with the fibre wall densities obtained from the CT analyses (see Section 3.1).

3.3. Lignin Inter-Diffusion Affecting Wet Tensile Strength

The dependence of wet tensile strength index (wet tensile strength divided by the grammage) on pressing temperature seems to be defined by the activation energy for the inter-diffusion of lignin between fibre surfaces. The inter-diffusion is expected to be proportional to $\exp\left(-\frac{E_a}{RT}\right)$ [10], where E_a is the activation energy, T is temperature, and R is gas constant. We obtained E_a/R by plotting ln(Wet tensile strength index) vs. 1/Temperature ($1/T$) and taking the slope of the linear fitting line. In Figure 6, this is done first for TMP only (Figure 6a) and then for the whole data (Figure 6b) with different furnishes at temperatures exceeding 150 °C. The relationship between ln(Wet tensile strength index) and $1/T$ seems quite linear in the range of 150–270 °C for all pulps. This is striking taking into account that the press type and associated nip dwelling time are different below (cylinder press) and above (steel belt press) 200 °C for the data. The above exponential temperature-dependence of lignin diffusion rate thus dominates over other factors when the level of wet tensile strength of pressed material is set by these processes.

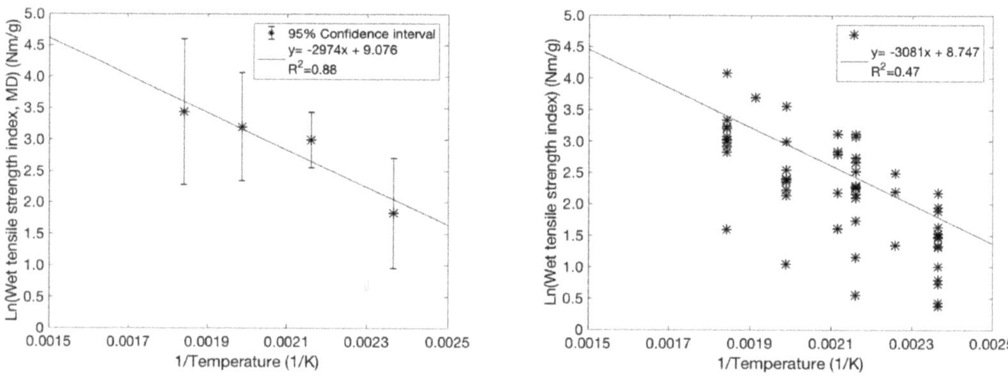

(a) (b)

Figure 6. The logarithm of wet tensile strength index plotted against 1/Temperature when pressed with either cylinder press (T = 150 °C, 190 °C; 6 MPa) or steel belt press (T = 230 °C, 270 °C; 8 MPa): (**a**) TMP sheets with preferred MD fibre orientation pressed at an initial solids content of 61%. The points represent an average of 10 data points and their confidence intervals. (**b**) Varied pulps and pressing conditions for standard laboratory sheets with uniform fibre orientation. Solids content varies in the range of 50–65%. The overall trend is described by a similar activation energy of 26 kJ mol^{-1} as in (**a**).

The temperature behaviour of TMP (Figure 6a) is quantitatively similar to that of the whole data (Figure 6b) with $E_a/R \approx 3080$ K, i.e., $E_a \approx 26$ kJ mol^{-1}. This value is close to the value of 29 kJ mol^{-1} obtained earlier for the diffusion of dissolved lignin from the interior of the chip to the bulk liquor, during the kraft pulping of Eucalyptus globulus wood [25]. Thus, the diffusion rate does not seem to be very sensitive to the type of lignin.

We studied the effect of lignin content of fibres by making similar plots for different pulps separately. A systematic increase in lignin content in the range of 0–12% was obtained for chemical kraft pulps by varying the cooking time. The results for these pulps were compared with similar data for CTMP (lignin content 27%) and TMP (lignin content 28%). Figure 7 shows both estimated E_a and extrapolated wet strength at $1/T = 0$ for the different cases. Here the $1/T = 0$ limit, plotted on a logarithmic axis, describes the order of magnitude of wet strength achievable in hot-pressing. On the other hand, a low E_a value seen for the smallest lignin contents indicates a relatively weak temperature dependence, which is generally coupled with a low $1/T = 0$ limit as well. It seems that at least c.a. 7% of lignin in kraft fibres is required to raise wet strength to a similar level as for the other pulps. On the other hand, lignin content of fibres higher than 12% does not seem to improve wet strength further, as both the activation energy and the $1/T = 0$ limit saturate in Figure 7. In other words, the main improvement on wet strength is achieved already for moderate lignin content of fibres. This suggests that a fairly thin surface layer of diffused lignin is sufficient to provide the maximal bonding between wet fibres.

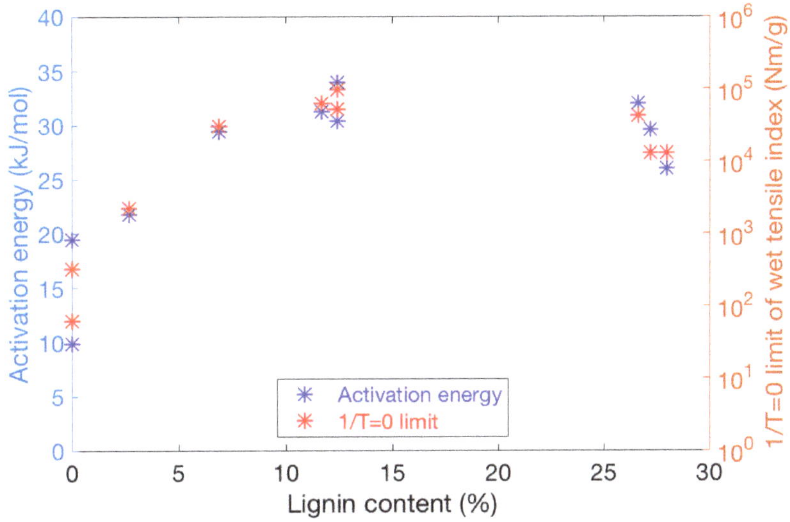

Figure 7. The apparent activation energy (left vertical axis) and extrapolated ln(Wet tensile strength index) at $1/T = 0$ (right logarithmic axis) for different pulps with varied lignin content. The extrapolation omits the degradation of fibre-wall polymers and therefore does not describe the true high-temperature limit of wet tensile index. The points up to 12% lignin content describe kraft pulps with varied cooking times in pulping. These results are compared with similar data for CTMP and TMP with lignin content of 26–28%.

In addition to the primary effects of pressing temperature and lignin content (i.e., pulp type) mentioned above, it is interesting to consider other parameters. Wet strength appears to have a similar level and temperature behaviour for sheets with uniform (Figure 6b) and non-uniform (Figure 6a) fibre orientation. This further suggests that the effective bonding of the contacting inter-fibre surfaces is more important for wet strength than the geometry of the fibre network. This idea is also supported by the observation that wet strength is surprisingly insensitive to nip pressure. When studying heat-treated sheets with and without applied nip pressure, we found no correlation between measured wet strength and average sheet density. On contrary, wet strength and solids content before pressing are correlated as shown in Figure 8. However, the total variation here is much smaller than that for varied temperatures. One possible reason for the correlation between the wet strength index and solids content could be the higher sheet temperature achieved when pressing a drier sheet, which accelerates the lignin inter-diffusion and thus enhances bonding.

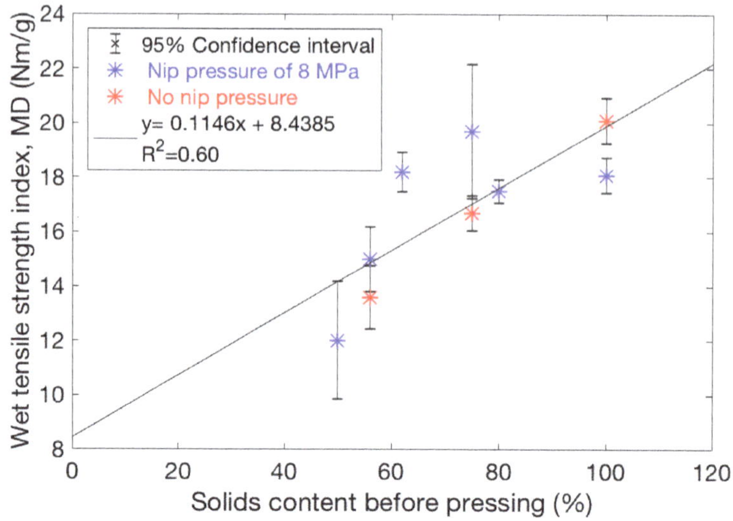

Figure 8. Wet tensile strength index of TMP sheets improves when the pressing is done on a dried web. Here the initial solids content is varied for the constant pressing temperature of 230 °C, keeping other process conditions fixed for all trial points. The points represent an average of 10 data points and their 95% confidence intervals. The applied pressure is 8 MPa for the blue markers, and 0 MPa for the red markers. Wet tensile strength seems not to be very dependent on pressure or density. The applied pressure (0.15 MPa over 23 s) exerted by the steel-belt after the pressing nip seems sufficient to improve wet tensile strength to the same level as when using the 8 MPa nip pressure.

3.4. Network Stiffness and Dry Tensile Strength

The same varying solids contents prior to pressing as in Figure 8 were used for the data in Figure 9, where elastic modulus and dry tensile strength index (dry tensile strength divided by the grammage) are compared against sheet density. The elastic modulus for oriented TMP sheets in MD increases with density (Figure 9a). This is expected as density generally determines the relative bond area for random fibre networks [26]. Nevertheless, the correlation between dry strength and density is rather poor for this particular fibre type (Figure 9b), suggesting that the inelastic behaviour after yielding of the fibre network is important for dry strength. All in all, it seems that the mechanisms underlying dry strength are much more complex than the inter-diffusion mechanism previously discussed in the case of wet strength. For example, in Figure 9b, there is a much higher dry strength value for a particular pressing condition corresponding to 62% solids content. In this case, the parallel measurements have very good reproducibility. Curiously, the elastic modulus, calculated from the same measurement curves, does not differ from the general trend observed for other conditions, as shown in Figure 9a. It is possible that the dry strength (and associated inelastic straining) is more sensitive to overheating of the fibre polymers than the wet strength. Thus, maximising dry strength may require a delicate balance of temperature and process dynamics for optimal moisture removal during hot-pressing. Generally, inelastic deformations are controlled by amorphous fibre components such as hemicelluloses, whose mechanical behaviour changes dramatically with varying moisture content and temperature [27–29].

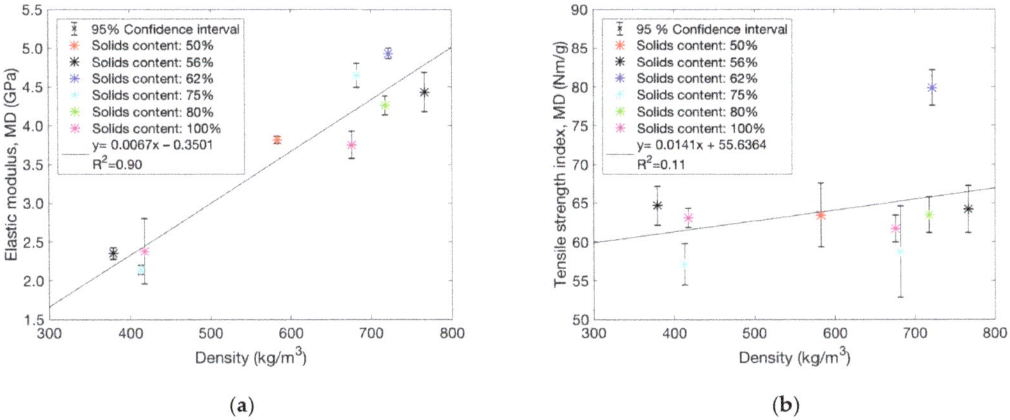

Figure 9. (**a**) Elastic modulus (MD) of hot-pressed sheets with and without nip pressure follows the density as expected. (**b**) Correlation between tensile strength index (MD) and density is still rather weak. The solids contents prior to pressing are indicated in the figures. Note the highest value, which appears like an outlier here, comes from the same measurement as the corresponding point in (**a**). The points represent an average of 10 data points and their 95% confidence intervals.

The deterioration temperatures of cellulose and hemicellulose differ slightly, and some differences for the high-temperature behaviour of elastic modulus and dry tensile strength could be expected. The hemicelluloses degrade at 230–315 °C, whereas lignin decomposes over a broader temperature range of 200–500 °C [30,31]. However, as seen in Figure 10 for varied pulp types, both elastic modulus and dry strength peak around 150–200 °C, followed up by a decrease for most cases when further increasing the temperature. In other words, the above slight differences in polymer degradation do not seem to change the big picture concerning the mechanical behaviour of materials with different pulp types. The only exceptions are a few kraft pulp samples, which contain some lignin that might shield hemicelluloses, and for which a similar decrease of mechanical properties beyond 200 °C is not observed.

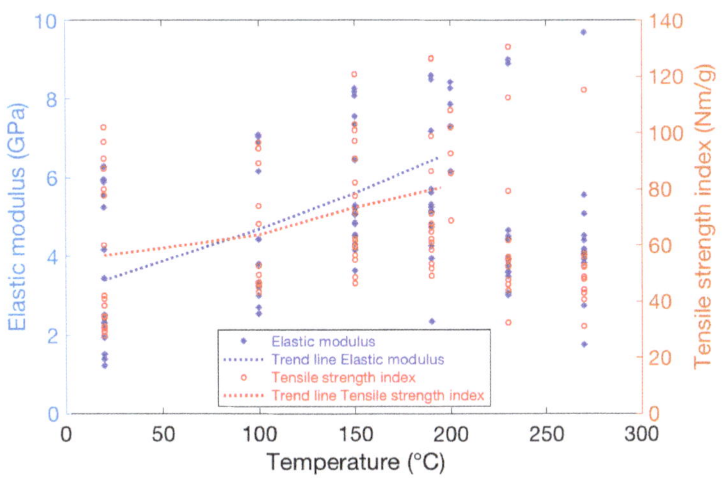

Figure 10. Elastic modulus and dry tensile strength index for a wide data set of sheets with uniform fibre orientation and different furnishes. One should notice that samples above 200 °C have been pressed using the steel-belt press, and those in the range of 20–200 °C have been pressed with cylinder press. The dashed trend lines describe the average behaviour for varied pressing temperatures.

Standard carbohydrate analysis (performed with SCAN-CM 71:09) of some of the samples pressed at 20 °C and 270 °C showed a small reduction in hemicellulose content (calculated according to KA 10.314) caused by hot-pressing. The samples containing nearly 100% carbohydrate showed the largest decrease in hemicelluloses (less than 10% decrease), perhaps due to the lack of "protecting" lignin. However, in general, the changes are rather small in all cases, which is in agreement with earlier findings in wood welding studies [32]. This means that it is not possible to explain the decrease in mechanical properties caused by hot-pressing, as seen in Figure 10, solely by observing the changes in hemicellulose content.

4. Discussion

Hot-pressing often produces a significant increase in the wet tensile strength of paper webs. This effect is strongest when the lignin content of the fibres exceeds 7–12% and the pressing temperature is as high as possible. It is possible that there is an optimum range of lignin contents for wet strength. The relative change in wet strength with temperature is similar for very different pulps, which is explained by a similar lignin inter-diffusion mechanism that strengthens inter-fibre bonding under wet conditions. The strong exponential temperature dependence of the diffusion rate appears to overrule other factors such as pressing time or changes in network density (affected by nip pressure) in determining wet strength levels. This suggests that the wet-strengthening mechanism described above is not sensitive to the amount of lignin diffusing into the bond region between fibres. Even a very thin layer of lignin is sufficient to glue fibre surfaces together so that the bond formed is water resistant. However, it appears that a lignin content of at least 7% is required to cover the surfaces well enough for the wet strength improvement to reach its full potential.

When investigating wet strength for TMP papers with varying initial solids content, the best results were obtained when pressing an initially dry web, which is also expected to have the highest web temperature. It should be noted that the theoretical intra-fibre vapour pressure can become very high, several tens of bars, when the temperature is 200 °C or higher and the solids content is below 80% (see Appendix A). This high internal pressure in the fibre walls does not appear to accelerate lignin transfer, at least when considering the observed changes in wet strength at different solids contents. In other words, lignin and water transport mechanisms seem to decouple from one another. This is very interesting since it is known that the presence of water critically affects the softening (i.e., Tg) of lignin.

The dry tensile strength of hot-pressed TMP paper shows a very complex behaviour under varying process conditions. Rather than having clear trends with varying solids content or temperature, certain conditions appear to be more optimal than others in unexpected ways. This behaviour differs depending on the pulp being pressed, e.g., mechanical pulp or chemical pulp, so it is difficult to draw general conclusions. However, it seems that softening, e.g., by water or sulfonation [13], is important for dry strength. It should be noted that dry strength is quite high even without pressing, so the relative changes are smaller than for wet strength. In addition, polymer degradation can degrade strength at high temperatures. Therefore, evaporation of water in the fibre walls can appropriately control the temperature rise and prevent polymer degradation. Perhaps the best conditions consist of pressing times and temperatures that are just sufficient to evaporate most of the water from the fibres but do not cause over-drying or heating of the fibres that degrades their strength properties.

5. Conclusions

The main findings in this study are highlighted below:

- Hot-pressing does not cause a noticeable change in density in the fibre walls, despite a large reduction in network porosity and mean pore size.
- The wet strength increases with increasing pressing temperature. The stronger bonding of the wet fibres can be explained by inter-diffusion of lignin macromolecules (with an activation energy around 26 kJ mol^{-1}) after lignin softening. The associated

exponential acceleration of diffusion with temperature dominates the setting of wet strength over other factors such as process dynamics or final material density.

- The highest solids content before hot-pressing for the TMP sheets was found to give the highest values for wet strength. A possible explanation for this is that when a drier sheet is pressed, a higher temperature is reached, which accelerates the inter-diffusion of the lignin and thus enhance bonding. No significant correlation was observed between the varied solids content before pressing and dry strength.
- The elastic modulus increases with the increasing density of the sheets after hot-pressing, as expected. On the other hand, the dry strength does not show the same trend, indicating that the inelastic behaviour after yielding is responsible for the observed differences among the trial points.
- For dry strength and elastic modulus, the optimum pressing temperature is lower than for wet strength due to the degradation of hemicelluloses.

Author Contributions: Conceptualisation, A.M. (Amanda Mattsson) and J.A.K.; methodology, A.M. (Amanda Mattsson), T.J., A.M. (Arttu Miettinen) and J.A.K.; formal analysis, A.M. (Amanda Mattsson), T.J., A.M. (Arttu Miettinen) and J.A.K.; investigation, A.M. (Amanda Mattsson) and J.A.K.; writing—original draft preparation, A.M. (Amanda Mattsson), T.J., A.M. (Arttu Miettinen) and J.A.K.; writing—review and editing, A.M. (Amanda Mattsson), T.J., A.M. (Arttu Miettinen), J.A.K. and P.E.; visualisation, A.M. (Amanda Mattsson), A.M. (Arttu Miettinen) and J.A.K.; funding acquisition, G.P. and P.E. All authors have read and agreed to the published version of the manuscript.

Funding: This research was funded by The Kamprad Family Foundation for Entrepreneurship, Research & Charity (grant number 20180234), The KK foundation, and Kempestiftelserna.

Institutional Review Board Statement: Not applicable.

Informed Consent Statement: Not applicable.

Data Availability Statement: The data in this study are available on request from the corresponding author.

Acknowledgments: The authors thank Licentiate of Engineering Javier Brugés for the help with editing the figures.

Conflicts of Interest: The authors declare no conflict of interest.

Appendix A

The vapour pressure P_v inside a fibre wall at elevated temperatures can be estimated based on the theory developed by Flory and Huggins [33]. We assume that all water is bound to the (hemi)cellulose gel. In a binary polymeric solution (w water, c cellulose) [33], the vapour pressure P_v is given by

$$ln\left(\frac{P_v}{P_0}\right) = \ln(\phi_w) + \left(1 - \frac{v_w}{v_c}\right)\phi_c + \chi\phi_c^2 \quad \text{(A1)}$$

where P_0 is the vapour pressure of pure solvent, approximated by the Antoine equation

$$log_{10}P_0 = A - \frac{B}{C+T};\ A = 8.071,\ B = 1731\ ^\circ\text{C},\ C = 233.4\ ^\circ\text{C}\ (P_0\ in\ Torr) \quad \text{(A2)}$$

In Equation (A1), ϕ_i is volume fraction, v_i is molar volume (molar mass divided by mass density), and χ is Flory–Huggins interaction parameter (for cellulose at moderate water contents). According to reference [34], the specific volume (or density) of cellulose does not vary much with temperature up to 190 °C if the pressure remains below 20 MPa. Assuming that the lignin content of fibre is c_l, the cellulose content is $1 - c_l$. The volumes of the different components are (assuming bulk phase dominates the volume for both components when the moisture content is high).

$$V_i = \frac{M_i}{\rho_i} \tag{A3}$$

For cellulose, we obtain the volume

$$V_c = \frac{(1-c_l)M_f}{\rho_c} \tag{A4}$$

where M_f is dry fibre mass and ρ_c is cellulose density. The water mass in terms of moisture content mc and dry fibre mass M_f becomes

$$M_w = \frac{mc\ M_f}{100 - mc} \tag{A5}$$

leading to the water volume

$$V_w = \frac{mc\ M_f}{(100-mc)\rho_W} \tag{A6}$$

where ρ_w is water density. The total volume of the water-cellulose gel can be approximated as

$$V_T \approx V_c + V_w = \left[\frac{1-c_l}{\rho_c} + \frac{mc}{(100-mc)\rho_w}\right]M_f = \frac{(1-c_l)(100-mc)\rho_w + mc\rho_c}{\rho_c(100-mc)\rho_w}M_f \tag{A7}$$

Thus, we can write volume fractions as

$$\phi_w = \frac{V_w}{V_T} \approx \frac{mc\ \rho_c}{(1-c_l)(100-mc)\rho_w + mc\rho_c} \tag{A8}$$

$$\phi_c = \frac{V_c}{V_T} \approx \frac{(1-c_l)(100-mc)\ \rho_w}{(1-c_l)(100-mc)\rho_w + mc\rho_c} \tag{A9}$$

In Figure A1 we show the behaviour of vapour pressure for varied temperatures and solids contents. The used values of the parameters are presented in Table A1.

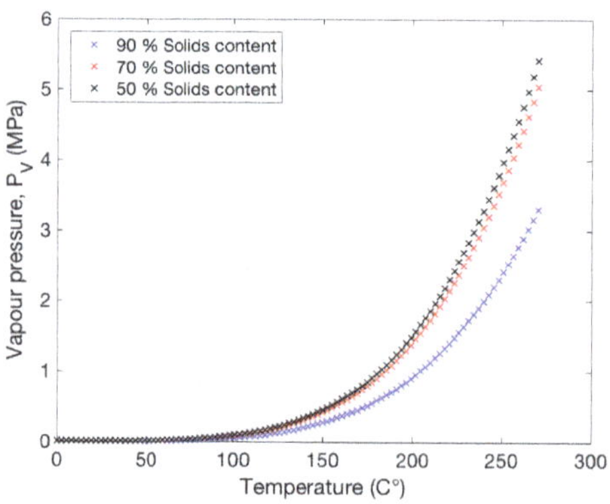

(a)

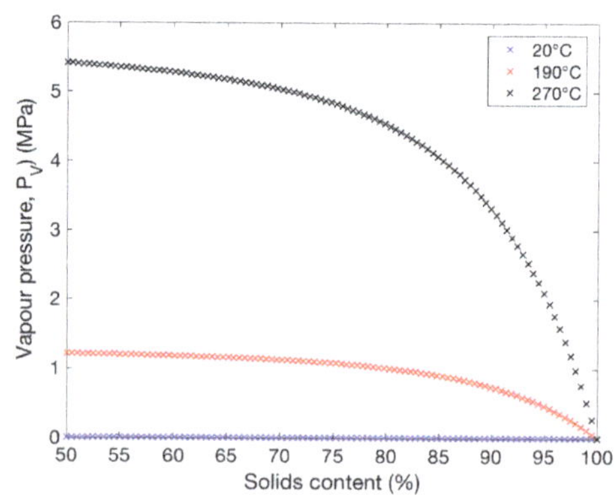

(b)

Figure A1. (**a**) Vapour pressure as a function of temperature for three different levels of solids content, 90%, 70%, and 50%. (**b**) Vapour pressure as a function of solids content for three different temperature levels, 20 °C, 190 °C, and 270 °C.

Table A1. The values of the parameters used for plotting the vapour pressure as a function of temperature and solids content in Figure A1.

Parameter	Value
Molar volume of water (v_w)	18.02 cm^3/mol
Molar volume of cellulose (v_c)	101.3 cm^3/mol
Flory–Huggins interaction parameter (χ)	0.67 [35]
Density of water (ρ_w)	1000 kg/m^3
Density of (crystal) cellulose (ρ_c)	1600 kg/m^3
Lignin content of fibre (c_l)	25%

References

1. Ourworldindata. Available online: https://ourworldindata.org/plastic-pollution (accessed on 3 March 2021).
2. Östlund, S.; Niskanen, K. *Mechanics of Paper Products*; De Gruyter: Berlin, Germany, 2021; ISBN 9783110617412.
3. Karlsson, M.; Paltakari, J. *Papermaking Science and Technology*; Karlsson, M., Ed.; Papermakin; Fapet Oy: Atlanta, GA, USA, 2008.
4. Norgren, S.; Pettersson, G.; Höglund, H. Strong paper from spruce CTMP—Part II: Effect of pressing at nip press temperatures above the lignin softening temperature. *Nord. Pulp Pap. Res. J.* **2018**, *33*, 142–149. [CrossRef]
5. Joelsson, T.; Pettersson, G.; Norgren, S.; Svedberg, A.; Höglund, H.; Engstrand, P. High strength paper from high yield pulps by means of hot-pressing. *Nord. Pulp Pap. Res. J.* **2020**, *35*, 195–204. [CrossRef]
6. Joelsson, T.; Pettersson, G.; Norgren, S.; Svedberg, A.; Höglund, H.; Engstrand, P. Improving paper wet-strength by increasing lignin content and hot-pressing temperature. *Tappi J.* **2020**, *19*, 487–499. [CrossRef]
7. Good, R.J. *Contact Angles and the Surface Free Energy of Solids BT—Surface and Colloid Science: Volume 11: Experimental Methods*; Good, R.J., Stromberg, R.R., Eds.; Springer: Boston, MA, USA, 1979; pp. 1–29, ISBN 978-1-4615-7969-4.
8. Vaziri, M.; Karlsson, O.; Abrahamsson, L.; Lin, C.F.; Sandberg, D. Wettability of welded wood-joints investigated by the Wilhelmy method: Part 1. Determination of apparent contact angles, swelling, and water sorption. *Holzforschung* **2021**, *75*, 65–74. [CrossRef]
9. Karlsson, O.; Torniainen, P.; Dagbro, O.; Granlund, K.; Morén, T. Presence of water-soluble compounds in thermally modified wood: Carbohydrates and furfurals. *BioResources* **2012**, *7*, 3679–3689. [CrossRef]
10. Gupta, P.R.; Rezanowich, A.; Goring, D.A.I. The adhesive properties of lignin. *Pulp Pap. Mag. Can* **1962**, *63*, 21–30.
11. Goring, D.A.I. Thermal softening of lignin, hemicelluolose and cellulose. *Pulp Pap* **1963**, *64*, T517–T527.
12. Back, E.L.; Salmen, N.L. Glass Transitions of Wood Components Hold Implications for Molding and Pulping Processes. *Tappi* **1982**, *65*, 107–110.
13. Joelsson, T.; Persson, E.; Pettersson, G.; Norgren, S.; Svedberg, A.; Engstrand, P. The impact of sulfonation and hot-pressing of low-energy high temperature chemi-thermomechanical pulp. *Holzforsch* **2021**, submitted.
14. Thakur, V.K.; Thakur, M.K.; Raghavan, P.; Kessler, M.R. Progress in green polymer composites from lignin for multifunctional applications: A review. *ACS Sustain. Chem. Eng.* **2014**, *2*, 1072–1092. [CrossRef]
15. Bajpai, P. *Pulp and Paper Production Processes and Energy Overview*; Elsevier: Amsterdam, The Netherlands, 2016; ISBN 9780128034118.
16. Pettersson, G.; Norgren, S.; Engstrand, P.; Rundlöf, M.; Höglund, H. Aspects on bond strength in sheet structures from TMP and CTMP—A review. *Nord. Pulp Pap. Res. J.* **2021**, *36*, 177–213. [CrossRef]
17. Joelsson, T.; Svedberg, A.; Norgren, S.; Pettersson, G.; Berg, J.-E.; Garcia-Lindgren, C.; Engstrand, P. Unique steel belt press technology for high strength papers from high yield pulp. *SN Appl. Sci.* **2021**, *3*, 561. [CrossRef]
18. Tomasi, C.; Manduchi, R. Bilateral filtering for gray and color images. In Proceedings of the Sixth International Conference on Computer Vision (IEEE Cat. No. 98CH36271), Bombay, India, 7 January 1998; pp. 839–846.
19. Otsu, N. A Threshold Selection Method from Gray-Level Histograms. *IEEE Trans. Syst. Man Cybern.* **1979**, *9*, 62–66. [CrossRef]
20. Turpeinen, T.; Myllys, M.; Kekalainen, P.; Timonen, J. Interface Detection Using a Quenched-Noise Version of the Edwards-Wilkinson Equation. *IEEE Trans. Image Process.* **2015**, *24*, 5696–5705. [CrossRef] [PubMed]
21. Hildebrand, T.; Rüegsegger, P. A new method for the model-independent assessment of thickness in three-dimensional images. *J. Microsc.* **1997**, *185*, 67–75. [CrossRef]
22. Ali, C.M.; Jean-Francis, B.; Elodie, B.; Patrice, M. 3D synchrotron X-ray microtomography for paper structure characterization of z-structured paper by introducing micro nanofibrillated cellulose. *Nord. Pulp Pap. Res. J.* **2016**, *31*, 219–224. [CrossRef]
23. Kellogg, R.M.; Wangaard, F.F. Variation in the cell-wall density of wood. *Wood Fiber Sci.* **1969**, *1*, 180–204.
24. Li, K.; Wang, S.; Chen, H.; Yang, X.; Berglund, L.A.; Zhou, Q. Self-Densification of Highly Mesoporous Wood Structure into a Strong and Transparent Film. *Adv. Mater.* **2020**, *32*, 2003653. [CrossRef]
25. Simão, J.P.F.; Carvalho, M.G.V.S.; Baptista, C.M.S.G. Heterogeneous studies in pulping of wood: Modelling mass transfer of dissolved lignin. *Chem. Eng. J.* **2011**, *170*, 264–269. [CrossRef]
26. Kaarlo, N. *Paper Physics*; Paperi ja Puu Oy: Atlanta, GA, USA, 2008; ISBN 9525216292.
27. Niskanen, K. *Mechanics of Paper Products*; De Gruyter: Berlin, Germany, 2011; ISBN 9783110254617.
28. Lucenius, J.; Valle-Delgado, J.J.; Parikka, K.; Österberg, M. Understanding hemicellulose-cellulose interactions in cellulose nanofibril-based composites. *J. Colloid Interface Sci.* **2019**, *555*, 104–114. [CrossRef]

29. Sorieul, M.; Dickson, A.; Hill, S.J.; Pearson, H. Plant fibre: Molecular structure and biomechanical properties, of a complex living material, influencing its deconstruction towards a biobased composite. *Materials* **2016**, *9*, 618. [CrossRef]
30. Börcsök, Z.; Pásztory, Z. The role of lignin in wood working processes using elevated temperatures: An abbreviated literature survey. *Eur. J. Wood Wood Prod.* **2021**, *79*, 511–526. [CrossRef]
31. Hubbe, M.A.; Pizzi, A.; Zhang, H.; Halis, R. Critical Links Governing Performance of Self-binding and Natural Binders for Hot-pressed Reconstituted Lignocellulosic Board without Added Formaldehyde: A Review. *BioResources* **2017**, *13*, 2049–2115. [CrossRef]
32. Delmotte, L.; Ganne-Chédeville, C.; Leban, J.-M.; Pizzi, A.; Pichelin, F. CP-MAS 13C NMR and FT-IR investigation of the degradation reactions of polymer constituents in wood welding. *Polym. Degrad. Stab.* **2008**, *93*, 406–412. [CrossRef]
33. Khansary, M.A. Vapor pressure and Flory-Huggins interaction parameters in binary polymeric solutions. *Korean J. Chem. Eng.* **2016**, *33*, 1402–1407. [CrossRef]
34. Jallabert, B.; Vaca-Medina, G.; Cazalbou, S.; Rouilly, A. The pressure–volume–temperature relationship of cellulose. *Cellulose* **2013**, *20*, 2279–2289. [CrossRef]
35. Hakalahti, M.; Faustini, M.; Boissière, C.; Kontturi, E.; Tammelin, T. Interfacial mechanisms of water vapor sorption into cellulose nanofibril films as revealed by quantitative models. *Biomacromolecules* **2017**, *18*, 2951–2958. [CrossRef]

polymers

MDPI

Article

Improvements of Arboblend V2 Nature Characteristics through Depositing Thin Ceramic Layers

Simona-Nicoleta Mazurchevici [1], Alina Marguta [1], Bogdan Istrate [2], Marcelin Benchea [2], Mihai Boca [1] and Dumitru Nedelcu [1,3,*]

1 Department of Machine, Manufacturing Technology, "Gheorghe Asachi" Technical University of Iasi, 700050 Iasi, Romania; simona0nikoleta@gmail.com (S.-N.M.); alinamarguta@yahoo.com (A.M.); mihaitzaboca@yahoo.com (M.B.)
2 Department of Mechanical Engineering, Mechatronics and Robotics, Faculty of Mechanical Engineering, "Gheorghe Asachi" Technical University of Iasi, 700050 Iasi, Romania; bogdan.istrate@tuiasi.ro (B.I.); marcelin.benchea@tuiasi.ro (M.B.)
3 Mechanical Engineering Department, Technical Sciences Academy of Romania, 030167 Bucharest, Romania
* Correspondence: dnedelcu@tuiasi.ro

Abstract: The paper aims to investigate the behavior of Arboblend V2 Nature biopolymer samples covered with three ceramic powders, Amdry 6420 (Cr_2O_3), Metco 143 (ZrO_2 $18TiO_2$ $10Y_2O_3$) and Metco 136F (Cr_2O_3-$xSiO_2$-$yTiO_2$). The coated samples were obtained by injection molding, and the micropowder deposition was achieved by using the Atmospheric Plasma Spray (APS) method, with varied thickness layers. The present study will only describe the results for nine-layer deposition because, as the number of layers' increases, the surface quality and mechanical/thermal characteristics such as wear, hardness and thermal resistance are also increased. The followed determinations were conducted: the adhesion strength, hardness on a microscopic scale by micro-indentation, thermal analysis and structural and morphological analysis. The structural analysis has highlighted a uniform deposition for the ZrO_2 $18TiO_2$ $10Y_2O_3$ layer, but for the layers that contained Cr_2O_3 ceramic microparticles, the deposition was not completely uniform. The thermal analysis revealed structural stability up to a temperature of 230 °C, the major degradation of the biopolymer matrix taking place at a temperature around 344 °C. The samples' crystalline structure as well as the presence of the Cr_2O_3 compound significantly influenced the micro-indentation and scratch analysis responses. The novelty of this study is given by itself the coating of the Arboblend V2 Nature biopolymer (as base material), with ceramic microparticles as the micropowder coating material. Following the undertaken study, the increase in the mechanical, tribological and thermal characteristics of the samples recommend all three coated biopolymer samples as suitable for operating in harsh conditions, such as the automotive industry, in order to replace plastic materials.

Keywords: biodegradable thermoplastic; coating; Amdry 6420; Metco 143; Metco 136F; hardness; adhesion; structure

Citation: Mazurchevici, S.-N.; Marguta, A.; Istrate, B.; Benchea, M.; Boca, M.; Nedelcu, D. Improvements of Arboblend V2 Nature Characteristics through Depositing Thin Ceramic Layers. *Polymers* **2021**, *13*, 3765. https://doi.org/10.3390/polym13213765

Academic Editors: Domenico Acierno and Antonella Patti

Received: 7 October 2021
Accepted: 27 October 2021
Published: 30 October 2021

1. Introduction

The thermal coating process finds its applicability in areas such as the automotive, aerospace and naval industries in order to improve the corrosion resistance, wear resistance and lifetime of the equipment part as follows: thermal barrier coatings for components that operate in severe thermal conditions (turbine blades, fuel parts, vanes) with the role of increasing their life by improving resistance to oxidation, heat and corrosion [1]; hard metal coatings such as tungsten and chromium carbide are commonly used to increase the wear and corrosion resistance of the parts surfaces [2]; corrosion resistance after heat treatment of a Cr_3C_2–NiCr coating showed significant improvement due to both the microstructural changes and the presence of a metallurgical bond at the coating–substrate interface [3];

titanium plasma spraying (TPS) used on polymers as polyethylene/polyamide highlighted good mechanical properties and increased hardness [4].

One of the most common thermal deposition techniques is Atmospheric Plasma Spray (APS), and the coatings obtained by zirconium and chromium ceramic layer depositions are widely spread materials in mechanical applications [5]. The main drawbacks of these coatings are given by the micro-cracks, isolation inhomogeneity and residual stresses (appears during cooling process) [6].

The Al_2O_3 (aluminium oxide (alumina)) powder with high purity used for coating, usually applied by the plasma spray method, presents electrical insulation in terms of thermal conductivity and dielectric characteristics. Furthermore, the coatings are wear-resistant, have higher hardness, higher temperature stability and are chemically inert. The ceramic coating is suitable for electrical, electronic and semiconductor manufacturing tool applications, electrostatic chucks and capacitors, vacuum chamber lines, etc. The high purity of the material ensures that it will not contaminate the semi-conductor components [7]. Nickel, zirconium, aluminium, molybdenum and chromium, as powder elements, are a good choice for applications that require high toughness with moderate resistance to fretting, erosion and scuffing. Zinc-based coatings are widespread in applications that require increased corrosion resistance, but if this chemical element is not alloyed with other elements such as tungsten carbide or nickel, it has low mechanical characteristics [8]. The degradation of the Zn–polymer interface has been studied by the authors of the paper [9], and they point out that the delamination can be inhibited by CO_2 gas in a humid environment. Inhibition depends very much on the polymer matrix's affinity for carbon dioxide. Other researchers such as [10] have used nano-zinc-oxide (5 wt %) and epoxy acrylate in order to obtain a corrosion protection coating for mild steel panels.

It is well known that the addition of ceramic particles, metals particles and biopolymers as a substrate provides a combination of properties of all three types of materials: biopolymer matrices and metallic and ceramic reinforcement components. This may result in the improvement of the physical and mechanical properties of the composite [11].

In the literature, the applications of function coatings refer mainly to corrosion protection characteristic, and there is a limited emphasis on equally important properties—for example, mechanical robustness, which is significant for determining the applicability of developed coatings. However, there are discussions on the principal working theories, procedures for preparation, performance investigations and applications of superhydrophobic coatings, for instance: a viable preventative method for controlling metal corrosion due to their mechanical stability and durability, and the shortcoming consists in the ability of maintaining this characteristic for a prolonged period of time [12]; corrosion inhibition of metallic materials with the help of smart coatings involves difficulties in achieving some features as thermal stability, resistances to scratching and strong chemical acids, high optical transmission, in situ healing, etc. [13]; coating of stainless steel and titanium bipolar plates in order to improve the corrosion resistance and electrical conductivity in PEMFC (proton-exchange membrane fuel cells), which has, for the moment as the main impediment in the large-scale spread of the product, a high manufacturing cost [14]; excellent anticorrosion durability obtained by coating the polyaniline–graphene oxide composite with zinc-based waterborne [15]; anticorrosive coatings in the marine field face the degradation, loss of adhesion and failure of coating systems [16]; metallic substrates (aluminum alloy) coated with active corrosion protection systems (with self-healing ability) such as silica−zirconia nanoparticles highlighted long-term corrosion protection and the ability to self-heal defects, and these characteristics are obtained by rigorous control of the inhibitor regular release at the moment when the corrosion process begins to arise [17]. A comparative review that accounts for all factors, including the durability and other mechanical properties, is essential for understanding the applicability of an advanced coating at a practical level.

In addition, another area targeted by coating researchers is related to composites based on polymer matrices with the inclusion of magnetic nano-sized particles: Polydimethylsiloxane (PDMS) coated with different concentrations of nanosized Ni@C core-shell [18];

Ni-silicone elastomagnetic composites [19]; polyacrylamide-based hydrogels coated with Ni ferrite [20]; polyetherurethane (TFX) and a biodegradable multiblock copolymer (PDC) with poly(p-dioxanone) as hard segment and poly(ε-caprolactone) as soft segment were investigated as matrix component, coated with iron oxide particles [21]; and oligo(ε-caprolactone)dimethacrylate/butyl acrylate, coated with Fe_3O_4 [22]; carbon-fiber–epoxy composites coated with two thermal ceramic particles (lass flakes and aluminum titanate) in order to create a thermal barrier for the substrate [23]. The interest of researchers was focused mainly on elastomagnetic effects [19] and the wide prospects of applications as follows: Ni ferrite with a highly organized structure as humidity sensors [20]; magnetic nanoparticles for practical applications which involve sensors and biosensors [24]; magnetoresistive sensors for applications where the ultimate field detection limits are required or as readers in hard disk drives [25]; Mg substitution on Ni-ferrite ceramics with applications in biomedicine, gas detection, heterogeneous catalysis, adsorption, etc. [26]; shape-memory materials through the inductive heating of magnetic nanoparticles in thermoplastic polymers [21]; and the incorporation of surface-modified superparamagnetic nanoparticles into a polymer matrix [22]. These materials also demonstrate prospects for biomedicine: drug delivery, hyperthermia, magnetic resonance imaging contrast enhancement [27] and the manipulation of cell membranes [28]; recording media and high-frequency applications—electromagnetic-wave-absorption materials [29], microwave absorption [30] and gigahertz microwave absorption [31].

In this paper, the authors have analysed three types of plasma jet coatings for intermediate layers (micrometallic powder) and one type of ceramic coating (aluminium oxide). The coatings were made on Arboblend V2 Nature substrate materials. The samples were obtained using injection moulding, and the method used for coating was atmospheric plasma spraying.

The registered trademark Arboblend®, developed by scientists and engineers from the German company Tecnaro in collaboration with those from the Fraunhofer Institute for Chemical Technology, is a 100% biodegradable biopolymer [32], and a part of the current research group has investigated the behavior of samples covered with metallic intermediate layer and ceramic final layer in the past but not at such a deep level and not following an experimental plan [33].

The aim of the manuscript was to obtain a new material with improved properties that can then be used successfully as a substitute for synthetic plastics in the automotive industry. Given this objective, the biodegradable material Arboblend V2 Naure was chosen, for which the research team previously studied the properties. The next step was to realize the coating with ceramic microparticles by using the APS method, after which the same characteristics were studied. The present study is not found in the research activity of other authors, so the proposed research through both technology and experimental results is constituted as a novelty element.

The introduction section provides a coatings overview. This chapter is followed by a description of the materials and methods used in the analysis of the coated samples. The results of the experimental research are presented and commented on in Section 3.4, and the conclusions part suggestively presents general comments on the main obtained results.

2. Materials and Methods

The thermoplastic material selected to be coated with ceramic microparticles was Arboblend V2 Nature. According to the information provided by the producers but also to some studies from the specialized literature [32,34,35], the basic matrix of the polymer is lignin, this being extracted from annual vegetable plants, so it is not necessary to use wood raw materials that require dozens of years or even longer to reach maturity and be used in the forestry and paper industry. This is important to mention because another source of lignin used for this material comes from its extraction from the paper industry waste. In addition, the Arboblend V2 Nature structure can contain a significant amount of polylactic acid (PLA—also biodegradable polyester) and other constituents such as bio-polyamides (bio-PA), cellulose, natural vegetable fibers and, for processing in good conditions, contains a small amount of natural additives (resins, waxes, shellac) [33,36,37].

Obtaining the necessary samples for coating with ceramic layers was realized by injection in the mold using the SZ-600H equipment (SHEN ZHOU, Zhangjiagang, China). The dimensions of the samples were (70 × 50 × 10) mm^3. The following technological parameters were used for injection: material melting temperature—165 °C; injection pressure—100 MPa; injection speed—80 m/min; cooling time—30 s.

The preparation of the rectangular biopolymer samples consisted in the fixed adhesion between metal strips, followed by the blasting and removal of impurities in order to obtain a surface roughness that was as low as possible. The final stage of preparation was degreasing with ethyl alcohol.

Atmospheric plasma spray (APS) technology (SPRAYWIZARD-9MCE, Sultzer-Metco, Westbury, New Yorkxz, USA/9MBspraying gun) was used to cover the injected samples. The technological coverage parameters used are shown in Table 1.

The deposition rate of the microparticles was constant. The thickness of the deposited ceramic layer was of the micrometers order and for the thermal control of the samples' melting temperature, a laser pyrometer was used throughout the process.

Three ceramic powders were used for the coating: Amdry 6420 (Cr_2O_3), Metco 143 (Cr_2O_3-xSiO_2-yTiO_2) and Metco 136F (ZrO_2 18TiO_2 10Y_2O_3). The three micropowders were deposited on three samples injected from Arboblend V2 Nature. On each sample, a distinct number of passes was made, namely, 5, 7 and 9 passes, in order to study the improvement or not of the mechanical characteristics with the increase in the deposited ceramic layer. However, this manuscript will only present the results for the samples obtained by performing 9 passes, because the objective of the paper is to highlight the uniformity and homogeneity of the deposited layers, which is not entirely revealed by the samples obtained with 3 and 5 passes. The experimental plan used to cover the samples with ceramic micropowders is highlighted in Table 2.

Table 1. Technological parameters used during the counting process.

Powder	Gun Type	N_2		H_2		Electric		9MP Powder Dispenser		Amount (g/min)	Spray Distance (mm)
		Pressure (Bar)	Gas Flow (NLPM)	Pressure (Bar)	Gas Flow (NLPM)	DC (A)	DC (V)	Carrier Gas Flow (NLPM)	Air Pressure (Bar)		
ZrO_2 18TiO_2 10Y_2O_3	9MB	3.4	44	3.4	6.6	400	70–80	5.3	1.4	144	137
Cr_2O_3		3.6	39	3.6	6.6	400	70–80	5.1	1.4	126	145
Cr_2O_3-xSiO_2-yTiO_2		3.7	42	3.7	6.6	400	70–80	5.1	1.4	132	145

Table 2. Experimental plan used to cover the samples with ceramic layer.

No.crt.	Sample Number	Powder Type	Number of Passes
1	1	143	5
2	2	143	7
3	3	143	9
4	4	6420	5
5	5	6420	7
6	6	6420	9
7	7	136	5
8	8	136	7
9	9	136	9

The ceramic micro powders were purchased from the Oerlikon Metco manufacturer (Bella Vista, New South Wales, Australia). The microparticles dimensions and shapes are varied as follows [38]:

- Chromium Oxide Thermal Spray Powder, Amdry 6420 (Cr_2O_3): angular, blocky morphology, (10–105) μm;
- Chromia–Silica composite powder, Metco 136F (Cr_2O_3-xSiO_2-yTiO_2): irregular or angular/blocky morphology, (9–110) μm, Cr_2O_3—Balance; SiO_2—(3.0–4.5)%; TiO_2 < 4.0%;
- Zirconia–Titania–Yttria Composite Powder, Metco™ 143 (ZrO_2 18TiO_2 10Y_2O_3): spheroidal morphology; typically particle size between (3–40) μm;
- The microindentation and scratch tests used the CETR UMT-2 microtribometer (Universal Materials Tester, CETR®, Campbell, SUA). Test conditions were as follows:
- Scratch analysis: a blade with a tip of 0.4 mm (radius at the tip) was used, the samples were fixed on the table and during the test samples were pressed with a vertical force of 10 N NVIDIA blade, moving the table over a distance of 10 mm in 60 s, and the test speed was 0.167 mm/s. The software performed the automatic test and recorded the following parameters: vertical force F_z, horizontal force F_x, time and distance of movement in the horizontal direction Y (of the fixing mass of the sample).
- Microindentation test: the Rockwell type indenter was used (cone with diamond tip having an angle of 120° and a radius at the peak of 200 microns), the samples were fixed on the table and during the test samples were pressed with a vertical force of 10 N (with steps/times described in work). Three samples of each type of powder were tested in order to be able to calculate the parameters statistically with the highest possible accuracy (hardness and Young's modulus). The software performed the automatic test and recorded the following parameters: vertical force F_z, time and vertical travel distance C of the indenter (with capacitive sensor). Other process parameters were loading time—30 s; holding time—15 s; unloading time—30 s; and sensor of (0.2–20) N. Microindentation tests involved testing three samples for each type of ceramic powder in order to confirm experimental repeatability. The average values were obtained by calculating the arithmetic average, and the standard deviation highlights the variation in a set of numbers compared to the calculated average value.

In order to determine the thermal, structural and morphological behavior, only the samples with 9 successive passes were selected for this study: sample 3 coated with Zirconia–Titania–Yttria Composite Powder, Metco™ 143, further noted with P3–143–9 passes; sample 6 coated with Chromium Oxide powder, Amdry 6420, noted with P6–6420–9 passes; sample 9 coated with Chromia–Silica composite powder, Metco 136F, noted with P9–136–9 passes. The used equipments for these analyzes were as follows:

Differential scanning calorimetry *(DSC)* was performed on a DSC 200 F3 Maia differential scanning calorimeter (NETZSCH-Gerätebau GmbH, Selb, Germany). The calibration of the device was realised in accordance with mercury (Hg), zinc (Zn), Indium (In), Tin (Sn) and Bismuth (Bi) standards. The mass of the analyzed samples was less than 30 mg. The experiments were analyzed in the atmosphere of inert gas (Ar). In this experiment, a sample and a reference (an empty crucible) were subjected to the same temperature program. The temperature program consisted of heating from room temperature (RT $\approx$ 20 °C) to 200 °C, then cooling from this temperature to RT. The heating and cooling speed used was 10 K/min. During the experiment, the reference and sample temperatures were measured and the temperature difference recorded between the two was converted into heat flux. The recorded thermograms were then evaluated using Proteus software (provided by NETZSCH). The tangent method was used to determine the transition temperatures. The area was determined by using a rectilinear baseline. Temperature of the transformation beginning (T_{onset}), temperature assigned to the peak (T_{peak}), temperature at the end of the transformation (T_{end}) and the amount of absorbed or dissipated heat were determined.

Thermogravimetric curves (*TG*), derived thermogravimetric curves (*DTG*) and differential thermal analyzes (*DTA*), were determined using Mettler Toledo TGA/SDTA 851 equipment. The mass of samples subjected to thermal decomposition was between 2.9 and 3.9 mg. It was worked in an air atmosphere with a flow rate of 20 cm^3/min. The study was realized in the temperature range of 25–700 °C using a heating rate of 10 °C/min. The processing of thermogravimetric curves was performed with the STARe SW 9.10 software from Mettler Toledo (Columbus, OH, USA). As in the case of DSC analysis, the beginning (T_{onset}), maximum (T_{peak}) and end (T_{end}) temperatures of each thermal degradation stage were determined. For each identified stage, it is indicated the loss mass percentage residue (W%).

SEM structural analysis (Scanning Electron Microscopy) was performed on the QUANTA 200 3D electron microscope (FEI Company, Fremont, CA, USA). Micrographic maps of the samples coated with ceramic micropowders were made on their surface to observe mainly the uniformity of the deposition. The main parameters considered were the following: the pressure inside the microscope chamber—60 Pa; detector—(Large Field Detector) for the analysis of non-conductive samples such as polymers, textile fibers, powders, etc.; tilt angle—0°; secondary electron acceleration voltage—20 Kv; working distance—15 mm; magnification power—500×–2000×.

To determine the chemical elements that appeared with the deposition of the ceramic layers, an Energy-dispersive X-ray spectroscopy (EDX) together with a SEM was performed. The SEM equipment was VegaTescan LMHII (TESCAN ORSAY HOLDING, Kohoutovice, Czech Republic) with EDX detector X Flash 6I10 from Bruker, Germany, using Esprit 2.2 software. The type of EDX analysis was in-line in order to capture as accurately as possible the difference in chemical composition between the resulted microceramic layer and substrate material.

X-ray diffraction analysis (XRD) was performed with the X'Pert Pro MRD X-ray diffractometer, which has a Cu kα anode X-ray tube, λ–1.54 Å, Panalytical equipment (PANalytical, Almelo, the Netherlands), on which a voltage of 45 kV was applied, the variation of the diffraction angle (2θ) being between 10 and 90°. Two X'Pert Data Collector programs were used to process the data and make the diagrams, namely X'Pert High Score Plus version number 3 and X'Pert Data Viewer version number 2.2 g (Malvern Panalytical, Malvern, UK). This analysis aimed to identify the existence of crystallization phases specific to ceramic micropowders deposited on the surface of the samples from Arboblend V2 Nature. The identification of the crystallization phases was performed by comparing the obtained data with those from the scientific literature.

The chemical composition analyses were performed in five distinct points, and an overall average composition was made with the help of Minitab software.

3. Results and Discussion

3.1. DSC Analyse

In order to establish the physical transformations that take place during the gradual heating of the Arboblend V2 Nature samples coated with ceramic micropowders, a DSC analysis was performed. Three distinct samples were used, one from each type of powder: P3–143–9 passes; P6–6420–9 passes; and P9–136–9 passes. The sample size was less than 5 mm; their mass was less than 20 mg.

During the heating of the three selected samples, three transformations were highlighted, two of them endothermic (I^{st} and IIIrd) and one exothermic (IInd), the same thermal behavior being highlighted for the sample injected from Arboblend V2 Nature but not covered with a ceramic layer [34,35].

The variation in the heat flow in relation to the recorded temperature for the three phase transformations of each analyzed sample is shown in Figure 1. In order to highlight the possible different thermal compartments of the coated samples, the three signals were overlapped.

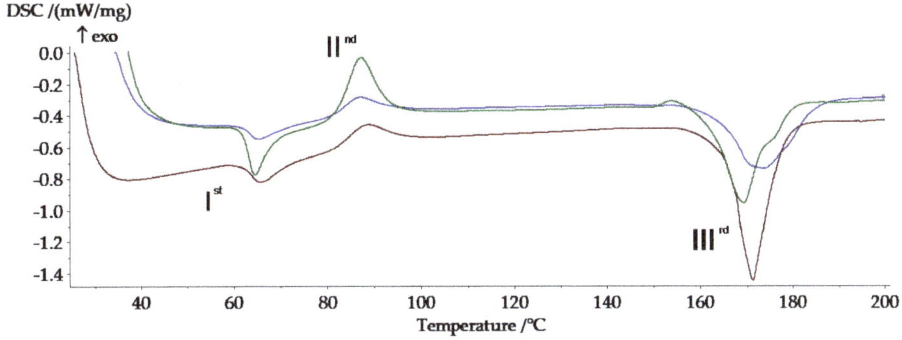

Figure 1. Highlighting the thermal behaviors of the tested samples: Ist—first transformation; IInd—second transformation; IIIrd—third transformation; P3–143–9 passes (green), P6–6420–9 passes (blue), P9–136–9 passes (red).

Both Figure 1 and Table 3 reveal changes in the temperatures at which the phase transformations take place. This aspect can be attributed to the small but still existing mass difference between the three analyzed samples (P3–143–9 passes 12.6 mg, P6–6420–9 passes 16.7 mg, P9–136–9 14.5 mg). At the same time, the amount of heat absorbed or dissipated is slightly different [39]. Another reason that could have generated this difference is the thickness of the deposited layer with the completion of the nine passes. As the dimensions of the microparticles differ in the case of the three ceramic powders, it is expected that the powder, whose microparticles are larger, will form a thicker layer, and thus the phase transitions will take place at slightly higher temperatures and the amount of absorbed or dissipated heat will be lower.

The critical temperatures of the three transformations are as follows: T_{onset} is the starting temperature; T_{peak} is the middle temperature; T_{end} is the finishing temperature (determined using the tangent method); and $\Delta H/m$ is the amount of dissipated/absorbed heat (using a rectilinear baseline).

Analyzing the first phase transformation, it is observed that the three samples register an endothermic maximum around 65 °C, a slightly lower transition temperature (64.7 °C) that the P3–143–9 passes sample (Zirconia–Titania–Yttria Composite Powder), whose powder has a smaller granulation. The first peak can be associated with a slow monotropic transformation of the solid-solid type and of some metastable crystals [40], which takes place with reduced heat absorption, −8.81 kJ/kg for P3–143–9 passes, −4.29 kJ/kg for the P6–6420–9 passes and −5.57 kJ/kg in the case of the P9–136–9 passes. The variation in the absorbed heat can be attributed, first of all, to the thickness of the deposited layers but also to the mass difference of the analyzed samples.

Table 3. Calorimetric characterization of samples coated with ceramic layers.

Sample	Transformation	T_{onset} [°C]	T_{peak} [°C]	T_{end} [°C]	$\Delta H/m$ [kJ/kg]
P3–143–9 passes	Ist	62.0	64.7	68.5	−8.81
	IInd	81.9	87.2	93.0	19.36
	IIIrd	162.7	169.3	175.3	−40.1
P6–6420–9 passes	Ist	62.1	65.3	71.2	−4.29
	IInd	81.4	87.2	94.2	7.16
	IIIrd	164.0	173.7	185.3	−38.14
P9–136–9 passes	Ist	61.8	65.8	71.8	−5.57
	IInd	81.7	88.7	95.6	8.66
	IIIrd	166.0	171.4	178.0	−48.38

The second peak takes place around the temperature of 86 °C, the powder with the highest granulation, the P9–136–9 passes, registering an increase in the transformation with 1.5 °C higher than the other two covered samples. The exothermic peak can be associated with the base biopolymer crystallization or with the reticular reorganization of lignin, the basic matrix of the biopolymer [39].

The third peak occurs with considerable heat absorption, in the case of all analyzed samples: −40.1 kJ/kg for the sample with smaller microparticles (P3–143–9 passes) and −48.38 kJ/kg for the sample with higher granulation (P9–136–9 passes). The endothermic transformation is attributed to the melting of the Arboblend V2 Nature biopolymer around the temperature of 170 °C, with small variations depending on the size of the ceramic microparticles.

3.2. TG Analyses

Knowing the thermal stability of Arboblend V2 Nature samples coated with ceramic powders is essential because their use in applications that require operation in severe working conditions, whether it is wear resistance or thermal resistance, requires the study of thermogravimetric behavior. It is desirable that the coating with ceramic microparticles increases the mechanical characteristics but also the thermal stability. Figure 2 compares the thermogravimetric (TG), derived thermogravimetric (DTG) and differential thermal (DTA) curves for the three samples coated with ceramic layers made from nine successive passes.

The main thermogravimetric characteristics of the P3–143–9 passes, P6–6420–9 passes and P9–136–9 passes samples are presented in Table 4.

Table 4. Thermogravimetric characteristics of the samples covered with ceramic micropowders.

Sample	Stage	T_{onset} [°C]	T_{peak} [°C]	T_{end} [°C]	W [%]	DTA Characteristic	Residue [%]
P3–143–9 passes	I	289	341	369	84.98	exo	3.81
	II	413	423	436	11.21	exo	
P6–6420–9 passes	I	282	346	373	84.61	exo	6.60
	II	413	426	438	8.79	exo	
P9–136–9 passes	I	281	347	367	88.06	exo	1.68
	II	415	426	438	10.26	exo	

T_{onset}, the temperature at which thermal degradation begins at each stage; T_{end}, the temperature at which the thermal degradation ends at each stage; T_{peak}, the temperature at which the degradation rate at each stage is maximum; W%, percentage mass loss at each stage; residue, the amount of degraded sample remaining at a temperature above 700 °C.

The three coated samples with different ceramic layers highlight two decomposition stages, the first recorded being around a temperature of 345 °C, with a significant mass loss of over 85%: this decomposition is attributed to the structural degradation of the basic constituent of the material, lignin. This stage consists of the formation of aromatic hydrocarbons, guaiacyl-/syringyl-type and hydroxy-phenolic compounds and more [40]. According to the manufacturer [32], another constituent contained by the analyzed biopolymer is PLA, which decomposes in considerable proportions in this temperature range [41,42]. According to the literature [43,44], PLA and pure lignin degrade completely up to a temperature of 500 °C.

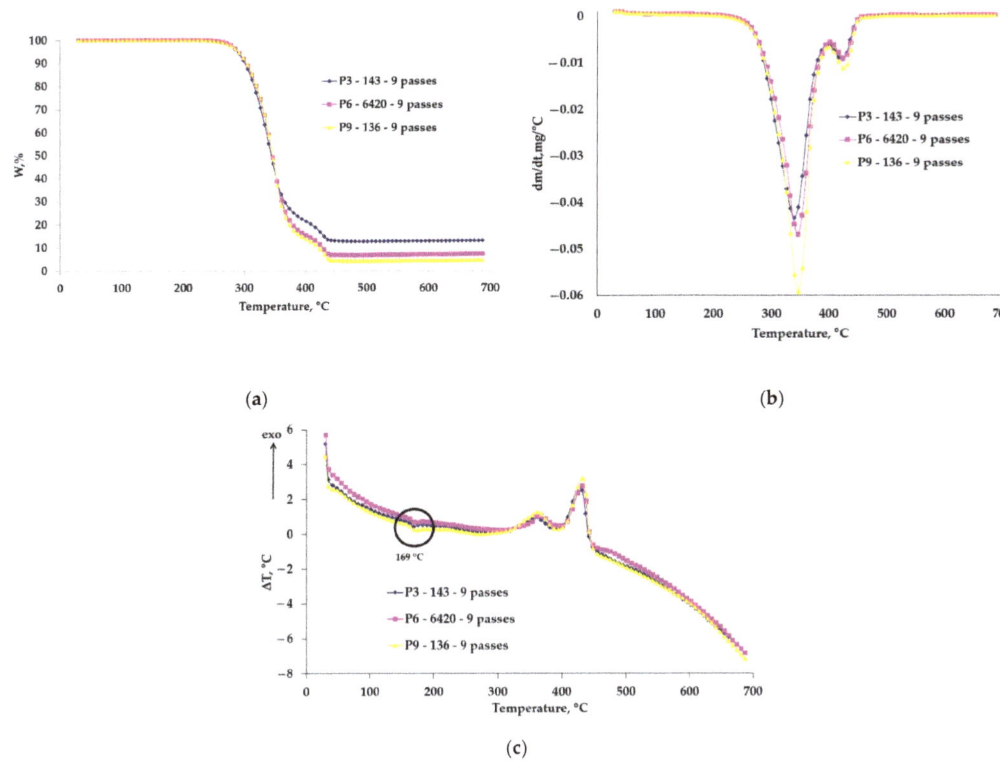

Figure 2. Thermogravimetric curves of the coated samples: (**a**) TG, (**b**) DTG and (**c**) DTA.

The curves obtained from the TG analysis were overlapped to highlight their difference in behavior, and the thermal stability of the coated samples is similar (Figure 2).

In the second stage with a T_{peak} around 425 °C, there is a mass loss in a percentage much lower than 10%, attributed to the thermal oxidation of the carbonic residue that appeared from the pyrolysis of lignin and/or PLA but also of another biodegradable constituent of the biopolymer that was introduced by the manufacturer as a binder (resin, wax, shellac, etc.) [33]. At a temperature of 700 °C, depending on the used ceramic powder type, a certain amount of residual mass is found. It is observed that the sample P6–6420–9 passes has the highest percentage of residue, 6.6%, most likely due to the higher amount of microparticles than, for example, in the case of the P3–143–9 passes sample, where the amount of ceramic powder deposited is much lower. The ceramic powders at the end of the analysis temperature have not yet reached the melting point of approximately 2500 ° C, their working temperature varying from 540 °C (P6–6420–9, P9–136–9 passes) to 980 °C (P3–143–9 passes) [38]. In addition, inorganic substances that are found in the composition of the biopolymer are very likely to be part of the residual mass [45].

Figure 2c shows the DTA curves where the melting temperature of Arboblend V2 Nature can be observed, 169 °C, very close to the values obtained by calorimetric analysis, especially in the case of the P3–143–9 passes sample.

3.3. Surface and Structure Analysis of Coated Samples

3.3.1. SEM Analysis

Figure 3 shows the morphological aspect of the Arboblend V2 Nature material coated with Metco ™ 143 (ZrO_2 $18TiO_2$ $10Y_2O_3$). A uniform coating of the biopolymer mass is observed. The coating consists of spherical component particles having a dimensions variation between 1 and 27 µm. The particles retain their spherical shape due to the very rapid cooling on contact with the base matrix. They do not flatten in the form of splats as is conventional in the case of coatings on metal substrate [46]. The fact that the basic matrix contains various particles in shape and size, in large quantities and evenly distributed, leads to an increase in mechanical properties.

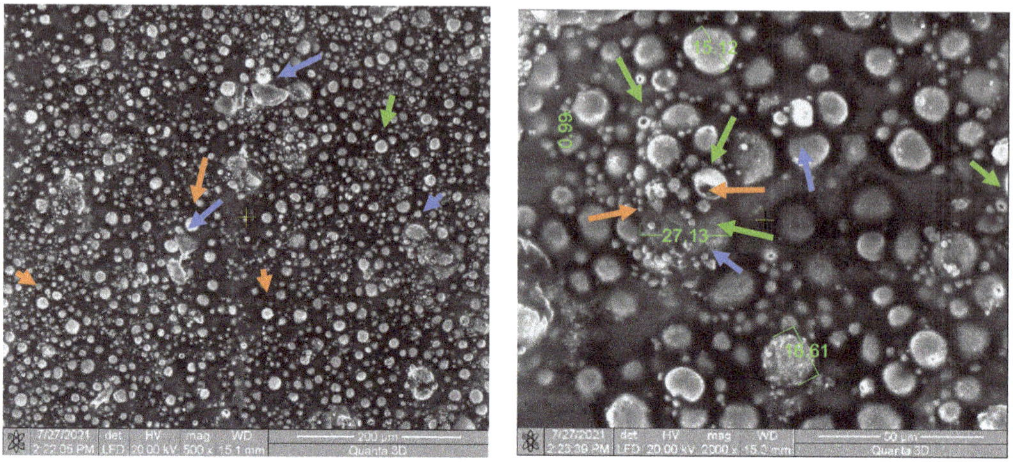

500×　　　　　2000×

Figure 3. SEM analysis of the P3–143–9 passes samples: Zirconium dioxide (bearing balls)—orange arrow; Titanium dioxide—blue arrow; Yttrium oxide—green arrow.

Yttrium oxide (green arrow, Figure 3) shows a porous spherical morphology in the form of a sintered agglomerate.

Figure 4 shows the polymer matrix containing particles from the coating formed by chromium oxide. Some of these particles are heterogeneously distributed, and another part is embedded in the polymeric structure. Their size varies from 18 µm to 30 µm, and they have rectangular shapes specific to chromium oxide. The spherical microparticles can be attributed to the presence of Fe_2O_3 and SiO_2, which are released in small quantities in the structure of the Amdry 6420 powder, maximum 0.4% and 0.45%, respectively. Compared to the P3–143–9 passes sample, the material incorporated a smaller amount of powder appearance, which represents the lower capacity of the coating and embedding chromium oxide in the polymeric structure.

The coating of chromium oxide, silicon oxide and titanium oxide (Cr_2O_3-xSiO_2-yTiO_2, Figure 5) highlights a relatively uneven distribution of microparticles, similar to the situation presented in Figure 4. The particles are of different shapes with mostly polyhedral appearance (TiO_2—green arrow) but there are also spherical (SiO_2—blue arrow) and rectangular (Cr_2O_3—orange arrow) structures, their dimensions varying between 1 and 63 µm. The particles are embedded in the polymer mass. Both Figures 4 and 5 show chromium oxide, and this compound does not have a better adhesion compared to the sample coated with P3–143–9 passes (Figure 3), which does not contain chromium oxide.

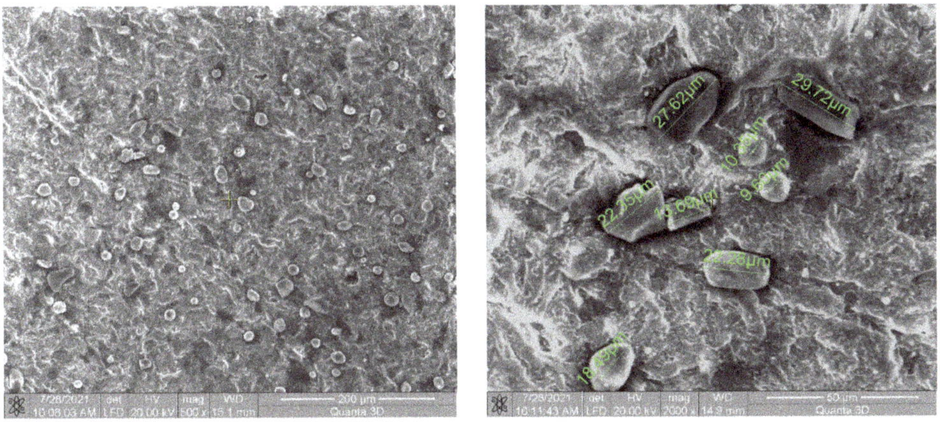

500×
2000×

Figure 4. SEM analysis of the P6–6420–9 passes samples.

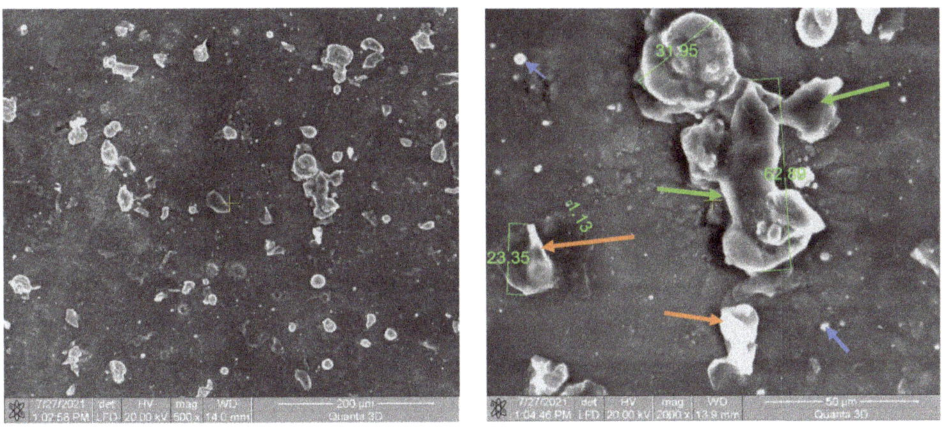

500×
2000×

Figure 5. SEM analysis of the P9–136–9 passes samples.

3.3.2. EDX Analysis

In order to observe the presence of the deposited ceramic layers, the line EDX analysis was performed together with the SEM analysis of the P3–143–9 passes, P6–6420–9 passes and P9–136–9 passes samples. This analysis series had as the point of study the edge area of the samples in order to highlight as accurately as possible the presence, distribution and concentration of the chemical elements which form the ceramic microlayers and also the biopolymeric support.

The EDX in-line analysis (Figure 6—yellow arrow) of all samples reflects the abundant presence of two chemical elements, carbon and oxygen. Their existence is closely related to the chemical composition of the biopolymer matrix which presents, according to the previous determinations, C and O in similar mass proportions, 48 $\pm$ 0.02% and 52 $\pm$ 0.02%, respectively [35]. The acquisition of data is a difficult one because the polymer matrix is not an electrical conductor, so the electrons coming out of the material are of much lower intensity than, for example, in the case of an analyzed metallic material.

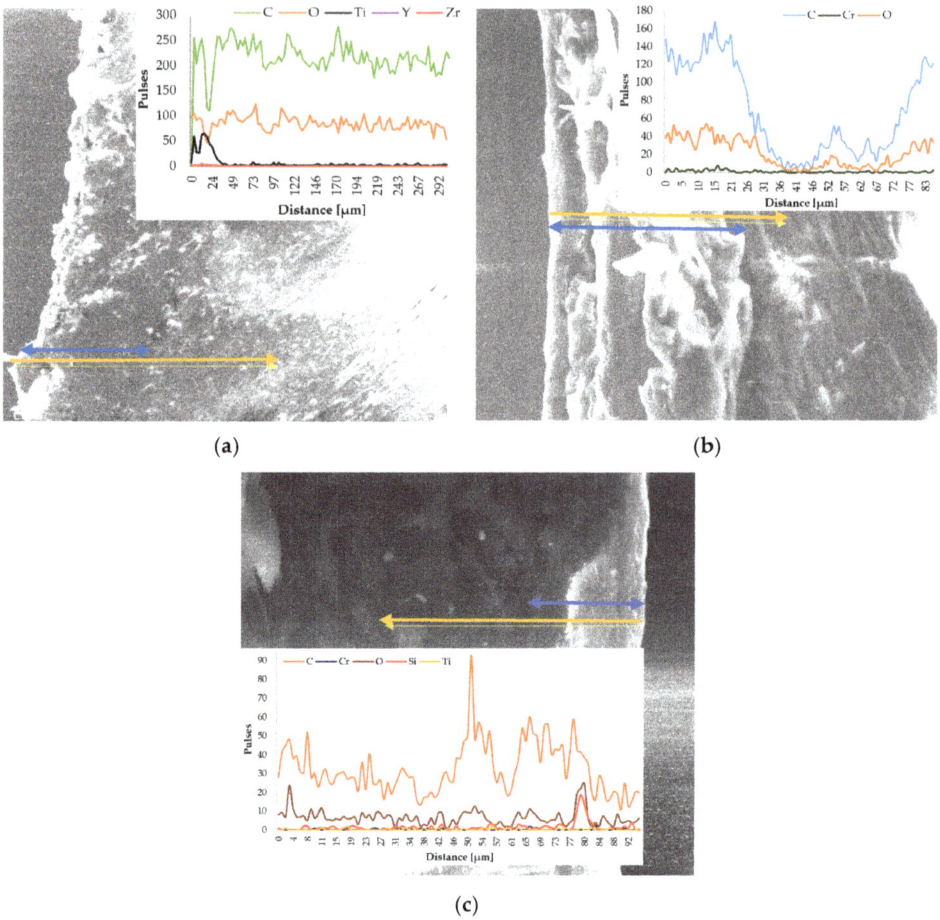

Figure 6. In line EDX analysis of the coated samples. (**a**) P3–143–9 passes, (**b**) P6–6420–9 passes, (**c**) P9–136–9 passes.

For the sample covered with Zirconia–Titania–Yttria Composite Powder (Figure 6a), the presence of ceramic microparticles in the first 20 μm is observed, an area that we consider to coincide with the thickness of the deposited ±0.02 ceramic layer (blue arrow). Titanium is the chemical element found in the most significant amount in the area of the ceramic layer, but the other two elements are highlighted by the graphic. The presence of a small amount of Zr and Ti during the entire test distance is due to the microparticles that came off of the brittle ceramic layer in the moment of EDX preparation. The higher amount of carbon than oxygen is attributed to the carbides that form during the embedding of molten microparticles in the solid (cold) polymeric matrix, carbides that provide surface hardness. Oxides are also formed but in a much smaller amount.

Chromium being a hard metallic element during the preparation of the P6–6420–9 passes sample, this one is visible over the entire analyzed surface; however, it can be observed that in the first part of the analyzed distance (first 20 μm), the amount of chromium is higher (Figure 6b). As in the case of the P3–143–9 passes sample, the amount of carbon is higher than that of oxygen. In the 35–75 μm interval, the graph highlights a steep decrease in the amount of elements reflected by the sample, but most likely this can be attributed to the transition between the deposited layer and the biopolymer support.

For the P9–136–9 passes sample (Figure 6c), the in–line analysis reveals the presence of the ceramic layer on the right side of the SEM image (last 15 μm), and the graph reflects the existence of the micropowder chemical elements. Silicon being a semiconductor chemical element reflects its presence in the structure of the ceramic layer much more than the than the other constituents. The Cr, Ti and O of the composite micropowders were highlighted by the analyzed sample.

3.3.3. XRD Analysis

The main purpose of the XRD analysis was to determine the structure of the samples made of Arboblend V2 Nature and coated with ceramic micropowders, Amdry 6420 (Cr_2O_3), Metco 143 (ZrO_2 $18TiO_2$ $10Y_2O_3$) and Metco 136F (Cr_2O_3-$xSiO_2$-$yTiO_2$), but also to identify possible crystal phases.

Figure 7 shows the phase diffractograms for the three samples with distinct ceramic coatings: P3–143–9 passes, P6–6420–9 passes and P9–136–9 passes. It can be seen that two of the three samples have a crystalline structure (P3–143–9 passes, P9–136–9 passes) highlighted by specific peaks, and the third (P6–6420–9 passes green diffractogram) has a semi- crystalline structure with the presence of small peaks of chromium oxide at four distinct 2θ angles, 24.25°, 33.39°, 35.88° and 54.61°, respectively, of low intensity, 1036 to 1356 [47–49]. The major peak registered at 16.73°, with a diffraction intensity of 2643, which, according to the literature, may be associated with the presence of polylactic acid in its the chemical composition (($C_3H_4O_2$)n) [50,51]. The presence of this compound is not accidental because it is due to the fact that the thickness of the deposited layer is very thin (small microparticles), thus, the equipment is detecting one of the basic material constituents.

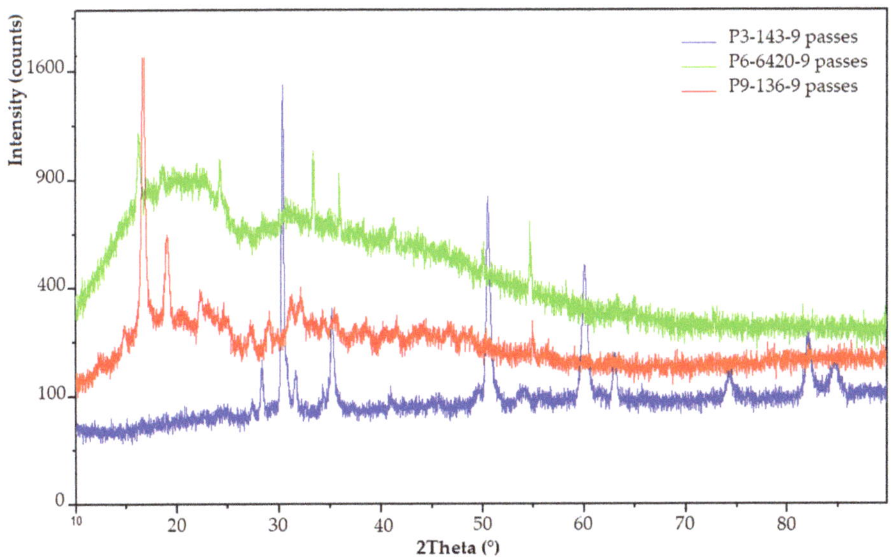

Figure 7. XRD analysis for ceramic coated samples: P3–143–9 passes (blue), P6–6420–9 passes (green), P9–136–9 passes (red).

In the case of the other two coatings, the presence of predominant peaks is observed, which correspond to the crystallization of certain compounds as follows:

- The P3–143–9 passes sample (blue diffractogram) has a strong crystalline aspect due to the large amount of zirconium dioxide particles present in the structure of the ceramic powder. Thus, to the ZrO_2 compound can be assigned the peaks from 2θ = 31.14°, 38.43°, 60°, 82° and 84.78° [46,52,53]. Titanium dioxide, as in the case of the P9–136–9 passes sample, is found at angles of 27.43°, 28.34°, 63.02° and 74.41° [54,55]. However, the intensity of the diffraction peaks is quite low, the highest being registered in the case of the 2θ = 28.34° angle. The low angle from 2-theta = 43°, was identified as the specific angle of Y_2O_3 crystallization [56].

- The P9–136–9 passes sample (red diffractogram) shows diffraction maxima associated with the presence of the polymer matrix, which has in its structure polylactic acid (16.73°) and lignin or natural fibers (19.04°) [53,57]. Diffraction angles corresponding to the coating with ceramic micropowder are also visible: The specific peaks to Cr_2O_3 crystallization at 2-theta angles of low intensity are 30.35°, 31.70°, 35.16°, 50.48° and 54.12° [47–49,58]. For SiO_2 microspheres, a peak located at about 2θ = 22.5 is observed [59]. No other diffraction peaks can be detected for this compound. According to the literature [54,55], the diffraction angles that can be attributed to the titanium dioxide (TiO_2) present are at 27.33° and 32.13° in the case of Metco 136F micropowder.

3.4. Scratch Analysis

The scratch test was performed in order to evaluate the adhesion of the hard (ceramic) coatings made on the surface of the Arboblend V2 Nature biopolymeric material.

Analyzing the curves presented in Figure 8, it is observed that one of them, the green curve (P6–6420–9 passes), shows a sudden and gradual transition of the apparent friction coefficient (A-COF), which means that the adhesion between the deposited thin layer and the polymeric material is better than in the case of the other two coatings due to the presence of chromium oxide. The other two tests have a good scratching behavior, but the P3–143–9 passes test (blue curve), recorded higher A-COF values than in the case of the P9–136–9 passes test.

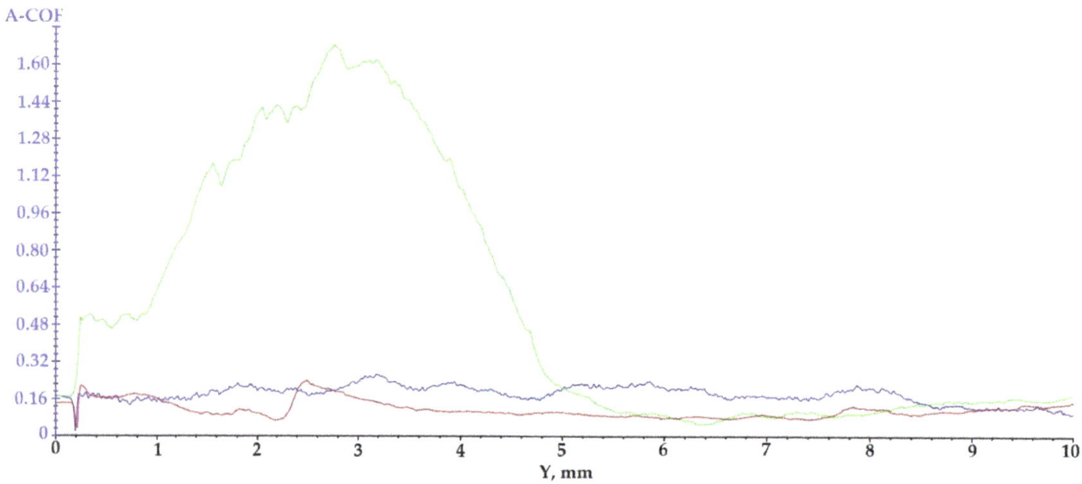

Figure 8. Results regarding the scratching behavior of the samples coated with ceramic layers: blue curve—P3–143–9 passes; green curve—P6–6420–9 passes; red curve—P9–136–9 passes.

The more peaks that appear in the variation of the apparent friction coefficient, the better the adhesion between the deposited layer and the polymeric material is.

For the P6–6420–9 passes sample, a high amplitude peak of A-COF is registered at the beginning of the test; the explanation for this would be related to the deposition granulation. It is very possible that the tip of the cutting tool (pin) has hung an area of deposited material with a larger granulation.

The samples injected from Arboblend V2 Nature and coated with ceramic micropowders showed the following behavior during the 60 s of testing (Figure 9): For sample P3–143–9 passes, blue curve, an increase in A-COF is observed in the first 4 s, after which its value begins to decrease sharply until the 16 s when it registers an increase followed quickly by a decrease. Starting with at 19 s, the average value of A-COF increases and begins to stabilize, reaching a maximum value of 0.53 at 50 s. The mean value of A-COF was 0.29 ± 0.16. The behavior of the sample coated with Zirconia–Titania–Yttria composite powder is a typical one, and the coatings in the first part of the test register variations of A-COF in order to later stabilize. Sample P6–6420–9 passes, green curve, reflects a completely different behavior from the first test. In the first part of scratching, the first 3 s, it reaches the maximum value of A-COF at 1.62, followed by a sudden decrease until 6 s. Next, the sample registers two fluctuations, and starting at 40 s, the A-COF value begins to increase, at the end of the 60 s reaching the value of 1.37. These fluctuations can be attributed to the variable dimensions (9–30 μm) of the microparticles that constitute the ceramic powder. The progressive increase recorded in the last 20 s of testing reflects the fact that the test pin detached ceramic microparticles from the sample surface, thus gradually becoming more and more rough. The mean value of A-COF for this sample, 0.56 ± 0.42, is the highest compared to the other two tested samples. The last test subjected to tribological determination, P9–136–9 passes, red curve, as well as the previous test, records fluctuations throughout the test with the A-COF value at the end of the test reaching a maximum of 0.37. The average value of A-COF for this sample is the lowest, 0.18 ± 0.08. The value registered is very close to that of the injected samples and not covered with ceramic layer, 0.16 for rotational determinations and 0.13 for oscillation ones [60]. This similarity comes from the fact that the coverage of the sample was not uniform, the adhesion and incorporation of Chromia–Silica composite powder, as it could be observed in the SEM images being very low, had a very high percentage both in the P6–6420–9 passes test and in the P9–136–9 passes test. This lack of deposition is due to the thermal behavior of chromium oxide, which is found in a very high percentage both in the P6–6420–9 passes and P9–136–9 passes samples.

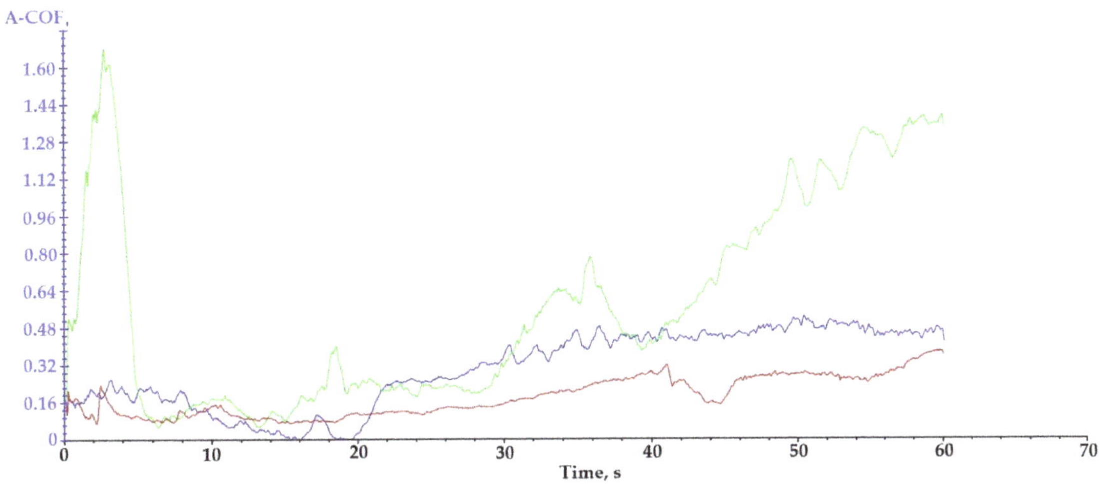

Figure 9. A-COF variation with test time for samples coated with ceramic micropowders: (blue) P3–143–9 passes; (green) P6–6420–9 passes; (red) P9–136–9 passes.

Table 5 shows the results of scratch testing in the case of the three samples covered with ceramic micropowders.

Table 5. Values of A-COF recorded by samples covered with ceramic layers.

Sample	A-COF Medium Value	A-COF Maximum	Time of A-COF Maximum[s]
P3–143–9 passes	0.29 ± 0.16	0.53	50
P6–6420–9 passes	0.56 ± 0.42	1.62/1.37	3.0/60
P9–136–9 passes	0.18 ± 0.08	0.37	60

3.5. Microindentation Test

To conduct the microindentation test, three samples were tested for each type of ceramic powder used to coat the Arboblend V2 Nature biopolymeric material. Repeated testing was to confirm the experimental stability. Figure 10 shows the evolution of the force as a function of the depth of penetration for all three samples subjected to analysis. The software package used (UMT Test Viewer, 2.16) allowed the reading of both the microhardness values and the Young's modulus. These values are presented in Table 6.

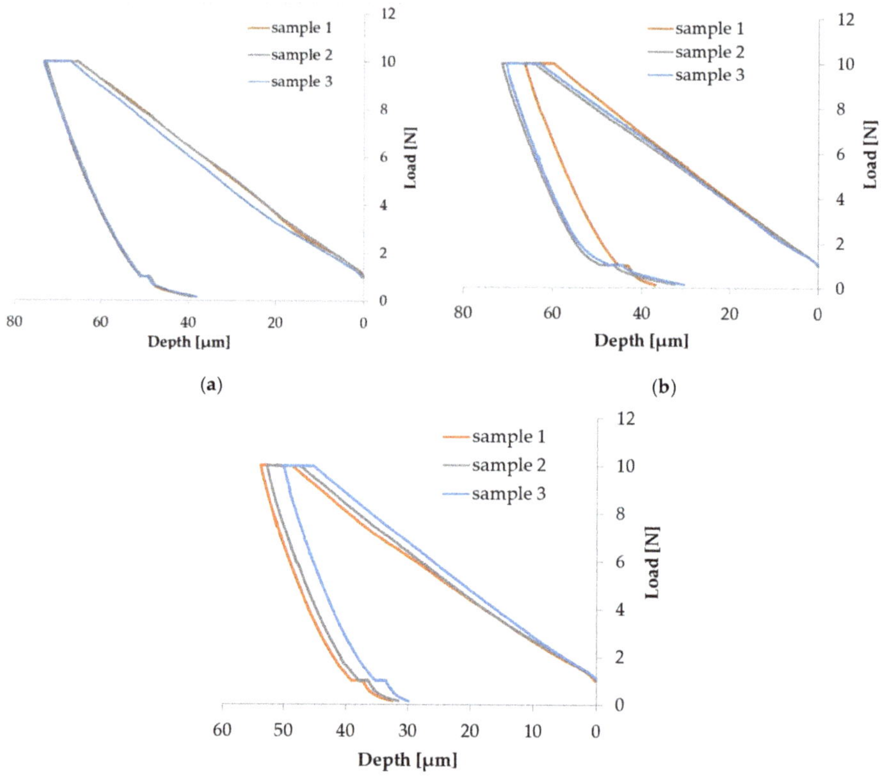

Figure 10. Results of microindentation tests for samples coated with ceramic micropowders: (**a**) P3–143–9 passes; (**b**) P6–6420–9 passes; (**c**) P9–136–9 passes.

The force applied to the indenter increases constantly during the charging phase and is maintained at the maximum value of 10 N. This phase is called creep, after which there is a decrease to zero in the discharge phase.

Table 6. Results obtained by microindenting the samples coated with ceramic micropowders.

Sample	Test	Max Load (N)	Max Depth (μm)	Young's Modulus (GPa)	Micro Hardness (GPa)
P3–143–9 passes	1	8.99	73.55	1.52	0.11
	2	8.7	72.78	1.69	0.11
	3	8.98	73.02	1.48	0.11
Average		8.98 ± 0.01	73.12 ± 0.4	1.56 ± 0.11	0.11 ± 0.00
P6–6420–9 passes	1	8.99	66.64	1.69	0.13
	2	8.98	71.70	1.91	0.11
	3	8.98	71.02	2.53	0.10
Average		8.99 ± 0.01	69.79 ± 2.75	2.04 ± 0.44	0.12 ± 0.01
P9–136–9 passes	1	8.99	53.77	2.92	0.16
	2	8.99	53.08	3	0.16
	3	8.97	50.42	2.9	0.17
Average		8.99 ± 0.01	52.42 ± 1.77	2.94 ± 0.05	0.17 ± 0.01

According to the data obtained from the micro-indentation, the P9–136–9 passes sample, although it does not have a uniform coating, presents the best values of microdurity, (0.17 ± 0.01 GPa), the maximum indentation depth being 52.42 ± 1.77 μm. Compared to the zirconium-based ceramic coating (P3–143–9 passes), the chromium oxide coatings have a higher hardness (P6–6420–9 passes, P9–136–9 passes) [35]. Another justification of the results comes from the reduced layer thickness due to the small dimension of microparticles (P3–143–9 passes) than in the case of the other two types of ceramic micropowders.

The lowest dispersion of the results was obtained in the case of the P3–143–9 passes sample, most likely due to the fact that the deposited ceramic layer was uniform. In addition, the other samples tested did not show large differences.

4. Conclusions

The coating of bio-based polymers has gained the attention of researchers worldwide. The purpose of coatings with various layers, be they ceramic, metal, etc., as well as in the case of reinforcements, is to increase the characteristics of the base material so that it responds better to certain industrial applications and can even replace a certain material, such as metal. The coatings have been realized in order to increase the mechanical, tribological and thermal characteristics of the samples (wear, hardness and increase in thermal resistance), thus becoming suitable in applications that require harsh working conditions, especially in the automotive industry. The coating of sample with ceramic microlayers led to the following results:

- A slight increase in the melting point, from 172 °C (base material [35,61]) up to 174 °C, varying depending on the thickness of the deposited layer. The thickness value is closely influenced by the microparticles size that constitute the ceramic powder. Thus, the larger the size of the microparticles, the higher the thermal resistance of the coated material is.
- No significant changes were visible both in terms of temperature range and the amount of mass lost during the thermal degradation, and the differences were also attributed in this situation to the size of the ceramic microparticles.
- Regarding the coatings' uniformity, the SEM surface analysis indicated a good and uniform incorporation of microparticles in the case of composite powder based on zirconium oxide. The other two ceramic micropowders in contact with the polymer matrix did not reveal a good adhesion due to the lower working temperature than the melting point (2435 °C).
- XRD and EDX analyses highlighted the presence of microceramic layers. Their crystalline/semi-crystalline structure confers hardness; thus, they are suitable for applications that require this feature. The coating with Cr_2O_3-xSiO_2-yTiO_2 (sample

9) formed the hardest layer (0.17 ± 0.01 GPa), which was demonstrated during the microindentation test. Analyzing the obtained results from the SEM and scratch analyses point of view, it can be concluded that the deposition was not uniform due to the fact that the adhesion between the microparticles of chromium oxide, silicon oxide and titanium oxide is lower than in the case of the other two samples, the average value of the apparent friction coefficient (A-COF) being 0.18 ± 0.08.

According to the obtained results regarding the adhesion of the ceramic layers on the polymer surface, it can be stated that the samples showed strong chemical bonds at the interface between the thin layers and Arboblend V2 Nature bio-based polymer. Thus, these coated materials can be used in industrial applications that require high surface hardness and thermal resistance. They can also successfully replace various non-biodegradable polymeric materials used in various applications such as those in the automotive and electronics industry (telephone covers, housings, worm wheels, car wiper system, etc.).

Author Contributions: Conceptualization, D.N.; data curation, B.I. and M.B. (Marcelin Benchea); formal analysis, S.-N.M. and A.M.; methodology, M.B. (Mihai Boca); supervision, D.N. and B.I.; validation, D.N.; writing—original draft, S.-N.M. and A.M.; writing—review and editing, S.-N.M.; funding acquisition, D.N. All authors have read and agreed to the published version of the manuscript.

Funding: This work was supported by a publication grant of the TUIASI, project number GI/P21/2021.

Institutional Review Board Statement: Not applicable.

Informed Consent Statement: Not applicable.

Data Availability Statement: The data presented in this study are available on request from the corresponding author.

Acknowledgments: This work was supported by a publication grant of the TUIASI, project number GI/P21/2021.

Conflicts of Interest: The authors declare no conflict of interest.

Nomenclature

TPS	Titanium Plasma Spraying
APS	Atmospheric Plasma Spray
PDMS	Polydimethylsiloxane
PDC	biodegradable multiblock copolymer
TFX	Polyetherurethane
PLA	polylactic acid
NLPM	normal litre per minute
DC	Direct Current
DSC	Differential scanning calorimetry
RT	room temperature
TG	Thermogravimetric curves
DTG	derived thermogravimetric curves
DTA	differential thermal analyzes
T_{onset}	starting temperature
T_{peak}	middle temperature
T_{end}	finish temperature
SEM	Scanning Electron Microscopy
LFD	Large Field Detector
EDX	Energy-dispersive X-ray spectroscopy
XRD	X-ray diffraction analysis
W%	percentage mass loss at each stage
A-COF	apparent friction coefficient
Y	testing distance (mm)

References

1. Lu, Z.; Myoung, S.W.; Kim, E.H.; Lee, J.; Hand Jung, Y.G. Microstructure Evolution and thermal durability with coating thickness in APS thermal barrier coatings. *Mater. Today Proc.* **2014**, *1*, 35–43. [CrossRef]
2. Matikainen, V.; Bolelli, G.; Koivuluoto, H.; Sassatelli, P.; Lusvarghi, L.; Vuoristo, P. Sliding wear behaviour of HVOF and HVAF sprayed Cr3C2-based coatings. *Wear* **2017**, *388–389*, 57–71. [CrossRef]
3. Suarez, M.; Bellayer, S.; Traisnel, M.; Gonzalez, W.; Chicot, D.; Lesage, J.; Puchi-Cabrera, E.S.; Staia, M.H. Corrosion behavior of Cr_3C_2–NiCr vacuum plasma sprayed coatings. *Surf. Coat. Technol.* **2008**, *202*, 4566–4571. [CrossRef]
4. Wypych, A.; Siwak, P.; Andrzejewski, D.; Jakubowicz, J. Titanium Plasma-sprayed coatings on polymers for hard tissue applications. *Materials* **2018**, *11*, 2536. [CrossRef]
5. Gariboldi, E.; Rovatti, L.; Lecis, N.; Mondora, L.; Mondora, G.A. Tribological and mechanical behaviour of Cr3C2–NiCr thermally sprayed coatings after prolonged aging. *Surf. Coat. Technol.* **2016**, *305*, 83–92. [CrossRef]
6. Tsui, Y.C.; Clyne, T.W. An analytical model for predicting residual stresses in progressively deposited coatings Part 1: Planar geometry. *Thin Solid Films* **1997**, *306*, 23–33. [CrossRef]
7. Thermal Spray Materials Guide. Available online: https://www.oerlikon.com/ecomaXL/files/oerlikon_BRO-0001.17_TS_MaterialGuide_EN.pdf (accessed on 4 August 2020).
8. Channagiri Mohankumar Praveen, K.; Avinash, L.; Manjunath Patel Gowdru, C.; Pimenov, D.Y.; Khaled, G. Electrodeposition based preparation of Zn–Ni alloy and Zn–Ni–WC nano-composite coatings for corrosion-resistant applications. *Coatings* **2021**, *11*, 712. [CrossRef]
9. Shkirskiy, V.; Uebel, M.; Maltseva, A.; Lefèvre, G.; Volovitch, P.; Rohwerder, M. Cathodic driven coating delamination suppressed by inhibition of cation migration along Zn polymer interface in atmospheric CO_2. *npj Mater. Degrad.* **2019**, *3*, 2. [CrossRef]
10. Aung, M.M.; Wong, J.L.; Hong, N.L. Improvement of anticorrosion coating properties in bio-based polymer epoxy acrylate incorporated with nano zinc oxide particles. *Ind. Eng. Chem. Res.* **2020**, *59*, 1753–1763. [CrossRef]
11. Rajkovic, A.; Uyttendaele, M.; Deley, W.; Van Soom, A.; Rijsselaere, T.; Debevere, J. Dynamics of boar semen motility inhibition as a semi-quantitative measurement of Bacillus cereus emetic toxin (Cereulide). *J. Microbiol. Methods* **2006**, *65*, 525–534. [CrossRef]
12. Divine, S.; Chun-Wei, Y.; Ian, L. Mechanical durability of engineered superhydrophobic surfaces for anti-corrosion. *Coatings* **2018**, *8*, 162. [CrossRef]
13. Ulaeto, S.; Rajan, R.; Pancrecious, J.; Rajan, T.; Pai, B. Developments in smart anticorrosive coatings with multifunctional characteristics. *Prog. Org. Coat.* **2017**, *111*, 294–314. [CrossRef]
14. Asri, N.; Husaini, T.; Sulong, A.; Majlan, E.; Daud, W. Coating of stainless steel and titanium bipolar plates for anticorrosion in PEMFC: A review. *Int. J. Hydrogen Energy* **2017**, *42*, 9135–9148. [CrossRef]
15. Fengjuan, X.; Cheng, Q.; Mingyang, G.; Junzhong, W.; Xinrui, Y.; Hongli, L.; Lin, Y. Anticorrosive durability of zinc-based waterborne coatings enhanced by highly dispersed and conductive polyaniline/graphene oxide composite. *Prog. Org. Coat.* **2018**, *125*, 79–88. [CrossRef]
16. Sorensen, P.; Kiil, S.; Dam-Johansen, K.; Weinell, C. Anticorrosive coatings: A review. *J. Coat. Technol. Res.* **2009**, *6*, 135–176. [CrossRef]
17. Zheludkevich, M.; Shchukin, D.; Yasakau, K.; Möhwald, H.; Ferreira, M. Anticorrosion coatings with self-healing effect based on nanocontainers impregnated with corrosion inhibitor. *Chem. Mater.* **2007**, *19*, 402–411. [CrossRef]
18. Plyushch, A.; Macutkevic, J.; Banys, J.; Kuzhir, P.; Kalanda, N.; Petrov, A.; Silvestre, C.; Uimin, M.A.; Yermakov, A.Y.; Shenderova, O. Carbon-coated nickel nanoparticles: Effect on the magnetic and electric properties of composite materials. *Coatings* **2018**, *8*, 165. [CrossRef]
19. Ausanio, G.; Hison, C.L.; Iannotti, V.; Lanotte, L.; Lanotte, L. Magneto-piezoresistance in elastomagnetic composites. *J. Appl. Phys.* **2011**, *110*, 063903. [CrossRef]
20. Dumitrescu, M.; Lisa Iordan, A.R.; Tudorache, F.; Petrila, I.; Borhana, A.I.; Palamaru, M.N.; Mihailescu, C.; Leontie, L.; Munteanu, C. Ni ferrite highly organized as humidity sensors. *Mater. Chem. Phys.* **2015**, *156*, 170–179. [CrossRef]
21. Mohr, R.; Kratz, K.; Weigel, T.; Lucka-Gabor, M.; Moneke, M.; Lendlein, A. Initiation of shape-memory effect by inductive heating of magnetic nanoparticles in thermoplastic polymers. *Proc. Nat. Acad. Sci. USA* **2006**, *103*, 3540–3545. [CrossRef]
22. Schmidt, A.M. Electromagnetic activation of shape memory polymer networks containing magnetic nanoparticles. *Macromol. Rapid Commun.* **2006**, *27*, 1168–1172. [CrossRef]
23. Baljinder, K.; Forkan, S.; Piyanuch, L.; Myler, P. Thermal protection of carbon fiber-reinforced composites by ceramic particles. *Coatings* **2016**, *6*, 22. [CrossRef]
24. Rocha-Santos, T.A. Sensors and biosensors based on magnetic nanoparticles. *TrAC Trends Anal. Chem.* **2014**, *62*, 28–36. [CrossRef]
25. Freitas, P.; Ferreira, R.; Cardoso, S.; Cardoso, F. Magnetoresistive sensors. *J. Phys. Condens. Matter* **2007**, *19*, 165221. [CrossRef]
26. Mocanu, Z.; Airimioaei, M.; Ciomaga, C.; Curecheriu, L.; Tudorache, F.; Tascu, S.; Iordan, A.; Palamaru, N.; Mitoseriu, L. Investigation of the functional properties of MgxNi1-xFe_2O_4 ceramics. *J. Mater. Sci.* **2014**, *49*, 3276–3286. [CrossRef]
27. Pankhurst, Q.A.; Connolly, J.; Jones, S.; Dobson, J. Applications of magnetic nanoparticles in biomedicine. *J. Phys. D Appl. Phys.* **2003**, *36*, R167. [CrossRef]
28. Berry, C.C.; Curtis, A.S. Functionalisation of magnetic nanoparticles for applications in biomedicine. *J. Phys. D Appl. Phys.* **2003**, *36*, R198. [CrossRef]

29. Liu, X.; Li, B.; Geng, D.; Cui, W.; Yang, F.; Xie, Z.; Kang, D.; Zhang, Z. (Fe, Ni)/C nanocapsules for electromagnetic-wave-absorber in the whole Ku-band. *Carbon* **2009**, *47*, 470–474. [CrossRef]
30. Wu, N.; Liu, X.; Zhao, C.; Cui, C.; Xia, A. Effects of particle size on the magnetic and microwave absorption properties of carbon-coated nickel nanocapsules. *J. Alloys Compd.* **2016**, *656*, 628–634. [CrossRef]
31. Li, Y.; Wang, J.; Liu, R.; Zhao, X.; Wang, X.; Zhang, X.; Qin, G. Dependence of gigahertz microwave absorption on the mass fraction of Co@C nanocapsules in composite. *J. Alloys Compd.* **2017**, *724*, 1023–1029. [CrossRef]
32. TECNARO–The Biopolymer Company. Available online: https://www.tecnaro.de/en/ (accessed on 20 August 2020).
33. Nedelcu, D.; Marguta, A.; Mazurchevici, S.; Munteanu, C.; Istrate, B. Micro-structural and morphological analyses of coated 'liquid wood' samples by ceramic particles. *Mater. Res. Express* **2019**, *6*, 085326. [CrossRef]
34. Mazurchevici, S.-N.; Mazurchevici, A.-D.; Nedelcu, D. Dynamical mechanical and thermal analyses of biodegradable raw materials for additive manufacturing. *Materials* **2020**, *13*, 1819. [CrossRef] [PubMed]
35. Mazurchevici, S.-N.; Motaș, J.G.; Diaconu, M.; Lisa, G.; Lohan, N.M.; Glod, M.; Nedelcu, D. Nanocomposite biopolymer arboblend V2 nature AgNPs. *Polymers* **2021**, *13*, 2932. [CrossRef]
36. Broitman, E.; Nedelcu, D.; Mazurchevici, S.; Glenat, H.; Grillo, S. Tribological and nanomechanical behavior of liquid wood. *J. Tribol.* **2019**, *141*, 022001. [CrossRef]
37. Nedelcu, D.; Santo, L.; Santo, A.G.; Plavanescu Mazurchevici, S. Mechanical behaviour evaluation of arboform material samples by bending deflection test. *Mater. Plast.* **2015**, *52*, 423–426.
38. Oerlikon Metco–Materials and Surface Solutions. Available online: https://www.oerlikon.com/metco/en/ (accessed on 12 June 2020).
39. Wagner, M. Thermal analysis in practice, fundamental aspects. In *Thermal Analysis in Practice*; Wagner, M., Ed.; Hanser: München, Germany, 2018; pp. 1–9. ISBN 9781569906439.
40. Brebu, M.; Vasile, C. Thermal degradation of lignin—A review. *Cell Chem. Technol.* **2010**, *44*, 353–363.
41. Cai, Y.H.; Xie, Y.C.; Tang, Y.; Zhao, L.S. Thermal decomposition kinetics of Poly(L-lactic acid) after heat treatment. *Open Mater. Sci. J.* **2015**, *9*, 28–32. [CrossRef]
42. Mofokeng, J.P.; Luyt, A.S.; Tabi, T.; Kovacs, J. Comparison of injection moulded, natural fibre-reinforced composites with PP and PLA as matrices. *J. Thermoplast. Compos. Mater.* **2011**, *25*, 927–948. [CrossRef]
43. Prieur, B.; Meub, M.; Wittemann, M.; Klein, R.; Bellayer, S.; Fontaine, G.; Bourbigot, S. Phosphorylation of lignin: Characteri-zation and investigation of the thermal decomposition. *RSC Adv.* **2017**, *7*, 16866–16877. [CrossRef]
44. Chien, Y.C.; Liang, C.; Liu, S.H.; Yang, S.H. Combustion kinetics and emission characteristics of polycyclic aromatic hydrocarbons from polylactic acid combustion. *J. Air Waste Manag. Assoc.* **2012**, *60*, 849–855. [CrossRef] [PubMed]
45. Sameni, J.; Krigstin, S.; Rosa, D.d.S.; Leao, A.; Sain, M. Thermal characteristics of lignin residue from industrial processes. *BioResources* **2014**, *9*, 725–737. [CrossRef]
46. Istrate, B.; Rau, J.V.; Munteanu, C.; Antoniac, I.V.; Saceleanu, S. Properties and in vitro assessment of ZrO_2-based coatings obtained by atmospheric plasma jet spraying on biodegradable Mg-Ca and Mg-Ca-Zr alloys. *Ceram. Int.* **2020**, *46*, 15897–15906. [CrossRef]
47. Fatemeh, T.; Alireza, S.; Hossein, M. Synthesis of Cr_2O_3/TiO_2 nanocomposite and its application as the blocking layer in solar cells. *J. Environ. Anal. Chem.* **2018**, *5*, 1000231. [CrossRef]
48. Bhagyashri, K.; Mahesh, N.; Garadkar, K.M.; Rahul, B.M.; Kiran Kumar, K.S.; Balu, D.A.; Shivaji, T. Ionic liquid assisted synthesis of chromium oxide (Cr_2O_3) nanoparticles and their application in glucose sensing. *J. Mater. Sci. Mater. Electron.* **2019**, *30*, 13984–13993. [CrossRef]
49. Sone, B.T.; Manikandan, E.; Gurib-Fakim, A.; Maaza, M. Single-phase α-Cr_2O_3 nanoparticles' green synthesis using Callistemon viminalis' red flower extract. *Green Chem. Lett. Rev.* **2016**, *9*, 85–90. [CrossRef]
50. Benwood, C.; Anstey, A.; Andrzejewski, J.; Misra, M.; Mohanty, A.K. Improving the impact strength and heat resistance of 3d printed models: Structure, property, and processing correlationships during fused deposition modeling (FDM) of Poly(Lactic Acid). *ACS Omega* **2018**, *3*, 4400–4411. [CrossRef]
51. Teixeira, E.d.M.; Campos, A.d.; Marconcini, J.M.; Bondancia, T.J.; Wood, D.; Klamczynski, A.; Mattosoa, L.H.C.; Glenn, G.M. Starch/fiber/poly(lactic acid) foam and compressed foam composites. *RSC Adv.* **2014**, *4*, 6616–6623. [CrossRef]
52. Wu, C.-S. Analysis of mechanical, thermal, and morphological behavior of polycaprolactone/wood flour blends. *J. Appl. Polym. Sci.* **2004**, *94*, 1000–1006. [CrossRef]
53. Onkar, M.; Savita, R. Monoclinic zirconium oxide nanostructures having tunable band gap synthesized under extremely non-equilibrium plasma conditions. *Proceedings* **2019**, *3*, 10. [CrossRef]
54. Sagadevan, S.; Jiban, P.; Isha, D. Hydrothermal synthesis of zirconium oxide nanoparticles and its characterization. *J. Mater. Sci. Mater. Electron.* **2016**, *27*, 5622–5627. [CrossRef]
55. Suning, L.; Qian, W.; Tao, C.; Zhihua, Z.; Ying, W.; Jiajun, F. Study on cerium-doped nano-TiO_2 coatings for corrosion protection of 316 L stainless steel. *Nanoscale Res. Lett.* **2012**, *7*, 227. [CrossRef]
56. Vorrada, L.; Natnapin, J.; Thirawich, C.; Achanai, B. The photocatalytic reduction of hexavalent chromiumby controllable mesoporous anatase TiO_2 nanoparticles. *Adv. Mater. Sci. Eng.* **2014**, *2014*, 348427. [CrossRef]
57. Tamrakar, R.K.; Upadhyay, K.; Bisen, D.P. Gamma ray induced thermoluminescence studies of yttrium (III) oxide nanopowders doped with gadolinium. *J. Radiat. Res. Appl. Sci.* **2014**, *7*, 526–531. [CrossRef]

58. Gupta, A.K.; Mohanty, S.; Nayak, S.K. Preparation and characterization of lignin nanofibre by electrospinnig technique. *Int. J. Sci. Eng. Appl. Sci.* **2015**, *1*, 184–190, ISSN:2395-3470.
59. Adnan, A.; Shahid, A.; Murtaza, S.; Shahid, M.R.; Shahzad, N.; Saadat, A.S. Structural and magnetic phase transition of sol–gel-synthesized Cr_2O_3 and $MnCr_2O_4$ nanoparticles. *J. Sol-Gel Sci. Technol.* **2016**, *80*, 96–102. [CrossRef]
60. Ferreira, C.S.; Santos, P.L.; Bonacin, J.A.; Passos, R.R.; Aparecido Pocrifka, L. Rice husk reuse in the preparation of SnO_2/SiO_2 nanocomposite. *Mater. Res.* **2015**, *18*, 639–643. [CrossRef]
61. Nedelcu, D.; Comaneci, R. Microstructure, mechanical properties and technology of samples obtained by injection from arboblend. *Indian J. Eng. Mater. Sci.* **2014**, *21*, 272–276.

polymers

MDPI

Article

Physical, Mechanical, and Morphological Properties of Treated Sugar Palm/Glass Reinforced Poly(Lactic Acid) Hybrid Composites

S. F. K. Sherwani [1], E. S. Zainudin [1,2,*], S. M. Sapuan [1,2,*], Z. Leman [1,2] and A. Khalina [2]

[1] Advanced Engineering Materials and Composites Research Centre (AEMC), Department of Mechanical and Manufacturing Engineering, Universiti Putra Malaysia (UPM), Serdang 43400, Malaysia; faisalsherwani786@gmail.com (S.F.K.S.); zleman@upm.edu.my (Z.L.)

[2] Laboratory of Biocomposite Technology, Institute of Tropical Forestry and Forest Products (INTROP), Universiti Putra Malaysia (UPM), Serdang 43400, Malaysia; khalina@upm.edu.my

* Correspondence: edisyam@upm.edu.my (E.S.Z.); sapuan@upm.edu.my (S.M.S.)

Abstract: This research was performed to evaluate the physical, mechanical, and morphological properties of treated sugar palm fiber (SPF)/glass fiber (GF) reinforced poly(lactic acid) (PLA) hybrid composites. Morphological investigations of tensile and flexural fractured samples of composites were conducted with the help of scanning electron microscopy (SEM). Alkaline and benzoyl chloride (BC) treatments of SPFs were performed. A constant weight fraction of 30% total fiber loading and 70% poly(lactic acid) were considered. The composites were initially prepared by a Brabender Plastograph, followed by a hot-pressing machine. The results reported that the best tensile and flexural strengths of 26.3 MPa and 27.3 MPa were recorded after alkaline treatment of SPF, while the highest values of tensile and flexural moduli of 607 MPa and 1847 MPa were recorded after BC treatment of SPF for SPF/GF/PLA hybrid composites. The novel SPF/GF/PLA hybrid composites could be suitable for fabricating automotive components.

Keywords: mechanical properties; sugar palm fiber; poly(lactic acid); alkaline treatment; benzoyl chloride treatment; hybrid composites

Citation: Sherwani, S.F.K.; Zainudin, E.S.; Sapuan, S.M.; Leman, Z.; Khalina, A. Physical, Mechanical, and Morphological Properties of Treated Sugar Palm/Glass Reinforced Poly(Lactic Acid) Hybrid Composites. *Polymers* **2021**, *13*, 3620. https://doi.org/10.3390/polym13213620

Academic Editors: Andrea Sorrentino, Domenico Acierno and Antonella Patti

Received: 22 September 2021
Accepted: 18 October 2021
Published: 20 October 2021

1. Introduction

Due to environmentally friendly customers' desires to save the earth, recently there has been a growing interest in using renewable resources and biodegradable products. For the development of polymer composites, many major industries have focused on using natural fibers. This is owing to the benefits offered by natural fibers (e.g., they are cost-effective, dense, easy to obtain, environmentally safe, non-toxic, durable, reusable, biodegradable, abrasion-resistant, have high strength and modulus, and are simple to process) [1,2]. Natural fibers' biodegradability makes them ideal for reinforcement in polymer composites [3,4]. Both natural and synthetic fibers can be used to create hybrid composites. This combination demonstrates outstanding structural and mechanical properties [5,6]. Recently, sugar palm has been considered by many studies as a desirable natural fiber due to its easy cellulose separation from other components [7]. Sugar palm (*Arenga pinnata Wurmb. Merr*) is found abundantly in Malaysia, Indonesia, India, and Thailand [8]. SPF has the advantages of having low density, biodegradable, lack of toxicity, low cost, non-abrasive nature, and long-lasting [9,10].

In the world of biodegradability, efforts in commercializing PLA polymers have been taking place since the last decade. PLA is a thermoplastic bio-based product obtained from fermentation processes of natural agricultural raw materials (starch and sugar) [11,12]. PLA is recyclable, readily decomposable, has cheap manufacturing costs, and is commercially available [13]. These features enable PLA to replace petroleum-based polymers for several biomedical, textile, plastic, 3D printing materials, and packaging applications [14,15]. However, some drawbacks of PLA include its high price and water sensitivity, low crystallinity rate, and fragility [16]. Many researchers aim towards overcoming these limitations, commonly by mixing PLA with natural fibers [17].

Moisture absorption is a major problem with natural fibers that negatively affects the strong interfacial bonding between a fiber and a matrix. This problem can be solved by fiber pre-treatment, which eliminates lignin and other related materials, improving the interfacial bonding between the fiber and the matrix [18,19]. Treating the lignocellulosic fiber for improved adhesiveness and good stress transfer from the matrix to the rigid fiber will increase the performance of different hybrid composites. Since hydrophilic fiber does not tightly bond with a hydrophobic polymer, weak fiber-matrix bonding degrades composite strength [8,18]. Treatment of the fiber is needed to enhance bonding between the fiber and the matrix. After the treatment, the fiber has reduced moisture absorption, increased bonding, and, most significantly, both enhanced physical and mechanical properties. Alkaline treatment is one of the most effective chemical treatments for fiber surface modification [18–20]. An alkaline treatment procedure has proven to be a successful treatment for extracting waxy substances and impurities [20,21]. As a result, the fiber's surface becomes rough, allowing for greater adhesion with the polymer. This treatment affects the fiber in two ways. Firstly, hydrogen bonding in the network structure is disrupted, enhancing the surface roughness. Secondly, lignin, wax, and oils are eliminated from the surface, enhancing the exposure of cellulose to the fiber surface. This results in increasing the number of possible reaction sites [22]. However, alkaline treatments have the added problem of fiber degradation at high concentrations, which may be addressed with a moderate chemical treatment such as benzoyl chloride.

The fiber treatment with BC decreases the hydrophilic nature of natural fibers and strengthens fiber attachment to the matrix, improving the bio-composites' strength [23,24]. Alkali pre-treatment is used during the benzoylation process. When further treated with BC, these alkaline pretreated fibers lead to the —OH groups of the cellulose fibers overtaken by benzoyl groups and render them hydrophobic [22]. Extractable materials such as waxes, lignin, and oil-coating materials are removed at this stage, exposing more volatile hydroxyl groups on the fiber surface. The fiber's —OH groups are substituted by the benzoyl groups, which binds to the cellulose backbone [25]. Prabhu et al. [24] addressed that the benzoylation of fiber improved adhesion to the fiber-matrix, significantly improved composite strength and decreased water absorption. The treatment improved the wettability of *Impomea pes-caprae* fiber and epoxy composites, resulting in stronger bonding and increased overall composite strength [26]. Palmyra palm-leaf stalk fiber composites were also treated with BC, which improved tensile strength and modulus by 60% [27]. The BC treatment strengthened fiber and matrix adhesion, improved strength, and reduced the water absorption character of the whole composite by reducing SPF's hydrophilic nature and strengthening the blend with epoxy resin matrix [25,28]. Currently, Mohd Izwan et al. [29] investigated SPF benzoylation treatment and found that the highest tensile strength was achieved after 15 min of soaking time, indicating good SPF properties for use as reinforcement in composites.

Several studies were conducted on the hybridization of natural–natural fibers, natural–synthetic fibers and synthetic–synthetic fibers in a single matrix, showing that hybrid composites have been promising to enhance mechanical properties [30–33]. However, in hybrid composites, fiber loading is typically limited to a maximum of 50% [34]. Atiqah et al. [35] studied the mechanical properties of kenaf-glass reinforced unsaturated polyester hybrid composites. They found that combining these fibers improved the tensile, flexural, and impact properties. Afzaluddin et al. [36] developed SPF/GF reinforced TPU hybrid composites and studied their physical, tensile, flexural, and impact properties. They reported that tensile and impact properties of the hybrid composites were increased with the increase of the content of SPF (%) compared to GF reinforced composites. Still, the flexural properties were improved when the content of GF was increased. Atiqah et al. [37]

studied the effects of silane and alkaline treatment of fiber on the tensile, flexural, and impact properties of SPF/GF/TPU hybrid composites and found that treated SPF can increase the mechanical properties of hybrid composites. In another study, Atiqah et al. [38] reported that physical properties could be improved by alkaline treatment of SPF in fabricating the SPF/GF/TPU hybrid composites. Recently, Radzi et al. [20] determined the physical and mechanical properties of roselle/SPF reinforced thermoplastic polyurethane hybrid composites and revealed that the alkaline treatment on RF/SPF hybrid composite has improved the mechanical properties of hybrid composite and proven the RF/SPF composite suitability for making automotive parts.

To the best of our knowledge, no study on treated SPF/GF/PLA hybrid composites has been reported in the literature. Novel combinations of treated SPF/GF are selected as effective reinforcements PLA hybrid composite due to their positive effects on improving physical, mechanical, and morphological properties. Due to the presence of SPF and PLA bio-degradable plastic manufactured from plant-based resources such as corn starch or sugarcane, this hybrid composite is environmentally friendly. The hybrid composites had a constant SPF/GFs weight fraction of 30%. Sherwani et al. [39] demonstrated that a 70:30 (PLA/SPF) ratio exhibits excellent mechanical properties, especially tensile and flexural properties. Hence, a 70% PLA to 30% fiber ratio is considered in this research. The effect of alkaline and benzoyl chloride treatment of SPFs on PLA were also determined by Sherwani et al. [40] and reported that the optimum percentage is 6% alkaline and the best soak duration is 15 min for enhancing the mechanical properties of SPF/PLA composites. As a result, a 6% alkaline and 15-min benzoyl chloride treatment is considered here. The aim of this paper is to evaluate the physical, tensile, flexural, and impact properties as well as the post-tensile and flexural morphological tests, aiming to propose the best natural/green hybrid composite formulation for engineering material which can be used for the applications of automotive (especially motorcycle) components manufacturing. With the help of this research, an environmentally friendly biodegradable material may be developed for possible application in the manufacture of motorcycle components.

2. Materials and Methods

2.1. Materials

Sugar palm fiber (length of fiber up to 1.19 m, average diameter 0.5 mm, the density of raw SPF 1.2–1.3 gm/cm^3, and tensile strength 15.5 MPa [41]) were purchased from Kampung Kuala Jempol, Negeri Sembilan, Malaysia. The poly(lactic acid) (NatureWork 2003D) (density 1.25 gm/cm^3 at 21.5 °C, yield tensile strength of 52 MPa, and melting point of 170 °C), BC with reagent plus 99%, ethanol, and E-glass fiber (properties of E-glass is shown in Table 1) were delivered by Mecha Solve Engineering, Petaling Jaya, Selangor, Malaysia. Sodium hydroxide (NaOH) pellet was delivered by Evergreen Engineering and Services, Taman Semenyih Sentral, Selangor, Malaysia.

Table 1. Some typical properties of E-glass fiber [42].

Liquidus Temp. (°C)	1140
Fiberising temp. (°C)	1200
Tensile strength at 25 °C (GPa)	3.7
Tensile modulus (GPa)	76
Density (g/cm^3)	2.53
Refractive index	1.550
Volume resistivity (Ωcm)	10^{15}
Dielectric constant at 25 °C and 10^6 Hz	6.6
Loss tangent at 25 °C and 10^{10} Hz (10^{-3})	3.9

The electrical resistivity of E-glasses at room temperature is exceptionally high. E-glass compositions by weight percentage are SiO_2-60; Al_2O_3-9; MgO-4; CaO-27.

2.2. Preparation of Sugar Palm Fiber

A crusher machine was used for crushing a bundle of SPF. The dry SPF was graded to a length of 10 mm to 15 mm using a crusher. The fiber was then cleaned several times with water to remove impurities. The SPF was left outdoor for 24 h before being dried in an air-circulating oven at 60 °C.

2.3. Chemical Treatments

2.3.1. Alkaline Treatment

Alkaline treatment of natural fibers was used to eliminate surface impurities as well as hemicelluloses within the fibers [20]. In our study, 50 g of SPFs was soaked in a 6% *w/v* alkaline solution of 1000 mL for 1 h at 25 °C. After that, they were immersed with an acetic acid solution till a neutral pH value was obtained prior to washing with distilled water and oven-drying for 24 h at 60 °C. The dried SPF was placed into zipper plastic storage bags. Table 2 shows the chemical composition of SPF and alkaline treated SPF.

Table 2. Composition of alkaline treated SPF.

	Cellulose (%)	Hemi-Cellulose (%)	Holo-Cellulose (%)	Lignin (%)	Extractive (%)	Ash (%)	References
SPF	43.88	7.24	51.12	33.24	2.73	1.01	[9]
Alkaline treated SPF (5% *w/v* NaOH solution)	82.33	3.97	86.3	0.06	-	0.72	[43]

2.3.2. Benzoyl Chloride Treatment

50 g SPF was immersed in 18% NaOH solution for 30 min, then washed SPF twice with tap water. The SPF was suspended in a 10% NaOH solution and vigorously agitated for 15 min in 50 mL BC. Once again, SPF was washed in water, filtered, and dried at 25 °C. SPFs is normally immersed in ethanol for an hour before being washed, filtered, and dried in a 60 °C oven for 24 h [28,29]. The reaction between the cellulosic —OH groups of fiber and BC is given in Scheme 1.

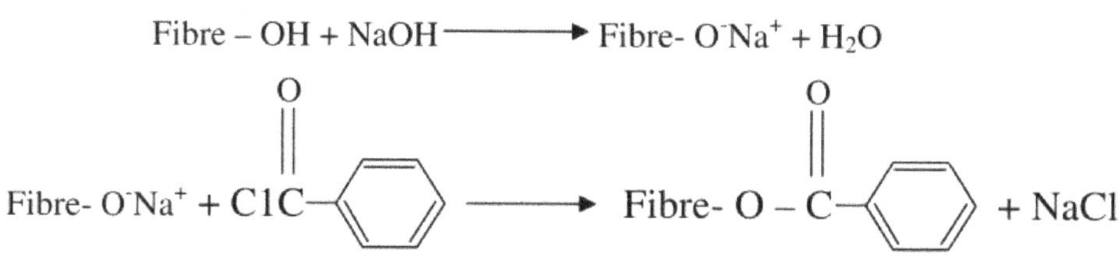

Scheme 1. The reaction between the cellulosic —OH groups of fiber and BC. Alkaline pretreated fibers, when further treated with BC, the OH groups of the cellulose fibers are replaced by benzoyl groups which reduced the hydrophilic character of fiber.

2.4. Fabrication of SPF/GF Reinforced PLA Hybrid Composites

The melting compounding and hot press moulding methods were used to prepare the SPF/GF reinforced PLA hybrid composites. The 10–15 mm sugar palm fiber, 12.5 mm chopped E glass fiber, and PLA pellets were dried at a temperature of 60 °C in electric ovens for 48 h. Nine sets of SPF/GF composites (30/0, 20/10, 15/15, 10/20, and 0/30) wt % reinforced PLA were developed as seen in Table 3. Based on past research [34,36,44], such a ratio was considered. In a Brabender Plastograph (co-rotating twin-screw extruder), untreated/treated SPF and chopped E-glass fibers reinforced PLA were mixed for 10 min at 160 °C with a rate of 50 rpm to ensure consistent mixing. These samples were then crushed using a crushing unit. After Brabender mixing, to reduce voids or gaps, the composite samples must be placed in an electric oven for 24 h at 60 °C before the hot press.

A compression moulding Techno Vation machine model 40 tons was used for hot-press moulding. These samples were pre-heated for 7 min at 170 °C before being completely pressed for 6 min. There were three vent cycles to remove voids in the composites. During the final cycle, the cold-pressed time was 6 min at 25 °C. Figure 1 describes the detailed methodology of this research.

Table 3. Formulation of non-hybrid and hybrid composites.

No. of Samples	Matrix	Reinforcement		
	PLA (wt %)	SPF		GF (wt %)
		Treatment	(wt %)	
S1	70	-	30	0
S2	70	-	0	30
S3	70	-	15	15
S4	70	6% NaOH	10	20
S5	70	6% NaOH	15	15
S6	70	6% NaOH	20	10
S7	70	15 min BC	10	20
S8	70	15 min BC	15	15
S9	70	15 min BC	20	10

BC—Benzoyl Chloride.

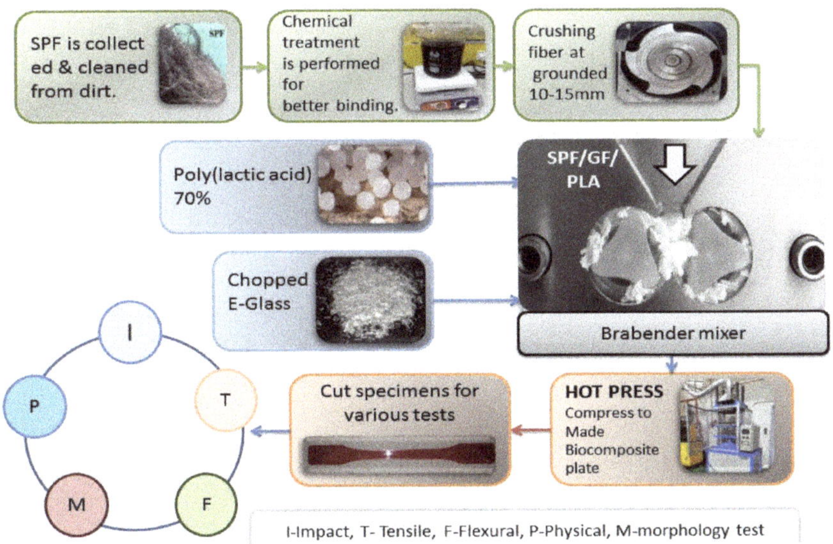

Figure 1. Detailed description of the methodology flow diagram.

3. Characteristics of SPF/PLA/GF Hybrid Composites

3.1. Density

Non-hybrid and hybrid composites' experimental densities were measured using the Mettler Toledo XS205 electronic densitometer as per the ASTM D792 standard [45]. Equation (1) was used to evaluate the theoretical density of the composite,

$$\text{Theoretical density } \rho_{ct} = \frac{1}{\left(\frac{w_{sf}}{\rho_{sf}}\right) + \left(\frac{w_{gf}}{\rho_{gf}}\right) + \left(\frac{w_m}{\rho_m}\right)} \tag{1}$$

where w_{sf}, w_{gf}, and w_m represent the weight fraction of the SPFs, glass fibers, and matrix, respectively, while ρ_{sf} , ρ_{gf} , and ρ_m denote the density of the SPF, glass fibers, and matrix, respectively. Subsequently, the volume fraction of voids in the percentage of the composites is calculated by using Equation (2) [18].

$$\text{Volume fraction of void } V_{void} = \frac{\rho_{ct} - \rho_{comp.\ exp.}}{\rho_{ct}} \times 100 \quad (2)$$

$\rho_{comp.\ exp.}$—Composite experimental density.

3.2. Moisture Content

All of the non-hybrid and hybrid composites were tested for moisture content analysis. In an oven, the composites were heated to 100 °C for 24 h. To calculate the moisture content, the weights of the composites were determined before (M_{bh}, gram) and after (M_{ah}, gram) being placed into the oven. The following Equation (3) was used:

$$\text{Moisture content } (\%) = \frac{M_{bh} - M_{ah}}{M_{bh}} \times 100 \quad (3)$$

3.3. Water Absorption Test

Non-hybrid and hybrid composites were tested for water absorption (*WA*) using the ASTM D570 standard [46]. From the composite plate, a rectangular geometry sample measuring 10 mm × 10 mm × 3 mm was removed. The weight of the composite was calculated as an initial mean before it was immersed in water, W_{iw}, and W_{fw} as a final stage mean after a day of immersion in water at about 25 °C. The test was conducted for eight days. The following Equation (4) was used for calculating water absorption:

$$\text{Water Absorption } (\%) = \frac{\left(W_{fw} - W_{iw}\right)}{W_{iw}} \times 100 \quad (4)$$

3.4. Thickness Swelling

The thickness swelling (*TS*) of all non-hybrid and hybrid composites with measurements of 10 mm × 10 mm × 3 mm was evaluated. Using a digital vernier caliper, the thickness was calculated as T_{1b} (before) and T_{2a} (after) they were immersed in water [20]. The *TS* reading was taken for eight days and was determined using Equation (5).

$$\text{Thickness Swelling } (\%) = \frac{T_{2a} - T_{1b}}{T_{1b}} \times 100 \quad (5)$$

3.5. Tensile Test

An Instron3366 universal testing machine (UTM, University Ave Norwood, Norwood, MA, USA) was used to conduct a tensile test in accordance with ASTM Standard D638-10 [47]. The gauge length of the non-hybrid and hybrid composites was 80 mm, and the crosshead velocity was 2 mm/min, using a 5 KN load cell. Five samples measuring 150 mm × 25 mm × 3 mm were tested. The average of the five samples provided the final result.

3.6. Flexural Test

Using an Instron 3365 dual column tabletop UTM with a span length of 50 mm and a crosshead speed of 12 mm/min, the flexural properties of non-hybrid and hybrid composites were tested according to the ASTM D790 standard [48]. Five composite samples measuring 127 mm × 12.7 mm × 3 mm were obtained from the composite plate. The average of the five samples provided the final result.

3.7. Impact Test

ASTM D256 (2010) [49] Standard Izod impact test samples measuring 65 mm × 15 mm × 3 mm were taken out of the non-hybrid and hybrid composite plates. Rayran RR/IMT/178 Izod impact testers were used for the impact study. For each sample, the average of the three readings was calculated for the final result of the impact test. Five identical samples were rigidly placed in a vertical position with each type of composite and struck in the center of the instruments with a pendulum with a force of 10 J. The impact had 2.75 J energy and a velocity of 3.46 m/s.

3.8. Morphological Investigations

SEM (Coxem-EM-30AX+) was used to examine broken surfaces of tensile and flexural samples for morphological analysis. Scanning electron microscopy (SEM) with a working distance of 14.7 mm, a 58 A emission current, and a 20.0 kV acceleration voltage was used. The samples were coated with a fine gold film for electrical conductivity, improving the image resolution appreciably.

3.9. Fourier Transform Infrared (FTIR)

The FTIR spectra were used to examine the presence of functional groups existing in non-hybrid and hybrid composites. The samples were segregated into 10 mm × 10 mm × 3 mm squares and then analyzed with a power FTIR spectrometer with attenuated total reflectance (ATR) (Nexus Analytics-Isio 713601, Petaling Jaya, Selangor Darul Ehsan, Malaysia). The wavenumber spectrum for the spectra was 4000 cm^{-1} to 500 cm^{-1}, and the measurements were taken with a resolution of 4.0 cm^{-1} and 16 scans per sample.

3.10. Statistical Analysis

SPSS software was used to perform an analysis of variance (ANOVA) on the obtained experimental results. 'Duncan's test was employed to conduct a mean comparison at a 0.05 level of significance ($p \leq 0.05$).

4. Results and Discussion

4.1. Density

Figure 2 illustrates the density ($\rho_{comp.\ exp.}$) value measurement for non-hybrid and hybrid SPF/GF/PLA composites. The density result showed that the density was improved by adding GF to the SP/PLA composites from 1.21 to 1.32 gm/cm^3. Comparing S1, –S3 composites, the density of S3 exhibited the highest value of 1.32 gm/cm^3, concluding that the hybridization of SPF/PLA composite increases the density of the composite. A similar finding was also revealed from a previous work done by Afzaluddin et al. [36] that density increment resulted from the addition of glass fiber into SPF/GF/TPU hybrid composites. This was due to the fact that GF has a higher density than SPF. The densities of untreated S3 and treated S5, S8 SP/GF hybrid composites were 1.32, 1.31, and 1.19 gm/cm^3. The density slightly decreased after treatment of SPF. In previous studies by Atiqah et al. [38] and Merlini et al. [50], density reductions of treated fibers were also recorded. Comparing alkaline treated S4–S6, and BC treated S7–S9 hybrid composites, density results showed that the density decreased slightly as the SPF content was increased, irrespective of whether the SPF has been alkaline or BC treated. As SPF wt % increased, the hydrophilic nature of the SPF made it more difficult to develop the composite properly. This led to voids in the hybrid composites, decreasing the composite density. A similar finding was supported by Safri et al. [51], where the density was measured for SPF/GF/epoxy hybrid composites.

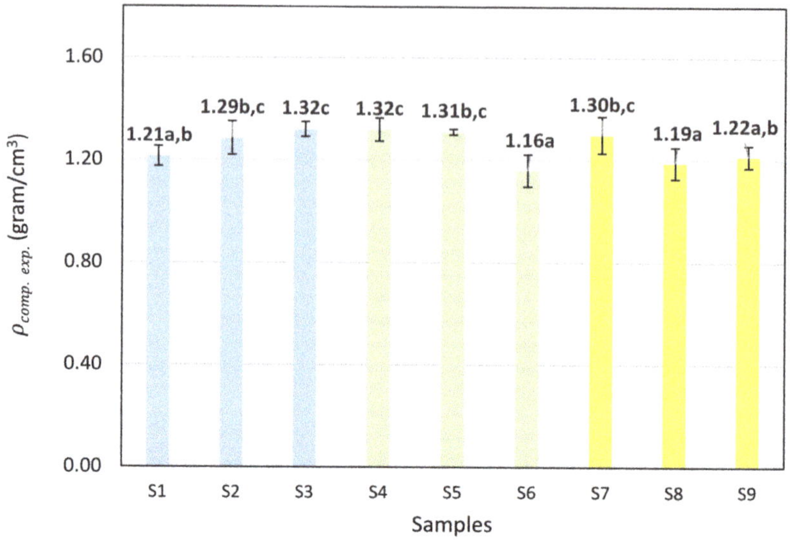

Figure 2. Composite experimental density $\rho_{comp.\ exp.}$ vs. samples for non-hybrid and hybrid SPF/GF/PLA composites. * Values with different letters in the figures are significantly different ($p < 0.05$).

4.2. Moisture and Void Contents

Table 4 shows moisture contents, theoretical/experimental composites density, and void contents of non-hybrid and hybrid SPF/GF/PLA composites. In general, the experimental and theoretical densities differed from each other due to a significant influence of voids and pores in the composite towards the behavior of the composite [40]. It is clear from Table 4 that, with the help of glass hybridization, the percentage of voids decreased, as confirmed by SEM images in the morphological investigation. The finding was also supported by the results from work by Radzi et al. [20]. The S4–S6 (alkaline treated) hybrid composites exhibited lower %voids than S7–S9 (benzoyl treated) hybrid composites. This was due to the good compatibility of SPF and GF with the PLA matrix. According to Jawaid et al. [36], voids formation is due to the incompatibility of natural fiber with matrix to displace all the trapped air which is entrained during fabrication of hybrid composite and incomplete wetting out of the fibers by the matrix. This research also reported that after Brabender mixing, the composite samples need to be placed in an electric oven for 24 h at 60 °C before the hot press. Otherwise, more voids are visible due to their moist content.

Table 4. Moisture contents, densities, and void contents of non-hybrid and hybrid SPF/GF/PLA composites.

Composite Specimens	Moisture Contents	ρ_{ct} (gr/cm^3)	$\rho_{comp.\ exp.}$ (gr/cm^3)	Void Contents V_{void}
S1	1.79%	1.23	1.21 ± 0.039 ^{a,b}	1.94%
S2	0.75%	1.40	1.29 ± 0.066 ^{b,c}	7.9%
S3	0%	1.34	1.32 ± 0.029 ^{c}	1.63%
S4	1.24%	1.38	1.32 ± 0.045 ^{c}	3.93%
S5	2.09%	1.34	1.31 ± 0.011 ^{b,c}	2.45%
S6	1.07%	1.30	1.16 ± 0.062 ^{a}	1.5%
S7	1.29%	1.38	1.30 ± 0.072 ^{b,c}	7.5%
S8	0%	1.34	1.19 ± 0.062 ^{a}	9.7%
S9	0%	1.30	1.22 ± 0.044 ^{a,b}	6.5%

* Values with different letters in the same column are significantly different ($p < 0.05$).

4.3. Water Absorption Analysis

Figure 3 shows the effect of treatments on the values of water absorption (*WA*) of non-hybrid and hybrid SP/GF/PLA composites. The higher SPF content (S1 composite) resulted in a higher *WA* value. The S2 (GF/PLA) composite was the lowest *WA* compared to other non-hybrid or hybrid composites. Among the SPF/GF/PLA hybrid composites, S5 and S8 composites had the lowest *WA* values as both composites exhibited the same 15 wt % SPF. S5 composite was alkaline treated, indicating that alkali hydroxyl (—OH) groups inside the molecules were broken down and when reacted with H_2O molecules and left the fiber structure. The other reactive molecules produce fiber-cell-O-Na groups in the molecular cellulose chains. This reduced hydrophilic hydroxyl groups and increased the water absorption resistance of the fiber. A considerable number of hemicelluloses, lignin, pectin, wax, and oil were also extracted after alkaline treatment [25], while BC treated S8 composite indicated that the fibers were treated with BC after pre-alkaline treatment. The —OH fiber groups were substituted by the benzoyl group and bonded to the backbone of the cellulose. This resulted in the increased hydrophobicity of the fiber and enhanced matrix adhesion. The following is the order of the decreasing value of hybrid and non-hybrid SPF/GF/PLA composites' water absorption S2< S8< S5< S7< S9< S6< S4< S3< S1. After the 6th day, no change was observed in *WA*. For hybrid composite, the higher loading of natural fibers was considered to have more *WA*, whereas the higher GF loading was related to lower *WA*. This was also consistent with the past work of Afzaluddin et al. [36], who found that water resistance increased with the addition of GF in the SPF/TPU composites.

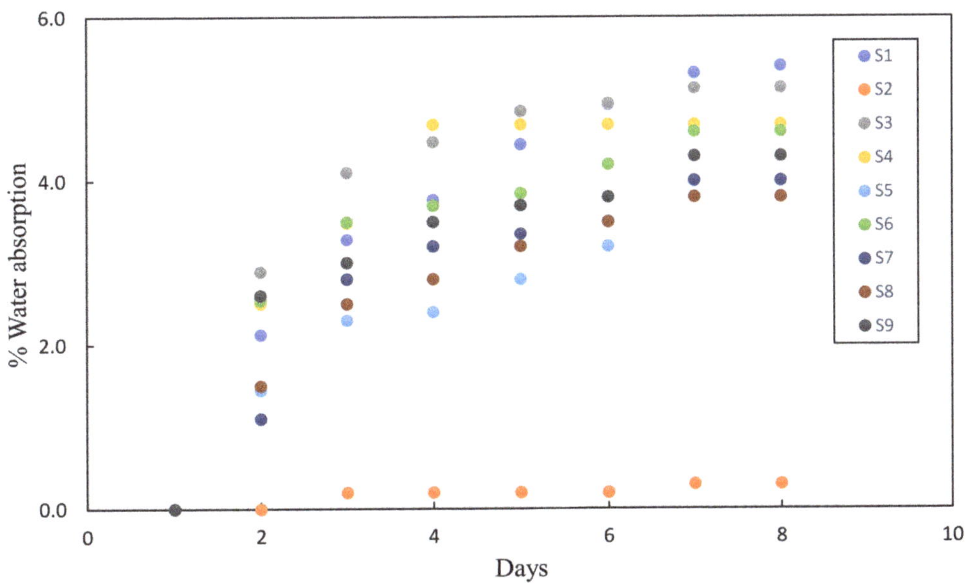

Figure 3. % Water absorption (*WA*) versus samples for non-hybrid and hybrid SPF/GF/PLA composites.

4.4. Thickness Swelling

Figure 4 presents the effect of treatments on the values of thickness swelling (*TS*) of non-hybrid and hybrid SP/GF/PLA composites. With the increase in *WA*, the thickness swellings of all the composites were increased. The higher the percentage of GF, the lower the *TS* value, while the higher the SPF content, the higher the *TS* value. This was due to the fact that GF possessed water resistance ability while SPF has hydrophilic nature [52]. The highest *TS* value was shown by untreated S1 (SPF/PLA) composite, having maximum content of SPF. In the S2 composite, no *TS* value was observed since it contained only

GF and PLA. Among the hybrid composite, the maximum *TS* value was observed in the S3 (untreated SPF/GF/PLA) hybrid composite. The thickness swelling was due to the prolonged immersion duration, where more water molecules were bonded to the hydrogen bonds of fiber. This *TS* value was randomly decreased after alkaline treatment due to the reduction in the micropores and collapsing capillaries, as well as the removal of wax and impurities at the fiber surface after the treatment. This predicted reduced water retention, as indicated by lessening the amount of water absorbed by the fiber. This reduced the *TS* value of the hybrid composite. Due to this cause, as the alkaline treated SPF content was increased, the *TS* value decreased. A similar finding was reported by N.B.M. Hafidz et al. [53] after alkaline treatment of palm oil fiber composites as well as kenaf fiber composites.

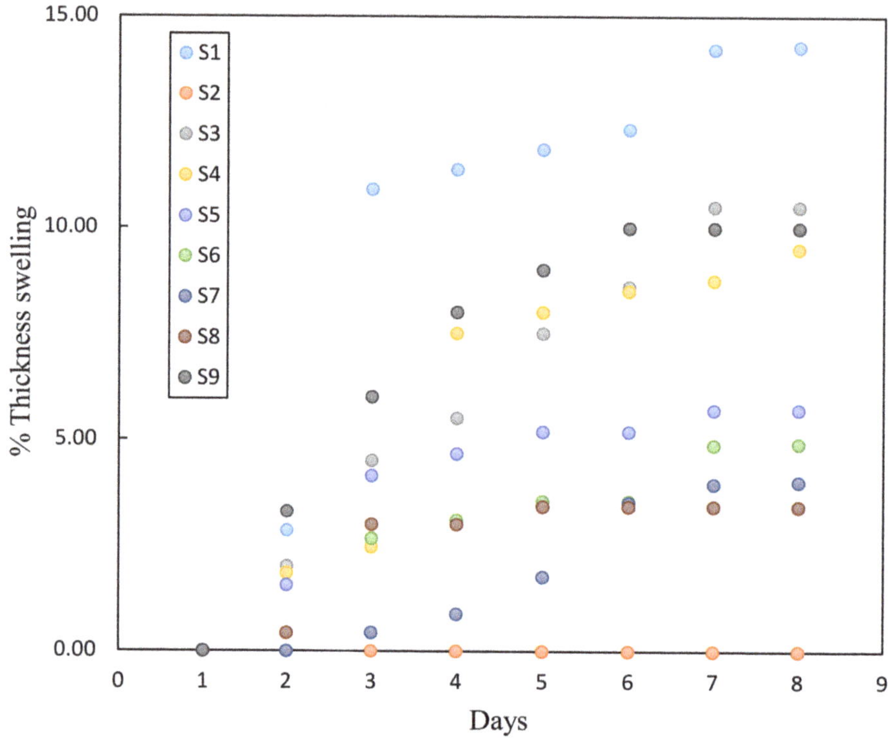

Figure 4. % Thickness swelling (*TS*) versus samples for non-hybrid and hybrid SPF/GF/PLA composites.

The reduction of *TS* was also observed after BC treatment of SPF for hybrid composite. S8 composite showed the lowest *TS* value among all hybrid composite. For the S9 composite, as the benzoyl treated SPF content was increased to 20 wt %, the *TS* value increased due to the disruption of lignin and polysaccharides during treatment that enhanced cellulose concentration. The cellulose chemical structure is composed of hydroxyl that is accessible to water. This directly stimulated the *WA*, thus increased the *TS* [51].

4.5. Tensile Testing

Figure 5a,b shows the effect of treatment on tensile strength and modulus of non-hybrid and hybrid SPF/GF/PLA composites. Hybrid composites were used for various weight percentages of SPFs and GFs. Hybrid composites had a combined weight percentage concentration of fibers (SPF/GF) fixed at 30%, whereas PLA was at 70%. Figure 5a confirmed that by incorporating the GFs, the tensile strength of the PLA composite was increased significantly. From Table 1, the tensile strength of GF was more than SPF, while the tensile strengths of S1 (SPF only) and S2 (GF only) reinforced PLA composite were 16.0 MPa and 23.7 MPa, which indicated that the tensile strength on the addition of GFs was better than untreated SPF. It was observed that the addition of alkali-treated SPF increased the tensile strength of the hybrid SPF/GF reinforced PLA composites. The highest tensile strength of 26.3 MPa was shown by the S6 composite among all hybrid composites. According to Afzaluddin et al. [36], the tensile strength of SPF/GF reinforced TPU hybrid composites could be increased by adding SPF. Among S4–S6 (alkaline treated) hybrid composites, S6 exhibited a maximum tensile strength of 26.3 MPa, while S5 and S4 demonstrated only 14 MPa and 18.7 MPa, respectively. This proves that the interfacial bonding between SPF and PLA matrix had improved after alkaline treatment. S6 hybrid composite exhibited maximum tensile strength that might be due to GF (10% loaded); therefore, SPF can effectively transfer the load from the GF on this particular composition [30]. When alkaline treated SPF/GF/PLA hybrid composites were compared with benzoyl treated SPF/GF/PLA hybrid composites, it is clearly shown in Figure 5a that the alkaline treatment was the best treatment approach used for the improvement of tensile strength. This was due to the improvement in the interface bonding by giving rise to additional sites of mechanical interlocking, facilitating the interpretation of fiber-matrix at the interface. The alkaline treatment of fiber increased the binding properties of the surface by eliminating natural and artificial impurities, which created rough surface topography [30]. This finding was also supported by the morphological investigation of this research.

S3, S5, and S8 had the same composition of 15/15/70 wt % SPF/GF/PLA. The only difference they possessed was the untreated or treated SPF. The analysis showed that the tensile strength of S5 was improved by 27% compared to the S3 hybrid composite. The tensile strength of the alkaline treated SPF/GF/PLA was higher than untreated hybrid composites due to the strong interfacial bonding between treated fiber with PLA matrix. A similar study was carried out by Atiqah et al. [3], that the surface treatment of fiber improved the tensile properties of SP/GF reinforced TPU hybrid composites. The minimum value of tensile strength was 9.3 MPa, which was recorded for S8. The inclusion of BC treated SPF in SPF/GF/PLA hybrid composites caused a remarkable decrease in tensile strength. One of the investigations by Swain et al. [54] demonstrated that the tensile strength only decreased when there was weak bonding between fiber and the matrix.

In general, the stiffness property of the SPF/GF reinforced PLA hybrid composites was also determined. Material stiffness property is mainly indicated by tensile modulus. In general, the value of tensile modulus was increased with an increase in wt % of GFs. This statement was valid for both untreated as well as treated SPF/GF reinforced PLA hybrid composite. The tensile moduli for BC treated SPF/GF/PLA hybrid composites were increased by the GF percentage, i.e., for 10, 15, and 20 wt % GF in S9, S8, and S7 composites, the value of tensile moduli were 500, 505, and 607 MPa.

Comparing untreated SPF of S1–S3 composites indicated increasing percentage of GFs improved the tensile modulus of composites. The highest tensile modulus was shown by S2, having 30 wt % of GFs as depicts in Figure 5b. This is since the tensile modulus and strength of GF was higher than SPF. The increment of tensile modulus denoted the improvement of load-bearing capacity.

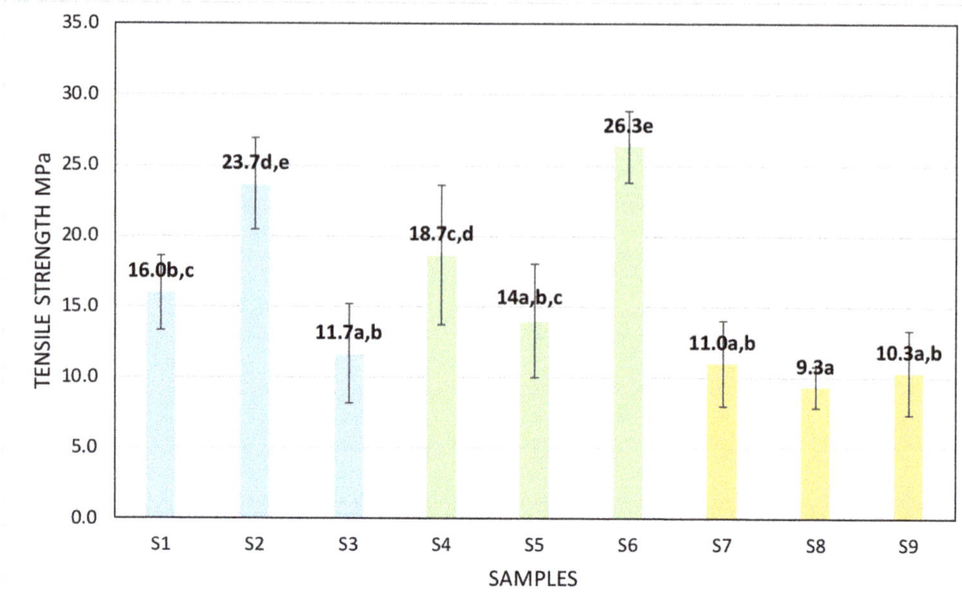

(a)

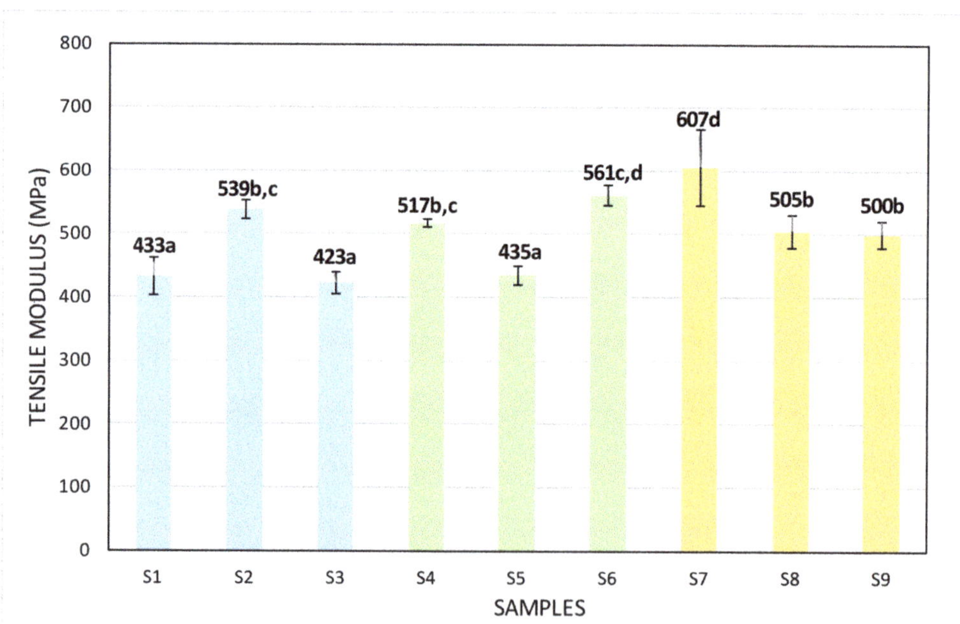

(b)

Figure 5. (**a**) Tensile strength versus samples of non-hybrid and hybrid SPF/GF/PLA composite. (**b**) Tensile modulus versus samples of non-hybrid and hybrid SPF/GF/PLA composite. * Values with different letters in the figures are significantly different ($p < 0.05$).

On comparing the alkaline treated SPF/GF/PLA composites, the analysis showed that the S6 hybrid composite exhibited the maximum tensile modulus of 561 MPa. The S6 hybrid composite had a maximum of 20 wt % of SPF, indicating improvement in the interfacial bonding between SPF and PLA matrix after the alkaline treatment of SPF that consequently resulted in the improved tensile modulus [20]. In addition, after alkaline treatment, rougher topography might cause the qualitative interface between fiber and matrix due to the removal of waxy and impurity substances. A similar increment of tensile modulus after alkaline was revealed by Mukhtar et al. [18] for SPF/GF-reinforced polypropylene composites. Comparing the alkaline, BC treated, and untreated SPF/GF/PLA hybrid composites, it might be noted that BC treatment increased the tensile modulus of the same wt % composite, i.e., S3, S5, and S8. Among these three composites having 15/15/70 wt % (SPF/GF/PLA), the highest value of tensile modulus 505 MPa was shown by S8, followed by 435 MPa and 423 MPa, for S5 and S3 composites, respectively. Previous studies [29,51,55] indicated that the treatment of fibers with BC enhanced the mechanical contact surface with the matrix, which enhanced the interfacial bonding between fiber and matrix and increased stress transfer and tensile modulus of the entire composite. The maximum tensile modulus was shown by the S7 (BC treated) hybrid composite. The tensile properties in terms of the tensile stress–strain curve of various composites is shown in Figure 6.

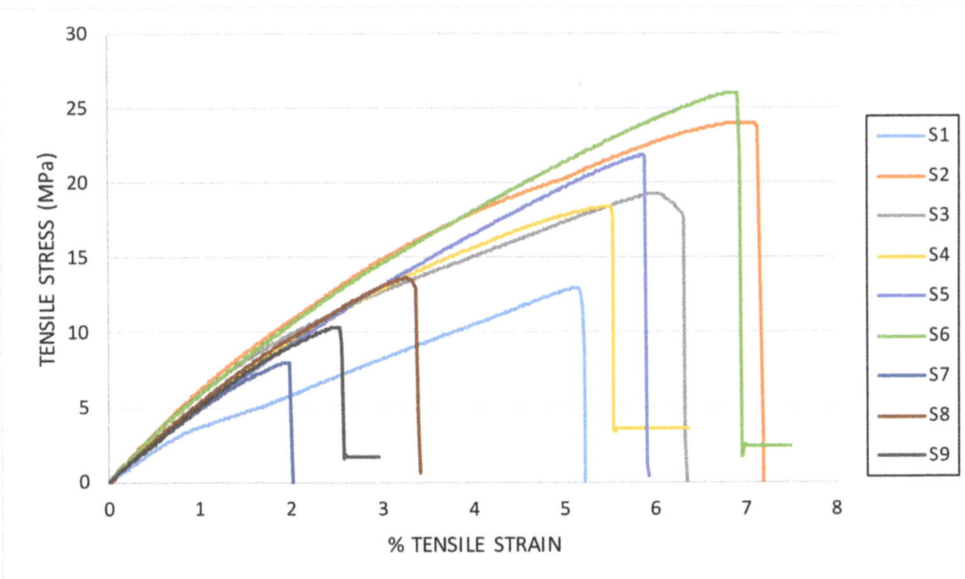

Figure 6. Tensile stress (MPa) versus % tensile strain for various composites.

4.6. *Flexural Testing*

Figure 7a,b shows the effect of treatment on flexural strength and modulus of non-hybrid and hybrid SPF/GF/PLA composites. The bending results analysis confirmed the flexural strength of the composite. In the S5 hybrid composite, the alkaline treatment of SPF enhanced the flexural strength. Comparing S3, S5, and S8 hybrid composite having the same wt % (SPF/GF/PLA-15/15/70), the S5 (alkaline treated) hybrid composite showed maximum flexural strength of 27.3 MPa among all the non-hybrid or hybrid composites. The incorporation of alkaline treated 15 wt % of SPF led to improved flexural strength of the S5 hybrid composite. After the alkaline treatment, about 17% of flexural strength was increased compared with the S3 (untreated) hybrid composite. The increasing of flexural strength after the alkaline treatment was due to partial elimination of hemicellulose, wax, as well as disruption —OH bonding on the fiber, which ensured a better adhesion bonding between the PLA matrix and SPF, whereas the S3 hybrid composite showed low flexural strength caused by poor interfacial bonding between the PLA matrix and SP fiber. This finding was comparable with the SPF/GF-reinforced polypropylene investigated by Mukhtar et al. [18]. Another research highlighted that the fiber treatment plays an important role for flexural strength value. The alkaline treatment of fiber can reduce the cell wall thickening, leading to improved adhesion between fiber and matrix [3]. Similar research reported the effect of alkaline treatment on mechanical properties for roselle/SPF reinforced TPU hybrid composites [20]. Atiqah et al. [3] reported that the flexural strength of SPF hybrid composites also depends on types of surface treatment. The contrast results of alkaline treatment with BC treatment for SPF/GF/PLA hybrid composites are shown. The value of flexural strengths for all three hybrid composites treated with BC were less than the untreated ones.

For non-hybrid S1 composite (only 30 wt % untreated SPF and 70 wt % PLA), the flexural strength was 26.3 MPa, which was supported by the study of Sherwani et al. [40] for flexural analysis of different ratio SPF/PLA composite. The analysis showed that the hybridization of GF did not significantly affect the flexural strength of composites. In the case of S2, it decreased by means of bending ability decrease on the addition of GF. Comparing with untreated SPF/GF, it was clear that the hybridization of the composite resulted in a good flexural modulus of 1811 MPa (S3 hybrid composite), while the flexural moduli for non-hybrid S1 and S2 composite were very low of 1317 MPa and 1564 MPa, respectively. From the comparison of the flexural modulus after alkaline and BC treatment for same wt % hybrid composites (i.e., S5 with S8), it was observed that after BC treatment, the flexural modulus was increased from 1336 MPa to 1536 MPa. BC treatment reduces the diameter of SPF as well as eliminate the lignin and wax layer on fiber. The compatibility between fiber and matrix increased due to the benzene rings' availability in the benzoyl group attached to the fibers.

Safri et al. [28] also reported the same reason for an increment of flexural modulus after BC treatment of SPF/GF/epoxy hybrid composites. On comparing S6 and S9 having the same wt % ratios (20/10/70) SPF/GF/PLA hybrid composites, almost the same flexural modulus about 1450 MPa was observed in all composites. This value is maximum among alkaline treated SPF, which was proved by Radzi et al. [20] that the alkaline treatment of SPF would improve flexural modulus due to better wetting and good interfacial bonding between matrix and the treated fiber. S7 exhibited 1847 MPa, the highest value of flexural modulus. This was due to various changes at the surface of SPF after BC treated that increased the adhesion between treated fiber and matrix and mechanical interlocking by distributing several small voids on the fiber surface and creating extra fiber interpenetration to the interface. A similar reason was reported for studying the effect of 6% alkaline treatment of SPF for roselle/SPF reinforced TPU hybrid composites [20].

Safri et al. [28] also reported the same reason for an increment of flexural modulus after BC treatment of SPF/GF/epoxy hybrid composites. On comparing S6 and S9 having the same wt.% ratios (20/10/70) SPF/GF/PLA hybrid composites, almost the same flexural modulus about 1450MPa was observed in all composites. This value is maximum among alkaline treated SPF, which was proved by Radzi et al. [20] that the alkaline treatment of SPF would improve flexural modulus because of better wetting and good interfacial bonding between matrix and the treated fibre. S7 exhibited 1847MPa, the highest value of flexural modulus. This was due to various changes at the surface of SPF after BC treated that increased the adhesion between treated fibre and matrix and mechanical interlocking by distributing several small voids on the fibre surface and creating extra fibre interpenetration to the interface. A similar reason was reported for studying the effect of 6% alkaline treatment of SPF for roselle/SPF reinforced TPU hybrid composites [20].

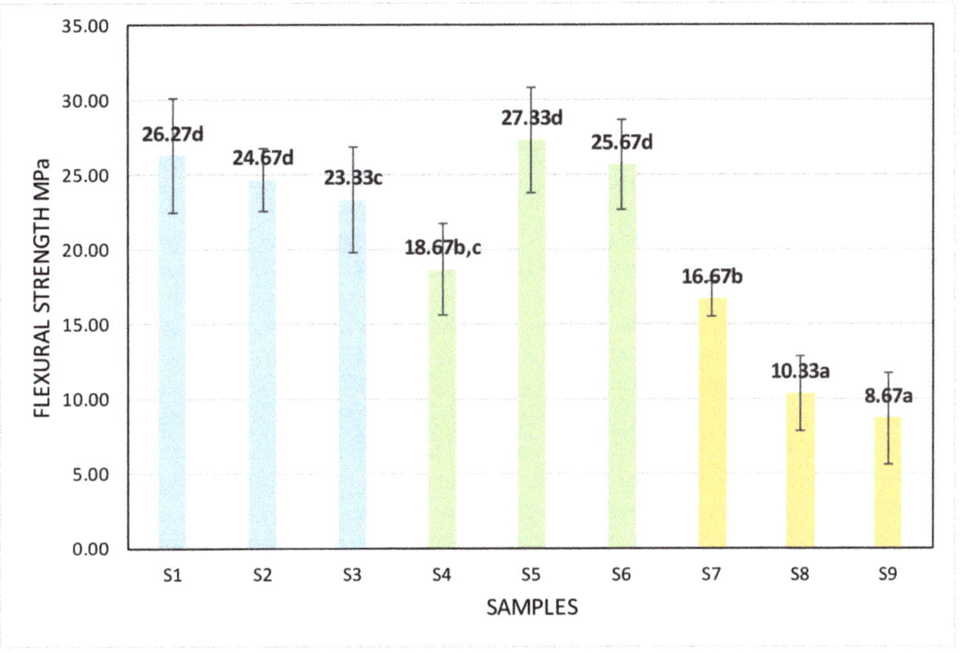

(a)

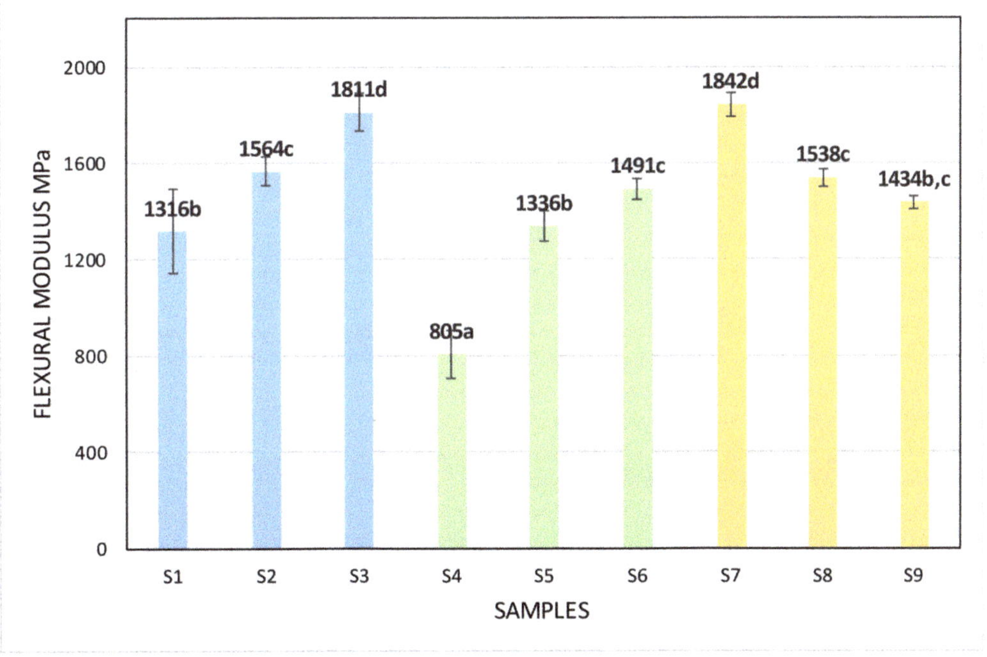

(b)

Figure 7. (**a**) Flexural strength versus samples of non-hybrid and hybrid SPF/GF/PLA composite. (**b**) Flexural modulus versus samples of non-hybrid and hybrid SPF/GF/PLA composite. * Values with different letters in the figures are significantly different ($p < 0.05$).

4.7. Morphological Investigations

The morphological analysis was conducted from two different perspectives: tensile and flexural testings of the fractured cross-sectional area, as shown in Figure 8. The analyses were carried out from 100× magnification to 300× magnification for both non-hybrid and hybrid SPF/GF/PLA composites from S1 to S9. In SEM morphology, S1 (untreated/non-hybrid) composites showed that the SPF demonstrated pull-outs, voids, were also visible that indicated poor compatibility with the PLA matrix. Due to the untreated SPF, some wax and lignin contents were visible, and these SEM images were identical to date palm leaf (DPL) images studied by Swain et al. [54], showing the untreated DPL's lignin and wax contents. Non-hybrid S2 composite revealed that GF was stretched when the tensile load was applied and fewer voids were observed, proving good tensile strength and modulus compared with S1 and S3 composites. Atiqah et al. [36] also mentioned that when the tensile load is applied to SPF/GF/TPU composites, GF stretching occurred. SEM images of the S3 hybrid composite revealed that SPF breakage began when analysing the flexural fracture surface, whereas the other two flexural fracture surfaces of S1 and S2 did not reveal any broken fibres. As a result, the flexural modulus of S3 hybrid composite was the highest among S1, S2, and S3 hybrid composites. SEM figures generally showed a good bonding between alkaline and BC treated fibres and PLA matrix compared with untreated SPF. S4, S5, and S6 (alkaline treated) hybrid composites possessed a rough surface than other hybrid composites that contributed to the enhancement of interfacial bonding between SPF and PLA matrix. The rough surface after alkaline treatment of SPF was because of the removal of hemicellulose, lignin, and waxy layer, where this part was visible on SEM images. Because of this reason, S4, S5, and S6 tensile and flexural fracture surfaces showed broken SPF, which indicated the increments in the tensile and flexural strengths after alkaline treatment. This phenomenon might be due to the rougher surface of SPF, the increased bonding strength between SPF and PLA matrix, and few visible voids on the surface. Breakage of the SPF indicated more energy dissipation when tensile or flexural loads were applied. SPF did not pull out, as proven by the breakage images of SPF in all three alkalines treated S4, S5, and S6 hybrid composites. Similar results were reported by Radzi et al. [20] for alkaline surface-treated roselle fibre/SPF hybrid composite that demonstrated better adhesion between treated fibre and matrix. The tensile and flexural properties of the hybrid composites were also enhanced. S7, S8, and S9 hybrid composites showed the big gaps/voids and low bonding strength between SPF and PLA matrix after BC treatment. SEM images also presented visible fibres break and dislocation, as well as voids/gaps, as aforementioned above. The weak interfacial adhesion caused the pull-outs in SPF. The fractured surface observed for S7, S8, and S9 (BC treated) hybrid composites showed broken/breakage SPF attributed to the increase of both flexural and tensile moduli of the hybrid composites. In previous studies, morphological analysis of BC treated fibre also reported that the wettability of treated fibre with matrix was increased [26,56].

Composite Specimens | **(a) Tensile Tests** | **(b) Flexural Tests**

S1

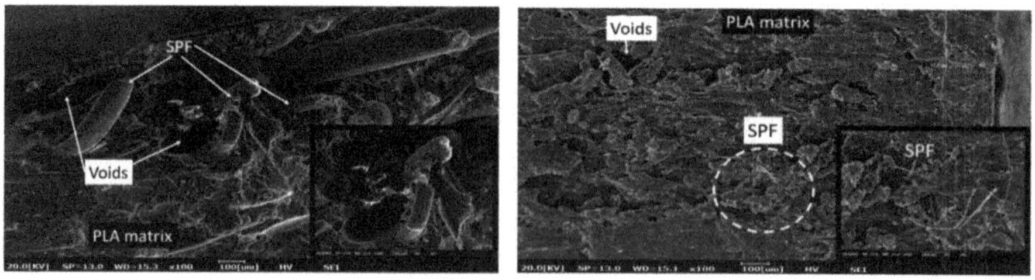

Figure 8. *Cont.*

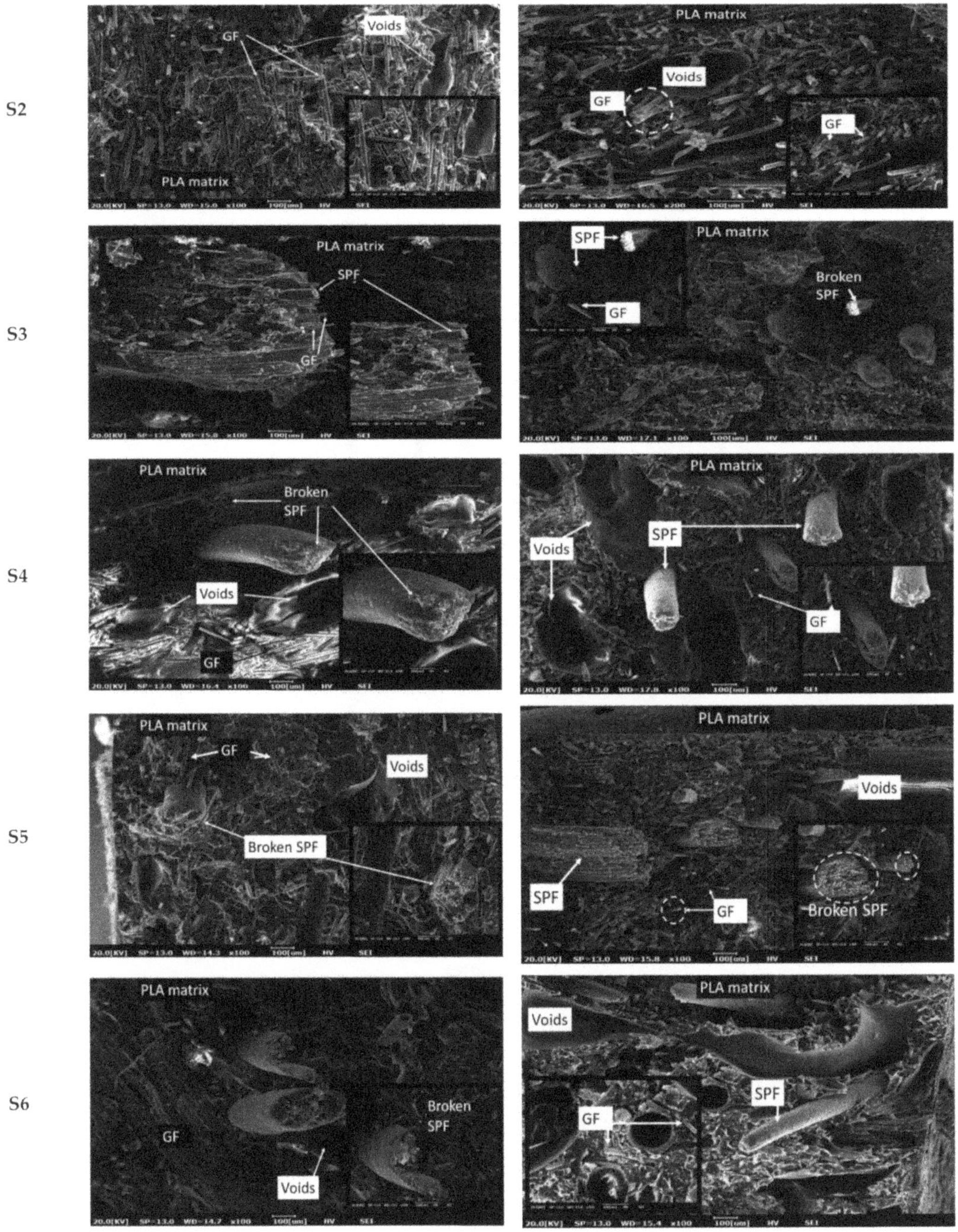

Figure 8. *Cont.*

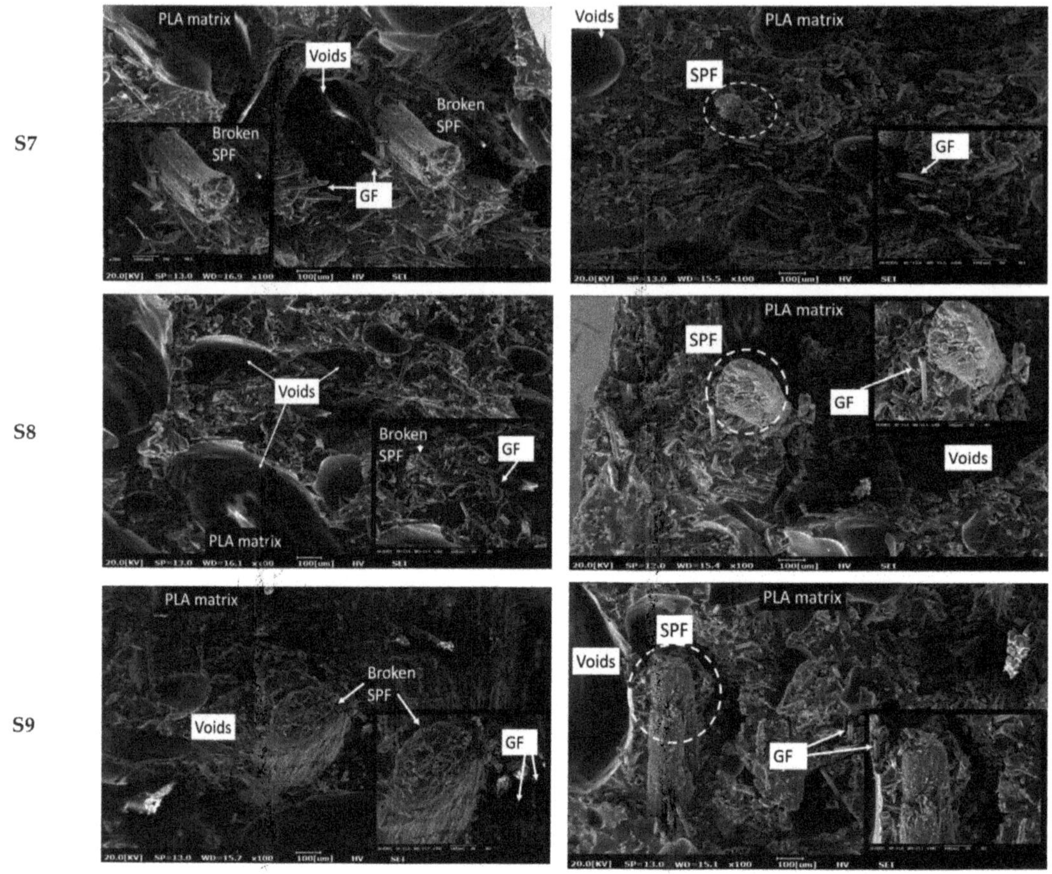

Figure 8. Morphological investigations for fractured surface analysis of (**a**) tensile and (**b**) flexural test. The SEM images clearly defined the presence of voids, breakage of SPF, GF, and PLA matrix.

SEM images revealed that composites of untreated SPF or BC treated had poor interfacial adhesion, as reported by fiber pull-outs and the presence of holes/voids/gaps. On the other side, alkaline treated S4, S5, and S6 hybrid composites were distinguished by fibers breakage that showed strong adhesion between SPF and PLA matrix. This morphological analysis showed that alkaline and BC treatment was able to modify the SPF surface for good adhesion, as reported in other studies [20,51]

4.8. Impact Testing

Table 5 shows tensile strength, tensile modulus, flexural strength, flexural modulus, and impact strength values of non-hybrid and hybrid SPF/GF/PLA composites. Impact strength is used to calculate the dissipation of total energy before ultimate fracture. Figure 9 shows the effect of treatment on the impact strength of non-hybrid and hybrid SPF/GF/PLA composites. Therefore, on hybridization, as wt % of GFs content was increased, the failure mechanism was GF fracture, not GF pull-out, due to the brittle nature of GF. Due to this reason, the composite can withstand a high-speed impact load at higher wt % GFs content. In general, the energy absorption mechanism was not in role, only the energy dissipated in frictional sliding of one fiber with the other due to the interaction of fibers. Moreover, impact strength was increased after hybridization, which increased the stress capabilities. The impact strength of S1 (untreated) composite was 2.09 kJ/m^2 that increased to 2.70 kJ/m^2 after the hybridization of GF for S3 hybrid composite. Non-hybrid S2 composite showed good impact strength of 3.07 kJ/m^2. S4 (alkaline treated SPF) hybrid composite exhibited the maximum impact strength value of 3.22 kJ/m^2 due to the removals of hemicellulose, lignin, and pectin, wax generation of moisture resistance, and the creation of rough fiber surface after alkaline treatment, which improved the adhesion between treated fiber and matrix. Comparing S4–S6 hybrid composites showed a decrease in impact strength as the wt % of glass content decreased. The impact strength directly depended upon the toughness of the entire composite. Fibers play a crucial role in impact resistance, where the combined effect of both fibers improved their impact strength. Uma Devi et al. [57] reported that as the percentage of GF increased, the impact strength increased for short pineapple fiber/GF/polyester hybrid composites.

Table 5. Tensile, flexural, and impact properties of non-hybrid and hybrid SPF/GF/PLA composites.

Composite Specimens	Tensile Strength (MPa)	Tensile Modulus (MPa)	Flexural Strength (MPa)	Flexural Modulus (MPa)	Impact Strength (kJ/m^2)	References
SP30	13.70	-	26.65	-	1.99	[40]
SA	17.83 ± 2.43	431 ± 43	32.34 ± 2.92	1209 ± 258	4.39	[41]
SB	7.01 ± 0.8	602 ± 7	6.07 ± 0.83	863 ± 106	2.39	[41]
S1	16.0 ± 2.64 ^{b,c}	433 ± 30 ^{a}	26.3 ± 3.85 ^{d}	1316 ± 175 ^{b}	2.09 ± 0.15 ^{a}	Current study
S2	23.7 ± 3.21 ^{d,e}	539 ± 14 ^{b,c}	24.7 ± 2.08 ^{d}	1564 ± 60 ^{c}	3.07 ± 0.2 ^{b}	Current study
S3	11.7 ± 3.51 ^{a,b}	423 ± 16 ^{a}	23.3 ± 3.51 ^{c}	1811 ± 80 ^{d}	2.70 ± 0.10 ^{b}	Current study
S4	18.7 ± 4.93 ^{c,d}	517 ± 7 ^{b,c}	18.7 ± 3.05 ^{b,c}	805 ± 100 ^{a}	3.22 ± 0.10 ^{b}	Current study
S5	14.0 ± 4.00 ^{a,b,c}	435 ± 15 ^{a}	27.3 ± 3.52 ^{d}	1336 ± 64 ^{b}	3.10 ± 0.14 ^{b}	Current study
S6	26.3 ± 2.51 ^{e}	561 ± 16 ^{c,d}	25.7 ± 3.05 ^{d}	1491 ± 45 ^{c}	2.79 ± 0.39 ^{b}	Current study
S7	11.0 ± 3.00 ^{a,b}	607 ± 60 ^{d}	16.7 ± 1.15 ^{b}	1842 ± 50 ^{d}	2.80 ± 0.25 ^{b}	Current study
S8	9.3 ± 1.52 ^{a}	505 ± 25 ^{b}	10.3 ± 2.51 ^{a}	1538 ± 35 ^{c}	2.78 ± 0.27 ^{b}	Current study
S9	10.3 ± 3.05 ^{b}	500 ± 20 ^{b}	8.7 ± 3.05 ^{a}	1434 ± 29 ^{b,c}	1.97 ± 0.22 ^{a}	Current study

* Values with different letters in the same column are significantly different ($p < 0.05$). SP30 defined for Untreated SPF (30%)/PLA (70%). SA defined for 6% conc. of NaOH treated SPF (30%)/PLA (70%). SB defined for 15 min soaking BC treated SPF (30%)/PLA (70%).

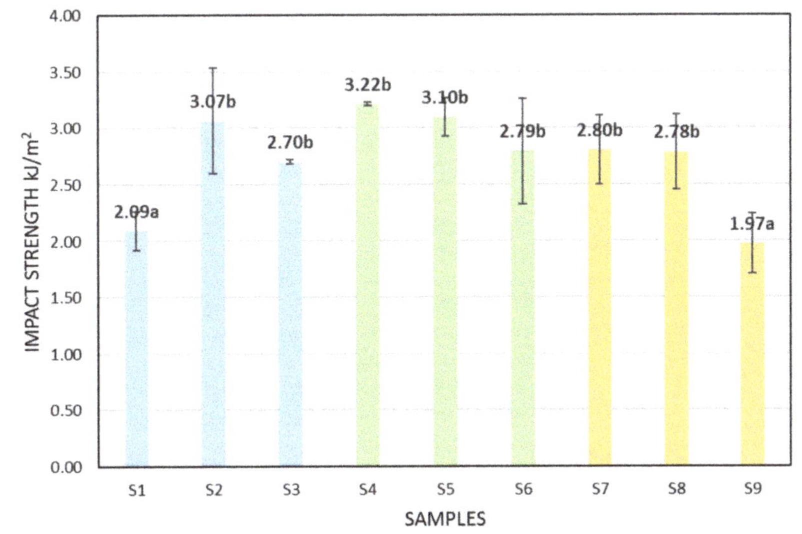

Figure 9. Impact strength versus samples of non-hybrid and hybrid SPF/GF/PLA composite. * Values with different letters in the figures are significantly different ($p < 0.05$).

Comparing the same wt % hybrid composites, i.e., S3, S5, and S8, S5 (alkaline treated) hybrid composite showed the highest impact value of 3.10 kJ/m^2 followed by 2.78 kJ/m^2 for S8 (BC treated) whereas S3 (untreated) exhibited the lowest impact strength value of 2.7 kJ/m^2. A similar improved impact strength value was reported by Atiqah et al. [3] after alkaline treatment of SPF for SPF/GF/TPU hybrid composites. After S4, the value of impact strength decreased as the SPF content was increased. With an increase in SPF content, the interspaces and stress concentration shoot up, which acted as crack propagation. The same trend of decreasing impact strength value with increasing fiber content was reported by Swain et al. [53] for date palm leaf/GF reinforced hybrid composite.

BC treated SPF also improved the impact strength from 2.70 kJ/m^2 for S3 (untreated) hybrid composite to 2.80 kJ/m^2 for S7 (BC treated) hybrid composite. This might be due to good interlocking between treated fiber and matrix, which allowed maximum energy absorption and stopped the crack propagation, enhancing the impact properties [58,59]. Thiruchitrambalam et al. [27] reported a 12% impact strength increment after BC treatment for palmyra palm leaf stalk fiber-polyester composites. Swain et al. [60] also revealed that after BC treatment of jute fiber, the impact strength increased for developing jute/epoxy composites. The lowest value of impact strength of 1.97 kJ/m^2 was shown by the S9 hybrid composite that might be due to the insufficient resistance to pull out fiber during impact fracture. Fracture of a matrix, fiber/matrix debonding, and fiber pull-out are three main causes for impact failure [60].

4.9. Fourier Transform Infrared (FTIR)

Fourier transform infrared (FTIR) spectroscopy is often used to verify the correct mixture of matrix-fiber ratio due to the interplay between components in polymer composites is complex [61–68]. Figure 10 shows the FTIR analysis for untreated, alkaline, BC treated of non-hybrid, and hybrid SPF/GF/PLA composites. From the figure, it is clear that all hybrid composite showed almost similar patterns. This analysis was used to determine the effect of alkaline and BC treatment on SPF and the chemical bonding nature between SPF, GF, and PLA. FTIR of untreated and treated SPF/GF/PLA composites revealed changes in the associated functional groups. Spectrum helps us to determine the presence of lignin, cellulose, and hemicellulose in (C-H rocking vibrations), 1180 cm^{-1} cellulose (C-O-C asymmetric valence vibration, 1316 cm^{-1} cellulose (C-H_2 rocking vibration), 1370 cm^{-1} cellulose (C-H_2 deformation vibration), 1424 cm^{-1} cellulose, 1227 cm^{-1} lignin (C-C plus C-O plus C=O stretch) [69,70].

In the range, 1300 cm^{-1} to 1160 cm^{-1} belonged to the C-C group (lignin in-ring stretch mode). The difference in peak heights of the untreated (S1, S3) and alkaline treated (S4–S6) hybrid composites was observed that resulted from the reduced amount of lignin and hemicelluloses in SPF after alkaline treatment. Significant changes at peaks 1180 cm^{-1} were observed, showing C-O in alkaline treated fiber concerning primary alcohol stretching, peak reduction compared to untreated fiber. Other studies have also confirmed this related disappearance of lignin and hemicelluloses after alkaline treatment that improved the adhesion between the fiber and the matrix [3,19,71]. The sharp peak at 1745 cm^{-1} for S1 composite was observed that was associated with the presence of hemicellulose C=O stretching vibration [71]. The intensity changes at the 1756 cm^{-1} peak showed the esterification among the —OH groups of SP fiber and -COOH terminal groups of PLA. A small peak in all nine samples at 2995 cm^{-1} was ascribed with the frequency of the O-H group [72,73]. This intensity was also decreased after treatment of SPF. After 6 wt % alkaline treatment of SPF, the —OH groups were substituted with –ONa groups. According to Bachtiar [71], at 4% alkaline treatment, the nature of SPF changes to hydrophilic since the cellulose I changed to Cellulose II.

The peak at 1450 cm^{-1} revealed C-C stretching in the aromatic ring, and the peak at 1719 cm^{-1} indicated the C=O stretching of the benzoyl carbonyl group in the benzoylated fiber. As the benzoyl group reacted with the —OH group of SPF, the hydrophilic character decreased by reducing the hydroxyl group, which was indicated at 2995 cm^{-1}. A similar FTIR result was also reported by Salisu et al. [58] after benzoylation of sisal fiber unsaturated polyester-reinforced composites.

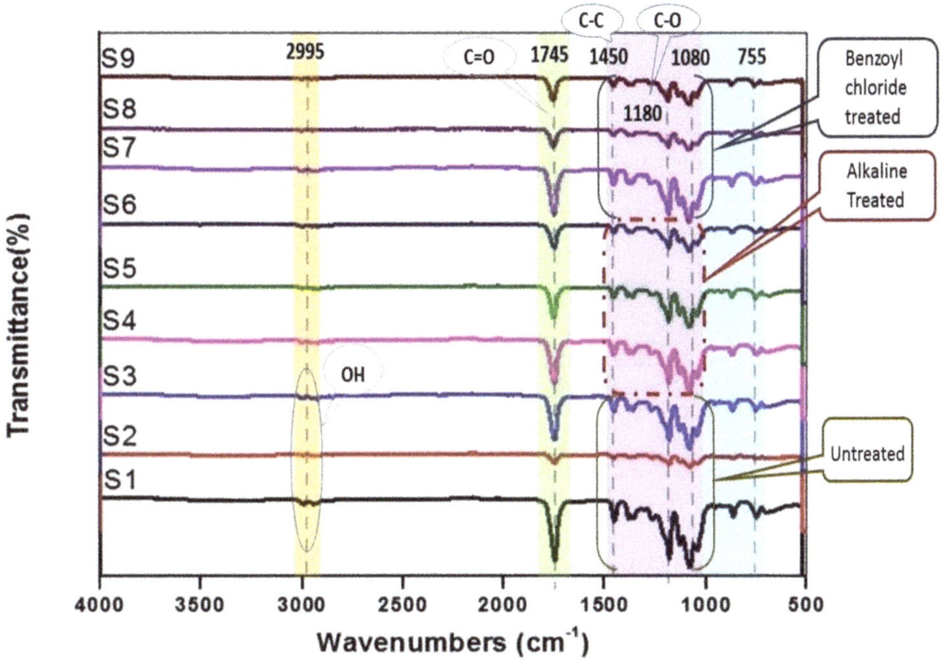

Figure 10. FTIR spectrum detailed data analysis of non-hybrid and hybrid SPF/GF/PLA composites.

5. Conclusions

The physical, mechanical, and morphological properties of treated SPF/GF reinforced PLA hybrid composites were investigated. This novel SPF/GF reinforced PLA hybrid composites exhibited lower densities after alkaline treatment of SPF, improved water absorption, and thickness swelling after both SPF treatments. It was observed that the incorporation of alkaline treated SPF/GF reinforced PLA matrix increased the tensile and flexural strengths. Both alkaline treated S6 and S5 hybrid composites exhibited the highest tensile strength of 26.3 MPa and flexural strength of 27.3 MPa. It was also found that BC treated SPF/GF reinforced PLA hybrid composites exhibited the highest tensile and flexural moduli. S7 hybrid composite recorded the highest tensile and flexural moduli of 607 MPa and 1847 MPa, respectively. Nevertheless, the incorporation of alkaline treated S4 hybrid composite showed the highest value of impact strength of 3.22 kJ/m^2. This value was reduced as the SPF content was increased. The morphological investigation revealed that alkaline treatment of SPF possessed better interfacial adhesion between SPF and PLA matrix. FTIR results also showed that after alkaline treatment, the adhesion between fiber and matrix was improved. This combination of alkaline and BC treated SPF/GF reinforced PLA hybrid composite resulted in good physical and mechanical properties. Therefore, this composite can be proposed for the fabrication of automotive components. Future research will primarily focus on replacing Acrylonitrile butadiene styrene (ABS) plastic motorcycle battery housing parts with these hybrid composites.

Author Contributions: Conceptualization, S.F.K.S., E.S.Z. and S.M.S.; methodology, S.M.S.; software, S.F.K.S.; validation, S.F.K.S. and E.S.Z.; formal analysis, S.F.K.S.; investigation, S.F.K.S. and S.M.S.; resources, S.F.K.S. and S.M.S.; data curation, S.F.K.S. and S.M.S.; writing—original draft preparation, S.F.K.S.; writing—review and editing, E.S.Z. and S.M.S.; visualization, S.F.K.S., E.S.Z. and S.M.S.; supervision, Z.L., A.K., E.S.Z. and S.M.S.; project administration, E.S.Z.; funding acquisition, E.S.Z. All authors have read and agreed to the published version of the manuscript.

Funding: This research was funded by Universiti Putra Malaysia for the financial support through the Grant Putra Berimpak UPM.RMC.800–3/3/1/GPB/2020/9694500 (vote number 9694500).

Institutional Review Board Statement: Not applicable.

Informed Consent Statement: Not applicable.

Data Availability Statement: The data that support the findings of this study are available from the corresponding author, upon reasonable request.

Acknowledgments: The authors are gratefully acknowledged to Universiti Putra Malaysia (UPM) for funding this research.

Conflicts of Interest: The authors declare no conflict of interest. The authors declare that they have no known competing financial interests or personal relationships that could have appeared to influence the work reported in this paper.

References

1. Sherwani, S.F.K.; Sapuan, S.M.; Leman, Z.; Zainuddin, E.S.; Ilyas, R.A. Chapter 15: Application of polymer composite materials in motorcycles: A comprehensive review. In *Biocomposite and Synthetic Composites for Automotive Applications*; Sapuan, S.M., Llyas, R.A., Eds.; Woodhead Publishing: Duxford, UK, 2021; pp. 401–426. [CrossRef]
2. Tarique, J.; Sapuan, S.M.; Khalina, A.; Sherwani, S.F.K.; Yusuf, J.; Ilyas, R.A. Recent developments in sustainable arrowroot (*Maranta arundinacea Linn*) starch biopolymers, fibres, biopolymer composites and their potential industrial applications: A review. *J. Mater. Res. Technol.* **2021**, *13*, 1191–1219. [CrossRef]
3. Atiqah, A.; Jawaid, M.; Sapuan, S.M.; Ishak, M.R. Effect of surface treatment on the mechanical properties of sugar palm/glass fibre-reinforced thermoplastic polyurethane hybrid composites. *BioResources* **2018**, *13*, 1174–1188. [CrossRef]
4. Hazrati, K.Z.; Sapuan, S.M.; Zuhri, M.Y.M.; Jumaidin, R. Extraction and characterization of potential biodegradable materials based on *Dioscorea hispida* tubers. *Polymers* **2021**, *13*, 584. [CrossRef]
5. Atiqah, A.; Maleque, M.A.; Jawaid, M.; Iqbal, M. Development of kenaf-glass reinforced unsaturated polyester hybrid composite for structural applications. *Compos. Part B Eng.* **2013**, *56*, 68–73. [CrossRef]
6. Nazrin, A.; Sapuan, S.M.; Zuhri, M.Y.M.; Ilyas, R.; Syafiq, R.; Sherwani, S.F.K. Nanocellulose reinforced Thermoplastic Starch (TPS), Polylactic Acid (PLA), and Polybutylene Succinate (PBS) for food packaging applications. *Front. Chem.* **2020**, *8*, 213. [CrossRef] [PubMed]
7. Siregar, J.; Zalinawati, M.; Cionita, T.; Rejab, M.; Mawarnie, I.; Jaafar, J.; Hamdan, M. Mechanical properties of hybrid sugar palm/ramie fibre reinforced epoxy composites. *Mater. Today Proc.* **2020**, *46*, 1729–1734. [CrossRef]
8. Mukhtar, I.; Leman, Z.; Ishak, M.R.; Zainudin, E.S. Sugar palm fibre and its composites: A review of recent developments. *BioResources* **2016**, *11*, 10756–10782. [CrossRef]
9. Ilyas, R.A.; Sapuan, S.M.; Ibrahim, R.; Abral, H.; Ishak, M.; Zainudin, E.; Asrofi, M.; Atikah, M.S.N.; Huzaifah, M.R.M.; Radzi, A.M.; et al. Sugar palm (*Arenga pinnata (Wurmb.) Merr*) cellulosic fibre hierarchy: A comprehensive approach from macro to nano scale. *J. Mater. Res. Technol.* **2019**, *8*, 2753–2766. [CrossRef]
10. Ilyas, R.; Sapuan, S.; Atikah, M.; Asyraf, M.; Rafiqah, S.A.; Aisyah, H.; Nurazzi, N.M.; Norrrahim, M. Effect of hydrolysis time on the morphological, physical, chemical, and thermal behavior of sugar palm nanocrystalline cellulose (*Arenga pinnata (Wurmb.) Merr*). *Text. Res. J.* **2021**, *91*, 152–167. [CrossRef]
11. Mukaffa, H.; Asrofi, M.; Sujito; Asnawi; Hermawan, Y.; Sumarji; Qoryah, R.D.H.; Sapuan, S.; Ilyas, R.; Atiqah, A. Effect of alkali treatment of piper betle fiber on tensile properties as biocomposite based polylactic acid: Solvent cast-film method. *Mater. Today Proc.* **2021**. [CrossRef]
12. Siva, R.; Valarmathi, T.; Samrot, A.V.; Jeevahan, J.J. Surface-modified and untreated *Cissus quadrangularis* reinforced polylactic composite. *Curr. Res. Green Sustain. Chem.* **2021**, *4*, 100121. [CrossRef]
13. Bendourou, F.E.; Suresh, G.; Laadila, M.A.; Kumar, P.; Rouissi, T.; Dhillon, G.S.; Zied, K.; Brar, S.K.; Galvez, R. Feasibility of the use of different types of enzymatically treated cellulosic fibres for polylactic acid (PLA) recycling. *Waste Manag.* **2020**, *121*, 237–247. [CrossRef] [PubMed]
14. Zhang, K.; Chen, Z.; Smith, L.M.; Hong, G.; Song, W.; Zhang, S. Polypyrrole-modified bamboo fiber/polylactic acid with enhanced mechanical, the antistatic properties and thermal stability. *Ind. Crop. Prod.* **2021**, *162*, 113227. [CrossRef]

15. Ali, W.; Mehboob, A.; Han, M.-G.; Chang, S.-H. Novel biodegradable hybrid composite of polylactic acid (PLA) matrix reinforced by bioactive glass (BG) fibres and magnesium (Mg) wires for orthopaedic application. *Compos. Struct.* **2020**, *245*, 112322. [CrossRef]
16. Rizvi, S.H.A.; Che, J.; Mehboob, A.; Zaheer, U.; Chang, S.-H. Experimental study on magnesium wire–polylactic acid biodegradable composite implants under in vitro material degradation and fatigue loading conditions. *Compos. Struct.* **2021**, *272*, 114267. [CrossRef]
17. Azlin, M.N.M.; Sapuan, S.M.; Zainudin, E.S.; Zuhri, M.Y.M.; Ilyas, R.A. Natural polylactic acid based fibre composites: A review. In *Advanced Processing, Properties and Applications of Starch and Other Bio-Based Polymers*; Faris, M., L-Oqla, M.A., Sapuan, S.M., Eds.; Elsevier Inc.: Amsterdam, The Netherlands, 2020; pp. 21–34. [CrossRef]
18. Mukhtar, I.; Leman, Z.; Zainudin, E.S.; Ishak, M.R. Hybrid and nonhybrid laminate composites of sugar palm and glass fibre-reinforced polypropylene: Effect of alkali and sodium bicarbonate treatments. *Int. J. Polym. Sci.* **2019**, *2019*, 1–12. [CrossRef]
19. Asim, M.; Jawaid, M.; Abdan, K.; Ishak, M.R. Effect of alkali and silane treatments on mechanical and fibre-matrix bond strength of kenaf and pineapple leaf fibres. *J. Bionic Eng.* **2016**, *13*, 426–435. [CrossRef]
20. Radzi, A.M.; Sapuan, S.M.; Jawaid, M.; Mansor, M.R. Effect of alkaline treatment on mechanical, physical and thermal properties of roselle/sugar palm fiber reinforced thermoplastic polyurethane hybrid composites. *Fibers Polym.* **2019**, *20*, 847–855. [CrossRef]
21. Shukor, F.; Hassan, A.; Hasan, M.; Islam, S.; Mokhtar, M. PLA/Kenaf/APP Biocomposites: Effect of alkali treatment and Ammonium Polyphosphate (APP) on dynamic mechanical and morphological properties. *Polym. Technol. Eng.* **2014**, *53*, 760–766. [CrossRef]
22. Nayak, S.; Mohanty, J. Erosion wear behavior of benzoyl chloride modified areca sheath fiber reinforced polymer composites. *Compos. Commun.* **2020**, *18*, 19–25. [CrossRef]
23. Siakeng, R.; Jawaid, M.; Ariffin, H.; Sapuan, S.M.; Asim, M.; Saba, N. Natural fiber reinforced polylactic acid composites: A review. *Polym. Compos.* **2018**, *40*, 446–463. [CrossRef]
24. Prabhu, L.; Krishnaraj, V.; Sathish, S.; Gokulkumar, S.; Karthi, N.; Rajeshkumar, L.; Balaji, D.; Vigneshkumar, N.; Elango, K. A review on natural fiber reinforced hybrid composites: Chemical treatments, manufacturing methods and potential applications. *Mater. Today Proc.* **2021**, *45*, 8080–8085. [CrossRef]
25. Kabir, M.M.; Wang, H.; Lau, K.T.; Cardona, F. Chemical treatments on plant-based natural fibre reinforced polymer composites: An overview. *Compos. Part B Eng.* **2012**, *43*, 2883–2892. [CrossRef]
26. Vijay, R.; Manoharan, S.; A, V.; Singaravelu, D.L.; Sanjay, M.R.; Siengchin, S. Characterization of raw and benzoyl chloride treated Impomea pes-caprae fibers and its epoxy composites. *Mater. Res. Express* **2019**, *6*, 095307. [CrossRef]
27. Thiruchitrambalam, M.; Shanmugam, D. Influence of pre-treatments on the mechanical properties of palmyra palm leaf stalk fiber–polyester composites. *J. Reinf. Plast. Compos.* **2012**, *31*, 1400–1414. [CrossRef]
28. Safri, S.N.; Sultan, M.T.; Saba, N.; Jawaid, M. Effect of benzoyl treatment on flexural and compressive properties of sugar palm/glass fibres/epoxy hybrid composites. *Polym. Test.* **2018**, *71*, 362–369. [CrossRef]
29. Izwan, S.M.; Sapuan, S.; Zuhri, M.; Mohamed, A. Effects of benzoyl treatment on NaOH treated sugar palm fiber: Tensile, thermal, and morphological properties. *J. Mater. Res. Technol.* **2020**, *9*, 5805–5814. [CrossRef]
30. Mishra, S.; Mohanty, A.; Drzal, L.; Misra, M.; Parija, S.; Nayak, S.; Tripathy, S. Studies on mechanical performance of biofibre/glass reinforced polyester hybrid composites. *Compos. Sci. Technol.* **2003**, *63*, 1377–1385. [CrossRef]
31. Darshan, S.; Suresha, B. Effect of basalt fiber hybridization on mechanical properties of silk fiber reinforced epoxy composites. *Mater. Today Proc.* **2020**, *43*, 986–994. [CrossRef]
32. Waghmare, P.M.; Bedmutha, P.G.; Sollapur, S.B. Investigation of effect of hybridization and layering patterns on mechanical properties of banana and kenaf fibers reinforced epoxy biocomposite. *Mater. Today Proc.* **2021**, 1–5. [CrossRef]
33. Venkatarajan, S.; Subbu, C.; Athijayamani, A.; Muthuraja, R. Mechanical properties of natural cellulose fibers reinforced polymer composites—2015–2020: A review. *Mater. Today Proc.* **2021**, *47*, 1017–1024. [CrossRef]
34. Nurazzi, N.M.; Khalina, A.; Sapuan, S.; Ilyas, R.; Rafiqah, S.A.; Hanafee, Z. Thermal properties of treated sugar palm yarn/glass fiber reinforced unsaturated polyester hybrid composites. *J. Mater. Res. Technol.* **2019**, *9*, 1606–1618. [CrossRef]
35. Afdzaluddin, A.; Maleque, A.; Iqbal, M. Synergistic effect on flexural properties of kenaf-glass hybrid composite. *Adv. Mater. Res.* **2013**, *626*, 989–992. [CrossRef]
36. Afzaluddin, A.; Jawaid, M.; Salit, M.S.; Ishak, M.R. Physical and mechanical properties of sugar palm/glass fiber reinforced thermoplastic polyurethane hybrid composites. *J. Mater. Res. Technol.* **2019**, *8*, 950–959. [CrossRef]
37. Atiqah, A.; Jawaid, M.; Sapuan, S.M.; Ishak, M.R. Mechanical and thermal properties of sugar palm fiber reinforced thermoplastic polyurethane composites: Effect of silane treatment and fiber loading. *J. Renew. Mater.* **2017**, 1–16. [CrossRef]
38. Atiqah, A.; Jawaid, M.; Sapuan, S.; Ishak, M.; Ansari, M.; Ilyas, R. Physical and thermal properties of treated sugar palm/glass fibre reinforced thermoplastic polyurethane hybrid composites. *J. Mater. Res. Technol.* **2019**, *8*, 3726–3732. [CrossRef]
39. Sherwani, S.F.K.; Sapuan, S.M.; Leman, Z.; Zainudin, E.S.; Khalina, A. Physical, mechanical and morphological properties of sugar palm fiber reinforced polylactic acid composites. *Fibers Polym.* **2021**, 1–11. [CrossRef]
40. Sherwani, S.F.K.; Sapuan, S.; Leman, Z.; Zainudin, E.; Khalina, A. Effect of alkaline and benzoyl chloride treatments on the mechanical and morphological properties of sugar palm fiber-reinforced poly(lactic acid) composites. *Text. Res. J.* **2021**. [CrossRef]
41. Ilyas, R.; Sapuan, S.; Ishak, M.; Zainudin, E.S. Sugar palm nanofibrillated cellulose (*Arenga pinnata (Wurmb.) Merr*): Effect of cycles on their yield, physic-chemical, morphological and thermal behavior. *Int. J. Biol. Macromol.* **2019**, *123*, 379–388. [CrossRef]

42. Jones, F.; Huff, N. Structure and properties of glass fibres. In *Handbook of Tensile Properties of Textile and Technical Fibres*; Chapter 15; Bunsell, A.R., Ed.; Woodhead Publishing Ltd., CRC Press: Washington, DC, USA, 2009; pp. 529–573. [CrossRef]
43. Ilyas, R.; Sapuan, S.; Ishak, M. Isolation and characterization of nanocrystalline cellulose from sugar palm fibres (*Arenga Pinnata*). *Carbohydr. Polym.* **2018**, *181*, 1038–1051. [CrossRef]
44. Ray, K.; Patra, H.; Swain, A.K.; Parida, B.; Mahapatra, S.; Sahu, A.; Rana, S. Glass/jute/sisal fiber reinforced hybrid polypropylene polymer composites: Fabrication and analysis of mechanical and water absorption properties. *Mater. Today Proc.* **2020**, *33*, 5273–5278. [CrossRef]
45. ASTM International. *ASTM. D792—13: Standard Test Methods for Density and Specific Gravity (Relative Density) of Plastics by Displacement*; ASTM International: West Conshohocken, PA, USA, 2013; Volume 15, pp. 145–149. [CrossRef]
46. ASTM International. *ASTM D570 Standard Test Method for Water Absorption of Plastics*; ASTM International: West Conshohocken, PA, USA, 2014; Volume 98, pp. 25–28. [CrossRef]
47. ASTM International. *ASTM Standard D638. Standard Test Method for Tensile Properties of Plastics*; ASTM International: West Conshohocken, PA, USA, 2010. [CrossRef]
48. ASTM International. *ASTM. D790-03. Standard Test Method for Flexural Properties of Unreinforced and Reinforced Plastics and Electrical Insulation Materials*; ASTM International: West Conshohocken, PA, USA, 2015; Volume 08.01, pp. 1–11. [CrossRef]
49. ASTM International. *ASTM Standard D256. Determining the Izod Pendulum Impact Resistance of Plastics*; ASTM International: West Conshohocken, PA, USA, 2010; pp. 1–20. [CrossRef]
50. Merlini, C.; Soldi, V.; Barra, G.M.D.O. Influence of fiber surface treatment and length on physico-chemical properties of short random banana fiber-reinforced castor oil polyurethane composites. *Polym. Test.* **2011**, *30*, 833–840. [CrossRef]
51. Safri, S.N.A.; Sultan, M.T.H.; Shah, A.U.M. Characterization of benzoyl treated sugar palm/glass fibre hybrid composites. *J. Mater. Res. Technol.* **2020**, *9*, 11563–11573. [CrossRef]
52. Radzi, A.; Sapuan, S.; Jawaid, M.; Mansor, M. Water absorption, thickness swelling and thermal properties of roselle/sugar palm fibre reinforced thermoplastic polyurethane hybrid composites. *J. Mater. Res. Technol.* **2019**, *8*, 3988–3994. [CrossRef]
53. Hafidz, N.B.M.; Rehan, M.B.M.; Mokhtar, H.B. Effect of alkaline treatment on water absorption and thickness swelling of natural fibre reinforced unsaturated polyester composites. *Mater. Today Proc.* **2021**. [CrossRef]
54. Swain, P.T.R.; Das, S.N.; Jena, S.P. Manufacturing and study of thermo-mechanical behaviour of surface modified date palm leaf/glass fiber reinforced hybrid composite. *Mater. Today Proc.* **2018**, *5*, 18332–18341. [CrossRef]
55. Majid, R.; Ismail, H.; Taib, R. Benzoyl chloride treatment of kenaf core powder: The effects on mechanical and morphological properties of PVC/ENR/kenaf core powder composites. *Procedia Chem.* **2016**, *19*, 803–809. [CrossRef]
56. Majid, R.A.; Ismail, H.; Taib, R.M. Processing, tensile, and thermal studies of poly(Vinyl Chloride)/epoxidized natural rubber/kenaf core powder composites with benzoyl chloride treatment. *Polym. Technol. Eng.* **2018**, *57*, 1507–1517. [CrossRef]
57. Devi, L.U.; Bhagawan, S.; Thomas, S. Polyester composites of short pineapple fiber and glass fiber: Tensile and impact properties. *Polym. Compos.* **2012**, *16*, 1–6. [CrossRef]
58. Salisu, A.A.; Musa, H.; Yakasai, M.Y.; Aujara, K.M. Effects of chemical surface treatment on mechanical properties of sisal fibre unsaturated polyester reinforced composites. *ChemSearch J.* **2015**, *6*, 8–13.
59. Sherwani, S.F.K.; Zainudin, E.S.; Sapuan, S.M.; Leman, Z.; Abdan, K. Mechanical properties of sugar palm (*Arenga pinnata Wurmb. Merr*)/glass fiber-reinforced poly(lactic acid) hybrid composites for potential use in motorcycle components. *Polymers* **2021**, *13*, 3061.
60. Swain, P.T.R.; Biswas, S. Influence of fiber surface treatments on physico-mechanical behaviour of jute/epoxy composites impregnated with aluminium oxide filler. *J. Compos. Mater.* **2017**, *51*, 3909–3922. [CrossRef]
61. Ilyas, R.; Sapuan, S.; Asyraf, M.; Dayana, D.; Amelia, J.; Rani, M.; Norrrahim, M.; Nurazzi, N.; Aisyah, H.; Sharma, S.; et al. Polymer composites filled with metal derivatives: A review of flame retardants. *Polymers* **2021**, *13*, 1701. [CrossRef]
62. Ilyas, R.; Sapuan, S.; Harussani, M.; Hakimi, M.; Haziq, M.; Atikah, M.; Asyraf, M.; Ishak, M.; Razman, M.; Nurazzi, N.; et al. Polylactic Acid (PLA) biocomposite: Processing, additive manufacturing and advanced applications. *Polymers* **2021**, *13*, 1326. [CrossRef]
63. Omran, A.A.B.; Mohammed, A.A.B.A.; Sapuan, S.M.; Ilyas, R.A.; Asyraf, M.R.M.; Koloor, S.S.R.; Petrů, M. Micro- and nanocellulose in polymer composite materials: A Review. *Polymers* **2021**, *13*, 231. [CrossRef] [PubMed]
64. Nurazzi, N.; Asyraf, M.; Khalina, A.; Abdullah, N.; Aisyah, H.; Rafiqah, S.; Sabaruddin, F.; Kamarudin, S.; Norrrahim, M.; Ilyas, R.; et al. A review on natural fiber reinforced polymer composite for bullet proof and ballistic applications. *Polymers* **2021**, *13*, 646. [CrossRef]
65. Nurazzi, N.M.; Asyraf, M.; Khalina, A.; Abdullah, N.; Sabaruddin, F.; Kamarudin, S.; Ahmad, S.; Mahat, A.; Lee, C.; Aisyah, H.; et al. Fabrication, functionalization, and application of carbon nanotube-reinforced polymer composite: An overview. *Polymers* **2021**, *13*, 1047. [CrossRef]
66. Suriani, M.; Rapi, H.; Ilyas, R.; Petrů, M.; Sapuan, S. Delamination and manufacturing defects in natural fiber-reinforced hybrid composite: A review. *Polymers* **2021**, *13*, 1323. [CrossRef]
67. Jumaidin, R.; Diah, N.; Ilyas, R.; Alamjuri, R.; Yusof, F. Processing and characterisation of banana leaf fibre reinforced thermoplastic cassava starch composites. *Polymers* **2021**, *13*, 1420. [CrossRef] [PubMed]

68. Suriani, M.; Radzi, F.; Ilyas, R.; Petrů, M.; Sapuan, S.; Ruzaidi, C. Flammability, tensile, and morphological properties of oil palm empty fruit bunches fiber/pet yarn-reinforced epoxy fire retardant hybrid polymer composites. *Polymers* **2021**, *13*, 1282. [CrossRef] [PubMed]
69. Ilyas, R.; Sapuan, S.M.; Ishak, M.R.; Zainudin, E.S. Effect of delignification on the physical, thermal, chemical, and structural properties of sugar palm fibre. *BioResources* **2017**, *12*, 8734–8754. [CrossRef]
70. Hazrati, K.; Sapuan, S.; Zuhri, M.; Jumaidin, R. Effect of plasticizers on physical, thermal, and tensile properties of thermoplastic films based on *Dioscorea hispida* starch. *Int. J. Biol. Macromol.* **2021**, *185*, 219–228. [CrossRef] [PubMed]
71. Bachtiar, D.; Salit, M.S.; Zainuddin, E.; Abdan, K.; Dahlan, K.Z.H.M. Effects of alkaline treatment and a compatibilizing agent on tensile properties of sugar palm fibrereinforced high impact polystyrene composites. *BioResources* **2011**, *6*, 4815–4823. [CrossRef]
72. Bakar, N.H.; Hyie, K.M.; Jumahat, A.; Kalam, A.; Salleh, Z.; Aidah, J. Effect of alkaline treatment on tensile and impact strength of kenaf/kevlar hybrid composites. *Appl. Mech. Mater.* **2015**, *763*, 3–8. [CrossRef]
73. Syafiq, R.M.O.; Sapuan, S.M.; Zuhri, M.R.M. Effect of cinnamon essential oil on morphological, flammability and thermal properties of nanocellulose fibre–reinforced starch biopolymer composites. *Nanotechnol. Rev.* **2020**, *9*, 1147–1159. [CrossRef]

polymers

MDPI

Article

Interfacial Adhesion and Mechanical Properties of Wood-Polymer Hybrid Composites Prepared by Injection Molding

Alexander Stadlmann [1,*], Andreas Mautner [2], Maximilian Pramreiter [1], Alexander Bismarck [2,3] and Ulrich Müller [1]

1 Department of Material Science and Process Engineering, Institute of Wood Technology and Renewable Materials, University of Natural Resources and Life Sciences Vienna, Austria (BOKU), Konrad Lorenz Strasse 24, 3430 Tulln an der Donau, Austria; maximilian.pramreiter@boku.ac.at (M.P.); ulrich.mueller@boku.ac.at (U.M.)

2 Polymer and Composite Engineering (PaCE) Group, Faculty of Chemistry, Institute of Material Chemistry and Research, University of Vienna, Währinger Strasse 42, 1090 Vienna, Austria; andreas.mautner@univie.ac.at (A.M.); alexander.bismarck@univie.ac.at (A.B.)

3 Department of Chemical Engineering, South Kensington Campus, Imperial College London, London SW7 2AZ, UK

* Correspondence: alexander.stadlmann@boku.ac.at; Tel.: +43-1-47654-89151

Citation: Stadlmann, A.; Mautner, A.; Pramreiter, M.; Bismarck, A.; Müller, U. Interfacial Adhesion and Mechanical Properties of Wood-Polymer Hybrid Composites Prepared by Injection Molding. *Polymers* **2021**, *13*, 2849. https://doi.org/10.3390/polym13172849

Academic Editors: Domenico Acierno and Antonella Patti

Received: 19 July 2021
Accepted: 16 August 2021
Published: 25 August 2021

Publisher's Note: MDPI stays neutral with regard to jurisdictional claims in published maps and institutional affiliations.

Abstract: Birch (*Betula pendula* Roth.) and beech (*Fagus sylvatica* L.) solid wood and plywood were overmolded with polyamide 6 (PA 6) and polypropylene (PP) to investigate their mechanical properties and interfacial adhesion. In the case of PA 6, maximum tensile shear strengths values of more than 8 to 9 MPa were obtained for birch and beech, respectively. The values are comparable to bond strengths of commercial joints bonded with formaldehyde-containing amino-plastics. Perpendicular to the wood elements, bond strength values of 3 MPa was achieved for PA 6. The penetration of the polymers into the wood structure results in a non-densified interphase and subsequent plastic deformation of the wood structure beyond the interphase. These compressed areas influenced the interfacial adhesion and mechanical interlocking. SEM and XPS analysis revealed different interpenetration behavior of the polymers into the wood structure, with chemical interaction confirmed only for wood and PA 6 but not PP.

Keywords: interfacial bond strength; wood-polymer composites; wood-polymer interface; XPS

1. Introduction

The mobility sector of the European Union is responsible for 26% of its total CO_2 emissions. Around one fourth of these emissions are driven by the weight of the vehicle [1], increasing the importance of lightweight materials for automotive parts like wood [2]. Manufacturing technologies such as milling, cutting, gluing, molding, etc. for the production of wood-based products are well studied and established in the wood industry. In order to introduce wood and wood-based materials in these new areas of application such as the automotive industry, it is necessary to consider new production technologies during the design of wood-based hybrid components. Mair-Bauernfeind et al. [3] investigated the sustainability of wood and wood-based materials compared to other materials such as steel, where wood showed environmental, economic and social advantages. In addition, wood has also been increasingly used in multi-story buildings in the form of wood-concrete hybrid construction for several years. As claimed by Franzini et al. [4], the bio-based material wood also offers better indoor air quality, lower carbon dioxide emissions and competitive costs compared to concrete. Due to its sustainable nature and comparably low density, in recent years there has been an increasing demand for the utilization of wood and renewable materials in the mobility sector [5–9]. Besides utilizing sustainable and renewable products, formaldehyde-free bonding and joining of wood get more and more importance. The total amount of adhesive in plywood production can reach levels up to 20% for continuous bond lines that require high loading [10,11]. Kohl et al. [7] presented the environmental impact of urea-formaldehyde bonded beech plywood. Commonly, manufactured structural components in this field are made from steel, aluminum, magnesium, polymers or polymer composites by means of pressing, deep drawing, casting and molding [12]. Due to their low resistance against various media—e.g., salts—metals are usually coated using synthetic coating systems [13]. In contrast to metals, components made of polymers that are stabilized with suitable additives exhibit better durability and weathering behavior [14]. However, low mechanical properties and poor creep behavior of polymer-based components without fiber reinforcement (e.g., dashboards, claddings or wheel cases) negatively affect the applicability for load-bearing components in automotive parts. In addition to the properties of wood in terms of load-bearing capacity and lower density compared to most polymers, wood also provides good resistance to accelerated weathering in salty environment [15]. Therefore, back-injection molding or overmolding is a suitable technique for combining the good properties of wood with the durability, weather resistance and elasticity of polymers [16] to create wood-polymer hybrid composites. In addition, functional parts, such as brackets or mounting aids, can be easily fabricated by injection molding.

Wood as a bio-based reinforcement in polymers, i.e., in wood-polymer composites (WPCs), is well studied and widely used in applications such as furniture, decking, automotive and building components [17,18]. The mechanical interlocking and the mechanical adhesion as well as the effect of different wood species on the wood-polymer interaction of WPCs have already been investigated [19,20]. Gacitua et al. [19] observed that molten polymer (high density polyethylene, HD-PE) penetrates into the wood micro-structure resulting in a mechanical interaction between polymer and wood. Furthermore, the viscosity of the polymer melt also influences the penetration behavior [20]. Further research was carried out by Sretenovic et al. [21] to better understand the micro-mechanical behavior of wood plastic composites (WPC), demonstrating stress transfer from the wood to an LDPE plastic matrix caused by mechanical interlocking.

The modification of wood fibers for WPC production with various coupling agents aiming to improve interfacial adhesion, thus increasing strength and impact properties to a large extent, is well studied. Keener et al. [22] investigated the interaction of different coupling agents, i.e., maleic anhydride, polyolefins and peroxides in agrofiber polypropylene (PP) and polyethylene (PE) composites. In PE composites those coupling agents triple the impact bending strength and double the tensile strength, whereas the strength of PP composites increased by more than 60%. Correa et al. [23] aimed to improve the adhesion of wood-flour PP composites using maleated coupling agents and observed an increased interfacial adhesion between the matrix and fibers, which led to an improved load transfer and thus increased mechanical properties.

Polymers, used as adhesive in plywood fabrication have been investigated by Fang et al. [24] and Chang et al. [25] using HD-PE to bond poplar veneers by hot-pressing. The influence of the moisture content (MC) of the veneers, pressing temperature and pressure as well the quantity of the HDPE films on the mechanical and physical properties was investigated and compared with conventional urea-formaldehyde adhesive bonded plywood. They found that the MC of the veneers affected HDPE penetration. With increasing MC, the penetration depth of the polymer melt into the vessels (pores) of the wood structure decreased, which resulted in lower mechanical properties. Increased pressing temperature and pressure increased the bond strength as more polymer melt did penetrate into the vessels, thereby improving the mechanical properties. Furthermore, the dimensional stability could be improved when using HDPE of a higher quality. Surface modification using silane agents to improve the performance of wood-polymer plywood was also investigated by Fang et al. [26], Liu et al. [27] and Bekhta et al. [28], which results in a significant increasing tensile shear strength and a reduction of water uptake and lead to an improved dimension stability. Regarding particleboard production, there are several formaldehyde-free synthetic and renewable adhesive systems available, with the main drawback being availability and higher costs [29]. Overall, the mechanical performance of pure wood adhesives was also well discussed by Stoeckl et al. [30], where a wide range in stiffness was found. Briefly, the commonly used adhesives in engineered wood products and wood composites [29–32], as well as so called wood-plastic composites (i.e., extrusion and injection molding of wood fibers and particles with different kinds of plastics) [33], and improvements with various coupling agents is well investigated so far. However, almost no research was found on wood-polymer hybrid composites produced by means of injection molding. However, the realization of directly overmolded wood will help to reduce production time and costs, number of production steps, formaldehyde emission and carbon footprint.

The present study aims to investigate mechanical properties and in particular the interfacial adhesion of wood-polymer hybrid composites prepared by injection molding. A frequently used wood-based material for non-structural automotive is plywood. Therefore, birch (*Betula pendula* Roth.) and beech (*Fagus sylvatica* L.) plywood boards were overmolded with PP at varying injection temperatures to investigate the effect of the injection temperature on the mechanical properties. A novel test setup was established to evaluate the tensile shear strength, the tensile strength perpendicular to the plane of the board and the tensile strength perpendicular to the edge. In addition to overmolded plywood, the tensile shear strength of birch and beech solid wood specimens overmolded with PP and polyamide 6 (PA 6) was investigated. To investigate the penetration depth of the polymer-melt into the wood micro-structure and the polymer wood adhesion, X-ray photoelectron spectroscopy (XPS) was performed. Furthermore, the interphase between the wood and the polymer was investigated by means of scanning electron microscopy (SEM). The main research questions of this study were as follows:

Q1. Does the polymer melt penetrate into the wood structure and what determines the adhesion between the polymer and wood, including chemical bonds?

Q2. Is the interfacial adhesion of an overmolded wood-polymer hybrid composites as strong as bonded wood products with commercial adhesives?

2. Materials and Methods

2.1. Materials

Industrially manufactured birch and beech plywood, with a thickness of 10 mm, and solid wood were sourced from Frischeis GmbH (Stockerau, Austria). Polypropylene (Daplen KSR 4525) and polyamide 6 (Grilon BZ 3) were provided by Borealis (Vienna, Austria) and EMS-Chemie AG (Domat, Switzerland), respectively.

2.2. Plywood Composites

Sixteen boards of each species, birch and beech, respectively, with a dimension of 297 × 146 mm were prepared using a circular saw and overmolded with PP as shown in Figure 1a. Injection molding was performed with an injection molding machine (Engel ES 1350/200 HL-V, Schwertberg, Austria) with a screw diameter of 70 mm. Each species was overmolded at three different injection temperatures to investigate the effect of the injection temperature on the interfacial adhesion and the mechanical properties of the wood-polymer composites. For this, cylinder temperatures were set to an average value of 220 °C, 240 °C and 260 °C, respectively. The volumetric flow rate was 15 cm^3/s at an injection pressure of 170 bar, injection time of 2 s and the cycle time was 80 s. The way point of the feed screw was set to 35 mm, which corresponds to a changeover point at 95 cm^3 after an injection time of 10 s and a holding pressure of 40 bar.

In order to investigate mechanical properties under three different load conditions, specimens were cut using a circular saw (Figure 1a–d). For tensile tests perpendicular to the plane of the board (σ_P), specimens with a dimension of 50 × 20 mm (Figure 1b), and for tensile tests perpendicular to the edge (σ_E) specimens with a dimension of 120 × 20 × 10 mm were produced (Figure 1c), with one specimen consisting of 2 individual parts. These two parts were welded together with a welding mirror. For this purpose the temperature of the welding mirror was set to 200 °C and the welding time was 2 s. Specimens for tensile shear (σ_S) tests, performed according to DIN EN 302-1, had a dimension of 120 × 20 × 4 mm, with the overlap length of the overmolded areas being 10 mm [34] (Figure 1d). In total 121 birch and 118 beech samples were prepared for this study.

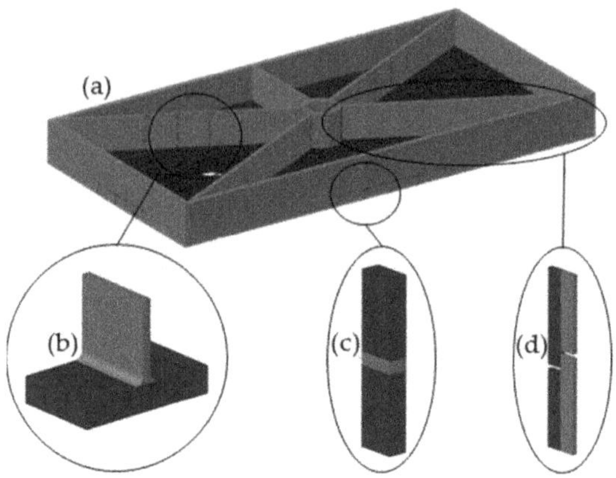

Figure 1. Schematic of the overmolded plywood boards (**a**) and schematic representation of the specimens for the tensile strength perpendicular to the plane of the board σ_P (**b**), tensile strength perpendicular to the edge σ_E (**c**) and tensile shear strength σ_S (**d**).

2.3. *Solid Wood Composites*

Forty board specimens were cut with a dimension of 140 × 140 mm from each solid wood species, planed to a thickness of 4 mm and overmolded with PP or PA 6, respectively, to a final thickness of 8 mm (Figure 2a). Injection molding was performed with an injection molding machine (Wittmann Battendfeld Smart Power 120/750 B 8) having a screw diameter of 70 mm.

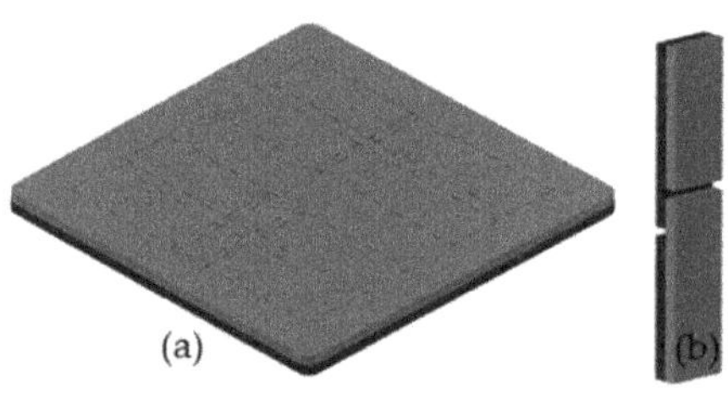

Figure 2. Schematic of the overmolded solid wood (**a**) and representation of the specimens for the tensile shear strength σ_S (**b**) according to ÖNORM EN 302-1 [34].

For PA 6, the cylinder temperature was set to an average value of 260 °C. The volumetric flow rate was 40 cm^3/s at an injection pressure of 360 bar, the injection time was 2 s and the cycle time was 64 s. The changeover point was set to 12 cm^3, with a holding pressure and time of 100 bar and 2 s, respectively.

For PP, the cylinder temperature was set to an average value of 260 °C, the volumetric flow rate to 15 cm^3/s, the injection pressure was 170 bar, the cycle time 66 s and injection time 5 s. The changeover point was set at 12 cm^3, with a holding pressure and time of 100 bar and 1 s, respectively. In total 108 specimens with a dimension of 110 × 20 × 8 mm were produced using a circular saw (Figure 2b). In total 50 birch and 58 beech samples were prepared.

2.4. Mechanical Properties of Wood-Polymer Composites

Mechanical tests were performed using a universal testing machine (Zwick/Roell Z20, Ulm, Germany). Prior to mechanical tests, all samples were stored under standard climate conditions (20 °C ± 2 °C, 65% ± 5% relative humidity) according to standard ISO 554 [35] until constant mass was reached. All tests were stopped after a 30% load reduction of the maximum force (F_{max}) was reached or failure occurred within 90 ± 30 s.

For plywood composites tensile tests perpendicular to the plane of the board (σ_P) were performed with clamps originally designed for testing internal bond strength of particle and fiber boards, [36] which were used to attach the wooden part to the testing machine as shown in Figure 3a. A pre-force of 10 N was applied before testing at a constant crosshead speed of 1 mm/min. σ_P was calculated according to DIN 52 188 [37], by dividing F_{max} through the calculated interface area. σ_E was determined following DIN 52 188 [37], depicted in Figure 3b. After a pre-force of 10 N was applied, the specimens were loaded at a constant speed of 0.3 mm/min. σ_S was determined following DIN EN 302-1 [34] (Figure 3c) with an applied pre-force of 20 N at a constant speed of 0.4 mm/min.

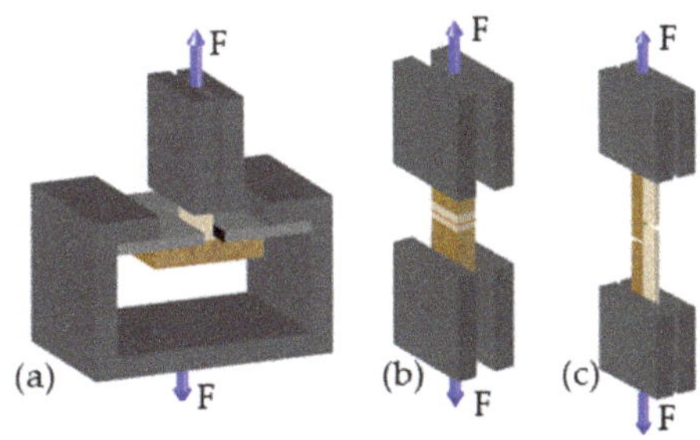

Figure 3. Schematic of the test setup for overmolded plywood specimens for strength measurements according to DIN 52 188 [37] and DIN EN 302-1 [34]: (**a**) test setup for tensile strength perpendicular to the plane of the board σ_P, (**b**) test setup for tensile strength perpendicular to the edge σ_E, and (**c**) test setup for tensile shear strength σ_S.

For solid wood composites, the tensile shear strength (σ_S) of birch and beech wood composites was assessed according to DIN EN 302-2 [34]. The samples were loaded with a pre-force of 10 N and tested at a constant crosshead speed of 0.6 mm/min.

2.5. Scanning Electron Microscopy (SEM) and X-ray Photoelectron Spectroscopy (XPS)

To investigate the penetration of the polymer into the wood structure on a microscopic level, two samples per combination were analyzed by means of SEM (Hitachi TM3030, Tokyo, Japan). To analyze the interphase of a cross-section of the overmolded samples, specimens with a dimension of about 3 × 8 mm were cut using a double-bladed circular saw. To obtain a smooth surface without any disturbing artefacts, the area of interest was cut with a razor blade.

XPS spectra were recorded to determine the penetration depth of the polymer into the microstructure of wood as well as chemical interactions between polymer and wood and to gain a deeper understanding into interfacial adhesion. Six solid wood specimens of each combination having a cross section of 8 × 4 mm were cut. The analysis was performed using an XPS system (Nexsa, Thermo-Scientific, Waltham, MA, USA) using an Al K_α radiation source operating at 72 W and an integrated flood gun. A pass energy of 200 eV, "Standard Lens Mode", CAE Analyzer Mode and an energy step size of 1 eV for the survey spectrum were used. The diameter of the X-ray beam was 100 μm. A line scan was performed where four analysis points were placed in the wood-polymer interphase (Figure 4, −2 to +1). Starting from the first analysis point (Figure 4, 0), two spots were placed in the wood direction (Figure 4, −1 to −2) and one measuring spot in the polymer direction (Figure 4, +1) at a distance of 200 μm, respectively. As a reference, additional analysis points were placed in the wood substrate and in the polymer bulk, respectively. Prior to analysis the surface was cleaned by sputtering with Ar-clusters (1000 atoms, 6000 eV, 1 mm raster size) for 60 s. High-resolution spectra of C_{1s}, N_{1s} and O_{1s} of 6 specimens were examined, acquired with 50 passes at a pass energy of 50 eV and an energy step size of 0.1 eV. These were analyzed using software package Thermo Avantage (v5.9914, Build 06617) with Smart background and Simplex Fitting algorithm by using Gauss-Lorentz Product. Peak profiles of C_{1s} and O_{1s} were deconvoluted.

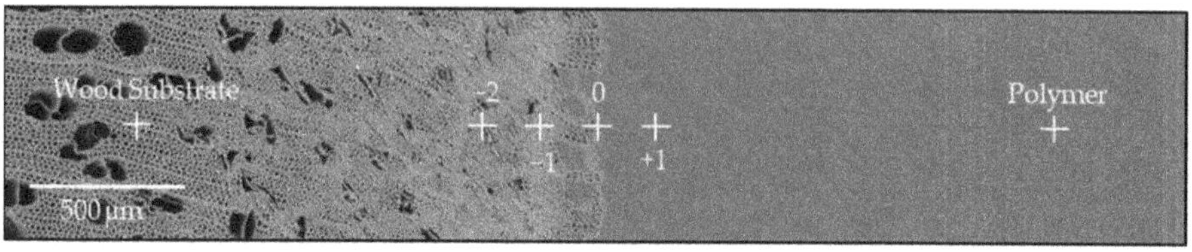

Figure 4. Micrograph of the overmolded solid wood specimens for X-ray photoelectron spectroscopy analyses (XPS) used to determine the elementary distribution of C, O and N and for high-resolution deconvoluted XPS spectra within the samples cross section, indicating the position of the X-ray beam.

2.6. Statistics

In this study a one-way analysis of variance with an error level of 0.05 was calculated using Excel 2016 (Microsoft, Redmond, WA, USA) to statistically evaluate the effect of the injection temperature and the wood species on the mechanical properties.

3. Results and Discussion

3.1. Mechanical Properties of Wood-Polymer Composites

3.1.1. Plywood-Polymer Composites

Figure 5 displays the results of all test configurations at the different injection temperatures for birch and beech plywood, respectively, overmolded with PP. Regarding the influence of the injection temperature on the mechanical properties of these wood-polymer composites no statistically significant effect was present, with the exception of σ_E for the overmolded birch plywood specimens, for which higher injection temperatures resulted in higher σ_E. Chang et al. [25] reported that the hot-pressing temperature and pressure exhibit an inflexion point, i.e., a certain pressure and temperature, at which penetration of the polymer into the wood structure with partly damaged cells and cracks and into the vessels is highest, thus resulting in the highest strength. Furthermore, no significant difference between birch and beech plywood for all test configurations was observed.

σ_P was on average (over all three injection temperatures) 3.16 ± 0.91 MPa and 2.89 ± 0.68 MPa for birch and for beech plywood, respectively. The highest σ_P for birch and beech plywood specimens was observed at an injection temperature of 240 °C (3.31 ± 1.15 MPa and 3.02 ± 0.49 MPa), similar to results reported by Chang et al. [25]. The tensile strength perpendicular to the edge was on average 5.08 ± 1.44 MPa and 4.78 ± 1.01 MPa for birch and beech plywood, respectively, with the highest values achieved at 260 °C for birch and 240 °C for beech plywood specimens (6.04 ± 1.04 MPa and 5.14 ± 1.24 MPa). Liu et al. [38] investigated the surface bond strength of engineered plywood in a similar fashion. Poplar veneers (*Populus tomentosa* Carriére) were bonded with chlorinated PP films on a wood fiber PP composite core layer (80% wood fiber and 20% PP) using a hot-pressing procedure. They observed surface bonding strength values of the veneers on the composite core layer, which is comparable with σ_P, of approx.

1.75 MPa, which was significantly lower compared to our study. The higher values were attributed to two reasons. On the one hand, the poplar veneers used have a significantly lower tensile strength perpendicular to the grain of about 1.7–2.8 MPa compared to birch (~7.0 MPa) and beech (~7.0–10.7 MPa) wood [39]. On the other hand, they prepared the PP-bonded plywood using a hot-pressing process, in which the pressure applied was about 5 MPa at a temperature of 110 °C, which is much lower compared to those used in our study. Improved penetration of PP into the wood at 170 bar (17 MPa) and 360 bar (36 MPa) for PA 6, respectively, the pressures used in this study, could eradicate the damage caused in the wood structure due to compressive failure of the top layers.

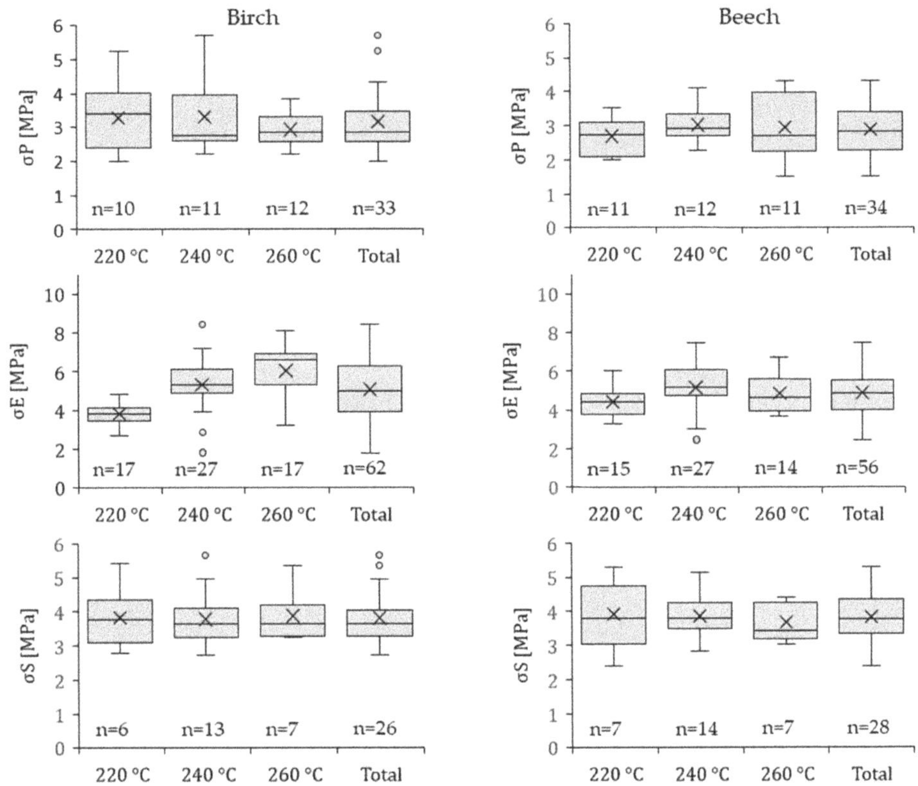

Figure 5. Strength properties of 121 birch plywood specimens and 118 beech plywood specimens overmolded with PP at three different injection temperatures. σ_P is the average tensile strength perpendicular to the plane of the board, σ_E is the average tensile strength perpendicular to the edge, σ_S is the average tensile shear strength and n is the number of the samples tested. The whiskers show minimum and maximum values. X is the mean value. ° indicate values of statistical outliers and—is the median.

σ_S of birch and beech plywood samples were on average 3.81 ± 0.76 MPa and 3.83 ± 0.69 MPa, respectively, with highest values observed at an injection temperature of 260 °C for birch and 220 °C for beech of 3.87 ± 0.68 MPa and 3.92 ± 0.94 MPa, respectively. Bekhta et al. [28] reported shear strength values for birch and beech plywood bonded with PA 6 and polyethylene (PE) higher than 3 MPa and 1.7 MPa, respectively. However, PP overmolded birch and beech plywood showed higher σ_S, which is thought to be mainly influenced by the different process parameters used in this study.

3.1.2. Solid Wood-Polymer Composites

Figure 6 summarizes the mean values, standard deviation and sample number for birch and beech solid wood, overmolded with PA 6 and PP, respectively. There was no statistically significant difference between the birch and beech solid wood composites. σ_S for birch-PA 6 were on average 5.71 $\pm$ 1.13 MPa, while σ_S for beech-PA 6 was slightly higher (6.36 $\pm$ 1.47 MPa). For PP-composites a slightly lower tensile shear strength was observed for birch solid wood compared to beech (2.33 $\pm$ 0.44 MPa and 2.54 $\pm$ 0.83 MPa, respectively). The observed maximum values for σ_S for the PA 6-composites were 8.65 MPa and 9.74 MPa and for PP-composites 2.98 MPa and 4.19 MPa for birch and beech, respectively. Compared to literature, the measured maximum values were similar, with tensile shear strengths of 9 MPa and 3.5 MPa reported for beech wood rods overmolded with PA 6 and PP, respectively [16]. However, a perfectly aligned longitudinal fiber orientation of the specimens results in fewer cut vessels and fibers and thus fewer open lumens into which polymer could penetrate, which in turn results in less mechanical interlocking and thus in a lower average tensile shear strength. For comparison, typical values for bonded birch and beech wood specimens (melamine-urea-formaldehyde (MUF), polyurethane (PU) and phenol-resorcinol-formaldehyde (PRF)) with commercially adhesives do exceed 10 to 11 MPa [31,32].

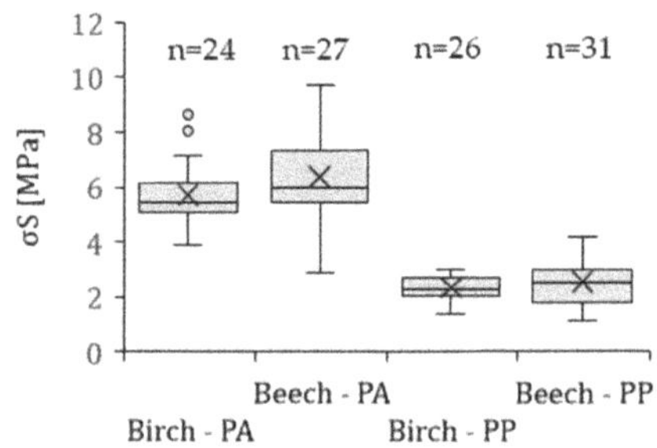

Figure 6. Average tensile shear strength σ_S of birch solid wood specimens and beech solid wood overmolded with PA 6 and PP, respectively; n is the number of the samples. The whiskers show minimum and maximum values. X is the mean value. ° indicate values of statistical outliers and—is the median.

The differences between PA 6 and PP composites can be explained by the different polarity of both polymers. It is supposed that the polar PA 6 [40] exhibits a good adhesion and/or sound bonding with the wood surface, which promotes higher strength values. In addition, the high temperature during the molding process (260 °C) degrades free hydrophilic groups of wood polymers, mainly the hemicelluloses [41,42]. The effects of thermally modified wood fibers on the adhesion to thermoplastics were also reported by Follrich et al. [43]. It can be assumed, that exposure to elevated temperatures leads to a more hydrophobic character of the wood surface enhancing the interfacial compatibility to hydrophobic polymers, which results in improved interfacial interactions. Furthermore, the higher strength for PA 6-composites in contrast to PP-composites can also be explained by the higher cohesive strength of PA 6.

3.2. Wood-Polymer Interfaces

3.2.1. Morphology of Wood Polymer Composites by SEM

Figure 7 shows cross sections of representative birch and beech solid wood composites. During the overmolding process, the melted polymers penetrated into the wood structure through the sliced vessels and fibers. For the specimens overmolded with PA 6, in comparison to PP-composites only minor penetration of the melt into the wood substrate was observed. Due to the high pressure used (360 bar) during the injection process, the melt flow was mainly directed in radial direction (Figure 7(1a–2b)). PP composites were fabricated at much lower injection pressure (170 bar), hence, the melt penetrated the wood cells in both directions, for birch and beech, respectively (Figure 7(3a–4b)). Furthermore, it was observed that the outer cellular structure (approximately 100 μm up to 200 μm) is stabilized by the polymer that penetrated the wood by filling the lumens of vessels and tracheids. Additionally, only a few micro gaps along the interface between the wood and the polymer were observed, which is interpreted as an indicator of good adhesion between the materials. In addition, wood rays and also the transition zone from early to late wood have a structurally reinforcing effect. As Mattheck and Kubler [44] presented, the many rays oriented perpendicular to the grain behave like beams, that lead to an increasing compressive strength of the wood structure. These compressed areas generate an increased interface and thus improved mechanical interlocking between the polymer and the wood surface. According to Sretenovic et al. [21] the mechanical interlocking influences the stress transfer from the polymer to the wood structure in wood fiber composites. In addition, Smith et al. [45] reported that both the porosity of the wood structure and the processing parameters are influencing mechanical interlocking. Furthermore, Gupta et al. [46] showed that there are strong correlations between surface roughness and interfacial adhesion.

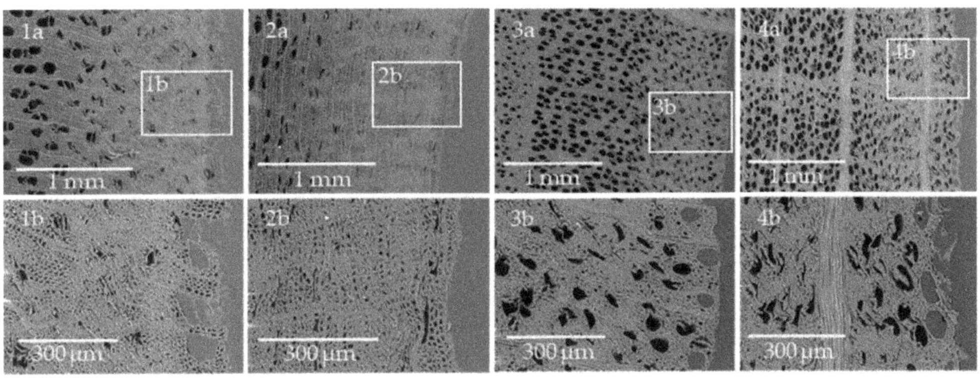

Figure 7. Representative SEM micrographs recorded on the cross-section of the overmolded specimens for birch solid wood (**1a–2b**) and beech solid wood (**3a–4b**); b represents always the detailed inset of a. (**1a**), (**1b**), (**3a**) and (**3b**) show samples overmolded in radial direction and (**2a**), (**2b**), (**4a**) and (**4b**) represents samples overmolded in tangential direction.

Beyond the stabilized interface the wood structure exhibits a zone of compressive failure, caused by plastic deformation of the wood structure during injection molding (Figure 7(1a–4a)). These compressed zones range approximately 0.4 to 1.1 mm into the wood structure depending on the species. Müller et al. [47] reported that a pressure of about 12 MPa is required to densify a diffuse-porous wood structure (e.g., birch and beech) perpendicular to the grain. Corresponding to the thickness of the compressed zone, the overmolding procedure causes almost homogeneous densification across the overmolded surface for both birch and beech. However, the compression zone of birch wood is much larger than the compression zone of beech wood, which is caused by its lower compression strength perpendicular to the grain [39,48,49]. A lower ratio of strength perpendicular to the grain to the injection pressure, leads to higher densification of the wood substrate, causing the formation of a so-called weak boundary layer, which influences the strength of wood polymer composites. In birch wood samples, failure occurs mainly in the weak boundary layer, which corresponds to the results of the mechanical tests, both for solid wood and plywood overmolded with PA 6 and PP as well as findings, as reported by Chang et al. [25]. Furthermore, it is clearly shown, that mainly vessels are compressed in this range. In contrast, tracheids and wood fibers are compressed mainly in the peripheral areas up to a depth of 200 to 600 μm.

3.2.2. Elemental Composition and Chemistry of the Interface in Wood-Polymer Composites

From XP spectra information regarding the penetration of polymer into the wood structure and their interaction can be derived. The elemental composition (C, N and O) at various positions within the wood polymer interphase extracted from scans is shown in Table 1. Additionally, as an initial indicator of the presence of polymer in the wood structure and vice versa, the atomic ratio O/C and N/C for PA 6 composites and the O/C ratio for PP composites were calculated. For all composites, the O/C ratio significantly decreased from the wood substrate through the interphase towards the bulk polymer. Specifically, the birch-PA 6composites exhibited a constant decrease of the O/C ratio from the wood towards the polymer, whereas for beech wood samples this ratio significantly decreases in the interphase (Table 1, −1 to +1), after that the O/C ratio remained constant. Correspondingly, the N/C ratio significantly increased in the interphase to polymer direction (Table 1, −1 to +1). PP composites exhibited a similar trend regarding the O/C ratio; it significantly decreased from the wood substrate towards the interphase, both for birch and beech composites (Table 1, −1 to +1). These results confirm that penetration of the polymer melt into the wood cell wall structure takes place and not only through the cut vessels and fibers as determined in the SEM analyses.

Table 1. Elemental composition of C, N and O over the sample cross-section for birch and beech solid wood overmolded with PA 6 and PP from XPS analysis. The position of the measuring spots placed on the samples are shown in Figure 4. In addition, the atomic O/C and N/C ratios determined by XPS analysis are presented.

Composite	Elemental	Wood Substrate	−2	−1	0	1	Polymer
Birch-PA 6	C at [%]	81.85	86.37	87.65	87.74	91.11	90.00
	N at [%]	0.19	0.14	0.15	2.05	4.28	5.02
	O at [%]	17.95	13.50	12.21	10.22	4.61	4.98
	O/C	0.219	0.156	0.139	0.116	0.051	0.055
	N/C	0.002	0.002	0.002	0.023	0.047	0.056
Beech-PA 6	C at [%]	67.77	67.76	67.37	73.07	89.24	87.74
	N at [%]	0.38	0.34	0.29	1.52	5.82	5.56
	O at [%]	31.85	31.90	32.34	25.41	4.95	6.71
	O/C	0.470	0.471	0.480	0.348	0.055	0.076
	N/C	0.006	0.005	0.004	0.021	0.065	0.063
Birch-PP	C at [%]	72.61	71.46	70.40	85.30	98.68	98.32
	N at [%]	0.45	0.27	0.33	0.35	0.41	0.40
	O at [%]	26.95	28.27	29.27	14.35	0.92	1.29
	O/C	0.371	0.396	0.416	0.168	0.009	0.013
Beech-PP	C at [%]	66.66	65.70	65.96	83.45	99.29	99.11
	N at [%]	0.32	0.28	0.24	0.38	0.30	0.31
	O at [%]	33.02	34.03	33.81	16.17	0.42	0.59
	O/C	0.495	0.518	0.513	0.194	0.004	0.006

Carbon is the dominant element in both wood and polymer. For this reason, the carbon peak from high resolution spectra was deconvoluted into four components. With regard to wood, the C 1 peak (C-C or C-H) at approx. 284 eV corresponds to carbon-carbon or carbon-hydrogen bonds and is predominant in lignin or polymers such as PP. The C 2 peak at approx. 286 eV corresponds to carbon-non-carbonyl oxygen bonds (C-O), a major moiety in cellulose. The C 3 peak at approx. 287 eV is assigned to carbon atoms bound to two non-carbonyl oxygens (O-C-O) or to one carbonyl oxygen (C=O), while the C 4 peak at approx. 289 eV represents carboxylic groups (O-C=O) [50,51]. For PA 6 the C peak was deconvoluted into three components according to the literature [52], with the C 1 peak at approx. 284 eV corresponding to the aliphatic carbon atoms CH_2 (**C-C** (C=O)-N-C), the C 2 at approx. 286 eV representing the carbon atoms linked to the amide nitrogen (C-C (C=O)-N-C) and the C 3 at approx. 287 eV representing the amide carbonyl group (C-C (**C=O)-N**-C). For aliphatic PP the C peaks were deconvoluted into two main components: C 1 peak (C-H or C-C) at approximately 284.5 eV and the C 2 peak (C-O) at approx. 286 eV [53]. To obtain information about the penetration as well the distribution of the chemical components, C 1 and C 3 peaks for specimens overmolded PA 6 and C 1 and C 2 peaks for specimens overmolded with PP were selected (Figure 8a–d and Figure A1).

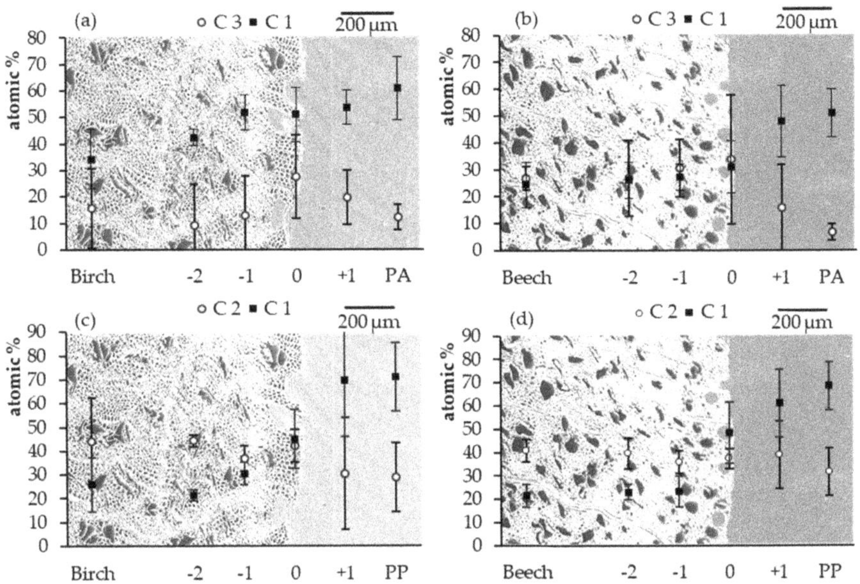

Figure 8. Mean values and standard deviation of the XPS results: Carbon peak components at C 1 and C 3 for birch solid wood (**a**) and beech solid wood (**b**) overmolded with PA 6 and the C 1 and C 2 for birch solid wood (**c**) and beech solid wood (**d**) overmolded with PP.

In general, an increase of the measured atomic percentage of the C 1 components from the interface in polymer direction and at the same time a decrease in wood direction for all samples was found. For birch-PA 6 composites (Figure 8a), the amount of the C 1 component decreased significantly from the polymer bulk towards the interphase from 61.0 at.% to 51.0 at.% (Figure 8a, PA to 0), while the amount of C 1 remained almost constant through the interphase but decreased towards the wood substrate. For the C 3 component, the highest value of 27.5 at.% was observed at the interphase (Figure 8a, 0), significantly decreasing in both directions. For beech-PA 6 (Figure 8b) almost similar results were observed. The amount of the C 1 component decreased from the PA 6 bulk to the wood substrate from 51.0 at.% to 24.4 at.%, with the interphase area having a constant value. The highest amount of C 3 was again observed in the interphase of 33.9 at.% (Figure 8b, 0), decreasing toward the wood and polymer direction. The results of C 1 and C 3 indicate that the polymer melt penetrated the wood structure during injection molding up to about ~400 µm (Figure 8a,b, −2). Additionally, a nonlinear trend of ratios between C 1 and C 3 was observed indicating that chemical reactions, e.g., transamidation occurs influencing the ratio beyond the trend expected from pure mixing of polymer and wood [54].

Figure 8c,d shows the elemental distribution across the cross section of PP composites. Similar results were observed for birch and beech wood. Highest values of C 1 were determined in the polymer bulk, as expected (71.1 at.% and 68.4 at.% for birch and beech specimens, respectively). These amounts then constantly decreased towards to point

0 within the interphase of 44.9 at.% and 48.2 at.% (Figure 8c,d, 0) for birch and beech, respectively. Beyond point 0 a significant decrease of the C 1 component was observed towards wood (Figure 8c,d, −1). Between point −1 and the wood substrate no significant difference between C 1 for birch-PP and beech-PP can be observed. For C 2, the elemental composition also does not differ significantly from each other. PP does not interact with the wood structure beyond van der Waals interactions as PA 6 does, based on polar groups present in PA 6 and being absent in PP. Furthermore, PP only penetrated the wood structure up until 200 μm (Figure 8c,d, −1).

Results of the XPS analyses assist in explaining the results of the mechanical tests. Higher strength and stiffness of wood-PA 6 composites can be explained by PA 6 penetrating the wood structure on a macro- but also microscopic level, which corresponds to previous findings [16] where beech wood rods were overmolded with different polymer materials. In addition, chemical interaction of PA 6 with wood takes place due to the polar nature of the material, resulting in better interfacial adhesion as compared to PP and thus improved mechanical properties of wood-PA 6 composites produced by injection molding. However, sound bonding to the wood occurs, both for PA 6 as well as for the more hydrophobic material PP due to formation of an interphase by polymer penetration into the wood.

Based on the presented results, the initially proposed research questions can be answered as follows: The used polymers penetrate into the peripheral porous structure through the sliced vessels and fibers thus forming an interphase which contributes to adhesion by mechanical interlocking. In case of PA 6 wood composites, additional chemical interactions do seem to contribute to improved adhesion. Additionally, the mechanical properties of the produced (unmodified) wood-polymer composites can compete with commercially bonded wood-wood composites and, therefore, this technology is suitable to manufacture wood polymer hybrid composites for structural applications for instance for the automotive sector.

4. Conclusions

Solid wood and plywood were overmolded with PP and PA 6 in order to investigate the influence of process parameters on interfacial adhesion between wood and polymer and the mechanical properties of wood-polymer composites. The mechanical properties of these composites are influenced only to a small extent by the processing temperature used. Temperature effects on the wood substrate are of minor importance. However, SEM and XPS analysis showed that substantial amounts of molten polymer penetrated into the wood substrate. A weak boundary layer of compacted cells formed in the wood substrate, extending from the wood-polymer interphase to a depth of 1 mm. Due to the high-pressure during injection molding, a weak boundary layer consisting of heavy densified cells formed in the wood substrate, extending from the wood-polymer interphase to a depth of 1 mm. The weak boundary layer, which varied between birch and beech wood, lowered the mechanical properties of the wood-polymer composites. However, polymer interpenetrated the wood substrate through vessels, which led to the formation of a stabilized interphase and improved mechanical properties. Due to its polar character, PA 6 interacts chemically with the wood substrate, resulting in the highest tensile shear strength observed, ranging from 8 MPa to 9 MPa for birch and beech, respectively. Nevertheless, sufficient bonding and mechanical interlocking of PP was also observed for both wood species.

Author Contributions: Conceptualization, A.S. and U.M.; methodology, A.S.; validation, A.S., M.P. and A.M.; formal analysis, A.S.; investigation, A.S. and A.M.; resources, A.S., A.M., M.P., A.B. and U.M.; data curation, A.S. and A.M.; writing—original draft preparation, A.S.; writing—review and editing, A.M., A.B. and U.M.; visualization, A.S.; supervision, U.M. All authors have read and agreed to the published version of the manuscript.

Funding: The results presented in this study are part of the research project "WoodC.A.R. (http://www.woodcar.eu/index.html accessed on 20 August 2021)" (FFG Project No.: 861.421). Financial support by the Austrian Research Promotion Agency (FFG), Styrian Business Promotion Agency (SFG), Standortagentur Tirol, and the companies Collano AG, DOKA GmbH, DYNAmore GmbH,

EJOT Austria GmbH, Forst-Holz-Papier, Holzcluster Steiermark GmbH, IB STEINER, Lean Management Consulting GmbH, Magna Steyr Engineering GmbH & Co KG, MAN Truck & Bus AG, Mattro Mobility Revolutions GmbH, Volkswagen AG and Weitzer Parkett GmbH & Co KG.

Informed Consent Statement: Not applicable.

Data Availability Statement: The raw/processed data required to reproduce these findings cannot be shared at this time due to legal or ethical reasons.

Acknowledgments: The outlined results in this study are part of the project "WoodC.A.R. (http://www.woodcar.eu/ accessed on 20 August 2021)" (Project No.: 861.421). Financial support by the Austrian Research Promotion Agency (FFG), Styrian Business Promotion Agency (SFG), Standortagentur Tirol, and the companies Collano AG, DOKA GmbH, DYNAmore GmbH, EJOT Austria GmbH, Forst-Holz-Papier, Holzcluster Steiermark GmbH, IB STEINER, Lean Management Consulting GmbH, Magna Steyr Engineering GmbH & Co KG, MAN Truck & Bus AG, Mattro Mobility Revolutions GmbH, Volkswagen AG, and Weitzer Parkett GmbH & Co KG is gratefully acknowledged.

Conflicts of Interest: The authors declare no conflict of interest.

Appendix A

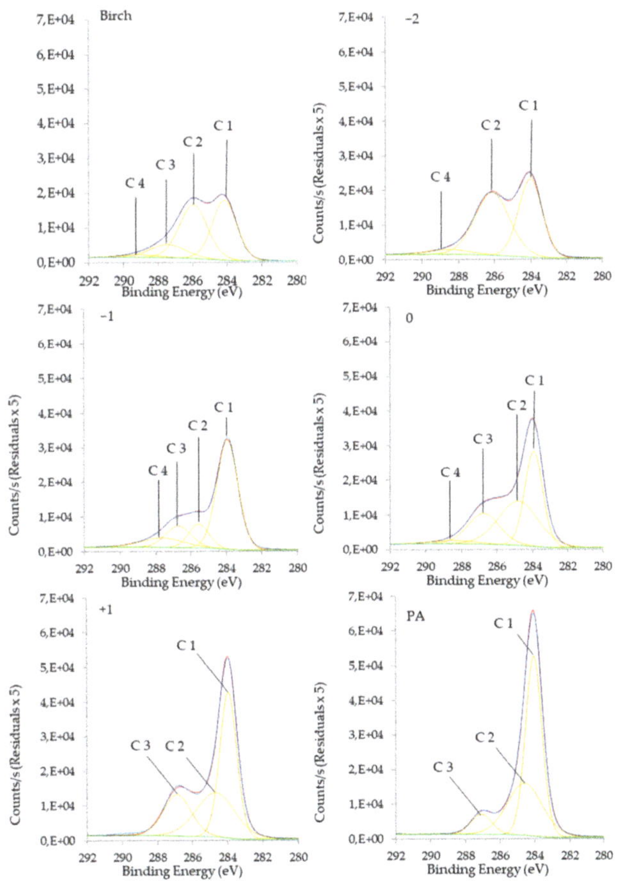

Figure A1. Representative deconvoluted XPS spectrum with relative areas of the C 1, C 2, C 3 and C 4 binding energy regions of the birch solid wood specimens, overmolded with PA 6, within the cross section of a sample.

References

1. Leitgeb, W.; Kerschbichler, S.; Jost, T.; Mayrhofer, P.; Wagner, W.; Müller, U. Holz im strukturellen Fahrzeugbau. In *IHD, Konferenzproceedings—2*; Interdisziplinäres Fahrzeugkolloquium: Dresden, Germany, 2016; pp. 1–118.
2. Statistisches Bundesamt. Available online: https://www.destatis.de/Europa/DE/Thema/Umwelt-Energie/CO2_Strassenverkehr.html (accessed on 16 August 2021).
3. Mair-Bauernfeind, C.; Zimek, M.; Asada, R.; Bauernfeind, D.; Baumgartner, R.J.; Stern, T. Prospective sustainability assessment: The case of wood in automotive applications. *Int. J. Life Cycle Assess.* **2020**, *25*, 2027–2049. [CrossRef]
4. Franzini, F.; Berghäll, S.; Toppinen, A.; Toivonen, R. Comparing wood versus concrete: An explorative study of municipal civil servants' beliefs about multistory building materials in Finland. *For. Prod. J.* **2021**, *71*, 65–76. [CrossRef]
5. Asada, R.; Cardellini, G.; Mair-Bauernfeind, C.; Wenger, J.; Haas, V.; Holzer, D.; Stern, T. Effective bioeconomy? A MRIO-based socioeconomic and environmental impact assessment of generic sectoral innovations. *Technol. Forecast. Soc. Change* **2020**, *153*, 119946. [CrossRef]
6. Mair, C.; Zimek, M.; Stern, T.; Baumgartner, R.J. Sustainability performance of wood-based materials: A systemic assessment of lightweight materials for the mobility sector. In Proceedings of the 24th Annual ISDRS Conference, Messina, Italy, 13–15 June 2018.
7. Kohl, D.; Link, P.; Böhm, S. Wood as a Technical Material for Structural Vehicle Components. *Procedia CIRP* **2016**, *40*, 557–561. [CrossRef]
8. Müller, U.; Feist, F.; Jost, T. Wood composites in the automotive industry of the future? Assumed crash tests and simulation of wood materials. *Holztechnologie* **2019**, 5–15. Available online: https://www.woodcar.eu/wa_files/04_jost%20et%20al.%202018_final.pdf (accessed on 20 August 2021).
9. Boland, C.S.; De Kleine, R.; Keoleian, G.A.; Lee, E.C.; Kim, H.C.; Wallington, T.J. Life cycle impacts of natural fiber composites for automotive applications: Effects of renewable energy content and lightweighting. *J. Ind. Ecol.* **2016**, *20*, 179–189. [CrossRef]
10. De Melo, R.R.; Del Menezzi, C.H.S. Influence of veneer thickness on the properties of LVL from Paricá (*Schizolobium amazonicum*) plantation trees. *Eur. J. Wood Wood Prod.* **2014**, *72*, 191–198. [CrossRef]
11. Dunky, M.; Niemz, P. *Holzwerkstoffe und Leime*; Springer: Berlin/Heidelberg, Germany, 2002; ISBN 9783642624759.
12. Friedrich, H.E. *Leichtbau in der Fahrzeugtechnik*; Springer: Stuttgart, Germany, 2017; ISBN 3658122951.
13. Luckeneder, G.; Autengruber, R.; Stellnberger, K.-H.; Kurz, J.F.T. Corrosion Protection of Galvanized Press-Hardening Steel: Main Influencing Factors and Mechanisms. In Proceedings of the Galvatech 2015 Conference Proceedings, Toronto, ON, Canada, 31 May–4 June 2015; pp. 868–875.
14. Pritchard, G. *Reinforced Plastics Durability*; CRC Press: Cambridge, UK, 1999; ISBN 0849305470.
15. Stadlmann, A.; Pramreiter, M.; Stingl, R.; Kurzböck, C.; Jost, T.; Müller, U. Durability of wood exposed to alternating climate test and natural weathering. *Forests* **2020**, *11*, 953. [CrossRef]
16. Stadlmann, A.; Veigel, S.; Dornik, F.; Pramreiter, M.; Steiner, G.; Müller, U. Bond strength of different wood-plastic hybrid components prepared through back injection moulding. *BioResources* **2020**, *15*, 1050–1061. [CrossRef]
17. Teaca, C.A.; Tanasa, F.; Zanoaga, M. Multi-component Polymer Systems Comprising Wood as Bio-based Component and Thermoplastic Polymer Matrices—An Overview. *BioResources* **2018**, *13*, 4728–4769. [CrossRef]
18. Gurunathan, T.; Mohanty, S.; Nayak, S.K. A review of the recent developments in biocomposites based on natural fibres and their application perspectives. *Compos. Part A Appl. Sci. Manuf.* **2015**, *77*, 1–25. [CrossRef]
19. Gacitua, E.W.; Wolcott, M.P. Morphology of Wood Species Affecting Wood-Plastic Interaction: Mechanical Interlocking and Mechanical Properties. In Proceedings of the 9th International Conference on Wood and Biofiber Plastic Composites, Madison, WI, USA, 21–23 May 2007.
20. Gacitua, W.; Wolcott, M. Morphology of wood species affecting wood-thermoplastic interaction: Microstructure and mechanical adhesion. *Maderas. Cienc. Tecnol.* **2009**, *11*, 217–231. [CrossRef]
21. Sretenovic, A.; Müller, U.; Gindl, W. Mechanism of stress transfer in a single wood fibre-LDPE composite by means of electronic laser speckle interferometry. *Compos. Part A Appl. Sci. Manuf.* **2006**, *37*, 1406–1412. [CrossRef]
22. Keener, T.J.; Stuart, R.K.; Brown, T.K. Maleated coupling agents for natural fibre composites. *Compos. Part A Appl. Sci. Manuf.* **2004**, *35*, 357–362. [CrossRef]
23. Correa, C.A.; Razzino, C.A.; Hage, E. Role of maleated coupling agents on the interface adhesion of polypropylene-wood composites. *J. Thermoplast. Compos. Mater.* **2007**, *20*, 323–339. [CrossRef]
24. Fang, L.; Chang, L.; Guo, W.J.; Ren, Y.P.; Wang, Z. Preparation and characterization of wood-plastic plywood bonded with high density polyethylene film. *Eur. J. Wood Wood Prod.* **2013**, *71*, 739–746. [CrossRef]
25. Chang, L.; Tang, Q.; Gao, L.; Fang, L.; Wang, Z.; Guo, W. Fabrication and characterization of HDPE resins as adhesives in plywood. *Eur. J. Wood Wood Prod.* **2018**, *76*, 325–335. [CrossRef]
26. Fang, L.; Chang, L.; Guo, W.J.; Chen, Y.; Wang, Z. Influence of silane surface modification of veneer on interfacial adhesion of wood-plastic plywood. *Appl. Surf. Sci.* **2014**, *288*, 682–689. [CrossRef]
27. Liu, Y.; Guo, L.; Wang, W.; Sun, Y.; Wang, H. Modifying wood veneer with silane coupling agent for decorating wood fiber/high-density polyethylene composite. *Constr. Build. Mater.* **2019**, *224*, 691–699. [CrossRef]
28. Bekhta, P.; Müller, M.; Hunko, I. Properties of thermoplastic-bonded plywood: Effects of the wood species and types of the thermoplastic films. *Polymers* **2020**, *12*, 2582. [CrossRef]

29. Solt, P.; Konnerth, J.; Gindl-Altmutter, W.; Kantner, W.; Moser, J.; Mitter, R.; van Herwijnen, H.W.G. Technological performance of formaldehyde-free adhesive alternatives for particleboard industry. *Int. J. Adhes. Adhes.* **2019**, *94*, 99–131. [CrossRef]
30. Stoeckel, F.; Konnerth, J.; Gindl-Altmutter, W. Mechanical properties of adhesives for bonding wood—A review. *Int. J. Adhes. Adhes.* **2013**, *45*, 32–41. [CrossRef]
31. Konnerth, J.; Gindl, W.; Harm, M.; Müller, U. Comparing dry bond strength of spruce and beech wood glued with different adhesives by means of scarf- and lap joint testing method. *Holz Roh Werkst.* **2006**, *64*, 269–271. [CrossRef]
32. Konnerth, J.; Kluge, M.; Schweizer, G.; Miljković, M.; Gindl-Altmutter, W. Survey of selected adhesive bonding properties of nine European softwood and hardwood species. *Eur. J. Wood Wood Prod.* **2016**, *74*, 809–819. [CrossRef]
33. Wolcott, M.P.; Englund, K. A technology review of WPC. In Proceedings of the 33rd International Particleboard/Composite Materials Symposium, Washingtonm DC, USA, 13–15 April 1999; pp. 103–111.
34. ÖNORM EN 302-1. *Klebstoffe für tragende Holzbauteile—Prüfverfahren—Teil 1: Bestimmung der Längszugscherfestigkeit*; Austrian Standards International: Vienna, Austria, 2013.
35. ISO 554. *Normalklimate für die Konditionierung und/oder Prüfung: Anforderungen International Organization of Standardization*; ISO: Geneva, Switzerland, 1976.
36. ÖNORM EN 319:1993. *Particleboards and Fibreboards—Determination of Tensile Strength Perpendicular to the Plane of the Board*; Austrian Standards International: Vienna, Austria, 2003.
37. DIN 52188. *Prüfung von Holz—Bestimmung der Zugfestigkeit Paralell zur Faser*; German Institute for Standardization: Berlin, Germany, 1979.
38. Liu, Y.; Sun, Y.; Hao, J.; Wang, W.; Song, Y.; Zhou, Z. Interface Bonding Properties and Mechanism of Poplar Board-Veneered Wood Fiber/Polypropylene Composites with Chlorinated Polypropylene Films as an Intermediate Layer. *Langmuir* **2019**, *35*, 13934–13941. [CrossRef] [PubMed]
39. Wagenführ, R.; Scheiber, C. *Holzatlas*; Fachbuchverlag Leibzig im Carl Hanser Verlag: Leibzig, Germany, 2007; ISBN 3446406492.
40. Adriaensens, P.; Pollaris, A.; Carleer, R.; Vanderzande, D.; Gelan, J.; Litvinov, V.M.; Tijssen, J. Quantitative magnetic resonance imaging study of water uptake by polyamide 4,6. *Polymer* **2001**, *42*, 7943–7952. [CrossRef]
41. Hill, C.A.S. *Wood Modification: Chemical, Thermal and Other Processes*; John Wiley & Sons: Hoboken, NJ, USA, 2006; Volume 5, ISBN 9780470021729.
42. Burmester, A. Zur Dimensionsstabilisierung von Holz. *Holz Roh Werkst.* **1975**, *33*, 333–335. [CrossRef]
43. Follrich, J.; Müller, U.; Gindl, W.; Mundigler, N. Effects of long-term storage on the mechanical characteristics of wood plastic composites produced from thermally modified wood fibers. *J. Thermoplast. Compos. Mater.* **2010**, *23*, 845–853. [CrossRef]
44. Mattheck, C.H.K. *Wood—The Internal Optimization of Trees*; Springer: Berlin/Heidelberg, Germany, 1997; ISBN 9783540620198.
45. Smith, M.J.; Dai, H.; Ramani, K. Wood–thermoplastic adhesive interface—Method of characterization and results. *Int. J. Adhes. Adhes.* **2002**, *22*, 197–204. [CrossRef]
46. Gupta, B.S.; Reiniati, I.; Laborie, M.P.G. Surface properties and adhesion of wood fiber reinforced thermoplastic composites. *Colloids Surfaces A Physicochem. Eng. Asp.* **2007**, *302*, 388–395. [CrossRef]
47. Müller, U.; Gindl, W.; Teischinger, A. Effects of cell anatomy on the plastic and elastic behaviour of different wood species loaded perpendicular to grain. *IAWA J.* **2003**, *24*, 117–128. [CrossRef]
48. Vorreiter, L. *Holztechnologisches Handbuch*; Verlag Georg Fromme und Co: Wien, Austria, 1949.
49. Bosshard, H.H. *Holzkunde. Band 1, Mikroskopie und Makroskopie des Holzes*; Birkhäuser Verlag: Basel, Switzerland, 1982; ISBN 978-3-0348-5323-1.
50. Popescu, C.M.; Tibirna, C.M.; Vasile, C. XPS characterization of naturally aged wood. *Appl. Surf. Sci.* **2009**, *256*, 1355–1360. [CrossRef]
51. Nzokou, P.; Kamdem, D.P. X-ray photoelectron spectroscopy study of red oak- (*Quercus rubra*), black cherry- (*Prunus serotina*) and red pine- (*Pinus resinosa*) extracted wood surfaces. *Surf. Interface Anal.* **2005**, *37*, 689–694. [CrossRef]
52. Oliveira, F.R.; Zille, A.; Souto, A.P. Dyeing mechanism and optimization of polyamide 6,6 functionalized with double barrier discharge (DBD) plasma in air. *Appl. Surf. Sci.* **2014**, *293*, 177–186. [CrossRef]
53. Morent, R.; De Geyter, N.; Leys, C.; Gengembre, L.; Payen, E. Comparison between XPS- And FTIR-analysis of plasma-treated polypropylene film surfaces. *Surf. Interface Anal.* **2008**, *40*, 597–600. [CrossRef]
54. Kashani Rahimi, S.; Otaigbe, J.U. The role of particle surface functionality and microstructure development in isothermal and non-isothermal crystallization behavior of polyamide 6/cellulose nanocrystals nanocomposites. *Polymer* **2016**, *107*, 316–331. [CrossRef]

polymers

MDPI

Article

Evaluation of Particleboards Made from Giant Reed (*Arundo donax* L.) Bonded with Cement and Potato Starch

Aranzazu Alejandra Ferrandez-García, Teresa Garcia Ortuño, Manuel Ferrandez-Villena *, Antonio Ferrandez-Garcia and Maria Teresa Ferrandez-García

Department of Engineering, Universidad Miguel Hernández, 03300 Orihuela, Spain; aranfer2@gmail.com (A.A.F.-G.); tgarcia@umh.es (T.G.O.); antonio.ferrandezg@umh.es (A.F.-G.); mt.ferrandez@umh.es (M.T.F.-G.)

* Correspondence: m.ferrandez@umh.es; Tel.: +34-966-749-716

Abstract: There is a general concern about the rationalization of resources and the management of waste. Plant residues can contribute to the development of new non-polluting construction materials. The objective of this study was to valorize a plant residue such as the giant reed and obtain a particleboard with cement using potato starch as a plasticizer in a manufacturing process involving compression and heat. The influence of cement and starch in different proportions and its stability over time were analyzed. Finally, their physical and mechanical properties were evaluated and compared to European Standards. High-quality sustainable particleboards (boards with high structural performance) were obtained and can be classified as P6 according to European Standards. Mechanical properties were improved by increasing the starch content and pressing time, whereas greater resistance to water was obtained by increasing the cement content. Giant reed particles seem to tolerate the alkalinity of the cement since there was no sign of degradation of its fibers. The use of these residues in the manufacture of construction materials offers a very attractive alternative in terms of price, technology and sustainability.

Keywords: composite; MOR; MOE; IB; panel

Citation: Ferrandez-García, A.A.; Ortuño, T.G.; Ferrandez-Villena, M.; Ferrandez-Garcia, A.; Ferrandez-García, M.T. Evaluation of Particleboards Made from Giant Reed (*Arundo donax* L.) Bonded with Cement and Potato Starch. *Polymers* **2022**, *14*, 111. https://doi.org/10.3390/polym14010111

Academic Editors: Domenico Acierno and Antonella Patti

Received: 28 November 2021
Accepted: 24 December 2021
Published: 29 December 2021

1. Introduction

Environmental awareness is increasing in society and with it, public concern about the rationalization of resources and the management of waste. One way to collaborate in solving both problems is by increasing the use of construction materials composed of plant fibers, given these products are easily recyclable and are not aggressive with the environment. The recovery of this waste is also in agreement with European policies related to the environment [1–3]. The manufacturing of products from plant residues avoids their elimination through incineration, reducing greenhouse gas emissions by fixing carbon during their life cycle. This could help EU member states fulfill their agreements ensuring that accounted greenhouse gas emissions from land use, land use change or forestry (LULUCF) are balanced by at least an equivalent accounted removal of CO_2 from the atmosphere in the period 2021 to 2030 [4].

Cement-bonded wood composite panels are not a novel concept, having been on the market during the past century [5]. Cement traditionally has been used to strengthen wood composites improving their mechanical performance [6,7], fire retardance, water resistance and insulation [8]. These products are currently used by the construction industry in applications such as walls, roof sheathing and tiles, floor, fences and sound barriers [9–11]. However, due to a decreasing availability of wood, there has been a deterioration in the mechanical properties of these commercial composites.

Studies are currently oriented towards the production of new generation composite boards with lignocellulosic residues of agricultural origin. The development of bio-renewable materials mixed with cement provides added value for the agricultural waste market. The main advantages of using lignocellulosic materials as reinforcement in cement are their low density, low cost, biodegradability, availability of large varieties of fibers throughout the world and the promotion of a new agricultural economy [12,13].

A variety of investigations of plant fibers with cements have been studied: rattan [14], bamboo [15], rice husk [16], sisal [17–19], coconut [20–26], sugar cane bagasse [27–29], babaçu [30], banana [31,32], coconut and abaca [33], oil palm [34–36], canary palm [37], arhar [38], agave lecheguilla [39], date palm [40], hemp [41–45], kenaf [46], giant reed [47], hazelnut shell, wood and tea [48], cork [49], jute [50,51] and corn [52].

Composites made of natural fibers absorb a large amount of water causing cracking due to the swelling of the fibers. The initial curing process of cement compounds with plant biomass is problematic due to the loss of water absorbed by the fibers. However, later internal curing is favored by the release of part of the water that the biomass had captured. Natural fibers with cement composites are more susceptible to a lack, rather than an excess, of water. High amounts of water in the cement–fiber mix cause the segregation of the mixture, whereas small amounts of water make compaction of the mixture difficult and favors the presence of heterogeneities.

Another issue is that plant fibers suffer degradation when in contact with cement. Some investigations have focused on the modification of the surface of the fibers in order to prevent degradation [40,46,53–56]. The properties of these composites are limited by a combination of factors such as heterogeneity, wettability and chemical compatibility of the natural fibers with cement. Composites reinforced with lignocellulosic material present a great variability in their mechanical properties in cement boards due to the deterioration of their properties because of alkaline degradation (hydrolysis) and fiber mineralization [57]. These mechanisms produce changes in the chemical composition of the fibers that cause a reduction in strength and degradation of the polymeric matrix, of the fiber/polymer matrix interfacial bond [57] and also of plant fibers resulting in a delay in the cement hydration [31].

Various components of the biomass, such as soluble sugars or low molar mass hemicelluloses, have adverse effects on the preparation and performance of concrete [58]. Therefore, selecting biomass sources with a low content of these compounds would minimize these drawbacks. Research on starch-based lightweight concrete has investigated the effect of its polysaccharides on cement [59,60], the retardant properties of the starch for cement setting [61], the dispersion mechanism of sulfonated starch as a water-reducing agent for cement [62], jute–cement panels with different proportions of starch [63] and particles of palm tree with cement and starch [37].

Different manufacture parameters of panels made of cement with lignocellulosic materials have been analyzed, concluding that cemented panels are high-strength construction materials. However, their industrial production requires a high investment and further research is needed to reduce cost manufacturing [64] and to evaluate the effects of fiber pretreatment methods and alternative curing methods for the long-term performance of these composites [12,65].

Giant reed (*Arundo donax* L.) is one of the largest species of grass growing in Mediterranean regions. It is a wild perennial plant to which no genetic improvement or genotype selection has been made. It grows annually, with average heights of 4 m and a mean thickness of 4 cm. Reed has traditionally been used in many Mediterranean countries. In the southeast of Spain, it was used in construction as part of the roof and floor up to the beginning of the 20th century. However, it is now in disuse and has become an environmental problem since it forms dense reed beds along river banks. When the water level rises, the reeds are uprooted and carried away on the current, forming large masses that block watercourses, causing flooding and sweeping away any structure that gets in their way. In the Segura River in Spain, the authorities are forced to make significant economic investments in order to keep reeds controlled, hence they are cleaned and processed in authorized landfills [66].

Several authors used giant reed biomass to obtain particle-boards with different adhesives. A study on multilayer panels of oriented particles [67] found that these boards have good mechanical behavior and were therefore suitable for use in load-bearing structures according to European Standards [68]. With urea formaldehyde (UF), Garcia Ortuño et al. [69] obtained panels with good properties of modulus of rupture (17.67 N/mm^2), modulus of elasticity (3025.90 N/mm^2) and internal bounding strength (1.31 N/mm^2) that could be commercialized. Giant reed particleboards have also been manufactured with unmodified starches [70] and presented good mechanical properties (modulus of rupture of 16.20 N/mm^2, modulus of elasticity of 2520.97 N/mm^2 and internal bounding strength of 0.39 N/mm^2) through a cyclical process of humidification, heat and pressure. The best properties were found when potato starch was used as an adhesive. In other tests of giant reed with cement mortars, Shon [71] observed that the thermal conductivity and density of the concrete decreased.

Plant residues can contribute to the developing of new non-polluting construction materials. The objective of this research was to valorize a plant residue such as the giant reed and obtain a particleboard, adding a very small amount of cement in comparison with the wood–cement composites and using potato starch as a plasticizer in a manufacturing process involving compression and heat. The influence of cement and starch in different proportions and its stability over time were analyzed. Finally, the particleboard's physical and mechanical properties were evaluated and compared to European Standards [68] in order to verify whether it could be used as a building material.

2. Materials and Methods

2.1. Materials

The materials used were residues of giant reed (*Arundo donax* L.) and different proportions of CEM II/B-LL 32.5 N Portland cement, potato starch (*Solanum tuberosum*) and water.

The giant reed biomass (Figure 1) was obtained from clearing the banks of the Segura River in southeast Spain. The reeds were laid out in a vertical position to dry outdoors for 12 months until their relative humidity was 8.2 ± 0.4%. They were then cut and shredded in a blade mill. The particles were collected in a vibrating sieve and only those that passed through the 0.25 mm sieve were selected.

Figure 1. Giant reed used in the manufacturing of the panels.

Potato starch from the food industry, with 90% purity, was used as a plasticizer. Chemically, starch is a mixture of two similar polysaccharides: amylose and amylopectin.

Potato starch typically contains large oval granules and gels at a temperature of 58–65 °C. Water was taken directly from the mains water supply, with an average temperature of 20 °C.

2.2. Manufacturing Process

The manufacturing process consisted of combining dry reed particles with cement in different proportions in weight (0, 10 and 20%) and starch (0, 5 and 10%). Then, 10% water was sprayed onto the mixture before stirring it for 15 min in a blender (LGB100, Imal s.r.l., Modena, Italy) until homogenized.

The mat, formed in a mold of dimensions 600 mm × 400 mm, was then placed in a hot press with a force of 2.6 MPa, a temperature of 100 °C and four different times (1, 2, 3 and 4 h). Subsequently, the boards were cooled to room temperature. A total of 132 panels were made, comprising seven types with four different classes (the 28 different configurations are shown in Tables 1 and 2).

Table 1. Types of panels manufactured.

Type	Number of Panels	Weight Dosage (%)	
		Starch	Cement
A	20	10	20
B	20	5	20
C	20	10	10
D	20	-	10
E	20	-	20
F	16	5	-
G	16	10	-

Table 2. Division of panels in classes according to their pressing time.

Pressing Time (h)	Class	Number of Panels
1	1	4
2	2	8 [1]
3	3	4
4	4	4

[1] Only for types A to E. For types F and G, the number of panels was four.

Twenty-eight days after manufacture, four specimens of each type (A to G) and class (1 to 4) were cut to the appropriate dimensions as indicated in European Standards [72] in order to carry out the tests needed to characterize the mechanical, physical and thermal properties of each of the boards being studied (Figure 2). Three hundred and sixty-five days after manufacture, four boards of class 2 from types A to E were cut and also tested.

Figure 2. Giant reed–cement–starch panels.

2.3. Methods

The method followed was experimental. The tests were conducted in the Materials Strength Laboratory of the Higher Technical College of Orihuela at Universidad Miguel Hernández, Elche. The values were determined according to European Standards established for wood particleboards [73].

After they were manufactured and cut, density [74], thickness swelling (TS) and water absorption (WA) after 2 and 24 h immersed in water [75], internal bonding strength (IB) [76], modulus of elasticity (MOE) and modulus of rupture (MOR) [77] were measured (Table 3). Later, the boards were evaluated according to European Standards [68]. In order to assess the resistance of the reed particles to the alkalinity of the cement, MOR and MOE tests were performed after 365 days on four class 2 panels of types A to E.

Table 3. Characteristics of the tests performed.

Test	N of Replicates (Per Panel)	N of Replicates (Total)	Size of the Specimens	Equipment Used
Relative Humidity	3	396	20 g	Model UM2000, Imal s.r.l.
Density	6	792	50 mm × 50 mm	Model IB700, Imal s.r.l.
Thickness Swelling (TS)	3	396	50 mm × 50 mm	Model 76-B0066/B Water Bath, Controls S.A. Model UM2000, Imal s.r.l.
Water Absorption (WA)	3	396	50 mm × 50 mm	Model 76-B0066/B Water Bath, Controls S.A. Model UM2000, Imal s.r.l.
Modulus of Rupture (MOR)	6	792	150 mm × 50 mm	Model UM2000, Imal s.r.l.
Modulus of Elasticity (MOE)	6	792	150 mm × 50 mm	Model UM2000, Imal s.r.l.
Internal Bonding Strength (IB)	3	396	50 mm × 50 mm	Model UM2000, Imal s.r.l.

The morphology of the inside of the experimental panels was examined using a scanning electron microscope (SEM) (Hitachi model S3000N, Hitachi, Ltd., Tokyo, Japan) equipped with an X-ray detector (Bruker XFlash 3001, Billerica, MA, USA). For the observations, images of fractured 5 mm × 5 mm cross-sections of the panels were taken.

The moisture content of the material was measured with a laboratory moisture meter (model UM2000, Imal s.r.l., Modena, Italy). The water immersion test for the panels was carried out in a heated tank (Model 76-B0066/B Water Bath, Equipos de Ensayo Controls S.A., Toledo, Spain).

The mechanical tests and density were performed with the universal testing machine (model IB700, Imal s.r.l., Modena, Italy), which complies with the velocity of 5 mm·min^{-1} for the bending test and 2 mm·min^{-1} for internal bonding strength.

For the statistical analysis, SPSS v. 26.0 software (IBM, Chicago, IL, USA) was used. Analysis of variance (ANOVA) was performed for a significance level of $\alpha < 0.05$. The standard deviation was obtained for the mean values of the tests.

3. Results

3.1. Physical Properties

Average density is showed in Figure 3, with all boards considered to have medium-high density. Boards made with starch alone (F, G) had a density range from 866 to 988 kg/m^3, those with cement alone (D, E) were from 942 to 1026 kg/m^3 and those with starch and cement (A, B, C) were from 990 to 1116 kg/m^3. Boards manufactured with a mix of cement and starch have higher density than using these two components separately. The boards with the highest density were B4 class, with a composition of 5% potato starch and 20% cement. They were in the hot plate press for 4 h, reaching an average density of 1116 kg/m^3.

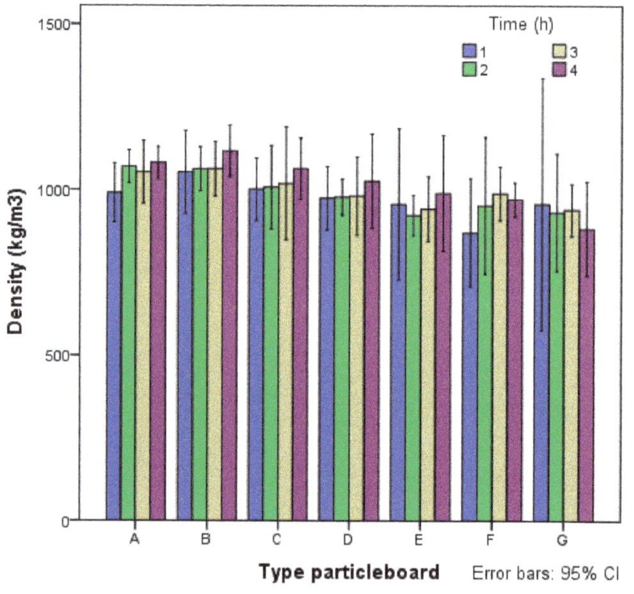

Figure 3. Mean density of type A–G particleboards.

Figure 4 shows the mean values of the thickness swelling tests (TS) and water absorption (WA) after 24 h of immersion in water. The boards with starch alone (F, G) had a TS from 37.2 to 56.3%, with cement (D, E) ranging from 26.7 to 58.5% and starch and cement (A, B, C) from 11.5 to 23.8%. WA follows the same tendency of TS. Type B, made with 5% starch and 20% cement, had the lowest WA value between 37.27 and 43.85%. Panels made of cement and starch achieved a great result and could be classified as structural boards.

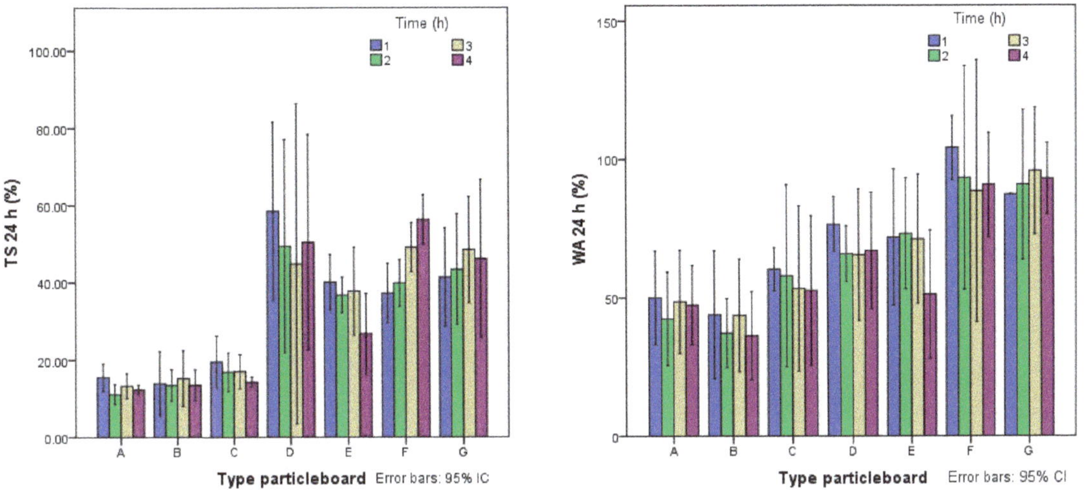

(**a**) (**b**)

Figure 4. (**a**) Thickness swelling (TS) after 24 h and (**b**) water absorption (WA) after 24 h.

3.2. Mechanical Properties

Modulus of rupture (MOR) and modulus of elasticity (MOE) results follow a similar pattern, as shown in Figure 5. MOR performance of the boards made with starch (F, G) ranged from 7.89 to 18.53 N/mm^2, with cement (D, E) ranging from 6.21 to 18.04 N/mm^2 and starch and cement (A, B, C) from 12.59 to 27.26 N/mm^2. MOE values with starch (F, G) ranged between 1663.3 and 2763.7 N/mm^2, with cement (D, E) ranging from 1234.8 to 3080.1 N/mm^2 and starch and cement (A, B, C) from 1846.1 to 4287.4 N/mm^2.

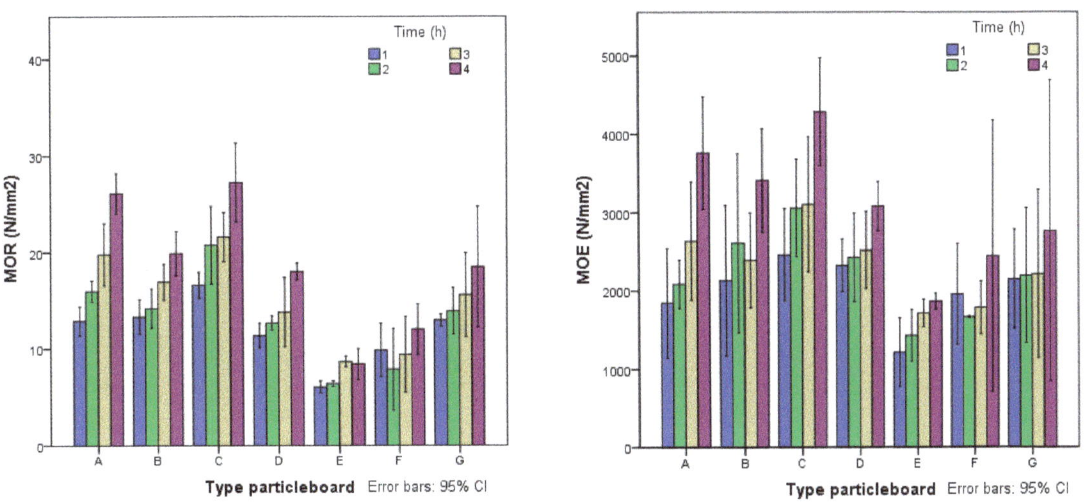

(**a**) (**b**)

Figure 5. (**a**) Modulus of rupture (MOR) and (**b**) and modulus of elasticity (MOE).

Type A and C boards, using starch and cement and a press time of 1 h, can be classified as P2. After being in the press for 4 h, these panels achieved the P7 requirements. The mechanical results obtained were very high, showing the good mechanical properties that can be achieved with boards made of giant reed–cement–starch.

The average values obtained for internal bonding strength (IB) (Figure 6) varied between 0.20 and 0.28 N/mm^2, with starch (F, G) with cement (D, E) ranging between 0.15 and 0.36 N/mm^2 and starch and cement (A, B, C) between 0.36 and 0.89 N/mm^2.

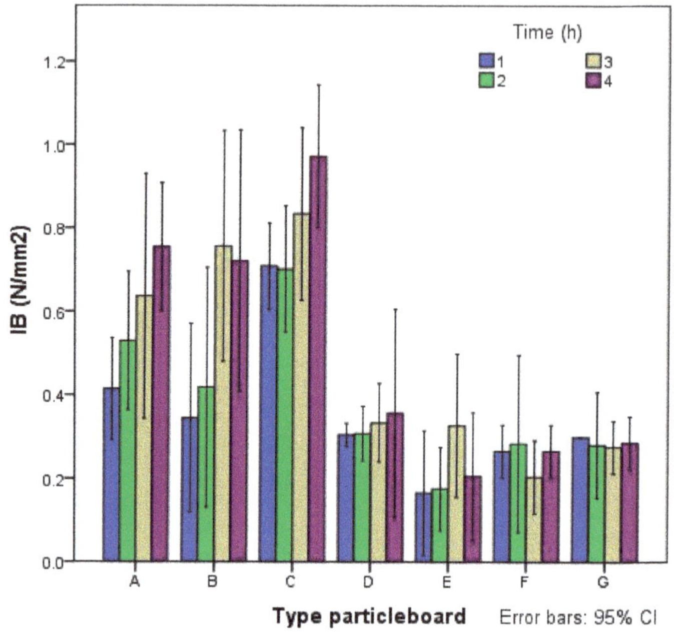

Figure 6. Internal bounding strength.

The results of the test indicate that the contribution of starch improved the properties of the boards, as was observed in the palm–cement boards [37]. Board types A, B and C could be classified as P7, whereas if only cement or starch was used, the boards only met the P1 requirements.

3.3. *Assessment of the Durability of the Fibers*

To verify that fibers were not degraded in contact with cement, four extra boards of classes A2, B2, C2, D2 and E2 (the types containing cement) were made. These extra specimens were cut for the mechanical tests that were carried out 365 days after manufacture and compared to the others specimens from the same class. The results obtained from MOR and MOE are shown in Figure 7. It can be concluded that boards made with starch and cement (classes A2, B2 and C2) had a significant increase in the MOR and MOE values at 365 days of setting. The boards that only contained cement (classes D2 and E2) presented similar values.

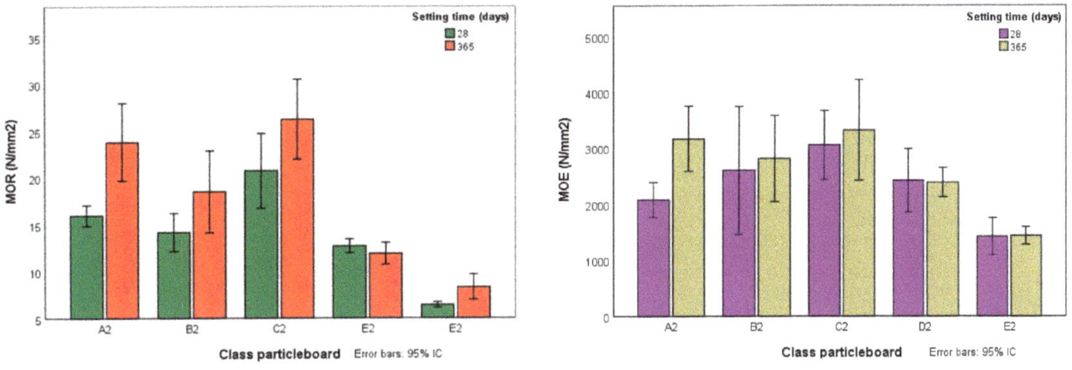

(a) (b)

Figure 7. (**a**) Modulus of rupture (MOR) and (**b**) modulus of elasticity (MOE) after 28 and 365 days of cement hydration.

This demonstrates that giant reed particles are not degraded by the alkalinity of the cement, as no deterioration was observed in their mechanical properties. Boards made of reed with cement and starch showed a notable improvement over time that may be justified by the beneficial effect that starch could exert on the setting of the cement. It is possible that the water absorbed by the starch at the beginning of the manufacturing process was gradually transferred to the cement, favoring its subsequent hydration.

3.4. Statistical Analysis

The ANOVA shown in Table 4 indicated that density depends on the type of board, whereas the time in the press has no influence. Density also depends on the amount of starch used but not the cement. TS and WA are dependent on the type of board and the amount of cement and starch in the mix. Pressing time is not influential in the physical properties. MOR, MOE and IB are influenced by all factors: type of board, pressing time and amount of cement and starch added to the mix.

Table 4. ANOVA of the test results.

Factor	Properties	Sum of Squares	d.f.	Half Quadratic	F	Sig
Type of panel	Density (kg/m^3)	246,938.733	6	41,156.455	8.851	0.000
	TS 24 h (%)	24,117.087	6	4019.514	79.859	0.000
	WA 24 h (%)	30,972.016	6	5162.003	24.673	0.000
	MOR (N/mm^2)	2203.595	6	367.266	22.757	0.000
	MOE (N/mm^2)	2.572×10^7	6	4.286×10^6	8290	0.000
	IB (N/mm^2)	3.947	6	0.658	10.880	0.000
Pressing time	Density (kg/m^3)	44,648.253	3	14,882.751	2.285	0.083
	TS 24 h (%)	421.488	3	140.496	0.496	0.686
	WA 24 h (%)	1060.883	3	353.628	0.708	0.549
	MOR (N/mm^2)	1081.134	3	360.378	13.461	0.000
	MOE (N/mm^2)	2.348×10^7	3	7.827×10^6	14.943	0.000
	IB (N/mm^2)	0.995	3	0.332	3.771	0.013

Table 4. *Cont.*

Factor	Properties	Sum of Squares	d.f.	Half Quadratic	F	Sig
Starch (%)	Density (kg/m^3)	225,752.148	2	112,876.074	24.920	0.000
	TS 24 h (%)	22,366.496	2	11,183.248	170.680	0.000
	WA 24 h (%)	28,956.715	2	14,478.357	71.060	0.000
	MOR (N/mm^2)	1686.149	2	843.074	40.973	0.000
	MOE (N/mm^2)	14,861,771.141	2	7,430,885.571	12.256	0.000
	IB (N/mm^2)	3.682	2	1.841	50.932	0.000
Cement (%)	Density (kg/m^3)	7589.261	1	7589.261	1.391	0.243
	TS 24 h (%)	80.590	1	80.590	7.165	0.010
	WA 24 h (%)	1418.878	1	1418.878	5.653	0.021
	MOR (N/mm^2)	307.374	1	307.374	15.118	0.000
	MOE (N/mm^2)	8,513,888.803	1	8,513,888.803	12.613	0.001
	IB (N/mm^2)	0.773	1	0.773	16.647	0.000

d.f.: degrees of freedom. F: Fisher–Snedecor distribution. Sig: significance.

3.5. SEM Observations

Scanning electron microscope (SEM) observations were made to investigate the interaction between the three components of the experimental boards. Using a standard method [78], the mineralogical evolution of cement during hydration of the boards was analyzed. Micrographs were obtained at 8 and 28 days after the manufacture of the boards of class 2, which contained 20% cement and 10% starch and had been pressed for 2 h. Samples for SEM were polished and plated with gold.

Figure 8a shows that starch has gelled and enveloped the cement grains that were being hydrated. Figure 8b shows how cement grains were hydrating eight days after panels were made. It can be differentiated that alite (calcium silicate) and belite (dicalcium silicate) were hydrating, forming tobermorite grains. Hydration cracks were also found.

(**a**)

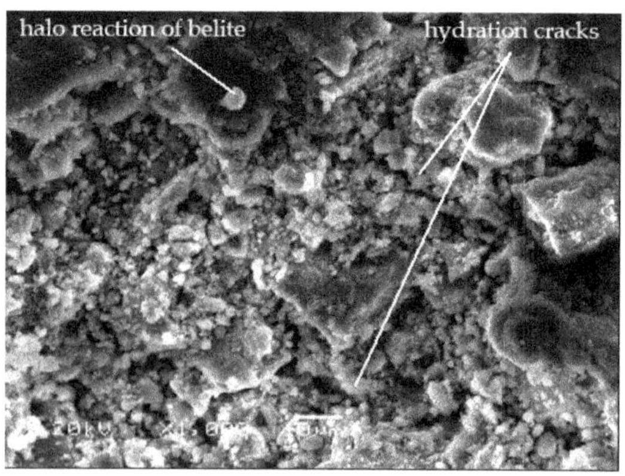

(**b**)

Figure 8. (**a**) Starch gel enveloping cement grains 3 days after hydration and (**b**) cement grains 8 days after hydration.

Figure 9a shows a cube of gypsum (calcium sulfate) that was added to the clinker to decrease the solubility of the aluminate. This cube is surrounded by crystals of tricalcium silicate (C_3S) transformed into tobermorite gel (hydrated calcium silicates), whose grains began to coalesce. In Figure 9b, tobermorite is seen over an organic gel, indicating that the starch has gelled.

(**a**) (**b**)

Figure 9. (**a**) Gypsum cube surrounded by tobermorite gel 3 days after hydration and (**b**) tobermorite over an organic gel 8 days after hydration.

Figure 10a shows that tobermorite gel has bonded, forming a continuous matrix, while in Figure 10b some tobermorite grains have coalesced forming a continuous layered matrix. Some thin hexagonal plates can be seen, indicating the presence of portlandite [$Ca(OH)_2$].

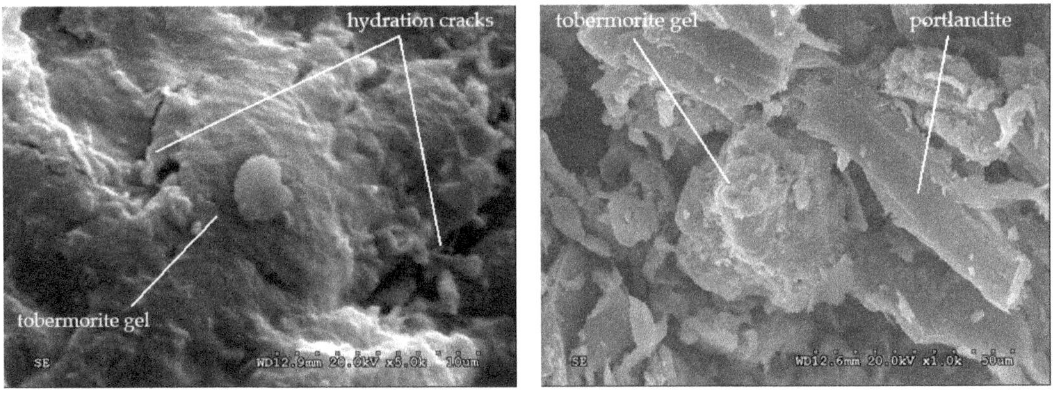

(**a**) (**b**)

Figure 10. (**a**) Tobermorite gel 8 days after hydration and (**b**) tobermorite and crystalized portlandite 8 days after hydration.

Figure 11a shows the halo reaction of belite (dicalcium silicate). In Figure 11b, alite crystals (calcium silicate) in the hydration process are seen, forming tobermorite grains which were forming a gel.

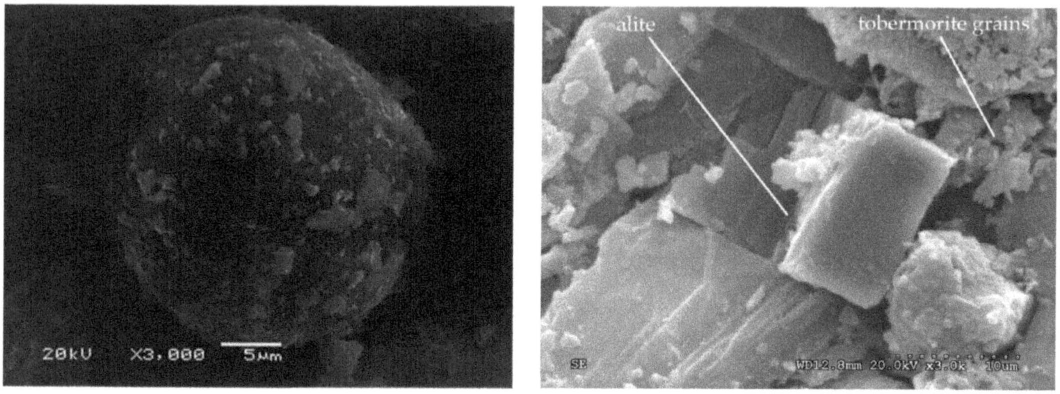

(a) (b)

Figure 11. (**a**) Belite 8 days after hydration and (**b**) tobermorite, alite and portlandite 8 days after hydration.

Portlandite (calcium hydroxide) over tobermorite (hydrated calcium silicates) is observed in Figure 12a, with no ettringite found. Figure 12b shows giant reed fibers glued together, with no degradation due to the alkalinity of the cement observed.

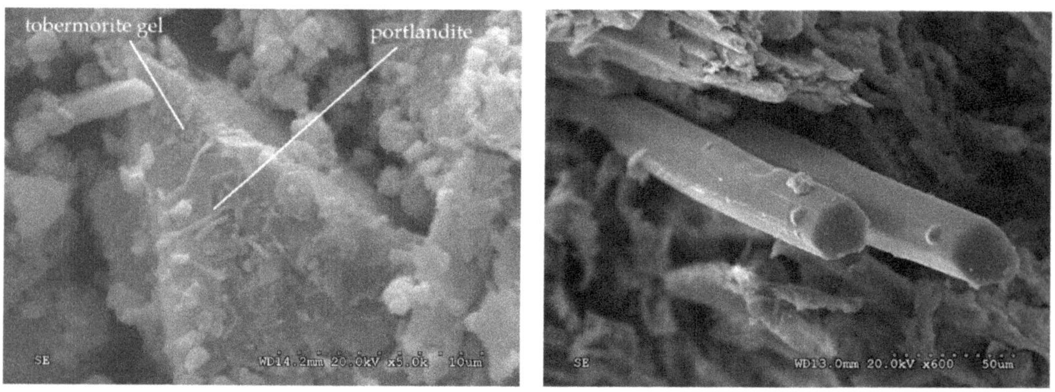

(a) (b)

Figure 12. (**a**) Portlandite over tobermorite 8 days after hydration and (**b**) giant reed fibers 8 days after hydration.

There is no visible sign of microcracks on the surface of the boards 8 or 28 days after setting in Figure 13. Even though other authors observed degradation of plant fibers in contact with cement by alkaline hydrolysis, the hydration process of the experimental boards of this study was optimal.

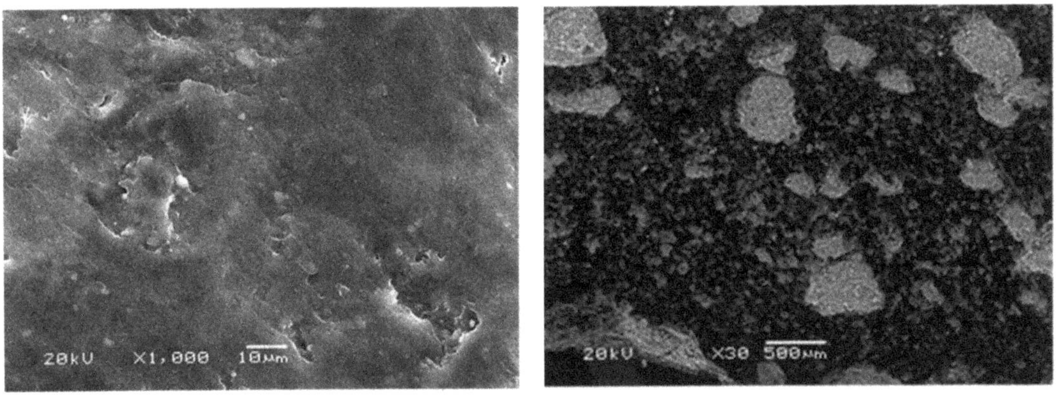

(a) (b)

Figure 13. Surface of the board (**a**) 8 days after hydration and (**b**) 28 days after hydration.

To elucidate the mechanism involved in why boards with starch (types A, B and C) have better properties than boards without starch (D and E), a micrograph of board type E2 (no starch, 20% cement and 2 h in the press) after 28 days of setting is shown in Figure 14.

Figure 14. Cross-sectional micrograph of a board type E2 28 days after hydration.

It shows cement components without hydration. This indicates that there was a water shortage that prevented a correct absorption of water by the cement. It is possible that water contained in the mixture before hot pressing was attracted by the reed particles and then evaporated when pressed. Boards made with starch could have retained this water and later released it to the cement to hydrate it properly, thereby improving the mechanical performance of the board.

4. Discussion

Particleboards are classified based on their mechanical properties and dimensional stability by European Standards [68], which establishes minimum requirements of their properties. Their grades range from P1 to P7 (Table 5).

Table 5. Classification of experimental panels based on European Standards [68].

Grade	Definition
P1	Boards for general use in dry conditions
P2	Boards for indoor application (including furniture) in dry conditions
P3	Non-structural boards for use in humid conditions
P4	Structural boards for use in dry conditions
P5	Structural boards for use in humid conditions
P6	High performance structural boards for use in dry conditions
P7	High performance structural boards for use in humid conditions

According to Table 6, boards whose only binder was 20% cement (type E) had lower mechanical properties, which indicates that hydration of the cement was not adequate and the water initially contained in the mixture was absorbed by the giant reed particles. When pressure and heat were applied to these boards, water evaporated, preventing the correct setting of the cement.

Table 6. Physical and mechanical properties of the experimental panels.

Type of Panel	Classification EN 312 [68]	MOR (N/mm^2)	MOE (N/mm^2)	IB (N/mm^2)	TS 24 h (%)
A1	P2	12.89 (1.44)	1846.04 (368.20)	0.41 (0.13)	15.54 (3.37)
A2	P3	15.98 (1.19)	2083.50 (335.15)	0.53 (0.22)	11.12 (2.85)
A3	P4	19.79 (3.06)	2636.88 (521.43)	0.66 (0.27)	13.33 (3.08)
A4	P6	26.11 (2.28)	3759.76 (377.99)	0.76 (0.21)	12.38 (1.35)
B1	P1	13.33 (1.13)	2130.78 (304.21)	0.34 (0.15)	13.96 (5.22)
B2	P2	14.19 (1.63)	2610.01 (520.34)	0.42 (0.23)	13.48 (3.28)
B3	P4	16.94 (1.49)	2392.63 (488.00)	0.79 (0.21)	15.25 (5.77)
B4	P4	19.88 (0.91)	3076.93 (431.62)	0.74 (0.14)	13.49 (1.59)
C1	P2	12.56 (6.07)	2021.89 (461.01)	0.54 (0.19)	24.22 (8.13)
C2	P3	20.76 (2.51)	3058.56 (389.95)	0.71 (0.09)	16.83 (3.09)
C3	P3	21.60 (1.60)	3104.68 (541.19)	0.84 (0.24)	16.95 (2.75)
C4	P6	27.26 (3.29)	4287.44 (554.31)	0.95 (0.33)	14.04 (1.03)
D1	P1	11.41 (0.81)	2325.87 (211.76)	0.31 (0.01)	58.50 (14.50)
D2	P1	12.72 (0.46)	2426.60 (356.39)	0.31 (0.04)	49.42 (0.17)
D3	P1	13.82 (1.44)	2519.48 (198.45)	0.34 (0.03)	44.83 (16.6)
D4	P1	18.04 (0.35)	3080.12 (126.14)	0.36 (0.10)	50.41 (11.20)
E1	-	6.22 (0.22)	1234.77 (131.71)	0.16 (0.11)	42.24 (1.56)
E2	-	6.40 (0.24)	1426.89 (167.47)	0.16 (0.06)	36.79 (3.71)
E3	-	8.71 (0.34)	1713.74 (11.04)	0.33 (0.10)	37.79 (7.15)
E4	-	9.45 (1.02)	1865.42 (64.53)	0.21 (0.09)	26.73 (6.59)
F1	-	9.90 (0.30)	1958.16 (71.76)	0.27 (0.01)	37.28 (1.85)
F2	-	7.90 (0.47)	1663.29 (1.38)	0.28 (0.02)	39.89 (0.67)
F3	-	9.46 (0.43)	1784.03 (37.78)	0.21 (0.01)	49.11 (0.71)
F4	-	12.06 (0.29)	2443.87 (51.51)	0.27 (0.01)	56.34 (0.71)
G1	P1	13.00 (0.07)	2148.25 (70.71)	0.30 (0.01)	41.41 (1.41)
G2	P1	14.04 (0.27)	2192.85 (95.95)	0.28 (0.01)	43.44 (1.59)
G3	P1	15.63 (0.48)	2211.76 (119.50)	0.28 (0.01)	48.49 (1.52)
G4	P1	18.53 (0.69)	2763.68 (72.18)	0.29 (0.01)	46.18 (2.28)

Table 6. *Cont.*

Type of Panel	Classification EN 312 [68]	MOR (N/mm^2)	MOE (N/mm^2)	IB (N/mm^2)	TS 24 h (%)
Thickness 6–13 mm [68]	P1	10.50	-	0.28	-
	P2	11.00	1800.00	0.40	-
	P3	15.00	2050.00	0.45	17.00
Thickness 6–10 mm [68]	P4	16.00	2300.00	0.40	16.00
	P5	18.00	2550.00	0.45	13.00
	P6	20.00	3150.00	0.60	16.00
	P7	22.00	3350.00	0.75	10.00

() standard deviation.

The boards with the best performance were types A4 and C4, which met European Standards [68] and could be classified as P6 (high performance structural load-bearing particleboard to use in dry conditions). In general, boards with the best properties are those in which a mixture of cement and starch has been used as a binder.

Higher proportions of cement produced lower TH, whereas higher proportions of starch resulted in better mechanical performances. Starch could have retained part of the water previously added onto the mixture to later transfer it to the cement, favoring its correct hydration. This confirmed a study that suggested that in order to improve the properties of cementitious compounds, an additive of cellulose and starch was added because it behaves as a good water-retention agent [62].

The two manufacturing variables considered in this study were the type of binder and the time in the hot plate press. The type of binder is the most influential parameter since all physical and mechanical properties varied according to it (density, TS, WA MOR, MOE and IB). The pressing time affected half of the properties tested (MOR, MOE and IB).

Several authors noticed that during the setting time of cement composites with plant fibers, there was a delay in the formation of hydration products [15,28,79]. They attributed it to different fiber components, especially sugars. A different study indicated that pectins contained in the jute fibers combined with the cement functioned as a growth inhibitor of calcium silicate hydrate (CSH) [41].

SEM micrographs showed calcium compounds at eight days of setting and therefore the hydration process had not been delayed. In addition, micrographs showed how the gelled starch enveloped the cement grains, which would make difficult its dilution and the interaction of the reed components with the cement.

The hydration of the cement in the experimental boards is in accordance with the study carried out with cement without plant fibers at seven days [78]. Some research [57] affirms that plant fibers degrade in contact with cement by alkaline hydrolysis. However, in this work, no signs of degradation of giant reed fibers were observed at 28 days. Another investigation [31] indicated that composites made of cement and plant fibers decrease performance in MOR and MOE after one year of setting due to carbonation of the matrix followed by lixiviation and progressive microcracking. In this paper, results appear to differ, with the MOR and MOE behavior of the panels even increasing, indicating that there was no incompatibility between the three components (giant reed, cement and starch).

Other researchers obtained composites made of cement and wood or plant fibers. Their results are shown in Table 7. Pine wood–cement composites have lower performance than the boards of this study. It is possible that this type of wood requires large amounts of cement to comply with European Standards [68]. With eucalypt and rubberwood or with pretreated coconut coir, the results of TS are better than in this study. Nonetheless, one of the uses of these composites is in humid conditions such as terraces or façades. It is possible that the pretreatment of the coconut or the amount of cement could improve the TS of the panels of this study. The behavior of canary palm composites with amounts of cement similar to this investigation is sufficient to be commercialized as P2 [68]. However, their properties are behind the giant reed–cement composites of this study.

Table 7. Properties obtained with cement composites.

Source	Material	Ratio Cement/Material	MOR (N/mm^2)	MOE (N/mm^2)	IB (N/mm^2)	TS 24 h (%)
[5]	Pine	3/1	15.8	5495	-	18.5
[9]	50% Eucalypt and 50% rubberwood	8/1	6.40	4090	0.34	1.80
[21]	Coconut coir	2/1	19.94	5315	0.73	3.64
[37]	Canary palm	1/5	15.76	1872	0.68	26.70
This work (C4)	Giant reed	1/10	27.26	4287.44	0.95	14.04

Industrial boards are normally manufactured with a wood/cement ratio of 1/5.5. The boards in this work have been made with a small amount of cement, with the ratio in weight of giant reed/cement being 1/0.2 maximum. Pressing time was a maximum of 4 h at 100 °C, while industrial boards need 8 h at 75 °C and drying times of 8 h at 90 °C. The experimental boards described in this study have been made with less time and with a much lower proportion of cement than in the production of industrial wood–cement boards. This represents significant energy savings in their manufacture and makes them potentially a more ecological product.

5. Conclusions

The effects of the binder (starch–cement) on the physical and mechanical properties of boards made of giant reed (*Arundo donax* L.) particles were investigated.

High-quality sustainable particleboards (boards with high structural performance) were obtained using giant reed and can be classified as P6 according to European Standards. Mechanical properties were improved by increasing the starch content, whereas greater resistance to water was obtained by increasing the cement content.

The pressing time of the boards in the hot plate press at 100 °C had a great influence in MOR, MOE and IB, which increased performance. Further tests are needed to determine the optimum pressing temperature and time for the manufacturing process.

Cement is mineralized in the giant reed–cement–starch particleboards following the same hydration process that occurred without fibers or additives. Giant reed particles seem to tolerate the alkalinity of the cement since there was no sign of degradation of its fibers, with MOR and MOE even increasing at 365 days of setting.

Using giant reed particles, cementitious composites can be made which could be used in the manufacture of various construction components. Therefore, the use of these residues in the manufacture of construction materials offers a very attractive alternative in terms of price, technology and sustainability.

Author Contributions: Idea and methodology, A.A.F.-G. and M.T.F.-G.; Experiments, M.T.F.-G. and A.A.F.-G.; Resources, T.G.O.; Statistics, M.F.-V. and A.F.-G.; Project administration, A.F.-G.; Supervision, M.F.-V.; Writing, A.A.F.-G. and A.F.-G.; Review, T.G.O. All authors have read and agreed to the published version of the manuscript.

Funding: This research was funded thanks to Agreement No. 4/20 between the company Aitana, Actividades de Construcciones y Servicios, S.L., and Universidad Miguel Hernandez, Elche.

Institutional Review Board Statement: Not applicable.

Informed Consent Statement: Not applicable.

Data Availability Statement: The data presented in this study are available within the article.

Acknowledgments: The authors would like to thank the company Aitana, Actividades de Construcciones y Servicios, S.L., for its support by signing Agreement No. 4/20 with Universidad Miguel Hernández, Elche, on 20 December 2019.

Conflicts of Interest: The authors declare no conflict of interest.

References

1. Directive 2008/98/EC of The European Parliament and of The Council on Waste and Repealing Certain Directives. *Off. J. Eur. Union* **2008**, *L312*, 3–30, Current consolidated version: 5 July 2018.
2. Directive 2010/75/EU of the European Parliament and of the Council of 24 November 2010 on industrial emissions (integrated pollution prevention and control). *Off. J. Eur. Union* **2010**, *L334*, 159–261, Current consolidated version: 6 January 2011.
3. Directive (EU) 2018/850 of the European Parliament and of the Council of 30 May 2018 amending Directive 1999/31/EC on the landfill of waste. *Off. J. Eur. Union* **2018**, *L150*, 100–108.
4. Regulation (EU) 2018/841 of the European Parliament and of the Council of 30 May 2018 on the inclusion of greenhouse gas emissions and removals from land use, land use change and forestry in the 2030 climate and energy framework, and amending Regulation (EU) No 525/2013 and Decision No 529/2013/EU. *Off. J. Eur. Union* **2018**, *L156*, 1–25.
5. Moslemi, A.A.; Pfister, S.C. The influence of cement/wood ratio and cement type on bending strength and dimensional stability of wood-cement composite panels. *Wood Fiber Sci.* **1987**, *19*, 165–175.
6. Ferrier, E.; Agbossou, A.; Michel, L. Mechanical behaviour of ultra-high-performance fibrous-concrete wood panels reinforced by FRP bars. *Compos. Part B Eng.* **2014**, *60*, 663–672. [CrossRef]
7. Li, M.; Khelifa, M.; Khennane, A.; El Ganaoui, M. Structural response of cement-bonded wood composite panels as permanent formwork. *Compos. Struct.* **2019**, *209*, 13–22. [CrossRef]
8. Brahmia, F.Z.; Horváth, P.G.; Alpár, T.L. Effect of pre-treatments and additives on the improvement of cement wood composite: A review. *BioResources* **2020**, *15*, 7288–7308. [CrossRef]
9. Okino, E.Y.; De Souza, M.R.; Santana, M.A.; da S Alves, M.V.; de Sousa, M.E.; Teixeira, D.E. Cement-bonded wood particleboard with a mixture of eucalypt and rubberwood. *Cem. Concr. Compos.* **2004**, *26*, 729–734. [CrossRef]
10. Tittelein, P.; Cloutier, A.; Bissonnette, B. Design of a low-density wood-cement particleboard for interior wall finish. *Cem. Concr. Compos.* **2012**, *34*, 218–222. [CrossRef]
11. Fuwape, J.A.; Oyagade, A.O. Bending strength and dimensional stability of tropical wood-cement particleboard. *Bioresour. Technol.* **1993**, *44*, 77–79. [CrossRef]
12. Karade, S.R. Cement-bonded composites from lignocellulosic wastes. *Constr. Build. Mater.* **2010**, *24*, 1323–1330. [CrossRef]
13. Ashori, A.; Tabarsa, T.; Valizadeh, I. Fiber reinforced cement boards made from recycled newsprint paper. *Mater. Sci. Eng. A* **2011**, *528*, 7801–7804. [CrossRef]
14. Olorunnisola, A.O.; Adefisan, O.O. Trial production and testing of cement-bonded particleboard from rattan furniture waste. *Wood Fiber Sci.* **2002**, *34*, 116–124.
15. Sudin, R.; Swamy, N. Bamboo and wood fibre cement composites for sustainable infrastructure regeneration. *J. Mater. Sci.* **2006**, *41*, 6917–6924. [CrossRef]
16. Torkaman, J.; Ashori, A.; Momtazi, A.S. Using wood fiber waste, rice husk ash, and limestone powder waste as cement replacement materials for lightweight concrete blocks. *Constr. Build. Mater.* **2014**, *50*, 432–436. [CrossRef]
17. Wei, J.; Meyer, C. Sisal fiber-reinforced cement composite with Portland cement substitution by a combination of metakaolin and nanoclay. *J. Mater. Sci.* **2014**, *49*, 7604–7619. [CrossRef]
18. Frazão, C.; Barros, J.; Toledo Filho, R.; Ferreira, S.; Gonçalves, D. Development of sandwich panels combining sisal fiber-cement composites and fiber-reinforced lightweight. *Concr. Cem. Concr. Comp.* **2018**, *86*, 206–223. [CrossRef]
19. Thomas, B.C.; Jose, Y.S. Impact of sisal fiber reinforced concrete and its performance analysis: A review. *Evolut. Intell.* **2019**, 1–11. [CrossRef]
20. Aggarwal, L.K. Studies on cement bonded coir fibre boards. *Cem. Concr. Compos.* **1992**, *14*, 63–69. [CrossRef]
21. Asasutjarit, C.; Hirunlabh, J.; Khedari, J.; Charoenvai, S.; Zeghmati, B.; Cheul Shin, U. Development of coconut coir-based lightweight cement board. *Constr. Build. Mater.* **2007**, *21*, 277–288. [CrossRef]
22. Silva, F.A.; Toledo Filho, R.D.; Melo Filho, J.A.; Fairbairn, E.M.R. Physical and mechanical properties of durable sisal fiber-cement composites. *Constr. Build. Mater.* **2010**, *24*, 777–785. [CrossRef]
23. Wang, W.; Chouw, N. The behaviour of coconut fibre reinforced concrete (CFRC) under impact loading. *Constr. Build. Mater.* **2017**, *134*, 452–461. [CrossRef]
24. Da Silva, E.J.; Marques, M.L.; Velasco, F.G.; Junior, C.F.; Luzardo, F.M.; Tashima, M.M. A new treatment for coconut fibers to improve the properties of cement-based composites—Combined effect of natural latex/pozzolanic materials. *Sustain. Mater. Technol.* **2017**, *12*, 44–51. [CrossRef]
25. Sathiparan, N.; Rupasinghe, M.N.; Pavithra, B.H. Performance of coconut coir reinforced hydraulic cement mortar for surface plastering application. *Constr. Build. Mater.* **2017**, *142*, 23–30. [CrossRef]
26. Farias Machado, N.A.; Barros Furtado, M.; Parra-Serrano, L.J.; de Oliveira Maia Parente, M.; Fiorelli, J.; Savastano Júnior, H. Agglomerated panels made from babaçu coconut residues. *Rev. Bras. Cienc. Agrár.* **2017**, *12*, 202–209. [CrossRef]
27. Aggarwal, L.K. Bagasse-reinforced cement composites. *Cem. Concr. Compos.* **1995**, *17*, 107–112. [CrossRef]
28. Bilba, K.; Arsène, M.A.; Ouensanga, A. Sugar cane bagasse fibre reinforcedcement composites. Part, I. Influence of the botanical components of bagasse on the setting of bagasse/cement composite. *Cem. Concr. Compos.* **2003**, *25*, 91–96. [CrossRef]
29. Cabral, M.R.; Nakanishi, E.Y.; Dos Santos, V.; Palacios, J.H.; Godbout, S.; Junior, H.S.; Fiorelli, J. Evaluation of pre-treatment efficiency on sugarcane bagasse fibers for the production of cement composites. *Arch. Civ. Mech. Eng.* **2018**, *18*, 1092–1102. [CrossRef]

30. Almeida, R.R.; Del Menezzi, C.H.S.; Teixeira, D.E. Utilization of the coconut shell of babaçu (*Orbignya* sp.) to produce cement-bonded particleboard. *Bioresour. Technol.* **2002**, *85*, 159–163. [CrossRef]
31. Savastano, H., Jr.; Warden, P.G.; Coutts, R.S. Potential of alternative fibre cements as building materials for developing areas. *Cem. Concr. Compos.* **2003**, *25*, 585–592. [CrossRef]
32. Teixeira, J.N.; Silva, D.W.; Vilela, A.P.; Junior, H.S.; de Siqueira Brandão, L.E.V.; Mendes, R.F. Lignocellulosic materials for fiber cement production. *Waste Biomass Valoriz.* **2020**, *11*, 2193–2200. [CrossRef]
33. Cabalo, V.B. Utilization of coconut coir and abaca stripping waste fibers for cement bonded board. *J. Sci. Eng. Technol.* **2015**, *3*, 183–188.
34. Basri, H.B.; Mannan, M.A.; Zain, M.F.M. Concrete using waste oil palm shells as aggregate. *Cem. Concr. Res.* **1999**, *29*, 619–622. [CrossRef]
35. Hermawan, D.; Subiyanto, B.; Kawai, S. Manufacture and properties of oil palm frond cement-bonded board. *J. Wood Sci.* **2001**, *47*, 208–213. [CrossRef]
36. Mannan, M.A.; Alexander, J.; Ganapathy, C.; Teo, D.C.L. Quality improvement of oil palm shell (OPS) as coarse aggregate in lightweight concrete. *Build. Environ.* **2006**, *41*, 1239–1242. [CrossRef]
37. Ferrandez-Villena, M.; Ferrandez-Garcia, C.E.; Garcia-Ortuño, T.; Ferrandez-Garcia, A.; Ferrandez-Garcia, M.T. Properties of cement-bonded particleboards made from Canary Islands palm (Phoenix canariensis Ch.) trunks and different amounts of potato starch. *Forests* **2020**, *11*, 560. [CrossRef]
38. Aggarwal, L.K.; Agrawal, S.P.; Thapliyal, P.C.; Karade, S.R. Cement-bonded composite boards with arhar stalks. *Cem. Concr. Compos.* **2008**, *30*, 44–51. [CrossRef]
39. Juarez, C.; Duran, A.; Valdez, P.; Fajardo, G. Performance of "Agave lecheguilla" natural fiber in Portland cement composites exposed to severe environment conditions. *Build. Environ.* **2007**, *42*, 1151–1157. [CrossRef]
40. Kriker, A.; Bali, A.; Debicki, G.; Bouziane, M.; Chabannet, M. Durability of date palm fibres and their use as reinforcement in hot dry climates. *Cem. Concr. Compos.* **2008**, *30*, 639–648. [CrossRef]
41. Sedan, D.; Pagnoux, C.; Smith, A.; Chotard, T. Mechanical properties of hemp fibre reinforced cement: Influence of the fibre/matrix interaction. *J. Eur. Ceram. Soc.* **2008**, *28*, 183–192. [CrossRef]
42. Diquélou, Y.; Gourlay, E.; Arnaud, L.; Kurek, B. Impact of hemp shiv on cement setting and hardening: Influence of the extracted components from the aggregates and study of the interfaces with the inorganic matrix. *Cem. Concr. Compos.* **2015**, *55*, 112–121. [CrossRef]
43. Çomak, B.; Bideci, A.; Bideci, Ö.S. Effects of hemp fibers on characteristics of cement based mortar. *Constr. Build. Mater.* **2018**, *169*, 794–799. [CrossRef]
44. Nguyen-Sy, T.; Tran-Le, A.D.; Nguyen-Thoi, T.; Langlet, T. A multi-scale homogenization approach for the effective thermal conductivity of dry lime–hemp concrete. *J. Build. Perform. Simul.* **2018**, *11*, 179–189. [CrossRef]
45. Fernea, R.; Manea, D.L.; Plesa, L.; Iernuțan, R.; Dumitran, M. Acoustic and thermal properties of hemp-cement building materials. *Procedia Manuf.* **2019**, *32*, 208–215. [CrossRef]
46. Mahjoub, R.; Yatim, J.M.; Mohd Sam, A.R.; Hashemi, S.H. Tensile properties of kenaf fiber due to various conditions of chemical fiber surface modifications. *Constr. Build. Mater.* **2014**, *55*, 103–113. [CrossRef]
47. Ferrández-García, C.E.; García-Ortuño, T.; Andreu-Rodríguez, J.; Ferrández-Villena, M.; Ferrández-García, M.T. Mechanical properties of a cement mortar reinforced with *Arundo donax* L. reeds. *Int. J. Civ. Eng.* **2015**, *2*, 305–308.
48. Demirbas, A.; Aslan, A. Effects of ground hazelnut shell, wood and tea waste on the mechanical properties of cement. *Cem. Concr. Res.* **1998**, *28*, 1101–1104. [CrossRef]
49. Karade, S.R.; Irle, M.A.; Maher, K. Influence of granule properties and concentration on cork–cement compatibility. *HolzRohWerkst* **2006**, *64*, 281–286. [CrossRef]
50. Ferrandez-García, M.T.; Ferrandez-Garcia, C.E.; Garcia-Ortuño, T.; Ferrandez-Garcia, A.; Ferrandez-Villena, M. Study of waste jute fibre panels (*corchorus capsularis* L.) agglomerated with Portland cement and starch. *Polymers* **2020**, *12*, 599. [CrossRef]
51. Ferreira, S.R.; Pepe, M.; Martinelli, E.; de Andrade Silva, F.; Toledo Filho, R.D. Influence of natural fibers characteristics on the interface mechanics with cement based matrices. *Compos. Part B Eng.* **2018**, *140*, 183–196. [CrossRef]
52. Ahmad, M.R.; Chen, B.; Oderji, S.Y.; Mohsan, M. Development of a new bio-composite for building insulation and structural purpose using corn stalk and magnesium phosphate cement. *Energy Build.* **2018**, *173*, 719–733. [CrossRef]
53. Zadorecki, P.; Flodin, P. Surface modification of cellulose fibers. I. Spectroscopic characterization of surface-modified cellulose fibers and their copolymerization with styrene. *J. Appl. Polym. Sci.* **1985**, *30*, 2419–2429. [CrossRef]
54. Zadorecki, P.; Flodin, P. Surface modification of cellulose fibers. II. The effect of cellulose fiber treatment on the performance of cellulose-polyester composites. *J. Appl. Polym. Sci.* **1985**, *30*, 3971–3983. [CrossRef]
55. Bilba, K.; Arsene, M.A. Silane treatment of bagasse fiber for reinforcement of cementitious composites. *Compos. Part A* **2008**, *39*, 1488–1495. [CrossRef]
56. Chakraborty, S.; Kundu, S.P.; Roy, A.; Adhikari, B.; Majumder, S.B. Polymer modified jute fibre as reinforcing agent controlling the physical and mechanical characteristics of cement mortar. *Constr. Build. Mater.* **2013**, *49*, 214–222. [CrossRef]
57. Yan, L.; Kasal, B.; Huang, L. A review of recent research on the use of cellulosic fibres, their fibre fabric reinforced cementitious, geo-polymer and polymer composites in civil engineering. *Compos. Part B Eng.* **2016**, *92*, 94–132. [CrossRef]

58. Vo, L.T.; Navard, P. Treatments of plant biomass for cementitious building materials. *Constr. Build. Mater.* **2016**, *12*, 161–176. [CrossRef]
59. Peschard, A.; Govin, A.; Grosseau, P.; Guilhot, B.; Guyonnet, R. Effect of polysaccharides on the hydration of cement paste at early ages. *Cem. Concr. Res.* **2004**, *34*, 2153–2158. [CrossRef]
60. Peschard, A.; Govina, A.; Pourchez, J.; Fredon, E.; Bertrand, L.; Maximilien, S.; Guilhot, B. Effect of polysaccharides on the hydration of cement suspension. *J. Eur. Ceram. Soc.* **2006**, *26*, 1439–1445. [CrossRef]
61. Dewacker, D.R.; Mcad, B. Cement Mortar Systems Using Blends of Polysaccharides and Cold-Water-Soluble Unmodified Starches. U.S. Patent 5575840, 19 November 1996.
62. Zhang, D.F.; Ju, B.Z.; Zhang, S.F.; He, L.; Yang, J.Z. The study on the dispersing mechanism of starch sulfonate as a water-reducing agent for cement. *Carbohydr. Polym.* **2007**, *70*, 363–368. [CrossRef]
63. Ferrández-García, M.T.; García-Ortuño, T.; Ferrández-Villena, M.; Andreu-Rodríguez, J.; Ferrández-García, C.E. Study of the properties of cement boards improved with starch. In Proceedings of the 8th Iberian Congress of Agroengineering, Libro de Actas: Retos de la Nueva Agricultura Mediterránea, Orihuela, Spain, 1–3 June 2015; Universidad Miguel Hernández de Elche: Elche, Spain, 2015; pp. 387–392, ISBN 978-84-16024-30-8.
64. Islam, M.R.; Rahman, F.; Islam, M.N.; Rana, M.N.; Nath, S.K.; Ashaduzzaman, M.; Shams, M.I. Cement-bonded lignocellulosic panel (CLP): A promising environmental friendly construction material for conservation of forest resources. In *Handbook of Ecomaterials*; Springer: New York, NY, USA, 2018; pp. 1–13. [CrossRef]
65. Onuaguluchi, O.; Banthia, N. Plant-based natural fibre reinforced cement composites: A review. *Cem. Concr. Compos.* **2016**, *68*, 96–108. [CrossRef]
66. Deltoro Torró, V.; Jiménez Ruiz, J.; Vilán Fragueiro, X.M. *Bases para el Manejo y Control de Arundo donax* L. *(Caña común). Colección Manuales Técnicos de Biodiversidad*; Conselleria d'Infraestructures, Territori i Medi Ambient. Generalitat Valenciana: Valencia, Spain, 2012; ISBN 978-84-482-5777-4.
67. Ferrández García, M.T. Análisis y Desarrollo de Tableros Multicapa de Caña Común (*Arundo donax* L.). Ph.D. Thesis, Universidad Miguel Hernández, Elche, Spain, 28 July 2008.
68. EN 312. In *Particleboards—Specifications*; European Committee for Standardization: Brussels, Belgium, 2010.
69. García-Ortuño, T.; Rodríguez, J.; Ferrández-García, M.T.; Ferrández-Villena, M.; Ferrández-García, C.E. Evaluation of the physical and mechanical properties of particleboard made from giant reed (*Arundo donax* L.). *BioResources* **2011**, *6*, 477–486. [CrossRef]
70. Ferrández-García, C.E.; Andreu-Rodríguez, J.; Ferrández-García, M.T.; Ferrández-Villena, M.; García-Ortuño, T. Panels made from giant reed bonded with non-modified starches. *BioResources* **2012**, *7*, 5904–5916. [CrossRef]
71. Shon, C.S.; Mukashev, T.; Lee, D.; Zhang, D.; Kim, J.R. Can common reed fiber become an effective construction material? Physical, mechanical, and thermal properties of mortar mixture containing common reed fiber. *Sustainability* **2019**, *11*, 903. [CrossRef]
72. EN 326: Wood-based panels, Cutting and Inspection. In *Part 1: Sampling and Cutting of Test Pieces and Expression of Test*; European Committee for Standardization: Brussels, Belgium, 1994.
73. EN 309: Particleboards. In *Definitions and Classification*; European Committee for Standardization: Brussels, Belgium, 2005.
74. EN 323: Wood-based panels. In *Determination of Density*; European Committee for Standardization: Brussels, Belgium, 1993.
75. EN 317: Particleboards and fiberboards. In *Determination of Swelling in Thickness after Immersion in Water*; European Committee for Standardization: Brussels, Belgium, 1993.
76. EN 319: Particleboards and fiberboards. In *Determination of Tensile Strength Perpendicular to the Plane of de Board*; European Committee for Standardization: Brussels, Belgium, 1993.
77. EN 310: Wood-based panels. In *Determination of Modulus of Elasticity in Bending and of Bending Strength*; European Committee for Standardization: Brussels, Belgium, 1993.
78. Giraldo, M.A.; Tobón, J.I. Mineralogical evolution of Portland cement during hydratation process. *Dyna* **2006**, *73*, 69–81.
79. Fan, M.; Ndikontar, M.K.; Zhou, X.; Ngamveng, J.N. Cement-bonded composites made from tropical woods: Compatibility of wood and cement. *Constr. Build. Mater.* **2012**, *36*, 135–140. [CrossRef]

polymers

MDPI

Article

Influence of Different Percentages of Binders on the Physico-Mechanical Properties of *Rhizophora* spp. Particleboard as Natural-Based Tissue-Equivalent Phantom for Radiation Dosimetry Applications

Siti Hajar Zuber [1], Nurul Ab. Aziz Hashikin [1,*], Mohd Fahmi Mohd Yusof [2], Mohd Zahri Abdul Aziz [3] and Rokiah Hashim [4]

1 School of Physics, Universiti Sains Malaysia, Penang 11800, Malaysia; hajarzuber@student.usm.my
2 School of Health Sciences, Universiti Sains Malaysia, Kelantan 16150, Malaysia; mfahmi@usm.my
3 Advanced Medical and Dental Institute, Universiti Sains Malaysia, Penang 13200, Malaysia; mohdzahri@usm.my
4 School of Industrial Technology, Universiti Sains Malaysia, Penang 11800, Malaysia; hrokiah@usm.my
* Correspondence: hashikin@usm.my; Tel.: +60-4-653-3670

Abstract: *Rhizophora* spp. particleboard with the incorporation of lignin and soy flour as binders were fabricated and the influence of different percentages of lignin and soy flour (0%, 6% and 12%) on the physico-mechanical properties of the particleboard were studied. The samples were characterised by Fourier transform infrared spectroscopy (FTIR), X-ray diffraction (XRD), scanning electron microscopy (SEM), energy dispersive X-ray (EDX), X-ray fluorescence (XRF) and internal bonding. The results stipulated that the addition of binders in the fabrication of the particleboard did not change the functional groups according to the FTIR spectrum. For XRD, addition of binders did not reveal any major transformation within the composites. SEM and EDX analyses for all percentages of binders added showed no apparent disparity; however, it is important to note that the incorporation of binders allows better bonding between the molecules. In XRF analysis, lower percentage of chlorine in the adhesive-bonded samples may be advantageous in maintaining the natural properties of the particleboard. In internal bonding, increased internal bond strength in samples with binders may indicate better structural integrity and physico-mechanical strength. In conclusion, the incorporation of lignin and soy flour as binders may potentially strengthen and fortify the particleboard, thus, can be a reliable phantom in radiation dosimetry applications.

Keywords: characterisation; particleboard; natural fiber composites; *Rhizophora* spp.

Citation: Zuber, S.H.; Hashikin, N.A.A.; Yusof, M.F.M.; Aziz, M.Z.A.; Hashim, R. Influence of Different Percentages of Binders on the Physico-Mechanical Properties of *Rhizophora* spp. Particleboard as Natural-Based Tissue-Equivalent Phantom for Radiation Dosimetry Applications. *Polymers* **2021**, *13*, 1868. https://doi.org/10.3390/polym13111868

Academic Editor: Domenico Acierno

Received: 29 April 2021
Accepted: 1 June 2021
Published: 4 June 2021

1. Introduction

In medical physics, phantoms are used as a substitute for humans for various experiments and treatment verification purposes. Since ionising radiation is usually involved, irradiation of human subjects is not allowed, as radiation poses health risks for developing cancers in the future. Phantoms are materials comprising of elemental composition and density close to human tissue, and possess equivalent radiation properties (i.e., they generate similar interactions when irradiated). Acrylic (poly(methyl methacrylate)), a type of thermoplastic, is the most common, commercially available and widely used tissue-equivalent phantom material. Although it is biocompatible and is recyclable, it is not biodegradable and recycling this material usually leads to environmentally harmful by-products. Recent statistic shows that more than 380 million tonnes of plastic are produced every year, and due to their non-biodegradable properties, these plastic wastes may end up as pollutants and affect the natural environment and oceans [1].

Wood is an example of naturally occurring, bio-degradable composite materials that can be considered to replace acrylic as a phantom material. Wood is made up of fibrous chains of cellulose molecules in a matrix of organic polymer lignin. Cellulose fibres from wood are part of natural fibres, and these fibres are among the most abundant and renewable resources, eco-friendly and low cost. Composite materials are often used in constructions, especially in building reinforced structures, bridges and composites products such as particleboard, fibre-reinforced polymer or fiberglass [2,3]. Natural fiber composites are also readily used for interior finishes such as, interior door panels and floor mats [4]. Composite materials are advantageous in terms of their flexibility in design, light weight, physical strength, durability and good resistance toward corrosion [5,6]. Particleboard is a wood-based composite panel product consisting of cellulosic particles of various sizes that may be bonded together with any binders, under the influence of heat and pressure. *Rhizophora* spp. was investigated and show potential as durable particleboard in the construction of furniture, interior applications and as tools in medical applications [7–10]. In radiation dosimetry applications, *Rhizophora* spp. has been studied for their potential as a phantom material and several studies have proved their properties as a tissue-equivalent material [10–12]. The composition and electrical properties of the tissue-mimicking phantom material are also important for better modelling of the phantom, adhering to the body structure represented by the phantom [13]. In order to develop composite phantom materials with tailored properties, one should be able to predict a property of a composite based on the properties of its constituents [14].

Adhesives, also known as binders, may be found naturally or produced synthetically [15]. Wood binder is used in most wood products including particleboards, boards, plywood and many other applications. Over the years, the development of binders in the wood industry shows significant changes, with higher production numbers, and with the emergence of synthetic binders, the range of binder formulations exploded. Commercial synthetic binders are commonly used in the wood industry, which include urea formaldehyde (UF), phenol formaldehyde (PF) and phenol resorcinol formaldehyde (PRF) [16,17]. This is due to their properties, which include excellent water resistance, and they provide better performance than a natural binder [18].

Despite popular use of synthetic resins in the wood industry, several harmful environmental effects were discovered [19,20]. In 2008, the Environmental Protection Agency (EPA) declared formaldehyde a carcinogenic material which brings about environmental and health concerns [20–23]. Among the side effects of formaldehyde towards humans are inflammation of the eyes, nose, throat, asthma-like respiratory allergy, asthma attack with shortness of breath, wheezing and coughing. In the environmental aspect, formaldehyde decomposes in air and forms formic acid and carbon monoxide. Animals that are affected by formaldehyde may experience chronic effects which include, shortened lifespan, reproductive problems and lower fertility. As a result, several attempts have been made to improve environmentally friendly binders in an effort to substitute formaldehyde as a binder in the wood industry.

Soy protein is one of the natural resources to produce natural wood adhesive [19]. Soy products are preferred as binders due to several factors which include, availability, being economical and that presence of carbohydrate in soy products serve as an inert dilute, which is an excellent addition in adhesive materials. They are also safe and perform well as a dominant bonding portion [24]. Other than that, soy protein can withstand hot or cold conditions during the fabrication process, and more efforts have been done to study soy protein as a binder in order to improve the wood bond strength [24–29]. Soy protein is also acknowledged for its good adhesive properties, due to its structures and several modifications. The structure of soy protein can be modified to adhere to the requirement, which may improve its properties. Hydrogen bond presented in the soy protein structure prevents it from interacting with the cellulosic structure of the wood; however, this disadvantage can be overcome by several modifications. The molecular structures of the soy protein can be modified by using physical, chemical or enzymatic means. The modifications result in the breaking down of hydrogen bond, allowing the soy protein structure to be able to react with the wood material with increased adhesion properties.

Lignin is part of the components that hold the plant fibers together and this property roused interests for it to be utilized as a suitable wood binder, which is low-cost, non-toxic, easily available, renewable and environmental-friendly. Lignin consists of approximately 40% of wood's mass and is among the most abundant natural polymers on earth, accounting for 12% to 33% of lignocellulosic biomass [30,31]. Lignin is often studied for the production of carbon fibers, thermoplastics and binders [32–35]. It has been studied as a binder for hundreds of years but the development of a binding system, mainly from lignin, has yet to be carried out [36]. Kraft lignin and lignosulfonates from hardwoods and agricultural residues are among the lignin products that show promising potentials as good wood binders, and more efforts have been done to make binder blends based on these two products [37–40].

There are many studies devoted to the characterisation of natural based composites and polymer as a phantom in radiation dosimetry applications; however, the study of the effect of different percentages of binders towards the characterisation outcomes and physico-mechanical properties are yet to be discussed. In this study, different percentages of lignin and soy flour were formulated as binders in the fabrication of *Rhizophora* spp. particleboard, to study their effects towards different analyses: i.e. Fourier transform infrared (FTIR), X-ray diffraction (XRD), scanning electron microscopy (SEM), energy dispersive X-ray (EDX) and X-ray fluorescence (XRF). Internal bonding analysis was also performed.

2. Materials and Methods

2.1. Sample Preparation of Rhizophora spp. Particleboard Bonded with Lignin and Soy Flour

Rhizophora spp. raw wood trunks were obtained from a coal factory in Perak and underwent several steps, which include drying, debarking, grinding and sieving, to prepare wood particles at approximately 0 μm to 103 μm particle sizes. Lignin (Sigma Aldrich, Merck, QRec, Petaling Jaya, Malaysia) and soy flour (Sigma Aldrich, Merck, QRec, Petaling Jaya, Malaysia) were utilised as binders for the fabrication of the particleboard. Different percentages of binders, i.e. 0%, 6% (3:1 soy flour to lignin) and 12% (3:1 soy flour to lignin) binders, were pre-determined to analyse their effects towards the characterisation and physico-mechanical properties. Bonded particleboards were fabricated at a target density of 1.0 g.cm^{-3}, while maintaining the *Rhizophora* spp. moisture content at approximately 1.39% L to 6.47% L. Moisture content of the sample is best being below 10% L to maintain its relative humidity, and may also be affected by type, size and geometry of the sample [15]. The fabrication process involved hot pressing by using a hot press machine at approximately 200 °C, with a pressure of 20 MPa for approximately 17 min to 20 min.

2.2. Fourier Transform Infrared (FTIR) Analysis

Analysis of the structure of the molecules and functional groups present in the samples were determined using an FTIR spectroscopy, to investigate the chemical bonding with the addition of soy flour and lignin as binders. The samples were ground into approximately 1.0 mg of powder, mixed with approximately <100 mg of potassium bromide (KBr), and was then compressed to form 1.0 mm thickness of pellet. In this analysis, FTIR spectrophotometer (IRPrestige21, Shimadzu, Japan) at the School of Industrial Technology, Universiti Sains Malaysia was utilized.

2.3. X-ray Diffraction (XRD) Analysis

XRD analysis was carried out using Bruker's D8 Advance X-Ray Diffractometer at the Center of Global Archeological Research, Universiti Sains Malaysia, to estimate the degree of crystallinity of the fabricated samples. On the sample holder, the powdered samples were hydraulically pressed into 25 mm diameter circular discs. The samples were radiated at energy of 40 kVp. The source used was Cu-K$_{\alpha 1}$, with wavelength λ of 1.54060 Å, scanning range of approximately 5° 2θ to 70° 2θ and scanning speed of 0.02° 2θ/s.

2.4. Analysis of Scanning Electron Microscopy (SEM) and Energy Dispersive X-ray (EDX)

The microstructures and elemental compositions of the bonded *Rhizophora* spp. samples were obtained and studied using a SEM (FEI Quanta FEG-650, Eindhoven, The Netherlands) and EDX at the Centre for Global Archeological Research, Earth Material Characterisation Laboratory, Universiti Sains Malaysia. The samples with the measurement of (5.0 × 5.0 × 0.5) cm^3 were taped to a specimen holder with two-sided adhesive tape, and coated with gold by a sputter coater (Quorum Q150T S, Quantum Design GmbH, Darmstadt, Germany). The images were taken at a magnification of 4000×. For EDX analysis, the elements were preset to focus only on carbon, nitrogen and oxygen.

2.5. X-ray Fluorescence (XRF) Analysis Using Omnian Analysis Software

The samples were analysed using a 4 kW wavelength dispersive XRF machine (Axios Max, Panalytical, Almelo, Holland, The Netherlands). This machine has a maximum voltage of up to 60 kV and 160 microamps (μA), and is equipped with several crystals such as, PE002, LiF200, PX1 and Ge 111. A representative portion of each sample was ground into 50 μm grain size using a motorised grinding machine, and was further ground manually to a finer grain size of 20 μm, suitable for XRF analysis. The specimen for the XRF analysis was made by igniting approximately 0.5 g of sample and 5.0 g of spectroflux, at approximately 1100 °C (for a duration of 20 min), before it was casted into a glass disc with 32 mm in diameter. The specimen was analysed for 10 major elements using a fully automated XRF spectrometer (Axios Max, Panalytical, Almelo, Holland, The Netherlands), with a standard elemental setup. The calibration technique was employed. The 10 element curves were constructed using 30 high quality international standard reference materials, comparable in composition to the unknown samples. For minor and trace elements, each sample was formed into pressed-powder pellet, using approximately 1 g of sample baked with approximately 6.0 g of Boric acid (in a 32 mm diameter round shaped disc, with the sample placed at the center, and Boric acid as a binder around it). The samples were pressed with a hydraulic press machine at 15 tonnes for 2 min. The XRF analysis was done by scanning for the presence of elemental peaks using the Omnian software. The weight percentages of the compound presented in the samples were recorded in the results upon the completion of the analysis.

2.6. Internal Bonding Analysis

Internal bond analysis was performed for the fabricated *Rhizophora* spp. bonded with different percentages of binders [8]. Metal blocks with the dimension of approximately (5.0 × 5.0) cm^2 were utilised in this analysis, and a hot melt glue was used to bind the samples to the metal blocks. Internal bond strength was determined by using a mechanical testing machine (Model UTM-5582; Instron, Norwood, MA, USA) with a load capacity of approximately 1000 kg, adapted from a previous study by Zuber et al. [8].

3. Results and Discussion

3.1. Evaluation of Functional Group using FTIR Analysis

The analyses for the *Rhizophora* spp. particleboard bonded with different percentages of binders were recorded between wavenumbers of 4000 cm^{-1} and 400 cm^{-1}, with a resolution of 4 cm^{-1}. Different relative transmittance values were determined by FTIR spectral analysis, as shown in Figures 1–3. Based on the overview of the FTIR values for cellulose, hemicellulose, saccharides and lignin, the broad peak at 3385.07 cm^{-1} in Figure 1 may represent cellulose and hemicellulose related to O-H stretching, in relation to hydrogen bond of hydroxyl groups [41,42]. The peak at 2904.80 cm^{-1} may indicate cellulose in relation to C-H stretching, whereas the peak at 1735.93 cm^{-1} may represent the softwood or hardwood with C=O stretching [41]. All the samples displayed almost the same spectra, as indicated by all the peaks presented in the Figures 1–3. Compared to the spectrum of binderless *Rhizophora* spp. sample, the spectra of samples incorporated with lignin and soy flour as binders showed a small increase in the peak intensity, as shown in peak 3402.43 cm^{-1} and 2809.80 cm^{-1}. However, no new peak appeared in the spectrum of *Rhizophora* spp. samples bonded with the binders, indicating that the introduction of lignin and soy flour did not change the functional groups according to the FTIR spectrum.

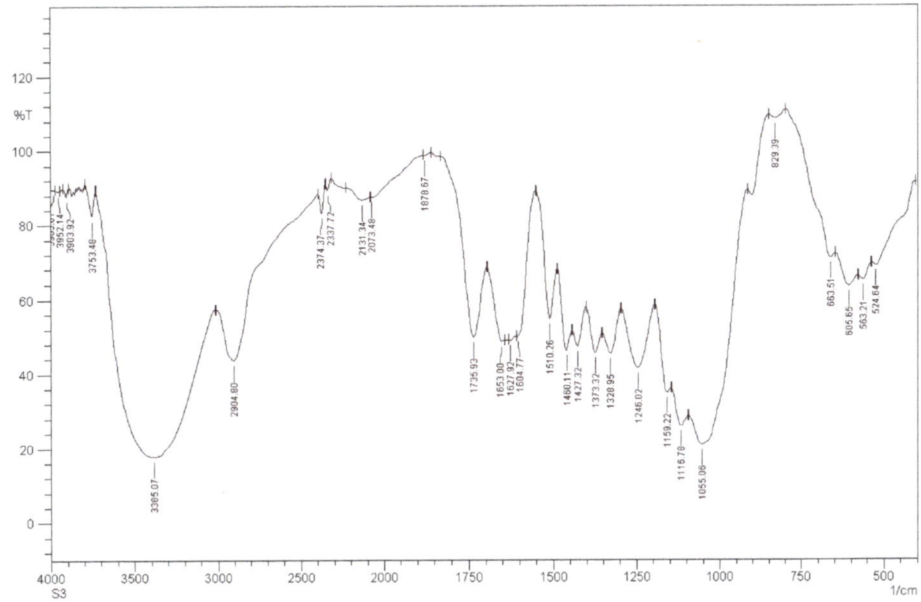

Figure 1. Transmittance spectrum of *Rhizophora* spp. particleboard bonded with 0% binder.

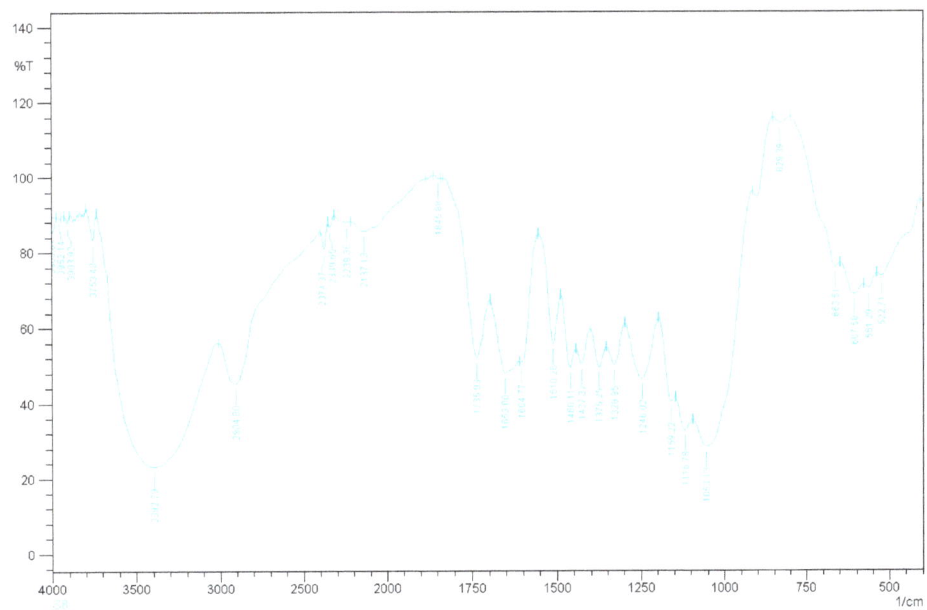

Figure 2. Transmittance spectrum of *Rhizophora* spp. particleboard bonded with 6% binders.

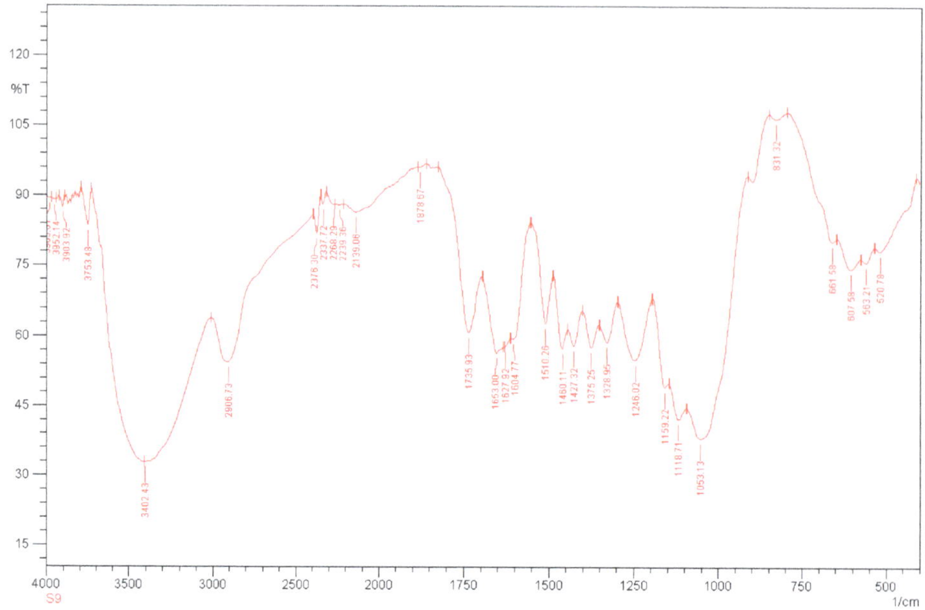

Figure 3. Transmittance spectrum of *Rhizophora* spp. particleboard bonded with 12% binders. Reprinted with permission from ref. [7].

3.2. Evaluation of XRD Spectrum

XRD spectrums of the *Rhizophora* spp. bonded with different percentages of binders are depicted in Figures 4–6. A peak intensity at around $2\theta = 22°$ was observed in the diffractogram of the sample bonded with 0% binders, whereas for samples with 6% and 12% binders, the peak intensity can be seen at $2\theta = 22.20°$, which corresponds to the crystalline properties of the composites [43]. The addition of lignin and soy flour led to a small increment of peak intensity at approximately $2\theta = 22.20°$, suggesting that the introduction of binders may indicate the changes in the structural order of the molecules, which may also influence the crystalline structure of the composites [44]. This may be due to the forces from molecular chain entanglements between the *Rhizophora* spp. wood particles and the binders, as they diffuse across the joint interface. Cross-linking may occur during the hot pressing, while interfacial diffusion during bonding is enhanced by chain scission [45]. However, no apparent disparity can be observed in all the spectrums, indicating that no major structural transformation can be clearly stipulated in this analysis. To conclude, incorporation of lignin and soy flour in the fabrication of *Rhizophora* spp. particleboard did not reveal apparent crystalline and amorphous transformations within the composites.

3.3. Evaluation of SEM Analysis

The micrographs of the particleboards with different percentages of binders are depicted in Figures 7–9. For the sample with 6% binders, the void spaces between the molecules were reduced when compared to the binderless sample, demonstrating a much smoother appearance which can be attributed to the hot pressing that allows the auto-condensation process between the wood particles and the binders, binding them together [7]. For the *Rhizophora* spp. sample with 12% binders as shown in Figure 9, the appearance of the molecules revealed a less smooth appearance, which may be due to insufficient bonding between the wood particles and the binders, causing an increase in lumen and gaps. Increased lignin content can affect the composites' compactness as cross-linking of lignin with the cell wall components may occur, which in return, reduce cellulose and hemicellulose accessibility to microbial enzymes, resulting in lower biomass digestibility [46]. Specks of bright appearance in all the figures may be due to the charging effect of electron or ion irradiations, which usually happens for a non-conductive specimen being examined [47]. The interactions between *Rhizophora* spp. and the binders were also influenced by the intermolecular forces that mediate the interaction, and the forces of attraction and repulsion between the molecules [25,48–50]. These attractive forces may allow better adhesion between the binders in their molecular forms, together with the raw wood particles. The natural properties of the raw wood itself are an advantage due to its cell cavities, which may allow the binders to infiltrates and provide better bonding with the help of thermal pressing. Data adapted from Zuber et al. revealed the thermogravimetric analysis (TGA) results of *Rhizophora* spp., soy flour and lignin. The mass degradation of *Rhizophora* spp., soy flour and lignin occurred at 303.35 °C, 236.11 °C and 279.64 °C, respectively, where the particles started to decompose at these temperatures. The mass degradation temperature should not be exceeded in the compression process of the particleboard, in order to improve the efficiency between the fibre chain of the particles and the binders [7]. The process of auto-condensation during the hot pressing had bound the molecules of the binders together with the *Rhizophora* spp. fibre; however, depending on the formulation of the composites, the microscopic appearances of the samples may slightly differ. In this study, the distribution of the *Rhizophora* spp. particles and binders was influenced by shrinking and compounding of the composites to a specified thickness by hot pressing at approximately 200 °C, leading to good interfacial bonding. Although there is no apparent disparity between the micrographs shown, it is important to note that incorporation of lignin and soy flour in the fabrication of the particleboard allows better bonding between the molecules, which will further improve the physical and mechanical strength of the particleboard [8].

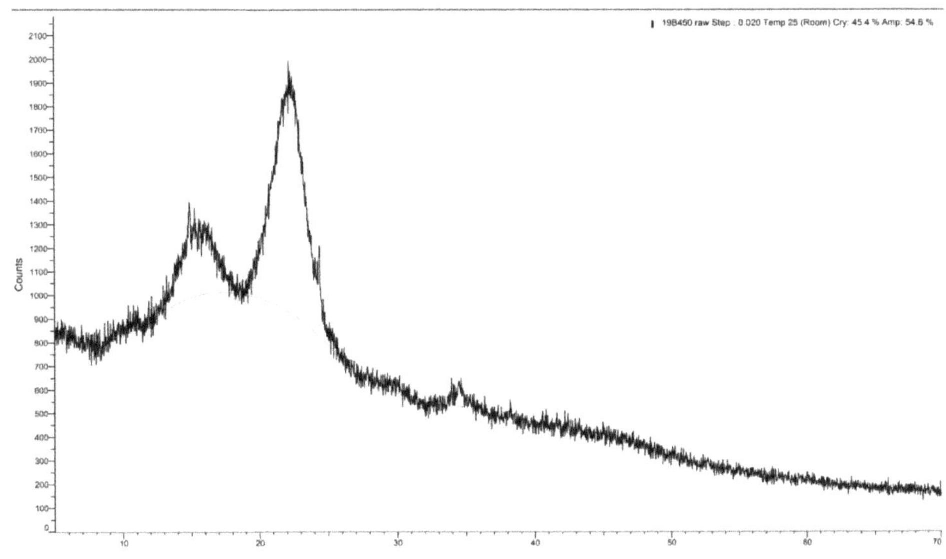

Figure 4. X-ray diffraction spectrum for *Rhizophora* spp. particleboard bonded with 0% binder.

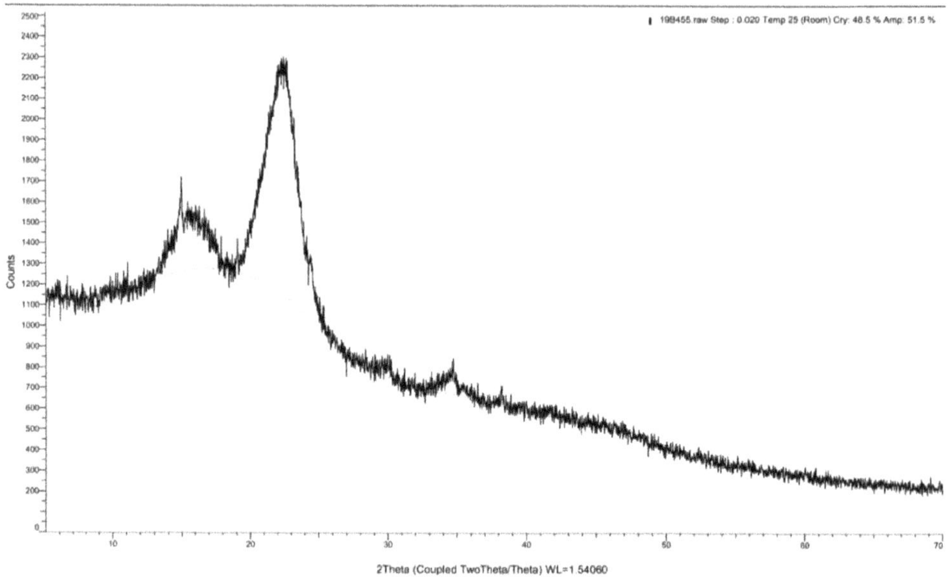

Figure 5. X-ray diffraction spectrum for *Rhizophora* spp. particleboard bonded with 6% binders.

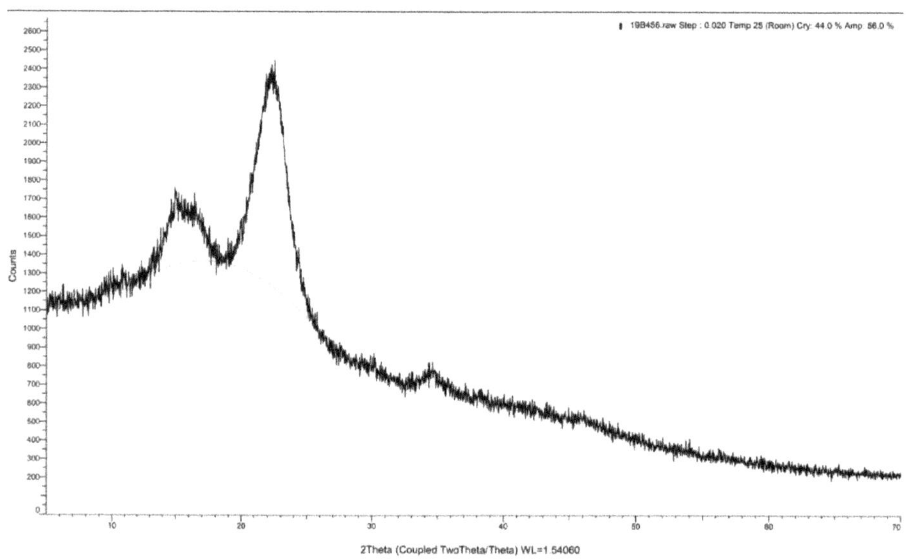

Figure 6. X-ray diffraction spectrum for *Rhizophora* spp. particleboard bonded with 12% binders Reprinted with permission from ref. [7].

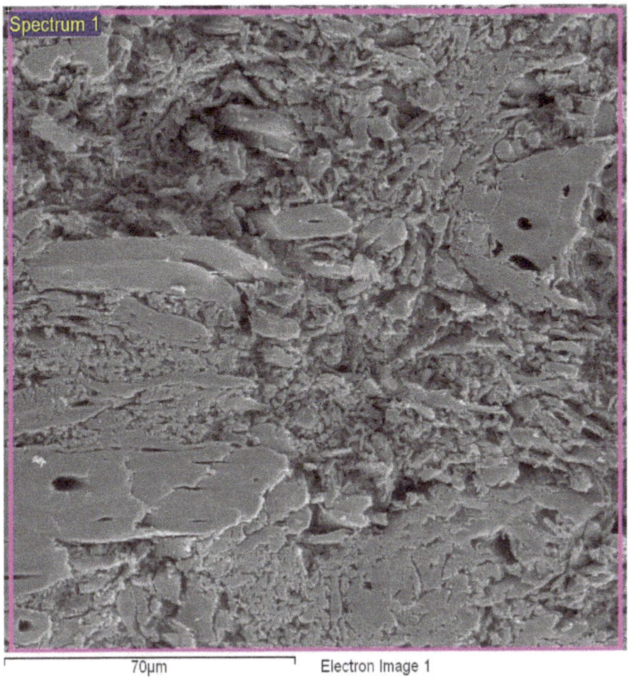

Figure 7. Micrograph of *Rhizophora* spp. particleboard bonded with 0% binder.

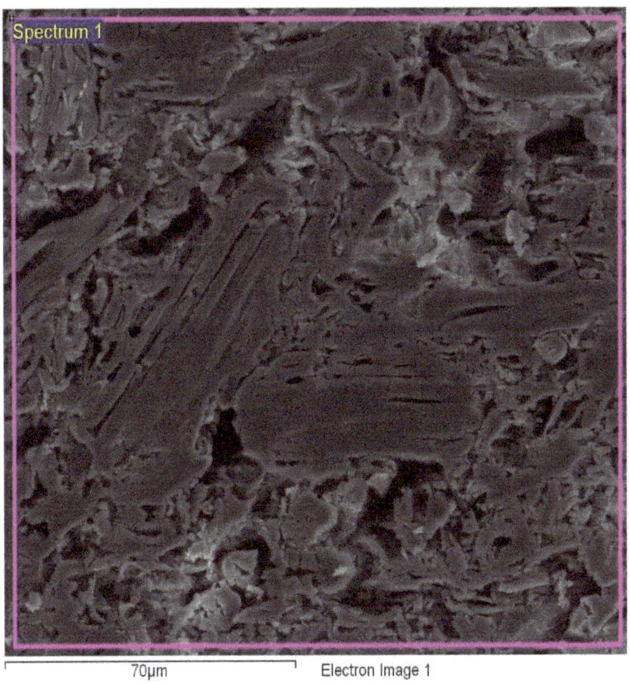

Figure 8. Micrograph of *Rhizophora* spp. particleboard bonded with 6% binders.

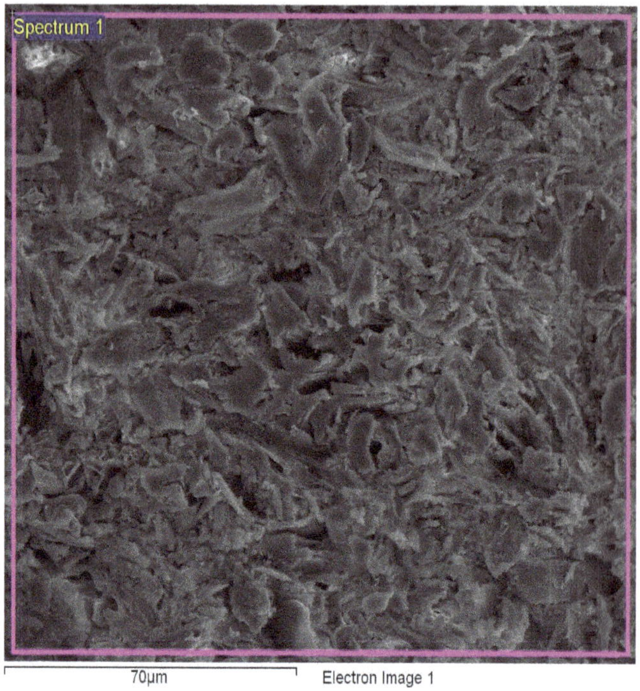

Figure 9. Micrograph of *Rhizophora* spp. particleboard bonded with 12% binders.

3.4. Evaluation of EDX Analysis

Figures 10–12 show EDX spectrums of the *Rhizophora* spp. particleboard at 0%, 6% and 12% binders. All the EDX spectrums of *Rhizophora* spp. particleboard bonded with different percentages of binders demonstrated visible carbon and oxygen signals, which confirmed the discernible presence of the carbon and oxygen in the composites [51,52]. Another peak in the spectrums may represent other elements such as gold, as the samples were coated with gold for better conductivity. Based on the figures, all spectrums display no discernible disparity; thus, the incorporation of binders in the fabrication of *Rhizophora* spp. particleboard did not greatly affect the percentages of elements presented in the composites.

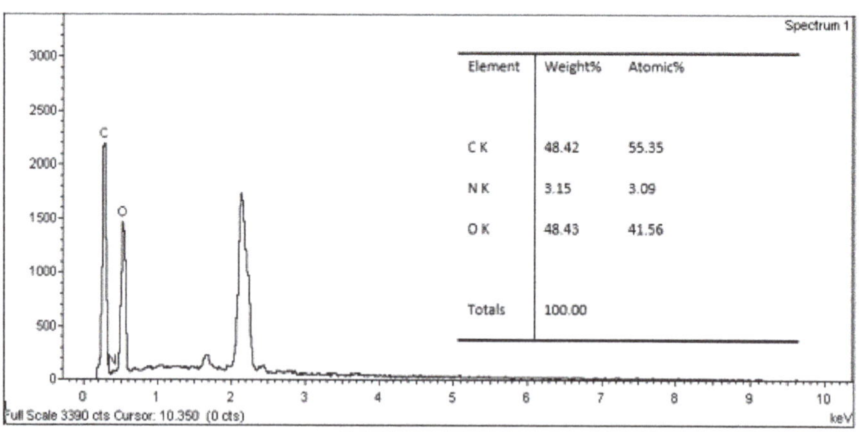

Figure 10. Energy dispersive X-ray spectrum for *Rhizophora* spp. particleboard bonded with 0% binders.

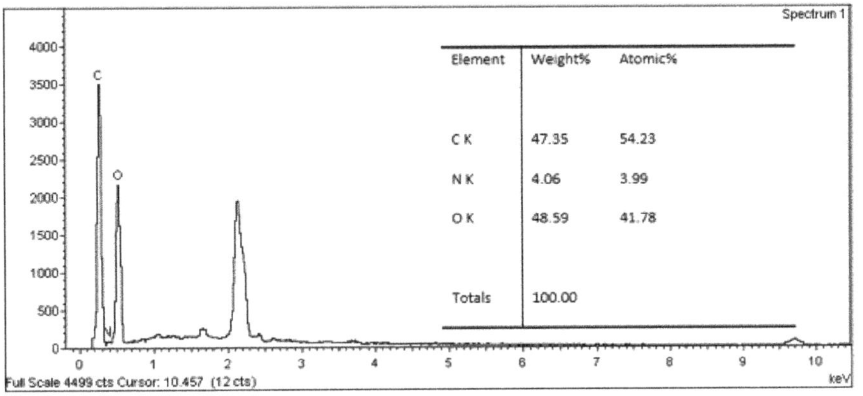

Figure 11. Energy dispersive X-ray spectrum for *Rhizophora* spp. particleboard bonded with 6% binders.

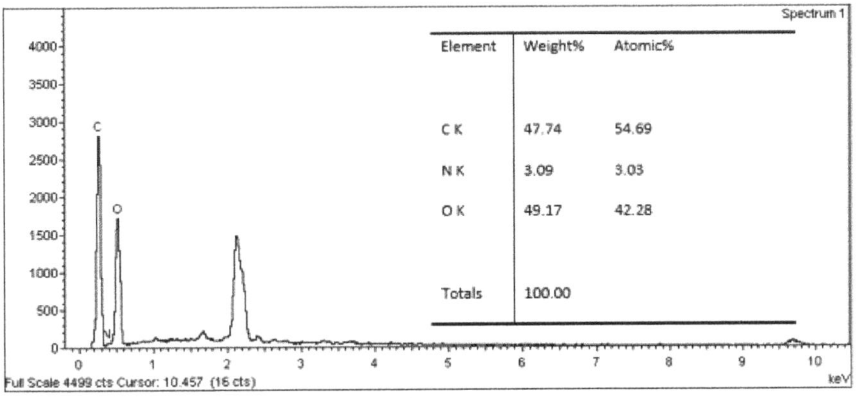

Figure 12. Energy dispersive X-ray spectrum for *Rhizophora* spp. particleboard bonded with 12% binders.

3.5. Evaluation of the XRF Analysis

The XRF method is the best way to identify major and trace elements within the natural composition of the particleboard. The XRF analysis of major elements is recorded in Table 1. Calcium oxide showed the highest weight percentage in all samples in the range of 40.319% to 52.744%. For *Rhizophora* spp. samples with the addition of lignin and soy flour, the dry weight of potassium oxide recorded the second highest concentration in the range of 14.883% to 18.399%, while chlorine percentage is between 11.623% to 12.947%. However, for the binderless particleboard (0% soy flour and lignin), chlorine recorded the second highest percentage after calcium oxide at 17.705%, followed by sodium oxide at 6.355%. The high percentage of chlorine in the wood-based sample may be due to combustion performed in a laboratory scale spectrometer. Nevertheless, the lower percentage of chlorine in the bonded samples may be advantageous to the overall particleboard outcomes, especially in maintaining the natural properties of the particleboard.

Table 1. X-ray fluorescence analysis of major and trace element in weight percentages using Omnian method with pressed powder pellet.

Samples	Calcium Oxide (CaO)	Potassium Oxide (K_2O)	Chlorine (Cl)	Sodium Oxide (Na_2O)	Others
0% binder	52.744	5.241	17.705	6.355	17.955
6% binders	42.740	14.883	12.947	5.036	24.394
12% binders	40.319	18.399	11.623	4.708	24.951

3.6. Internal Bond Analysis

The analysis of internal bonding was performed and the result is shown in Figure 13 [8].

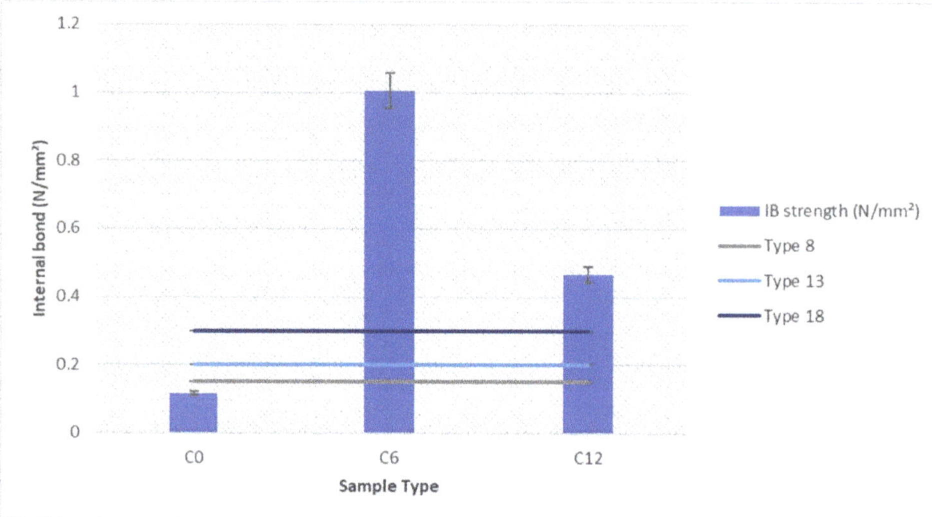

Figure 13. Internal bonding of soy-lignin bonded *Rhizophora* spp. particleboards at different percentages. C = 0 µm to 103 µm particle sizes; 0 = 0% soy flour and lignin, 6 = 4.5% soy flour and 1.5% lignin, 12 = 9% soy flour and 3% lignin.

The minimum requirement according to JIS A-5908 includes classification of Type 8 (0.15 $N.mm^{-2}$), Type 13 (0.2 N.mm$-^2$) and Type 18 (0.3 $N.mm^{-2}$) [53]. The *Rhizophora* spp. sample with binders satisfy all three classifications by JIS, whereas the sample with 0% binder did not satisfy any of the requirements. Increased internal bonding strength displayed by samples with binders indicates improved structural stability and durability of the composites [44]. Hot pressing allows the high temperature to be imparted on the formulation of *Rhizophora* spp. with binders, which may create strong bonding with reduced lumen voids between the particles in the composites. The strength of the particleboard may be achieved due to the vessel element and parenchymatous cell of the *Rhizophora* spp. that are closely attached under high pressure condition. In this study, internal bond represents the mechanical properties of the composites, and based on the result, the addition of lignin and soy flour increased the internal bond strength, which may improve its mechanical strength and structural integrity.

4. Conclusions

Evidence from the FTIR, XRD, SEM, EDX, XRF and internal bonding revealed the potential use of lignin and soy flour as binders in the fabrication of *Rhizophora* spp. particleboard, as a phantom material in radiation dosimetry applications. Different percentages of binders used in the fabrication of the particleboard did not greatly affect the properties of the particleboard as a phantom material; however, lower percentages of chlorine and increased internal bonding strength in the sample with binders may be advantageous in maintaining the natural properties of the particleboard, and improve the mechanical strength of the samples, respectively.

Author Contributions: Conceptualisation, S.H.Z., N.A.A.H., M.F.M.Y., M.Z.A.A. and R.H.; methodology, S.H.Z., M.F.M.Y., M.Z.A.A. and R.H.; validation, N.A.A.H., M.F.M.Y. and R.H.; formal analysis, S.H.Z.; investigation, S.H.Z., M.F.M.Y. and M.Z.A.A.; resources, N.A.A.H., M.F.M.Y. and R.H.; data curation, S.H.Z.; writing—original draft, S.H.Z.; writing—review & editing, S.H.Z. and N.A.A.H.; visualisation, S.H.Z.; supervision, N.A.A.H., M.F.M.Y., M.Z.A.A. and R.H.; project administration, N.A.A.H.; funding acquisition, N.A.A.H., M.F.M.Y. and R.H. All authors have read and agreed to the published version of the manuscript.

Funding: This research was funded by Universiti Sains Malaysia Short-Term Grant (304/PFIZIK/6315322), the School of Industrial Technology Grant (1001/PTEKIND/8014083) and the Universiti Sains Malaysia Bridging Grant (304.PPSK.6316324).

Institutional Review Board Statement: Not applicable.

Informed Consent Statement: Not applicable.

Data Availability Statement: The data presented in this study are available on request from the corresponding author.

Acknowledgments: The authors thanked the School of Physics, School of Industrial Technology and Advanced Medical and Dental Institute, Universiti Sains Malaysia for allowing this research to be conducted in the respective schools/institute. The first author of this paper is financially sponsored by the UTM Academic Fellow Scheme (SLAM) (2019–2021) and the author would like to thank UTMLead and the Faculty of Science, Universiti Teknologi Malaysia, Johor for making this study possible.

Conflicts of Interest: The authors declare no conflict of interest.

References

1. Ritchie, H.; Roser, M. Plastic pollution. Our World Data. 2018. Available online: https://ourworldindata.org/plastic-pollution (accessed on 2 June 2021).
2. Gay, D.; Hoa, S.V.S.; Tsai, W. *Composite Materials: Design and Applications*; CRC Press: Boca Raton, FL, USA, 2002.
3. Mathur, V.K. Composite materials from local resources. *Constr. Build. Mater.* **2006**, *20*, 470–477. [CrossRef]
4. Jawaid, M.; Thariq, M. *Sustainable Composites for Aerospace Applications*; Woodhead Publishing: Cambridge, UK, 2018.
5. Balakrishnan, P.; John, M.J.; Pothen, L.; Sreekala, M.S.; Thomas, S. Natural fibre and polymer matrix composites and their applications in aerospace engineering. In *Advanced Composite Materials for Aerospace Engineering*; Rana, S., Fangueiro, E., Eds.; Woodhead Publishing: Cambridge, UK, 2016; pp. 365–383.
6. Rajak, D.K.; Pagar, D.D.; Kumar, R.; Pruncu, C.I. Recent progress of reinforcement materials: A comprehensive overview of composite materials. *J. Mater. Res. Technol.* **2019**, *8*, 6354–6374. [CrossRef]
7. Zuber, S.H.; Hashikin, N.; Mohd Yusof, M.F.; Hashim, R. Lignin and soy flour as adhesive materials in the fabrication of *Rhizophora* spp. particleboard for medical physics applications. *J. Adhes.* **2020**. [CrossRef]
8. Zuber, S.H.; Hashikin, N.A.A.; Mohd Yusof, M.F.; Hashim, R. Physical and mechanical properties of soy-lignin bonded *Rhizophora* spp. particleboard as a tissue-equivalent phantom material. *Bioresources* **2020**, *15*, 3. [CrossRef]
9. Yusof, M.F.M.; Hashim, R.; Tajuddin, A.; Bauk, A.S.; Sulaiman, O. Characterization of tannin-added *Rhizophora* spp. particleboards as phantom materials for photon beams. *Ind. Crops. Prod.* **2017**, *95*, 467–474. [CrossRef]
10. Samson, D.; Shukri, A.; Jafri, M.Z.M.; Hashim, R.; Aziz, M.Z.A.; Yusof, M.F.M. Tissue equivalent materials from SPC-SPI/NAOH/IA-PAE bonded mangrove wood characterized for radiation therapy dosimetry. *Sci. Proc. Ser.* **2020**, *2*, 115–120.
11. Yusof, M.F.M.; Khatijah Abd Hamid, P.N.; Tajuddin, A.A.; Abdullah, R.; Hashim, R.; Bauk, S.; Isa, N.M.; Md Isa, M.J. Characterization and attenuation study on tannin-added *Rhizophora* spp. particleboard at high energy photon and electron. In Proceedings of the AIP Conference 2017, Selangor, Malaysia, 8–10 August 2016; Volume 1799, p. 40002.
12. Samson, D.O.; Jafri, M.M.; Hashim, R.; Sulaiman, O.; Aziz, M.Z.A.; Yusof, M.F.M.; Shukri, A. *Rhizophora* SPP particleboards incorporating defatted soy flour bonded with NaOH/IA-PAE: Towards a water equivalent phantom material. *Radiat. Phys. Chem.* **2020**, *176*, 109057. [CrossRef]
13. Said, M.; Seman, N.; Jaafar, H. Characterization of human head phantom based on its dielectric properties for wideband microwave Imaging application. *J. Teknol.* **2015**, *73*. [CrossRef]
14. Dinulović, M.; Rašuo, B. Dielectric modeling of multiphase composites. *Compos. Struct.* **2011**, *93*, 3209–3215. [CrossRef]

15. Tousi, E.T.; Hashim, R.; Bauk, S.; Jaafar, M.S.; Abuarra, A.M.H.; Ababneh, B. Some properties of particleboards produced from *Rhizophora* spp. as a tissue-equivalent phantom material bonded with *Eremurus* spp. *Measurement* **2014**, *54*, 14–21. [CrossRef]
16. Pizzi, A. Wood products and green chemistry. *Ann. For. Sci.* **2016**, *73*, 185–203. [CrossRef]
17. Kent, J.A. *Kent and Riegel's Handbook of Industrial Chemistry and Biotechnology*; Springer: Berlin/Heidelberg, Germany, 2007; Volume 1665.
18. Frihart, C.R. Potential for Biobased adhesives in wood bonding. In Proceedings of the 59th International Convention of Society of Wood Science and Technology, Curitiba, Brazil, 6–10 March 2016; pp. 84–91.
19. Ferdosian, F.; Pan, Z.; Gao, G.; Zhao, B. Bio-based adhesives and evaluation for wood composites application. *Polymers* **2017**, *9*, 70. [CrossRef] [PubMed]
20. Hashim, R.; Nadhari, W.N.A.W.; Sulaiman, O.; Kawamura, F.; Hiziroglu, S.; Sato, M.; Tanaka, R. Characterization of raw materials and manufactured binderless particleboard from oil palm biomass. *Mater. Des.* **2011**, *32*, 246–254. [CrossRef]
21. Bertaud, F.; Tapin-Lingua, S.; Pizzi, A.; Navarrete, P.; Petit-Conil, M. Development of green adhesives for fibreboard manufacturing, using tannins and lignin from pulp mill residues. *Cellul. Chem. Technol.* **2012**, *46*, 449–455.
22. Bosetti, C.; McLaughlin, J.K.; Tarone, R.E.; Pira, E.; La Vecchia, C. Formaldehyde and cancer risk: A quantitative review of cohort studies through 2006. *Ann. Oncol.* **2008**, *19*, 29–43. [CrossRef]
23. Dongre, P.; Driscoll, M.; Amidon, T.; Bujanovic, B. Lignin-furfural based adhesives. *Energies* **2015**, *8*, 7897–7914. [CrossRef]
24. Frihart, C.R.; Lorenz, L. Protein modifiers generally provide limited improvement in wood bond strength of soy flour adhesives. *For. Prod.* **2013**, *63*, 138–142. [CrossRef]
25. Frihart, C.R.; Hunt, C.G. Adhesives with wood materials-bond formation and performance. In *Wood Handbook: Wood as an Engineering Material*; Forest Products Laboratory: Madison, WI, USA, 2010.
26. Frihart, C.R.; Satori, H. Soy flour dispersibility and performance as wood adhesive. *J. Adhes. Sci. Technol.* **2013**, *27*, 2043–2052. [CrossRef]
27. Frihart, C.R.; Wescott, J.M. Improved water resistance of bio-based adhesives for wood bonding. In Proceedings of the ICECFOP1-1st International Conference on Environmentally-Compatible Forest Products: Fernando Pessoa University, Oporto, Portugal, 22–24 September 2004; Edicoes Universidade Fernando Pessoa: Porto, Portugal, 2004; pp. 293–302.
28. Hojilla-Evangelista, M.P. Adhesive qualities of soybean protein-based foamed plywood glues. *J. Am. Oil Chem. Soc.* **2002**, *79*, 1145–1149. [CrossRef]
29. Khosravi, S.; Khabbaz, F.; Nordqvist, P.; Johansson, M. Protein-based adhesives for particleboards. *Ind. Crops Prod.* **2010**, *32*, 275–283. [CrossRef]
30. Sjöström, E.; Alén, R. *Analytical Methods in Wood Chemistry, Pulping, and Papermaking*; Springer: Berlin/Heidelberg, Germany, 2013.
31. Hu, F.; Ragauskas, A. Pretreatment and lignocellulosic chemistry. *Bioenergy Res.* **2012**, *5*, 1043–1066. [CrossRef]
32. Kubo, S.; Kadla, J.F. Poly (ethylene oxide)/organosolv lignin blends: Relationship between thermal properties, chemical structure, and blend behavior. *Macromolecules* **2004**, *37*, 6904–6911. [CrossRef]
33. Kubo, S.; Kadla, J.F. Lignin-based carbon fibers: Effect of synthetic polymer blending on fiber properties. *J. Polym. Environ.* **2005**, *13*, 97–105. [CrossRef]
34. Pizzi, A.; Salvadó, J. Lignin-based wood panel adhesives without formaldehyd. *Holz Roh Werkst.* **2007**, *65*, 65.
35. Montané, D.; Torné-Fernández, V.; Fierro, V. Activated carbons from lignin: Kinetic modeling of the pyrolysis of Kraft lignin activated with phosphoric acid. *Chem. Eng. J.* **2005**, *106*, 1–12. [CrossRef]
36. Hemmilä, V.; Trischler, J.; Sandberg, D. Lignin: An adhesive raw material of the future or waste of research energy? In Proceedings of the Northern European Network for Wood Science and Engineering, Hannover, Germany, 11–12 September 2013; pp. 98–103.
37. Çetin, N.S.; Özmen, N. Studies on lignin-based adhesives for particleboard panels. *Turk. J. Agric. For.* **2003**, *27*, 183–189.
38. Gosselink, R.J.; Van Dam, J.E.; De Jong, E.; Gellerstedt, G.; Scott, E.L.; Sanders, J.P. Effect of periodate on lignin for wood adhesive application. *Holzforschung* **2011**, *65*, 155–162. [CrossRef]
39. Martínez, C.; Sedano, M.; Munro, A.; López, P.; Pizzi, A. Evaluation of some synthetic oligolignols as adhesives: A molecular docking study. *J. Adhes. Sci. Technol.* **2010**, *24*, 1739–1751. [CrossRef]
40. Zhang, W.; Ma, Y.; Wang, C.; Li, S.; Zhang, M.; Chu, F. Preparation and properties of lignin–phenol–formaldehyde resins based on different biorefinery residues of agricultural biomass. *Ind. Crops Prod.* **2013**, *43*, 326–333. [CrossRef]
41. Abraham, E.; Deepa, B.; Pothan, L.; Jacob, M.; Thomas, S.; Cvelbar, U.; Anandjiwala, R. Extraction of nanocellulose fibrils from lignocellulosic fibres: A novel approach. *Carbohydr. Polym.* **2011**, *86*, 1468–1475. [CrossRef]
42. Zhao, J.; Zhang, W.; Zhang, X.; Zhang, X.; Lu, C.; Deng, Y. Extraction of cellulose nanofibrils from dry softwood pulp using high shear homogenization. *Carbohydr. Polym.* **2013**, *97*, 695–702. [CrossRef]
43. Khan, A.; Khan, R.A.; Salmieri, S.; Le Tien, C.; Riedl, B.; Bouchard, J.; Lacroix, M. Mechanical and barrier properties of nanocrystalline cellulose reinforced chitosan based nanocomposite films. *Carbohydr. Polym.* **2012**, *90*, 1601–1608. [CrossRef]
44. Samson, D.O.; Shukri, A.; Jafri, M.Z.M.; Hashim, R.; Sulaiman, O.; Aziz, M.Z.A.; Yusof, M.F.M. Characterization of *Rhizophora* SPP. Particleboards with SOY protein isolate modified with NaOH/IA-PAE adhesive for use as phantom material at photon energies of 16.59–25.26 keV. *Nucl. Eng. Technol.* **2020**, *53*, 216–233. [CrossRef]
45. *Adhesive Bonding*; William Andrew Publishing: Boston, MA, USA, 2009; pp. 145–173.
46. Sarkanen, K.V.; Ludwig, C.H. *Liguins. Occurrence, Formation, Structure, and Reactions*; John Wiley & Sons: Hoboken, NJ, USA, 1971.

47. Kim, K.H.; Akase, Z.; Suzuki, T.; Shindo, D. Charging effects on SEM/SIM contrast of metal/insulator system in various metallic coating conditions. *Mater. Trans.* **2010**, *10*, 1080–1083. [CrossRef]
48. Hansen, C.M. 50 Years with solubility parameters—Past and future. *Prog. Org. Coat.* **2004**, *51*, 77–84. [CrossRef]
49. Yemele, M.C.N.; Cloutier, A.; Diouf, P.N.; Koubaa, A.; Blanchet, P.; Stefanovic, T. Physical and mechanical properties of particleboard made from extracted black spruce and trembling aspen bark. *For. Prod. J.* **2008**, *58*, 38–46.
50. Hansen, C.M. *Hansen Solubility Parameters: A User's Handbook*; CRC Press: Boca Raton, FL, USA, 2007.
51. Cwa, C.W.S. Kayu Tropika Sebagai Bahantara Setaraan Tisu Untuk Kajian Dosmteri. Master's Thesis, Universiti Sains Malaysia, Kuala Lumpur, Malaysia, 1993. (In Malay).
52. Marashdeh, M.W.; Hashim, R.; Tajuddin, A.A.; Bauk, S.; Sulaiman, O. Effect of particle size on the characterization of binderless particleboard made from *Rhizophora* SPP. Mangrove wood for use as phantom material. *BioResources* **2011**, *6*, 4028–4044.
53. Japan Standard Association. *Japanese Industrial Standard for Particle Board JIS A 5908*; Japanese Standard Association: Jepang, Japan, 2003.

polymers

MDPI

Article

Characterization of 3D Printing on Jute Fabrics

Edgar Adrián Franco-Urquiza [1,*], Yael Ramírez Escamilla [2] and Perla Itzel Alcántara Llanas [2]

[1] National Council for Science and Technology (CONACYT—CIDESI), Center for Engineering and Industrial Development, Carretera Estatal 200, Querétaro 76265, Mexico

[2] Department of Mechatronics, Center for Engineering and Industrial Development (CIDESI), Carretera Estatal 200, Querétaro 76265, Mexico; c.yael.r.escamilla@gmail.com (Y.R.E.); perla.alcantara@cidesi.edu.mx (P.I.A.L.)

* Correspondence: edgar.franco@cidesi.edu.mx

Abstract: This work evaluates the feasibility to manufacture polylactic acid (PLA) composites using jute fiber fabrics. For characterization, PLA-fused filament was successfully deposed onto jute fabrics to print dog-bone tensile specimens (Type I specimen from ASTM D638). The jute fabrics were chemically modified, treated with flame retardant additives, and sprayed with aerosol adhesive to improve the mechanical properties of PLA/Jute fabric composites. The elastic modulus and the strength of PLA were higher than PLA composites, and the plastic deformation of the PLA composites was slightly lower than PLA. Tomography scans revealed the fabrics were well oriented and some adherence between jute fabrics and PLA. Viscoelastic properties of PLA composites resulted in the reduction in storage modulus and the reduction in intensity in the damping factor attributed to segmental motions with no variations in the glass transition temperature. Flame retardant and spray adhesive on jute fabrics promoted better response to time of burning than PLA and PLA with modified fibers. The results presented in this work lead to the need for a more detailed investigation of the effect of plant fiber fabrics as reinforcement of 3D printed objects for industrial applications.

Keywords: jute fabrics; 3D printing; mechanical properties; eco-friendly composites

Citation: Franco-Urquiza, E.A.; Escamilla, Y.R.; Alcántara Llanas, P.I. Characterization of 3D Printing on Jute Fabrics. *Polymers* **2021**, *13*, 3202. https://doi.org/10.3390/polym13193202

Academic Editors: Domenico Acierno and Antonella Patti

Received: 6 August 2021
Accepted: 6 September 2021
Published: 22 September 2021

1. Introduction

Additive manufacturing (AM) of polymers is an automated process for producing three-dimensional objects from computer-aided design (CAD) data, and it is mostly used for prototyping that cannot manufacture one-piece products. The relevance of this technology has been constantly evolving over the years, and it is standardized by the common standards ISO/ASTM 52900:2015.

FFF, also known as fused deposition modeling (FDM), is the 3D printing of polymers based on the extrusion process. The object is built by depositing melted thermoplastic layer-by-layer through a heated nozzle onto the platform or over previously printed layers until the designed element is completed. Conventional-fused filaments, such as polyamide, acrylonitrile–butadiene–styrene (ABS), polyether ether ketone (PEEK), and PLA, are commercially available for domestic users. PLA is an aliphatic polyester corn starch-based thermoplastic and has been extensively examined in the literature as the most popular biodegradable material used for AM [1–10].

It is possible to find extensive literature, including several reviews, related to developing new experimental thermoplastic filaments prepared by the extrusion process to enhance the mechanical properties of 3D printed objects [1–3,11–14]. Part of this extensive research is aimed to find alternatives to recycled and biodegradable filaments for the sustainability of 3D printing [3,14,15].

Santana et al. [16] developed an exciting work of unifying the technology of textile concrete and additive manufacturing to develop composites of geopolymer matrix reinforced with printed polyethylene terephthalate glycol, commonly known as PETG, mesh. The composites were subjected to the notched prism bending test. The homogeneous reinforcement (volumetric polymer content of 4.75%) and the graduated (volumetric polymer content of 3.34%) produced an increase in toughness of 47 and 52 times, respectively, in addition to conserving the maximum load supported and reducing the volumetric content of the reinforcing material without compromising the mechanical performance of the composites.

Other works are focused on polymer modification and blends [4,5,17]. However, the most attractive seems to be the reinforced fused filaments with metal particles [18–20], clay minerals [6,21–23], graphene [8,24–26], glass [27], or carbon chopped fibers [28,29], and, more recently, continuous fibers [9,30,31].

A particular interest for several industrial applications is the use of continuous natural fibers as a substitute for glass fibers because of their mechanical and acoustic properties in combination with their end-of-life management and positive alternative to reduce carbon footprint. Ecological 3D printed objects are currently produced by combining long natural fibers and melted plastic, which are embedded in the hot block and deposed in a simultaneous manner [12,13,32–34]. Cellulosic fibers are widely available in most countries and are cost-effective with low density. They are biodegradable, renewable, non-hazardous, and non-abrasive. Furthermore, its specific mechanical properties are comparable to glass fibers. The purpose of adding these monofilament cellulosic fibers is to improve the mechanical properties of composite materials, including the construction industry, to improve the ductility and post-crack toughness of the composites [35]. Besides many advantageous properties of the natural fiber reinforced ecological composites, there are also some drawbacks, such as incompatibility with hydrophobic matrices, high water absorption, lower processing window, and bad surface appearance.

Numerous researchers have investigated PLA/natural fiber composites. Recently, Wis et al. developed over-molded jute / PLA fabric composites (OMC) on a laboratory scale. In that work, the authors developed hybrid organic composites prepared with thermoplastic composite technology and obtained lightweight composite components for structural parts. In this process, a reinforcing sheet composed of continuous glass or thermoplastic carbon fiber, called an organo-sheet, is over-molded using a thermoplastic polymer in an injection molding process. The composite sheets obtained are rigid, high-strength, and, at the same time, still have a detailed shape. The results obtained showed that the flexural modulus and the strength of OMC improved compared to pure PLA. Dynamic mechanical analysis showed that the thermomechanical resistance of PLA was improved for OMC [36]. Jerpdal et al. investigated the influence of overmolded temperature on tensile modulus, shrinkage, and strain for an insert made of self-reinforced polyethylene terephthalate (PET). The authors observed that a temperature above the glass transition temperature leads to relaxation of residual stresses and reduction in tensile modulus up to 18%. The study shows fascinating results, which may lead to new areas of application for self-reinforced PET [37].

Despite the extensive research and development effort of plant fiber-embedded polymers for 3D printing applications, there are no publications that deal with the use of fused filament deposition onto plant fiber fabrics; this is the research gap this work pretends to fill.

Among all-natural fibers, jute fiber seems to be a promising fiber with relevant research fields due to its good mechanical properties compared to other natural fibers, such as kenaf, sisal, and hemp [38,39]. The bag industry is the largest consumer of jute fibers because they represent an ecological option. However, many jute bags at the end of their lives are wasted and gone to the landfill every year.

This work aims to prepare new green composites through 3D printing PLA onto jute fabrics to evaluate the mechanical performance that allows discovering multiple industrial applications.

2. Experimental Section

2.1. Materials

The commercial PLA-based filament with a diameter of 1.75 mm and a nominal density of 1.27 g/cm^3 from 3D MARKET® (Querétaro, Mexico) was used in this work. According to the supplier, this PLA filament has a tensile strength break between 55–65 MPa and a modulus of elasticity of 0.42 GPa. Natural jute fabrics with plain weave configuration and thickness of 0.91 mm (Figure 1) were obtained from bolsas publicitarias® (Yucatán, Mexico) The fiber contains approximately 70 threads count, elastic modulus of 11 GPa, and tensile strength of 44 MPa. The mechanical properties were previously calculated following the methodology of the ASTM C1557.

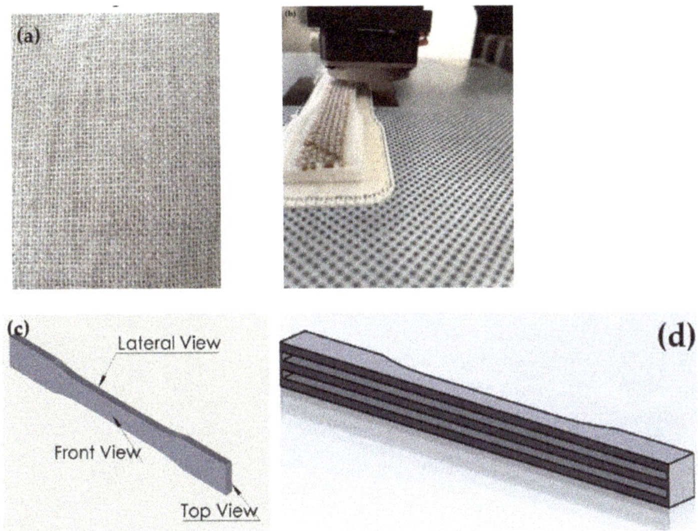

Figure 1. Photographs and schematic representations of: (**a**) jute fabrics, (**b**) 3D printing on jute fabrics, (**c**) tensile specimens' solid-like configuration, and (**d**) specimens with gaps to place the jute fabrics. The solid-like picture also presents the viewing directions for tomography scans.

2.2. 3D Printing Fabrication

A Zortrax M200 desktop 3D printer(Zortrax, Olsztyn, Poland) was used to print ASTM D638 Type I tensile specimens, previously modeled using SolidWorks software and exported to the 3D printing software as an STL file. Two different tensile specimen configurations were modeled. The first one was a solid-like specimen printed just with PLA. The second specimen included two longitudinal gaps of 0.91 mm, corresponding to the space to place the jute fabric, as schematized in Figure 1.

PLA was fused through a 0.4-mm-diameter nozzle at 200 °C and a printing speed of 50 mm/s over a bed platform heated at 50 °C. The specimens were built with 0.14-mm-layer thickness in a flat orientation with rectilinear pattern and an infill density of 90%. Jute fabrics dog bone geometry was cut using regular scissors and placed in the 0/90 direction (parallel to the uniaxial tension). The solid-like specimen was continuously printed, whereas the composite specimens required interrupting the 3D printing process to place the jute fiber fabric, as presented in Figure 1b. All composite specimens contained two jute fabric layers.

In this work, various strategies were used to evaluate the feasibility of increasing the mechanical properties of 3D printed composites. In this way, the characterization of these materials was carried out using the materials listed in Table 1.

Table 1. Materials description and their reference used in this work.

Reference	Description
PLA	PLA
PLA/J-M	PLA/jute fiber modified
PLA/J-R	PLA/jute fiber with flame retardant
PLA/J-MR	PLA/jute fiber modified and flame retardant
PLA/J-A	PLA/jute fiber with adhesive
PLA/J-RA	PLA/jute fiber with flame retardant and adhesive

Firstly, jute fabrics (J-M) were washed in an ionized water bath at 75 °C for 2 h, and dried at 85 °C for 2 h in an air convection oven. Afterward, jute fabrics were chemically treated with 5% NaOH and diazonium salt at alkali, acidic, and neutral media to increase compatibility with PLA.

Jute fabrics (J-R) were treated using a commercial flame retardant Flamebar S3 from Bollom fire protection. According to the supplier, the jute fibers were immersed for 12 h in a stainless steel container, having at least 70% of the solution, calculated on the submerged jute fabrics' weight. Afterward, the jute fabrics were dried at 85 °C for 24 h in an air convection oven.

Jute fabrics (J-A) were sprayed with Hi-Tack 71 from 3M™, which is a mist aerosol adhesive recommended by 3M for its use for the manufacturing composites, including infusion and dry lamination. Spraying was carried out at a 45° angle before fiber placement during 3D printing.

The jute fabrics (J-MR) were firstly modified and subsequently treated with flame retardant. For the case of the J-MA, the J-M fabrics were treated and stored; then, the adhesive was applied to the fabric just a few minutes before to place it on the PLA during the 3D printing process.

2.3. Methods

Uniaxial tensile tests were performed according to ASTM D638-14 using a universal testing machine Instron® 647(Instron®, Norwood, MA, USA) with a load cell of 10 kN. Ten specimens of each material were tested at room temperature (23 °C ± 2 °C) and at a 5-mm/min crosshead speed, and the curves showed in the results and discussion section are the representative curves based on the average behaviour revelaed during the tensile tests. Young's modulus (E) and yield strength (σ_y) were obtained from the engineering stress versus strain curves, and the elastic deformation was measured using a video extensometer MTS Advantage video extensometer (AVX) with 25 mm lens. The video-extensometer recognizes patterns on surfaces to acquire measurement data for strain calculations processed by MTS TestSuite™. Photographs of the failure zone after tensile tests were taken using a Zeiss stereomicroscope Discovery V8.

Three-dimensional computed tomography (CT) scans were performed in a GE phoenix v|tome|x instrument to visualize the inner of the 3D printed tensile specimens and detect adherence between PLA and jute fabrics. The analysis was conducted using X-Ray at 160 kV and 240 μA.

The viscoelastic behavior was evaluated in a Dynamical Mechanical Analysis DMA Discovery 850 from TA Instruments (Waters Corporation, Cary, NC, USA). The tests were performed in a single cantilever configuration at a frequency of 1 Hz and an amplitude of 30 μm. The specimens with 50 × 12 × 4 mm nominal dimensions were tested from 20 °C to 145 °C using a heating ramp of 5 °C/min. The curves of storage modulus (E′) and damping factor (tan δ) were obtained following the ASTM-D7028.

Flammability is highly interesting to analyze in ecological composite systems since it has become a crucial parameter in several industrial applications such as aeronautics, automotive, construction, or textile clothing. The flame retardant behavior of PLA and PLA composites was evaluated according to the methodology of chapter 3 of the Aircraft Materials Fire Test Handbook based on the FAR 25 Appendix F part III. The methodology proposed in this regulation can be used to predict the behavior of plastic or textile materials for diverse industrial sectors and aerospace applications. The test allows determining the burning speed of the specimens supported horizontally in a stainless steel cabin with air inlets on the top. Then, the flame (using methane gas) burns the specimen for 15 s. Subsequently, the ignition source is removed, and the test specimen is observed for time and extent of burning. An average burning rate is reported. Flammability tests were conducted in a multipurpose flammability chamber Deatak model MP-1 using high-purity methane gas (99%). Distance and time measurements were made with calibrated equipment, including Mitutoyo rulers with a resolution of 0.5 mm and a chronometer Control Company model 1025MX with a resolution of 0.1 s. The specimens with 50 mm × 13 mm × 4 mm nominal dimensions were conditioned at 21 °C and 55% humidity for at least 24 h before testing.

3. Results and Discussion

The representative tensile engineering stress versus strain curves for PLA and PLA composites are shown in Figure 2.

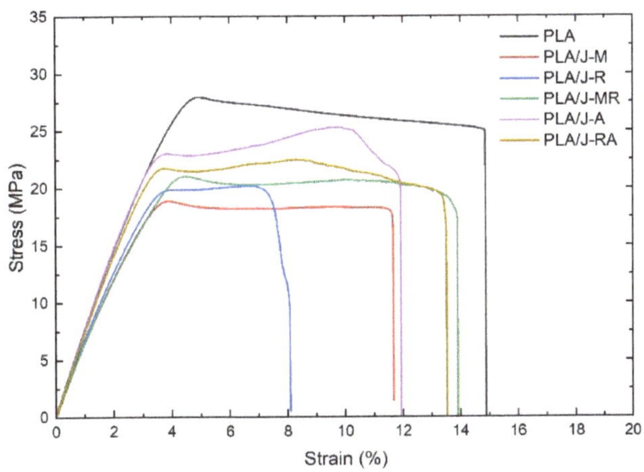

Figure 2. Stress vs strain curves corresponding to PLA and PLA composites.

It is possible to appreciate similar behavior for all materials evaluated in this work. The curves show a linear elastic region followed by diffusive necking and relatively low deformation until failure. It is worth noticing the presence of a shoulder in the plastic region developed for the composites. The adherence between jute fibers and PLA matrix requires a higher level of stress before failure, and it is the cause of the shoulder presence. The shoulder was evident in the PLA/J-A and PLA/J-RA composites.

The tensile parameters like E, σ_y, and deformation at break (ε_b) are summarized in Table 2.

Table 2. Mechanical parameters of PLA and PLA composites.

Material	E (GPa)	σ_y (MPa)	σ_b (MPa)	ε_b (%)
PLA	1.98 ± 0.02	27.93 ± 1.25	25.13 ± 1.16	14.76 ± 0.98
PLA/J-M	1.22 ± 0.23	18.81 ± 3.76	18.24 ± 1.98	11.48 ± 1.25
PLA/J-R	1.41 ± 0.13	19.84 ± 1.86	19.72 ± 1.98	7.51 ± 2.68
PLA/J-MR	1.26 ± 0.46	21.11 ± 2.33	19.41 ± 1.98	13.68 ± 1.26
PLA/J-A	1.83 ± 0.19	22.97 ± 2.16	21.88 ± 1.98	11.76 ± 1.89
PLA/J-RA	1.62 ± 0.16	21.78 ± 1.89	19.76 ± 1.98	13.36 ± 1.36

The elastic modulus and strength of PLA are notably higher than PLA composites. The PLA/J-A and PLA/J-RA composites presented intermediate stiffness and strength values, and the rest had low mechanical properties.

On the other hand, PLA and PLA composites showed similar deformation values, except for the PLA/J-R, which showed low ductility and sudden failure. In general, the PLA specimens did not develop necking nor whitening. On the contrary, the specimens presented a homogeneous deformation with a brittle-like break during the tensile test.

It results show that the jute fiber is not compatible with PLA, which is confirmed by the low mechanical performance observed by the composite containing jute fiber with flame retardant (PLA/J-R). On the other hand, the modified treatment applied to the jute fabrics does not seem to influence the fiber-matrix interaction. Nonetheless, the combination of modified fibers followed by the flame retardant application seems to influence the molecular compatibility, which favors the mechanical properties of the PLA/J-MR composite. According to the mechanical results obtained, spray adhesive could prove to be the best strategy to achieve a better interaction between jute fabrics and PLA.

Ductility decreases because the fabrics restrict plastic deformation, although an increase in stiffness and strength is usually expected because the fibers promote the reinforcing effect of the polymeric matrix. Lack of compatibility, lack of adhesion, and distortion of fabrics also affect the mechanical performance of fiber-reinforced polymer systems.

CT scan tomography is a powerful non-destructive testing tool for observing the disposition of the natural fiber fabrics into the 3D printing PLA specimens (Figure 3). We used CT scans to detect fabrics' inner disruptions and reveal possible adhesion between fibers and PLA in this work.

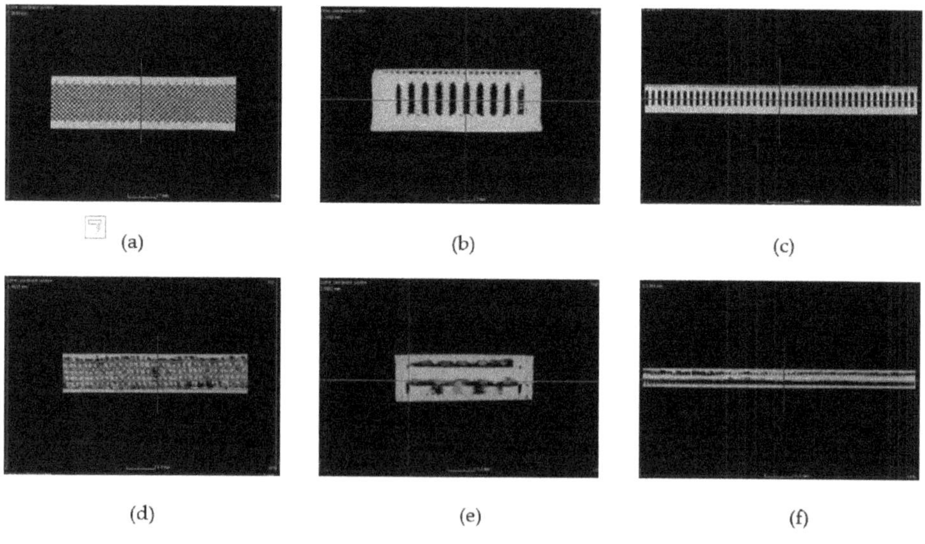

Figure 3. Tomography scans corresponding to PLA (upper) and PLA composite (bottom): (**a**) front view, (**b**) top view, (**c**) lateral view, (**d**) front view, (**e**) top view, (**f**) lateral view.

Because of densities, PLA looks white, jute fiber appears grey, and the air is black. The inspection was performed in the gage zone of the tensile specimens before the test (Figure 1c).

The front view of PLA (Figure 3a) shows the rectilinear pattern with an infill density of 90%. This pattern develops the configuration of a stacking sequence similar to bridge pillars-like, as appreciated in the top and lateral views of the PLA specimens (Figure 3b,c, respectively). On the other hand, tomography of the PLA composite reveals that the jute fabrics are well-aligned, without distortions or fiber displacements, and are easily identifiable, as observed in the front view of Figure 3d. Furthermore, some interaction between fibers and PLA seems to occur, as observed in the top and lateral views (Figure 3e,f, respectively), although it is necessary to underline that the rectilinear pattern is altered or distorted when the fused filament is printed onto the fabrics. The previous promotes the shell-like appearance, compared to the top and lateral views of PLA and PLA composites in Figure 3. Considering that the jute fabric configuration is a conventional woven (not spread-tow), the fused filament is placed over the weft and warp surface with a different texture than PLA filament, which could be altering the 3D printing pattern and favoring some adherence between fibers and PLA that restrict the ductility during the tensile tests.

After the tensile test, the specimens broke suddenly due to the failure of the PLA matrix, keeping part of the jute fiber fabric together until its complete breakage, as can be seen in Figure 4.

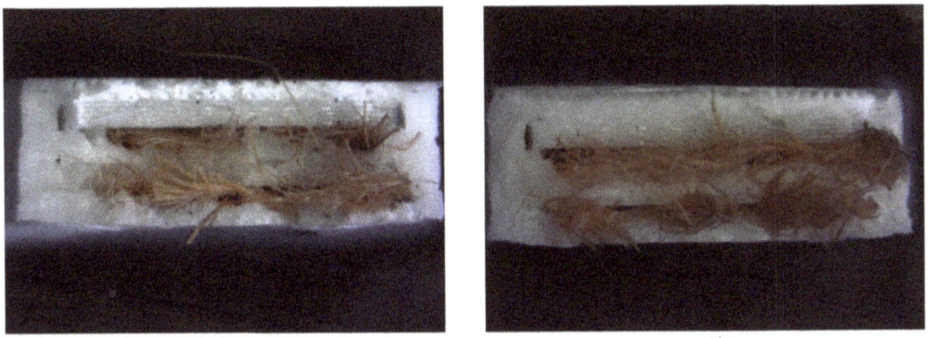

(a) (b)

Figure 4. Photographs of the failure surface corresponding to: (**a**) PLA/J-M, (**b**) PLA/J-RA. All specimens failed similarly.

DMA provides relevant information on the viscoelastic behavior of polymers and composites. Storage modulus measures stress stored in the specimens as mechanical energy, while the damping factor is a typical measure of energy dissipation. DMA curves corresponding to storage modulus and damping factor are presented in Figure 5.

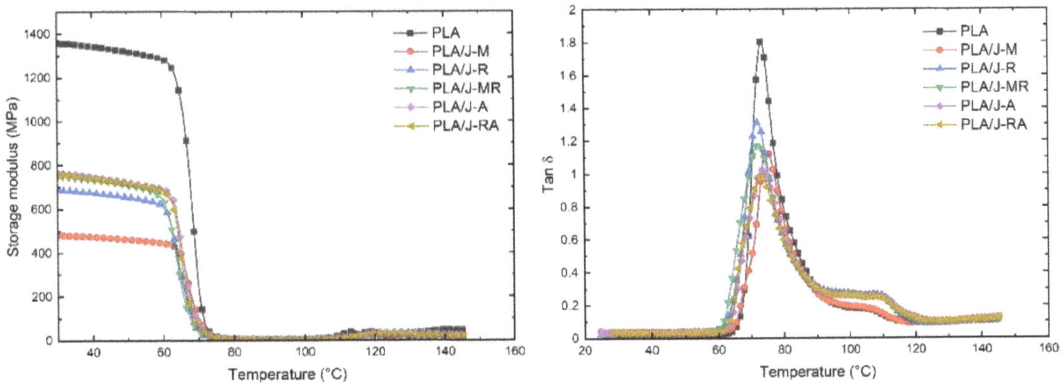

(a) (b)

Figure 5. DMA curves corresponding to PLA and PLA composite: (**a**) storage modulus, (**b**) tang δ.

The storage modulus curve contains the glassy region, the leathery zone, and the rubbery plateau. In this work, the shape of the DMA curves was similar between PLA and PLA composites (Figure 5a). All materials show that the storage modulus remains practically constant until the onset temperature (T_{onset}) indicates the leathery region's starting point, where the modulus drops abruptly, followed by the rubbery plateau. Then, the storage modulus increases above 115 °C, which is associated with the cold crystallization of the amorphous part of PLA. The T_{onset} of PLA and some PLA composites is similar (between 60 °C and 62 °C). The PLA/J-M shows the higher Tonset (64 °C).

From 20 °C to 60 °C, the PLA presents a storage modulus of 1358 MPa, which is notoriously higher than the value obtained for the PLA composites (~757 MPa) at the same temperature range. The previous should imply that the jute fabrics do not act as reinforcement in the glassy region.

The damping factor or Tan δ curves of PLA and PLA composite are similar (72.8 to 73.2 °C). The height of the tan δ peak is associated with the chain mobility of the amorphous region in the polymer composites. In this study, the peak position at approximately 73 °C indicates that the glass transition temperature of PLA is not altered by the addition of jute fabrics but affected the chain mobility in the amorphous region due to the confinement effect resulting in the reduction in tan δ peak height.

According to the methodology of chapter 3 of the Aircraft Materials Fire Test Handbook, the material meets the approval criteria if three specimens of the same material show a burning rate of less than 63.5 mm/min, considering that the thickness of specimens must be less than or equal to 13.0 mm. In this work, the specimens were out of specifications. Nonetheless, the burn rates were obtained using a calibrated distance of 38.1 mm. A horizontal burning test was carried out for the burning time and burning rate of the PLA and PLA composites for the flammability properties at room temperature. Table 3 presents the results obtained.

Table 3. Flammability test data report for PLA and PLA composites.

Material	Length (mm)	Width (mm)	Thickness (mm)	Burning Distance (mm)	Time of Burning (s)	Rate of Burning (mm/min)
PLA-1	51.3	13.23	4.77	38.1	49.55	46.14
PLA-2	50.49	13.27	4.75	38.1	41.65	54.89
PLA-3	50.66	13.24	4.74	38.1	40.33	56.68
PLA/J-M-1	50.36	13.69	4.94	38.1	35.08	65.17
PLA/J-M-2	51.16	13.76	5.07	38.1	48.88	46.77
PLA/J-M-3	51.59	14.03	5.13	38.1	50.83	44.97
PLA/J-R-1	51.75	13.67	4.84	38.1	64.57	35.40
PLA/J-R-2	50.72	13.53	4.81	38.1	63.62	35.93
PLA/J-R-3	49.95	13.54	5.16	38.1	70.03	32.64
PLA/J-MR-1	51.52	13.73	5.24	38.1	67.5	33.87
PLA/J-MR-2	51.7	13.41	5.11	38.1	63.29	36.12
PLA/J-MR-3	50.64	13.71	5.24	38.1	69.99	32.66
PLA/J-A-1	50.21	13.38	4.99	38.1	67.46	33.89
PLA/J-A-2	50.09	13.65	5.08	38.1	57.54	39.73
PLA/J-A-3	50.07	13.5	4.98	38.1	63.95	35.75
PLA/J-RA-1	50.38	13.45	5.01	38.1	62.5	36.58
PLA/J-RA-2	48.16	13.39	4.97	38.1	62.21	36.75
PLA/J-RA-3	50.16	13.4	5.28	38.1	60.11	38.03

It can be seen that the rate of burning of PLA and PLA composites meet the acceptance criteria of the reference standard, except for one value obtained for the PLA/J-M-1). Considering that this value is out of specifications and atypical data, it could be discarded. Figure 6 shows the time and rate of burning obtained for PLA and PLA composites.

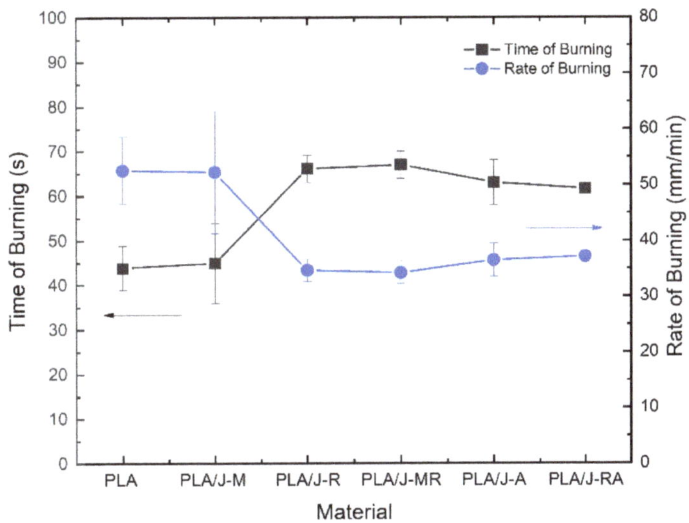

Figure 6. Time and rate of burning obtained for PLA and PLA composites.

It is possible to appreciate that the PLA and PLA composites containing modified jute fibers present low burning times (PLA/J-M). It means that the retardant treatment employed in this work induces a practical function on the flammability of PLA because the increase in burning time implies improvements in flame retardancy. Similarly, it appears that the spray adhesive exhibits some flame retardant effect since the time of burning exhibited for the PLA/J-A and PLA/J-RA displayed similar results to the PLA/J-R.

4. Conclusions

The results obtained in this work confirm the feasibility of producing 3D printing objects using plant fiber fabrics as filler. PLA-fused filament was successfully deposed onto natural fiber fabrics to print dog bone tensile specimens for characterization. In the first instance, the mechanical properties obtained for the PLA composites are not superior to PLA. However, viscoelastic properties allowed to identify the jute fabrics hinder the molecular motions, and the glass transition temperature remains.

Flame retardant and spray adhesive on jute fabrics promoted better composites' responses than PLA and PLA with modified fibers. One characteristic feature observed is the flame retardancy which increases because of the combined effect between jute fabrics and PLA, forming dense char which further resists the flame propagation.

The results obtained allow us to visualize the potential use of flame retardants for this composite material and the vast flammability analysis that can be deepened through cone calorimeter or limited oxygen index tests.

The use of plant fiber fabrics as reinforcement of 3D printed objects is a vast field of research. This work confirms the feasibility that plant fiber fabrics can be used as effective reinforcement. Other advantages of re-using this ecological waste material are that natural fibers are much less dense than synthetic fibers and polymers, which leads to developing lightweight composites and less raw material consumption for 3D printing parts and components. The results presented in this work lead to the need for a more detailed investigation of the effect of plant fiber fabrics as reinforcement of 3D printed objects for industrial applications. There are many questions about these new ecological materials, such has how the addition of distinct flame retardants, the fabric configuration, or different treated fibers can affect their mechanical performance. Some of these questions are being studied by our research group.

Author Contributions: Conceptualization, E.A.F.-U. Funding acquisition, E.A.F.-U. Writing—original draft, E.A.F.-U. and Y.R.E. Formal analysis, P.I.A.L. and Y.R.E. Validation, P.I.A.L. and Y.R.E. Investigation, P.I.A.L. and Y.R.E., Methodology, E.A.F.-U. All authors have read and agreed to the published version of the manuscript.

Funding: This work was supported by the Secretary of Public Education (SEP) and National Council of Science and Technology (CONACYT) through the Basic Science Fund (SEP-CONACYT Ciencia Basica) (grant number 257458) and funded by CONACYT through the Institutional Fund for Regional Promotion for Scientific, Technological and Innovation Development (FORDECYT), grant number 297708. The APC was funded by FORDECYT, grant number 297708.

Data Availability Statement: This study did not report any data.

Acknowledgments: Edgar Franco conveys special appreciation to the "CONACYT Researchers Program (Catedras CONACYT)".

Conflicts of Interest: The author declares no conflict of interest.

References

1. Cuan-Urquizo, E.; Barocio, E.; Tejada-Ortigoza, V.; Pipes, R.B.; Rodriguez, C.A.; Roman-Flores, A. Characterization of the mechanical properties of FFF structures and materials: A review on the experimental, computational and theoretical approaches. *Materials* **2019**, *16*, 895. [CrossRef]
2. Ming, Y.; Zhang, S.; Han, W.; Wang, B.; Duan, Y.; Xiao, H. Investigation on process parameters of 3D printed continuous carbon fiber-reinforced thermosetting epoxy composites. *Addit. Manuf.* **2020**, *33*, 101184. [CrossRef]
3. Anderson, I. Mechanical Properties of Specimens 3D Printed with Virgin and Recycled Polylactic Acid. 3D Print. *Addit. Manuf.* **2017**, *4*, 110–115. [CrossRef]
4. Qahtani, M.; Wu, F.; Misra, M.; Gregori, S.; Mielewski, D.F.; Mohanty, A.K. Experimental Design of Sustainable 3D-Printed Poly(Lactic Acid)/Biobased Poly(Butylene Succinate) Blends via Fused Deposition Modeling. *ACS Sustain. Chem. Eng.* **2019**, *7*, 14460–14470. [CrossRef]
5. Prasong, W.; Muanchan, P.; Ishigami, A.; Thumsorn, S.; Kurose, T.; Ito, H. Properties of 3D Printable Poly(lactic acid)/Poly(butylene adipate-co-terephthalate) Blends and Nano Talc Composites. *J. Nanomater.* **2020**, 8040517. [CrossRef]
6. Paspali, A.; Bao, Y.; Gawne, D.T.; Piestert, F.; Reinelt, S. The influence of nanostructure on the mechanical properties of 3D printed polylactide/nanoclay composites. *Compos. Part B Eng.* **2018**, *152*, 160–168. [CrossRef]
7. Farah, S.; Anderson, D.G.; Langer, R. Physical and mechanical properties of PLA, and their functions in widespread applications —A comprehensive review. *Adv. Drug Deliv. Rev.* **2016**, *107*, 367–392. [CrossRef] [PubMed]
8. Caminero, M.Á.; Chacón, J.M.; García-Plaza, E.; Núñez, P.J.; Reverte, J.M.; Becar, J.P. Additive manufacturing of PLA-based composites using fused filament fabrication: Effect of graphene nanoplatelet reinforcement on mechanical properties, dimensional accuracy and texture. *Polymers* **2019**, *11*, 799. [CrossRef] [PubMed]
9. Tian, X.; Liu, T.; Yang, C.; Wang, Q.; Li, D. Interface and performance of 3D printed continuous carbon fiber reinforced PLA composites. *Compos. Part A Appl. Sci. Manuf.* **2016**, *88*, 198–205. [CrossRef]
10. Pakkanen, J.; Manfredi, D.; Minetola, P.; Iuliano, L. *About the Use of Recycled or Biodegradable Filaments for Sustainability of 3D Printing BT, Sustainable Design and Manufacturing 2017*; Campana, G., Howlett, R.J., Setchi, R., Cimatti, B., Eds.; Springer International Publishing: Cham, Switzerland; Heidelberg, Germany, 2017; pp. 776–785.
11. Blok, L.G.; Longana, M.L.; Yu, H.; Woods, B.K.S. An investigation into 3D printing of fibre reinforced thermoplastic composites. *Addit. Manuf.* **2018**, *22*, 176–186. [CrossRef]
12. Le Duigou, A.; Correa, D.; Ueda, M.; Matsuzaki, R.; Castro, M. A review of 3D and 4D printing of natural fibre biocomposites. *Mater. Des.* **2020**, *194*, 108911. [CrossRef]
13. Sekar, V.; Fouladi, M.H.; Namasivayam, S.N.; Sivanesan, S. Additive Manufacturing: A Novel Method for Developing an Acoustic Panel Made of Natural Fiber-Reinforced Composites with Enhanced Mechanical and Acoustical Properties. *J. Eng.* **2019**. [CrossRef]
14. Woern, A.L.; Byard, D.J.; Oakley, R.B.; Fiedler, M.J.; Snabes, S.L.; Pearce, J.M. Fused particle fabrication 3-D printing: Recycled materials' optimization and mechanical properties. *Materials* **2018**, *11*, 1413. [CrossRef] [PubMed]
15. Hopewell, J.; Dvorak, R.; Kosior, E. Plastics recycling: Challenges and opportunities. *Philos. Trans. R. Soc. Lond. B. Biol. Sci.* **2009**, *364*, 2115–2126. [CrossRef]
16. Santana, H.A.; Amorim Júnior, N.S.; Ribeiro, D.V.; Cilla, M.S.; Dias, C.M.R. 3D printed mesh reinforced geopolymer: Notched prism bending. *Cem. Concr. Compos.* **2021**, *116*, 103892. [CrossRef]
17. Marchewka, J.; Laska, J. Processing of poly-l-lactide and poly(l-lactide-co-trimethylene carbonate) blends by fused filament fabrication and fused granulate fabrication using RepRap 3D printer. *Int. J. Adv. Manuf. Technol.* **2020**, *106*, 4933–4944. [CrossRef]
18. Fafenrot, S.; Grimmelsmann, N.; Wortmann, M.; Ehrmann, A. Three-dimensional (3D) printing of polymer-metal hybrid materials by fused deposition modeling. *Materials* **2017**, *10*, 1199. [CrossRef]

19. Khan, I.; Kamma-Lorger, C.S.; Mohan, S.D.; Mateus, A.; Mitchell, G.R. *The Exploitation of Polymer Based Nanocomposites for Additive Manufacturing: A Prospective Review*; Trans Tech Publications Ltd.: Bäch, Switzerland, 2019; Volume 890, ISBN 1139574620.
20. Podsiadły, B.; Skalski, A.; Wałpuski, B.; Słoma, M. Heterophase materials for fused filament fabrication of structural electronics. *J. Mater. Sci. Mater. Electron.* **2019**, *30*, 1236–1245. [CrossRef]
21. Salavati, M.; Yousefi, A.A. Polypropylene–clay micro/nanocomposites as fused deposition modeling filament: Effect of polypropylene-g-maleic anhydride and organo-nanoclay as chemical and physical compatibilizers. *Iran. Polym. J.* **2019**, *28*, 611–620. [CrossRef]
22. Herrero, M.; Peng, F.; Núñez Carrero, K.C.; Merino, J.C.; Vogt, B.D. Renewable Nanocomposites for Additive Manufacturing Using Fused Filament Fabrication. ACS Sustain. *Chem. Eng.* **2018**, *6*, 12393–12402. [CrossRef]
23. Aumnate, C.; Limpanart, S.; Soatthiyanon, N.; Khunton, S. PP/organoclay nanocomposites for fused filament fabrication (FFF) 3D printing. *Express Polym. Lett.* **2019**, *13*, 898–909. [CrossRef]
24. Guo, H.; Lv, R.; Bai, S. Recent advances on 3D printing graphene-based composites. *Nano Mater. Sci.* **2019**, *1*, 101–115. [CrossRef]
25. Aumnate, C.; Pongwisuthiruchte, A.; Pattananuwat, P.; Potiyaraj, P. Fabrication of ABS/Graphene Oxide Composite Filament for Fused Filament Fabrication (FFF) 3D Printing. *Adv. Mater. Sci. Eng.* **2018**, *2018*, 2830437. [CrossRef]
26. Gnanasekaran, K.; Heijmans, T.; van Bennekom, S.; Woldhuis, H.; Wijnia, S.; de With, G.; Friedrich, H. 3D printing of CNT- and graphene-based conductive polymer nanocomposites by fused deposition modeling. *Appl. Mater. Today* **2017**, *9*, 21–28. [CrossRef]
27. Sodeifian, G.; Ghaseminejad, S.; Yousefi, A.A. Preparation of polypropylene/short glass fiber composite as Fused Deposition Modeling (FDM) filament. *Results Phys.* **2019**, *12*, 205–222. [CrossRef]
28. Spoerk, M.; Savandaiah, C.; Arbeiter, F.; Traxler, G.; Cardon, L.; Holzer, C.; Sapkota, J. Anisotropic properties of oriented short carbon fibre filled polypropylene parts fabricated by extrusion-based additive manufacturing. *Compos. Part A Appl. Sci. Manuf.* **2018**, *113*, 95–104. [CrossRef]
29. Ivey, M.; Melenka, G.W.; Carey, J.P.; Ayranci, C. Characterizing short-fiber-reinforced composites produced using additive manufacturing. *Adv. Manuf. Polym. Compos. Sci.* **2017**, *3*, 81–91. [CrossRef]
30. Wang, X.; Jiang, M.; Zhou, Z.; Gou, J.; Hui, D. 3D printing of polymer matrix composites: A review and prospective. *Compos. Part B Eng.* **2017**, *110*, 442–458. [CrossRef]
31. Matsuzaki, R.; Ueda, M.; Namiki, M.; Jeong, T.K.; Asahara, H.; Horiguchi, K.; Nakamura, T.; Todoroki, A.; Hirano, Y. Three-dimensional printing of continuous-fiber composites by in-nozzle impregnation. *Sci. Rep.* **2016**, *6*, 1–7. [CrossRef]
32. Balla, V.K.; Kate, K.H.; Satyavolu, J.; Singh, P.; Tadimeti, J.G.D. Additive manufacturing of natural fiber reinforced polymer composites: Processing and prospects. *Compos. Part B Eng.* **2019**, *174*, 106956. [CrossRef]
33. Alkbir, M.F.M.; Sapuan, S.M.; Nuraini, A.A.; Ishak, M.R. Fibre properties and crashworthiness parameters of natural fibre-reinforced composite structure: A literature review. *Compos. Struct.* **2016**, *148*, 59–73. [CrossRef]
34. Shekar, H.S.S.; Ramachandra, M. Green Composites: A Review. *Mater. Today Proc.* **2018**, *5*, 2518–2526. [CrossRef]
35. Yan, L.; Kasal, B.; Huang, L. A review of recent research on the use of cellulosic fibres, their fibre fabric reinforced cementitious, geo-polymer and polymer composites in civil engineering. *Compos. Part B Eng.* **2016**, *92*, 94–132. [CrossRef]
36. Wis, A.A.; Kodal, M.; Ozturk, S.; Ozkoc, G. Overmolded polylactide/jute-mat eco-composites: A new method to enhance the properties of natural fiber biodegradable composites. *J. Appl. Polym. Sci.* **2020**, *137*, 48692. [CrossRef]
37. Jerpdal, L.; Schuette, P.; Ståhlberg, D.; åkermo, M. Influence of temperature during overmolding on the tensile modulus of self-reinforced poly(ethylene terephthalate) insert. *J. Appl. Polym. Sci.* **2020**, *137*, 48334. [CrossRef]
38. Hill, C.A.S.; Norton, A.; Newman, G. The water vapor sorption behavior of natural fibers. *J. Appl. Polym. Sci.* **2009**, *112*, 1524–1537. [CrossRef]
39. Defoirdt, N.; Biswas, S.; De Vriese, L.; Tran, L.Q.N.; Van Acker, J.; Ahsan, Q.; Gorbatikh, L.; Van Vuure, A.; Verpoest, I. Assessment of the tensile properties of coir, bamboo and jute fibre. *Compos. Part A Appl. Sci. Manuf.* **2010**, *41*, 588–595. [CrossRef]

polymers

MDPI

Article

Rotational Rheology of Wood Flour Composites Based on Recycled Polyethylene

Antonella Patti [1,*], Gianluca Cicala [1] and Stefano Acierno [2,*]

[1] Department of Civil Engineering and Architecture (DICAr), University of Catania, Viale Andrea Doria 6, 95125 Catania, Italy; gianluca.cicala@unict.it
[2] Department of Engineering, University of Sannio, Piazza Roma 21, 82100 Benevento, Italy
* Correspondence: antonella.patti@unict.it (A.P.); stefano.acierno@unisannio.it (S.A.)

Abstract: In this paper, we study the effect of the addition of wood flour as a filler in a recycled polyethylene (r-PE) in view of its potential applications in 3D printing. The composites, prepared by melt mixing, are characterized with torque measurements performed during the compounding, dynamic rotational rheology, and infrared spectroscopy. Data show that the introduction of wood results in increased viscosity and in sensible viscous heating during the compounding. The r-PE appear to be stable at temperatures up to 180 °C while at higher temperatures the material shows a rheological response characterized by time-increasing viscoelastic moduli that suggests a thermal degradation governed by crosslinking reactions. The compounds (with wood loading up to 50% in wt.) also shows thermal stability at temperatures up to 180 °C. The viscoelastic behavior and the infrared spectra of the r-PE matrix suggests the presence of branches in the macromolecular structure due to the process. Although the addition of wood particles determines increased viscoelastic moduli, a solid-like viscoelastic response is not shown even for the highest wood concentrations. This behavior, due to a poor compatibility and weak interfacial adhesion between the two phases, is however promising in view of common processing technologies as extrusion or injection molding.

Keywords: wood-polymer composites (WPC); recycled thermoplastics; torque measurements; rheological properties; dynamic mechanical analyses

Citation: Patti, A.; Cicala, G.; Acierno, S. Rotational Rheology of Wood Flour Composites Based on Recycled Polyethylene. *Polymers* **2021**, *13*, 2226. https://doi.org/10.3390/polym13142226

Academic Editor: José António Covas

Received: 17 June 2021
Accepted: 2 July 2021
Published: 6 July 2021

1. Introduction

Plastics are one of the main components of products of everyday life and industrial activities, such as packaging, agriculture, automotive, and biomedical applications, becoming an essential element for the way of life. As a consequence, an increase in plastic production and plastic products has been verified in recent times. Conventional plastics are materials with (relatively) high strength and durability and requiring hundreds of years to break down under normal ambient conditions. This represents an important disadvantage from the perspective of environmental impact and aspects of pollution [1].

As reported by Singh and Sharma [2], over 300 million metric tons of plastic are produced each year and half of that amount is spent on disposal applications, i.e., activities lasting less than one year: the product is used and thrown away one year after purchase. The accumulation of solid plastic waste in the environment has become an increasingly important worldwide problem to consider and deal with [3].

Over the past decades, environmentalists have devoted more and more effort to the impact of chemical and industrial processes and, as a consequence, in a number of countries, governments have promoted rules and laws to protect the quality of the environment for the future [4]. In this context, chemical industries have been pushed to adopt non-polluting chemical processes and materials, reduce the use of hazard chemicals, efficiently use raw materials, and reduce emissions and wastes. This has translated into a growing interest in biodegradable renewable systems, such as composites reinforced with plant fibers [5].

Natural or lignocellulosic materials, such as jute, coir, rice husk, bagasse, or sisal have been taken into account for the production of "green" composites, by replacing the traditional synthetic (glass or carbon or aramid-based fibers), or common fillers (calcium carbonate, silica, or talc), in widespread used thermoplastics (polyethylene, polypropylene, and polystyrene) [6].

As cost-effective natural filler, wood has often adopted in the plastics industry to improve the strength or stiffness of thermoplastics, to reduce the costs, to enhance recyclability and eco-compatibility, or diminish the wear on processing equipment. In this perspective, wood-plastic composites were early used for exterior non-structural or semi-structural building products in the civil engineering (such as decking, fencing, siding, window frames, and roof tiles [7]) or also actualized for various industries, including automotive, household items, packaging, and consumer goods or living supplies as yard products [8].

A wood-plastic composite (WPC) refers consists of wood-based elements (such as lumber, veneer, fibers, or particles) embedded in a polymer matrix. Typically, extrusion, thermoforming, compressive or injection molding have been adopted, as processing techniques, in the compounds preparation. Unfortunately, wood start to degrade at approximately 200 °C and this represents a limiting condition to the choice of the polymeric matrix in WPCs. Typical polymers, such as polypropylene (PP), polyethylene (PE), polystyrene (PS), and poly vinyl chloride (PVC), have been adopted in wood-polymer compounds [9].

For example, in work by Wang et al. [10], polypropylene (PP) composites reinforced with 15, 30, and 45 wt.% wood powder (WP) were prepared by injection molding and characterized in terms of thermal, mechanical, and dynamic mechanical properties. Experimental results show that the presence of wood in polypropylene determined on the one hand an increment of strength and rigidity but, on the other hand, the toughness of the compounds did not improve. However, the authors affirmed the potentiality of WP-PP composites to replace fiberglass-reinforced composites in construction, sports facilities, and the automotive industry.

The influence of five types of plastics, high density polyethylene (HDPE), low density polyethylene (LDPE), polypropylene (PP), polyvinylchloride (PVC), and polystyrene (PS), on the mechanical properties of wood-composites have been studied by Ratanawilai et al. [11]. According to their data, WPCs based on PS and PP achieved the highest performance in terms of strength and weathering compared to the other matrices.

Additionally, recycled plastics were used for the preparation of wood-based composites; for example the effects of recycling operations on the mechanical and rheological properties of wood-flour in a blend of low and high density polyethylenes has been investigated by Habibi et al. [12]. A pre-mixing of the two polymers was found to increase flexural and tensile properties of the developed compounds. An increment in complex viscosity and storage modulus was measured in prepared samples by melt-premixing compared to those obtained by simultaneous compounding.

Despite the interesting features of WDCs systems, some difficulties came from the shaping through the traditional methods of extrusion and injection molding. Rheological characterization has been proven to be a very useful technique for the understanding of the manufacturing processes and to optimize process conditions and formulations in order to achieve good quality of the final products [13].

In recent years, the interest on the additive manufacturing (AM), also called 3D printing or rapid prototyping, has markedly concerned several fields [14]. The most popular procedure among 3D printing techniques is the so-called fused deposition modeling (FDM), during which extruded filaments are printed one on top the others following a geometric pattern into a three-dimensional structure, according to layer-by-layer deposition. On the one hand, the versatility offered by 3D printers, linked to a great reduction in equipment costs, have promoted a wide spread of this technology. On the other hand, critical issues regarding the environmental impact, coming from energy and material consumption, and wastes production, have arisen. Possible actions in order to improve the sustainability of additive manufacturing consist in the development of sustainable (eventually from recycling operation of waste products) for 3D printing [15]. A wide range of performance in 3D printed materials can be achieved with the right combination of polymer, filler, and additives. The commonly used thermoplastics available for FDM are acrylonitrile-butadiene-styrene (ABS), polylactic acid (PLA), and polyamide (PA), but frequently also polyamide or nylon (PA), polycarbonate (PC), poly-methylmethylacrylate (PMMA), polyethylene (PE), and polypropylene (PP). Nowadays, most researches are focused on composites and biocomposites, i.e., at least one component derived from biological or natural sources. The use of natural materials such as wood helps to reduce the use of petroleum-based plastics and, thus, the environmental impact. Different amounts of wood flour (up to 50 wt.%) were used as reinforcing filler in polylactide acid (PLA) by testing physical, mechanical, and rheological properties of the 3D printed composites [16]. The effect of extrusion temperature on the physical properties of the printed wood/PLA samples such as moisture content, surface roughness, water absorption rate, and thickness swelling rate, were studied in the work of Yang [17]. The 3D printable composite filaments have been made from cardboard dust and high density polyethylene (HDPE) [18]. Waste products have been proposed for 3D printing applications: post-consumer textile waste and polyethylene terephthalate (PET) water bottles were melt compounded to form a monofilament feedstock for extrusion-based 3D printing platforms [19].

In this framework, the purpose of this work is to explore the characteristics of a polymer filament based on recycled polyethylene in view of its potential applications as raw material in extrusion and 3D printed parts. The inclusion of wood flour in the chosen matrix is proposed in order increase the eco-compatibility and the overall performance of final products. At regard, rotational rheological tests were conducted on samples to investigate aspects related to the thermal stability of obtained compounds, and effects of powder content on polymer viscoelasticity. Torque and temperature measurements were also reported to examine the processability of formulations. Finally, spectroscopic analysis was used to confirm the thermal degradation of the tested materials.

2. Materials and Methods

A commercial low-density recycled polyethylene (r-PE), obtained by regenerating films coming from the greenhouse cover, supplied in pellets by ILAP (Ragusa, Italy) was used as matrix. A wood flour, micrometric in size, supplied from wood processing factory, was used as filler.

Composites, containing recycled polyethylene (r-PE) and 25, 35, and 50 wt./wt.% of wood (WD) flour as filler, were prepared by combining the two components in a batch mixer (Brabender Plastograph EC-Brabender GmbH and Co. KG, Germany). The mixing equipment was connected to a drive unit (torque rheometer) that allowed the collection of torque data and the thermal control of the mixing chamber during the compound preparation. The temperature of the mixing chamber measured through four thermocouples located in different points: in the center of the rotating screws (bulk temperature), near the rear wall, front wall, and mixer bowl. The bulk temperature and torque as a function of mixing time are provided by the software; when the set-point temperature is achieved by all the thermocouples, a torque calibration is done, and the chamber is filled (to about the 80% of its volume) pouring polymer pellets and wood flour. The compounding phase was conducted at a screw rotational speed of 30 RPM, a temperature of 180 °C, and a mixing time of 15 min. Before the composite preparation, the wood flour was dried overnight under vacuum in an oven at a temperature of 50 °C. The material that exited the mixer was reduced to fragments and extruded at 180 °C in a twin screw extruder (mod. KETSE 20/40 D, Brabender) to form a filament.

Viscoelastic characterization of molten composites was conducted using a strain-controlled rotational rheometer (mod. ARES-G2, by TA Instruments, New Castle, DE, USA), equipped with parallel plates (25 mm in diameter) geometry and a forced-convection oven for the temperature control. Preliminary strain sweep tests were performed in order to identify the linear viscoelastic region of the material. The materials were subjected to small–amplitude oscillations at a frequency of 1 rad/s and a strain amplitude of 1% for a duration of 1500 s (also called "Time sweep tests") at four different temperatures, ranging from 160 °C to 220 °C, in order to verify the thermal stability of matrix and of the compounds. The linear viscoelastic characterization of the samples in the frequency range from 0.1 to 100 rad/s (the test are also called "Frequency sweep tests") was performed at 160 °C, 170 °C, 180 °C. All the experiments were carried out on specimens under air atmosphere in order to assess the material behavior under real working conditions.

An infrared spectrometer (mod. Spectrum 65 FT IR, produced by Perkin Elmer, Waltham, MA, USA), was used in attenuated total reflection (ATR) modality to acquire the spectra of r-PE and r-PE-WD compounds. The samples were analyzed before and after a heat treatment conducted in an oven at 180 °C for 1 h. Afterward, the same samples were subjected to further a heat treatment at 220 °C for 1 h and IR spectra were collected again. During the examination, a range of wavenumber equal to 400–4000 cm^{-1}, a resolution of 4 cm^{-1}, and 16 scans were adopted. For each sample, the base line correction and advanced ATR correction related to the specific used crystal of diamond were carried out by Omnic Software.

A schematic of the experimental procedures is presented in Figure 1.

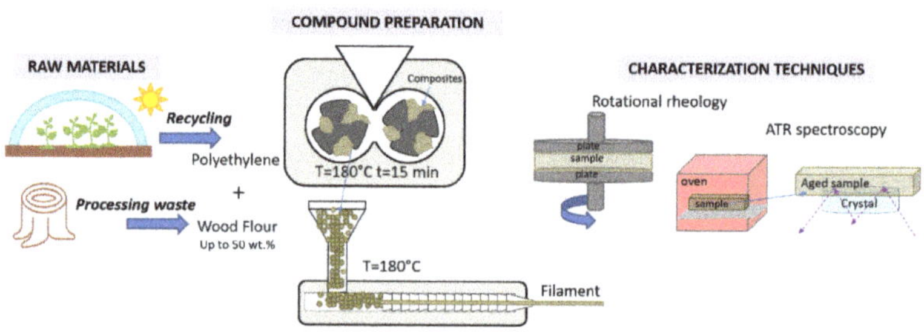

Figure 1. Schematic representation of compounding phase and characterization techniques.

3. Results and Discussion

3.1. Torque and Temperature Measurements during Compounding Phase

During the preparation of the compounds, the torque, and the temperature were measured as a function of mixing time and the corresponding data are reported, respectively, in Figure 2a,b.

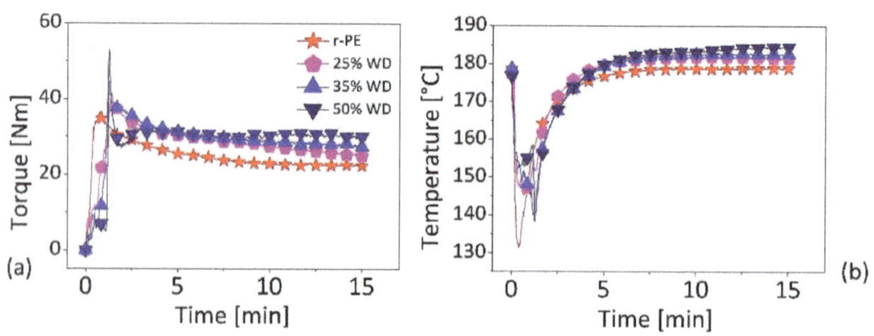

Figure 2. Torque and temperature measurement as a function of mixing time during the compounding phase. (Legend in Figure 2b as in Figure 2a).

The torque (see Figure 2a) shows a non-monotonous behavior with a local maximum, due to the fact that the polymer (fed as a solid) takes a few minutes to melt, followed by decline toward a steady-state value. For all the materials the torque stabilizes after about 7–10 min and subsequently does not show any sensible decrease until the end of the compounding (i.e., 15 min), thus suggesting that mixing has taken place and no degradation of the materials is occurring.

As visible from Figure 2b the temperature of the mixing chamber, which starts from the set-point value of 180 °C, shows a sudden drop in correspondence of the loading of the materials, which are poured at room temperature. When the material are fed to the mixing chamber, they are heated and compacted, a void-free state is created, and the melting of the polymer starts at the interface with the metallic parts, which are above the melting temperature of polyethylene (140 °C–150 °C, [20]), [21,22]. The melting is an endothermic phenomenon, consisting into the breakage of crystalline regions, requiring energy. For this reason, after an initial decrease, the temperature rises to the nominal set-point value (180 °C). It should be mentioned that, while for the r-PE the initial decline in temperature is almost instantaneous, reaching approximately 130 °C, in the case of composites this temperature drop takes longer and a higher minimum (around 140 °C) is attained. This behavior is due to a lower quantity of polymer to be melted and to reduced thermal diffusivity and specific heat of the r-PE/WOOD systems which leads to a slow-down the heat transmission dynamics and to reduced energy request for the melting process [23].

All the systems containing the wood flour (WD) show similar behaviors of temperature evolution. It should be mentioned that the composites at the end of mixing time of 15 min achieve temperatures that are slightly higher than the set-point of 180 °C (see Table 1 for more details). This overshoot in the temperature is particularly evident for the composite containing at 50 wt.% of WD, which reaches a temperature approximately 5 °C higher with respect to the set-point. This behavior, as the higher values of the torque (also reported in Table 1), is consistent with increased viscosities of these systems (as we will see in the next paragraph) due to the friction between the wood particles in motion within the melted matrix. In other words, when the filler loading is increased, less quantity of polymer is available for incorporating particles, the inter-particles distance is reduced, and the frequency of particle-to-particle collisions increases. As the viscosity of the system increases higher torques are required for the mixing, and an increased viscous heating takes place [24]. This effect becomes more evident at high WD loadings, given an increased shear stress during the kneading of materials in the chamber and the low thermal conductivity of wood particles (0.09–0.19 W/mK) [25] compared to that of polyethylene (0.30–0.44 W/mK) [26]. In this condition, the system are less able to dissipate the energy causing mixture overheating.

Table 1. Processing parameters during the compounding phase: M is the average torque during the last 10 min of mixing; T is the temperature reached at the end of mixing period, and TME is the total mechanical energy ($2\pi N \int Mdt$; N = 30 rpm).

Material	M [Nm]	T [°C]	TME [kJ]
r-PE	23	179	70
25% WD	27	181	79
35% WD	28	183	81
50% WD	30	185	80

According to previous a work on the compounding [27], processing variables, such as torque (M) and total mechanical energy (TME), shown in Table 1, provide useful information about the dispersion of the filler within the matrix. A comparison of the values corresponding to the matrix (r-PE alone) and to the composites show for the total mechanical energy (TME) an increase of about 14% for the composites independently of the wood content. Such a behavior suggests a poor dispersion of filler within the matrix.

3.2. *Dynamic Rotational Rheology*

3.2.1. Time Sweep Tests

Thermal stability of the matrix (recycled polyethylene, r-PE) was studied through time sweep test, performed at four different temperatures 160 °C, 180 °C, 200 °C, 220 °C. Experimental results are shown in Figure 3 in terms of storage (G′) and loss (G″) moduli as a function of time (1500 s).

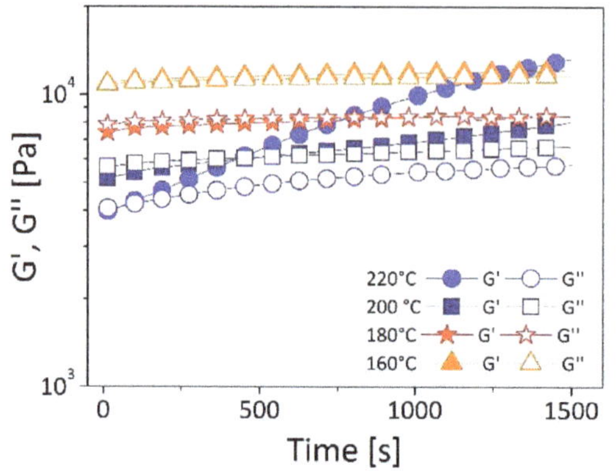

Figure 3. Storage (G′) and Loss (G″) modulus as function of time and temperature for recycled polyethylene. (r-PE) Measurements were conducted at an oscillatory frequency of 1 rad/s and a strain amplitude of 1%.

By heating from 160 °C to 220 °C, a strong change in rheological response of the resin is observed. More in details, while at 160 °C and 180 °C both G′ and G″ are almost time-independent over the entire experiment; at the temperatures of 200 °C and 220 °C, both G′ and G″ show and increasing behavior with G′ having faster kinetics compared to G″.

This behavior can be attributed to a thermal degradation of the polymer that seems to be negligible at 160 °C and 180 °C and relevant at 200 °C and 220 °C. These temperatures are consistent with data from literature for both low- and high-density polyethylene. In particular Dordinejad et al. [28], who studied the degradation of low-density polyethylene by time sweep rheometry, found that the growth rates of viscoelastic moduli increase with temperature and attributed this behavior to crosslinking reactions dominating in the thermal degradation mechanism. Mariani et al. [29], who studied the correlation between processability and properties of a high density polyethylene with a rheological approach, found increasing viscoelastic moduli with time and temperature; they attributed this behavior to an increase in molecular weight of the polymer due to the creation of long-chain branching originating from the reaction between the alkyl radical, formed during polymer processing, and vinyl groups of unsaturated ends of HDPE.

As our matrix has proved to be stable at temperatures as high as 180 °C, the thermal stability of the composites was studied at the temperature of 180 °C. The viscoelastic properties, in terms of storage modulus (G′) and complex viscosity (η^*), are reported as a function of time in Figure 4.

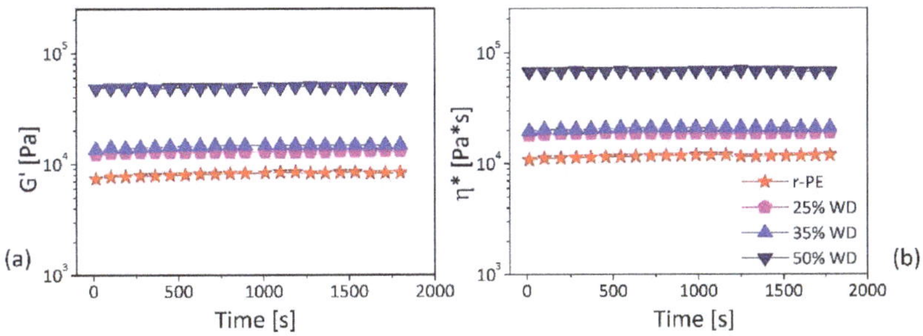

Figure 4. Time sweep tests performed at 180 °C (at an oscillatory frequency $\omega = 1$ rad/s and a strain amplitude $\gamma = 1\%$) for the developed formulations: (**a**) G′ and (**b**) η^*.

All the materials, also at highest filler loading (50 wt.%), show rheological properties almost time-independent for the entire duration of the experiments, thus confirming the thermal stability of the materials at temperatures as high as 180 °C. A similar behavior was observed by Mazzanti et al. [30] in the case of wood-polypropylene composites, who related the stable moduli to absence of wood degradation at a concentration of 70 wt.%. It should be mentioned that the good stability of our composites is in line with observations of Sliwa et al. [31], who found a spectacular improvement of thermal stability of wood-polymer composites due to the formation of a protective barrier of char which hinders thermo-oxidation process. Furthermore, this behavior is consistent with the formation of tortuous pathways of dispersed fillers in the polymer that slows the mass diffusion of degradation products from the polymer's bulk to the gas phase, reduces overall the thermal conductivity, and acts as a protection against thermal decomposition observed for composites [32].

3.2.2. Frequency Sweep Tests

In Figure 5, the storage and the loss moduli at 180 °C are plotted, for the different wood contents, as a function of the oscillatory frequency (ω). All the materials show the typical viscoelastic behavior of entangled polymeric liquid, with the loss modulus (G″) dominating at low frequencies, the storage modulus (G′) dominating at higher frequency, and a crossover (where G′ = G′) at intermediate frequencies.

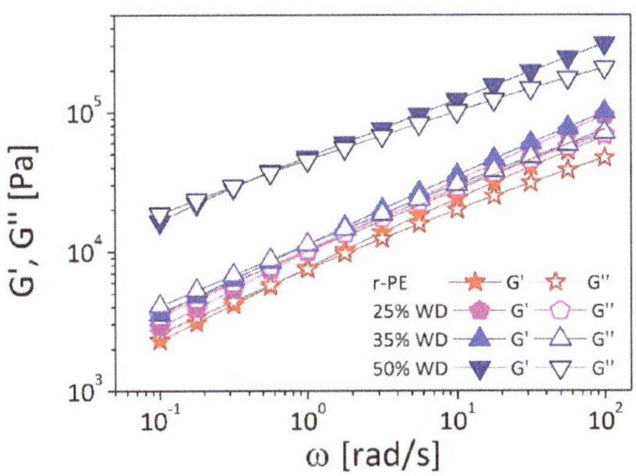

Figure 5. Storage (G′) and Loss (G″) modulus vs oscillatory frequency at temperature of 180 °C for composites.

From the observation of Figure 5, we note that as the amount of wood is increased the viscoelastic moduli move towards higher values and the crossover shifts towards lower frequencies with respect to the r-PE thus evidencing the reinforcing effect of the filler.

More in details, it should be noticed that even for the formulation containing the highest amount of wood loadings (50% WD) a low-frequency region with a frequency-independent elastic modulus G′, typical of interconnected networks, cannot be observed. This behavior indicates poor compatibility and weak interfacial adhesion between the two phases that prevent the interactions responsible for a rheological percolation and the formation of a three-dimensional network [33,34]. On the contrary, the fact that the crossover frequency decreases as a function of the filler loadings, indicates that, despite the weak affinity between the two phases, a greater amount of particles limits the motion and slows down relaxation dynamics of polymeric chains. These results are in line with other works of wood-plastic composites [35], where an increase in G′ and G″, an attitude of material to elastic behavior, and a crossover point moving towards lower frequency were observed with increasing filler concentration. Literature data suggest for our composites a possible improvement of melt strength [36] which could be very useful in view of the utilization of these material for the production of printed objects.

The temperature dependency of viscoelastic behavior of the wood–plastic composites is shown in Figure 6, where the storage modulus (G′) and the tangent of the phase shift angle (tan δ) are plotted as a function of the oscillatory frequency for system filled with 50 wt.% of wood (Figure 6a,b).

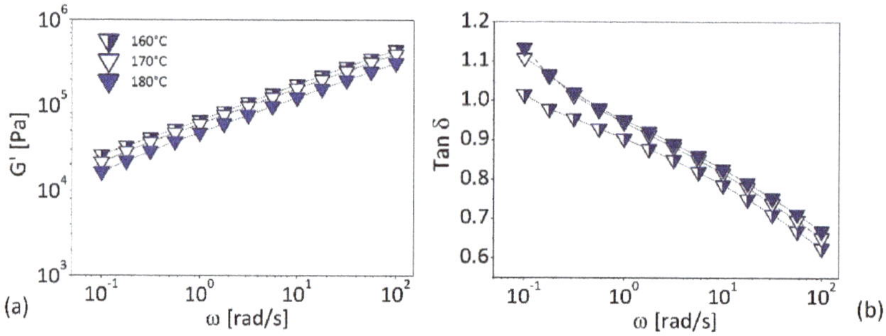

Figure 6. Effect of temperature on storage modulus (**a**) and dissipation factor (**b**) of composites at 50% in wt. of wood flour.

The data reported in Figure 6 and, in particular, the similarity of the G′ against ω curves (on double logarithmic scale) collected at the different temperatures, suggest that for the elastic modulus, and more generally for any viscoelastic function, the effect of the temperature can be separated by that of the frequency, and can be described by a temperature-shift factor (a_T). The temperature dependence of a_T represent the effect of temperature on viscoelastic properties; the values of a_T are determined empirically by making tan δ curves (i.e., a density independent viscoelastic function) superimpose on the one chosen as reference. In our case, we have chosen a reference temperature (at which the shift factor is equal to unity) to be T_0 = 180 °C. If the complex viscosity (η*) data are plotted against the reduce frequency $\omega \cdot a_T$ the master curves reported in Figure 7 are obtained. The master curves shown in Figure 6 represent the frequency dependence that would have been obtained measuring the complex viscosity over a wider frequency range at the (reference) temperature of 180 °C.

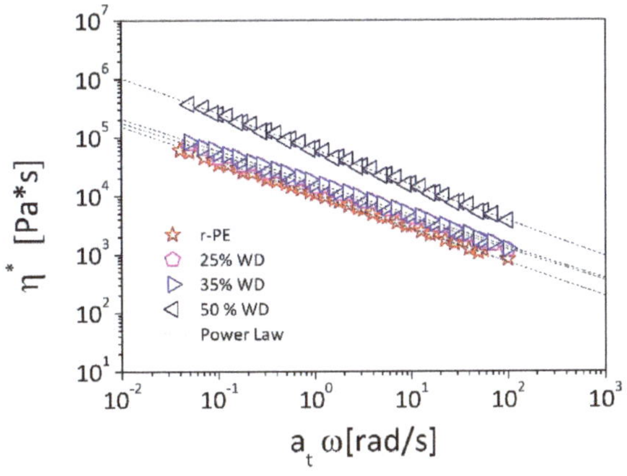

Figure 7. Complex viscosity against reduce frequency master-curves of composites. The dotted lines represent fittings with a power-law model (Equation (1)).

A knowledge, and a modeling, of the viscoelastic behavior of materials as a function of deformation rates and of the temperature is essential in order to adjust processing parameters and obtain a final product with good properties [37]; we derive in the following section analytical description of both variations.

The complex viscosities of our materials (see Figure 7) show a decreasing behavior upon the oscillatory frequency, a so-called shear-thinning or pseudoplastic behavior, that can be attributed to molecular alignments and disentanglements of the polymer chains at faster deformation rated. The most simple model that can be used to capture such a behavior is the power-law, or Ostwald-de Waele, model:

$$\eta^*(\omega) = k\omega^{n-1} \tag{1}$$

where k is the flow consistency (Pa*s^n) and n is the flow index.

By fitting the experimental data reported in Figure 7 with Equation (1), consistency and flow index are obtained as a function of the wood concentration (Figure 8a). It can be observed that the consistency increases as a strong function of the filler loading with a 6-fold increase for the system containing 50% of wood with respect to the base r-PE. On the contrary, the flow index shows a weak dependency upon the wood contents with values around 0.4 thus suggesting the shear thinning behavior depends on the polymeric matrix and not on the filler. These results are in good agreement with the studies of Mazzanti et al. [30] and of Laufer et al. [38] who studied, respectively, of wood-PP and wood-LDPE and observed the similar variations of consistency and flow index as a function of the amount of the particles in the matrix.

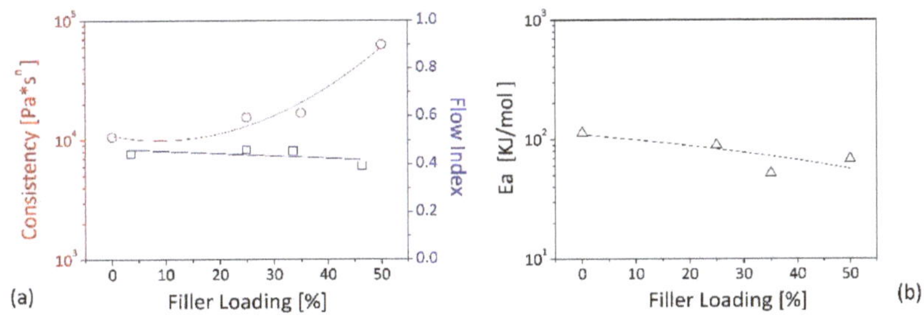

Figure 8. Consistency and index flow (**a**), and activation energy (**b**) against filler loading. The lines in figure are to guide the eye.

The temperature dependence of viscoelastic properties is described by the shift factor a_T, which at temperatures well above the glass transition temperature, can be represented by a simple Arrhenius-type equation:

$$a_T = exp\left[\frac{E_a}{R}\left(\frac{1}{T} - \frac{1}{T_0}\right)\right] \tag{2}$$

where T_0 is the reference temperature (in our case set to 180 °C), E_A is the activation energy for the process, and R is the universal gas constant.

The activation energy is plotted in Figure 8b as a function the filler content for the developed compounds. It should be noted that for the matrix we have determined a value of the order of 100 kJ/mol while for the higher wood loadings we have observed values of the order of 50 kJ/mol for. As lower values of the activation energy indicate a lower sensitivity of viscoelastic properties to temperature changes, it can be affirmed that the wood-filled systems behave more in a solid-like manner also from this point of view. This behavior is consistent with literature data, even though lower values are reported for homo-polyethylene (~26 kJ/mol) and for short-chain branched PE (~34 kJ/mol). However, as it is known that long-chain branching provokes an increase in the activation energy of polyethylene, the data suggest the presence of branching in our r-PE. As stated by Park [39], the free volume is not the main aspect that dominates the macromolecular motion. At testing temperatures higher than the glass transition temperature (T_g), the free volume is larger than that available at T_g. In this condition, the primary fact governing the chain mobility is the thermal activation. The branching increased the activation energy of polyethylene due to the hindrance of segmental motion for interaction of long-chain branches. In other words, the higher the activation energy observed for the r-PE suggests a possible presence long-chain branches formed during processing.

The variation of complex viscosity with deformation rates, described by power-law equation (Equation (1)), allows for the calculation of viscosity at processing deformation rates, which are higher than those experimentally explored. Equation (1) can be used to adjust the processing condition for each formulation. For example, typical shear rates at gate and in the mold are in a range between 100 and 1000 s^{-1}, and under such conditions, a shear-viscosity value below 1000 Pa·s is suggested for suitable compositions [40]. Use of Equation (1) suggests that while the r-PE respects such a condition at the temperature of 180 °C while the compounds at 50 wt.% of WD requires different condition in terms of temperature (>180 °C) or shear rate (>10^3 s^{-1}).

3.3. Infrared Spectroscopy

Infrared spectroscopy is widely used as a qualitative technique to monitor the thermal degradation of polyethylene samples. A comparison among ATR spectra of the matrix (r-PE) at room temperatures, and after heating at 180 °C and 220 °C for 1800 s, respectively, are reported in Figure 9.

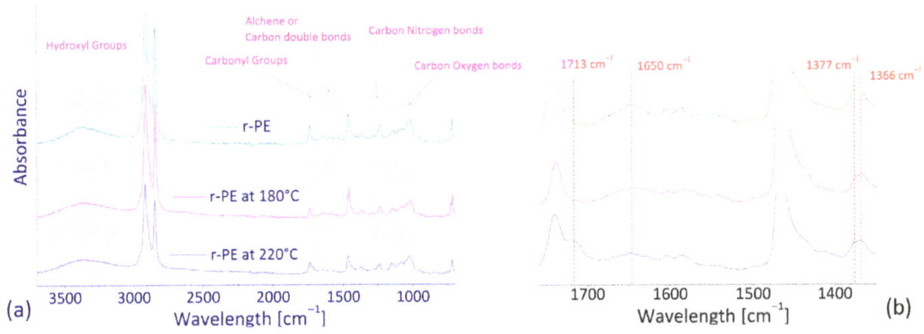

Figure 9. ATR spectra of recycled polyethylene (r-PE), and heated samples at 180 and 220 °C for 1800 s. (**a**) Frequency range of 800–3700 cm^{-1}; and (**b**) zoom in the range 1350–1760 cm^{-1}.

For all the samples, the characteristic peaks around 2900 cm^{-1} (C–H stretching), 1460 cm^{-1} (CH_2 deformation bending), and 720 cm^{-1} (CH rocking bending), commonly attributed to polyethylene homopolymer [41], are clearly visible. Different changes in IR spectra attributed to sample aging can be identified according to previous literature: (i) broad peaks from 3100 to 3700 cm^{-1} ascribed to hydroxyl groups; (ii) alkenes, or carbon double bonds in the range of 1600 to 1680 cm^{-1}; (iii) region of 1690–1810 cm^{-1} associated to carbonyls; (iv) region at 1000–1200 cm^{-1} representative of carbon-oxygen bonds; and (v) carbon–nitrogen bond in the range of 1200 to 1280 cm^{-1}. According to work of Brandon [42], usually pure polyethylene should not have nitrogen in the chemical structure, however nitrogen is present in the air and can be contained in plastic additives.

Generally, in presence of oxygen and high temperatures, the oxidation process of polyethylene starts from weak bonds close to terminal vinyl groups and generates oxygen-containing groups, such as alcohol, ketone, acid, and so on [43]. In the samples treated at 220 °C compared to pristine r-PE, it can be noted: (i) the formation of a shoulder in correspondence of 1713 cm^{-1} (close to the peak at 1720 cm^{-1}, typical of carbonyl groups), which can be correlated to the presence of ketones; (ii) the reduction in the peak at 1650 cm^{-1}, which can be linked to a decrease in double bonds [41]; (iii) a small increment in intensity at 1377 cm^{-1}, which can be due to the occurrence of branching. Indeed, as attested by Gulmine et al. [44], a characteristic peak at 1377 cm^{-1}, typical of CH_3 symmetric deformation, is present in LDPE and LLDPE spectra, together with peak at 1366 cm^{-1}. This latter, attributed to the CH_2 wagging deformation, is present alone in HDPE spectra. The typical peak around 909 cm^{-1} for representing changes in terminal vinyl groups was too weak to be detected accurately, as verified by Brandon et al. [42].

Infrared spectrum of wood flour, reported in literature [45,46], showed the predominance of hydroxyl band at 3100 to 3700 cm^{-1} and ether species (C–O–C) at 1000–1150 cm^{-1}, then weak bands at 1800–1680 cm^{-1} of carbonyl groups, at 1650–1500 cm^{-1} of vinyl (C=C) groups, at 1450 and 1350 cm^{-1} due to methyl species (CH_3).

By comparing spectra of the composite containing 50 wt.% of wood flour (see Figure 10) a strong decrease in intensities in the fingerprint region can be observed after the heat treatment at 220 °C. This result confirms the occurrence of thermal decomposition mechanism in wood-based compounds at temperature above 180 °C and their stability at lower temperature. Such a result is consistent with the rheological data show in the previous section.

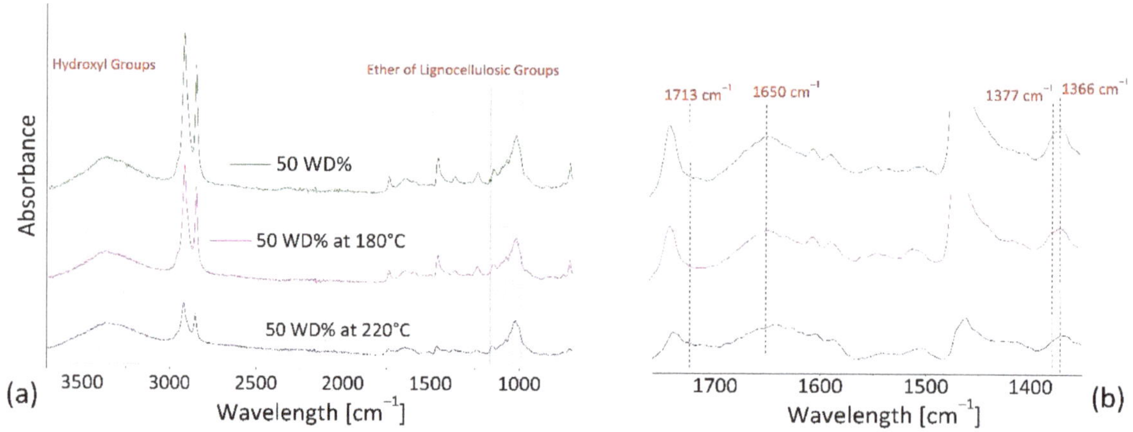

Figure 10. ATR spectra of composites at 50% in wt. of WD flour, and heated samples at 180 °C and 220 °C for 1800 s. (**a**) Frequency range of 800 to 3700 cm^{-1} and (**b**) zoom in the range of 1350 to 1760 cm^{-1}.

4. Conclusions

In this study, we have prepared by melt blending a series of composites incorporating wood flour (25, 35, and 50% in wt.) as a filler in a recycled LDPE. The composites were characterized in terms of torque measurements performed during the mixing process, dynamic rotational rheology performed on the molten sample, and ATR spectroscopy performed on processed solid samples. Torque data have shown that the introduction of wood particles within the polymeric resin produced an increase in the overall viscosity of systems that led to sensible viscous heating phenomena during compounding. The thermal stability of the r-PE (the matrix) was verified at temperatures of 180 °C for processing times up to of 1500 s, while at higher temperatures the materials showed a rheological response characterized by time-increasing viscoelastic moduli. This behavior suggested the occurrence of crosslinking reactions governing the thermal degradation mechanism. The compounds, realized with wood loading up to 50% in wt., have shown time-independent viscoelastic properties at 180 °C, thus confirming their thermal stability at these conditions. The viscoelastic behavior of the r-PE matrix (investigated in the temperature window from 160 °C to 180 °C) suggested the presence of (short or long) branches in the macromolecular structure that were confirmed by the ATR spectra. The addition of wood particles in the r-PE matrix determined an increment in both the storage (G′) and loss (G″) moduli, and even for the highest WD concentrations (50 wt. %) the viscoelastic spectra did not show the presence of interconnected structures. This behavior was attributed to a poor compatibility and weak interfacial adhesion between the two phases. However, the improving the elastic component in the composites due to the presence of wood flour was considered promising in view of applications to printing, and extrusion or molding technologies. Indeed, PE-wood composites seem capable of achieving good mechanical properties, processability, and durability without the use of harmful chemicals.

Author Contributions: Conceptualization, G.C., data curation, A.P., S.A.; writing—original draft preparation, A.P.; writing—review and editing G.C., S.A.; supervision, S.A. All authors have read and agreed to the published version of the manuscript.

Funding: This research received no external funding.

Institutional Review Board Statement: Not Applicable.

Informed Consent Statement: Not Applicable.

Data Availability Statement: The data presented in this study are available on request from the corresponding author.

Acknowledgments: A. Patti wishes to thank the Italian Ministry of Education, Universities and Research (MIUR) in the framework of Action 1.2 "Researcher Mobility" of The Axis I of PON R&I 2014-2020 under the call "AIM- Attrazione e Mobilità Internazionale". S. Acierno and G. Cicala acknowledge the support of the Italian Ministry of University, project PRIN 2017, 20179SWLKA "Multiple Advanced Materials Manufactured by Additive technologies (MAMMA)".

Conflicts of Interest: The authors declare no conflict of interest.

References

1. Singh, N.; Hui, D.; Singh, R.; Ahuja, I.P.S.; Feo, L.; Fraternali, F. Recycling of plastic solid waste: A state of art review and future applications. *Compos. Part B Eng.* **2017**, *115*, 409–422. [CrossRef]
2. Singh, P.; Sharma, V.P. Integrated Plastic Waste Management: Environmental and Improved Health Approaches. *Procedia Environ. Sci.* **2016**, *35*, 692–700. [CrossRef]
3. Lebreton, L.; Andrady, A. Future scenarios of global plastic waste generation and disposal. *Palgrave Commun.* **2019**, *5*, 6. [CrossRef]
4. Patti, A.; Cicala, G.; Acierno, D. Eco-Sustainability of the Textile Production: Waste Recovery and Current Recycling in the Composites World. *Polymers* **2020**, *13*, 134. [CrossRef] [PubMed]
5. Ashori, A. Wood-plastic composites as promising green-composites for automotive industries! *Bioresour. Technol.* **2008**, *99*, 4661–4667. [CrossRef]
6. Dinesh, S.; Kumaran, P.; Mohanamurugan, S.; Vijay, R.; Singaravelu, D.L.; Vinod, A.; Sanjay, M.R.; Siengchin, S.; Bhat, K.S. Influence of wood dust fillers on the mechanical, thermal, water absorption and biodegradation characteristics of jute fiber epoxy composites. *J. Polym. Res.* **2019**, *27*, 1–13. [CrossRef]
7. Xiong, C.; Qi, R.; Wang, Y. Wood-thermoplastic composites from wood flour and high-density polyethylene. *J. Appl. Polym. Sci.* **2009**, *114*, 1160–1168. [CrossRef]
8. Schwarzkopf, M.J.; Burnard, M.D. Wood-plastic composites—Performance and environmental impacts. In *Environmental Footprints and Eco-Design of Products and Processes*; Springer: Berlin/Heidelberg, Germany, 2016; pp. 19–43.
9. Gardner, D.J.; Han, Y.; Wang, L. Wood-Plastic Composite Technology. *Curr. For. Rep.* **2015**, *1*, 139–150. [CrossRef]
10. Wang, W.; Liu, L.; Ding, N.; Zhang, R.; Yu, J. Mechanical and thermal behavior analysis of wood–polypropylene composites. *Text. Res. J.* **2021**, *91*, 347–357. [CrossRef]
11. Ratanawilai, T.; Taneerat, K. Alternative polymeric matrices for wood-plastic composites: Effects on mechanical properties and resistance to natural weathering. *Constr. Build. Mater.* **2018**, *172*, 349–357. [CrossRef]
12. Habibi, M.; Kazemi Najafi, S.; Ghasemi, I. Rheological and mechanical properties of composites made from wood flour and recycled LDPE/HDPE blend. *Iran. Polym. J. (Engl. Ed.)* **2017**, *26*, 949–956. [CrossRef]
13. Mazzanti, V.; Mollica, F. A Review of Wood Polymer Composites Rheology and Its Implications for Processing. *Polymers* **2020**, *12*, 2304. [CrossRef] [PubMed]
14. Saroia, J.; Wang, Y.; Wei, Q.; Lei, M.; Li, X.; Guo, Y.; Zhang, K. A review on 3D printed matrix polymer composites: Its potential and future challenges. *Int. J. Adv. Manuf. Technol.* **2020**, *106*, 1695–1721. [CrossRef]
15. Suárez, L.; Domínguez, M. Sustainability and environmental impact of fused deposition modelling (FDM) technologies. *Int. J. Adv. Manuf. Technol.* **2020**, *106*, 1267–1279. [CrossRef]
16. Kariz, M.; Sernek, M.; Obućina, M.; Kuzman, M.K. Effect of wood content in FDM filament on properties of 3D printed parts. *Mater. Today Commun.* **2018**, *14*, 135–140. [CrossRef]
17. Yang, T.C. Effect of extrusion temperature on the physico-mechanical properties of unidirectional wood fiber-reinforced polylactic acid composite (WFRPC) components using fused depositionmodeling. *Polymers* **2018**, *10*, 976. [CrossRef]
18. Gregor-Svetec, D.; Leskovšek, M.; Vrabič Brodnjak, U.; Stankovič Elesini, U.; Muck, D.; Urbas, R. Characteristics of HDPE/cardboard dust 3D printable composite filaments. *J. Mater. Process. Technol.* **2020**, *276*, 116379. [CrossRef]
19. Carrete, I.A.; Quiñonez, P.A.; Bermudez, D.; Roberson, D.A. Incorporating Textile-Derived Cellulose Fibers for the Strengthening of Recycled Polyethylene Terephthalate for 3D Printing Feedstock Materials. *J. Polym. Environ.* **2021**, *29*, 662–671. [CrossRef]
20. Gopalan, M.; Mandelkern, L. The effect of crystallization temperature and molecular weight on the melting temperature of linear polyethylene. *J. Phys. Chem.* **1967**, *71*, 3833–3841. [CrossRef]
21. Chen, C.-H.; Mao, C.-F.; Lo, Y.-W. Study of fusion percolation thresholds of rigid PVC compounds. *J. Appl. Polym. Sci.* **2001**, *81*, 3022–3029. [CrossRef]
22. Russo, P.; Patti, A.; Petrarca, C.; Acierno, S. Thermal conductivity and dielectric properties of polypropylene-based hybrid compounds containing multiwalled carbon nanotubes. *J. Appl. Polym. Sci.* **2018**, *135*, 46470–46479. [CrossRef]
23. Tomaszewska, J.; Sterzyński, T.; Zajchowski, S. Thermal and structural effects of poly(vinyl chloride)/(wood flour) compound gelation in the Brabender mixer. *J. Vinyl Addit. Technol.* **2011**, *17*, 239–244. [CrossRef]
24. Hidalgo, J.; Jiménez-Morales, A.; Torralba, J.M. Torque rheology of zircon feedstocks for powder injection moulding. *J. Eur. Ceram. Soc.* **2012**, *32*, 4063–4072. [CrossRef]
25. Çavuş, V.; Şahin, S.; Esteves, B.; Ayata, Ü. Determination of Thermal Conductivity Properties in Some Wood Species Obtained from Turkey. *Bioresources* **2019**, *14*, 6709–6715.
26. Ngo, I.L.; Jeon, S.; Byon, C. Thermal conductivity of transparent and flexible polymers containing fillers: A literature review. *Int. J. Heat Mass Transf.* **2016**, *98*, 219–226. [CrossRef]

27. Patti, A.; Acierno, D.; Latteri, A.; Tosto, C.; Pergolizzi, E.; Recca, G.; Cristaudo, M.; Cicala, G. Influence of the processing conditions on the mechanical performance of sustainable bio-based PLA compounds. *Polymers* **2020**, *12*, 2197. [CrossRef] [PubMed]
28. Dordinejad, A.K.; Sharif, F.; Ebrahimi, M.; Rashedi, R. Time-sweep rheometry for evaluating polyethylene degradation behavior: Effect of formulation and process conditions. *Polym. Test.* **2018**, *70*, 39–46. [CrossRef]
29. Mariani, P.; Carianni, G.; Menconi, F.; La Mantia, F.P. Correlation between processability and properties of a high density polyethylene by a rheological approach. *Macromol. Chem. Phys.* **2002**, *203*, 1602–1605. [CrossRef]
30. Mazzanti, V.; Mollica, F.; El Kissi, N. Rheological and mechanical characterization of polypropylene-based wood plastic composites. *Polym. Compos.* **2016**, *37*, 3460–3473. [CrossRef]
31. Sliwa, F.; El Bounia, N.E.; Marin, G.; Charrier, F.; Malet, F. A new generation of wood polymer composite with improved thermal stability. *Polym. Degrad. Stab.* **2012**, *97*, 496–503. [CrossRef]
32. Patti, A.; Nele, L.; Zarrelli, M.; Graziosi, L.; Acierno, D. A Comparative Analysis on the Processing Aspects of Basalt and Glass Fibers Reinforced Composites. *Fibers Polym.* **2021**, *22*, 1449–1459. [CrossRef]
33. Kaseem, M.; Hamad, K.; Deri, F.; Ko, Y.G. Material properties of polyethylene/wood composites: A review of recent works. *Polym. Sci. Ser. A* **2015**, *57*, 689–703. [CrossRef]
34. Patti, A.; Russo, P.; Acierno, D.; Acierno, S. The effect of filler functionalization on dispersion and thermal conductivity of polypropylene/multi wall carbon nanotubes composites. *Compos. Part B Eng.* **2016**, *94*, 350–359. [CrossRef]
35. Godard, F.; Vincent, M.; Agassant, J.F.; Vergnes, B. Rheological behavior and mechanical properties of sawdust/polyethylene composites. *J. Appl. Polym. Sci.* **2009**, *112*, 2559–2566. [CrossRef]
36. Lozano, K.; Bonilla-Rios, J.; Barrera, E.V. A study on nanofiber-reinforced thermoplastic composites (II): Investigation of the mixing rheology and conduction properties. *J. Appl. Polym. Sci.* **2001**, *80*, 1162–1172. [CrossRef]
37. Tazi, M.; Erchiqui, F.; Godard, F.; Kaddami, H.; Ajji, A. Characterization of rheological and thermophysical properties of HDPE-wood composite. *J. Appl. Polym. Sci.* **2014**, *131*. [CrossRef]
38. Laufer, N.; Hansmann, H.; Koch, M. Rheological Characterisation of the Flow Behaviour of Wood Plastic Composites in Consideration of Different Volume Fractions of Wood. *Proc. J. Phys. Conf. Ser.* **2017**, *790*, 012017. [CrossRef]
39. Park, H.E.; Dealy, J.; Münstedt, H. Influence of long-chain branching on time-pressure and time-temperature shift factors for polystyrene and polyethylene. *Rheol. Acta* **2006**, *46*, 153–159. [CrossRef]
40. Liu, W.; Xie, Z.; Bo, T.; Yang, X. Injection molding of surface modified powders with high solid loadings: A case for fabrication of translucent alumina ceramics. *J. Eur. Ceram. Soc.* **2011**, *31*, 1611–1617. [CrossRef]
41. Weon, J. Il Effects of thermal ageing on mechanical and thermal behaviors of linear low density polyethylene pipe. *Polym. Degrad. Stab.* **2010**, *95*, 14–20. [CrossRef]
42. Brandon, J.; Goldstein, M.; Ohman, M.D. Long-term aging and degradation of microplastic particles: Comparing in situ oceanic and experimental weathering patterns. *Mar. Pollut. Bull.* **2016**, *110*, 299–308. [CrossRef] [PubMed]
43. Huang, H.-J.; Wang, Q.-B.; Xie, B.-H.; Yang, W.; Yang, M.-B. Thermal Oxidation and Structural Changes of Degraded Polyethylene in an Oxygen Atmosphere. *J. Macromol. Sci. Part B* **2011**, *50*, 1376–1387. [CrossRef]
44. Gulmine, J.V.; Janissek, P.R.; Heise, H.M.; Akcelrud, L. Polyethylene characterization by FTIR. *Polym. Test.* **2002**, *21*, 557–563. [CrossRef]
45. Soccalingame, L.; Perrin, D.; Bénézet, J.C.; Mani, S.; Coiffier, F.; Richaud, E.; Bergeret, A. Reprocessing of artificial UV-weathered wood flour reinforced polypropylene composites. *Polym. Degrad. Stab.* **2015**, *120*, 313–327. [CrossRef]
46. Andrzejewski, J.; Barczewski, M.; Szostak, M. Injection Molding of Highly Filled Polypropylene-based Biocomposites. Buckwheat Husk and Wood Flour Filler: A Comparison of Agricultural and Wood Industry Waste Utilization. *Polymers* **2019**, *11*, 1881. [CrossRef] [PubMed]

polymers

MDPI

Article

Flow Characteristics, Mechanical, Thermal, and Thermomechanical Properties, and 3D Printability of Biodegradable Polylactide Containing Boehmite at Different Loadings

Dimakatso Makwakwa [1,2], Vincent Ojijo [1], Jayita Bandyopadhyay [1] and Suprakas Sinha Ray [1,2,*]

1 DST-CSIR National Centre for Nanostructured Materials, Council for Scientific and Industrial Research, Pretoria 0001, South Africa; DMakwakwa@csir.co.za (D.M.); vojijo@csir.co.za (V.O.); jbandyopadhyay@csir.co.za (J.B.)
2 Department of Chemical Sciences, University of Johannesburg, Doornfontein, Johannesburg 2028, South Africa
* Correspondence: rsuprakas@csir.co.za

Citation: Makwakwa, D.; Ojijo, V.; Bandyopadhyay, J.; Ray, S.S. Flow Characteristics, Mechanical, Thermal, and Thermomechanical Properties, and 3D Printability of Biodegradable Polylactide Containing Boehmite at Different Loadings. *Polymers* **2021**, *13*, 2019. https://doi.org/10.3390/polym13122019

Academic Editors: Antonella Patti and Domenico Acierno

Received: 7 May 2021
Accepted: 2 June 2021
Published: 21 June 2021

Abstract: This work investigates the effects of modification of polylactide (PLA) using dicumyl peroxide (DCP) as a crosslinker and Joncryl as a chain extender on boehmite distribution. The PLA/boehmite (PLA/BA) composites at various concentrations were prepared via a twin-screw extruder. Transmission electron microscopy showed more agglomerations of BA particles when Joncryl and DCP were added individually to the PLA matrix, with lesser agglomeration upon simultaneous addition of DCP and Joncryl, which led to an enhancement of 10.7% of the heat distortion temperature and 8.8% of the modulus. The existence of fine dispersed BA particles in the BA3 sample improved the cold crystallization by 4 °C. Moreover, the maximum reinforcing effect in increasing the storage modulus of the prepared system was observed upon concurrent addition of DCP and Joncryl, with minimum reinforcing effect upon individual addition of DCP and Joncryl. In general, a bio-based PLA composite base BA with enhanced properties was successfully prepared for various applications.

Keywords: mechanical properties; biodegradable; thermomechanical properties; thermal properties; dicumyl peroxide; Joncryl

1. Introduction

Interest in polylactide (PLA) has grown both industrially and academically due to its biocompatibility, biodegradability, and transparency [1]. Consequently, it has found application in the automotive, packaging, and medical industries, amongst others [2]. Regardless of its excellent properties, it also has limitations, such as poor thermal stability, brittleness, and slow crystallization during processing [3–6].

Generally, there are ways of improving the properties of PLA, including the addition of fillers or blending with other polymers [3,7–10]. For instance, Nieddu et al. [11] reported that 5 wt.% nanoclay increased the modulus of neat PLA. Other fillers, such as nano-silicon dioxide (SiO_2) [12] and boehmite (BA) [10], have been found to improve the thermal stability, crystallizability, and mechanical and electrical conductivity of the polymers. Amongst these, BA has attracted more research interest due to its low cost, high flammability, high surface area, noble dispensability, and thermal stability [10,13]. Moreover, BA particles in nanometric form have greater reactivity than in micrometric form, owing to the increase of their surface area in nanometric dimensions [2]. A known challenge is the distribution of the hydrophilic BA in the hydrophobic PLA matrix, which leads to poor interfacial bonding between the polymer and the filler.

Several studies have been conducted on the modification of BA to improve the interfacial bonding between the polymer and the filler [10,13–18]. For instance, Malwela et al. [10] reported the impact of BA surface modification with *p*-toluene sulfonic acid on the mechanical, thermal, and rheological properties of PP/PS blends. Their results revealed that the addition of 1, 3, 5, and 7 wt.% modified BA particles in the PP/PS blend improved the modulus by 5.3, 13.5, 20.3, and 27.5%, respectively, due to the uniformly dispersed agglomerations of BA particles within the PS phase, whereas in the case of untreated BA, slight improvements of 0.2, 11.4, 17.8, and 25.7%, respectively, were observed, attributed to the agglomerations within the edges of the PS phase. On the other hand, the inclusion of 1 wt.% untreated BA particles had no effect on the crystallization temperature, while the crystallization temperature peaks moved from 116.90 to 119.0, 126.6, and 127.7 °C for the 3, 5 and 7 wt.%, respectively. Khumalo et al. [17] incorporated BA particles in low-density and high-density polyethylene (LDPE and HDPE). The authors revealed that 2 wt.% BA particles improved the resistance of the resulting composite to thermo-oxidative degradation from 0.5 to 2%. In 10 wt.% BA the improvement was from 1.5 to 3%, due to the presence of BA being reported for polyoxymethylene. Das et al. [2] reported the mechanical, thermal, and fire properties of biodegradable PLA/BA composites. Their results revealed that the incorporation of 3 wt.% BA nanoparticles increased the tensile strength of PLA by 57%, and the cold crystallization was observed in the range of 120–125 °C. However, this kind of modification includes a solvent, which has drawbacks, such as high cost and environmental problems.

One of the most common approaches to improving the distribution of the filler in the polymer matrix is through polymer modification [14,16]. Several studies on the interfacial modification of different polymers have been reported [1,19,20]. These modifications include Joncryl as a chain extender and DCP as an initiator. Joncryl is a polymeric chain extender with a low epoxy equivalent weight that reacts with the chain ends of polycondensates and effectively increases their melt viscosity. Additionally, it is a multifunctional oligomer chain extender that was intended to reverse the degradation in PLA; it has epoxy functional groups, which would react with the carboxyl and hydroxyl groups [1,20–24]. Joncryl has shown improvement in the interaction between carboxyl groups of PLA and the reactive functional epoxy group of the chain extender, and consequently, the enhancement of the crystallizability and mechanical properties of PLA [25]. Generally, DCP is a free radical generator in a polymer system, with the possibility of crosslinking or chain branching. Similarly, Joncryl helps with chain branching and extension in polyesters, such as PLA.

This study focuses on producing the PLA/BA composite in order to demonstrate the 3D printing of the samples. This study aimed to use the novel system to study the mechanical, flow, and thermal properties of PLA. We used DCP as a free radical generating agent in the reactive extrusion in order to enhance the amount of macroradicals that could introduce long-chain branching in the PLA chain. Joncryl, a patented, multifunctional, reactive polymer with improved thermal stability/chain extenders for specific food-contact applications, and polycondensation polymers including poly(ethylene terephthalate), was used as a crosslinking agent to extend the PLA chain and to improve the crystallizability of the PLA. Then, we demonstrated the 3D-orientability of the sample, with good distribution. This study aims to improve the BA particles' distribution in the PLA matrix using chain extension and branching, which will lead to enhanced mechanical properties.

2. Materials and Methods

2.1. Materials

The PLA used in this work was a commercial grade (PLA 4032D) purchased from NatureWorks LLC (Minnetonka, MN, USA), with a melt flow index (MFI) of 6 g/10 min (2.16 kg load) at 190 °C and a density of 1.23 g cm^{-3}, while the BA powder was of a commercial grade manufactured by SASOL, under the trade name Dispersal 40, containing 80% Al_2O_3, donated by SASOL Germany. DCP was obtained from Sigma-Aldrich (Johannesburg, South Africa), with a molecular weight of 270.37 g/mol, density of 1056 g/mL, vapor pressure of 15.4 mmHg, and a melting point between 39 and 41 °C, the chain extender Joncryl ADR 4368 CS was donated by BASF South Africa. The chemical structures of DCP and Joncryl are shown in Figure 1.

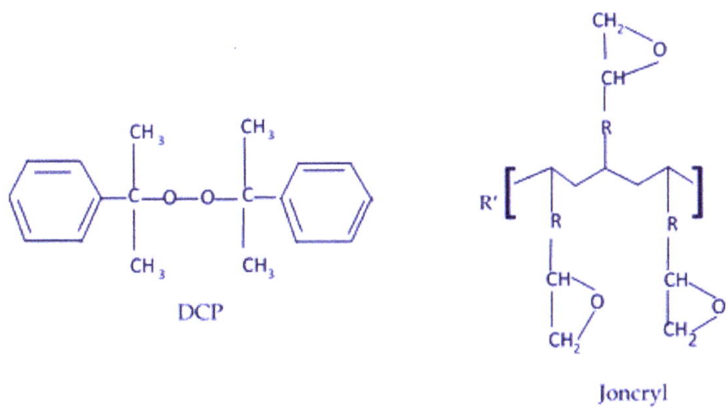

Figure 1. Chemical structure of dicumyl peroxide (DCP) and Joncryl.

2.2. Preparation of the Samples

Prior to processing, PLA was dried at 80 °C under vacuum for 12 h. The samples with different compositions (Table 1) were melt compounded in a co-rotating twin-screw extruder from Thermo Scientific, Waltham, MA, USA, with an L/D of 40. The extruder conditions were as follows: feeding rate 5.6 g/min; screw speed 202 rpm; and barrel temperatures from the hopper to the die were 140, 160, 180, 180, 180, 180, 180, 180, and 190 °C, respectively. The composites were then compression molded into different specimens using a Carver compression molder (Carver laboratory Model 973214A, Wabash, IN, USA) at a temperature of 190 °C and pressure of 1 MPa for 6 min, then cooled to room temperature.

Table 1. Sample names and compositions.

Sample Code	PLA	DCP	Joncryl	BA (wt.%)
PLA	100	-	-	-
PLA/DCP	99.95	0.05	-	-
PLA/J	99.40	-	0.6	-
PLA/DCP/J	99.35	0.05	0.6	-
PLA/BA	98.00	-	-	2
PLA/DCP/BA	97.95	0.05	-	2
PLA/J/BA	97.40	-	0.6	2
PLA/BA2/DCP/J	97.35	0.05	0.6	2
BA3	96.35	0.05	0.6	3
BA4	95.35	0.05	0.6	4
BA5	94.35	0.05	0.6	5
BA6	93.35	0.05	0.6	6
BA10	89.35	0.05	0.6	10
BA20	79.35	0.05	0.6	20

2.3. Characterization

A Fourier-transform infrared (FTIR) spectroscope from Perkin Elmer (Model: Spectrum 100, Branford, CT, USA) was used to verify the chemical interactions between PLA, DCP, Joncryl, and BA within the wavelength range of 500–4000 cm^{-1}. For all of the spectra, 32 scans were collected, with a resolution of 4 cm^{-1}.

The compression-molded disc specimens were used for XRD measurements using an X'pert PRO diffractometer from PANalytical (EA Almelo, The Netherlands). The operating voltage was 45 KV, and the current was 40 mA. The exposure time and the scanning rate used were 29 min 45 s and 0.011°/min, respectively.

The distribution of BA particles in the PLA matrix was investigated using transmission electron microscopy (TEM) (JOEL, JEM 2100, Tokyo, Japan) with an acceleration voltage of 200 kV. The samples were prepared using a Leica (Austria) EM FC6 cryo-ultramicrotome at −100 °C, a cutting speed of 3 mm, and a feed rate of 80 nm. The samples were sliced using a diamond knife.

Differential scanning calorimetry (DSC) measurements were studied using a DSC-Q2000 instrument from TA Instruments, New Castle, DE, USA. Pellets with a mass of approximately 4–5 mg were heated from −20 °C to 190 °C at a rate of 10 °C/min, and then maintained at that temperature for 5 min. Samples were cooled to −20 °C at a rate of 10 °C/min and kept constant for 5 min, then heated to 190 °C at a rate of 10 °C/min. The heating and cooling cycles were conducted under nitrogen as the purge gas, with a flow rate of 25 mL/min for all samples. The glass transition temperature (T_g), melting temperature (T_m), enthalpy of fusion (ΔH_m), crystallization temperature (T_c), cold crystallization temperature (T_{cc}), and enthalpy of cold crystallization (ΔH_{cc}) were obtained.

The dynamic mechanical analysis (DMA) was conducted using a PerkinElmer DMA (Model 8000, Branford, CT, USA) analyzer in dual cantilever-bending mode. The temperature was measured at a frequency of 1 Hz, strain amplitude of 0.01%, and heating rate of 2 °C/min in the temperature range of −80 to +115 °C.

Heat distortion temperatures (HDTs) were measured using a CEAST HDT-VICAT instrument, and the measurements were recorded using the following conditions: Oil bath preheated to 30 °C; T_{start} = 30 °C; heating rate (ϕ) = 120 °C/h; T_{max} = 100 °C; end = 0.34 mm; span = 64 mm; and stress = 450 kPa.

The melt state rheological properties of neat PLA and composites were investigated using an Anton Paar stress/strain-controlled rheometer Physica MCR501 (Garz, Austria) with parallel plates of 25 mm in diameter. The injection-molded disc samples were used for this test. Frequency sweep tests were carried out from 0.1 to 100 rad/s. Each sample was melted in a parallel plate at 190 °C for 5 min to remove the remaining thermal history, and a dynamic strain sweep was then performed in order to determine the common linear region. The melt flow rate (MFR) properties of the neat PLA and composites were investigated using a melt flow meter (multiweight). The pelletized samples were used for this test, weighing 4 g per sample.

Tensile tests were performed in order to determine the modulus, yield strength, and elongation at break of each material, using an Instron 5966 tester (Instron Engineering Corp., Norwood, MA USA) with a load cell of 10 kN, in accordance with ASTM 638D standards. The test was carried out under tension mode at a single strain rate of 5 mm/min at room temperature. The dog-bone-shaped specimens were analyzed.

3. Results

3.1. Flow Properties

The melt flow rates (MFRs) of the prepared samples are shown in Table 2. PLA shows a higher MFR, implying lower viscosity and ease of processability. The addition of DCP and Joncryl individually decreased the MFR of PLA by 32.3 and 50.2%, respectively. A decrease in the MFR of PLA/DCP could be attributed to chain–chain coupling, which results from the interaction of radicals generated by DCP on PLA chains, as shown in Supplementary Figure S1. On the other hand, Joncryl as a chain extender could have increased the molecular weight and viscosity of PLA due to the reaction between the epoxide groups of Joncryl and the carboxylic groups of PLA (see Supplementary Figure S2) [25]. The FTIR spectroscopy shows the reactions between the chain extenders, and will be discussed in the FTIR section. The simultaneous addition of both Joncryl and DCP did not significantly affect the MFR of PLA/DCP/J compared to the PLA/J system. Upon the addition of BA at 2 and 3 wt.% to PLA/DCP/J, the MFR decreased slightly, by 12.7 and 8.5%, respectively, suggesting the immobilization of PLA chains by the nanoparticles. However, with further increase in BA particles, the MFR started to increase, probably due to the separation and weakening of PLA chains by the rigid BA particles.

Table 2. The MFR of all samples.

Sample Name	MFR(g/10 min) Using 2.16 kg Load at 190 °C
PLA	15.49
PLA/DCP	10.49
PLA/J	7.71
PLA/DCP/J	7.66
PLA/BA	11.19
PLA/DCP/BA	9.94
PLA/J/BA	5.86
PLA/DCP/J/BA2	6.69
BA3	7.01
BA4	7.78
BA5	8.11
BA6	9.93
BA10	10.50
BA20	11.52

DCP and Joncryl have a significant influence on the structural properties of polymers. In particular, the molecular weight (M_W) and distribution (MWD) can be affected due to chain scission or chain–chain coupling in the presence of DCP. In addition, chain extenders such as Joncryl can also modify the structural properties of a polymer. Rheology is a powerful tool to elucidate the changes in the molecular structures of polymers in melt states. Figure 2 shows the plots of G′ and G″ against angular frequency from 0.1 to 100 rad/s. In these plots, the crossover point between G′ and G″ provides information about the changes in the M_W and MWD. The horizontal shift of the crossover frequency (G_ω) to lower values is related to an increase in M_W. In contrast, the vertical shift of the crossover modulus (G_m) indicates an increased broadening of the MWD. It is worth mentioning that the crossover point is determined within the tested range (0.1 to 100 rad/s). For neat PLA (Figure 2a), the crossover point could not be determined within the tested range. However, looking at the distance between G′ and G″ at 100 rad/s, it can be noticed that the G′ and G″ curves are closer, suggesting lower angular frequency for the PLA/DCP (Figure 2b) system compared to neat PLA.

Moreover, a dramatic decrease in G_ω was noticed in the PLA/J system (Figure 2c), indicating an increase in the M_W of PLA. A horizontal shift to higher values could be noticed in the PLA/DCP/J system (Figure 2d); this suggests that there was a reduction in the M_W of PLA, which could be attributed to chain scission when both DCP and Joncryl were added. The incorporation of BA did not significantly change the M_W of the PLA/DCP/J system (Figure 2e–i), as can be seen from the distance between G′ and G″ at 100 rad/s.

Figure 3a shows the complex viscosity of the prepared PLA/DCP/J-based composites against angular frequency. As evident from Figure 3, the flow behavior of PLA, PLA/DCP, PLA/J, and PLA/DCP/J followed a similar trend to that noted from the MFR analysis. However, PLA/J showed the highest viscosity due to an increase in the M_W when Joncryl was introduced to PLA. This observation is attributable to the branches formed by the Joncryl and the introduction of long-chain branching (LCB) in the PLA structure, conveying pseudosolid-like behavior. The viscosity of the composites was solely dependent on the distribution of BA particles in the PLA matrix. The BA3 system showed better distribution, and exhibited a higher viscosity than the other composites (Figure 3b). With the increase in filler concentration, the viscosity decreased, possibly due to the poor distribution and agglomeration of BA particles, forming the weak points in the matrix. Further increase in BA concentrations resulted in an increase in viscosity, due to the reinforcing effect of the particles.

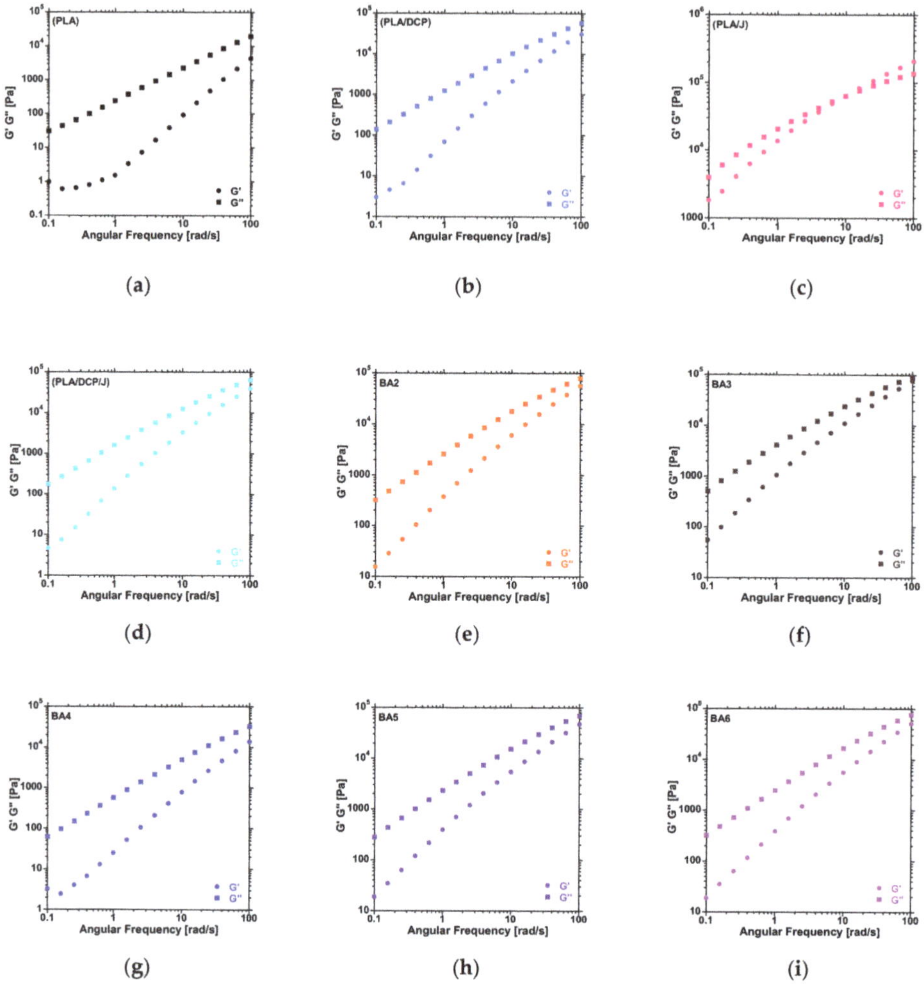

Figure 2. Storage and loss modulus curves of the PLA and BA composites at different contents.

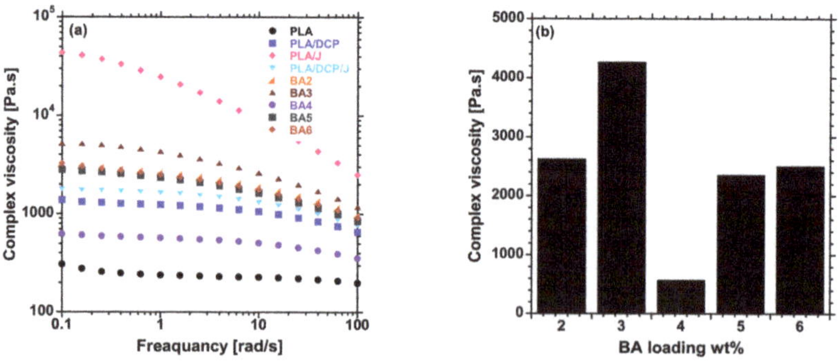

Figure 3. (**a**) Complex viscosity curves of neat PLA and BA composites at different loadings, and (**b**) viscosity at 1 frequency (rad/s) BA composites at different loadings.

3.2. Chain Extension

FTIR spectroscopy was used to compare the processed PLA/BA chain extender and crosslinking systems, in order to identify the reaction that may have transpired between the PLA/BA and chain extender/crosslinking. Figure 4a shows the FTIR spectra of neat PLA and composites. In the case of neat PLA, the peaks at 1754, 1455, and 1369 cm^{-1} are due to C=O stretching, C–H deformation, and C–O–H bands, respectively. Meanwhile, 1186 and 1080 cm^{-1} are assigned to –C–O stretching, 868 cm^{-1} is attributed to –C–C stretching, and 755 cm^{-1} is attributed to C–H bending. The H–O–H peak at 1630 cm^{-1} was not obtainable from the processed PLA, due to the existence of thermal chain scission at the C–O bond [26]. Supplementary Figure S3 shows the vibrations at 2852 and 3000 cm^{-1} assigned to the O–H stretching, and 2922 cm^{-1} due to the axial C–H stretching bond [27]. The FTIR spectra of neat BA (Supplementary Figure S4a) reveal that the vibrations at 3309 and 3095 cm^{-1} relate to the O–H stretching of BA [28]. The vibration at 1397 cm^{-1} is attributed to the amorphous surface structure that exists in crystalline BA [29]. In the case of neat DCP, as shown in Supplementary Figure S4b, vibrations are observed at 1727 cm^{-1} due to the C=O stretching, at 910 cm^{-1} due to C=C stretching, and at 762 cm^{-1} and 698 cm^{-1} due to C–H bending. Supplementary Figure S4c shows that in the FTIR spectra for neat Joncryl, vibrations are observed at 907 and 843 cm^{-1}, attributed to the symmetric and asymmetric ring deformation of cyclic epoxide [30,31].

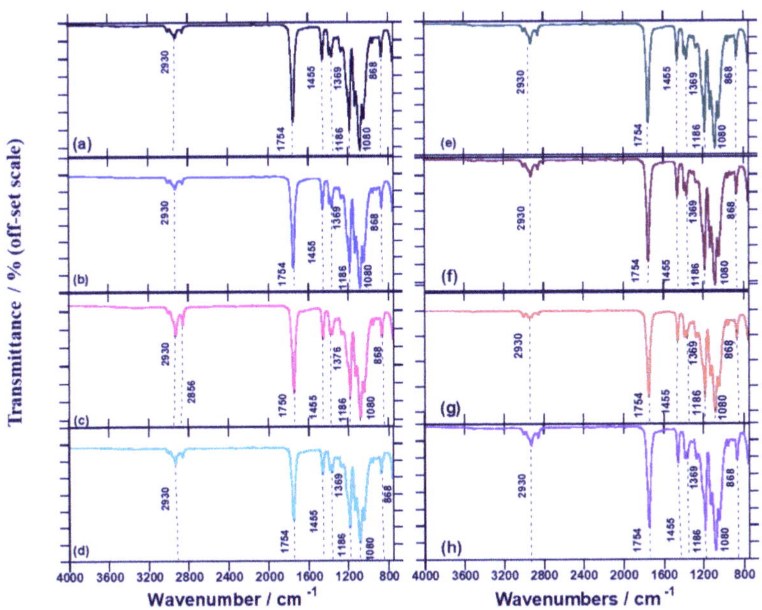

Figure 4. FTIR spectra for (**a**) neat PLA, (**b**) PLA/DCP, (c) PLA/J, (**d**) PLA/DCP/J, (**e**) PLA/BA, (f) PLA/BA/DCP, (**g**) PLA/BA, and (**h**) PLA/BA/DCP/J samples.

Figure 4e shows that there is no significant change to the PLA matrix upon addition of BA, due to the low concentration of BA embedded in the PLA matrix. Further, upon addition of DCP to the PLA/BA composite, as shown in Figure 4f, DCP undergoes homolytic cleavage when heat is applied, breaking down into free radicals and assisting in the removal of H's from the PLA chains in order to create free radicals on the backbone structure of PLA, as shown in Supplementary Figure S1. This phenomenon is attributed to the propagation of the radical reaction to form a crosslinked/branched structure of PLA [13]. Therefore, PLA produces free radicals on the tertiary C atoms, which become stabilized in reactive extrusion [4]. In the case of Joncryl, the 907 and 843 cm^{-1} peaks disappear, owing to the interaction of the epoxy groups with carboxyl groups on the PLA, suggesting that the reaction occurrs between the Joncryl epoxy and the PLA terminal functional group [32]. Surprisingly, there is a synergistic effect of peak decreases for Joncryl at 2922 cm^{-1}. These results relate to the decrease observed in the MFR results and the peak decrease shown in Supplementary Figure S2. Upon addition of all the components, the peak at 2922 cm^{-1} increases, suggesting that the initiator has created macroradicals after the chain extender has truly extended the PLA chain for the BA attachment. In all of the composites, vibrations at 1752, 1455, 1186, 1080, and 868 cm^{-1} remain unchanged. Inata and Matsumura [26] reported that the epoxides might react with carboxyl and hydroxyl end groups of polyesters, and the electrophilic group with the carboxyl end groups. It can be concluded that the chain extension/crosslinked/branched structures in the polymer composites play an important role in improving the properties of the reactive composites in a controlled way.

3.3. BA Distribution

To measure the distribution of BA particles in the PLA matrix, samples containing different BA concentrations were cryosectioned and viewed under TEM. Figure 5a illustrates that the particles of BA formed more agglomerates in the PLA matrix, and that the particles were not well distributed. Figure 5b shows that the addition of the initiator in the PLA matrix decreased the agglomerations of BA particles and distributed the particles better than in PLA/BA. This is attributed to the fact that the viscosity of the PLA matrix was increased by chain extension and/or branching, which assisted in breaking the BA agglomerates. On the other hand, better distribution of BA particles was observed in the presence of Joncryl, due to more chain branching that was created in the PLA structure. Furthermore, Figure 5d shows that the addition of all components at once produced a fair distribution of BA particles and strong intercomponent bonding. This observation correlates with the FTIR spectroscopy results in the next section, which show the interfacial bonding between all components.

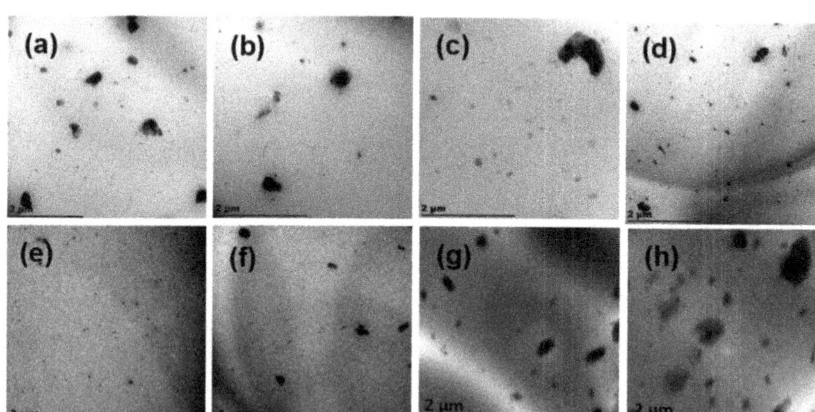

Figure 5. TEM images of (**a**) PLA/BA, (**b**) PLA/DCP/BA2, (**c**) PLA/J/BA2, (**d**) PLA/DCP/J/BA2, (**e**) BA3, (**f**) BA4, (**g**) B5, and (**h**) BA6.

Furthermore, it is evident from Figure 5e that upon the addition of 3 wt.% BA to the PLA matrix, the finest distribution was observed, showing an optimal distribution amongst all composites, owing to better intercomponent bonding amongst the neat PLA and BA particles. Das et al. [2] reported similar results in PLA/BA, revealing that the 3 wt.% BA loading in the PLA matrix was the optimal distribution. This led to improved mechanical properties, which were affected by the distribution of the BA particles in the PLA composites. On the other hand, the increase in BA concentration (i.e., 4, 5, 6, 10, and 20 wt.%) produced poorer distribution and more agglomerations of the BA particles in the PLA matrix.

3.4. Non-Isothermal Crystallization of the Modified PLA Systems

DSC was used to study the effects of BA, DCP, Joncryl, and the resultant structures on the crystallization and melting temperature of the PLA matrix. The DSC data are summarized in Table 3. The degree of crystallinity (χ_m) during cold crystallization, during heating (χ_{cc}), and total crystallinity (χ_c) were calculated using the following equations [2,23]:

$$X_m = \frac{\Delta H_m}{\varnothing_{PLA} \quad \Delta H_m^{\circ}} \tag{1}$$

or

$$X_{cc} = \frac{\Delta H_{cc}}{\varnothing_{PLA} \quad \Delta H_m^{\circ}}$$

$$X_c = X_m - X_{cc} \tag{2}$$

where ΔH_m is the melting enthalpy, ΔH_{cc} is the enthalpy of cold crystallization, $\varnothing PLA$ is the weight fraction of PLA, and $\Delta H^{\circ}{}_m$ is the enthalpy of fusion of 100% PLA, taken as 93.7 J/g [33]. Figure 6 shows the DSC thermograms from the second heating. PLA shows diverse transitions; the first transition is related to the PLA T_g at (60 °C), the second transition is allied with the T_{cc} at (110.24 °C), and the last transition is linked with the T_m at (169.11 °C). In addition, Figure 6a also shows various melting temperature peaks at 164.13 and 169.11 °C. Moreover, there was no T_c detected for PLA during the cooling cycle, because PLA crystallizes very slowly [34]. Upon addition of BA, and chain extension/branching by DCP/Joncryl, the T_m of the samples moved to the low side of the graph compared to neat PLA. This observation suggests that BA acted as a weak nucleating agent in the samples; meanwhile, 3 wt.% showed a profound shift to the lower side of the T_m peak (165.5 °C), suggesting a strong nucleating effect. Similar nucleating effects were reported by Malwela et al. [10] and Das et al. [2]. On the other hand, the T_{cc} temperatures were also affected by the addition of BA, due to the nucleating effect. Upon addition of BA, the T_c value of the composites decreased, moving towards the lower crystallization temperatures, indicating enhanced nucleation. Malwela et al. [10] reported a similar nucleating effect.

Additionally, BA limited the mobility of the PLA macromolecules, restricting their chain arrangement. When the molecular structure of PLA was altered by DCP and Joncryl, the T_{cc} was reduced, and the crystallinity increased. This phenomenon is related to PLA degradation [35–37]. However, when the BA content was increased, the crystallinity in 3 wt.% decreased due to the well-dispersed BA particles causing a physical barrier in the PLA matrix. The T_g of the samples remained unchanged regardless of incorporating BA or alterations to the molecular structure of PLA. This observation is related to the DMA results that will be discussed later. Overall, the T_m and T_c of the composites decreased compared to the neat PLA, confirming that BA, DCP, and Joncryl are good nucleating agents. Small loading of the nucleating agent assisted in forming the polymer crystals; meanwhile, high loading of the nucleating agent restricted the ordered arrangement of the molecular chain, leading to low crystallinity. Moreover, during heating, more crystals were formed; as a result, the PLA crystallinity was improved.

The XRD patterns of neat PLA, BA powder, and composites are shown in Figure 7. The diffraction patterns on the as-received BA powder were observed at 2θ = 13.98°, 28.12°, 38.36°, 49.46°, 55.11°, and 64.60°, attributed to the (20, 120, 031, 200, 002, and 151 crystallographic planes, respectively. In the case of neat PLA, the broad amorphous peak at 16.50° was observed and ascribed to the 200/100 crystallographic plane of PLA crystal, consistent with the features of PLA [38–41]. Upon the addition of various BA concentrations to PLA, the features of BA at the peak of interest (2¦È = 13.98°) were also recognized, signifying the presence of the filler in the composites. On the other hand, the XRD patterns of all of the composites show an intensive peak around 16.21°, and have slightly moved to a higher angle, suggesting that the crystal size of the composites has decreased due to the interaction and distribution of BA in the PLA matrix [2]. Chain branching contributes to the higher crystallinity of PLA.

Table 3. DSC measurements of neat PLA and composites.

Sample Code	T_{cc} (°C)	ΔH_{cc} (J/g)	X_{cc} (%)	ΔH_m (J/g)	T_m (°C)	X_m (%)	X_c (%)
Neat PLA	110.24 ± 0.8	31.21 ± 1.9	33.31	33.46 ± 0.7	169.11 ± 0.1	35.71	2.40
PLA/DCP	102.72 ± 0.1	26.72 ± 0.1	28.67	31.48 ± 0.1	167.49 ± 0.1	33.80	5.13
PLA/J	108.09 ± 0.1	21.72 ± 0.1	23.32	29.30 ± 0.1	165.78 ± 0.1	31.46	8.14
PLA/DCP/J	-	-	-	28.77 ± 0.1	164.00 ± 0.1	31.08	-
PLA/BA	104.98 ± 0.7	29.47 ± 0.02	32.09	32.56 ± 0.3	167.63 ± 0.7	35.46	3.37
PLA/DCP/BA	105.36 ± 0.3	19.81 ± 0.4	21.58	23.15 ± 0.9	167.28 ± 0.3	25.22	3.64
PLA/J/BA	107.02 ± 0.9	27.20 ± 1.8	29.80	29.35 ± 0.7	166.65 ± 0.5	32.16	2.36
PLA/BA2/DCP/J	105.98 ± 1.0	24.47 ± 0.5	26.98	30.37 ± 0.1	167.14 ± 0.3	33.48	6.50
BA3	106.0 ± 0.6	24.5 ± 0.9	32.0	31.4 ± 0.7	165.51 ± 0.4	33.3	2.8
BA4	102.9 ± 0.9	28.9 ± 0.4	32.6	32.5 ± 0.7	166.60 ± 1.1	34.8	3.8
BA5	105.6 ± 0.3	29.10 ± 0.6	31.2	32.0 ± 0.8	166.72 ± 0.8	36.4	5.0
BA6	106.3 ± 1.2	27.60 ± 0.9	30.4	30.8 ± 0.2	167.10 ± 0.5	35.2	4.8
BA10	108.8 ± 0.9	25.01 ± 0.8	29.87	32.47 ± 0.1	167.98 ± 0.6	38.78	8.91
BA20	110.1 ± 0.6	26.28 ± 1.2	35.34	32.38 ± 0.7	168.53 ± 1.5	43.55	8.21

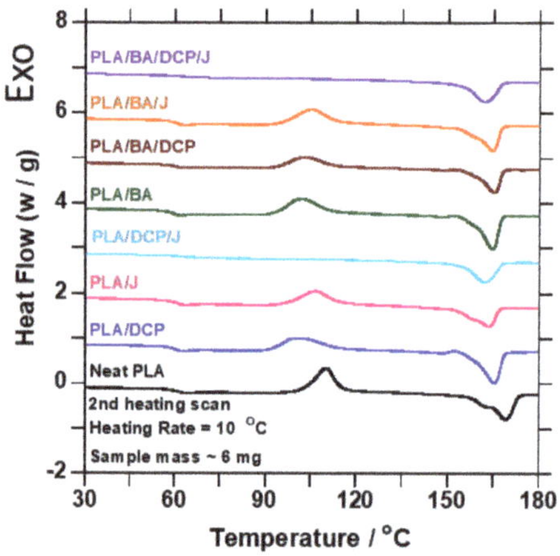

Figure 6. DSC traces of the second heating curve of neat PLA and samples containing DCP/Joncryl and BA.

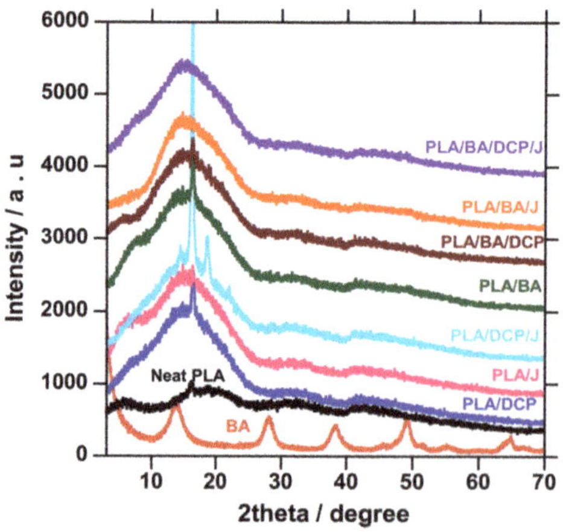

Figure 7. XRD patterns of the neat PLA and composite.

Similarly, this tends to reduce the crystal sizes. Therefore, it was important to calculate the crystallite size of the samples, using the Scherrer equation shown below (Equation (3)) as a mathematical expression of the relationship between full width at half maximum FWHM and the crystallite size. The results of the crystalline size for all samples are listed in Table 4.

$$FWHM = \frac{K\gamma}{L\cos\theta} \quad (3)$$

where $FWHM$ is the full width at half maximum attained from the instrument, λ is the wavelength of the X-ray that was used for the diffraction, L is the crystalline size, θ is the peak position ($2\theta/2$) in radians obtained from the instrument, and K is a shape factor constant with a value of 0.9 [42]. Neat BA shows a crystalline size of 39.59, and PLA 12.8, with a T_m of 169.11 °C from DSC curves. When a polymer is heated at the minimal T_m and the equilibrium T_m ($T_m{}^0$), the remaining well-ordered structures in the melt will significantly influence the crystallinity [43]. Upon alteration of the PLA structure, the crystal size decreased, and the addition of various BA loading further increased the crystal size with 5 wt.% as the threshold. The small crystalline size is due to the crystal growth of the polymer, which is attained by the extra addition of folded polymer chain segments, meaning that the sample has a lower T_m value [44]. Farid et al. [45] reported a similar observation. Further, the crystal size was not dependent on the BA concentration. In conclusion, we observed that the BA and the resultant branched structure acted as good nucleating agents, and this observation was consistent with the DSC analysis.

Table 4. Crystal sizes of neat PLA and composites.

Sample Name	2θ	θ	FWHM	Crystal Size (nm)
BA	13.91	6.96	0.038397	36.4
Neat PLA	16.29	8.15	0.119381	11.7
PLA/DCP	16.47	8.24	0.153414	9.1
PLA/J	15.29	7.65	0.213628	6.5
PLA/DCP/J	16.31	8.16	0.008029	174.4
PLA/BA	16.44	8.22	0.117286	12.0
PLA/DCP/BA	16.40	8.20	0.176802	8.0
PLA/J/BA	15.14	7.57	0.205251	6.8
PLA/BA2/DCP/J	14.90	7.45	0.186576	7.5
BA3	16.46	8.23	0.135961	11.2
BA4	16.45	8.22	0.122696	12.4
BA5	16.44	8.22	0.106988	14.3
BA6	14.82	7.41	0.199142	7.6
BA10	16.61	8.30	0.135786	11.2

3.5. HDT

The HDT results of neat PLA and composites are listed in Table 5. The HDT of neat PLA slightly improved from about 0.6 to 1.77 °C after the addition of BA and molecular structure alteration with DCP and Joncryl, which led to the enhancement of the mechanical properties of PLA. In the case of BA's inclusion in the PLA matrix, the results show an improvement of 1.7 °C; this increase is motivated by the higher degree of BA crystallinity, owing to its nucleating properties. This result is related to the DSC results shown in Figure 6. Upon the molecular structure alteration of PLA by DCP and Joncryl, the HDT increased by 2.0 and 2.7 °C, respectively, due to the chain extension and/or branching.

Further, upon the addition of different BA concentrations, the HDT increased with increasing loading. These results are consistent with the MFR test results reported in Table 2. Overall, the DSC and HDT results show that the incorporation of BA and the alteration of PLA's molecular structure enhance the distribution of BA particles and the mechanical properties of PLA.

Table 5. HDT values of neat PLA and composites.

Sample Name	HDT (°C)
Neat PLA	51.3 ± 0.21
PLA/DCP	51.9 ± 1.48
PLA/J	54.2 ± 0.40
PLA/DCP/J	53.4 ± 0.29
PLA/BA	53.0 ± 0.06
PLA/DCP/BA	53.3 ± 0.38
PLA/J/BA	54.0 ± 0.15
PLA/BA2/DCP/J	53.1 ± 0.55
BA3	56.8 ± 0.91
BA4	60.2 ± 0.88
BA5	62.2 ± 0.49
BA6	66.7 ± 0.90

3.6. Thermomechanical Properties

The effects of BA distribution on the thermomechanical properties of PLA were examined. Figure 8a shows the storage modulus (E′) of the samples as a function of temperature. The E′ of the samples is discussed at two different phases: glassy phase, below the T_g, where the polymer chains are highly restricted; and transition phase, at the T_g of PLA (60 °C). The glassy phase shows that neat PLA has a very low E′ compared to the composites. The reinforcing effect of BA in increasing the E′ of PLA was noted. However, the addition of DCP further increased the E′ of PLA due to the enhanced distribution of BA and the possibility of branched chains and/or crosslinking, which contributed to the rigidity of the PLA matrix. On the other hand, when Joncryl was added to PLA/BA, it further increased the E′ higher than DCP in PLA/BA, because of the chain extender used to extend the PLA chains.

Furthermore, the concurrent addition of DCP and Joncryl to PLA/BA further increased the E′. The presence of both DCP and Joncryl induces chemical bonding between PLA and BA, as shown in Figure 4; hence, better distribution of BA, as shown in Figure 5. This results in strong interfacial bonding between PLA and BA; hence, the E′ increases when both DCP and Joncryl are added. The increase in E′ can also be attributed to the increase in crystallinity, as shown in Figure 3. With an increase in temperature, the E′ decreased, as expected. Region 2 illustrates the T_g of all of the samples, as listed in Table 6 and shown in Figure 7; the T_g of all of the samples remained almost the same, indicating no effect of BA distribution on the T_g. The T_g of all of the samples examined from the tan delta curve (Figure 7) clearly shows no effect on the T_g, although it was expected that the T_g would move to higher temperatures due to the chain restriction in the presence of BA particles. Overall, it is evident that the storage modulus was dependent on the distribution of BA.

Table 6. Storage modulus and Tg of neat PLA and composites.

Sample Name	*Tg* °C	Storage Modulus [Pa]
PLA	69.44	2.03×10^9
PLA/DCP	67.36	1.50×10^9
PLA/J	70.13	1.99×10^9
PLA/DCP/J	70.83	2.62×10^9
PLA/BA	68.06	2.38×10^9
PLA/DCP/BA	68.75	2.45×10^9
PLA/J/BA	69.24	2.56×10^9
PLA/BA2/DCP/J	69.44	2.63×10^9
BA3	68.22	2.34×10^9
BA4	69.39	2.27×10^9
BA5	69.84	2.20×10^9
BA6	67.76	2.29×10^9

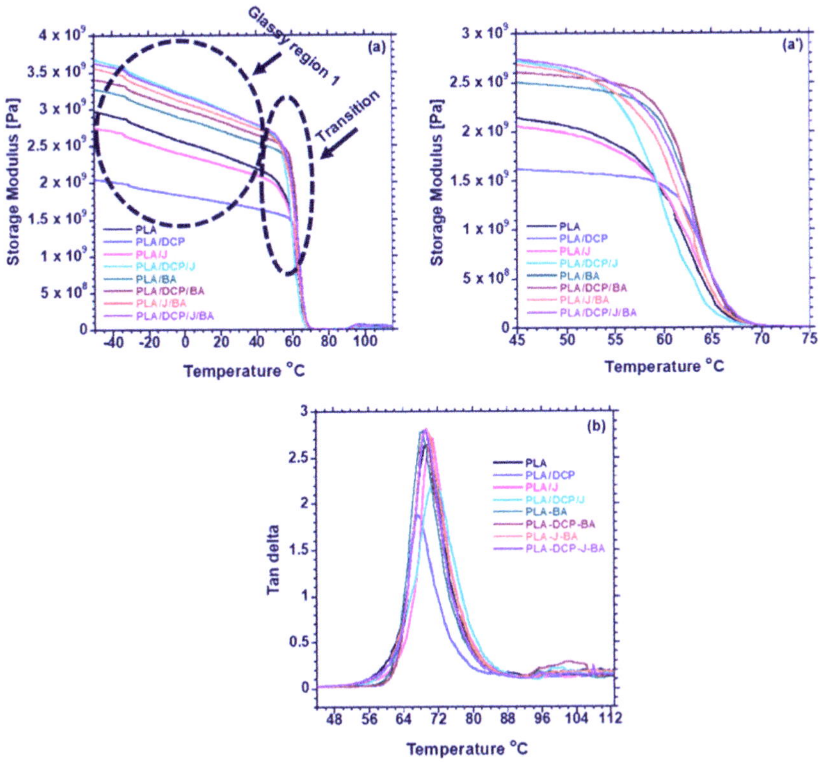

Figure 8. DMA plots of (**a**) storage modulus; (**a′**) storage modulus and (**b**) tan delta curve of neat PLA and composites.

3.7. Tensile

Figure 9 displays the tensile modulus (E′) and elongation at break (ε_{ba}) of the neat PLA, PLA/DCP, PLA/J, PLA/DCP/J/, and PLA/DCP/J/BA composites at various concentrations of BA. The neat rigid PLA shows a high E′ of 2040 MPa. Expectedly, PLA exhibits a low ε_{ba} of 4.8%. Upon the structural modification of PLA using DCP and Joncryl individually, the ε_{ba} did not change significantly. However, the E′ of PLA/DCP was higher than that of PLA and PLA/J. The entanglement of crosslinked structures could have caused this increase.

On the other hand, the structural modification of PLA using both DCP and Joncryl concurrently did not change the ε_{ba}, while the E′ slightly increased with respect to neat PLA. Upon the addition of 2–4 wt.% BA to the matrix, the E′ slightly decreased, steadily increasing as the BA concentration increased. At the same time, the ε_{ba} increased in low concentrations (2 and 3 wt.%), with a decrease at 4 and 5 wt.% loading. We believe that the good distribution of the filler and improved matrix interaction enhanced the stress transfer of the materials [2]. In this case, the better distribution observed in 3 wt.% loading, as shown in Figure 5, did not lead to the highest enhanced E′ and ε_{ba}. Overall, the structural modification of PLA and the incorporation of BA in all of the systems did not significantly affect the E′ and ε_{ba} of PLA.

3.8. 3DP

Figure 5e revealed that the BA3 sample showed a fair distribution of BA particles compared to other composites. It is noteworthy that the distribution of the filler particles and the polymer used play key roles in the 3DP process. Due to several extrusions involved in the 3DP process, improved distribution is required in order for heat to be well dissipated in the polymer matrix. Based on that, Supplementary Figure S5 shows the demonstration of the 3D-printed components for neat PLA and BA3 samples. The BA3 sample was chosen from amongst the other composites due to its fair distribution of BA particles in the PLA matrix. The fused deposition modelling (FDM) process based on extrusion technology was used as the technical basis for successfully printing biodegradable PLA/BA composites. A desktop printer (Wanhao Dupilcator i3 plus, Odessa, FL, USA) with a 0.4 mm nozzle was used to produce the 3D-printed specimen. The printing conditions were as follows: nozzle temperature 210 °C; bed temperature 50 °C; print speed of 60 mm/s; 1 perimeter wall; 2 top and bottom layers; and a layer height of 0.2 mm. Supplementary Figure S4 shows different shapes printed by the 3D printer for neat PLA and BA3 samples. A detailed paper concerning this process will follow.

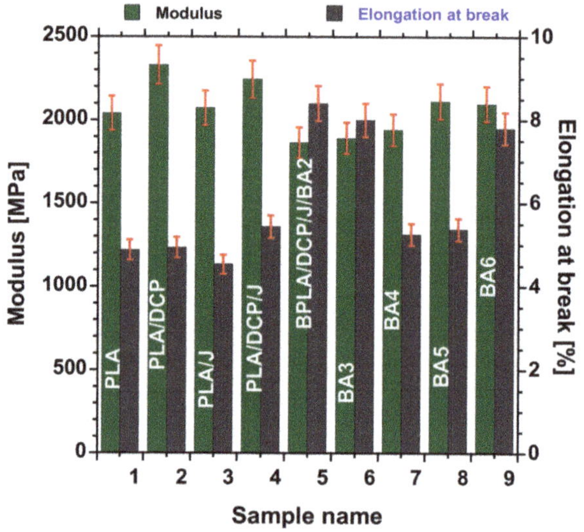

Figure 9. Tensile properties of neat PLA and BA composites with different loadings.

4. Conclusions

The present work investigated the effects of strategic modification with DCP and Joncryl on the PLA/BA composites, as well as the influence of BA distribution on the thermomechanical, mechanical, and flow properties. We found that the structural alteration of PLA by Joncryl and DCP had a significant effect on the flow properties and control of degradation. Based on the flow properties and FTIR spectroscopic analysis, the mechanism of stabilization is most likely chain extension. The chain extension leads to the long-chain branched structure in the sample containing Joncryl and DCP. The enhanced distribution of BA particles in the PLA matrix, and the interfacial bonding, were responsible for improving the PLA's properties. The mechanical properties of the PLA increased. The 3 wt.% BA showed the optimal distribution, and the PLA/DCP/J/BA3 system was chosen for further studies. The produced composite was indeed 3D printable. For future work, it will be interesting to investigate the influence of BA concentration on the PLA/DCP/J/BA system, in order to understand the effects of BA particle distribution on the thermal properties of the resulting system. In conclusion, the structural alteration with DCP and Joncryl successfully improved the BA particle distribution, leading to enhanced thermomechanical and HDT properties of PLA.

Supplementary Materials: The following are available online at https://www.mdpi.com/article/10.3390/polym13122019/s1, Figure S1: The mechanism of reaction between PLA and DCP, Figure S2: The mechanism of reaction between PLA and Joncryl, Figure S3: FTIR spectra for: (a) neat PLA, (b) PLA/DCP (c) PLA/J, (d) PLA/DCP/J, (e) PLA/BA, (f) PLA/BA/DCP, (g) PLA/BA, and (h) PLA/BA/DCP/J, Figure S4: FTIR spectra for neat materials of (a) BA, (b) DCP, and (c) Joncryl, Figure S5: The printability of: (a) neat PLA square shape, (b) BA3 square shape, (c) neat PLA CSIR logo, and (d) BA3 CSIR logo.

Author Contributions: Conceptualization, D.M.; writing—original draft preparation, D.M.; writing—review and editing, V.O., J.B., and S.S.R. All authors have read and agreed to the published version of the manuscript.

Funding: The authors thank the Department of Science and Innovation and Council for Scientific and Industrial Research, South Africa, for their financial support.

Informed Consent Statement: Not applicable.

Data Availability Statement: Not applicable.

Conflicts of Interest: The authors declare no conflict of interest.

References

1. Ojijo, V.; Ray, S.S. Super toughened biodegradable polylactide blends with non-linear copolymer interfacial architecture obtained via facile in-situ reactive compatibilization. *Polymer* **2015**, *80*, 1–17. [CrossRef]
2. Das, K.; Ray, S.S.; Chapple, S.; Wesley-Smith, J. Mechanical, thermal, and fire properties of biodegradable polylactide/boehmite alumina composites. *Ind. Eng. Chem. Res.* **2013**, *52*, 6083–6091. [CrossRef]
3. Kocsis, J.K. *Nano and Micro Mechanics of Polymer Blends and Composites*; Hanser: Munich, Germany, 2009; Volume 4, pp. 425–470.
4. Streller, R.C.; Thomann, R.; Torno, O.; Mülhaupt, R. Morphology, crystallization behavior, and mechanical properties of isotactic poly (propylene) nanocomposites based on organophilic boehmite. *Macromol. Mater. Eng.* **2009**, *294*, 380–388. [CrossRef]
5. Siengchin, S.; Karger, J.K.; Thoman, R. Nanofilled and/or toughened POM composites produced by water-mediated melt compounding: Structure and mechanical properties. *eXPRESS Polym. Lett.* **2008**, *2*, 746–756. [CrossRef]
6. Mallucelli, G.; Alongi, J.; Gioffredi, E.; Lazzari, M. Thermal, rheological, and barrier properties of waterborne acrylic nanocomposite coatings based on boehmite or organo-modified montmorillonite. *J. Therm. Anal. Calorim.* **2013**, *111*, 1303–1310. [CrossRef]
7. Najafi, N.; Heuzey, M.C.; Carreau, P.J. Polylactide (PLA)-clay nanocomposites prepared by melt compounding in the presence of a chain extender. *Compos. Sci. Technol.* **2012**, *72*, 608–615. [CrossRef]
8. Liu, M.; Pu, M.; Ma, H. Preparation structure and thermal properties of polylactide/speiolite nanocomposites with and without organic modifiers. *Compos. Sci. Technol.* **2012**, *72*, 1508–1514. [CrossRef]
9. Nijenhuis, A.J.; Colstee, E.; Grijpma, D.W.; Pennings, A.J. High molecular weight poly(l-lactide) and poly(ethylene oxide) blends: Thermal characterization and physical properties. *Polymer* **1996**, *37*, 5849–5857. [CrossRef]
10. Malwela, T.; Khumalo, V.M.; Selayian, R.; Ray, S.S. Characterization of polypropylene/polystyrene boehmite alumina nanocomposites: Impact of filler surface modification on the mechanical, thermal, and rheological properties. *J. Appl. Polym. Sci.* **2018**. [CrossRef]
11. Nieddu, E.; Mazzucco, L.; Gentile, P.; Benko, T.; Balbo, V.; Mandrile, R.; Ciardelli, G. Preparation and biodegradation of clay composites of PLA. *React. Funct. Polym.* **2009**, *69*, 371–379. [CrossRef]
12. Zhang, Q.; Yang, H.; Fu, Q. Kinetics-controlled compatibilization of immiscible polypropylene/polystyrene blends using nano-SiO2 particles. *Polymer* **2004**, *45*, 1913. [CrossRef]
13. Agwuncha, S.C.; Ray, S.S.; Jayaramudu, J.; Khoathane, C.; Sadiku, R. Influence of boehmite nanoparticle loading on the mechanical, thermal and rheological properties of biodegradable polylactite/poly (E-caprolactone) blends. *Macromol. Mater. Eng.* **2015**, *300*, 31–47. [CrossRef]
14. Adhikari, R.; Brostow, W.; Datashvili, T.; Henning, S.; Menard, B.; Menard, K.; Michler, G. Effect of surfactant treated boehmite nanoparticles on properties of block copolymers. *Mater. Res. Innov.* **2012**, *16*, 19. [CrossRef]
15. Bocchini, S.; Therias, S.M.; Gardette, J.L.; Camino, G. Influence of nanodispersed boehmite on polypropylene photooxidation. *Polym. Degrad. Stab.* **2007**, *92*, 1847–1856. [CrossRef]
16. Pedrazzoli, D.; Khumalo, V.M.; Kocsis, J.K.; Pegoretti, A. Thermal, viscoelastic and mechanical behavior of polypropylene with synthetic boehmite alumina nanoparticles. *Polym. Test.* **2014**, *35*, 92–100. [CrossRef]
17. Khumalo, V.M.; Kocsis, J.K.; Thomann, R. Polyethylene/synthetic boehmite alumina nanocomposites: Structure, mechanical, and perforation impact properties. *J. Mater. Sci.* **2011**, *46*, 422–428. [CrossRef]
18. Duangphet, S.; Szegda, D.; Song, J.; Tarverdi, K. The effect of epoxy-functionalized chain extender on crystal growth and morphology of poly 9 3- hydroxybutyrate-co-3-hydroxyvalerate) (PHBV). *KKU Eng. J.* **2016**, *43*, 470–473.
19. Tuba, F.; Khumalo, V.M.; Kocsis, J.K. Essential work of fractured of Poly caprolactone/boehmide alumina nanocomposites: Effect of surface coatin. *J. Appl. Polym. Sci.* **2013**. [CrossRef]

20. Najafi, N.; Heuzey, M.C.; Carreau, P.J.; Wood-Adams, P.M. Control of thermal degradation of polylactide(pla)-clay nanocomposites using chain extenders. *Polym. Degrad. Stab.* **2012**, *97*, 554–565. [CrossRef]
21. Villalobos, M.; Awojulu, A.; Greeley, T.; Turco, G.; Deeter, G. Oligomeric chain extenders for economic reprocessing and recycling of condensation plastics. *Energy* **2006**, *31*, 3227–3234. [CrossRef]
22. Pilla, S.; Kim, S.G.; Auer, G.K.; Gong, S.; Park, C.B. Microcellular extrusion foaming of polylactide with chain-extender. *Polym. Eng. Sci.* **2009**, *49*, 1653–1660. [CrossRef]
23. Arruda, L.C.; Magaton, M.; Bretas, R.E.S.; Ueki, M.M. Influence of chain extender on mechanical, thermal and morphological properties of blown films of PLA/PBAT blends. *Polym. Test.* **2015**, *42*, 27–37. [CrossRef]
24. Kong, Y.; Hay, J.N. The measurement of the crystallinity of polymers by DSC. *Polymer* **2002**, *43*, 3873–3878. [CrossRef]
25. Ahmed, A.S.M.; Saeed, H.A.M.; Eltahir, Y.A.; Xia, Y.M.; Wang, Y. Modification of PLA with chain extender. *Appl. Mech. Mater.* **2015**, *716–717*, 44–47.
26. Batakliev, T.; Doycheva, I.P.; Angelov, V.; Georgiev, V.; Ivanov, E.; Kotsilkova, R.; Casa, M.; Cirillo, C.; Adami, R.; Sarno, M.; et al. Effects of graphene nanoplatelets and multiwall carbon nanotube on the structure and mechanical properties of Polylactide composites. *Appl. Sci.* **2019**, *9*, 469. [CrossRef]
27. Boll, A.D.B.; Fiedle, B. Prediction of thermal exposure and mechanical behavior of epoxy resin using artficial neural networks and fourier transform infrared spectroscopy. *Polymers* **2019**, *11*, 363. [CrossRef]
28. Dickie, S.A.; McQuillan, A.J. In-situ infrared spectroscopic studies of adsorption processes on boehmite particle films: Exchange of surface hydroxyl groups observed upon chelation by acetylacetone. *Langmuir* **2004**, *20*, 111630–111636. [CrossRef]
29. Ram, S. Infrared spectral study of molecular vibrations in amorphous, nanocrystalline and AlO(OH) αH_2O bulk crystals. *Infrared Phys. Technol.* **2001**, *42*, 547–560. [CrossRef]
30. Inata, H.; Matsumura, S. Chain extenders for polyesters. I. addition-type chain extenders reactive with carboxyl end groups of polyesters. *J. Appl. Polym. Sci.* **1985**, *30*, 3325–3337. [CrossRef]
31. Aharoni, S.M.; Forbes, C.E.; Hammond, W.B.; Hindenlang, D.M.; Mares, F.; O'Brien, K.; Sedgwick, R.D. High-Temperature reactions of hydroxyl and carboxyl PET chain end groups in the presence of aromatic phosphite. *J. Polym. Sci. Part A Polym. Chem.* **1986**, *24*, 1281–1296. [CrossRef]
32. Wang, Y.; Chunhua, F.U.; Yongxiang, L.U.O.; Changshun, R.; Zang, Y.; Ya, F.U. Melt synthesis and characterization of poly(L-lactic Acid) chain linked by multifunctional epoxy compound. *J. Wuhan Univ. Technol. Mater. Sci. Ed.* **2010**, *25*. [CrossRef]
33. Zhai, W.; Ko, Y.; Zhu, W.; Wong, A.; Park, C.B. A study of the crystallization, melting, and foaming behaviors of polylactic acid in compressed CO_2. *Int. J. Mol. Sci.* **2009**, *10*, 5381–5397. [CrossRef] [PubMed]
34. Quero, E.; Muller, A.J.; Signori, F.; Coltelli, M.B.; Bronco, S. Isothermal cold-crystallization of PLA/PBAT blends with and without the addition of acetyl tributyl citrate. *Macromol. Chem. Phys* **2012**, *213*, 36–48. [CrossRef]
35. Ojijo, V.; Ray, S.S.; Sadiku, R. Role of specific interfacial area in controlling properties of immiscible blends of biodegradable polylactide and poly [(butylene succinate)-co-adipate]. *ACS Appl. Mater. Interfaces* **2012**, *12*, 6690–6701. [CrossRef]
36. Wang, L.; Jing, X.; Cheng, H.; Hu, X.; Yang, L.; Huang, Y. Blends of linear and long-chain branched poly (l-lactide) with high melt strength and fast crystallization rate. *Ind. Eng. Chem. Res.* **2012**, *51*, 10088–10099. [CrossRef]
37. Nofar, M.; Zhu, W.; Park, C.B.; Randall, J. Crystallization kinetics of linear and long-chain-branched polylactide. *Ind. Eng. Chem. Res.* **2011**, *50*, 13789–13798. [CrossRef]
38. Yeh, J.T.; Yang, M.G.; Wu, C.H.; Wu, X.; Wu, C.S. Study on the crystallization kinetic and characterization of poly (lactic acid) and poly(vinyl alcohol) blends. *Polym. Plast. Tech. Eng.* **2008**, *47*, 1289–1296. [CrossRef]
39. Tabatabaei, S.H.; Ajji, A. Crystal structure and orientation of uniaxially and biaxially oriented PLA and PP nanoclay composite films. *J. Appl. Polym. Sci.* **2011**, *124*, 4854–4863. [CrossRef]
40. Zhang, M.; Thomas, N.L. Blending polylactic acid with polyhydroxybutyrate: The effect on thermal, mechanical, and biodegradation properties. *Adv. Pol. Tech.* **2011**, *30*, 67–79. [CrossRef]
41. Bucci, D.Z.; Tavares, L.B.B.; Sell, I. PHB packaging for the storage of food products. *Polym. Test.* **2005**, *24*, 564–571. [CrossRef]
42. Suryanarayana, C.; Norton, M.G. *X-ray Diffraction: A Practical Approach*; Springer: Berlin/Heidelberg, Germany, 1998. [CrossRef]
43. Shen, J.; Zhou, Y.; Lu, Y.; Wang, B.; Shem, C.; Chen, J.; Zhang, B. Later stage melting of isotactic polypropylene. *Macromolecules* **2020**, *53*, 2136–2144. [CrossRef]
44. Becker, G.W.; Bottenbruch, L.; Binsack, R.; Braun, D. *Engineerinng Thermoplastics Polyamides*; Hanser Verlag: Munich, Germany, 1998; ISBN 3-446-16486-3.
45. Farid, T.; Herrera, V.N.; Kristiina, O. Investigation of crystalline structure of plasticized poly (lactic acid)/Banana nanofibers composites. *Mater. Sci. Eng.* **2018**, *369*, 012031. [CrossRef]

polymers

MDPI

Article

Optimization of the Sustainable Production of Resistant Starch in Rice Bran and Evaluation of Its Physicochemical and Technological Properties

Ruta Vaitkeviciene [1,]*, Joana Bendoraitiene [2], Rimgaile Degutyte [1], Mantas Svazas [1] and Daiva Zadeike [1,]*

[1] Department of Food Science and Technology, Faculty of Chemical Technology, Kaunas University of Technology, 50254 Kaunas, Lithuania

[2] Department of Polymer Chemistry and Technology, Faculty of Chemical Technology, Kaunas University of Technology, 50254 Kaunas, Lithuania

* Correspondence: ruta.vaitkeviciene@ktu.lt (R.V.); daiva.zadeike@ktu.lt (D.Z.)

Abstract: In this study, the optimization of ultrasound (US) (850 kHz, 120 W) processing parameters (temperature, time, and power) for the enhanced production of resistant starch (RS) in rice bran (RB) matrixes was performed. The effect of US cavitation at different temperatures on the morphology, physicochemical properties, and mechanical performance of RS was evaluated. Ultrasonication at 40–70 °C temperatures affected the chemical structure, reduced the crystallinity of RS from 23.85% to between 18.37 and 4.43%, and increased the mechanical and thermal stability of RS pastes, indicating a higher tendency to retrograde. US treatment significantly ($p < 0.05$) improved the oil (OAC) and water (WAC) absorption capacities, swelling power (SP), solubility (WS), and reduced the least-gelation concentration (LGC). The mathematical evaluation of the data indicated a significant effect ($p < 0.05$) of the US parameters on the production of RS. The largest increment of RS (13.46 g/100 g dw) was achieved with US cavitation at 1.8 W/cm^2 power, 40.2 °C temperature, and 18 min of processing time. The developed method and technology bring low-temperature US processing of rice milling waste to create a new sustainable food system based on modified rice bran biopolymers.

Keywords: rice bran; ultrasound cavitation; resistant starch microstructure; crystallinity; hydration properties; mechanical performance

Citation: Vaitkeviciene, R.; Bendoraitiene, J.; Degutyte, R.; Svazas, M.; Zadeike, D. Optimization of the Sustainable Production of Resistant Starch in Rice Bran and Evaluation of Its Physicochemical and Technological Properties. *Polymers* **2022**, *14*, 3662. https://doi.org/10.3390/polym14173662

Academic Editors: Domenico Acierno and Antonella Patti

Received: 10 August 2022
Accepted: 31 August 2022
Published: 3 September 2022

1. Introduction

Cereal processing by-products, being a low-cost source of nutritionally valuable components, such as dietary fibres, resistant starches, proteins, minerals, and bioactive compounds, are promising materials for the valorization of sustainable bio-based components for food.

Resistant starch (RS) is a biopolymer of plant origin that is resistant to digestion in the small intestine [1]. In terms of RS's chemical and structural features, five types of RS have been identified [2]. RS1 is a starch that is physically unavailable for digestion by enzymes due to its intact seed or grain structure. RS2 refers to the inherent resistance of starch granules (potato, banana, etc.) to digestion due to its native conformation. RS3 represents the largest RS fraction, which is primarily retrograded amylose. RS type 4 is a chemically modified starch, and RS5 is a helical-structured lipid–amylose complex.

Starch is composed of the two polysaccharides, amylose, a linear polysaccharide with α-1,4-glycosidic bonds, and amylopectin, which is a larger branched polysaccharide with α-1,4 and α-1,6 linkages [3,4]. The crystalline region consists of double helices of amylopectin, while the amorphous region is formed by amylose chains and branched segments of amylopectin. Starch retrogradation involves the recrystallization of amylose chains and the formation of tightly packed double helices stabilized by hydrogen bonds, and by the association of amylopectin chains into double helices, it is thus protected from enzyme attack [3]. Retrograded starch can seriously affect the functional properties of starch products, such as texture, stability, transparency, viscoelasticity, and digestibility [4]. The factors affecting starch retrogradation include the structure and the type of starch, the amylose to amylopectin ratio, storage conditions, and the starch modification process [4–6].

Nowadays, RS is characterized as a functional biopolymer with physiological effects comparable to those assigned to dietary fiber [2]. RS also brings a positive nutritional effect to food products as a functional ingredient and an exceptional prebiotic [7,8].

Rice processing by-products have the potential to be utilized as an RS source, but the natural RS content depends on the rice variety and can be relatively low [9]. Moreover, the better part of the conventional processes of food production decreases the undigestible fraction of starch in foods [7]. The content of RS can be enhanced during technological processes [10,11], and the physical or hydrothermal modification of starchy raw material [12,13]. The emerging food processing technologies, effectively changing the physicochemical properties and mechanical performance of food components, are of great interest nowadays [14,15].

The nutritional and technological properties of foods can be maintained through the use of preserving production conditions, leading to the increased interest in developing safe and cost-effective technologies. In the last decade, ultrasound (US) has been used extensively in food processing to enhance the physicochemical and functional properties of food components [16–20]. However, most of the research so far has focused on systems based on pure starch [21,22], although starch as such is not consumed. Future research, therefore, should be directed to RS within the context of an all-food matrix.

The technological and mechanical properties of starch depend on water absorption and starch formulation, processing, and retrogradation. Higher amylose/amylopectin ratios of starch generally give material of higher strength and plasticity, and also affects the gelatinization process, which generates the uniform amorphous thermoplastic structure by the action of temperature and mechanical force or ultrasound cavitation [14,23]. Quantifying these variables is important for the use of starch materials in different applications.

From this perspective, our work has been devoted to the development of a new sustainable food system based on rice bran biopolymers, in light of the potential application of US technology for the enhanced production of RS, by optimizing RS3 preparation conditions, thereby taking into account the effect of US processing temperature on the morphology, structural changes, mechanical performance, and technological properties of RS, and the possibility of the valorization and recovery of rice milling by-products.

2. Materials and Methods

2.1. Rice Bran Material

In this study, the analyzed food matrix was rice bran (RB) obtained from a local mill (SC "Ustukiu malunas", Pasvalys, Lithuania) after the brown rice milling process. A fraction with a particle size of 315 μm (protein 11.20%, starch 66.41%, carbohydrates 81.53%, fat 5.38%, ash 1.89% dry wight basis, dw) was used for the experiment. Commercial native rice starch (S-7260) was purchased from Sigma Aldrich (Saint Louis, MO, USA). Raw materials were stored in plastic jars at 4 °C temperature.

2.2. Ultrasonication Procedure

Parameters of the starch sonication were selected on the basis of procedures reported by Wang and Bai [24]. For the US treatment, the RB sample (10 g) and distilled water were mixed at a solid/liquid ratio of 1:3. The mixture was placed in a tightly closed plastic bag (sample thickness 15 mm), and the sample was treated with US in an ultrasonic high-power bath (Type 5/1575) connected to a 120W-HF-Generator (Meinhardt Ultraschalltechnik, Leipzig, Germany). Sonication was carried out at 850 kHz frequency and 50% US intensity for different times (15–35 min) at different temperatures (30–70 °C). After sonication, each sample was subsequently stored at 4 °C for 24 h, then dried by an oven at 50 °C and subjected to the isolation and determination of the starch. Each experiment was repeated four times.

2.3. Starch Content Measurement

The content of resistant starch (RS) was determined using a Megazyme Assay Kit (K-RSTAR, Megazyme Int., Wicklow, Ireland) for resistant starch determination based on the AOAC method (2002.02). The RB sample (100 ± 5 mg) was digested with a pancreatic α-amylase and amyloglucosidase (PAA/AMG) at 37 °C (pH 6.0) for 16 h with continual stirring, during which time the non-resistant starch was solubilized and hydrolyzed to D-glucose. The reaction was terminated by the addition of an equal volume of 95% ethanol and RS was recovered as a pellet through centrifugation (3000× *g*, 10 min); this was then washed twice with 50% (*v*/*v*) ethanol and centrifuged. RS in the pellet was dissolved in 2M KOH by vigorously stirring for 20 min in an ice-water bath. The solution was neutralized with acetate buffer and the starch was quantitatively hydrolyzed to glucose with AMG. D-Glucose was quantified with glucose oxidase/peroxidase reagent (GOPOD), and this was the measurement of the RS content in the sample. Digestible starch (DS) was determined by measuring free-glucose content with GOPOD in the original supernatant and the washings. The RS and rapid digestible starch (RDS) contents were calculated and expressed as grams per 100 g of the dry weight of raw material.

2.4. Amylose Content Determination

The amylose content in the RS samples was determined using a Megazyme Amylose/Amylopectin assay kit (Megazyme Ltd., Winklow, Ireland) according to the manufacturer instructions.

2.5. Starch Isolation

Starches were isolated from the RB material by the alkaline extraction before and after the enzymatic digestion step [10]. The starch isolation scheme is presented in Figure S1. The procedure of the isolation of starch comprises the mixing with 0.2% NaOH for 2 h and further centrifugation at 3200× *g* for 15 min. The extraction was repeated two times. After the protein extraction, the starch precipitates were washed two times with 150 mL of distilled water by centrifugation. Isolated starches were lyophilized and stored in a desiccator until further use.

2.6. Starch Morphology Evaluation

Granule morphology of starches was quantified using a scanning electron microscopy (SEM) (Quanta 200 FEG, FEI, Hillsboro, OR, USA). Each powdered sample was placed on an adhesive tape attached to an aluminum stub and randomly examined at 2500× magnification.

2.7. Crystallinity Analysis

Starch crystallinity was analyzed with a D8 Advance X-ray Diffractometer Bruker AXS (Bruker, Karlsruhe, Germany). Data were collected using CuK_α radiation (U = 40 kV and I = 40 mA), a 0.02 mm nickel filter, the Bragg-Brentano geometry, and a fast-counting detector Bruker LynxEye. The samples were scanned over the 2θ range of 5–30° at a speed of 6°/min. The relative crystallinity was determined by calculating the areas below and above the curve that correspond to the amorphous and crystalline regions, respectively [25]. The ratio of the amorphous area to the total area was taken as the relative crystallinity.

2.8. Fourier Transform Infrared Spectroscopy (FT-IR)

The FT-IR spectra of the samples were recorded using an FT-IR Spectrometer (Frontier, PerkinElmer Inc., Waltham, MA, USA) with a single reflectance horizontal ATR (Attenuated Total Reflectance) cell equipped with a diamond crystal. The data were recorded in the range from 4000 to 560 cm^{-1} with a resolution of 4 cm^{-1}.

2.9. Pasting Properties

The thermomechanical behavior of the RS paste (peak viscosity, pasting temperature, final viscosity) was analyzed by a Brabender Micro-Visco-Amylograph (Brabender GmbH, Duisburg, Germany) according to Yang et al. [18] with some modifications. Suspensions of each sample (10%, *w*/*w*) were heated from 25 to 95 °C at a rate of 6.5 °C/min, held at 95 °C for 15 min, cooled to 50 °C at 6.5 °C/min, and held at 50 °C for 2 min with a speed of rotating paddle of 250 rpm.

2.10. Specific Viscosity Measurement

The mechanical properties of RS slurry (0.4 mg/mL) in 1N KOH were investigated by an Ostwald's capillary viscometer (tube diameter 0.99 mm) held in a temperature-controlled water bath (Bath-Immersion Thermostat E5, medingLab Temperiertechnik, Hochdorf, Germany) [26]. Time of flow was measured four times, and the average values were taken for specific viscosity calculation.

2.11. Least Gelation Concentration

The least-gelation concentration (LGC) of starches was determined according to the method of Sathe and Salunkhe [27], analyzing the ability of starch suspension to form a gel after heating and cooling. For the experiment, test tubes containing 5 mL of 2–20% starch suspensions in distilled water were heated in an 80 °C water bath for 1 h, followed by rapid cooling in an ice-water bath, and further cooling at 4 °C for 2 h. The LGC was defined as the concentration allowing the gel to not slip from the tubes when inverted.

2.12. Experimental Design and Statistical Analysis

A Central Composite Design (CCD) consisting of 21 experiments and 3 center points for each power group was employed as the experimental design performed in a random order at four replications. The factors investigated included independent variables: temperature (T, °C), treatment time (t, min), and power (P, W/cm^2) at different levels. Response Surface Methodology (RSM) [28] was used to investigate the effect of the US-assisted treatment parameters on the resistant starch (RS) content. Furthermore, the optimization of US processing conditions was performed by developing the simplest possible mathematical model with a coefficient of determination higher than 80%. The response was the RS content exclusively obtained under the influence of US action (dY, g/100 g dw). Results were analyzed using Design-Expert trial version 13.0.1.0 software (StatEase Inc., Minneapolis, MN, USA). The analysis of variance (ANOVA) was conducted for the assessment of the suitability of the model using the coefficient of '*lack of fit*' and the Fisher value (F).

All chemical analyses were performed at least in triplicate. Statistical analysis of the data was performed using SPSS software (ver. 27.0, IBM, Chicago, IL, USA). The significant differences between the means were evaluated by one-way ANOVA at a significance level of 0.05.

3. Results

3.1. Resistant Starch Production Rate

Factors affecting the increment of RS due to the formation of RS type 3 in the rice bran matrix were studied at 50% US intensity (Figure 1A). The amount of RS3 measured at different US processing time points was compared to the initial content in the untreated RB (11.51 g/100 g d.w.).

Most of the RS was produced at the early period of ultrasonication (15–20 min); a longer US treatment time (25–35 min) enhanced the degradation of starch, significantly decreasing the content of RS. Sonication at higher temperatures (60–70 °C) caused a slight increase (up to 5.0%) in the RS content during 15–20 min of ultrasonication, but a decrease up to 73.5% at longer sonication times (25–35 min) was fixed (Figure 1A). Herewith, the rapidly digestible starch (RDS) content increased by 22.7–53.4% with increasing US time and temperature, increasing the total starch (TS) extraction yield (Figure 1B).

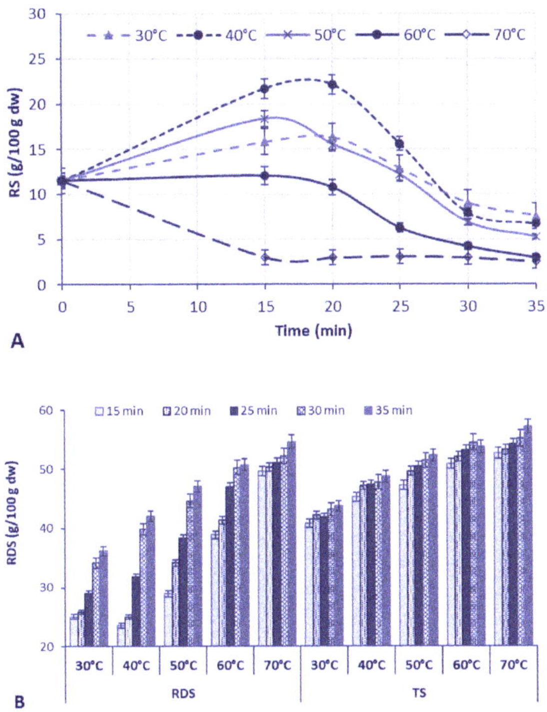

Figure 1. The effect of rice bran (*m/w* ratio 1:3) ultrasound treatment at different temperatures and times on resistant starch (RS) (**A**), rapid digestible starch (RDS), and total starch (TS) (**B**) contents.

The results are in agreement with other reports, showing that the yield of starch could be improved due to acoustic cavitation that disrupts the protein and fibre matrix, allowing the enhanced release of the starch granules [29]. In other cases, the US-initiated increase in the RS content at US temperatures of 40–50 °C may be due to the increased amylose content because of the cleavage of long chains and depolymerization of amylopectin due to the effect of US cavitation, contributing to the alignment or aggregation of the molecule, which leads to an enhanced yield of retrograded starch [18].

According to the literature [30], the US pre-treatment destroyed the amorphous areas of rice starch granules with a low impact on the crystalline pattern and chain length distribution of amylopectin [30]. Debranched starches can be involved in the formation of RS3, since the linear fragments of amylose with 13–30 glucose units of greater mobility and long amylopectin chains strongly contribute to RS3 formation [30]. Reduced chains of amylose have a tendency to form a more crystalline strongly bonded structure with a higher concentration of more stable and resistant double helices. The release of amylopectin side chains before retrogradation favored re-association leading to reduced starch digestibility [31].

At US temperatures above the rice starch gelatinization temperature, the layered structure of the starch became unstable, and that may be the reason for the reduced relative crystallinity and lower content of RS. The ANOVA of the initial experimental data showed that both the US temperature and time were the significant factors ($p < 0.0001$) for the RS production (Tables S1 and S2). The RS contents in all treatments at 30–50 °C significantly depended on the temperature ($p = 0.0001$, F = 267.36–749.01), and time ($p = 0.0001$, F = 21.63–576.92).

3.2. The Effect of Ultrasonication on Resistant Starch Increment and Mathematical Model Analysis

The effect of the US processing temperature (T), time (t), and power (P) on the RS yield was determined via optimization by CCD and RSM. The yield of RS varied significantly from 1.51 to 12.61 g/100 g d.w. when applying different T (30–50 °C), t (15–25 min), and P (1.3, 1.5, and 1.8 W/cm^2) combinations (Table 1).

Table 1. Experimental design and experimental results.

Experiment Number	Processing Variables			dY [a] (g/100 g d.w.)		
	t (min)	T (°C)	P (W/cm^2)	Experimental	Predicted	RE (%)
Low level (−1)	15	30	1.3			
Medium level (0)	20	40	1.5			
High level (+1)	25	50	1.8			
1	1	1	1	2.02 ± 0.08	2.00	0.85
2	1	−1	−1	1.51 ± 0.04	1.53	−1.13
3	−1	1	−1	6.98 ± 0.13	7.00	−0.28
4	1	−1	1	3.82 ± 0.07	3.80	0.46
5	−1	1	1	8.23 ± 0.18	8.21	0.30
6	1	1	−1	1.09 ± 0.02	1.11	−1.56
7	−1	−1	−1	4.42 ± 0.09	4.44	−0.44
8	−1	−1	1	5.94 ± 0.13	5.92	0.42
9	−1	0	1	12.34 ± 0.18	12.48	−1.12
10	−1	0	−1	10.32 ± 0.21	10.28	0.43
11	1	0	1	5.22 ± 0.10	5.26	−0.85
12	1	0	−1	4.21 ± 0.07	4.18	0.83
13	0	1	0	4.87 ± 0.08	4.83	0.87
14	0	−1	0	5.89 ± 0.11	5.93	−0.71
15	0	0	−1	10.75 ± 0.12	10.69	0.60
16	0	0	1	12.72 ± 0.04	12.61	0.84
17	1	0	0	4.76 ± 0.08	4.72	0.85
18	−1	0	0	11.34 ± 0.23	11.38	−0.35
19	0	0	0	11.72 ± 0.12	11.65	0.60
20	0	0	0	11.61 ± 0.19	11.65	−0.34
21	0	0	0	11.65 ± 0.14	11.65	0.00

[a] dY: resistant starch content; T: temperature; t: time; P: power; RE: relative error.

A quartic-order regression model (Table S3), describing the relationship between the RS content increment ($dY = Y - Y_0$), obtained under the influence of the US processing, and independent variables, was constructed in the following equation:

$$dY = Y - Y_0 = 11.65 - 3.33t - 0.5519T + 0.9644P - 0.8837tT - 0.28tP - 0.1419TP - 3.6t^2 - 6.27T^2 - 0.1387tTP + 0.8807t^2T - 0.1393t^2P + 1.19tT^2 - 0.023T^2P + 2.47t^2T^2 - 0.0643t^2TP + 0.3412tT^2P, \quad (1)$$

where dY is the RS content increment; Y is the total RS content; Y_0 is the RS content in raw material; t, T, and P are the values of temperature, time, and US power, respectively.

The ANOVA of the model confirmed that the sonication temperature, time, power, and their quadratic effects and linear interaction effects among the factors t and T significantly affected the RS formation ($p < 0.0001$) (Table S4). In addition, there were significant linear interaction effects among the factors of tP and TP ($p = 0.0002$ and $p = 0.0094$, respectively), and their quadratic effects on the RS content: T^2 (F = 13636.80) > t (F = 5168.84) > t^2 (F = 4055.14) > P (F = 1049.64) > T^2t^2 (F = 1001.51) > Tt (F = 728.10) > tT^2 (F = 437.29) > t^2T (F = 263.98) > T (F = 163.30).

Based on the statistical analysis, the model was significant for dY, showing that it was reasonably reproducible and well-fitting to the experimental data: Fisher value—2839.52 ($p < 0.0001$), variation coefficient 1.19 %, determination coefficient $R^2 = 0.9998$. The results showed that the p-value for *'lack-of-fit'* was not significant ($p = 0.1049$); hence, the model was satisfactory in explaining the obtained data at a 95% confidence level. The predicted

R^2 value (0.9736) for the experimental design was close to the adjusted R^2 value (0.9994), indicating that 97% of the data could be described by this regression model. Comparing the experimental data with the theoretical prediction, the relative error (RE) varied between 0.17 and 1.56% (Table 1), indicating that the model could be used to estimate the response for the optimization.

To our knowledge, there are no reports on the optimization of high-power US-assisted production of RS in rice bran. The optimization of the US process conditions to increase the RS content was previously demonstrated for pure pea starch and buckwheat starch as raw materials [24,32]. However, such process could have complicated applications on an industrial scale because of the high cost of pure starches. In the present work, the US processing was optimized by implication of a different approach, and rice milling by-products were used as the raw material instead of pure rice starch to withdraw the losses of bran nutrients while improving the RS content; these results can be preferable for the food industry.

3.3. Optimization and Prediction of Process Parameters

Based on the obtained model, a three-dimensional plot was constructed to predict the relationship between the independent variables. Predictive plotting of the optimum conditions based on the increase in RS content was carried out by varying the factors. Figure 2 presents the relationships between the response value (dY) and the independent variables (time, temperature, and power).

The elliptical form of the contours and a relatively sharp slope of the surface responses indicate a strong interaction between the analyzed factors and a strong influence of the sonication parameters on the response value. Based on RSM, the optimal conditions for the production of RS in RB were set as follows: US power of 1.8 W/cm^2, temperature 40.2 °C, and time 18 min. However, the optimal conditions proposed for the US power were located at the maximum level studied (+1), indicating that there is a possibility of exploring the production of RS outside the studied level, but this would be outside of the technical characteristics of the device used.

Under optimal conditions, the predicted RS increment (dY) shows at 13.46 g/100 g d.w. After three repeated tests, the average RS content was 13.44 g/100 g d.w. with a small relative error (RE = 0.42). The actual value was close to the theoretical value, indicating that the optimization results were reliable and practical.

US as a sustainable processing technique has the potential to produce RS with desirable properties and mechanical performance. Modification at different US processing temperatures can achieve functional properties not found in untreated starches, which may have specific applications in the food industry.

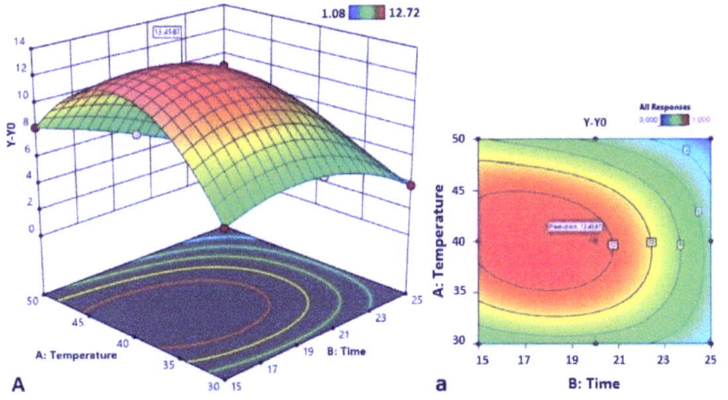

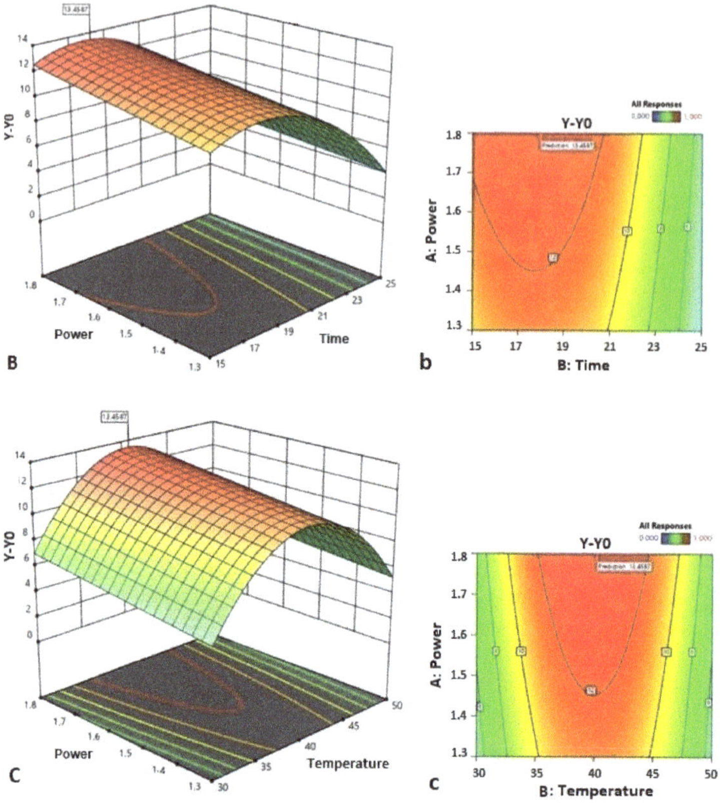

Figure 2. Response surface plots (**A**–**C**) and contour plots (**a**–**c**) of the interactions between ultrasound temperature and time (**A**), power and time (**B**), and power and temperature (**C**) for the RS increment ($dY = Y - Y_0$).

3.4. Morphological Characterization of Ultrasound-Modified Rice Bran Resistant Starch

Scanning electron micrographs (SEM) of RS isolated from sonicated samples at different temperatures of RB are presented in Figure 3. The micrographs of the native rice starch and untreated RS (RS_{UN}) showed typical irregular-shaped small-size granules (Figure 3A,B), corroborating the structure reported by Ashwar et al. [10] and Yang et al. [19]. The changes in the RS granule morphology became more visible when US was applied.

Ultrasonication for 20 min at 40 °C already induced the disintegration of the RS_{40C} granules and the formation of crushed particles (Figure 3C), while the granules of RS_{50C} (Figure 3D) already showed visible differences in shape and size (higher amount of crushed particles) compared to those treated at 40 °C. Elevation of the temperature to 60 °C resulted in an aggregation of small granules (Figure 3E); the formation of a film-like structure was noticed at a temperature of 70 °C (Figure 3F).

Similar tendencies were reported by Noor et al. [33], who indicated a notable difference in the microstructure between RS2 samples of US-treated and untreated lotus stem starch. As Sujka [17] reported, US processing (20 kHz, 170 W, 20 °C, 30 min) affected the average diameter and pore size distribution in rice, corn, wheat, and potato starches. According to Ding et al. [30], the morphological characteristics of retrograded starch (RS3) changed due to high-power US treatment (20 kHz, 100–600 W, 30 min), resulting in a more compact block-shaped structure.

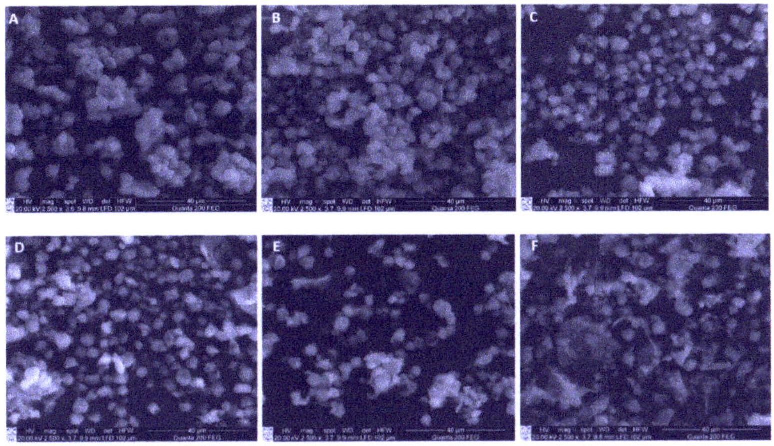

Figure 3. SEM images of native rice starch (**A**), resistant starch isolated from RB untreated (**B**) and ultrasonicated (1.8 W/cm^2, 20 min) at 40 °C (**C**), 50 °C (**D**), 60 °C (**E**), and 70 °C (**F**) temperatures (magnification ×2500).

In our study, the increase in the US temperature from 50 to 70 °C tended to a more intensive disruption of larger starch granules and the aggregation of small granules due to the combined effect of the US cavitation and temperature (Figure 3E–G). According to the literature, US cavitation induces high-frequency vibrations, which can break the rice starch and cause a substantial deformation of the granule structure at the starch gelatinization temperature (62–79 °C) [34]. Based on these observations, the change in the pasting behavior is explained in terms of the solubilization of the swollen starch granules and starch aggregates induced by sonication, as was reported by Zuo et al. [35]. Overall, our study showed the visible fragmentation of starch particles of RS isolated from US-treated RB material.

3.5. Crystallinity of Rice Resistant Starches

The results of X-ray diffraction (Figure 4) indicate that the untreated RS had the typical A-type crystallinity pattern (corresponding to tight-packed amylopectin double helices) with the characteristic peaks at diffraction angles (2θ) of 15°, 16.9°, 17.9°, a weak peak at 20° 2θ, and an additional peak at 23° 2θ, as is similar to the observations of other researchers [18,36].

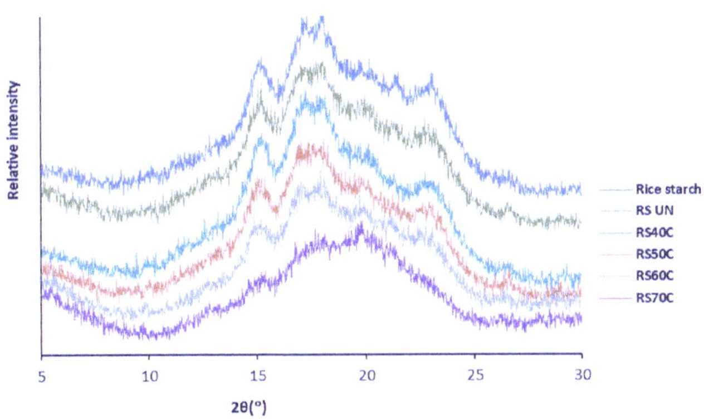

Figure 4. X-ray diffraction patterns of native rice starch and resistant starches (RS) isolated from rice bran that was untreated (UN) and ultrasonicated (1.8 W/cm^2, 20 min) at different temperatures.

According to the literature, the width of the peaks is assigned to the size of the crystallites: a discrete peak represents a higher crystallinity of the starch [25], while a broad diffraction peak indicates lower crystallinity and an amorphous character [37].

Ultrasonication under mild conditions (40 °C) resulted in a minor decrease in crystallinity at 20° and 23° 2θ, a reduced peak sharpness at 15°, and a doublet at 16.9° and 17.9° 2θ compared to the starches isolated from the untreated RB (RS_{UN} sample). The RS samples obtained after sonication at higher temperatures already showed lower crystallinity peaks. With a temperature of 70 °C, the disappearance of peaks at 15° and 23° 2θ, and a doublet at 16.9° and 17.9° 2θ was noticed.

The higher crystallinity of the native rice starch (26.31%) and RS_{UN} (23.85%) (Table 2) implies a stronger interaction between the double helices within the crystalline regions [37]. Similar to maize starch [14], ultrasonication at temperatures lower than the gelatinization temperature did not cause polymorphic changes in the rice RS diffraction pattern (A-type), but a reduction in the degree of crystallinity (up to 17.95–18.36%) was observed. Sonication at temperatures of 60 and 70 °C strongly reduced the degree of crystallinity to 14.40% and 4.43%, respectively.

Table 2. Amylose content and degree of crystallinity (%) of native rice starch and resistant starches (RS) isolated from rice bran that was untreated and sonicated at different temperatures.

Starch Samples	Amylose (%)	Crystallinity (%)	X-ray Diffraction Peak Intensity (r.u.)
Rice starch	25.35 ^{d}	26.31 ^{a}	2677 ^{a}
RS_{UN}	25.62 ^{d}	23.85 ^{b}	2663 ^{a}
RS_{40C}	37.47 ^{c}	18.36 ^{c}	2568 ^{b}
RS_{50C}	53.10 ^{a}	17.95 ^{c}	2385 ^{c}
RS_{60C}	51.56 ^{ab}	14.40 ^{d}	2325 ^{c}
RS_{70C}	49.97 ^{b}	4.02 ^{e}	1098 ^{d}

Data with different superscript letters within the column represent significant differences ($p < 0.05$). RS_{UN}: resistant starch isolated from untreated RB; RS_{40C}, RS_{50C}, RS_{60C}, and RS_{70C}: resistant starches isolated from RB sonicated at 40–70 °C.

Jiranuntakul et al. [38] also reported that the crystalline kind of starch in rice with an A-type diffraction design remains unaltered after heat-moisture treatment, with an average crystallite size varying from 4 to 20 nm.

Reduced crystallinity was reported for US treated waxy corn starch [17], rice starch [18], potato starch [39], and oat starch [40], and Noor et al. [33] also reported a decreased crystallinity and increased amorphous character of RS from lotus stem after 35 min of sonication (20 kHz, 100–400 W) under cold conditions. According to BeMiller and Huber [41], changes in the amorphous and crystalline regions, breaking, and destroying the double-helical order during mechanical treatment of starch were detected.

3.6. Chemical Structure Characterization

The FT-IR spectra of the rice starches are displayed in Figure 5. The chemical structure of the starch granules was analyzed after ultrasonication for 20 min at temperatures of 40–70 °C.

The intensity of absorption at 995 cm^{-1} (C–O stretching) and 1149 cm^{-1} (peak E; C–O–C asymmetric stretching), and the intensity of the peak at 1336 cm^{-1}, corresponding to C–H symmetric bending [37], was relatively stronger for the native rice starch and RS_{UN} than for the ultrasonicated RS. The RS samples showed C–H stretching at 2926 cm^{-1} (peak A2) and at 1646 cm^{-1} (peak C; C–O bending associated with –OH group) that may be due to the picking of amylose helices during the combined US and thermal treatment.

The stretching at 1514 cm^{-1} (peak D) could be assigned to the asymmetrical stretching of the carboxylate group (–COO). In addition, the bonding to the carbonyl group (C=O) displayed an absorption peak at the 1720 cm^{-1} (peak B) region as a result of a possible polysaccharide oxidation process, which was more noticeable after US treatment. The bands at 999 cm^{-1} (peak H) and 1076 cm^{-1} (peak F) characterize the crystalline structure, and the band at 1016 cm^{-1} (peak G) was attributed to the amorphous structure of the starch [33].

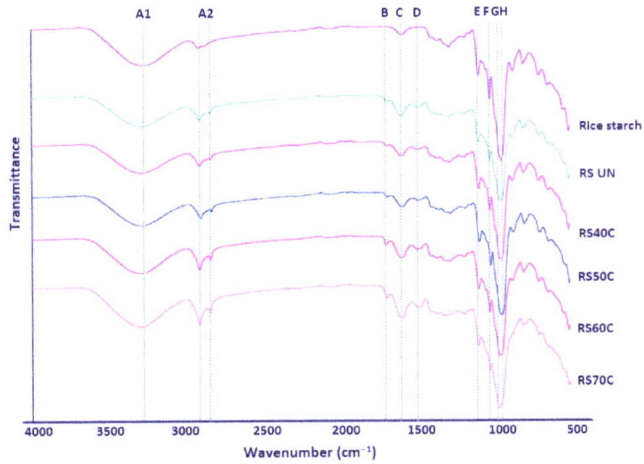

Figure 5. FT-IR spectra of native rice starch and resistant starches (RS) isolated from rice bran (RB) that was untreated (UN) and ultrasonicated (1.8 W/cm^2, 20 min) at different temperatures.

The observed differences in the FT-IR spectra of the RS samples may indicate a greater structural disorganization of the starch macromolecules, and the possible conformational and chemical changes (an increase in amylose content, hydrolysis, or oxidation) as a result of the action of the physical treatment and temperature [42]. The peaks H and F of the native rice starch exhibited the highest absorption, while in the case of the RS samples, the corresponding peaks were of lower intensity. The intensities of the latter bands of the US-treated RS decreased with increasing sonication temperature. The maximum intensity at peak G was reached for the RS samples ultrasonicated at temperatures of 60 and 70 °C, signifying a decreased crystallinity and increased amorphous character.

The results were similar to those reported by Young et al. [18], who reported the stretching of C–OH, C–C, and C–H groups in the case of rice starch modified by US (22 kHz, 25 °C, 20 min), and also with Ma et al. [37], who reported similar alteration of the crystalline and amorphous areas of RS isolated from Laird lentils by thermal treatment. Noor et al. [33] showed a decrease in absorbance at 995 and 1047 cm^{-1}, indicating a decreased crystallinity of RS from lotus stem due to US exposure.

In the case of amylose, the RS samples isolated from the US-treated RB had significantly higher ($p < 0.05$) contents of amylose (37.20–53.10%) than the untreated RS (Table 2). An increase of 46.2% of the amylose content was already determined when US was applied for 20 min at 40 °C; an average two-fold increase was observed with increasing sonication temperature at 50–60 °C compared to the RS_{UN} sample (25.62%).

The present results were in line with the study of Falsafi et al. [40], who reported that sonication increased the content of apparent amylose in oat starch, suggesting the depolymerization of amylopectin. US cavitation increased the amylose content in potato, mung bean, sago, and maize starches, indicating the decomposition of the molecular structures and formation of linear amylose fragments [43].

3.7. Pasting Properties

The pasting curve profiles of the US-treated RS samples were similar to that of the untreated RS (Figure 6), indicating that sonication had no significant influence on the profile shape, compatible with the studies by Yang et al. [18] and Bian and Chung [44].

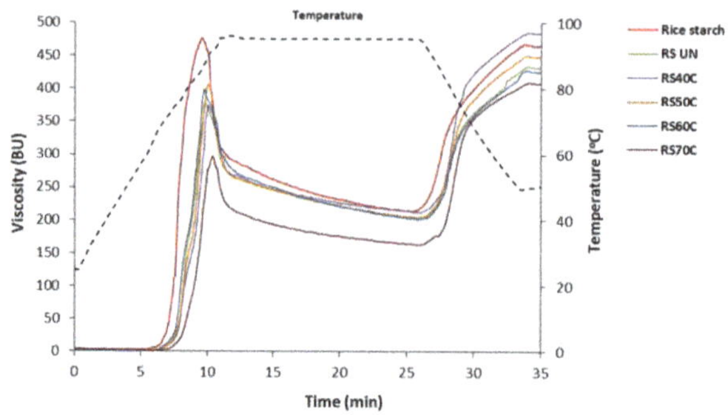

Figure 6. Pasting curves of native rice starch and resistant starches (RS) isolated from rice bran that was untreated (UN) and ultrasonicated (1.8 W/cm^2, 20 min) at different temperatures. BU: Brabender units.

In the case of the pasting properties (Table 3), the native rice starch had the lowest pasting temperature (67.90 °C) and the highest peak viscosity (474.02 BU), while the RS_{UN} showed a significantly higher ($p < 0.05$) pasting temperature (72.95 °C) and lower peak viscosity (370.63 BU) due to the decrease in the crystalline/amorphous ratio and the double helix content of starch that occurred during the milling of the rice grain [45]. The specific viscosity (η_s) of the RS slurry increased with a 40 °C US temperature and decreased at temperatures between 50 and 70 °C.

Table 3. Pasting properties and specific viscosity (η_s) of commercial rice starch and resistant starches (RS) isolated from untreated and sonicated rice bran.

Starch Samples	Pasting Temp. (°C)	Peak Viscosity [a]	Final Viscosity [a]	ηs(mPa·s)	LGC (%)
Rice starch	67.90 ^{d}	474.02 ^{a}	467.27 ^{a}	4.38 ^{b}	8
RS_{UN}	72.95 ^{b}	382.14 ^{bc}	428.06 ^{c}	4.27 ^{b}	14
RS_{40C}	73.47 ^{b}	370.63 ^{c}	485.34 ^{a}	5.78 ^{a}	12
RS_{50C}	73.93 ^{b}	403.74 ^{b}	451.62 ^{ab}	3.62 ^{c}	10
RS_{60C}	71.82 ^{bc}	398.32 ^{b}	426.18 ^{c}	3.09 ^{d}	10
RS_{70C}	75.07 ^{a}	298.51 ^{d}	408.37 ^{d}	2.86 ^{e}	8

Data with different superscript letters within the column represent significant differences ($p < 0.05$). RS_{UN}: resistant starch isolated from untreated RB; RS_{40C}, RS_{50C}, RS_{60C}, RS_{70C}: resistant starches isolated from RB sonicated for 20 min at 40–70 °C temperatures; [a] BU: Brabender units.

Ultrasonication at 40 °C slightly increased the pasting temperature of the RS_{40C} paste and reduced the peak viscosity compared to the RS_{UN} paste. The final viscosity value of the RS increased significantly at US temperatures of 40–50 °C but decreased compared to the untreated RS due to the disruption and dissolution of the starch granules. This indicates that US-assisted treatment increased the mechanical and thermal stability of the rice RS pastes and caused a higher tendency of the RS granules to retrograde.

With temperatures of 50–60 °C, the RS pastes exhibited a slightly higher peak viscosity ($p < 0.05$) compared to the RS_{40C} with no significant effect ($p \geq 0.05$) on pasting temperature (Table 3). The RS_{70C} that was isolated after sonication at 70 °C had the highest pasting temperature (75.07 °C) but a strongly reduced peak viscosity (298.51 BU), indicating that the starch granules swelled slowly.

This can be explained by the induced intermolecular interactions, making this starch more resistant to thermal treatment. The percentage content of amylose increased, possibly due to the destruction of starch polysaccharides. The branched structure of amylopectin macromolecules were broken down, depending on the ultrasound intensity, while amylose formed a more compact structure with greater intermolecular interactions, i.e., additional hydrogen bonds due to the increased content of carbonyl and carboxylate groups (Figure 5), increasing the gelatinization temperature of the starch granules. At high US temperatures, the maximum viscosity of potentially damaged macromolecules decreased, which characterizes the interaction of maximally swollen particles (the smaller the particles, the lower the viscosity). In this case, the final viscosity was also lower due to the destruction of the starch polysaccharides and smaller particles, as was confirmed by the higher amylose content and the additional B and D peaks appearing in the FT-IR spectrum. During gelatinization, all starch became amorphous, so there was no significant difference in the profile of the pasting curves. The profile depended more on the shape, size, intermolecular interactions (chemical composition), and molecular weight of the starch granules.

The obtained results agreed with other studies reporting that US treatment reduces the peak viscosity of rice and waxy corn starches [17,34]. The physical damage of starch granules reduces the paste viscosity due to increased water penetration and hydration [46]. For starch, mechanical pre-treatment reduces the granule size and increases the content of the amorphous phase, making it possible to alter the hydrating and paste-forming properties of modified starch [47]. Additionally, sonication, initiating the breakage of hydrogen bonds, thereby reducing the interaction between the starch granules, leads to a viscosity reduction, as could be explained by the profiles of the pasting curves [20].

3.8. Technological Properties

The technological properties of the RS isolated from the untreated and US-treated RB material are presented in Table 4. The results showed that sonication significantly ($p < 0.05$) improved the oil (OAC) and water (WAC) absorption capacities, swelling power (SP), solubility (WS), and reduced the least-gelation concentration (LGC) of the RS compared to the untreated sample.

Table 4. Amylose content and technological and mechanical properties of rice starch and resistant starch (RS) isolated from untreated and sonicated rice bran.

Sample	WAC (g/g)	OAC (g/g)	SP (%)	WS (%)
Rice starch	4.30 ^{e}	2.37 ^{d}	5.22 ^{e}	3.64 ^{e}
RS_{UN}	4.58 ^{d}	2.70 ^{c}	5.78 ^{d}	4.81 ^{d}
RS_{40C}	5.91 ^{b}	3.08 ^{b}	6.17 ^{c}	6.06 ^{c}
RS_{50C}	5.59 ^{c}	3.15 ^{ab}	6.62 ^{b}	6.67 ^{b}
RS_{60C}	6.17 ^{b}	3.27 ^{a}	9.32 ^{a}	16.45 ^{a}
RS_{70C}	7.54 ^{a}	3.49 ^{a}	9.58 ^{a}	17.01 ^{a}

Data with different superscript letters within the column represent significant differences ($p < 0.05$). Samples: RS_{UN}: resistant starch from untreated RB; RS_{40C}, RS_{50C}, RS_{60C}, RS_{70C}: resistant starch isolated from RB sonicated at an appropriate temperature and time. WAC: water absorption capacity; OAC: oil absorption capacity; SP: swelling power; WS: water solubility; LGC: least gelation concentration.

Among the samples, the sonicated RS absorbed the highest amounts of water and oil (5.91–7.54 g/g and 3.08–3.49 g/g, respectively) compared to the untreated RS (4.58 and 2.70 g/g, respectively) and native rice starch (4.30 and 2.37 g/g, respectively). The SP and WS of the sonicated RS samples increased significantly ($p < 0.05$) with increasing US temperature compared to the RS_{UN}.

The LGC can be related to the particle size; native rice starch granules form the firmest gel (after retrogradation) at the lowest concentration, while for RS_{70C}, the lowest LGC can be explained by the involvement of amylose macromolecules in retrogradation, possibly strengthening RS gels [48]. The native rice starch contained a higher proportion of long-chain amylose, which is less soluble (Table 4).

The increased solubility of the RS samples isolated from the US-treated RB could be explained by thermal/ultrasound disruption, inducing depolymerization and degradation of amylose and amylopectin [48], thus leading to fragments of shorter branched chains, a rise in amylose content, and damage generation on the surface of the starch particles, as was confirmed by SEM.

According to the literature, US cavitation can initiate the degradation of chains between amylopectin molecules, contributing to strengthened linear fractions that can influence the increase in swelling capacity and solubility of starch granules [38]. When the US treatment was performed longer or at higher temperatures, the internal ordered structure of the starch granules changed to a disintegrated form owing to the absorption of water by amylopectin. The degree of swelling is an indicative characteristic of amylopectin, and the ultrasound temperature can lead to a steady loss of crystallinity in amylopectin molecules, thus reducing the viscosity. According to Wang et al. [49], a loss of crystallinity in amylopectin was a result of the US-induced disintegration of potato starch granules. The study by Ding et al. [50] showed that the swelling and solubility of retrograded corn starch after sonication was also improved in comparison to native starch, indicating that the crystalline structure was damaged due to the changes in the interaction between chains within the crystalline and amorphous regions of the starch granule.

4. Conclusions

This study showed that US treatment induced the structural disorganization of rice bran resistant starch, changing its physicochemical characteristics and influencing its mechanical preferences and hydration properties. Modification by US processing achieved functional properties not found in untreated starches, which may have specific applications in the food industry. US treatment at different temperatures (40–70 °C) affected the chemical structure, reduced the crystallinity of the RS from 23.85% to between 18.37% and 4.43%, and increased the mechanical and thermal stability of the RS pastes, indicating a higher tendency of the RS granules to retrograde. The pasting results clearly showed that the most viscous paste formed in the case of the RS_{40C} sample, indicating a strong interaction between the colloidal particles and soluble polysaccharide molecules. This finding could partly explain the enhanced content of such starch and its resistance to digestive enzymes. The US processing significantly ($p < 0.05$) improved the oil and water absorption capacities, swelling power, solubility, and gelation properties.

So far, there has been no report on the improvement of RS content through US treatment of rice-milling by-products. The data evaluation indicated a significant effect ($p < 0.05$) of three studied factors (US temperature, time, and power). The regression analysis confirmed that 97% of the data could be described by the obtained mathematical model. Through the optimization using the RSM, optimal conditions for rice bran US treatment were obtained: a time of 18 min, a temperature of 40.2 °C, and a US power of 1.8 W/cm^2; under optimized conditions, the predicted RS increment was 13.46 g/100 g d.w.

The technological concept adapted in this study differs from the idea of using pure starch as the raw material to increase the RS content. The raw material used in this study was rice bran, which excludes the difficulty of drying the slurry of starch in the following technological stages. US treatment, resulting in a significant increment of resistant starch in the rice bran matrix, thereby increases its potential to be used as a functional bio-based component for food.

Supplementary Materials: The following supporting information can be downloaded at: https://www.mdpi.com/article/10.3390/polym14173662/s1, S.1: Chemical analysis; S.2. Determination of technological properties; Figure S1: Rice bran starch and resistant starch isolation and measurement scheme; Table S1: Multiple comparisons between means of different US (1.3 W/cm^2) temperature groups (significant at $p < 0.05$); Table S2: Multiple comparisons between means of different US (1.3 W/cm^2) time groups (significant at $p < 0.05$); Table S3: Significant coefficients of quartic model equation in terms of coded factors; Table S4: Analysis of variance of the regression parameters for a quartic model for the response factor. References [51–53] are cited in the supplementary materials.

Author Contributions: Methodology, formal analysis, validation, writing—original draft, writing—review & editing, R.V.; Methodology, data curation, writing—review & editing, J.B.; Methodology, formal analysis, software, M.S.; Data curation, writing—review & editing, R.D.; Conceptualization, resources, data curation, writing—review & editing, supervision, project administration, D.Z. All authors have read and agreed to the published version of the manuscript.

Funding: This research was funded by the European Regional Development Fund under a grant agreement No. 01.2.2-LMT-K-718-01-0062 with the Research Council of Lithuania (LMTLT).

Institutional Review Board Statement: Not applicable.

Informed Consent Statement: Not applicable.

Data Availability Statement: Not applicable.

Conflicts of Interest: The authors have no conflict of interest to declare.

References

1. Englyst, H.N.; Kingman, S.M.; Cummings, J.H. Classification and measurement of nutritionally important starch fractions. *Eur. J. Clin. Nutr.* **1992**, *46*, S33–S50. [PubMed]
2. Raigond, P.; Ezekiel, R.; Raigond, B. Resistant starch in food: A review. *J. Sci. Food Agric.* **2015**, *95*, 1968–1978. [CrossRef] [PubMed]
3. Mohamed, I.O. Effects of processing and additives on starch physicochemical and digestibility properties. *Carbohydr. Polym. Technol. Appl.* **2021**, *2*, 100039. [CrossRef]
4. Chang, Q.; Zheng, B.; Zhang, Y.; Zeng, H. A comprehensive review of the factors influencing the formation of retrograded starch. *Int. J. Biolog. Macromol.* **2021**, *186*, 163–173. [CrossRef] [PubMed]
5. Patel, H.; Royall, P.G.; Gaisford, S.; Williams, G.R.; Edwards, C.H.; Warren, F.J.; Flanagan, B.M.; Ellis, P.R.; Butterworth, P.J. Structural and enzyme kinetic studies of retrograded starch: Inhibition of α-amylase and consequences for intestinal digestion of starch. *Carbohydr Polym.* **2017**, *164*, 154–161. [CrossRef] [PubMed]
6. Zhou, Y.; Meng, S.; Chen, D.; Zhu, X.; Yuan, H. Structure characterization and hypoglycaemic effects of dual modified resistant starch from indica rice starch. *Carbohydr. Polym.* **2014**, *103*, 81–86. [CrossRef] [PubMed]
7. Homayouni, A.; Amini, A.; Khodavirdiv, A.K.; Mortazavian, A.M.; Esazadeh, K.; Pourmoradian, S. Resistant starch in food industry: A changing outlook for consumer and producer. *Starch-Stärke* **2013**, *65*, 102–114. [CrossRef]
8. Zaman, S.A.; Sarbini, S.R. The potential of resistant starch as a prebiotic. *Crit. Rev. Biotechnol.* **2016**, *36*, 578–584. [CrossRef]
9. Thachil, M.T.; Chouksey, M.K.; Gudipati, V. Amylose-lipid complex formation during extrusion cooking: Effect of added lipid type and amylose level on corn-based puffed snacks. *Int. J. Food Sci. Technol.* **2014**, *49*, 309–316. [CrossRef]
10. Ashwar, B.A.; Gani, A.; Wani, I.A.; Shah, A.; Masoodi, F.A.; Saxena, D.C. Production of resistant starch from rice by dual autoclaving-retrogradation treatment: In vitro digestibility, thermal and structural characterization. *Food Hydrocoll.* **2016**, *56*, 108–117. [CrossRef]
11. Zheng, Y.; Wei, Z.; Zhang, R.; Deng, Y.; Tang, X.; Zhang, Y.; Liu, G.; Liu, L.; Wang, J.; Liao, N.; et al. Optimization of the autoclave preparation process for improving resistant starch content in rice grains. *Food Sci. Nutr.* **2020**, *8*, 2383–2394. [CrossRef] [PubMed]
12. Cabrera-Ramírez, A.H.; Cervantes-Ramírez, E.; Morales-Sánchez, E.; Rodriguez-García, M.E.; de la Luz Reyes-Vega, M.; Gaytán-Martínez, M. Effect of extrusion on the crystalline structure of starch during RS5 formation. *Polysaccharides* **2021**, *2*, 187–201. [CrossRef]
13. Hung, P.V.; Ngo, L.V.; Nguyen, T.L.P. Resistant starch improvement of rice starches under a combination of acid and heat-moisture treatments. *Food Chem.* **2016**, *191*, 67–73. [CrossRef] [PubMed]
14. Maniglia, B.C.; Castanha, N.; Rojas, M.L.; Augusto, P.E.D. Emerging technologies to enhance starch performance. *Curr. Opin. Food Sci.* **2021**, *37*, 26–36. [CrossRef]
15. Patti, A.; Acierno, D.; Latteri, A.; Tosto, C.; Pergolizzi, E.; Recca, G.; Cristaudo, M.; Cicala, G. Influence of the processing conditions on the mechanical performance of sustainable bio-based PLA compounds. *Polymers* **2020**, *12*, 2197. [CrossRef]
16. Amini, A.M.; Razavi, S.M.A.; Mortazavi, S.A. Morphological, physicochemical, and viscoelastic properties of sonicated corn starch. *Carbohydr. Polym.* **2015**, *122*, 282–292. [CrossRef]
17. Sujka, M. Ultrasonic modification of starch-Impact on granules porosity. *Ultrason. Sonochem.* **2017**, *37*, 424–429. [CrossRef]
18. Yang, Q.-Y.; Lu, X.-X.; Chen, Y.-Z.; Luo, Z.-G.; Xiao, Z.-G. Fine structure, crystalline and physicochemical properties of waxy corn starch treated by ultrasound irradiation. *Ultrason. Sonochem.* **2019**, *51*, 350–358. [CrossRef]
19. Yang, W.; Kong, X.; Zheng, Y.; Sun, W.; Chen, S.; Liu, D.; Zhang, H.; Fang, H.; Tian, J.; Ye, X. Controlled ultrasound treatments modify the morphology and physical properties of rice starch rather than the fine structure. *Ultrason. Sonochem.* **2019**, *59*, 104709. [CrossRef]
20. Wang, H.; Xu, K.; Ma, Y.; Liang, Y.; Zhang, H.; Chen, L. Impact of ultrasonication on the aggregation structure and physicochemical characteristics of sweet potato starch. *Ultrason. Sonochem.* **2020**, *63*, 104868. [CrossRef]
21. Bonto, A.P.; Tiozon, R.N., Jr.; Sreenivasulu, N.; Camacho, D.H. Impact of ultrasonic treatment on rice starch and grain functional properties: A review. *Ultrason. Sonochem.* **2021**, *71*, 105383. [CrossRef] [PubMed]
22. Okonkwo, V.C.; Kwofie, E.M.; Mba, O.I.; Ngadi, M.O. Impact of thermo-sonication on quality indices of starch-based sauces. *Ultrason. Sonochem.* **2021**, *73*, 105473. [CrossRef] [PubMed]
23. Faisal, M.; Kou, T.; Zhong, Y.; Blennow, A. High amylose-based bio composites: Structures, functions and applications. *Polymers* **2022**, *14*, 1235. [CrossRef] [PubMed]
24. Wang, L.; Bai, X. The producing technology of resistant starch from buckwheat using ultrasonic treatment. *Agrotechnology* **2017**, *6*, 2.
25. Frost, K.; Kaminski, D.; Kirwan, G.; Lascaris, E.; Shanks, R. Crystallinity and structure of starch using wide angle X-ray scattering. *Carbohydr. Polym.* **2009**, *78*, 543–548. [CrossRef]

26. Abbas, K.A.; Abdulkarim, S.M.; Saleh, A.M.; Ebrahimian, M. Suitability of viscosity measurement methods for liquid food variety and applicability in food industry-a review. *J. Food Agric. Environ.* **2010**, *8*, 100–107.
27. Sathe, S.K.; Salunkhe, D.K. Isolation, partial characterization and modification of the great northern bean (*Phaseolus vulgaris* L.) starch. *J. Food Sci.* **1981**, *46*, 617–621. [CrossRef]
28. Khuri, A.I.; Mukhopadhyay, S. Response surface methodology. Wiley Interdiscip. *Rev. Comput. Stat.* **2010**, *2*, 128–149.
29. Benmoussa, M.; Hamaker, B.R. Rapid small scale starch isolation using a combination of ultrasonic sonication and sucrose density separation. *Starch-Stärke* **2011**, *63*, 333–339. [CrossRef]
30. Ding, Y.; Luo, F.; Lin, Q. Insights into the relations between the molecular structures and digestion properties of retrograded starch after ultrasonic treatment. *Food Chem.* **2019**, *294*, 248–259. [CrossRef]
31. Villas-Boas, F.; Facchinatto, W.M.; Colnago, L.A.; Volanti, D.P.; Franco, C.M.L. Effect of amylolysis on the formation, the molecular, crystalline and thermal characteristics and the digestibility of retrograded starches. *Int. J. Biol. Macromol.* **2020**, *163*, 1333–1343. [CrossRef] [PubMed]
32. You, Q.; Zhang, X.; Fang, X.; Yin, X.; Luo, C.; Wan, M. Ultrasonic-assisted preparation and characterization of RS3 from pea starch. *Food Bioprocess Technol.* **2019**, *12*, 1244–1249. [CrossRef]
33. Noor, N.; Gani, A.; Jhan, F.; Jenno, J.L.H.; Da, M.A. Resistant starch type 2 from lotus stem: Ultrasonic effect on physical and nutraceutical properties. *Ultrason. Sonochem.* **2021**, *76*, 105655. [CrossRef] [PubMed]
34. Park, D.J.; Han, J.A. Quality controlling of brown rice by ultrasound treatment and its effect on isolated starch. *Carbohydr. Polym.* **2016**, *10*, 30–38. [CrossRef] [PubMed]
35. Zuo, J.Y.; Knoerzer, K.; Mawson, R.; Kentish, S.; Ashokkumarm, M. The pasting properties of sonicated waxy rice starch suspensions. *Ultrason. Sonochem.* **2009**, *16*, 462–468. [CrossRef] [PubMed]
36. Man, J.; Yang, Y.; Zhang, C.; Zhang, F.; Wang, Y.; Gu, M.; Liu, Q.; Wei, C. Morphology and structural characterization of high-amylose rice starch residues hydrolyzed by porcine pancreatic α-amylase. *Food Hydrocoll.* **2013**, *31*, 195–203. [CrossRef]
37. Ma, Z.; Yin, X.; Hu, X.; Li, X.; Liu, L.; Joyce, I.; Boye, J.I. Structural characterization of resistant starch isolated from Laird lentils (*Lens culinaris*) seeds subjected to different processing treatments. *Food Chem.* **2018**, *263*, 163–170. [CrossRef]
38. Jiranuntakul, W.; Puttanlek, C.; Rungsardthong, V.; Puncha-Arnon, S.; Uttapap, D. Microstructural and physicochemical properties of heat-moisture treated waxy and normal starches. *J. Food Eng.* **2011**, *104*, 246–258. [CrossRef]
39. Nie, H.; Li, C.; Liu, P.H.; Lei, C.Y.; Li, J.B. Retrogradation, gel texture properties, intrinsic viscosity and degradation mechanism of potato starch paste under ultrasonic irradiation. *Food Hydrocoll.* **2019**, *95*, 590–600. [CrossRef]
40. Falsafi, S.R.; Maghsoudlou, Y.; Rostamabadi, H.; Rostamabadi, M.M.; Hamedi, H.; Hosseini, S.M.H. Preparation of physically modified oat starch with different sonication treatments. *Food Hydrocoll.* **2019**, *89*, 311–320. [CrossRef]
41. BeMiller, J.N.; Huber, K.C. Nonchemical modification of food starch functionalities. *Ann. Rev. Food Sci. Technol.* **2015**, *6*, 19–69. [CrossRef] [PubMed]
42. Nawaz, H.; Waheed, R.; Nawaz, M.; Shahwar, D. Physical and chemical modifications in starch structure and reactivity. In *Chemical Properties of Starch*; Emeje, M., Ed.; IntechOpen: London, UK, 2020; pp. 1–21.
43. Chan, H.-T.; Bhat, R.; Karim, A.A. Effects of sodium dodecyl sulphate and sonication treatment on physicochemical properties of starch. *Food Chem.* **2010**, *120*, 703–709. [CrossRef]
44. Bian, L.; Chung, H.-J. Molecular structure and physicochemical properties of starch isolated from hydrothermally treated brown rice flour. *Food Hydrocoll.* **2016**, *60*, 345–652. [CrossRef]
45. Liu, T.Y.; Ma, Y.; Yu, S.F.; Shi, J.; Xue, S. The effect of ball milling treatment on structure and porosity of maize starch granule. *Inn. Food Sci. Emer. Technol.* **2011**, *12*, 586–593. [CrossRef]
46. Cui, R.; Zhu, F. Effect of ultrasound on structural and physicochemical properties of sweet potato and wheat flours. *Ultrason. Sonochem.* **2020**, *66*, 105118. [CrossRef] [PubMed]
47. Dome, K.; Podgorbunskikh, E.; Bychkov, A.; Lomovsky, O. Changes in the crystallinity degree of starch having different types of crystal structure after mechanical pretreatment. *Polymers* **2020**, *12*, 641. [CrossRef]
48. Vela, A.J.; Villanueva, M.; Solaesa, Á.G.; Ronda, R. Impact of high-intensity ultrasound waves on structural, functional, thermal and rheological properties of rice flour and its biopolymers structural features. *Food Hydrocoll.* **2021**, *11*, 106480. [CrossRef]
49. Wei, B.; Pu, Q.; Wu, C.; Zou, J.; Shi, K.; Xu, B.; Ma, Y. The temperature induced disruption process of amylopectin under ultrasonic treatments. *Int. J. Biol. Macromol.* **2022**, *205*, 297–303. [CrossRef]
50. Ding, Y.; Xiao, Y.; Ouyang, Q.; Luo, F.; Lin, Q. Modulating the in vitro digestibility of chemically modified starch ingredient by a non-thermal processing technology of ultrasonic treatment. *Ultrason. Sonochem.* **2021**, *70*, 105350. [CrossRef]
51. The Association AACC. *Approved Methods of the AACC International*, 10th ed.; The Association AACC: Saint Paul, MN, USA, 2000.
52. Ahmad, M.; Gani, A.; Masoodi, F.A.; Rizvi, S.H. Influence of ball milling on the production of starch nanoparticles and its effect on structural, thermal, and functional properties. *Int. J. Biol. Macromol.* **2020**, *151*, 85–91. [CrossRef]
53. Li, A.; Gao, Q.; Ward, R. Physicochemical properties and in vitro digestibility of resistant starch from mung bean (*Phaseolus radiatus*). *Starch-Stärke* **2011**, *63*, 171–178. [CrossRef]

polymers MDPI

Article

Characterization of Polyhydroxybutyrate-Based Composites Prepared by Injection Molding

Marcos M. Hernandez, Nevin S. Gupta, Kwan-Soo Lee, Aaron C. Pital, Babetta L. Marrone, Carl N. Iverson and Joseph H. Dumont *

C-CDE Chemical Diagnostics and Engineering, Los Alamos National Laboratory, Los Alamos, NM 87545, USA; mhernandez@lanl.gov (M.M.H.); nevin5@lanl.gov (N.S.G.); kslee@lanl.gov (K.-S.L.); acpital@lanl.gov (A.C.P.); blm@lanl.gov (B.L.M.); iverson@lanl.gov (C.N.I.)

* Correspondence: joseph.dumont@lanl.gov

Abstract: The waste generated by single-use plastics is often non-recyclable and non-biodegradable, inevitably ending up in our landfills, ecosystems, and food chain. Through the introduction of biodegradable polymers as substitutes for common plastics, we can decrease our impact on the planet. In this study, we evaluate the changes in mechanical and thermal properties of polyhydroxybutyrate-based composites with various additives: Microspheres, carbon fibers or polyethylene glycol (2000, 10,000, and 20,000 MW). The mixtures were injection molded using an in-house mold attached to a commercial extruder. The resulting samples were characterized using microscopy and a series of spectroscopic, thermal, and mechanical techniques. We have shown that the addition of carbon fibers and microspheres had minimal impact on thermal stability, whereas polyethylene glycol showed slight improvements at higher molecular weights. All of the composite samples showed a decrease in hardness and compressibility. The findings described in this study will improve our understanding of polyhydroxybutyrate-based composites prepared by injection molding, enabling advancements in integrating biodegradable plastics into everyday products.

Keywords: biodegradable plastics; polyhydroxybutyrate; polymer composites; compounding; injection molding

Citation: Hernandez, M.M.; Gupta, N.S.; Lee, K.-S.; Pital, A.C.; Marrone, B.L.; Iverson, C.N.; Dumont, J.H. Characterization of Polyhydroxybutyrate-Based Composites Prepared by Injection Molding. *Polymers* **2021**, *13*, 4444. https://doi.org/10.3390/polym13244444

Academic Editors: Domenico Acierno and Antonella Patti

Received: 2 November 2021
Accepted: 13 December 2021
Published: 18 December 2021

1. Introduction

Single use plastics (SUPs) are fossil fuel-based materials commonly used in the food and beverage industry, designed to be disposed of immediately after use. These plastics, often composed of polypropylene, polystyrene, or polyethylene are historically challenging to recycle, and current waste collection systems do not have the capacity to safely and effectively dispose of our recycle waste plastic on a global scale [1]. As a result, these SUPs that end up in landfills slowly make their way into our ecosystems, oceans, and our food chain, contributing to society's growing plastic pollution problems [2–5]. Current forecasts show that by 2050, only 26% of the 33 billion tons of plastics produced worldwide will have been recycled. To reduce the alarming pollution caused by our plastics consumption, the development and transition to truly biodegradable plastics is crucial.

Polyhydroxybutyrates (PHBs) are polyesters, which can be biologically produced by microorganisms, such as cyanobacteria or synthetically produced [6,7]. While these materials were first discovered in the 1920s, they only recently have gained interest due to their biodegradable nature [8]. In fact, recent studies have shown that PHB anaerobically degrades more rapidly at moderate temperatures than other biomaterials [9]. PHBs are used in industries, such as biomedicine and agriculture [10–12]. Although PHBs are extremely attractive for their degradability, they are also known for possessing limited mechanical properties due to the presence of large crystals in the material, making it unsuitable for applications, such as flexible packaging films [13]. Furthermore, the manufacture of these materials using industrial film/packaging equipment is rendered more difficult by the proximity between their degradation and melting temperatures [14–16]. While efforts have been made to predict the material thermal behavior and find PHBs suitable for a wide range of applications, more experimental work and modifications are needed to improve the performance of the materials [17,18].

Compounding, i.e., incorporation of additives into polymers, is essential to the manufacturing of polymer composites and the modification of their chemical, thermal, and mechanical properties. For instance, carbon fibers have been used to improve the thermomechanical and electrical properties of polymers, as well as to strengthen the material [19–22]. Microspheres are hollow fillers that are often used as a blowing agent to reduce the density of a variety of materials and improve their compressibility with applications ranging from footwear to the space industry [23–25]. Finally, polyethylene glycol (PEG) is a biodegradable polyether, used for its plasticizing properties to decrease polymer rigidity, which increases the materials process ability, as well as to provide lubricating coatings and decrease brittleness [26,27].

While carbon fibers, microspheres, and PEG in other polymers have been extensively described in the literature, there is little to no information on their effects when incorporated in PHB. In fact, the thermal and mechanical properties of PHB composites have not been well explored and this study is an attempt to fill the gap in the literature. Here, we compound these additives with PHBs, using a single-screw extruder retrofitted with an in-house injection mold. The injection-molded samples are an attempt to produce parts using a state-of-the-art industrial technique to modify the mechanical properties without altering the intrinsic nature of the materials. The samples were characterized using a wide array of microscopy, spectroscopic, thermal, and mechanical techniques, such as Fourier-Transform InfraRed (FT-IR) spectroscopy, thermogravimetric analysis (TGA), hardness (Shore A), and compression testing.

Improving the chemical, thermal, and mechanical properties of PHBs, while retaining their biodegradable nature is essential to transition from petroleum-based to biodegradable plastics and reduce the impact of plastic pollution around the world. This work will help identify the additives and processing conditions for PHB biopolymers and composites that will accelerate their wider adoption.

2. Materials and Methods

2.1. Materials

PHB granules (BU396312) were purchased from Goodfellow Cambridge Ltd. located in Cambridge, United Kingdom. Carbon fibers (Pyrograff PR-19-XT-PS) were purchased from Applied Sciences, United States. Expancel 930DU120 microspheres were purchased from Nouryon Company located in Bohus Ale, Sweden. The 2000 molecular weight (MW) PEG was purchased from Alfa Aesar, located in Tewksbury, Massachusetts, USA. Ten thousand and 20,000 MW PEG were purchased from Sigma-Aldrich, located in St. Louis, MO, USA. An EX2 model extruder was purchased from Filabot (Barre, VT, USA). All of the materials were used as received.

2.2. Methods

Injection Molding: The EX2 was preheated to 200 °C and fitted with an injection mold. The EX2 is a single-screw extruder that is equipped with a compact 3 stage extrusion screw. The stainless steel screw is proprietary to Filabot, and designed to pressurize the material during processing. The heating element in the extruder, allows for a relatively consistent temperature along the extrusion screw. A custom injection mold was designed and manufactured to produce a 6.35 mm thickness disk at 25.4 mm diameter (Figure 1). This geometry was selected for its repeatability and ease of use in mechanical testing. The custom mold was attached at the nozzle end of the extruder and allowed to equilibrate with the preheated extruder. For sample production, the rate of extrusion was approximately 75% (18 RPM) of the full capability.

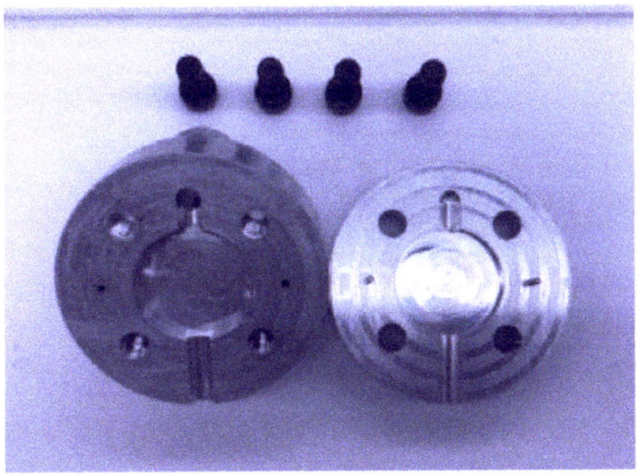

Figure 1. Custom split-die injection mold with a cavity size of 25.4 × 6.35 mm used to form samples for compression and hardness testing.

Samples were prepared by combining additives with PHB granules (*w*/*w*) in a zip lock bag and mixed by shaking. After the contents appeared to be well mixed, they were poured into the hopper, and extruded until the material exited the outlet of the injection mold. Then, the extruder was shut off, and the material was held at room temperature for 1 min in the mold. Thereafter, the mold was removed from the extruder and cooled in a water bath prior to sample removal.

The density was estimated by taking the sample mass (obtained by scale) and dividing it by the volume of the sample.

Hardness testing: A Shore A digital hardness probe and test stand manufactured by Zwick Roell Group (Ulm, Germany) was used to test the surface hardness of the discs. Each disc was probed at five different locations on the surface. The five probing sites were chosen uniformly across all of the samples as shown in Figure 2. Then, the collected values were averaged to find the hardness of each sample.

Figure 2. Shore A hardness probing sites along the sample surface.

Compression testing: A 3300 series single column testing system produced by Instron Co. (Norwood, MA USA) was used to test the compressibility of the discs. Instrom Co. compression platens (model 2501-083) were used as the compression attachment (refer to Figure 3. Samples were compressed to 1 MPa at a rate of 0.02 mm/s. Four compression cycles were completed, with the third cycle used to analyze the percent deformation due to the compression for a comparison between the formulations. This deformation is representative of the change in dimension of the sample thickness due to the compressibility of the material.

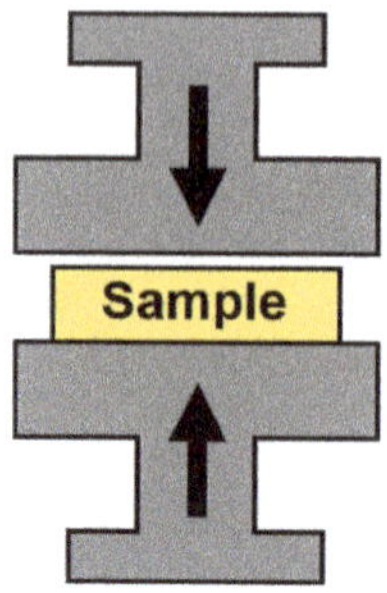

Figure 3. Compression testing configuration.

Fourier Transform Infrared (FT-IR) spectroscopy: A Nicolet iS50 FT-IR manufactured by Thermo Fisher Scientific (Waltham, MA USA), was used to analyze the chemical compounds present in the samples. Thirty-two scans were collected for each sample, with the background collected after every 120 min.

Thermogravimetric Analysis (TGA): A TGA 550 produced by TA Instruments (New Castle, DE, USA) was used to analyze the thermal degradation (T_d) of the samples. All of the tests were performed in a nitrogen filled test chamber. The following method was used: Equilibrate to 50 °C, heat 10 °C/min to 900 °C, end cycle. Trios software (TA instruments) was used to analyze the data. T_d was reported at the 5% wt loss across all of the samples.

Microscopy: A VHX 6000 digital microscope from KEYENCE (Osaka, Japan) was used to characterize the surface conditions of the samples. Images were taken at 20× and 100×, with full ring lighting. For microsphere composites, a partial ring was used to accent the pores present in the sample. Photographs and microscopy images were taken to evaluate the visual changes of the prepared samples and analyze changes in porosity, discoloration or surface imperfections. Cross sections were taken by cutting an outlet filament from the mold exit in half.

Scanning Electron Microscopy: Analysis of the sample morphology was performed using a FEI Quanta 200 Scanning Electron Microscope (SEM). PHB samples were sliced and mounted with carbon tape on standard $\frac{1}{2}$″ SEM posts. Samples were Au coated to an approximate thickness of (0.2 mA for 60 seconds), prior to testing to avoid charging on the surface, since the sample is non-conductive. The samples included a cross section, top, and bottom portion of the material. The micrographs were collected using the secondary electron imaging mode with an accelerating voltage of 5 kV, a spot size of 5.0, and a working distance of 10 mm. Images were processed in Python 3.9 via PIL and Matplotlib libraries to remove software artifacts and extract metadata from which the scale bars are produced.

3. Results and Discussion

The carbon fibers, microspheres, and PEG at concentrations ranging from 1–3%, 1–3%, and 5–25%, respectively, were separately combined with PHB granules in an extruder. The process involved the heating and compounding of the polymers and additives. Then, this mix was pushed through into the mold to form the test samples (Figure 4). The typical melting point for PHB is approximately 180 °C, and the initiation of decomposition is roughly above 220 °C [28]. Based on these properties, the material was processed at 200 °C to prevent degradation, while still having a fully melted polymer.

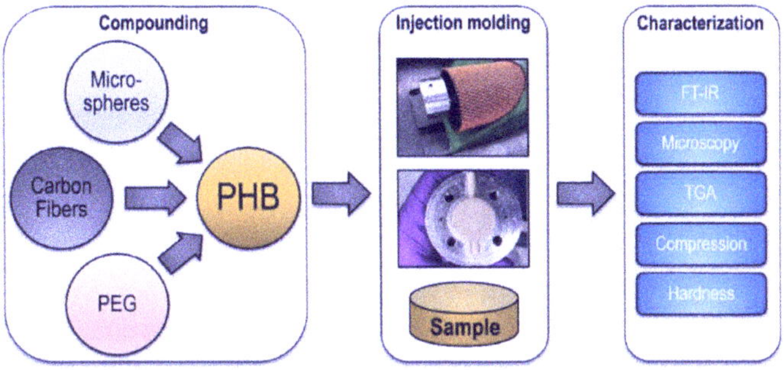

Figure 4. Sample process of preparing PHB based composites, through injection molding, to characterization.

3.1. Microscopy Analysis

Figure 5 shows the color variations across the samples. The microsphere/PHB sample remained visually similar to the control sample. In the PEG/PHB composites, darker areas of brown were seen along the surface, most noticeably for the 2000 MW sample. This discoloration, could have been the result of the degradation of PHB or PEG; or incomplete mixing between the polymer and additives [29]. Since the PHB was processed below the degradation temperature (T_d), the discoloration was likely due to the degradation of the PEG. In fact, previous studies have shown that PEG can start degrading at temperatures as low as 45 °C at relatively low MW values (6000) [30]. The PEG/PHB sample prepared with 20,000 MW PEG had a similar appearance with respect to the control sample. As expected, the carbon fiber/PHB composite showed a significant change in color from tan to black. Additional microscopy photographs can be found in Figure S1.

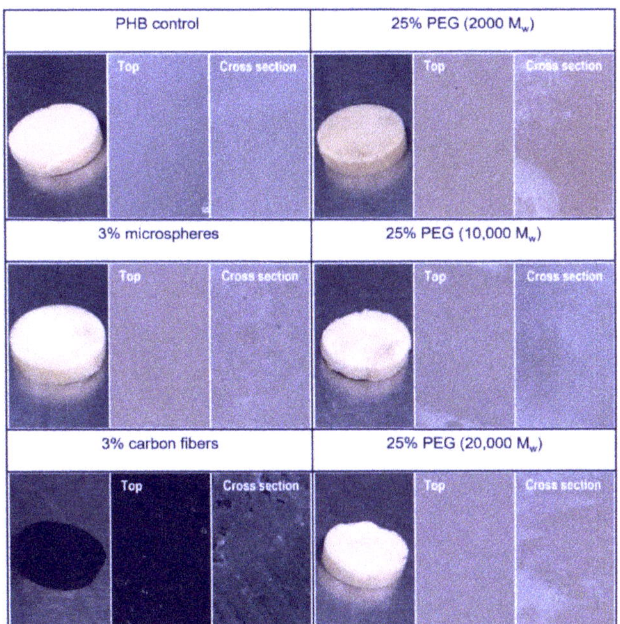

Figure 5. PHB composite sample surfaces photographed at 1×, 20×, and 100× for the analysis of surface changes. The 20× and 100× photographs were taken under full ring lighting conditions.

High-resolution confocal microscopy photographs highlight a fairly low number of distinguishable features from 500× to 2000×, as shown in Figure S2. The samples appear uniform, and no visible defect can be highlighted. The sample with 2000 MW PEG, as expected, shows a slight discoloration. The sample with PHB does show some porosity. However, at these resolutions, we can make no assessment of the distribution. High-definition images of the latter show that the pores are evenly distributed throughout the sample, as shown in Figure S3. Finally, SEM images of the same samples show samples that are uniform with no significant defects or multiphases to be distinguished, aside from the sample with microspheres that highlight well-dispersed pores ranging from a few microns to slightly over 100 μm, as shown in Figure S4.

Using confocal microscopy and the measurement tool, it was possible to see the porosity change in the material from the expansion of the microspheres, as shown in Figure 6. As the spheres underwent processing in the extruder at high temperature, the thermal expansion of the microspheres was initiated. Based on the manufacturer specification, the microspheres reached approximately 120 μm in its expanded form. From the measurement of the pores, it was found that some of the microspheres expanded from 25 μm to approximately 134 μm. The remaining unexpanded spheres, were likely due to the inadequate time under room temperature during processing. Despite only a fraction of spheres reaching full expansion, the density of the samples was decreased.

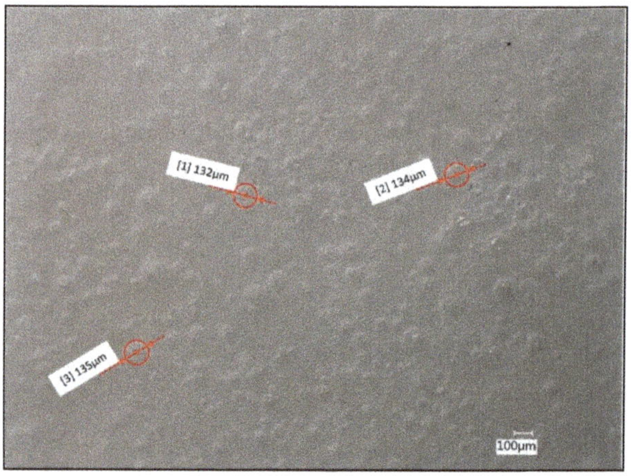

Figure 6. PHB/microsphere (3%) composite sample surface with measured pore diameters of fully formed microspheres. Image taken at 100× under partial-ring lighting conditions.

We were able to determine a rough estimate of the density of our samples given the dimensions of the discs and their weight, as shown in Table 1. The use of microspheres was aimed at reducing the density of our material, offering a lighter material. As we increased the concentration of hollow fillers, the density decreased, similar to previous studies [31]. We noted that while the density did decrease, it was by much less than expected with a reduction of only 11.5%. This was possibly due to the increased pressure from the geometry of the die, as well as an incomplete expansion of the microspheres.

Table 1. Density of PHB/microsphere composites.

Name	Density (g/cm^2)
PHB (control)	1.05
1% Microspheres	1.04
2% Microspheres	0.95
3% Microspheres	0.94

3.2. IR Spectroscopy Analysis

Infrared spectra of the PHB control and composites were collected, as shown in Figure 7. All of the spectra showed a strong absorption band at the 1720 and 1280 cm^{-1} peaks, corresponding to a C=O carbonyl group and C-O from the ester group, respectively, and are commonly observed when studying PHB [32]. The consistent presence of these peaks and similar peak ratios across the samples suggests that the PHB was mostly unchanged during the extrusion process. This absence of degradation phenomena is further confirmed, when compared against the IR spectra of the unprocessed PHB polymer (shown in Figure S6). The 2930 cm^{-1} peak corresponds to the alkane groups present in the PEG. Similarly, PHB has a presence of alkane groups at this peak. As a result, the groups for the two molecules overlap causing the peak to broaden [33]. Additional IR spectra that highlight different concentrations of additives can be found in Figure S5.

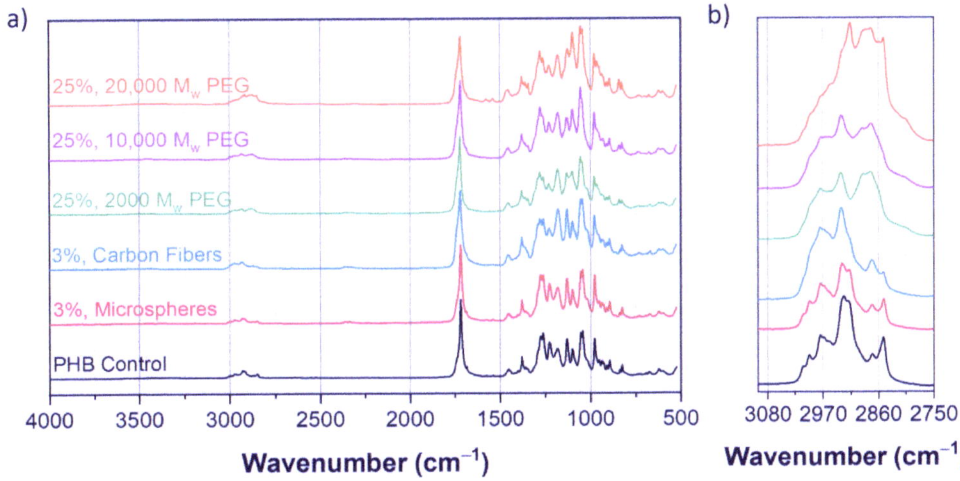

Figure 7. (**a**) IR spectra of PHB composites in the 500 to 4000 cm^{-1} frequency range. Absorption bands at 1280 and 1720 cm^{-1} are representative of the PHB content. (**b**) IR spectra in the 2750 to 3100 cm^{-1} range, for emphasizing on the effects of the PEG content at the 2930 cm^{-1} peak.

3.3. Thermal Analysis

Figure 8a shows the changes in degradation temperature across various samples. The T_d values reported correspond to a 5% weight loss in the material [34].

Figure 8b shows the baseline T_d value for PHB of 227 °C, corresponding to 5 wt% mass loss. The addition of 3% of the carbon fibers and microspheres shows a negligible change of 222 and 224 °C, respectively. The addition of 25% 2000, 10,000, and 20,000 MW PEG showed a change in T_d from 227 °C in the control samples without additives, to 227, 235, and 242 °C, respectively. As the MW of the PEG was increased, the thermal stability of the polymer was improved resulting in a higher temperature of degradation. We believe that this was due to the increasing length of the polymer chain, influencing the thermal stability of the material directly as MW increased [35]. Additional TGA curves for each additive at various concentrations can be found in Figure S7.

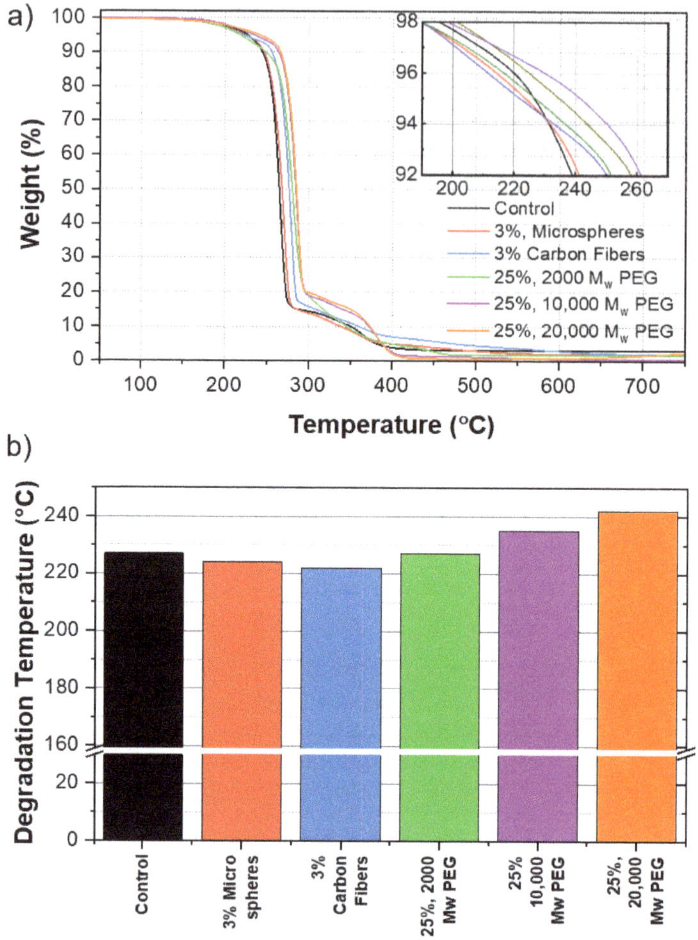

Figure 8. (**a**) TGA curves of PHB composites. The inset shows the degradation temperature, T_d across samples. (**b**) Comparison of T_d for PHB control against PHB composite samples.

3.4. Mechanical Characterization: Hardness

Material hardness is essential to understand how the material will resist plastic deformation and penetration. The enhancement of mechanical response of PHB composites through the compounding of additives could lead to new applications.

Figure 9 shows a Shore A value of 97 for the control sample. The addition of 3% carbon fibers and microspheres resulted in a decrease in Shore A values to 89 and 88, respectively. Similarly, the addition of 25% 2000, 10,000, and 20,000 MW PEG also reduced the hardness to 94, 88, and 89, respectively.

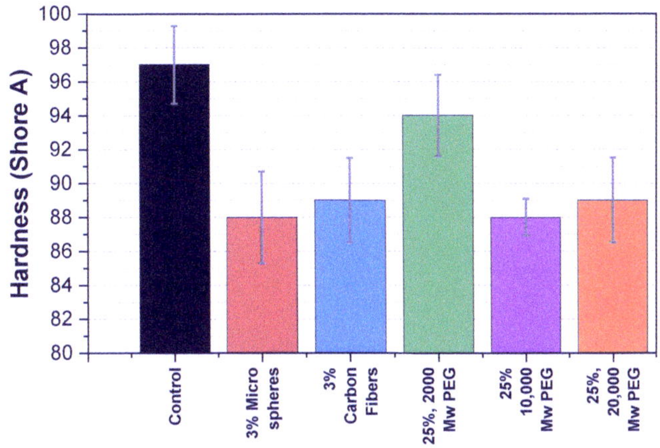

Figure 9. Shore A hardness comparison of PHB composite samples against the control sample. The bar shows the mean hardness (±standard deviation) of five measurements on each sample.

The PHB composites in this study exhibited a decrease in hardness with the introduction of each of the additives. A study by Gavali et al. on polylactic acid (PLA)/carbon fiber composites reported an 85% increase in hardness after adding 20% (w/w) carbon fibers [36]. Therefore, we expected to see a similar increase. However, as seen in Figure 9, we found a decrease in the hardness for the carbon fiber sample. In this paper, we propose that the carbon fibers disrupt the intramolecular and intermolecular interactions (hydrogen bonding and Van der Waals) of the PHB chains and therefore, weaken the lattice structure, making the samples softer. In addition, we extend this explanation to the microspheres sample.

PEG is often used as a plasticizer for lowering the rigidity of materials. We anticipated that this effect would lower the hardness of the composite. We found that the PEG did in fact lower the hardness of the material for the 10,000 and 20,000 MW samples. At higher molecular weights, we believe the PEG acts as an effective plasticizer by allowing the PHB chains to slide past each other when a force is applied. Additional hardness figures for each additive at various concentrations can be found in Figure S8.

3.5. Mechanical Characterization: Compressibility

Compression tests are used to understand the material performance under loading conditions, ductility, and compressibility. As we apply a load to the sample, we can further explore how the material will deform/displace.

Figure 10 shows a displacement of 6% for our control sample. The addition of 3% of the carbon fibers and microspheres showed a lowered displacement of 4.4 and 5.2%, respectively. Similarly, the addition of 25% 2000, 10,000, and 20,000 MW PEG showed a lowered displacement of 4.5, and 5%, respectively. Of note, all of the values have an experimental error of ±0.15% due to the instrument.

The PHB/microsphere composite showed a decrease in displacement from 6 to 5.2%. In previous work on macromolecular microsphere composite hydrogels, an increase in displacement was seen as a result of the microspheres [37]. This was opposite to our composite. Here, we suggest that the microsphere and carbon fiber additives occupied free space in the polymer matrix. As the samples were compressed, the chains were no longer able to rearrange as freely, which resulted in a decrease in the displacement. Unlike in the hardness tests that applied a force to a specific point, the compression tests applied a force to the entire sample creating a closed system that limited the mobility of the polymer chains.

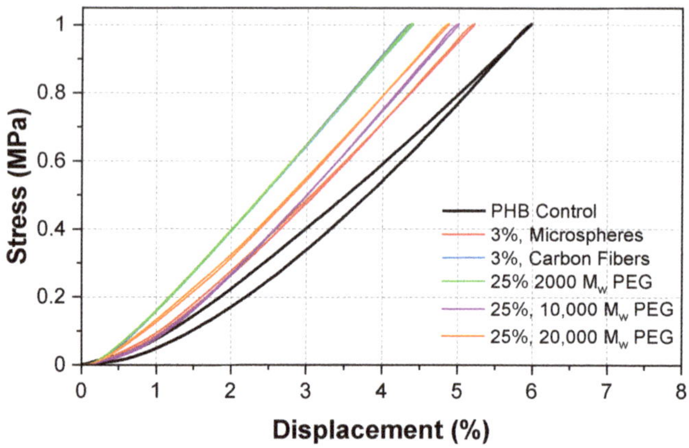

Figure 10. Stress/displacement diagram of PHB composites under 1 MPa compression for the analysis of maximum displacement.

A decrease in displacement was seen across all of the PEG/PHB composites. The compression properties of the composites depend on the polymer (PEG) chain length and how the material was prepared. At low densities and compression rates, a solid-like behavior occurs, which is most common in shorter chains. This could explain why the lowest displacement occurred at a MW of 2000. As the MW of the PEG in the sample was increased, the compressibility also increased, due to the longer chain length. This process could be interpreted as a polymer chain rearrangement, which develops in slow compression rates [38]. Additional stress/displacement curves can be found in Figure S9. In addition, a bar graph highlighting the changes in maximum displacement for the highest additive concentrations is shown in Figure S10.

The experiments conducted in this study were necessary to gain a baseline understanding of the effects of the selected additives on PHB. First, using both high-resolution confocal and SEM microscopy to analyze the morphology of the different PHB samples produced, we determined that despite the fact that only a single screw extruder was used to perform the compounding, which introduced the various additives into the polymer matrix, the samples appeared uniform, with no apparent segregation into multiple phases. The PHB sample with low MW PEG shows a slight discoloration/browning of the material, likely due to a premature degradation. Microscopy also highlighted an incomplete expansion in some of the microspheres that could stem from a pressure increase in the mold during processing. Second, IR spectroscopy has shown that PHB was mostly unaltered during the extrusion process. This phenomenon was further confirmed through a comparison between the composites and the IR spectra of the unprocessed PHB polymer (Figure S6). In addition, we found that the thermal stability of PHB/PEG composites was improved with higher molecular weights, possibly due to the interaction of the polymer chain with the PEG. Finally, mechanical testing showed a decrease in hardness, as well as a decrease in compressibility for all of the samples. The proposed mechanisms that could help explain these results are the intra- and intermolecular interactions from the carbon fibers, the reduced density from the microspheres, and the plasticizing effects of the PEG that overall disrupt the crystallinity of the material. Furthermore, the occupied free space of the microspheres and carbon fibers, as well as the solid-like behavior from the PEG could explain the unexpected lack of flexibility of these materials.

Compared to other bioplastics, such as PLA, PHB exhibits a lower thermal stability and mechanical properties that make it less desirable for mass market applications [39]. While PLA-PHB blends are promising optiona for the replacement of traditional plastics, focusing on PHB-only polymers is crucial for end-of-life considerations, such as in single use plastics [10]. Improving on the degradation temperature and expanding on the process temperature range, as was done in this study, is essential for broadening the single use applications of these materials. For instance, this allows the consideration of PHBs not only for single use packaging, such as cups, plates, and utensils, but also for the expansion to more challenging applications, such as traditional or microwave oven meals and hot beverages [40].

4. Conclusions

This work served as a preliminary study of the effect of carbon fibers, microspheres, PHB, and PEG on PHB composites, and contributed to understanding the thermal and mechanical behaviors of these materials. We compounded PHBs with various concentrations of additives using a single-screw extruder retrofitted with an in-house injection mold. The injection-molded samples were an attempt to use a state-of-the-art industrial technique to produce the parts. We investigated the changes of properties, which are induced by the addition of carbon fibers, microspheres or PEG additives using microscopy, spectroscopic, thermal, and mechanical techniques. We have shown that the PEG of higher MW had an improved thermal stability, slightly increasing the T_d of these composites. In addition, we noted that the percentage displacement of the PEG composites was slightly higher at higher MWs.

The addition of microspheres did not have noticeable effects on T_d, presumably due to an incomplete expansion of the microspheres. Both the carbon fibers and microsphere additives resulted in a decreased hardness of the composites, likely from the disruption of the interactions between the PHB chains. In this study, the PHB and additives were mixed under a relatively short mixing time in the extruder, as they underwent a single pass. If the material had been processed continuously, as in an industrial complex, we may have obtained a more homogenous mixture. While the effects of the additives used in this study were subtle, the results were repeatable. In future studies, additional aspects of injection molding processes may be adopted, in order to investigate the effects of injection molding beyond utilization as a sample formation method. Our results help in building a foundation for systematic composite manufacturing and characterization that will enable novel applications for PHBs and a wider adoption across industries. With single-use plastics contributing heavily towards global plastic pollution, the design of viable bioplastic composites will allow the development of new applications for these promising materials. Expanding the thermal and mechanical capabilities of biopolymer composites to reach higher temperatures and levels of mechanical stress/strain, will increase the viability of the materials as replacements for traditional non-biodegradable polymers.

With the improvements explored in this study, we continue to expand the possibilities of PHB composites. As the T_d is improved further, more demanding temperature applications, such as microwavable plastics, hot beverage products, and high temperature packaging are increasingly achievable. Similarly, as additional improvements are made on the mechanical properties of PHB, the applications in foams, appliance packaging, and electronic packaging are likely candidates. Future studies on the oxygen diffusion, water barrier, and elasticity properties of PHB will continue to expand the applications of this polymer.

Supplementary Materials: The following are available online at https://www.mdpi.com/article/10.3390/polym13244444/s1, Figure S1: PHB composite sample surface at 100× under full ring lighting, Figure S2: IR Spectra for PHB composite samples, Figure S3: TGA curves for PHB composites, Figure S4: Shore A hardness for PHB composites, Figure S5: Stress/strain diagrams for PHB composites, Figure S6: Maximum displacement diagram for PHB composites, Figure S7:TGA curves for PHB composites, Figure S8: Shore A hardness for PHB composites, Figure S9: Stress/strain diagrams for PHB composites, Figure S10: Maximum displacement diagram for PHB composites.

Author Contributions: Conceptualization, design, and supervision, J.H.D.; sample preparation, characterization, and analysis, M.M.H.; aided in the characterization and analysis, N.S.G.; compounding design, K.-S.L.; A.C.P.; characterization and analysis project guidance, B.L.M. and C.N.I. All authors contributed to literature collection, assessment, and manuscript writing, review and editing. All authors have read and agreed to the published version of the manuscript.

Funding: This research was funded by the Laboratory Directed Research and Development (LDRD) office, grant no. 20190001DR (BioManufacturing with Intelligent Adaptive Control—BioManIAC). M.M.H was supported by the NNSA Minority Serving Institutes (MSIPP) Partnership for Research and Education Consortium in Ceramics and Polymers (PRE-CCAP); grant DE-NA0003947. Los Alamos National Laboratory is operated by Triad National Security, LLC, for the National Nuclear Security Administration of the US Department of Energy under contract no. 89233218CNA000001.

Institutional Review Board Statement: Not applicable.

Informed Consent Statement: Not applicable.

Data Availability Statement: All of the data needed to evaluate the conclusions in the paper are present in the paper and/or the Supplementary Materials. Additional data related to this paper may be requested from the authors.

Acknowledgments: The authors wish to thank Leon Lopez for the machining of the molds. M.M.H. and J.H.D. wish to acknowledge the support of the LANL's Under Represented Minority Partnership Program: Tommy Rockward and Ulises Martinez and the NNSA MSIPP Program Managers: David Canty, Jennifer Kline, and Julie Spyres.

Conflicts of Interest: The authors declare no conflict of interest.

References

1. Williams, M.; Gower, R.; Green, J.; Whitebread, E.; Lenkiewicz, Z.; Schröder, P. *No Time to Waste: Tackling the Plastic Pollution Crisis before It's Too Late*; Institute of Development Studies: Falmer, UK, 2019.
2. Geyer, R.; Jambeck, J.R.; Law, K.L. Production, use, and fate of all plastics ever made. *Sci. Adv.* **2017**, *3*, e1700782. [CrossRef]
3. Eriksen, M.; Lebreton, L.C.; Carson, H.S.; Thiel, M.; Moore, C.J.; Borerro, J.C.; Galgani, F.; Ryan, P.G.; Reisser, J. Plastic pollution in the world's oceans: More than 5 trillion plastic pieces weighing over 250,000 tons afloat at sea. *PLoS ONE* **2014**, *9*, e111913.
4. Schnurr, R.E.; Alboiu, V.; Chaudhary, M.; Corbett, R.A.; Quanz, M.E.; Sankar, K.; Srain, H.S.; Thavarajah, V.; Xanthos, D.; Walker, T.R. Reducing marine pollution from single-use plastics (SUPs): A review. *Mar. Pollut. Bull.* **2018**, *137*, 157–171. [CrossRef] [PubMed]
5. Chen, Y.; Awasthi, A.K.; Wei, F.; Tan, Q.; Li, J. Single-use plastics: Production, usage, disposal, and adverse impacts. *Sci. Total Environ.* **2020**, *752*, 141772. [CrossRef] [PubMed]
6. Tang, X.; Chen, E.Y.-X. Chemical synthesis of perfectly isotactic and high melting bacterial poly (3-hydroxybutyrate) from bio-sourced racemic cyclic diolide. *Nat. Commun.* **2018**, *9*, 2345. [CrossRef]
7. DiCiccio, A.M.; Longo, J.M.; Rodríguez-Calero, G.G.; Coates, G.W. Development of highly active and regioselective catalysts for the copolymerization of epoxides with cyclic anhydrides: An unanticipated effect of electronic variation. *J. Am. Chem. Soc.* **2016**, *138*, 7107–7113. [CrossRef]
8. Lenz, R.W.; Marchessault, R.H. Bacterial polyesters: Biosynthesis, biodegradable plastics and biotechnology. *Biomacromolecules* **2005**, *6*, 1–8. [CrossRef]
9. Yagi, H.; Ninomiya, F.; Funabashi, M.; Kunioka, M. Mesophilic anaerobic biodegradation test and analysis of eubacteria and archaea involved in anaerobic biodegradation of four specified biodegradable polyesters. *Polym. Degrad. Stab.* **2014**, *110*, 278–283. [CrossRef]
10. Arrieta, M.P.; Samper, M.D.; Aldas, M.; López, J. On the use of PLA-PHB blends for sustainable food packaging applications. *Materials* **2017**, *10*, 1008. [CrossRef]
11. Daitx, T.S.; Giovanela, M.; Carli, L.N.; Mauler, R.S. Biodegradable polymer/clay systems for highly controlled release of NPK fertilizer. *Polym. Adv. Technol.* **2019**, *30*, 631–639. [CrossRef]
12. Yeo, J.C.C.; Muiruri, J.K.; Thitsartarn, W.; Li, Z.; He, C. Recent advances in the development of biodegradable PHB-based toughening materials: Approaches, advantages and applications. *Mater. Sci. Eng. C* **2018**, *92*, 1092–1116. [CrossRef]
13. Nosal, H.; Moser, K.; Warzała, M.; Holzer, A.; Stańczyk, D.; Sabura, E. Selected Fatty Acids Esters as Potential PHB-V Bioplasticizers: Effect on Mechanical Properties of the Polymer. *J. Polym. Environ.* **2021**, *29*, 38–53. [CrossRef]
14. Avella, M.; Martuscelli, E.; Raimo, M. Review Properties of blends and composites based on poly (3-hydroxy) butyrate (PHB) and poly (3-hydroxybutyrate-hydroxyvalerate)(PHBV) copolymers. *J. Mater. Sci.* **2000**, *35*, 523–545. [CrossRef]
15. Kunioka, M.; Doi, Y. Thermal degradation of microbial copolyesters: Poly (3-hydroxybutyrate-co-3-hydroxyvalerate) and poly (3-hydroxybutyrate-co-4-hydroxybutyrate). *Macromolecules* **1990**, *23*, 1933–1936. [CrossRef]
16. Lee, S.N.; Lee, M.Y.; Park, W.H. Thermal stabilization of poly (3-hydroxybutyrate) by poly (glycidyl methacrylate). *J. Appl. Polym. Sci.* **2002**, *83*, 2945–2952. [CrossRef]

17. Pilania, G.; Iverson, C.N.; Lookman, T.; Marrone, B.L. Machine-learning-based predictive modeling of glass transition temperatures: A case of polyhydroxyalkanoate homopolymers and copolymers. *J. Chem. Inf. Modeling* **2019**, *59*, 5013–5025. [CrossRef] [PubMed]
18. Bejagam, K.K.; Iverson, C.N.; Marrone, B.L.; Pilania, G. Molecular dynamics simulations for glass transition temperature predictions of polyhydroxyalkanoate biopolymers. *Phys. Chem. Chem. Phys.* **2020**, *22*, 17880–17889. [CrossRef] [PubMed]
19. Botelho, E.; Rezende, M.; Lauke, B. Mechanical behavior of carbon fiber reinforced polyamide composites. *Compos. Sci. Technol.* **2003**, *63*, 1843–1855. [CrossRef]
20. McAdam, B.; Fournet, B.M.; McDonald, P.; Mojicevic, M. Production of polyhydroxybutyrate (PHB) and factors impacting its chemical and mechanical characteristics. *Polymers* **2020**, *12*, 2908. [CrossRef]
21. Latiff, A.A.; Mohamad, N.; Nasir, M.H.M.; Rahmah, S.S.; Mahamood, M.A.; Abdullah, M.I.H.C.; Ab Maulod, H.E. Correlation of wear characteristics with hardness of recycled carbon fiber prepreg reinforced polypropylene composites. *J. Mater. Res.* **2016**, *31*, 1908–1913. [CrossRef]
22. Labouriau, A.; Robison, T.; Geller, D.; Cady, C.; Pacheco, A.; Stull, J.; Dumont, J.H. Coupled aging effects in nanofiber-reinforced siloxane foams. *Polym. Degrad. Stab.* **2018**, *149*, 19–27. [CrossRef]
23. Yuan, J.; An, Z.; Zhang, J. Effects of hollow microsphere surface property on the mechanical performance of high strength syntactic foams. *Compos. Sci. Technol.* **2020**, *199*, 108309. [CrossRef]
24. Kmetty, Á.; Litauszki, K. Development of poly (lactide acid) foams with thermally expandable microspheres. *Polymers* **2020**, *12*, 463. [CrossRef] [PubMed]
25. Dumont, J.H.; Murphy, E.; Lee, K.-S.; Labouriau, A. Insulating Boron-Containing PDMS-Based Materials for Space and Nuclear Applications. *Transactions* **2019**, *120*, 752–755.
26. Mohapatra, A.K.; Mohanty, S.; Nayak, S. Effect of PEG on PLA/PEG blend and its nanocomposites: A study of thermo-mechanical and morphological characterization. *Polym. Compos.* **2014**, *35*, 283–293. [CrossRef]
27. Tekin, D.; Birhan, D.; Kiziltas, H. Thermal, photocatalytic, and antibacterial properties of calcinated nano-TiO_2/polymer composites. *Mater. Chem. Phys.* **2020**, *251*, 123067. [CrossRef]
28. Janigová, I.; Lacík, I.; Chodak, I. Thermal degradation of plasticized poly (3-hydroxybutyrate) investigated by DSC. *Polym. Degrad. Stab.* **2002**, *77*, 35–41. [CrossRef]
29. Shang, S. What Makes Clear Polypropylene Discolor? *Med. Plast. Degrad. Resist. Fail. Anal.* **1998**, 79.
30. Han, S.; Kim, C.; Kwon, D. Thermal/oxidative degradation and stabilization of polyethylene glycol. *Polymer* **1997**, *38*, 317–323. [CrossRef]
31. Liu, Q.; Shao, L.; Xiang, H.; Zhen, D.; Zhao, N.; Yang, S.; Zhang, X.; Xu, J. Biomechanical characterization of a low density silicone elastomer filled with hollow microspheres for maxillofacial prostheses. *J. Biomater. Sci. Polym. Ed.* **2013**, *24*, 1378–1390. [CrossRef]
32. Wendlandt, K.-D.; Geyer, W.; Mirschel, G.; Hemidi, F.A.-H. Possibilities for controlling a PHB accumulation process using various analytical methods. *J. Biotechnol.* **2005**, *117*, 119–129. [CrossRef] [PubMed]
33. Shinzawa, H.; Uchimaru, T.; Mizukado, J.; Kazarian, S.G. Non-equilibrium behavior of polyethylene glycol (PEG)/polypropylene glycol (PPG) mixture studied by Fourier transform infrared (FTIR) spectroscopy. *Vib. Spectrosc.* **2017**, *88*, 49–55. [CrossRef]
34. Iulianelli, G.C.V.; David, G.d.S.; dos Santos, T.N.; Sebastião, P.J.O.; Tavares, M.I.B. Influence of TiO2 nanoparticle on the thermal, morphological and molecular characteristics of PHB matrix. *Polym. Test.* **2018**, *65*, 156–162. [CrossRef]
35. Treetharnmathurot, B.; Ovartlarnporn, C.; Wungsintaweekul, J.; Duncan, R.; Wiwattanapatapee, R. Effect of PEG molecular weight and linking chemistry on the biological activity and thermal stability of PEGylated trypsin. *Int. J. Pharm.* **2008**, *357*, 252–259. [CrossRef]
36. Gavali, V.C.; Kubade, P.R.; Kulkarni, H.B. Mechanical and thermo-mechanical properties of carbon fiber reinforced thermoplastic composite fabricated using fused deposition modeling method. *Mater. Today Proc.* **2020**, *22*, 1786–1795. [CrossRef]
37. Huang, T.; Xu, H.; Jiao, K.; Zhu, L.; Brown, H.R.; Wang, H. A novel hydrogel with high mechanical strength: A macromolecular microsphere composite hydrogel. *Adv. Mater.* **2007**, *19*, 1622–1626. [CrossRef]
38. Spigone, E.; Cho, G.-Y.; Fuller, G.G.; Cicuta, P. Surface rheology of a polymer monolayer: Effects of polymer chain length and compression rate. *Langmuir* **2009**, *25*, 7457–7464.
39. Abdelwahab, M.A.; Flynn, A.; Chiou, B.-S.; Imam, S.; Orts, W.; Chiellini, E. Thermal, mechanical and morphological characterization of plasticized PLA–PHB blends. *Polym. Degrad. Stab.* **2012**, *97*, 1822–1828. [CrossRef]
40. Dey, A.; Dhumal, C.V.; Sengupta, P.; Kumar, A.; Pramanik, N.K.; Alam, T. Challenges and possible solutions to mitigate the problems of single-use plastics used for packaging food items: A review. *J. Food Sci. Technol.* **2021**, *58*, 3251–3269. [CrossRef]

polymers

Review

Durability of Biodegradable Polymer Nanocomposites

Tatjana Glaskova-Kuzmina [1,*], Olesja Starkova [1], Sergejs Gaidukovs [2], Oskars Platnieks [2] and Gerda Gaidukova [3]

1 Institute for Mechanics of Materials, University of Latvia, LV-1004 Riga, Latvia; olesja.starkova@lu.lv
2 Institute of Polymer Materials, Faculty of Materials Science and Applied Chemistry, Riga Technical University, P.Valdena 3/7, LV-1048 Riga, Latvia; sergejs.gaidukovs@rtu.lv (S.G.); oplatnieks@gmail.com (O.P.)
3 Latvian Maritime Academy, Flotes 3-7, LV-1016 Riga, Latvia; gerda.gaidukova@rtu.lv
* Correspondence: tatjana.glaskova-kuzmina@lu.lv

Abstract: Biodegradable polymers (BP) are often regarded as the materials of the future, which address the rising environmental concerns. The advancement of biorefineries and sustainable technologies has yielded various BP with excellent properties comparable to commodity plastics. Water resistance, high dimensional stability, processability and excellent physicochemical properties limit the reviewed materials to biodegradable polyesters and modified compositions of starch and cellulose, both known for their abundance and relatively low price. The addition of different nanofillers and preparation of polymer nanocomposites can effectively improve BP with controlled functional properties and change the rate of degradation. The lack of data on the durability of biodegradable polymer nanocomposites (BPN) has been the motivation for the current review that summarizes recent literature data on environmental ageing of BPN and the role of nanofillers, their basic engineering properties and potential applications. Various durability tests discussed thermal ageing, photo-oxidative ageing, water absorption, hygrothermal ageing and creep testing. It was discussed that incorporating nanofillers into BP could attenuate the loss of mechanical properties and improve durability. Although, in the case of poor dispersion, the addition of the nanofillers can lead to even faster degradation, depending on the structural integrity and the state of interfacial adhesion. Selected models that describe the durability performance of BPN were considered in the review. These can be applied as a practical tool to design BPN with tailored property degradationand durability.

Keywords: biodegradable polymers; nanocomposites; durability; biodegradation; environmental ageing; creep; modelling

Citation: Glaskova-Kuzmina, T.; Starkova, O.; Gaidukovs, S.; Platnieks, O.; Gaidukova, G. Durability of Biodegradable Polymer Nanocomposites. *Polymers* **2021**, *13*, 3375. https://doi.org/10.3390/polym13193375

Academic Editors: Domenico Acierno and Antonella Patti

Received: 7 September 2021
Accepted: 28 September 2021
Published: 30 September 2021

1. Introduction

With an increasing global awareness of plastic wastes, there is a huge demand for environmentally friendly solutions such as biodegradable polymers (BP) [1,2]. Moreover, the development of alternative biodegradable materials is motivated due to reasonable limits and the depletion of petroleum resources and rising concerns over the increasing fossil CO_2 contents in the atmosphere [3]. Recently, many efforts are being made to improve these materials' quality and functionality, resulting in their applicability in food packaging, agriculture, furniture, construction, engineering and various smart applications [4–7]. The investigation of degradational processes of polymers and the ways to stabilize them is an extremely important area from the scientific and industrial point of view, and a better understanding of polymer degradation will ensure the long life of the products [8]. Therefore, considering the long-term aspects of such applications, the durability of biodegradable polymers and composites becomes crucial and should be investigated. Moreover, insufficient knowledge of mechanical properties, durability and long-term performance under environmental ageing restricts this new class of sustainable materials for advanced applications [9–11].

Generally, bioplastics could be classified into petroleum-based biodegradable polymers, renewable resource-based polymers and polymers from mixed sources (bio- and petroleum-based) as shown in Figure 1. According to the classification, the biodegradability of the polymers depends on the structure but not on the raw material source [12]. Therefore, biodegradable polymers may include both petroleum-based and bio-based polymers. We focus only on several cheap, abundant biodegradable biopolymers herein—polylactide (PLA), polycaprolactone (PCL), polybutylene succinate (PBS), polybutylene adipate-terephtalate (PBAT), polyhydroxyalkanoate (PHA) and thermoplastic starch (TPS).

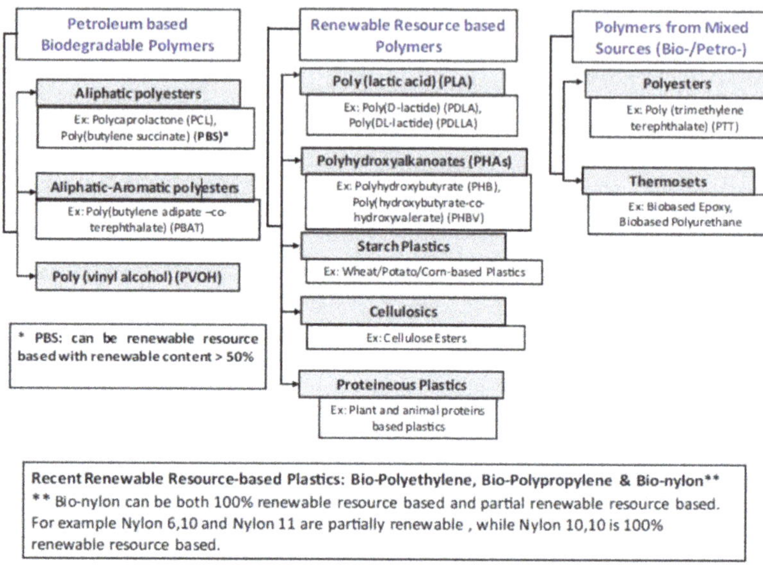

Figure 1. Classification of biopolymers. Reproduced with permission from [12]. Copyright © (2013). Elsevier Ltd. (licence No. 5142931044479).

The addition of nanofillers into BP effectively develops durable bioplastics with controlled functional properties and degradation rates [4–6,10–15]. Besides improvements in thermal, mechanical and barrier properties, some nanofillers can provide additional functionality to the polymer matrix, e.g., antimicrobial [16,17] and "smart" properties [18–20]. The main engineering properties of biodegradable polymer nanocomposites (BPN) were summarized in several recent reviews [21–25]. Some issues related to composites' preparation and mechanical behaviour with nano-sized reinforcement (i.e., silver nanoparticles, carbon nanofillers, nano-hydroxyapatite and cellulose nanocrystals) in comparison with composites with larger micron-sized inclusions were highlighted in [21]. The results on design, preparation and characterization of biodegradable polymer/layered silicate nanocomposites were reviewed in [22,23]. A comprehensive review of nanocellulose addition's impact on various synthetic and biopolymer composite materials was provided in [24]. Different properties and potential applications of bio-based poly(butylene succinate) (PBS) composites, including nanocomposites, were highlighted in [25].

Despite increasing interest in the research of BPN, most studies are based on their preparation techniques and the characterization of their fundamental structure–property relationships, while durability issues are rarely reported. The lack of research on the durability of bio-based and biodegradable polymers and composites and the emphasis on the need for this type of research was highlighted in several articles [8,26,27]. Thus, the potential of BPN under different environmental conditions should be thoroughly reviewed and understood to expand their applications to long-term and advanced solutions.

The main aim of the work is to provide insights into the BPN durability and estimate the role of different nanofillers on the overall performance and durability of BP. Recent literature results on the durability performance of BP and BPN were analyzed under environmental ageing and mechanical load conditions. Some existing models for BPN durability prediction were reviewed and discussed.

2. Biodegradable Polymers and their Basic Engineering Properties

Biodegradable polymers are abundant and obtainable from natural sources like cellulose, starch and chitosan. They have seen relative success in their applications, but they cannot replace the complete functionality of common fossil plastics like polyolefins, polystyrene, polyethylene terephthalate and others. Thus, commercial biodegradable alternatives to commodity plastics based on polyester structure have been developed and commercialized in the last decade [28]. Emerging bio-based and biodegradable synthetic plastics include polylactic acid (PLA), polycaprolactone (PCL), polybutylene succinate (PBS), polybutylene succinate adipate (PBSA) and polybutylene adipate terephthalate (PBAT). Polyester-produced microorganisms are known as polyhydroxyalkanoates (PHA), which can be further divided into polymer grades like polyhydroxybutyrate (PHB), polyhydroxy valerate (PHV) or their copolymer PHBV. In addition, few conventional fossil-based polymers like polyvinyl alcohol (PVOH) can biodegrade and should be included in this group of materials [29]. To achieve sustainability goals of reducing fossil CO_2 new more efficient bio-synthesis routes still need to be explored and optimized. In this regard, many advances have been made in the biorefinery field that has yielded most of modern bio-based plastics, but still, issues like relatively higher price and lack of legislation have delayed the transition to bio-based and biodegradable polymer materials worldwide [30].

Biopolymer materials could be differentiated depending on their interaction with water and structure as more hydrophilic and hydrophobic groups are present in the backbone. Natural bio-based polymers are usually hydrophilic; thus, their broad engineering applications are limited, while chemical modifications can change this property resulting in structures like cellulose acetate, nitrocellulose, etc. Thus, biopolyesters have emerged as a non-polar alternative for various applications that require contact with water, humidity and preservation of a sterile environment [31]. In addition, these polymers are often more thermally stable, melt-processed and easily modified with flame retardants for safety purposes [32]. While key characteristics of polymers are achieved with relatively high molecular weight. Especially polymers with molecular weight above 100 000 g/mol can have properties like high ductility, superelasticity and shape memory [33,34].

The characteristic properties of some of the most widely used biodegradable polymers are summarized in Table 1. Starch in the form of thermoplastic starch (TPS) is widely used for various packaging materials and other short life span products. TPS is obtained as a blend using plasticizers and/or other biodegradable polymers as additives. Thus, there is a significant disparity of properties for TPS, but the material itself is usually much more sensitive to water than biodegradable plastics.

Table 1. BP and their characteristic physical and mechanical properties [35–40].

	PLA	PCL	PBS	PBAT	PHA	TPS
Density, g/cm^3	1.21–1.30	1.11–1.15	1.22–1.26	1.26	1.18–1.26	0.85–1.00
Melting point, °C	165–170	58–65	110–115	89	160–190	100–160 *
Glass transition, °C	55–65	−65–60	−35–20	−30–20	10–40	−60–10
Tensile strength, MPa	30–60	20–45	20–35	15–25	30–50	0.5–50
Young's modulus, GPa	2–4	0.2–0.4	0.2–0.4	0.05–0.10	3–4	0.05–0.50
Elongation at break, %	2–10	300–1000	30–500	500–1100	4–12	10–300

* TPS does not melt but is processed at these temperatures.

Synthetic biopolymers can be sorted into two groups: relatively soft with large elongation values like PBS, PCL, PBAT and the second group with PLA and PHA with relatively high elastic modulus and low elongation values lead to brittleness without additives.

PCL has a relatively low melting temperature, limiting its applications and is commonly used for specific purposes like biomedicine. PBS and PBAT have great potential for film preparation required in packaging and agriculture [35,41]. In addition, they are an excellent matrix for the preparation of composite materials. Usually, incorporated particles in the matrix restrict polymer chain movements, resulting in elongation values, which are already low for PLA and PHA. The addition of plasticizers is common for PLA and PHA composite materials [42,43]. Studies indicate that PHA can degrade in various environments, including seawater, while PLA needs specific soil conditions [44]. The drawback of PHA is the relatively high cost of production. In addition, PHA and PLA have a relatively narrow range of thermal processing, while PBS and PBAT have been reported to be much more stable during melt processing [45,46].

3. Potential Nanofillers for Biodegradable Polymers

The main drawback of biopolymers is that most of them have poor mechanical and thermal properties limiting their use in structural applications. Both natural and synthetic nanofillers could be used to improve the physical-mechanical properties of biopolymers. In the case of both matrix and filler derived from renewable resources, a fully renewable and biodegradable nanocomposite could be produced [39].

Different nanofillers may introduce different properties to BPN resulting in specific applications [12]. Mostly, recent applications of BPN are limited to packaging, biomedical, antibacterial and smart applications. The scope of possible applications for different combinations of BP and nanofillers is reviewed in Table 2. Examples of smart applications of BPN include piezoresistive vapour sensors for PLA filled with multiwall carbon nanotubes (MWCNT), shape-memory applications for poly(d,l-lactide) filled with Fe_3O_4, and electrical/electromagnetic applications for PLA/PHBV filled with MWCNT [18–20]. The addition of electrically conductive fillers (e.g., carbon black, carbon nanotubes, nanofibres, graphene, Fe_3O_4) into biopolymers may result not only in improved nucleating, mechanical, thermal and fire-retardant properties, but also may introduce tailored electrical and thermal conductivity [15,47]. These composites can be promising as materials for manufacturing sensors with sensitivity to such factors as strain, temperature or organic solvents [7].

Table 2. Scope of applications for BPN.

Type of Application	Biopolymer	Nanofiller	References
Packaging	PLA	ZnO	[11,48–51]
	PLA	MMT	[17,49]
	PLA	Nanocellulose	[48,51–53]
	PBS	ZnO	[54]
	PBS	Nanocellulose	[48,55]
	Starch	Ag, ZnO, CuO	[56]
	Starch	Nanocellulose	[48]
	PCL	ZnO/nanocellulose	[57]
Biomedical applications	PLA	ZnO	[26]
	PLA	TiO_2	[58]
	PLA	Fe_3O_4	[47]
Antimicrobial applications	PLA	Ag	[49]
	PLA	MMT	[1,17]
	PBS	ZnO	[16,17]
	Cellulose acetate	Cu	[59]
Smart applications	PLA	MWCNT	[18]
	Poly(d,l-lactide)	Fe_3O_4	[19]
	PLA/PHBV	MWCNT	[20]

It should be noted that the addition of nanofillers could negatively affect biopolymer properties. For example, the advanced degradation of PLA chains resulting in reduced thermomechanical properties was observed upon the addition of some metal oxides such as calcium oxide (CaO), magnesium oxide (MgO) or other metallic compounds such as layered double hydroxides [47]. Similarly, the addition of untreated ZnO nanoparticles into PLA resulted in intense degradation at melt-processing temperature, described by the transesterification reactions and 'unzipping' depolymerization of PLA [11,26]. Nevertheless, the surface treatment of ZnO by using silanes may improve the physicochemical characteristics of PLA.

4. Biodegradation of BPN

The overall degradation process of biopolymers and biocomposites could be related to light, heat, moisture, chemical and microbial treatment on the bulk polymer material [60]. Biodegradation (i.e., biotic degradation) is a chemical degradation of materials (polymers) provoked by the action of microorganisms such as bacteria, fungi and algae. While a biodegradable polymer is a degradable polymer wherein the primary degradation mechanism is through the action of metabolism by microorganisms [61]. Different bacteria mainly guide the biodegradation of a macromolecular structure. Commonly, applying the complex factors of light, heat and microorganisms could significantly pronounce the intensity of polymers' physical and chemical changes, leading to a noticeable drop in the material's properties, partial disintegration and complete disappearance. For the efficient biological activity of bacteria, the polymer materials should have at least contact with soil and compost. In contrast, the full burial in the soil media of the polymer could facilitate the biodegradation process. In general, all biological functions, for example, bacterial biodegradation, are strongly dependent on the presence of water [62].

The biodegradation of polymer material could temporarily or permanently create small molecules that should be accumulated in the environment [63]. As reported, the formed oligomers, monomers and metabolic intermediates can interact with living organisms in the soil, adversely affecting the environment [64]. To that, environmental issues of the persistency and ecotoxicity of the developed compounds become very important in the biodegradation process investigations [65].

Many authors report that the polymer chain topology, macromolecular network structure, molecular chain weight and size can severely affect PBS, PBSA, PLA, PHA and other bio-based polymers biodegradation in soil [66–68]. The temperature, moisture, pH and the population of active microorganisms are essential factors to facilitate the polymers' biodegradation [69]. These conditions are broadly reviewed and reported in the literature; they depend on the soil characteristics, which vary from place to place and season to season. In comparison, the industrial composting conditions are easy to control due to several strictly physically/chemically controlled parameters and the standardized environment [70,71].

Several authors report that the biodegradation of the composites differs from the unfilled polymers [55,72,73] (e.g., see Figure 2). The degradation process depends on the nature, the chemical modification and the content of the used fillers [74,75]. Synthetic and natural fillers of different sizes and shapes are broadly used to control the biodegradable polymers performance properties [55,72,76]. Carbon, metallic, metallic oxide, cellulose and other micro-and nanoparticles have been very popular in the last decade [77–80]. Fillers could enormously change the overall degradation characteristics of the polymer materials [81]. For example, spent coffee particles significantly enhanced the tensile properties but strongly decreased the biodegradation time for biopolymers [82]. Similar biodegradation enhancement in the soil is observed for microcellulose and nanocellulose particles loaded biocomposites [83].

Figure 2. Photos of (**a**) PBS; (**b**) 40% NFC; (**c**) 7/3; (**d**) 5/5; (**e**) 3/7 and (**f**) 40% MCC films during biodegradation studies in soil burial test conducted in composting conditions [55].

The changed biodegradation mechanism of the biocomposites due to the nanoparticles' antibacterial properties was reported in [84,85]. It is possible to improve the antibacterial properties of bio-based polymers by adding the nanofillers having antibacterial properties (e.g., ZnO, Ag, MMT, etc.) [1,16,86]. Thus, the antimicrobial activity of ZnO-modified PBS films was proven to be effective against representative food spoilage bacteria (*S. aureus* and *E. coli*) at minimal content of 6 wt.% of ZnO [16]. Moreover, a synergistic effect in enhancing the antimicrobial properties against the bacteria, as mentioned above, was found by combinatorial use of Ag/ZnO/CuO nanofillers in the formulation of starch-based films [56].

The microorganisms multiply and prosper at mild temperatures in the presence of moisture and a source of carbon [54]. There is a significant concern to add antimicrobial properties to bio-based polymers to diminish the quantity and propagation of microbes (bacteria, fungi) by using antimicrobial agents. The antibacterial activity is analyzed mostly by transmission electron microscopy (TEM), scanning electron microscopy (SEM), Fourier-transform infrared spectroscopy (FTIR), nuclear magnetic resonance (NMR), zeta potential and dynamic light scattering (DLS) analyses [87].

Another way is to add the antibacterial agents such as, e.g., the bacteriocin (antibacterial peptides) to crystalline nanocellulose and incorporate such bacteriocin immobilized crystalline nanocellulose into bio-based polymers as antibacterial agents to have antibacterial properties with enhanced strength of the films and better biodegradability [88].

Still, the antimicrobial/antibacterial action mechanisms when the nanofillers are added to biopolymers are not fully understood [16,17,47,48]. However, the leading hypothesis is related to the photocatalytic generation of many reactive oxygen species to the formation of the ions [17], consequent leakage of intracellular substances, and lastly, the destruction of bacterial cells [16].

5. Durability Performance of BPN

According to a general definition provided in [9], material durability is related to the ability of a material to withstand a wide variety of physical processes and chemical degradation reactions from the exterior environment. The environmental factors can be solitary or combined action of moisture, oxygen and bacteria attacks, mechanical loading, wear and tear, and extreme temperature conditions. Basic durability tests include thermo- and photo-oxidative ageing, creep and fatigue, water absorption and hydrothermal ageing (Figure 3). Due to the breakdown of the macromolecules' structure from the water absorption and oxidation process–induced during the exposure to the environments, the functional properties of biopolymers and bio-based nanocomposites could deteriorate. For example, in [9] it was shown that the applied accelerated weather conditions did not cause significant changes in the mechanical properties of biocomposites made of flax fibres and epoxidized soybean oil–based thermosetting resin. An increase of hardness, tensile strength and modulus, and decrease of elongation at break and impact strength was attributed to the decreased chain mobility and increased crosslinking density after the tests.

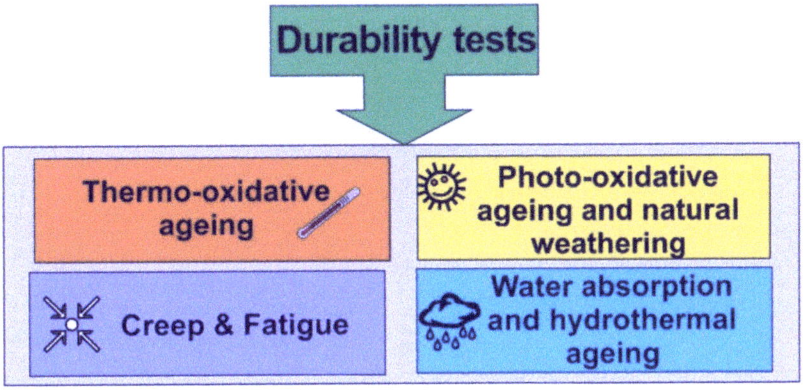

Figure 3. Basic types of durability tests for polymer composite materials.

Due to environmental degradation, different reversible and irreversible consequences may occur to the tested materials, such as decreased molecular weight (chain scissoring), reduced mechanical properties, embrittlement and cracks, colour fading and spots [1,8,89]. Moreover, it should be noted that the ageing behaviour and mechanism of the unfilled polymers are usually less complex than those for filled composite materials. This is due to the presence of different components in the composite such as fillers, fibres, additives, plasticizers, antioxidants, etc., each contributing to the environmental degradation of the composite as a whole [90,91].

Usually, ageing tests are rather long-term, lasting several years or even decades, and, therefore, accelerated ageing tests are applied to imitate specific environmental conditions at an increased rate. Moreover, these studies allow predicting the performance and investigate the degradation mechanisms of the materials and are very important to understand the material ageing behaviour for specific conditions and applications [92–95].

Environmental ageing in some cases causes an increase in the degree of crystallinity of the polymers. For instance, the rise in the degree of crystallinity by 50% was found for the neat PLA after accelerated weathering [27]. It was attributed to the relief of thermal stresses introduced due to the manufacturing process, which occurs under high temperature, and also re-aligning of broken chains due to chain scissoring into a more organized structure.

Ageing tests include different conditioning, e.g., in a climate chamber, exposure of natural weathering and UV irradiation, photo-oxidation, thermal-oxidation, water absorption, humidity, microbial, chemical degradation, thermal cycling/fatigue or a combination of these methods. The results of recent studies of the durability of BPN are summarized in Table 3.

Table 3. Recent studies on durability of various BPN *.

BP Matrix	Filler (Content)	Type of Durability Testing	Indicator	Reference
PLA	ZnO (0.1, 1 wt.%)	Thermal	Glass trans. temperature	[26]
	ZnO (1, 2, 3 wt%)	Water absorption	Diffusivity	[11]
	CaO, MgO (5 wt%)	Thermal	Pyrolysis	[47]
	MMT (5 wt.%)	Microbial	Molecular weight	[17]
	CNT (2, 5 wt%)	Thermal	Crystal. temperature	[15]
	CNF (1, 5 wt.%)	Hydrothermal	Glass trans. temperature	[13]
	CNC (1, 5 wt.%)	Water absorption	Hydrolytic degradation rate	[96]
	ZnO: Cu/Ag (0.5–1.5 wt%)	Microbial	SEM images	[50]
	Nanoclays (OMMT, HNT, Laponite®, 1, 5 wt.%)	Microbial	CO_2 evolution	[97]
	SiO_2	Creep tests	Creep resistance	[98]
PBS	ZnO (0.5, 1, 3 wt.%)	Photo-oxidative	Crystal. temperature	[54]
	ZnO (2–10 wt.%)	Microbial	Inhibition zone diameter	[16]
	MMT (0–10 wt.%)	Hydrothermal	Tensile strength and modulus	[95]
	GnP	Water absorption	Permeability	[99]
	CNF (12–40 wt.%)	Thermal	Crystal. temperature	[55]
PBSA	TiO_2 (0.5–1.5 wt.%)	Photo-oxidative	Crystal. temperature	[100]
Starch	CNF (5–20 wt.%)	Thermal	Creep resistance	[10]
	Ag, ZnO, CuO (0.66–3 wt%)	Microbial	SEM images	[56]
	Cellulose nanofibres (10 wt.%)	Thermal	Activation energy	[101]
	MWCNT (0.005–0.055 wt%))	Thermal	Glass trans. temperature	[102]
PHB	Bentonite (2–6 wt.%)	Thermal	Crystal. temperature	[103]
	nAg (0.25–1.25 mM)	Microbial, hydrolytic	SEM, glass trans. temperature	[104]
	MMT (1–10 wt.%)	Thermal	Glass trans. temperature	[87]
PCL	Nanoclay (6–26 wt.%)	Thermal	Glass trans. temperature	[105]
	Nanocellulose/ZnO (2–8 wt.%)	Thermal	Phase trans. temperature	[57]
	Bentonite (1.5, 3 wt.%)	Creep	Creep resistance	[22]
	MMT, MWCNT, SiO_2 (0.5–2.5 wt.%)	Thermal	Activation energy	[106]
	GO (0.1 wt%)	Creep	Creep resistance	[107]
Cellulose acetate	Cu (2, 6 mol.%)	Microbial	SEM images	[59]
	Ag/MMT (3, 5 wt.%)	Microbial, thermal	Inhibition reduction rate, glass trans. temperature	[86]
PVA	CNC/GO/Ag (0.5 wt.%)	Bacterial	Antibacterial efficiency	[108]
PLA/PHBV	TiO_2	Thermal	Activation energy	[109]
PLA/PBS	CNC (1–3wt.%)	Barrier	Permeability, oxygen transmission rate	[96]
PBAT	CNT (1–5wt.%)	Creep and stress relaxation	Creep resistance	[110]
PVA/ST/GL	HN (0.25–5 wt.%)	Water absorption	Water solubility, water contact angle	[111]

* designations according to the list of abbreviations.

5.1. Thermo-Oxidative Ageing

Generally, thermal degradation of polymers is a very complex phenomenon that involves physical, chemical and thermal processes [112]. During the manufacturing process and service life, polymers are generally exposed to thermo-oxidative degradation, which causes degradation of their performance, especially for long-term applications [90,113,114]. According to Table 3, the thermal degradation of PLA filled with different nanofillers (ZnO and CNT) caused the change in glass transition [26] and crystallization [15] temperatures. Similar results regarding the change in crystallization temperature were reported for PBS filled with CNF [55] and PHB filled with bentonite [103], and regarding the change in glass transition temperature due to thermal degradation for starch-filled with MWCNT [102], PHB filled with MMT [87] and PCL filled with nanoclay [105].

As reported in [115], for BP at elevated temperatures (above glass transition temperatures) a random chain scission mechanism occurs, determining a significant level of molecular degradation and polymer embrittlement. In addition, it was experimentally proven [116] that the oxidative degradation of PLA occurs at moderate temperatures (below PLA melting temperature) with a significant reduction of the polymer molar mass. According to Figure 4 the molar weight of PLA, aged at different constant temperatures (100, 130 and 150 °C), changed almost linearly as a function of temperature at different time-sections (indicated on the graph). An antioxidative degradational process could be minimized by adding the antioxidants to polymers, such as hindered phenols or amines and organophosphorus compounds [117].

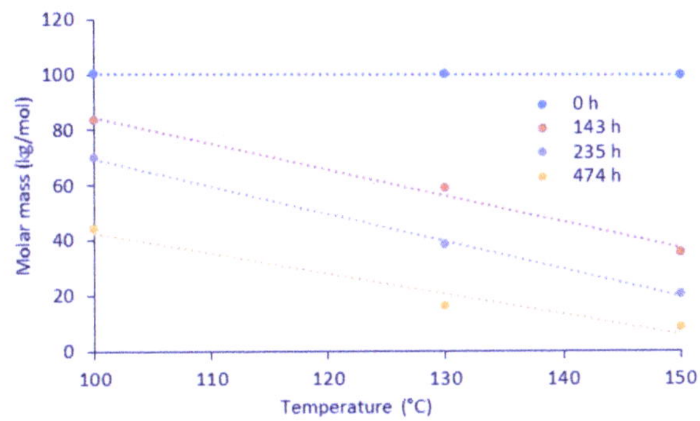

Figure 4. Molar weight of PLA during thermo-oxidative ageing as a function of exposure temperature. Dots: experimental data used from [116]; lines: linear approximations.

To study the kinetics of thermal degradation of BP and BPN different isoconversional methods could be applied [101,109,118–120]. The degradation rate for the isothermal process is given by a general relationship [101]:

$$\frac{d\alpha}{dt} = k(T)f(\alpha), \tag{1}$$

where $k(T)$ is the rate constant at temperature T, α is a specific degree of degradation or conversion (e.g., given by the mass loss in TGA tests) and $f(\alpha)$ is a function of the reaction model related to the degradation mechanism.

For non-isothermal measurements at a constant heating rate $\beta = \frac{dT}{dt}$ and the rate constant given by Arrhenius equation, Equation (1) takes the form:

$$\beta\frac{d\alpha}{dT} = A\exp\left(\frac{E_a}{RT}\right)f(\alpha), \tag{2}$$

where A is a constant, E_a is the activation energy and R is the gas constant.

Equation (1) is the basic equation used for the prediction of the degradation evolution. In general, determining the pre-exponential factor and the activation energy is challenging since both parameters could be interrelated conversion functions. The isoconversional methods provide simplified procedures for characterizing the degradation kinetics by presuming temperature independence of the pre-exponential factor and the activation energy in Equation (1). The latter could be evaluated without presuming any specific form of the degradation function $f(\alpha)$, while changes in E_a vs. α changes are assumed to be related to changes in the degradation mechanism. Isoconversional methods require a series of experiments with different temperature programs and obtaining E_a as a function of the conversion degree [118].

The activation energy can be calculated by various methods. Friedman's method is based on Equation (2) [101,109,118]:

$$\ln\frac{d\alpha}{dT} = \ln\frac{A}{\beta} + \ln f(\alpha) - \frac{E_a}{RT}, \tag{3}$$

It is seen from Equation (2) that if the function $f(\alpha)$ is constant for a particular value of α, the sum of the first two terms in Equation (3) also give a constant. Then, plotting $\ln(d\alpha/dT)$ vs. $1/T$ give straight lines with the slope $(-E_a/R)$.

In the Ozawa–Flynn–Wall method [109,118,120], it is assumed that the conversion function $f(\alpha)$ is invariant to the heating rate irrespective of the degree of conversion α. Equation (2) could be written as follows:

$$\ln\beta = \ln\left[\frac{Af(\alpha)}{d\alpha/dT}\right] - \frac{E_a}{RT} \quad (4)$$

The method involves measuring the temperatures corresponding to fixed values of α from tests at different heating rates β. The activation energy could be determined from the slope of $\ln\beta$ vs. 1/T straight lines according to Equation (4).

Titania nanoparticles incorporated into PLA/PHBV blends catalyzed the degradation process and inhibited the diffusion of the degradation volatiles out of the sample [118]. TGA tests were performed at different heating rates and the activation energy of degradation was calculated according to the Ozawa–Flynn–Wall model Equation (4). Alternative isoconversional methods for processing thermogravimetric data are highlighted in [101,118,120].

Chrissafis et al. have compared thermal degradation mechanisms of PCL and its nanocomposites containing different nanoparticles (pristine and modified MMT, MWCNT and fumed silica). Thermogravimetric analysis using non-isothermal conditions was performed at different heating rates and the activation energies were estimated using the Ozawa–Flynn–Wall Equation (4) and Friedman methods Equation (3). It was verified that nanoparticles did not affect the degradation mechanism but only the decomposition rate and thermal stability of PCL. Accelerated decomposition of PCL was observed for nanocomposites filled with modified MMT with quaternary ammonium salts and SiO_2 nanoparticles promoted by aminolysis and hydrolytic degradation due to the presence of the reactive groups on their surface. At the same time, unmodified MMT and MWCNT inhibited thermal degradation of PCL due to the shielding effect.

Nanoreinforcing is an effective way to improve the thermal stability of BP extending their high-performance applications. For instance, the results of DSC showed that the presence of MWCNT had a nucleating effect on both the melt crystallization and the cold crystallization of PLA [15]. Similarly, it was proven that ZnO acted as a disruptor of the PLA crystallization process, causing the degradation of PLA polymer chains during melt processing, and shifted the polymer glass transition temperature (T_g) to lower temperatures [26]. Moreover, it was shown that PBS polymer matrix could effectively shield the NFC nanofiller from thermal degradation resulting in a lower mass-loss rate and degradation over a wider and upper-temperature range [55]. Adding cellulose nanofibres to glycerol plasticized starch significantly enhanced the activation energy by 52% [101]. Meanwhile, for PHB/organically modified clay nanocomposites, the activation energy did not vary greatly with the degree of degradation, denoting degradation in one step with similar values for pure PHB and all nanocomposites [103].

5.2. Photo-Oxidative Ageing

Exposure to ultraviolet (UV) light can limit the scope of applications for BP as they can become fragile during storage, transportation and outdoor use [121]. The operational environment causes oxidation and cleavage of small molecular components, which leads to the deterioration of physical properties [122]. The addition of TiO_2 nanofillers improves UV resistance and the mechanical performance of BP and conventional petroleum-based polymers [100]. According to FTIR results indicated the degradation of the poly(butylene succinate-co-butylene adipate) (PBSA) matrix caused by high-energy UV light was significantly reduced with the addition of only 1.5 wt% of TiO_2 nanoparticles.

The results obtained on viscosity analysis indicated that TiO_2 nanoparticles inhibited the chain scission of PBSA matrix under irradiation and led to the reduced deterioration of their mechanical properties than that of unmodified PBSA films during the photoaging process [100]. According to Figure 5, the relative change of complex viscosity of PBSA filled with TiO_2 nanoparticles, after 360 h of UV irradiation was maximally reduced for PBSA with 1 wt.% of TiO_2. It can be attributed to the diminished dispersion at higher filler loadings leading to faster degradation, depending on the structural integrity and the state of interfacial adhesion. By using FTIR, pyrolysis gas chromatograph-mass spectrometer (PGC-MS), DSC and SEM, similar results were also obtained for the PBSA matrix filled with ZnO nanoparticles (0.5– wt.%), demonstrating that ZnO nanoparticles can hinder the photodegradation of PBSA [54].

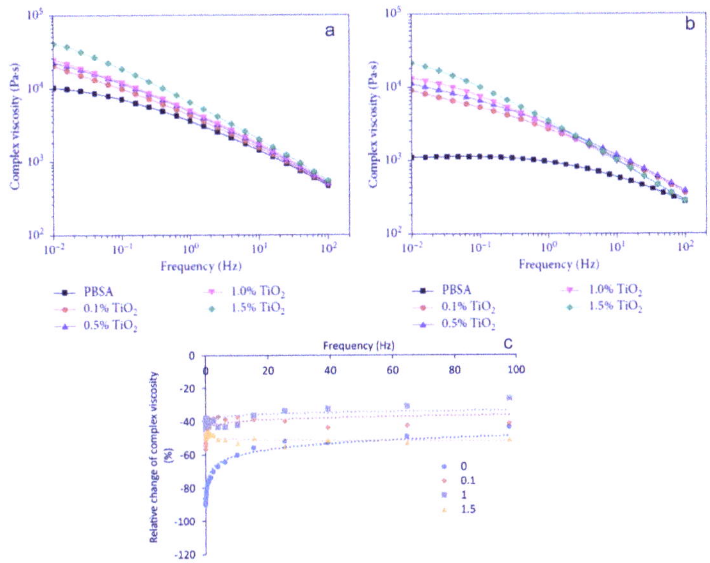

Figure 5. Complex viscosity of pure PBSA and PBSA/TiO_2 nanocomposites (**a**) before and (**b**) after 360 h of UV irradiation as a function of frequency at 140 °C (reproduced from [100], copyright © (2019). Hindawi, (**c**) relative change of complex viscosity of PBSA filled with TiO_2 (relative weight content is indicated on the graph) vs. frequency. Dots: experimental data used from [100]; lines: approximations by logarithmic functions.

The most appropriate and popular measurement of photodegradation is UV irradiation in a weatherometer [8]. This method allows outdoor accelerated exposure testing of plastics at the simulated desert and sub-tropical climatic conditions and applies to a range of polymer materials including films, sheets, laminates and extruded and moulded samples.

5.3. Water Absorption and Hygrothermal Ageing

Moisture or water affects hydrophilic constituents of BPN through immersion, cycles of spraying and condensation [123]. Water transport is governed by three mechanisms, i.e., the diffusion through micro-gaps between polymer chains, capillary transport into interfaces and transport through micro-gaps caused by swelling of hydrophilic constituents [90,124,125].

Some of the BP (e.g., PLA, PVA, starch, cellulose acetate) are well known to be able to absorb a considerable amount of water due to their amorphous nature that allows water molecules to penetrate more easily than into semi-crystalline polymers (e.g., PBS, PCL, etc.) [11,13,96,99]. To minimize the water absorption content and subsequent degradation of physical and mechanical properties of BP different nanofillers, such as ZnO [11] and CNC [96] could be added. The nanofillers act as crosslinking entanglements leading to lower water absorption in the nanocomposite than the neat polymers.

Thus, significant improvement in barrier properties of poly(D,L-lactide) (PDLLA), i.e., water absorption resistance, was obtained by the addition of the CNW (Figure 6) [13]. These results showed that even a small quantity of cellulose nanowhiskers (1 wt.%) inhibited water absorption and hence retarded the degradation, modifying the kinetics of the hydrolytic process in PDLLA polymers.

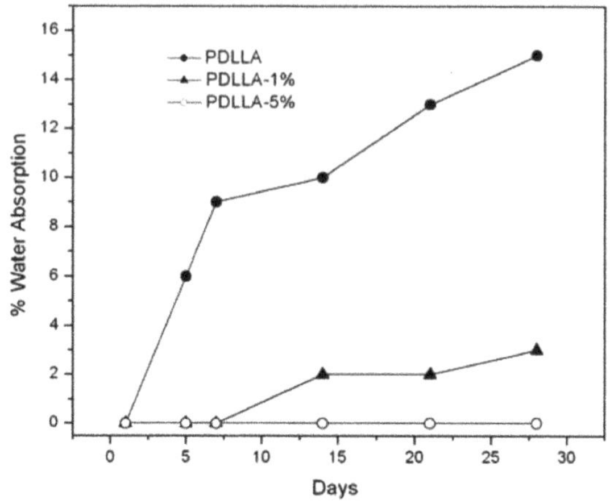

Figure 6. Kinetics of water absorption of poly(D,L-lactide) filled with cellulose nanowhiskers at different filler contents indicated on the graph. Reproduced with permission from [13]. Copyright © (2011). Elsevier Ltd. (licence No. 5135221438116).

According to the Nielsen model, the relative permeability coefficient is inverse proportional to the tortuosity factor [23,125]:

$$\frac{P}{P_0} = \frac{1-\varphi}{k} \tag{5}$$

where P and P_0 are the permeabilities of the composite and neat polymer, respectively.

The diffusion phenomena in polymers filled with filler particles could be associated with the tortuosity factor k, which is a function of the filler aspect ratio (α) and volume content (φ) [23,125–128]

$$k = 1 + \frac{\alpha}{2}\,\varphi. \tag{6}$$

For instance, PLA filled with ZnO [11,129], CaO or MgO [47], MMT [17,130], MWCNT [15] and CNW [13] are characterized by different tortuosity factors according to Equation (6). The tortuosity factor–filler volume fraction relationship is shown in Figure 6. The densities, relative weight fractions and aspect ratios for the nanofillers were taken from the data provided in the according papers.

From Figure 7 it is obvious that the tortuosity factor for 1D nanoparticles ZnO, MgO and CaO is close to unity, while 2D CNW only slightly contributes to its increasing at high filler volume fractions. MMT and MWCNT having a high aspect ratio (50 and 100, accordingly) improve the barrier properties of the polymers [90]. Hence, inhibited water absorption can retard the degradation, modifying the kinetics of the hydrolytic process in BP. Moreover, a good filler–matrix adhesion would reduce water molecules' penetration into BP to reduce the water absorption properties [131,132].

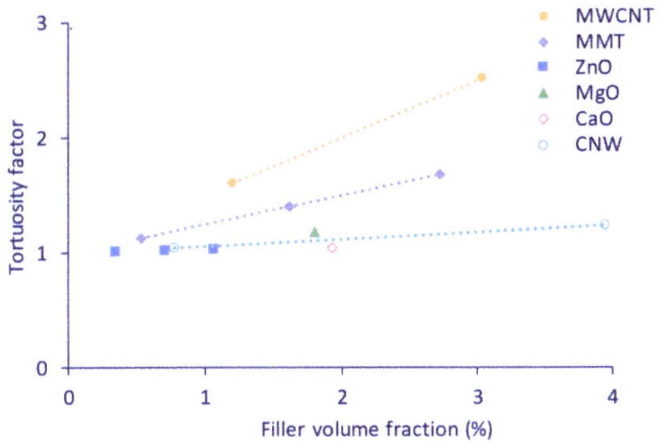

Figure 7. Tortuosity factor as a function of filler volume fraction for PLA filled with different nanofillers (indicated in the legend).

Bharadwaj [133] has modelled permeability in polymer-layered silicate nanocomposites and modified Nielsen's model by taking into account orientational effects. The Bharadwaj model is given by the following relation [99,133]:

$$\frac{P}{P_0} = \frac{1-\varphi}{1+\frac{\alpha\varphi}{2}\cdot\frac{2}{3}\cdot\left(S+\frac{1}{2}\right)} \tag{7}$$

where S is the orientation of fillers in the nanocomposites. S take the values of -0.5, 0 and 1 for fillers oriented perpendicularly, randomly and parallel to the membrane surface. It is seen that the Bharadwaj model (Equation (7)) reduces to Nielsen's model (Equation (5)) at $S = 1$.

Okamoto highlighted that relative permeability as a function is inverse proportional to the tortuosity factor for different biodegradable polymer/layered silicate nanocomposites [23].

The water absorption property of BP and BPN can also be determined by Fick's law that in some cases could be given by a simplified equation [131]:

$$\frac{M_t}{M_s} = k\cdot t^n \tag{8}$$

where M_t is the moisture content at time t, M_s is the moisture content at the saturated point and n are constants determined from the fitting curve of plot $\log(M_t/M_s)$ vs. $\log(t)$, accordingly. Thus, depending on the n-value, the moisture diffusion property of the composite can be divided into three cases: when $n = 0.5$ and the diffusion is Fickian, when $0.5 < n < 1$, and the diffusion is non-Fickian or anomalous; and when $n > 1$ [132].

Another parameter of Fick's model is the diffusion coefficient (D) which determines the ability of water molecules to diffuse and penetrate the composite structure. Its value is calculated from the slope of the plot of M_t/M_s vs. time ($t^{0.5}$) by the following equation [90,131,134]:

$$\frac{M_t}{M_s} = \left(\frac{4}{h}\right)\left(\frac{D}{\pi}\right)^{0.5}\cdot t^{0.5} \tag{9}$$

where h is the thickness of the specimen.

Cosquer et al. have studied the influence of graphene nanoplatelets (GnP) on water absorption kinetics of biodegradable PBS [99]. GnP, being hydrophobic nanofillers of a high aspect ratio, act as efficient impermeable barriers. The diffusivity of PBS decreased by about 40% compared to nanocomposites with 2wt.% GnP (Figure 8). The improvement was attributed to a purely geometric type phenomenon, i.e., with increasing the tortuosity. The tortuosity factor was estimated by the ratio of the diffusion coefficients of the neat polymer and nanocomposites by using a relation similar to Equation (5). The tortuosity factor was found to be independent of the water activity. The Bharadwaj model Equation (7) applied for fitting the experimental data on water and dioxygen permeability showed reasonable results.

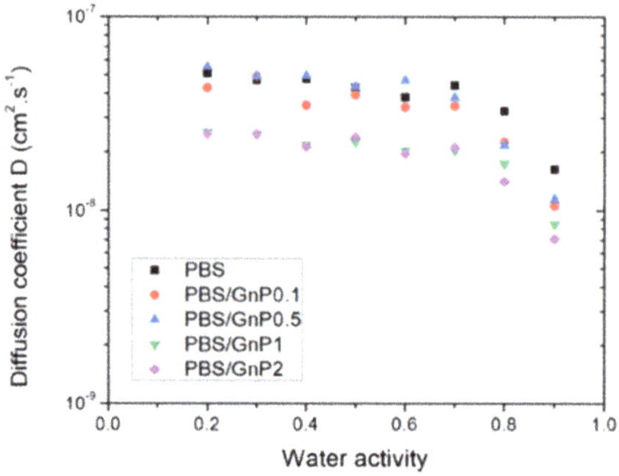

Figure 8. Diffusion coefficient as a function of water activity for neat PBS and PBS/GnP nanocomposites [99].

Hydrothermal ageing results in changes in physical (e.g., T_g) and mechanical properties. The inherent structure is deteriorated appearing in loss of interfacial adhesion and reinforcement efficiency. Thus, the durability of composites and the extent of degradation could be assessed by comparing these parameters in the reference and aged states.

The interfacial adhesion between the filler and polymer matrix plays an essential role in determining the mechanical properties of composites. Pukanszky's model is among the most widely used models for assessing the filler–matrix bond strength. Originally, the model was developed for particulate polyolefin-based composites [135,136], although later it has been successfully applied for other heterogeneous polymer systems, including bio-based and biodegradable blends [137] and nanocomposites [98]. According to Pukanszky's model, the composite strength σ_c and the polymer matrix strength σ_m are related by the equation [135–138]:

$$\sigma_c = \sigma_m \frac{1-\varphi}{1+2.5\varphi} \exp(B\varphi) \quad (10)$$

where B is the adhesion parameter: an empirical constant, which is dependent on the surface area of the particles, particles density and interfacial bonding energy. B value is 0 for very weak adhesion and can be increased, depending on the adhesion strength.

The interfacial adhesion of PLA filled with different types of fumed silica nanoparticles was estimated by Dorigato et al. [98]. The adhesion parameter B was dependent on the surface treatment of SiO_2 and varied from 3.8 to 2.5 with the highest value for pristine nanoparticles. An opposite effect of the improved interfacial adhesion with surface modifications of the filler is observed for PLA/sugarcane leaves fibre biofilms [131]. Bleaching treatment by H_2O_2 improved the interfacial adhesion between PLA and sugarcane leaves and thus enhanced biofilms' tensile strength, evidenced by the increased B factor from 6.6 to 7.5. Low adhesion factors of around 1.3 were found for PBS/wine lees [110] and PBS filled with microcrystalline cellulose particles [138]. The filler–polymer bond strength was enhanced by chemical modifications of the MCC surface [138]. Nanni and Messori have studied the strength properties of PHBH and PHBV composites filled with natural fillers [139]: the determined B values were in the range of 2.2–3.3 for pristine and 2.7–3.6 silane-treated fillers.

The interfacial adhesion is deteriorated due to ageing. By comparing B factors for pristine (unaged) and aged composites it is possible to estimate the extent of degradation on the mechanical properties quantitatively. For example, hydrothermal ageing of PBS/MCC composites [138] decreased the filler–matrix bond strength manifested in B drop from

1.37 to 0.78 for the reference and aged samples, respectively. Sugiman et al. reported about 1.4-fold decrease of B caused by water absorption in a polymer system filled with inorganic fillers [140].

The reinforcement efficiency is also affected by ageing. Nanofillers could act as reinforcement and contribute to the improvement of elastic properties of polymers. The reinforcement efficiency could be estimated in different ways. In a general case, the overall effectiveness of the reinforcement in a composite could be estimated by a simple empirical relationship [141–143]:

$$E_c = E_m(1 + r\varphi) \tag{11}$$

where r is the reinforcement efficiency factor; E_c and E_m are the elastic moduli of the composite and matrix, respectively. Platnieks et al. have studied the elastic properties of PBS/NFC composites processed by melt blending and solution casting [142]. By comparing r factors, the authors demonstrated the superior effectiveness of the former processing method. Hydrothermal ageing effects on the stiffness reduction of epoxy/graphene oxide nanocomposites appeared in the decrease of the reinforcement efficiency and r drop from 1.6 to 0.14 were highlighted in [141].

An alternative way to assess the filler contribution into the elastic properties of composites is based on an analysis of DMTA data and elastic moduli evolution when passing the glass transition. A so-called parameter C, relating the storage moduli in the glassy E'_g and rubbery E'_r, is given by the ratio [141–143]:

$$C = \frac{\left(E'_g/E'_r\right)_c}{\left(E'_g/E'_r\right)_m} \tag{12}$$

where subscripts c and m correspond to composite and matrix, respectively. For well-dispersed fillers and good compatibility with the polymer matrix, $C < 1$. The lower is C, the most efficient the reinforcement effect is. C factors of PBS filled with nanofibrillated cellulose (NFC) prepared by different processing routes were compared [142]. It was found that samples with 15 wt.% NFC processed by melt processing are characterized by higher $C = 0.69$ than those processed by solution casting. The reduced reinforcement efficiency and increase of C factors related to hydrothermal ageing effects were found in [141] by the example of epoxy/graphene oxide nanocomposites.

5.4. Creep

Viscoelastic properties of polymer-based composites have a critical role, especially in long-term applications, indicating the time-dependent deformation of materials as a function of temperature, stress and strain [144–147]. With increasing stress/temperature values, a creep compliance function becomes nonlinear on stress which can be described by various phenomenological models considering the creep of the composite and neat matrix. Different additives could be introduced to reduce creep deformations [10,98,120,139,145]. Nanofillers reduce the creep curve's elastic component and the viscous flow of the material with increment in viscoelastic deformations [145].

To estimate the effect of the fillers on the long-term deformability of BP, the creep parameters should be denoted. Figure 9 shows a schematic strain vs. time curve in a standard creep-recovery.

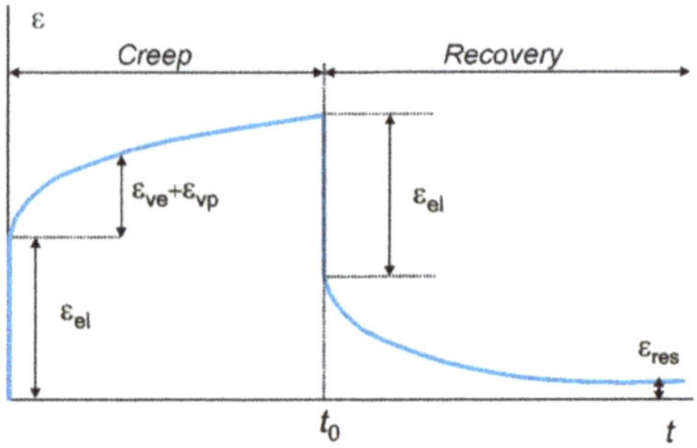

Figure 9. A schematic strain vs. time curve in a creep-recovery test.

The total creep strain ε is given by a sum of three components:

$$\varepsilon = \varepsilon_{el} + \varepsilon_{ve} + \varepsilon_{vp} \tag{13}$$

where subscripts *el*, *ve* and *vp* correspond to elastic, viscoelastic and viscoplastic strain components, respectively. The residual strain (ε_{res}) is defined as a permanent viscoplastic strain accumulated during the whole loading period and remaining after unloading after a time period longer than that of loading.

It was found [10] that the addition of cellulose nanofibrils (CNF) to starch-based nanocomposite films significantly decreased all creep deformations (viscoelastic and plastic, elastic and residual) shown in Figure 10. The concentration of CNF above 20 wt.% was found to accelerate the creep behaviour due to poor dispersion, whereas the nanocomposite films with CNF content of 15 wt.% revealed the lowest creep performance.

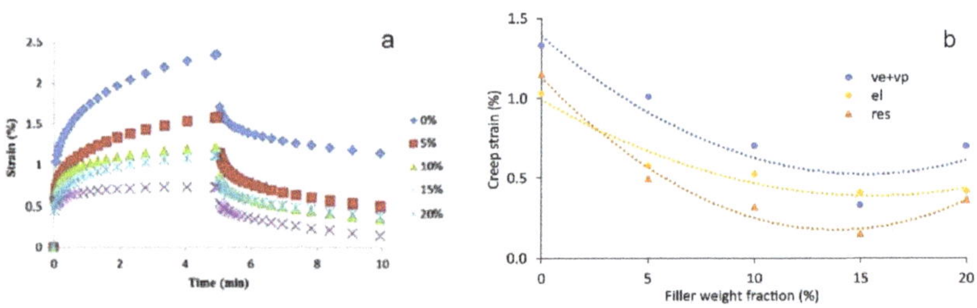

Figure 10. The (**a**) creep-recovery vs. time curves of starch only and starch–CNF composite films (reproduced with permission from [10], copyright © (2015), Elsevier Ltd., licence No. 5143001008321). (**b**) Viscoelastic and viscoplastic (ve+vp), elastic (el) and residual (res) strains of starch modified with cellulose nanofibrils vs. filler weight fraction. Dots: experimental data from [10]; lines: approximations by polynomial functions.

The most common model for creep description is the three-parameter Findley power law given by the following equation [110,139,148,149]:

$$\varepsilon(t) = \varepsilon_{el} + kt^{n} \tag{14}$$

where k and n ($0 < n < 1$) are material parameters. The power-law models are considered empirical models without attaching importance to a physical background.

The Burgers model is a combination of Maxwell and Kelvin–Voigt elements connected in series. According to Equation (13) and the Burgers model formulation, the creep strain is given by the following relation [10,98,139,150]:

$$\varepsilon(t) = \frac{\sigma}{E_M} + \frac{\sigma}{E_K}\left[1 - \exp\left(-\frac{t}{\tau}\right)\right] + \frac{\sigma}{\eta_M}t \tag{15}$$

where E_M and E_K are the elastic moduli of the Maxwell and Kelvin springs, $\tau = \eta_K/E_K$ is the retardation time of the Kelvin–Voigt element; η_K and η_M are the viscosities of the Kelvin and Maxwell dashpots, respectively.

The creep of glassy solids and semicrystalline polymers is described by the Kohlrausch–Williams–Watts (KWW) model. This is based on considerations that "viscoelastic changes in polymeric matrices occur because of incremental molecular jumps due to several segments chains jumps between different positions of relative stability". The creep strain is given by a Weibull-like function [120]:

$$\varepsilon(t) = \varepsilon_i + \varepsilon_c\left[1 - \exp\left(-\left(\frac{t}{t_c}\right)^{\beta_c}\right)\right] \tag{16}$$

where ε_i is the instantaneous elastic strain, ε_c is the limit viscous creep strain, t_c and β_c are the scale (characteristic time) and shape parameters, respectively. Expanding Equation (16) in a series and considering the first term only derives from the Findley model Equation (14).

The Weibull distribution equation is also applied to model the creep recovery behaviour. The recovery strain ε_{rec} is determined by the viscoelastic strain recovery ε_c and the residual strain ε_{res} caused by a viscous flow effect and is given by the following equation [110]:

$$\varepsilon_{rec}(t) = \varepsilon_c\left[\exp\left(-\left(\frac{t - t_0}{t_r}\right)^{\beta_r}\right)\right] + \varepsilon_{res}, \tag{17}$$

where t_0 is time of stress removal, t_r and β_r are the characteristic time and shape parameters, respectively.

Temperature growth results in accelerating relaxation processes in polymers and thus changing their viscoelastic response (e.g., creep compliance, relaxation modulus). This fact is widely applied to predict the long-term properties of polymers and their composites by using Time–Temperature Superposition Principle (TTSP) [146,147]. TTSP is based on the assumption that time and temperature are interrelated and interequivalent. A temperature increase leads to a parallel shift of the relaxation spectrum of a polymer, and this shift is characterised by so-called shift factors a_T. The long-term viscoelastic behaviour is predicted by shifting the short-term test data presented in logarithmic time axes to a generalized master curve for $\log a_T$ values. The lifetime of a polymer system t at an operating temperature T is determined by a ratio of the shift factors according to the relation [118]

$$t = \frac{a_{T_0}}{a_T} \cdot t_0 \tag{18}$$

where T_0 is the reference temperature, t_0 is the lifetime at T_0; a_{T_0} and a_T are the shift factors at corresponding temperatures. For simplicity, $a_{T_0} = 1$ is usually taken. TTSP has temperature limitations in terms of the shift function. The Williams–Landel–Ferry (WLF) equation is valid for the temperature range between T_g and T_g +100 °C [146,147]:

$$\log a_T = -\frac{C_1(T - T_0)}{C_2 + T - T_0} \tag{19}$$

where C_1 and C_2 are material parameters.

The Arrhenius equation is applied for a_T calculations at $T < T_g$:

$$\log a_T = \frac{E_a}{2.303R}\left(\frac{1}{T} - \frac{1}{T_0}\right) \tag{20}$$

where T_0 is taken in Kelvin; other designations are the same as in Equation (2).

Nanni and Messori [139] applied the Burger, KWW and Findley models to describe the nonlinear creep of PBS biocomposites filled with wine lees. By comparing parameters of the models representing elastic, viscoelastic and viscoplastic strain components, it has been quantitatively demonstrated that the addition of fillers into PBS resulted in the reduced creep of biocomposites. The earlier authors' study successfully applied the KWW model to describe creep PHBH- and PHBV-based biocomposites [120]. The master curves were generated by applying TTSP and the temperature shift factors were calculated according to the WLF model Equation (19). The long-term predictions for wine lees–filled biopolymers demonstrated much lower creep in the same time spans.

The creep and creep-recovery behaviour of starch-based nanocomposite films with CNF up to 20 wt.% have been studied by Li et al. [10]. The experimental data were effectively fitted by the Burgers model with parameters strongly dependent on the amount of the filler. TTSP was successfully applied for predicting the long-term creep behaviour of nanocomposites. The temperature shift factors were calculated according to the Arrhenius equation Equation (20).

Amiri et al. applied the Findley model to describe nonlinear creep of bio-based resin (methacrylated epoxidized sucrose soyate (MESS) reinforced with flax fibres [148]. Following TTSP and considering biocomposites as thermorheologically complex materials, the authors used horizontal and vertical shifts to generate the master curves. The long-term prediction agreed well with the experimental data.

In a recent study of Ollier et al. [149], creep of PCL reinforced with pristine and organo-modified bentonites up to 3 wt.% was investigated. The Findley and Burgers models were applied. Master curves were constructed using TTSP demonstrating substantial improvement in the creep resistance of nanocomposites for the long term.

The improved creep stability of PLA filled with fumed silica nanoparticles of different specific surface areas and surface functionalization was discovered by Dorigato et al. [98]. The authors applied the Burgers model and demonstrated that nanoparticles mainly contribute to the increased values of viscous components (η_K and η_M in Equation (15)).

Guedes et al. [150] and the Burgers model applied the modified three-element standard solid model of Kontou–Zacharatos [106] to describe the nonlinear viscoelastic behaviour of PLA-PCL fibres monitored in creep, stress-relaxation and quasi-static tensile tests. Dry and saturated in saline solution fibres were tested. Based on phenomenological considerations, the elastic spring is replaced by a nonlinear strain-dependent spring and the linear dashpot in the Maxwell element is replaced by an Eyring type one. Such modifications allowed authors to reduce the required fitting parameters and describe nonlinear viscoelastic–viscoplastic behaviour under moderate and large deformations, both in monotonic and cyclic loading.

Ding and coworkers have studied creep and stress relaxation of PBAT biocomposites containing CNT up to 5 wt.% [110]. The authors applied the Findley and Burgers models for creep description, while creep recovery was modelled by the Weibull distribution function given by Equation (17). PBAT/CNT nanocomposites possessed lower viscoelastic and viscoplastic strains that appeared in changed creep model parameters than the unfilled polymer.

Kontou et al. have studied the time-dependent behaviour of PBAT/PLA blends (commercial name Ecovio®) reinforced with different types of wood fibres up to 30 wt.% [106,151]. Highly nonlinear viscoelastic/viscoplastic behaviour observed in creep tests was modelled by the Burgers and Findley models [106]. A constitutive model presenting a combination of the transient network model, related to the viscoelasticity, with a plasticity theory has been developed in [115] describing the experimental data of stress-relaxation, monotonic loading and creep-recovery in a unified manner.

Qiu et al. have studied the time-dependent plastic failure of PLA/PBS blends in tensile tests at different strain rates [123]. The deformation behaviour of the blends with the improved ductility was modelled by the Chaboche viscoplastic model with nonlinear hardening variables.

5.5. Modelling of Mechanical Properties Accompanied by Biodegradation

The structure and properties of biodegradable polymers change in time comparable to non-biodegradable counterparts, with the test time and service life of materials. This fact should be considered when modelling such materials' time-dependent properties (e.g., creep, stress relaxation, fatigue, etc.). Parameters involved in traditional models need to be related to "inherent" degradation of the structure that could also be accelerated by an external mechanical load. In several studies for different biopolymers [148,152–156], degradation of mechanical properties (e.g., tensile strength) is directly related to molecular weight reduction.

The time-dependent behaviour of biodegradable PLA-PCL fibres during their hydrolytic degradation in a phosphate buffer medium has been studied by Viera et al. [152,153] (37 °C, 16 weeks). It was found that the decrease of tensile strength σ of the fibres follows the same trend as the decrease of molecular weight M_n. Modelling hydrolysis as a first-order kinetic mechanism, the hydrolytic damage d_h is defined as follows [152]:

$$d_h = 1 - \frac{\sigma_t}{\sigma_0} = 1 - \frac{M_{nt}}{M_{n0}} = 1 - e^{-ut} \quad (21)$$

where u is is the degradation rate of the material. Subscripts t and 0 are related to the corresponding parameters at current time t and initial (non-degraded) values. By incorporating Equation (21) into the constitutive models such as Neo-Hookean and Mooney–Rivlin hyperelastic models [152] and Bergström–Boyce viscoelastic model for polymer undergoing large deformations [153], stress–strain behaviour of PLA-PCL fibres for different degradation times was effectively predicted. The approach could be extended to other biodegradable polymers.

Singh and coauthors have studied the effect of hydrolytic degradation and strain rate on the tensile properties of PLA fibres [156]. The authors applied the modified three-element standard solid model (according to Khan [157]) introducing degradation-dependent stiffness parameters, while viscous (strain-dependent) parameters are assumed to be unaffected by degradation. Stiffness (Young modulus E) degradation was presented in a similar to Equation (21) way by using an exponential law:

$$\frac{E_t}{E_0} = e^{-ut} \quad (22)$$

The model Equation (22) parameters are determined experimentally by assessing E changes upon different degradation times.

Breche and coauthors have studied the evolution of the physical and mechanical properties of PLA-b-PEG-b-PLA biodegradable triblock copolymers caused by hydrolytic degradation [154]. Samples were immersed in phosphate buffer solution at 37 °C for up to 12 weeks. The stress relaxation was modelled by a linear viscoelastic model introducing a stiffness-related degradation variable. Similarly to [156], the viscous component was considered independent of degradation time, at least at the early stages. Then, following Viera et al.'s definition [152] and Equation (21), the degradation variable is defined as

$$d_h = 1 - \frac{\sigma_{max}(d_h)}{\sigma_{max}(d_h = 0)} = 1 - \frac{\sigma_{relax}(d_h, 0)}{\sigma_{relax}(d_h = 0, 0)} \quad (23)$$

where σ_{max} is the maximum stress reached at the end of load for a given degradation state d_h and corresponds to the time zero of relaxation, while σ_{relax} is the initial value of the relaxation stress for degraded and undegraded material. The degradation variable Equation (23) was linked to the molecular weight changes of the material according to the following dependencies [153]:

$$\begin{aligned} d_h &= c\cdot\left(1-\frac{M_{nt}}{M_{n0}}\right), && t < t_c \\ d_h &= a\cdot exp\left(-b\frac{M_{nt}}{M_{n0}}\right), && t \geq t_c \end{aligned} \tag{24}$$

where a, b and c are material parameters; t_c is a critical time, when the evolution of properties degradation is changed considerably (3 weeks for the PLA-b-PEG-b-PLA copolymers understudy).

Zhang et al. [158] proposed a semiempirical approach for predicting the strength of biodegradable medical polyesters, namely, PLA and polyglycolide (PGA) and their copolymers, during hydrolytic degradation. Three different phases in the mesoscopic-scale (amorphous, crystalline and vacancy phases) were defined and further integrated into the multiscale heterogeneous strength model. The strength of the amorphous and crystalline phases was related to molecular weight through power-law dependencies, while the cavity-related zones were considered as the zero-strength phase. The total strength is given by the equation:

$$\sigma(t) = \sigma_{A0}\cdot\alpha_A\left[\frac{M_{nA}(t)}{M_{n0}}\right]^{\beta_A} + \sigma_{C0}\cdot\alpha_C\left[\frac{M_{nC}(t)}{M_{n0}}\right]^{\beta_C} \tag{25}$$

where σ_0 and M_{n0} are the initial strength and molecular weight, respectively. $M_n(t)$ is the molecular weight at time t, α and β are material parameters. Subscripts A and C in Equation (25) are related to the amorphous and crystalline phases, respectively.

The considered models and methods may be adapted and used for other biodegradable polymers and nanocomposites. The incorporation of nanoparticles into BP contributed to the improved barrier properties and decreased degradation of polymers [13,14,111,141] that will appear in the decreased damage parameter d_h in Equations (21)–(25).

The models could be considered effective tools in designing biopolymer composites with tailored degradation and durability.

6. Conclusions

Environmental degradation mainly promotes a significant decrease in mechanical properties, particularly when the molecular weight of BP is low. Thus, incorporating nanofillers into BP could attenuate the loss of mechanical properties and improve durability. At the same time, in the case of poor dispersion, the nanofillers can lead to faster degradation, depending on the structural integrity and the state of interfacial adhesion. To fully understand and interrelate numerous factors (e.g., moisture, temperature, etc.) that can affect the degradation process of BPN, combined and comprehensive scientific investigations are required.

A correlation between outdoor and accelerated weathering should be established experimentally and particular analytical models should be developed. The degradation of mechanical properties could be modelled as a function of the duration of environmental ageing. The incorporation of nanoparticles into BP contributed to the improved barrier properties and decreased degradation of BP. The models considered in the review could be effective tools in designing biopolymer composites with tailored degradation and durability. Moreover, models could be developed to combine the effects of temperature and humidity to predict the durability of BP and BPN.

Author Contributions: Conceptualization, T.G.-K. and O.S.; writing—original draft preparation, T.G.-K., O.S., S.G., O.P. and G.G.; writing—review and editing, T.G.-K., O.S. and S.G.; funding acquisition, O.S. All authors have read and agreed to the published version of the manuscript.

Funding: This research was funded by Latvian Research Council: project No. lzp-2020/2-0207 "Durability of sustainable bio-based polymer nanocomposites under the combined effects of mechanical loading and environmental ageing". G.G. has been supported by the European Regional Development Fund within the Activity 1.1.1.2 "Post-doctoral Research Aid" of the Specific Aid Objective 1.1.1 "To increase the research and innovative capacity of scientific institutions of Latvia and the ability to attract external financing, investing in human resources and infrastructure" of the Operational Programme "Growth and Employment" (No. 1.1.1.2/VIAA/3/19/478).

Data Availability Statement: Not applicable.

Conflicts of Interest: The authors declare no conflict of interest.

Abbreviations

BP	biodegradable polymer
BPN	biodegradable polymer nanocomposites
CNC	cellulose nanocrystals
CNF	cellulose nanofibrils
CNT	carbon nanotubes
CNW	cellulose nanowhiskers
DLS	dynamic light scattering
DMTA	dynamic mechanical thermal analysis
DSC	differential scanning calorimetry
FTIR	Fourier-transform infrared spectroscopy
GL	glycerol
GnP	graphene nanoplatelets
GO	graphene oxide
HN	halloysite nanotubes
KWW	Kohlrausch–Williams–Watts
MCC	microcrystalline cellulose
MESS	methacrylated epoxidized sucrose soyate
MMT	montmorillonite
MWCNT	multiwall carbon nanotubes
NFC	nanofibrillated cellulose
NMR	nuclear magnetic resonance
PBAT	polybutylene adipate terephthalate
PBS	polybutylene succinate
PBSA	polybutylene succinate adipate
PEG	poly(ethylene glycol)
PCL	polycaprolactone
PDLLA	poly(D,L-lactide)
PLA	polylactic acid
PHA	polyhydroxyalkanoates
PHB	polyhydroxybutyrate
PHBV	poly(3-hydroxybutyrate-co-3-hydroxyvalerate)
PVOH	polyvinyl alcohol
PHV	polyhydroxy valerate
SEM	scanning electron microscopy
ST	starch
TEM	transmission electron microscopy
TGA	thermogravimetric analysis
TPS	thermoplastic starch
TTSP	time–temperature superposition principle
UV	ultraviolet
WLF	Williams–Landel–Ferry

References

1. Sharma, B.; Malik, P.; Jain, P. Biopolymer reinforced nanocomposites: A comprehensive review. *Mater. Today Commun.* **2018**, *16*, 353–363. [CrossRef]
2. Titone, V.; Correnti, A.; La Mantia, F.P. Effect of moisture content on a biodegradable polyester's processing and mechanical properties. *Polymers* **2021**, *13*, 1616. [CrossRef]
3. Ates, B.; Koytepe, S.; Ulu, A.; Gurses, C.; Thakur, V.K. Chemistry, structures, and advanced applications of nanocomposites from biorenewable resources. *Chem. Rev.* **2020**, *120*, 17, 9304–9362. [CrossRef]
4. Remiš, T.; Belský, P.; Kovárík, T.; Kadlec, J.; Azar, M.G.; Medlín, R.; Vavrunková, V.; Deshmukh, K.; Sadasivuni, K.K. Study on structure, thermal behavior, and viscoelastic properties of nanodiamond-reinforced poly (vinyl alcohol) nanocomposites. *Polymers* **2021**, *13*, 1426. [CrossRef]
5. Rouf, T.B.; Kokini, J.L. Biodegradable biopolymer–graphene nanocomposites. *J. Mater. Sci.* **2016**, *51*, 9915–9945. [CrossRef]
6. Fukushima, K.; Abbate, C.; Tabuani, D.; Gennari, M.; Camino, G. Biodegradation of poly(lactic acid) and its nanocomposites. *Polym. Degrad. Stab.* **2009**, *94*, 1646–1655. [CrossRef]
7. Murariu, M.; Dubois, P. PLA composites: From production to properties. *Adv. Drug Deliv. Rev.* **2016**, *107*, 17–46. [CrossRef]
8. Pandey, J.K.; Reddy, K.R.; Pratheep Kumar, A.; Singh, R.P. An overview on the degradability of polymer nanocomposites. *Polym. Degrad. Stab.* **2005**, *88*, 234–250. [CrossRef]
9. Chang, B.P.; Mohanty, A.K.; Misra, M. Studies on durability of sustainable biobased composites: A review. *RSC Adv.* **2020**, *10*, 17955. [CrossRef]
10. Li, M.; Li, D.; Wang, L.-J.; Adhikari, B. Creep behavior of starch-based nanocomposite films with cellulose nanofibrils. *Carbohydr. Polym.* **2015**, *117*, 957–963. [CrossRef]
11. Pantani, R.; Gorrasi, G.; Vigliotta, G.; Murariu, M.; Dubois, P. PLA-ZnO nanocomposite films: Water vapor barrier properties and specific end-use characteristics. *Eur. Polym. J.* **2013**, *49*, 3471–3482. [CrossRef]
12. Reddy, M.M.; Vivekanandhan, S.; Misra, M.; Bhatia, S.K.; Mohanty, A.K. Biobased plastics and bionanocomposites: Current status and future opportunities. *Prog. Polym. Sci.* **2013**, *38*, 1653–1689. [CrossRef]
13. De Paula, E.L.; Mano, V.; Pereira, F.V. Influence of cellulose nanowhiskers on the hydrolytic degradation behavior of poly(D,L-lactide). *Polym. Degrad. Stab.* **2011**, *96*, 1631–1638. [CrossRef]
14. De Paula, E.L.; Mano, V.; Duek, E.A.R.; Pereira, F.V. Hydrolytic degradation behavior of PLLA nanocomposites reinforced with modified cellulose nanocrystals. *Quím. Nova* **2015**, *38*, 1014–1020. [CrossRef]
15. Wu, D.; Wu, L.; Zhou, W.; Zhang, M.; Yang, T. Crystallization and biodegradation of polylactide/carbon nanotube composites. *Polym. Eng. Sci.* **2010**, *50*, 1721–1733. [CrossRef]
16. Petchwattana, N.; Covavisaruch, S.; Wibooranawong, S.; Naknaen, P. Antimicrobial food packaging prepared from poly(butylene succinate) and zinc oxide. *Measurement* **2016**, *93*, 442–448. [CrossRef]
17. Sharma, R.; Jafari, S.M.; Sharma, S. Antimicrobial bio-nanocomposites and their potential applications in food packaging. *Food Control* **2020**, *112*, 107086. [CrossRef]
18. Kumar, B.; Castro, M.; Feller, J.F. Poly(lactic acid)–multi-wall carbon nanotube conductive biopolymernanocomposite vapour sensors. *Sens. Actuat. B Chem.* **2012**, *161*, 621–628. [CrossRef]
19. Zheng, X.; Zhou, S.; Xiao, Y.; Yu, X.; Li, X.; Wu, P. Shape memory effect of poly(d,l-lactide)/Fe_3O_4 nanocomposites by inductive heating of magnetite particles. *Colloids Surf. B Biointerfaces* **2009**, *71*, 67–72. [CrossRef]
20. Silva, A.P.B.; Montagna, L.S.; Passador, F.R.; Rezende, M.C.; Lemes, A.P. Biodegradable nanocomposites based on PLA/PHBV blend reinforced with carbon nanotubes with potential for electrical and electromagnetic applications. *Express Polym. Lett.* **2021**, *15*, 987–1003. [CrossRef]
21. Armentano, I.; Puglia, D.; Luzi, F.; Arciola, C.R.; Morena, F.; Torre, S.M.; Torre, L. Nanocomposites based on biodegradable polymers. *Materials* **2018**, *11*, 795. [CrossRef]
22. Bari, S.S.; Chatterjee, A.; Mishra, S. Biodegradable polymer nanocomposites: An overview. *Polym. Rev.* **2016**, *56*, 287–328. [CrossRef]
23. Okamoto, M. Biodegradable Polymer/Layered Silicate Nanocomposites: A Review. In *Handbook of Biodegradable Polymeric Materials and Their Applications*; Mallapragada, S., Narasimhan, B., Eds.; Iowa State University: Ames, IO, USA, 2005; Volume 1, pp. 1–45. ISBN 1-58883-053-5.
24. Abdoulhdi, A.B.O.; Abdulrahman, A.B.A.M.; Sapuan, S.M.; Ilyas, R.A.; Asyraf, M.R.M.; Rahimian, K.S.S.; Petru, M. Micro- and Nanocellulose in Polymer Composite Materials: A Review. *Polymers* **2021**, *13*, 231. [CrossRef]
25. Rafiqah, S.A.; Khalina, A.; Harmaen, A.S.; Tawakkal, I.A.; Zaman, K.; Asim, M.; Nurrazi, M.N.; Lee, C.H. A Review on Properties and Application of Bio-Based Poly(Butylene Succinate). *Polymers* **2021**, *13*, 1436. [CrossRef]
26. Anžlovar, A.; Kržan, A.; Žaga, E. Degradation of PLA/ZnO and PHBV/ZnO composites prepared by melt processing. *Arab. J. Chem.* **2018**, *11*, 343–352. [CrossRef]
27. Sawpan, M.A.; Islam, M.R.; Beg, M.D.H.; Pickering, K. Effect of accelerated weathering on physico-mechanical properties of polylactide bio-composites. *J. Environ. Polym. Degrad.* **2019**, *27*, 942–955. [CrossRef]
28. Mtibe, A.; Motloung, M.P.; Bandyopadhyay, J.; Ray, S.S. Synthetic biopolymers and their composites: Advantages and limitations—an overview. *Macromol. Rapid Commun.* **2021**, *42*, 2100130. [CrossRef]

29. Chiellini, E.; Corti, A.; D'Antone, S.; Solaro, R. Biodegradation of poly (vinyl alcohol) based materials. *Prog. Polym. Sci.* **2003**, *28*, 963–1014. [CrossRef]
30. Esposti, D.M.; Morselli, D.; Fava, F.; Bertin, L.; Cavani, F.; Viaggi, D.; Fabbri, P. The role of biotechnology in the transition from plastics to bioplastics: An opportunity to reconnect global growth with sustainability. *FEBS Open Bio* **2021**, *11*, 967–983. [CrossRef]
31. Vroman, I.; Tighzert, L. Biodegradable Polymers. *Materials* **2009**, *2*, 307–344. [CrossRef]
32. Yiga, V.A.; Lubwama, M.; Pagel, S.; Benz, J.; Olupot, P.W.; Bonten, C. Flame retardancy and thermal stability of agriculturalresidue fiber-reinforc ed polylactic acid: A Review. *Polym. Compos.* **2021**, *42*, 15–44. [CrossRef]
33. Higuchi-Takeuchi, M.; Morisaki, K.; Toyooka, K.; Numata, K. Synthesis of high-molecular-weight polyhydroxyalkanoates by marine photosynthetic purple bacteria. *PLoS ONE* **2016**, *11*, e0160981. [CrossRef]
34. Xu, D.J.; Guo, B.-H. Poly(butylene succinate) and its copolymers: Research, development and industrialization. *Biotechnol. J.* **2010**, *5*, 1149–1163. [CrossRef]
35. Van de Velde, K.; Kiekens, P. Biopolymers: Overview of several properties and consequences on their applications. *Polym. Test.* **2002**, *21*, 433–442. [CrossRef]
36. Zhang, Y.; Rempel, C.; McLaren, D.C. 16—Thermoplastic Starch in Innovations. In *Food Packaging*, 2nd ed.; Han, J.H., Ed.; Academic Press: Cambridge, MA, USA, 2014; pp. 391–412. [CrossRef]
37. Lackner, M.; Ivanič, F.; Kováčová, M.; Chodák, I. Mechanical properties and structure of mixtures of poly(butylene-adipate-co-terephthalate) (PBAT) with thermoplastic starch (TPS). *Int. J. Biobased Plast.* **2021**, *3*, 126–138. [CrossRef]
38. Wang, X.; Peng, S.; Chen, H.; Yu, X.; Zhao, X. Mechanical properties, rheological behaviors, and phase morphologies of high-toughness PLA/PBAT blends by in-situ reactive compatibilization. *Compos. B Eng.* **2019**, *173*, 107028. [CrossRef]
39. Petersson, L.; Kvien, I.; Oksman, K. Structure and thermal properties of poly(lactic acid)/cellulose whiskers nanocomposite materials. *Compos. Sci. Technol.* **2007**, *67*, 2535–2544. [CrossRef]
40. Vytejčková, S.; Vápenkab, L.; Hradecký, J.; Dobiáš, J.; Hajšlová, J.; Loriot, C.; Vannini, L.; Poustka, J. Testing of polybutylene succinate based films for poultry meat packaging. *Polym. Test.* **2017**, *60*, 357–364. [CrossRef]
41. Wei, D.; Wang, H.; Ziaee, Z.; Chibante, F.; Zheg, A.; Xiao, H. Non-leaching antimicrobial biodegradable PBAT films through a facile and novel approach. *Mater. Sci. Eng. C* **2016**, *58*, 986–991. [CrossRef]
42. Darie-Niță, R.N.; Râpă, M.; Sivertsvik, M.; Rosnes, J.T.; Popa, E.E.; Dumitriu, R.P.; Marincaș, O.; Matei, E.; Predescu, C.; Vasile, C. PLA-based materials containing bio-plasticizers and chitosan modified with rosehip seed oil for ecological packaging. *Polymers* **2021**, *13*, 1610. [CrossRef]
43. Mekonnen, T.; Mussone, P.; Khalil, H.; Bressler, D. Progress in bio-based plastics and plasticizing modifications. *J. Mater. Chem. A* **2013**, *1*, 13379. [CrossRef]
44. Cho, J.Y.; Park, S.L.; Lee, H.-J.; Kim, S.H.; Suh, M.J.; Ham, S.; Bhatia, S.K.; Gurav, R.; Park, S.-H.; Park, K.; et al. Polyhydroxyalkanoates (PHAs) degradation by the newly isolated marine Bacillus sp. JY14. *Chemosphere* **2021**, *283*, 131172. [CrossRef]
45. Myung, J.; Flanagan, J.C.A.; Waymouth, R.M.; Criddle, C.S. Expanding the range of polyhydroxyalkanoates synthesized by methanotrophic bacteria through the utilization of omega-hydroxyalkanoate co-substrates. *AMB Express* **2017**, *7*, 118. [CrossRef]
46. Signori, F.; Coltelli, M.-B.; Bronco, S. Thermal degradation of poly(lactic acid) (PLA) and poly(butylene adipate-co-terephthalate) (PBAT) and their blends upon melt processing. *Polym. Degrad. Stab.* **2009**, *94*, 74–82. [CrossRef]
47. Fan, Y.; Nishida, H.; Moria, T.; Shirai, Y.; Endo, T. Thermal degradation of poly(L-lactide): Effect of alkali earth metal oxides for selective L,L-lactide formation. *Polymer* **2004**, *45*, 1197–1205. [CrossRef]
48. Nazrin, A.; Sapuan, S.M.; Zuhri, M.Y.M.; Ilyas, R.A.; Syafiq, R.; Sherwani, S.F.K. Nanocellulose re109124inforced thermoplastic starch (TPS), polylactic acid (PLA), and polybutylene succinate (PBS) for food packaging applications. *Front. Chem.* **2020**, *8*, 213. [CrossRef]
49. Raquez, J.-M.; Habibi, Y.; Murariu, M.; Dubois, P. Polylactide (PLA)-based nanocomposites. *Prog. Polym. Sci.* **2013**, *38*, 1504–1542. [CrossRef]
50. Vasile, C.; Râpă, M.; Ștefan, M.; Stan, M.; Macavei, S.; Darie-Niță, R.N.; Barbu-Tudoran, L.; Vodnar, D.C.; Popa, E.E.; Ștefan, R.; et al. New PLA/ZnO:Cu/Ag bionanocomposites for food packaging. *Express Polym. Lett.* **2017**, *11*, 531–544. [CrossRef]
51. Luzi, F.; Fortunati, E.; Jiménez, A.; Puglia, D.; Chiralt, A.; Torre, L. PLA nanocomposites reinforced with cellulose nanocrystals from posidonia oceanica and ZnO nanoparticles for packaging application. *J. Renew. Mater.* **2017**, *5*, 2. [CrossRef]
52. Sung, S.H.; Chang, Y.; Han, J. Development of polylactic acid nanocomposite films reinforced withcellulose nanocrystals derived from coffee silverskin. *Carbohydr. Polym.* **2017**, *169*, 495–503. [CrossRef] [PubMed]
53. Zinge, C.; Kandasubramanian, B. Nanocellulose based biodegradable polymers. *Eur. Polym. J.* **2020**, *133*, 109758. [CrossRef]
54. Zhang, Y.; Xu, J.; Guo, B. Photodegradation behavior of poly(butylene succinate-co-butyleneadipate)/ZnO nanocomposites. *Colloids Surf. A Physicochem.* **2016**, *489*, 173–181. [CrossRef]
55. Platnieks, O.; Gaidukovs, S.; Barkane, A.; Sereda, A.; Gaidukova, G.; Grase, L.; Thakur, V.K.; Filipova, I.; Fridrihsone, V.; Skute, M.; et al. Bio-based poly(butylene succinate)/microcrystalline cellulose/nanofibrillated cellulose-based sustainable polymer composites: Thermo-mechanical and biodegradation studies. *Polymers* **2020**, *12*, 1472. [CrossRef]
56. Peighambardoust, S.J.; Peighambardoust, S.H.; Pournasir, N.; Pakdel, P.M. Properties of active starch-based films incorporating a combination of Ag, ZnO and CuO nanoparticles for potential use in food packaging applications. *Food Packag. Shelf* **2019**, *22*, 100420. [CrossRef]

57. Gibril, M.E.; Ahmed, K.K.; Lekha, P.; Sithole, B.; Khosla, A.; Furukawa, H. Effect of nanocrystalline cellulose and zinc oxide hybrid organic–inorganic nanofiller on the physical properties of polycaprolactone nanocomposite films. *Microsyst. Technol.* **2019**. [CrossRef]
58. Chen, C.; Lv, G.; Pan, C.; Song, M.; Wu, C.; Guo, D.; Wang, X.; Chen, B.; Gu, Z. Poly(lactic acid) (PLA) based nanocomposites—A novel way of drug-releasing. *Biomed. Mater.* **2007**, *2*, L1–L4. [CrossRef] [PubMed]
59. Abou-Yousef, H.; Saber, E.; Abdel-Aziz, M.S.; Kamel, S. Efficient alternative of antimicrobial nanocomposites based on cellulose acetate/Cu-NPs. *Soft Mater.* **2018**, *16*, 141–150. [CrossRef]
60. Van der Zee, M. Biodegradability of Polymers Mechanisms and Evaluation Methods. In *Handbook of Biodegradable Polymers*; Bastioli, C., Ed.; RAPRA Technology Limited: Telford, UK, 2006.
61. Leja, K.; Lewandowicz, G. Polymer biodegradation and biodegradable polymers—A review. *Polish J. Environ. Stud.* **2010**, *19*, 255–266.
62. Müller, R.J. Biodegradation Behavior of Polymers in Liquid Environments. In *Handbook of Biodegradable Polymers*; Bastioli, C., Ed.; RAPRA Technology Limited: Telford, UK, 2006.
63. Lucas, N.; Bienaime, C.; Belloy, C.; Queneudec, M.; Silvestre, F.; Nava-Saucedo, J.-E. Polymer biodegradation: Mechanisms and estimation techniques. *Chemosphere* **2008**, *73*, 429–442. [CrossRef] [PubMed]
64. Walsh, J.H. Ecological considerations of biodeterioration. *Int. Biodeteter. Biodegrad.* **2001**, *48*, 16–25. [CrossRef]
65. Frits, J. Ecotoxicological Aspects in the Biodegradation Process of Polymers. In *Handbook of Biodegradable Polymers*; Bastioli, C., Ed.; RAPRA Technology Limited: Telford, UK, 2006.
66. El-Hadi, A.; Schnabel, R.; Straube, E.; Muller, G.; Henning, S. Correlation between degree of crystallinity, morphology, glass temperature, mechanical properties and biodegradation of poly(3-hydroxyalkanoate) PHAs and their blends. *Polym. Test.* **2002**, *21*, 665–674. [CrossRef]
67. Huang, S.J. Poly(lactic acid) and Copolyesters. In *Handbook of Biodegradable Polymers*; Bastioli, C., Ed.; RAPRA Technology Limited: Telford, UK, 2006.
68. Müller, R.J. Aliphatic-Aromatic Polyesters. In *Handbook of Biodegradable Polymers*; Bastioli, C., Ed.; RAPRA Technology Limited: Telford, UK, 2006.
69. Siracusa, V. Microbial Degradation of Synthetic Biopolymers Waste. *Polymers* **2019**, *11*, 1066. [CrossRef]
70. German Institute for Standardisation. *DIN 54900, Part 1–4, Testing of the Compostability of Plastics*; German Institute for Standardisation: Berlin, Germany, 1998.
71. ASTM. *ASTM D6002-96, Standard Guide for Assessing the Compostability of Environmentally Degradable Plastics*; ASTM INTERNATIONAL: West Conshohocken, PA, USA, 2002.
72. Abreu, A.S.; Oliveira, M.; de Sá, A.; Rodrigues, R.M.; Cerqueira, M.A.; Vicente, A.A.; Machado, A. Antimicrobial nanostructured starch based films for packaging. *Carbohydr. Polym.* **2015**, *129*, 127–134. [CrossRef] [PubMed]
73. Kim, S.H.; Kim, H.J.; Lee, J.W.; Choi, I.G. Biodegradability of bio-flour filled biodegradable poly(butylene succinate) bio-composites in natural and compost soil. *Polym. Degrad. Stab.* **2006**, *91*, 1117–1127. [CrossRef]
74. Neibolts, N.; Platnieks, O.; Gaidukovs, S.; Barkane, A.; Thakur, V.K.; Filipova, I.; Mihai, G.; Zelca, Z.; Yamaguchi, K.; Enachescu, M. Needle-free electrospinning of nanofibrillated cellulose and graphene nanoplatelets based sustainable poly (butylene succinate) nanofibers. *Mater. Today Chem.* **2020**, *17*, 100301. [CrossRef]
75. Platnieks, O.; Gaidukovs, S.; Barkane, A.; Gaidukova, G.; Grase, L.; Thakur, V.K.; Filipova, I.; Fridrihsone, V.; Skute, M.; Laka, M. Highly loaded cellulose/poly (butylene succinate) sustainable composites for woody-like advanced materials application. *Molecules* **2020**, *25*, 121. [CrossRef]
76. Al Awak, M.M.; Wang, P.; Wang, S.; Tang, Y.; Sun, Y.-P.; Yang, L. Correlation of carbon dots' light-activated antimicrobial activities and fluorescence quantum yield. *RSC Adv.* **2017**, *7*, 30177–30184. [CrossRef]
77. Amjadi, S.; Almasi, H.; Ghorbani, M.; Ramazani, S. Preparation and characterization of TiO2NPs and betanin loaded zein/sodium alginate nanofibers. *Food Packag. Shelf Life* **2020**, *24*, 100504. [CrossRef]
78. Arfat, Y.A.; Ahmed, J.; Ejaz, M.; Mullah, M. Polylactide/graphene oxide nanosheets/clove essential oil composite films for potential food packaging applications. *Int. J. Biol. Macromol.* **2018**, *107*, 194–203. [CrossRef]
79. Arfat, Y.A.; Ahmed, J.; Jacob, H. Preparation and characterization of agar-based nanocomposite films reinforced with bimetallic (Ag-Cu) alloy nanoparticles. *Carbohydr. Polym.* **2017**, *155*, 382–390. [CrossRef]
80. Platnieks, O.; Gaidukovs, S.; Neibolts, N.; Barkane, A.; Gaidukova, G.; Thakur, V.K. Poly(butylene succinate) and graphene nanoplatelet–based sustainable functional nanocomposite materials: Structure-properties relationship. *Mater. Today Chem.* **2020**, *18*, 100351. [CrossRef]
81. Iovino, R.; Zullo, R.; Rao, M.A.; Cassar, L.; Gianfreda, L. Biodegradation of poly(lactic acid)/starch/coir biocomposites under controlled composting conditions. *Polym. Degrd. Stab.* **2008**, *93*, 147–157. [CrossRef]
82. Gaidukova, G.; Platnieks, O.; Aunins, A.; Barkane, A.; Ingrao, C.; Gaidukovs, S. Spent coffee waste as a renewable source for the production of sustainable poly(butylene succinate) biocomposites from a circular economy perspective. *RSC Adv.* **2021**, *11*, 18580–18589. [CrossRef]
83. Platnieks, O.; Barkane, A.; Ijudina, N.; Gaidukova, G.; Thakur, V.K.; Gaidukovs, S. Sustainable tetra pak recycled cellulose / Poly(Butylene succinate) based woody-like composites for a circular economy. *J. Clean. Prod.* **2020**, *270*, 122321. [CrossRef]

84. Chowdhury, S.; Teoh, Y.L.; Ong, K.M.; Zaidi, N.S.R.; Mah, S.-K. Poly (vinyl) alcohol crosslinked composite packaging film containing gold nanoparticles on shelf life extension of banana. *Food Packag. Shelf Life* **2020**, *24*, 100463. [CrossRef]
85. Venkatesh, C.; Laurenti, M.; Bandeira, M.; Lanzagorta, E.; Lucherini, L.; Cauda, V.; Devine, D.M. Biodegradation and Antimicrobial Properties of Zinc Oxide–Polymer Composite Materials for Urinary Stent Applications. *Coatings* **2020**, *10*, 1002. [CrossRef]
86. Dairi, N.; Ferfera-Harrar, H.; Ramos, M.; Garrigós, M.C. Cellulose acetate/AgNPs-organoclay and/or thymol nano-biocomposite filmswith combined antimicrobial/antioxidant properties for active food packaging use. *Int. J. Biol. Macromol.* **2019**, *121*, 508–523. [CrossRef]
87. Achilias, D.S.; Panayotidou, E.; Zuburtikudis, I. Thermal degradation kinetics and isoconversional analysis of biodegradable poly(3-hydroxybutyrate)/organomodified montmorillonite nanocomposites. *Thermochim. Acta* **2011**, *514*, 58–66. [CrossRef]
88. Bagde, P.; Nadanathangam, V. Mechanical, antibacterial and biodegradable properties of starch film containing bacteriocin immobilized crystalline nanocellulose. *Carbohydr. Polym.* **2019**, *222*, 115021. [CrossRef]
89. Aniskevich, K.; Glaskova, T.; Janson, Y. Elastic and sorption characteristics of an epoxy binder in a composite during its moistening. *Mech. Compos. Mater.* **2005**, *41*, 341–350. [CrossRef]
90. Aniskevich, A.; Glaskova-Kuzmina, T.C. 3. Effect of moisture on elastic and viscoelastic properties of fibre reinforced plastics: Retrospective and current trends. In *Creep and Fatigue in Polymer Matrix Composites*, 2nd ed.; Guedes, R.M., Ed.; Woodhead Publishing: Oxford, UK, 2019; pp. 83–120. [CrossRef]
91. Glaskova-Kuzmina, T.; Zotti, A.; Borriello, A.; Zarrelli, M.; Aniskevich, A. Basalt fibre composite with carbon nanomodified epoxy matrix under hydrothermal ageing. *Polymers* **2021**, *13*, 532. [CrossRef] [PubMed]
92. Ogunsona, E.O.; Misra, M.; Mohanty, A.K. Accelerated hydrothermal aging of biocarbon reinforced nylon biocomposites. *Polym. Degrad. Stab.* **2017**, *139*, 76–88. [CrossRef]
93. Goliszek, M.; Podkościelna, B.; Sevastyanova, O.; Fila, K.; Chabros, A.; Pączkowski, P. Investigation of accelerated aging of lignin-containing polymermaterials. *Int. J. Biol. Macromol.* **2019**, *123*, 910–922. [CrossRef]
94. Sánchez, M.L.; Capote, G.; Patiño, J.P. Effect of surface treatment of fibers on the accelerated aging of biocomposites. *Constr. Build. Mater.* **2021**, *271*, 121875. [CrossRef]
95. Phua, Y.J.; Chow, W.S.; Mohd Ishak, Z.A. The hydrolytic effect of moisture and hygrothermal aging on poly(butylene succinate)/organo-montmorillonite nanocomposites. *Polym. Degrad. Stab.* **2011**, *96*, 1194–1203. [CrossRef]
96. Luzi, F.; Fortunati, E.; Jiménez, A.; Puglia, D.; Pezzolla, D.; Gigliotti, G.; Kenny, J.M.; Chiralt, A.; Torre, L. Production and characterization of PLA PBS biodegradable blends reinforced with cellulose nanocrystals extracted from hemp fibres. *Ind. Crop. Prod.* **2016**, *93*, 276–289. [CrossRef]
97. Castro-Aguirre, E.; Auras, R.; Selke, S.; Rubino, M.; Marsh, T. Impact of Nanoclays on the Biodegradation of Poly(Lactic Acid) Nanocomposites. *Polymers* **2018**, *10*, 202. [CrossRef]
98. Dorigato, A.; Sebastiani, M.; Pegoretti, A.; Fambri, L. Effect of Silica Nanoparticles on the Mechanical Performances of Poly(Lactic Acid). *J. Polym. Environ.* **2012**, *20*, 713–725. [CrossRef]
99. Cosquer, R.; Pruvost, S.; Gouanvé, F. Improvement of Barrier Properties of Biodegradable Polybutylene Succinate/Graphene Nanoplatelets Nanocomposites Prepared by Melt Process. *Membranes* **2021**, *11*, 151. [CrossRef]
100. Cai, L.; Qi, Z.; Xu, J.; Guo, B.; Huang, Z. Study on the photodegradation stability of poly(butylene succinate-co-butylene adipate)/TiO_2 nanocomposites. *J. Chem.* **2019**, 5036019. [CrossRef]
101. Ahuja, D.; Kumar, L.; Kaushik, A. Thermal stability of starch bionanocomposites films: Exploring the role of esterified cellulose nanofibers isolated from crop residue. *Carbohydr. Polym.* **2021**, *255*, 117466. [CrossRef]
102. Famá, L.; Rojoe, P.G.; Bernal, C.; Goyanes, S. Biodegradable starch based nanocomposites with low water vapor permeability and high storage modulus. *Carbohydr. Polym.* **2012**, *87*, 1989–1993. [CrossRef]
103. Ollier, R.P.; D'Amico, D.A.; Schroeder, W.F.; Cyras, V.P.; Alvarez, V.A. Effect of clay treatment on the thermal degradation of PHB based nanocomposites. *Appl. Clay Sci.* **2018**, *163*, 146–152. [CrossRef]
104. Jayakumar, A.; Prabhu, K.; Shah, L.; Radha, P. Biologically and environmentally benign approach for PHB-silver nanocomposite synthesis and its characterization. *Polym. Test.* **2020**, *81*, 106197. [CrossRef]
105. Nivedita, S.; Shiny, J. Effect of unmodified and modified montmorillonite on the properties of PCL based ultrafiltration membrane for water treatment applications. *J. Water Process. Eng.* **2018**, *21*, 61–68. [CrossRef]
106. Georgiopoulos, P.; Kontou, E.; Christopoulos, A. Short-term creep behavior of a biodegradable polymer reinforced with wood-fibers. *Composites* **2015**, *80*, 134–144. [CrossRef]
107. Ramazani, S.; Karimin, M. Study the molecular structure of poly(ε-caprolactone)/graphene oxide and graphene nanocomposite-nanofibers. *J. Mech. Behav. Biomed. Mater.* **2016**, *61*, 484–492. [CrossRef]
108. Bai, H.; Liang, Z.; Wang, D.; Guo, J.; Zhang, S.; Ma, P.; Dong, W. Biopolymer nanocomposites with customized mechanical property and exceptionally antibacterial performance. *Compos. Sci. Technol.* **2020**, *199*, 108338. [CrossRef]
109. Mofokeng, J.P.; Luyt, A.S. Morphology and thermal degradation studies of melt-mixed PLA/PHBV biodegradable polymer blend nanocomposites with TiO_2 as filler. *J. Appl. Polym. Sci.* **2015**, *50*, 10. [CrossRef]
110. Ding, K.; Wei, N.; Zhou, Y. Viscoelastic behavior and model simulations of poly(butylene adipateco-terephthalate) biocomposites with carbon nanotubes: Hierarchical structures and relaxation. *J. Compos. Mater.* **2016**, *50*, 1805–1816. [CrossRef]
111. Abdullah, Z.W.; Dong, Y. Biodegradable and Water Resistant Poly(vinyl) Alcohol (PVA)/Starch (ST)/Glycerol (GL)/Halloysite Nanotube (HNT) Nanocomposite Films for Sustainable Food Packaging. *Front Mater.* **2019**, *6*, 58. [CrossRef]

112. Liu, X.; Wang, Y.; Yu, L.; Tong, Z.; Chen, L.; Liu, H.; Li, X. Thermal degradation and stability of starch under different processing conditions. *Starch* **2013**, *65*, 48–60. [CrossRef]
113. Amaro, L.P.; Cicogna, F.; Passaglia, E.; Morici, E.; Oberhauser, W.; Al-Malaika, S.; Tzankova Dintcheva, N.; Coiai, S. Thermo-oxidative stabilization of poly(lactic acid) with antioxidant intercalated layered double hydroxides. *Polym. Degrad. Stab.* **2016**, *133*, 92–100. [CrossRef]
114. Ray, S.; Cooney, R.P. Ch. 9. Thermal degradation of polymer and polymer composites. In *Handbook of Environmental Degradation of Materials*, 3rd ed.; Kutz, M., Ed.; William Andrew Applied Science Publishers: Oxford, UK, 2018; pp. 185–206. [CrossRef]
115. Kopinke, F.D.; Remmler, M.; Mackenzie, K.; Moder, M.; Wachsen, O. Thermal decomposition of biodegradable polyesters—II. Poly(lactic acid). *Polym. Degrad. Stab.* **1996**, *53*, 329–342. [CrossRef]
116. Rasselet, D.; Ruellan, A.; Guinault, A.; Miquelard-Garnier, G.; Sollogoub, C.; Fayolle, B. Oxidative degradation of polylactide (PLA) and its effects on physical and mechanical properties. *Eur. Polym. J.* **2014**, *50*, 109–116. [CrossRef]
117. Murphy, J. *Additives for Plastics Handbook*, 3rd ed.; Elsevier Advanced Technology: Oxford, UK, 2001; 484p.
118. Plota, A.; Masek, A. Lifetime prediction methods for degradable polymeric materials—A short review. *Materials* **2020**, *13*, 4507. [CrossRef]
119. Barkane, A.; Platnieks, O.; Jurinovs, M.; Gaidukovs, S. Thermal stability of UV-cured vegetable oil epoxidized acrylate-based polymer system for 3D printing application. *Polym. Degrad. Stab.* **2020**, *181*, 109347. [CrossRef]
120. Chrissafis, K.; Antoniadis, G.; Paraskevopoulos, K.M.; Vassiliou, A.; Bikiaris, D.N. Comparative study of the effect of different nanoparticles onthe mechanical properties and thermal degradation mechanism ofin situ prepared poly(e-caprolactone) nanocomposites. *Compos. Sci. Technol.* **2007**, *67*, 2165–2174. [CrossRef]
121. Zhang, H.; Yu, J.; Wang, H.; Xue, L. Investigation of microstructures and ultraviolet aging properties of organomontmorillonite/SBS modified bitumen. *Mater. Chem. Phys.* **2011**, *129*, 769–776. [CrossRef]
122. Konstantinou, I.K.; Albanis, T.A. TiO_2-assisted photocatalytic degradation of azo dyes in aqueous solution: Kinetic and mechanistic investigations. *Appl. Catal. B* **2004**, *49*, 1–14. [CrossRef]
123. Gonzalez-Lopez, M.E.; del Campo, A.S.M.; Robledo-Ortíz, J.R.; Arellano, M.; Perez-Fonseca, A.A. Accelerated weathering of poly(lactic acid) and its biocomposites: A review. *Polym. Degrad. Stabil.* **2020**, *179*, 109290. [CrossRef]
124. Beg, M.D.H.; Pickering, K.L. Accelerated weathering of unbleached and bleached Kraft wood fibre reinforced polypropylene composites. *Polym. Degrad. Stabil.* **2008**, *93*, 1939–1946. [CrossRef]
125. Glaskova, T.I.; Guedes, R.M.; Morais, J.J.; Aniskevich, A.N. A comparative analysis of moisture transport models applied to epoxy binder. *Mech. Compos. Mater.* **2007**, *43*, 377–388. [CrossRef]
126. Choudalakis, G.; Gotsis, A.D. Free volume and mass transport in polymer nanocomposites. *Curr. Opin. Colloid Interface Sci.* **2012**, *17*, 132–140. [CrossRef]
127. Nielsen, L.E. Models for the Permeability of Filled Polymer Systems. *J. Macromol. Sci. Part A Chem.* **1967**, *1*, 929–942. [CrossRef]
128. Tan, B.; Thomas, N.L. Tortuosity model to predict the combined effects of crystallinity and nano-sized clay mineral on the water vapour barrier properties of polylactic acid. *Appl. Clay Sci.* **2017**, *141*, 46–54. [CrossRef]
129. Murariu, M.; Benali, S.; Paint, Y.; Dechief, A.-L.; Murariu, O.; Raquez, J.-M.; Dubois, P. Adding value in production of multifunctional polylactide (PLA)–ZnO nanocomposite films through alternative manufacturing methods. *Molecules* **2021**, *26*, 2043. [CrossRef] [PubMed]
130. Rostamzad, H.; Paighambari, S.Y.; Shabanpour, B.; Ojagh, S.M.; Mousavi, S.M. Improvement of fish protein film with nanoclay and transglutaminase for food packaging. *Food Packag. Shelf Life* **2016**, *7*, 1–7. [CrossRef]
131. Nanthakumar, K.; Yeng, C.M.; Chun, K.S. Tensile and water absorption properties of solvent cast biofilms of sugarcane leaves fibre-filled poly(lactic) acid. *J. Thermoplast. Compos. Mater.* **2020**, *33*, 289–304. [CrossRef]
132. Chan, M.Y.; Koay, S.C.; Husseinsyah, S.; San, S.T. Cross-linked chitosan/corn cob biocomposite films with salicylaldehyde on tensile, thermal, and biodegradable properties: A comparative study. *Adv. Polym. Technol.* **2016**, *34*, 1229–1239. [CrossRef]
133. Bharadwaj, R.K. Modeling the Barrier Properties of Polymer-Layered Silicate Nanocomposites. *Macromolecules* **2001**, *34*, 9189–9192. [CrossRef]
134. Crank, J. *The Mathematics of Diffusion*, 2nd ed.; Clarendon Press: Oxford, UK, 1975.
135. Pukanszky, B. Influence of interface interaction on the ultimate tensile properties of polymer composites. *Composites* **1990**, *21*, 255–262. [CrossRef]
136. Danyadi, L.; Janecska, T.; Szabo, Z.; Nagy, G.; Moczo, J.; Pukanszky, B. Wood flour filled PP composites: Compatibilization and adhesion. *Compos. Sci. Technol.* **2007**, *67*, 2838–2846. [CrossRef]
137. Imre, B.; Pukánszky, B. Compatibilization in bio-based and biodegradable polymer blends. *Eur. Polym. J.* **2013**, *49*, 1215–1233. [CrossRef]
138. Gaidukovs, S.; Platnieks, O.; Gaidukova, G.; Starkova, O.; Barkane, A.; Beluns, S.; Thakur. Understanding the Impact of Microcrystalline Cellulose Modification on Durability and Biodegradation of Highly Loaded Biocomposites for Woody like Materials Applications. *J. Polym. Environ.* **2021**. [CrossRef]
139. Nanni, A.; Messori, M. Effect of the wine lees wastes as cost-advantage and natural fillers on the thermal and mechanical properties of poly(3-hydroxybutyrate-cohydroxyhexanoate) (PHBH) and poly(3-hydroxybutyrate-cohydroxyvalerate) (PHBV). *J. Appl. Polym. Sci.* **2019**, *137*, 48869. [CrossRef]

140. Sugiman, S.; Salman, S.; Maryudi, M. Effects of volume fraction on water uptake and tensile properties of epoxy filled with inorganic fillers having different reactivity to water. *Mater. Today Commun.* **2020**, *24*, 101360. [CrossRef]
141. Starkova, O.; Gaidukovs, S.; Platnieks, O.; Barkane, A.; Garkusina, K.; Palitis, E.; Grase, L. Water absorption and hydrothermal ageing of epoxy adhesives reinforced with amino-functionalized graphene oxide nanoparticles. *Polym. Degrad. Stabil.* **2021**, *191*, 109670. [CrossRef]
142. Platnieks, O.; Sereda, A.; Gaidukovs, S.; Thakur, V.K.; Barkane, A.; Gaidukova, G.; Filipova, I.; Ogurcovs, A.; Fridrihsone, V. Adding value to poly (butylene succinate) and nanofibrillated cellulose-based sustainable nanocomposites by applying masterbatch process. *Ind. Crop. Prod.* **2021**, *169*, 113669. [CrossRef]
143. Jyoti, J.; Singh, P.B.; Arya, A.K.; Dhakate, S.R. Dynamic mechanical properties of multiwall carbon nanotube reinforced ABS composites and their correlation with entanglement density, adhesion, reinforcement and C factor. *RSC Adv.* **2016**, *6*, 3997. [CrossRef]
144. Glaskova, T.; Aniskevich, K.; Borisova, A. Modeling of creep for MWCNT/epoxy nanocomposite. *J. Appl. Polym. Sci.* **2013**, *129*, 3314–3324. [CrossRef]
145. Sadasivuni, K.K.; Saha, P.; Adhikari, J.; Deshmukh, K.; Ahamed, M.B.; Cabibihan, J.-J. Recent advances in mechanical properties of biopolymer composites: A review. *Polym. Compos.* **2019**, *41*, 32–59. [CrossRef]
146. Aniskevich, K.; Starkova, O.; Jansons, J.; Aniskevich, A. *Long-Term Deformability and Aging of Polymer Matrix Composites*; Nova Science Publishers: Hauppauge, NY, USA, 2011; ISBN 978-1-61470-406-5.
147. Brinson, H.F.; Brinson, L.C. *Polymer Engineering Science and Viscoelasticity. An Introduction*; Springer Science Business Media: Berlin, Germany, 2008; ISBN 978-0-387-73860-4.
148. Amiri, A.; Yu, A.; Webster, D.; Ulven, C. Bio-Based Resin Reinforced with Flax Fiber as Thermorheologically Complex Materials. *Polymers* **2016**, *8*, 153. [CrossRef]
149. Ollier, R.P.; Casado, U.; Nicolini, A.T.; Alvarez, V.A.; Pérez, C.J.; Ludueña, L.N. Improved creep performance of melt-extruded polycaprolactone/organo-bentonite nanocomposites. *J. Appl. Polym. Sci.* **2021**, *138*, e50961. [CrossRef]
150. Guedes, R.M.; Singh, A.; Pinto, V. Viscoelastic modelling of creep and stress relaxation behaviour in PLA-PCL fibres. *Fiber Polym.* **2017**, *18*, 2443–2453. [CrossRef]
151. Kontou, E.; Spathis, G.; Georgiopoulos, P. Modeling of nonlinear viscoelasticity-viscoplasticity of bio-based polymer composites. *Polym. Degrad. Stabil.* **2014**, *110*, 203–207. [CrossRef]
152. Vieira, A.C.; Vieira, J.C.; Ferra, J.M.; Magalhães, F.D.; Guedes, R.M.; Marques, A.T. Mechanical study of PLA–PCL fibers during in vitro degradation. *J. Mech. Behav. Biomed.* **2011**, *4*, 451–460. [CrossRef]
153. Vieira, A.C.; Guedes, R.M.; Tita, V. Constitutive modeling of biodegradable polymers: Hydrolytic degradation and time-dependent behavior. *Int. J. Solids Struct.* **2014**, *51*, 1164–1174. [CrossRef]
154. Breche, Q.; Chagnon, G.; Machado, G.; Girarda, E.; Nottelet, B.; Garricc, X.; Faviera, D. Mechanical behaviour's evolution of a PLA-b-PEG-b-PLA triblock copolymer during hydrolytic degradation. *J. Mech. Behav. Biomed.* **2016**, *60*, 288–300. [CrossRef]
155. Qiu, T.Y.; Song, M.; Zhao, L.G. Testing, characterization and modelling of mechanical behaviour of poly (lactic-acid) and poly (butylene succinate) blends. *Mech. Adv. Mater. Mod. Process.* **2016**, *2*, 7. [CrossRef]
156. Singh, A.; Guedes, R.M.; Paiva, D.; Magalhães, F.D. Experiment and modelling of the strain-rate-dependent response during in vitro degradation of PLA fbres. *SN Appl. Sci.* **2020**, *2*, 177. [CrossRef]
157. Khan, A.S.; Lopez-Pamies, O.; Kazmi, R. Thermo-mechanical large deformation response and constitutive modeling of viscoelastic polymers over a wide range of strain rates and temperatures. *Int. J. Plast.* **2006**, *22*, 581–601. [CrossRef]
158. Zhang, T.; Jin, G.; Han, X.; Gao, Y.; Zeng, Q.; Hou, B.; Zhang, D. Multiscale modelling for the heterogeneous strength of biodegradable polyesters. *J. Mech. Behav. Biomed.* **2019**, *90*, 337–349. [CrossRef] [PubMed]

polymers

MDPI

Article

Environmental Effects on Strength and Failure Strain Distributions of Sheep Wool Fibers

Olesja Starkova [1,*], Alisa Sabalina [1], Vanda Voikiva [2] and Agnese Osite [3]

1 Institute for Mechanics of Materials, University of Latvia, Jelgavas 3, LV-1004 Riga, Latvia; alisasabalina@gmail.com
2 Institute of Chemical Physics, University of Latvia, Jelgavas 1, LV-1004 Riga, Latvia; vanda.voikiva@lu.lv
3 Department of Analytical Chemistry, University of Latvia, Jelgavas 1, LV-1004 Riga, Latvia; agnese.osite@lu.lv
* Correspondence: olesja.starkova@lu.lv

Abstract: Sheep wool is an eco-friendly, renewable, and totally recyclable material increasingly used in textiles, filters, insulation, and building materials. Recently, wool fibers have become good alternatives for reinforcement of polymer composites and filaments for 3D printing. Wool fibers are susceptible to environmental degradation that could shorten their lifetime and limit applications. This study reports on the mechanical properties of sheep wool fibers under the impact of humid air and UV irradiation. The results of single fiber tensile tests showed a noticeable gauge length effect on the fibers' strength and failure strain. Long (50 mm) fibers possessed about 40% lower characteristics than short (10 mm) fibers. Environmental aging decreased the elastic modulus and strength of the fibers. Moisture-saturated fibers possessed up to 43% lower characteristics, while UV aging resulted in up to a twofold reduction of the strength. The most severe degradation effect is observed under the coupled influence of UVs and moisture. The two-parameter Weibull distribution was applied for the fiber strength and failure strain statistical assessment. The model well predicted the gauge length effects. Moisture-saturated and UV-aged fibers were characterized by less extensive strength dependences on the fiber length. The strength and failure strain distributions of aged fibers were horizontally shifted to lower values. The results will contribute to be reliable predictions of the environmental durability of sheep wool fibers and will extend their use in technical applications.

Keywords: single fiber test; tensile properties; Weibull distribution; gauge length; environmental degradation; UV aging; moisture; durability

Citation: Starkova, O.; Sabalina, A.; Voikiva, V.; Osite, A. Environmental Effects on Strength and Failure Strain Distributions of Sheep Wool Fibers. *Polymers* **2022**, *14*, 2651. https://doi.org/10.3390/polym14132651

Academic Editors: Domenico Acierno and Antonella Patti

Received: 8 June 2022
Accepted: 27 June 2022
Published: 29 June 2022

1. Introduction

Increased concerns on availability, climate neutrality, and sustainability of available resources set new standards for the design of novel eco-friendly materials, their practical use, and planning of the end-of-life of products. Owing to their abundance and biodegradability, natural fibers have become good candidates for substituting for fossil-based counterparts in some technical applications [1–3].

Animal fibers are the second most widely used natural fibers after plant fibers (e.g., flax, cotton, hemp, jute, kenaf, abaca, etc.). Sheep wool is the most used commercially, particularly in the apparel and textile industry, agriculture, filtering elements, and thermal and acoustic insulation materials [3–6]. For instance, around 450 tons of wool are produced annually in Latvia [7]. According to rough estimates, only about 30% of the collected wool is used to make yarn, while the rest of the wool mostly becomes an agricultural waste [8]. Along with its availability, cost-effectiveness, biodegradability, and sustainability, wool has a range of unique properties that can expand applications of this natural material to other sectors and convert waste to resources. High hygroscopicity, thermal and sound insulation, flame retardant, and antistatic properties are desirable for materials used in building and automotive sectors [9,10]. Wool fibers are durable and flexible; their mechanical characteristics are comparable to or exceed the values of other natural fibers, e.g., cotton and coir/husk [1,11]. Thus, wool is used as fiber reinforcement in mortars and concrete [9,12,13], polymer composites, and laminates [2,14–17], increasing the value of the wool fiber. Recently, wool has become an alternative material in some high-tech industrial sectors, e.g., it is used in reinforcing compounds for additive manufacturing [11,18].

Wool is the natural protein fiber obtained from sheep, goats, camels, rabbits, and other mammals. It comprises hair-like multicellular fibers which grow out of skin follicles at a rate of around 100 mm per year [9]. Wool is mainly composed of keratin; the fibers have a complex structure and consist of a cortex and a surrounding cuticle layer [5]. A scaly fiber surface makes it easier to spin the fleece, which is used in textile fabrics and contributes to enhanced adhesion, which is favorable in reinforcing polymer compounds. The structure and mechanical properties of wool fibers depend on the breed, sex, and age of the sheep [19,20]. Bouagga et al. showed that variations in the elastic modulus and strength of the Tunisian sheep wool fibers of different breeds and sexes are in the range of 5–20%, and they correlated these data with fiber diameters, crystallinity degree, and cysteine amount [19].

Like many natural fibers, wool fibers possess a complex internal structure and marked variability in geometrical characteristics and mechanical properties [21–23]. Unlike most man-made fibers, which are geometrically uniform, wool exhibits between-fiber and within-fiber diameter variations caused by, for example, changing growing conditions. Fiber flaws or morphological defects, which can be built-in or induced during processing, serve as initiation sites of fiber failure. The strength of natural fibers exhibits a substantial scatter on diameter and dependence on fiber length [23–25]. In the study of Guo et al. [25], the strength of short (10 mm) palm fibers is 20% higher than that of long (40 mm) fibers. Sia et al. [23] reported strength variations up to 16% for banana fibers of different lengths. Parlato et al. [13] studied low-quality wool of a Sicilian sheep breed and reported on strength decrease within the fiber diameter: from 200 MPa for 50 μm fibers down to 50 MPa for 90 μm fibers.

According to the weakest link concept, fiber fails at a point with an internal flaw or where fiber diameter is small, or a combination of both. The amount of such "weak links" and the probability of reaching their breaking limits increases with the growing length of fibers. Thus, longer fibers are typically characterized by lower strength and greater data scatter [22]. Due to the wide dispersion of mechanical properties, statistical analysis methods are required to evaluate the fibers' probabilistic strength and failure strain. Weibull distribution is commonly utilized to study the discrete fiber property [21,26,27]. The two-parameter Weibull distribution is among the most used models for assessing natural fibers' strength scatter characteristics [22–25]. For fibers possessing essential diameter variations, the modified Weibull model is applied to obtain accurate predictions [13,22,25].

Fibers of natural origin are generally susceptible to environmental degradation. This fact can essentially reduce the durability of natural fiber-based products and complicate the prediction of their long-term performance [28]. Heat, UV light, and atmospheric humidity are among common external factors affecting fiber appearance and performance during their growth, processing, and use. Wool fibers composed of keratin are more susceptible to chemical damage and unfavorable environmental conditions than the cellulose in the plant fibers [2]. Thermal and UV aging destroy wool fibers' surface composition, leading to their yellowing and up to a twofold reduction of their strength [29]. Wool is hydrophilic and absorbs water up to 30% of its weight [2,11]. Absorbed moisture plasticizes the fibers, increasing their stretching ability while reducing the elastic and strength characteristics [9,13]. In order to fully characterize the mechanisms of environmental impact on fibers, it is necessary to evaluate not only the actual change in their properties but also the change in the distribution parameters depending on fiber geometrical characteristics.

Most of the works reported in the literature focused on the characterization of specific sheep wool fibers and determining their strength, and rarely on failure strain and distribution [22,30,31]. A few works highlighted environmental impacts on the mechanical characteristics of sheep wool [9,13] and some plant fibers [26]. However, no systematic studies on the analysis of the strength and failure strain distributions of aged fibers with their length variations were found in the literature.

The present study is aimed to evaluate the strength and failure strain distributions of the sheep wool fibers under the effect of environmental factors. The single and coupled influence of moisture and UV irradiation on tensile properties of Latvian dark-headed sheep wool fibers is studied on fibers of different lengths. The total experimental campaign on around 300 virgin and environmentally aged fibers was carried out, and the results were statistically analyzed using Weibull distribution. The dependences of the strength and failure strain on the gauge length of fibers exposed to different environments were obtained and fitted by the two-parameter Weibull model. The obtained results contribute to the comprehension of the environmental durability of products made of wool fibers and reliable predictions of their long-term performance. This, in turn, will promote local development opportunities and waste management through extended use of natural and renewable local resources for the development of novel products of low carbon impact and energy-efficient properties.

2. Materials and Methods

2.1. Wool Fibers

The wool samples were collected according to Australian Standard AS/NZ 4492.1 from the Latvian dark-headed breed sheep of a similar age (2–4 years) and the same gender (female). Animals of this age and gender were chosen because they make up the majority of sheep flocks not only in Latvia but throughout the world. The wool is sheared regularly, which makes it more homogenous. Thus, these samples are more representative of the characteristics of wool fibers. The collected sheep wool fibers were washed with detergent (soap with surfactants) in warm water (50–60 °C) and rinsed several times in water. Then they were dried at room temperature, in air, for a day [6]. A photo of an original wool ball and an SEM micrograph of individual fibers are shown in Figure 1. Single fibers for testing were carefully pulled out of the woolball one by one.

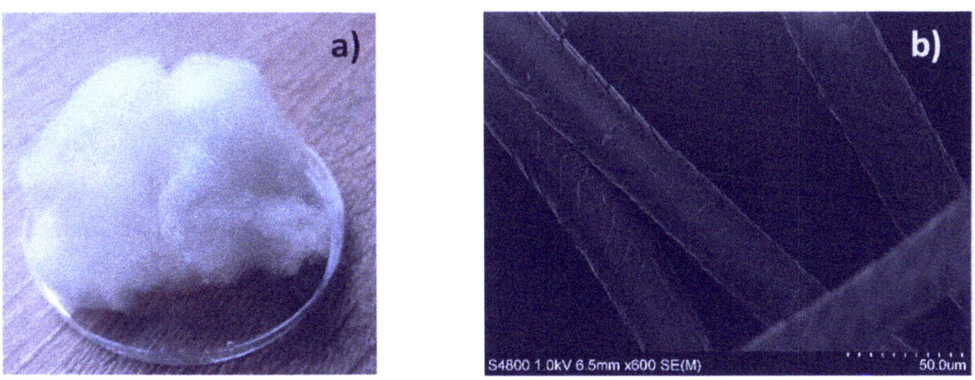

Figure 1. Photo of a wool ball (**a**) and individual fibers by SEM (**b**).

2.2. Single Fiber Tests

Quasistatic tensile tests were made according to ASTM D3379-75 by using the universal testing machine, Zwick, with a 100 N load cell. The wool fiber was mounted on a paper frame and additionally fixed with adhesive paper tape tabs (Figure 2). Fibers of different gauge lengths (LE) were studied: LE = 10, 30, and 50 mm (denoted in the text as LE10, LE30, and LE50, respectively). Scissor cuts were made on both sides of the paper frame tabs at the mid-gage just before the start of the test. The tests were performed at the crosshead speed rate of 1 mm/min under ambient conditions (22 ± 1 °C, RH = 37 ± 5%). The diameter of each fiber was measured by an optical microscope (equipped with a Moticam 2300 digital camera) with 8X magnification at five points along its length, and the average value was used to calculate the cross-sectional area. Between 20 and 40 tests were completed for each specific group of fibers conditioned under different environments, with different LE (Table 1).

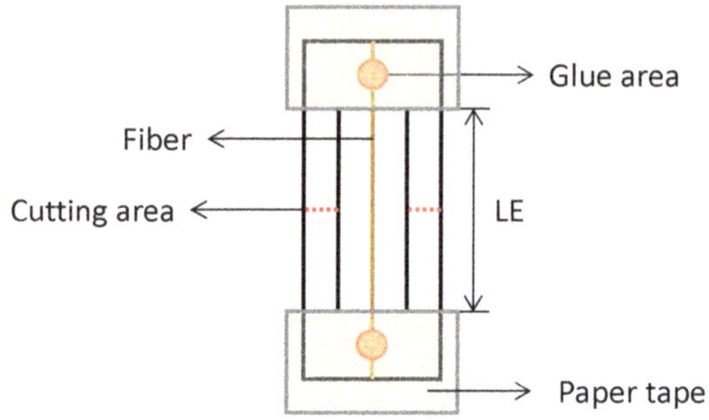

Figure 2. Schematic drawing of a fiber mounted in a paper frame for a tensile test.

2.3. Fiber Conditioning

Fibers mounted in paper frames were conditioned under different environments (Table 1). "As produced", samples were stored in a plastic box under ambient conditions (relative humidity RH = 37 ± 5% and temperature T = 22 ± 2 °C). These fibers were considered as the reference samples. A part of the reference fibers was placed in a desiccator under a saturated salt solution of K_2SO_4, giving an RH = 98%, under an ambient temperature. This group of samples was divided into two depending on the duration of fiber exposure in this environment: 1 week and 2 months for "RH98" and "RH98-ext", respectively ("ext" means extended time). Another part of the reference fibers was exposed to ultraviolet irradiation (denoted as "UV") for 24 h. Irradiation tests were made by using a high-pressure mercury-vapor discharge lamp, UV DRT230, as a UV source, giving a UV-A emission with the strongest peak at about 365 nm. The irradiation intensity was adjusted to 4 mW/cm^2. The temperature under the lamp was maintained at 35 ± 2 °C. In addition, UV irradiation effects were studied on fibers preliminary conditioned under a humid environment. This group of fibers is denoted as "UV-RH98".

Table 1. Details on fiber conditioning under different environments.

Notation	Environment	Duration	Gauge Length (LE), mm
Ref	Ambient: RH = 37 ± 5%, T = 22 ± 2 °C	-	10, 30, 50
RH98	RH = 98%, T = 22 ± 2 °C	1 week	10, 30, 50
RH98-ext	RH = 98%, T = 22 ± 2 °C	2 months	30
UV	I = 4 mW/cm^2, T = 35 ± 2 °C; "Ref" fibers	24 h	10, 30, 50
UV-RH98	I = 4 mW/cm^2, T = 35 ± 2 °C; "RH98" fibers	24 h	30

2.4. Scanning Electron Microscopy (SEM)

SEM images were obtained using a Hitachi S4800 Scanning electron microscope, with an operating voltage of 1.0 current kV, with 5–7 A. Wool fibers samples were electrically bonded to a sample analysis table with electrically conductive tape, and measurements were made. Reference and UV fibers were studied.

2.5. Optical Microscopy

Optical micrographs were taken using an inverted microscope (Olympus IX 71) in the PL mode using an Hg lamp light source (U-LH100HG) with a fluorescence filter set (U-MWU2), and in the micro-extinction spectroscopy (MExS) transmission mode, with a 10× objective lens (CPLNFLN 10XPH, NA 0.3).

3. Statistical Analysis with Weibull Distribution

Weibull distribution based on the weakest link theory for the failure strength, σ, states that the probability of failure $P(\sigma)$ of a material component of volume V is [25,32,33].

$$P(\sigma) = 1 - \exp\left[-\frac{V}{V_0}\left(\frac{\sigma}{\sigma_0}\right)^m\right] \quad (1)$$

where V_0 is the unit volume, m is the shape parameter (Weibull modulus), and σ_0 is the characteristic strength or scale parameter. The strength distribution with a lower m tends to perform a larger scatter and vice versa. Equation (1) is the most widely used formulation of Weibull distribution, called a two-parameter Weibull distribution function, used in failure analysis of various fibers.

The value P is estimated using a probability index, e.g.,

$$P = \frac{i - 0.3}{N + 0.4} \quad (2)$$

where i is the rank of the respective data points of strength placed in ascending order, and N is the total number of data points (tests/fibers). Alternative probability indices are given in [33].

When the cross-sectional area of all fibers is the same, the volume, V, in Equation (1) can be replaced by the gauge length, L. In a general case of fibers with geometrical irregularities, Equation (1) can be rewritten in the following form [13,21,22]:

$$P(\sigma) = 1 - \exp\left[-\left(\frac{L}{L_0}\right)^\alpha\left(\frac{\sigma}{\sigma_0}\right)^m\right] \quad (3)$$

where L is the gauge length of the fiber and L_0 is the unit length (normally $L_0 = 1$ for mathematical convenience). Parameter α ($0 < \alpha \leq 1$), known as the Gutans–Tamuzs [34] or Watson–Smith [35] parameter, is introduced in order to account for diameter variations [13,22,23]; $\alpha = 1$ for a constant within-fiber diameter.

Rearranging Equation (3) while taking the logarithm of both sides provides the following formula:

$$\ln(-\ln(1-P)) - \alpha\,\ln(L/L_0) = m\,\ln\sigma - m\,\ln\sigma_0 \quad (4)$$

As follows from Equation (4), by plotting $\ln(-\ln(1-P)) - \alpha\,\ln(L/L_0)$ against $\ln\sigma$, a linear graph is produced. Parameter m represents the slope of this line, while σ_0 is estimated from the intercept with the ordinate for the given L and α. The parameter α in Equations (3) and (4) is determined using the coefficient of variation of the diameter CV_d (i.e., the average within-fiber diameter variation of N fibers), and α represents the slope of $\ln(CV_d)$ versus $\ln(L)$ line [22,23,25,32]. The diameter variation among fibers can be neglected for long samples since their average diameter at each gauge length is close to each other [22].

Once the Weibull distribution parameters (m, σ_0) and diameter variation parameter, α, are determined, the average value of the strength is obtained [24,25]

$$\langle\sigma\rangle = \sigma_0\left(\frac{L}{L_0}\right)^{-\frac{\alpha}{m}}\Gamma\left(1 + \frac{1}{m}\right) \quad (5)$$

where $\Gamma(x)$ is the gamma function.

Analogously to the failure strength distribution in Equation (3), the two-parameter failure strain, ε, distribution can be written as [21,22]:

$$P(\varepsilon) = 1 - \exp\left[-\left(\frac{L}{L_0}\right)^{\alpha}\left(\frac{\varepsilon}{\varepsilon_0}\right)^{m_\varepsilon}\right] \tag{6}$$

where m_ε and ε_0 are the shape and scale parameters, respectively.

The mean failure strain is determined by the relationship similar to Equation (5):

$$\langle\varepsilon\rangle = \varepsilon_0\left(\frac{L}{L_0}\right)^{-\frac{\alpha}{m_\varepsilon}}\Gamma\left(1 + \frac{1}{m_\varepsilon}\right) \tag{7}$$

A procedure for the determination of m_ε and ε_0 is analogous to this described above for m and σ_0.

4. Results and Discussions

4.1. Diameter Variations

Wool fibers are traditionally positioned as fibers with geometrical irregularities for which diameters vary greatly, not only among fibers but also along the fiber length [22]. The distribution of the diameters measured for the whole population of the reference wool fibers (total of 625 measurements) is shown in Figure 3. As seen from the histogram, diameter values lie in the range of 17–73 μm with the mean value of 37.4 (±6.8) μm. These values are in accordance with literature data for other sheep wool fibers: 15–50 μm [11], 50–90 μm [13], and 25 μm [22].

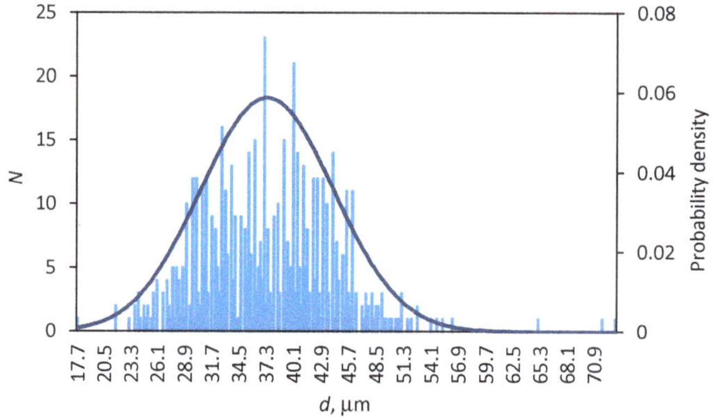

Figure 3. Distribution of the reference wool fiber diameters; the line is the probability density by a normal distribution.

After exposure of fibers to a humid environment and UV rays, their diameters did not change significantly, and d values remained within the distribution for the reference samples. The mean values of diameters for fibers of different gauge lengths in the reference and aged states are listed in Table 2. Figure 4 demonstrates an example of diameter variations of 24 different fibers due to their conditioning in a humid environment ("RH98-ext" samples, Table 1). Each point on the graph is the average from 5 measurements within the same fiber before and after its aging.

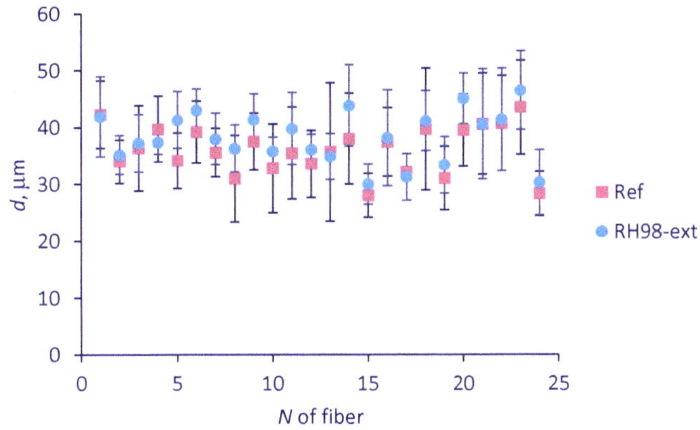

Figure 4. Diameter of fibers in the reference state and after exposure under an RH of 98%.

4.2. Tensile Properties of Wool Fibers

Representative stress-strain diagrams of the reference and aged fibers are shown in Figure 5. The wool fibers possess a highly non-linear viscoelastic–viscoplastic behavior typical for many natural fibers [13,21,22,25,26]. Generally, four regions can be distinguished on the stress-strain curve: (i) an almost linear part at low strains below 2–3%; (ii) a non-linear region with decreasing stress gradient ending with a yield; (iii) a steady-state phase with a constant stress gradient; (iv) a stress hardening region up to fiber breakage. The last phase is not always present, and failure occurs before the final stage (e.g., for UV-irradiated fibers). The elastic modulus was determined in the linear part of the stress-strain curve, i.e., in the region I.

The average values of the elastic modulus (E), strength (σ), and failure strain (ε) for all the groups of wool fibers are listed in Table 2. The reference fibers of a 30 mm gauge length are characterized by $\langle E \rangle = 3.93 \pm 0.61$ GPa, $\langle \sigma \rangle = 142.8 \pm 30.3$ MPa, and $\langle \varepsilon \rangle = 25.9 \pm 11.4\%$. These data compare well to those reported in the literature for other types of sheep wool [1,2,9,13,18], although they are somewhat smaller than those obtained with high-quality Merino wool [22] and some other natural fibers [1,11,36]. The specific strength and stiffness of the sheep wool fibers are comparable with some wood and plant fibers and synthetic polymer fibers. These values confirm the suitability of wool fibers as a reinforcement material for concrete and polymer composites [3,5,13].

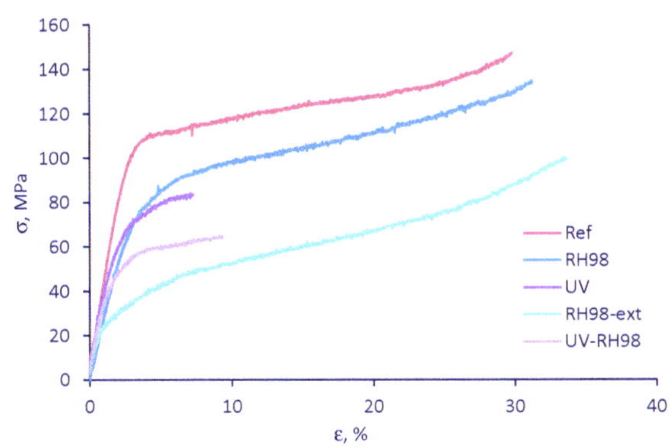

Figure 5. Representative stress-strain diagrams of wool fibers exposed to different environments; LE = 30 mm.

The presented data (Table 2) reveal the gauge length effect on the fiber strength that is compatible with the weakest link concept. The strength and failure strain of long (50 mm) fibers are about 40% lower than those for short (10 mm) fibers. A larger gauge length indicates a higher probability of defects, larger flaw numbers, and, thus, lower tensile strength [21,25]. Like all natural fibers, wool fibers inevitably develop internal defects during their growth. Fibers break in the defective or weakest parts; thus, each strength value in tests represents the strength of the weakest part of each fiber [24]. The strength of fibers reduces as the gauge length increases due to the increased number of flaws that appeared at a longer length. For instance, Zhang et al. [22] reported on the gauge length effect on the strength of merino wool fibers: 215 MPa for short (10 mm) and 200 Mpa for 100 mm long fibers. Similar effects are observed for various plant fibers [21,23,25,33].

Environmental aging significantly affected the mechanical performance of the wool fibers. Aged fibers possessed lower strength and stiffness compared to their pristine counterparts (Figure 5). Stretching ability drastically decreased for UV-aged fibers, while remaining unchanged or increased for moisture-saturated samples. At the same time, UV irradiation moderately affected the elastic modulus of the fibers. According to the data in Table 2 and comparing the average values for the reference and aged fibers of a 30 mm gauge length, the strength reduction is by 13%, 34%, 42%, and 53% for RH98, RH98-ext, UV, and UV-RH98 samples, respectively. For comparison with other studies, the strength of dry and wet (100% RH) wool fibers differ by about 16% (260 MPa and 190 MPa, respectively) [9] and 12% (86 MPa and 75 MPa) [13]. Similarly, the elastic modulus decreased by 22%, 43%, and 25% for RH98, RH98-ext, and UV-RH98 samples, respectively. It can be concluded that the combined action of moisture and the UV resulted in higher fiber degradation compared to that of single environmental factors. Long-term (2 months) conditioning under a humid environment affected the strength and elastic modulus of the fibers to a greater extent than their 1-week exposure. This fact cannot be solely related to the plasticization effect of absorbed moisture, but it is associated with moisture-induced structural degradation. The latter can, in turn, result in an additional moisture ingress into the fibers and increase the negative impact on the strength. According to the data of our previous study on moisture diffusion into the sheep wool (not shown here), the "Fickian" saturation was achieved in 3 days and reached the value of about 23%. Thus, RH98 fibers were assumed as fully saturated. In another study [13], it is reported that ten minutes is a sufficient time for wool fibers to reach saturation in distilled water. The moisture content of RH98-ext fibers was not evaluated in this study. At the same time, it should be noted that neither short-term nor long-term exposure of the fibers to a humid environment resulted in a noticeable change in their diameter (Section 4.1) and visual appearance (Section 4.5). It is assumed that two counterbalancing processes are taking place in moisture-saturated fibers. Absorbed moisture plays a role of a plasticizer and facilitates movements of macromolecular chain segments of keratin, the main structural part of sheep wool fibers. This results in increased failure strains. At the same time, water ingress results in swelling and hydrolytic degradation, leading to the development of additional defects in the fiber structure. Then, an opposite effect and decrease of the ultimate strains are expected.

Table 2. Mechanical characteristics and Weibull distribution parameters of the wool fibers.

Environment	LE, mm	Number of Tests	<d> [1], μm	<E>, GPa	<σ>, MPa	<ε>, %	m	σ_0 (R^2, %) [2]	m_ε	ε_0 (R^2, %)
Ref	10	20	37.63 (±4.50)	4.58 (±1.13)	178.0 (±37.2)	34.4 (±13.1)	4.80	311.6 (96%)	2.77	90.6 (94%)
	30	42	38.14 (±5.72)	3.93 (±0.61)	142.8 (±30.3)	25.9 (±11.4)	5.62	283.1 (91%)	1.85	190.4 (93%)
	50	25	37.14 (±4.66)	4.16 (±0.61)	105.8 (±31.7)	19.7 (±11.3)	3.59	353.43 (98%)	1.47	324.5 (96%)
RH98	10	25	37.29 (±4.55)	2.31 (±0.51)	136.0 (±28.3)	47.0 (±12.8)	5.33	226.9 (98%)	4.57	86.4 (95%)
	30	30	41.68 (±5.39)	3.06 (±0.59)	124.1 (±23.8)	25.7 (±9.8)	5.99	236.2 (94%)	2.55	111.0 (95%)
	50	25	36.97 (±3.86)	2.84 (±0.51)	110.3 (±26.7)	20.7 (±9.9)	4.65	279.3 (96%)	1.88	191.6 (95%)
RH98-ext	30	24	39.42 (±5.52)	2.23 (±0.53)	94.4 (±28.3)	30.6 (±8.6)	3.73	259.9 (98%)	3.66	86.6 (97%)
UV	10	26	35.97 (±3.80)	3.24 (±0.58)	104.5 (±26.8)	16.7 (±12.7)	3.95	207.1 (95%)	1.32	103.2 (94%)
	30	23	40.05 (±8.10)	3.95 (±1.23)	82.4 (±33.9)	8.4 (±8.0)	2.56	353.2 (98%)	1.45	94.8 (79%)
	50	25	34.83 (±5.00)	4.02 (±0.66)	88.8 (±16.9)	6.1 (±5.2)	6.06	182.2 (95%)	1.72	65.8 (77%)
UV-RH98	30	26	40.09 (±5.89)	2.95 (±0.76)	66.6 (±22.1)	9.0 (±7.0)	3.05	228.7 (98%)	1.66	77.9 (89%)

[1] The mean diameter is the average diameter of N fibers, while the diameter for each fiber is the average value from 5 measuring points along the fiber length. [2] R^2 is the correlation coefficient in the Weibull plots.

The mechanical properties of natural fibers are well known to be strongly dependent on their diameter [13,21,24]. Figure 6 demonstrates the elastic modulus and strength as functions of diameters of the reference and aged wool fibers of the gauge length LE = 30 mm. Each point on the graphs corresponds to the data from one test of a fiber with an average within-fiber diameter, d. A decreasing trend is observed for the elastic modulus: fibers of greater diameters possess lower stiffness in each sample group (Figure 6a). Similar trends were reported for flax [21] and "Valle del Belice" sheep wool [13] fibers. Strength exhibits high variability from fiber to fiber, although with no clear trend on their diameter (Figure 6b). Environmental aging decreased the mechanical characteristics of the fibers but did not affect the overall trends of their changes with diameter variations.

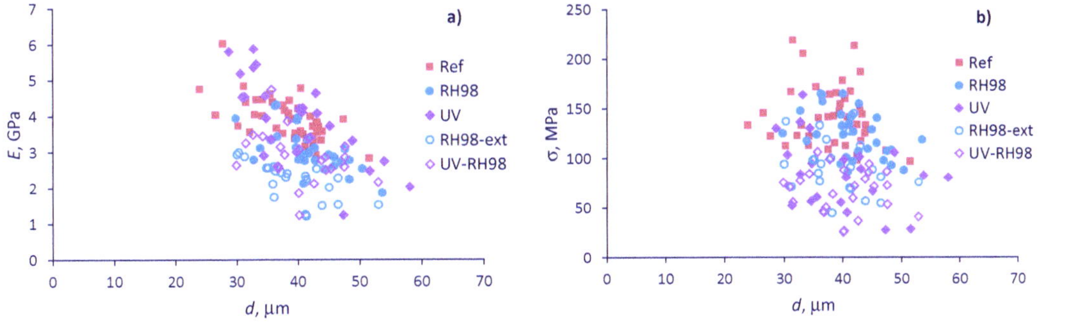

Figure 6. Elastic modulus (**a**) and strength (**b**) as functions of the diameter of the reference and aged wool fibers; LE = 30 mm.

According to literature data, the strength of fibers is assumed to be inversely proportional to their diameter [24]. However, this is not the case for the wool fibers under study. Figure 7a demonstrates the average strength, $<\sigma>$, as a function of the average fiber diameter, $<d>$, for all groups of samples. The strength values are within the data scatter range and do not indicate any specific relationship between $<\sigma>$ and $<d>$. It is also known from the literature that the variability of diameters, expressed through the coefficient of variation CV_d, increases with the length of fibers [22,23]. The linear log–log dependence between these two parameters gives the parameter α in Equation (3), which represents the slope of the line [25,32]. Figure 7b shows $\ln(CV_d)$ versus $\ln(L)$ dependence for the wool fibers under study. Contrary to the premises, no clear trend between these two parameters is observed. Thus, effects from diameter variation among fibers are ignored in further data analysis, and $\alpha = 1$ in Equations (3)–(7).

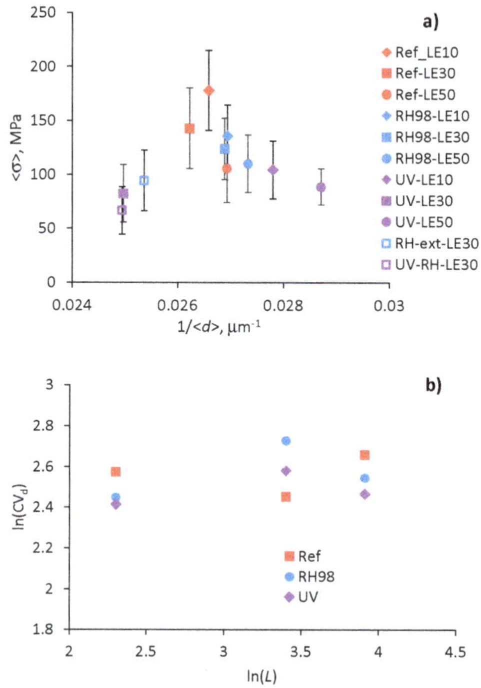

Figure 7. Average fiber strength as a function of fiber diameter (**a**) and relationship between the coefficient of variation of diameter and fiber length (**b**).

The axial strength and stiffness of fibers are determined by their internal structure; thus, these mechanical characteristics are usually correlated [18,21]. In addition, this relationship, although with some deviations, remains valid after the environmental aging of a material [37]. Figure 8 shows the strength versus the elastic modulus for all tested samples, i.e., for the reference and aged wool fibers of different gauge lengths. One point on the graph represents the data of a tensile test of a specific fiber. Despite noticeable data scatter, a definite trend is observed: fibers with a higher elastic modulus possess higher strength. Environmental aging results in a decrease of both mechanical characteristics. Longer fibers possess lower σ and E values within each group of aged samples (Figure 8 and Table 2).

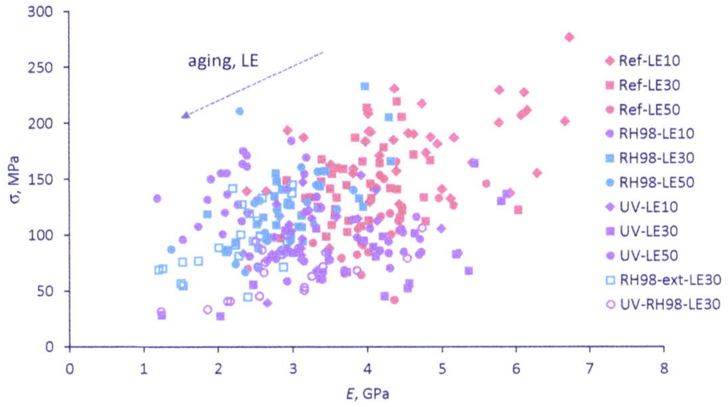

Figure 8. Strength versus the elastic modulus for wool fibers of different lengths (LE) from different environments.

4.3. Weibull Strength Distribution Analysis

Based on Equation (4), modified Weibull linear plots of $\ln(-\ln(1-P)) - \alpha\ln(L)$ vs. $\ln\sigma$ of the reference fibers at different gauge lengths are shown in Figure 9. Similar plots were obtained for environmentally aged fibers. Due to weak correlations between the ultimate properties and diameter of fibers (Figure 7), α was assumed to be equal to unity in all calculations. The correlation coefficients, R^2, are mostly in the range of 95%–98% (Table 2), indicating a reasonable degree of linearity between the linear regression of the fiber strength and experimental data.

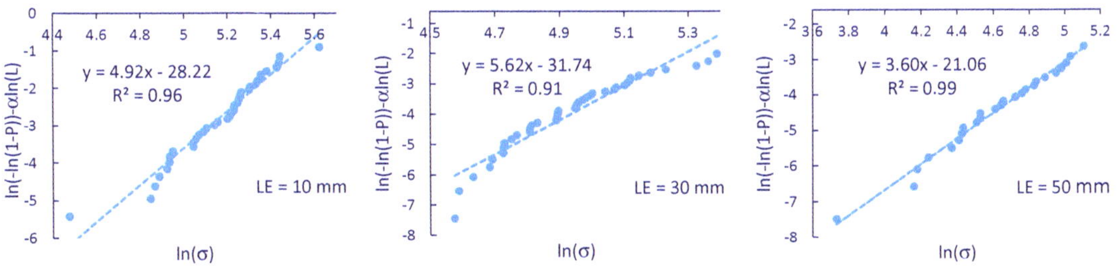

Figure 9. Weibull plots for fiber strength at different gauge lengths.

The strength distribution of the reference and environmentally aged fibers is shown in Figure 10. The data confirm the applicability of the Weibull distribution for the wool fiber tensile strength analysis both in the reference state and after their aging. An increase in the gauge length of fibers, as well as their exposure to a humid environment and UV irradiation, resulted in a horizontal shift of $P(\sigma)$ curves to lower σ values. Some deviations from the Weibull distribution can be associated with the weakest fibers and is related to their damage during the sample preparation process. The Weibull parameters m and σ_0 of the reference and aged fibers are listed in Table 2.

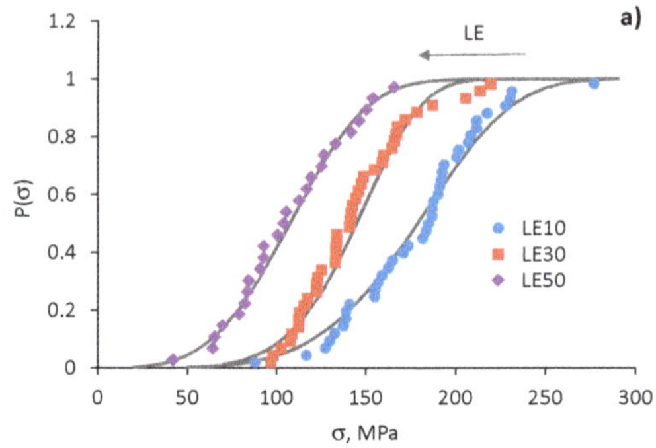

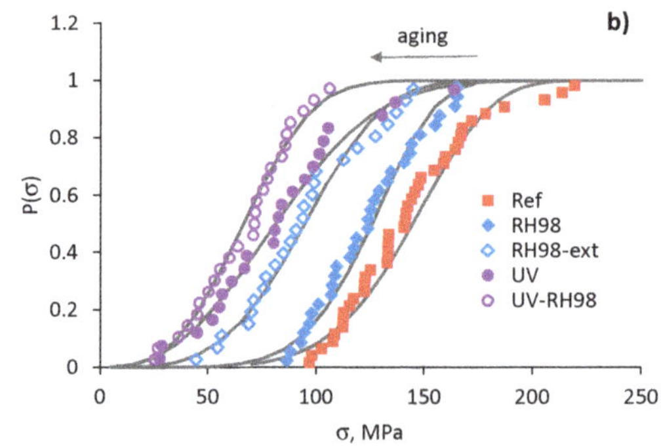

Figure 10. Strength distribution of the reference fibers of different gauge lengths (**a**) and fibers from different environments with LE = 30 mm (**b**). Lines are calculations by Equation (3).

Figure 11a represents the dependence of the average strength on the gauge length of fibers and related predictions using Equation (5). The Weibull distribution parameters used for calculations are shown near the curves. These values are of the same order but different from those obtained by linear regression of the Weibull plots, and are listed in Table 2. Similar notes were made in other studies considering flax fiber strength distribution [21,26]. At the same time, changes in σ_0 caused by fiber aging and determined by fitting Equation (5) correlate well with changes in the average strength (Section 4.2 and Table 2); the σ_0 of RH98 and UV samples is 14% and 44% lower than that of the reference fibers with a σ_0 = 360 MPa. The scale parameter σ_0, related to the characteristic strength of the fibers, reduces due to the increased number of flaws caused by aging effects and fiber degradation.

The Weibull shape parameter, m, of the reference wool fibers takes the values from 3.6 to 5.6 (Table 2) that correlate well with data for various natural fibers reported elsewhere [21,23,25,33]. Weak correlations between the parameter m and the gauge length of fibers were established within each group of samples, although the literature data often reveal that m decreases as the fiber length increases, i.e., longer gauge length results in greater strength scatter [23,25,30]. An increasing trend is noticed for m determined by fitting the average strength data by Equation (5). In this formulation and data presentation, according to Figure 11a, a higher m is associated with a lower dependence, $<\sigma>$, vs. LE. Thus, aging, particularly UV irradiation, resulted in the mitigation of the gauge length effect on the strength of the wool fibers. This fact indicates a leveling of the number of defects per fiber length after their exposure to harsh environments.

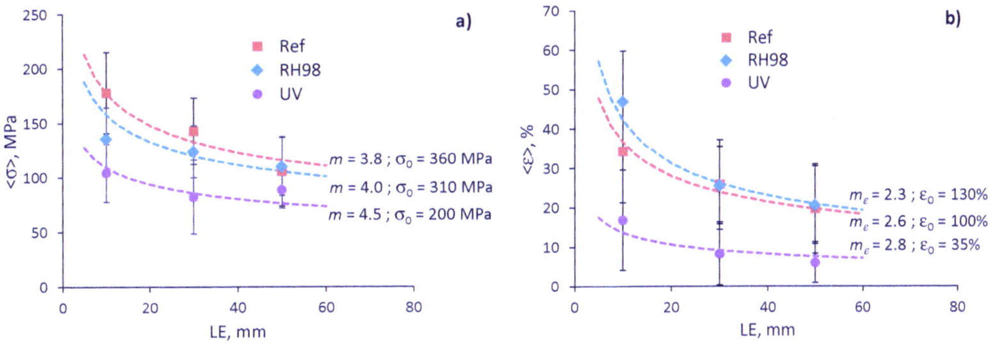

Figure 11. Strength (**a**) and failure strain (**b**) as functions of the gauge length of fibers from different environments. Solid lines are calculations by Equations (5) and (7), respectively; α = 1.

4.4. Weibull Failure Strain Distribution Analysis

Failure strain distribution analysis is similar to that done for the strength and described in Section 4.3. The modified Weibull plots $\ln(-\ln(1-P)) - \alpha\ln(L)$ vs. $\ln\varepsilon$ for the reference and aged fibers (RH98 and UV-RH98) of a 30 mm gauge length are shown in Figure 12. Similar plots were obtained for all groups of fibers. Weibull parameters m_ε and ε_0 according to Equation (6) are listed in Table 2. Due to weak correlations between the ultimate properties and diameter of the fibers, as mentioned above, it is assumed $\alpha = 1$.

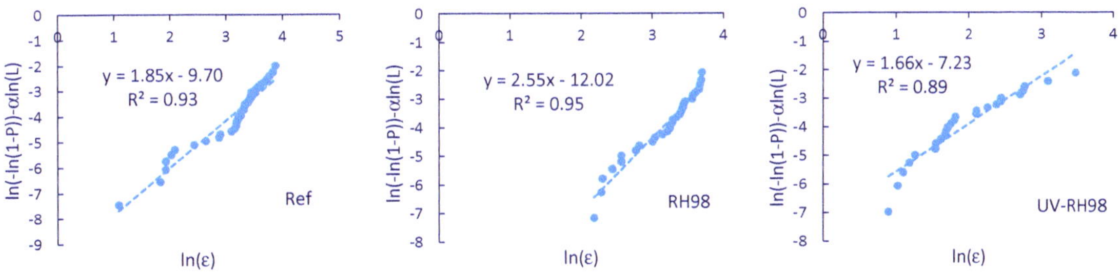

Figure 12. Weibull plots for failure strain of fibers in the reference and aged states at LE = 30 mm.

Figure 13 demonstrates the failure strain distribution of the reference and aged fibers. Overall, the data confirm the applicability of the two-parameter Weibull distribution for the failure strain analysis of wool fibers. However, the linear regression of the failure strain is characterized by lower correlation coefficients (R^2 = 77–95%) compared to those obtained at the strength analysis (Table 2). m_ε values are in the range of 1.3–4.6, which are also noticeably lower than m. These facts indicate a higher data dispersion for the failure strain compared to the strength, which can also be noticed by comparing the fitting data in Figures 10 and 13. Similar to the strength data, an increase in the gauge length of fibers resulted in a horizontal shift of $P(\varepsilon)$ curves to lower ε values (Figure 10a). The same effect is observed for UV-aged fibers and the opposite shift to a higher ε for moistened fibers.

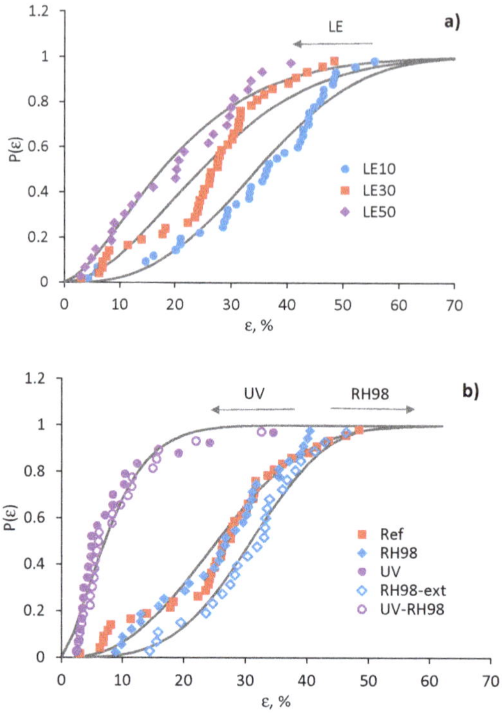

Figure 13. Failure strain distribution of the reference fibers of different gauge lengths (**a**) and fibers from different environments with an LE = 30 mm (**b**). Lines are calculations by Equation (5).

The average failure strain dependencies on the gauge length of the fibers for the reference and aged fibers are shown in Figure 11b. The data are finely fitted by Equation (7), with parameters shown near the curves. The characteristic failure strain, ε_0, of the UV-aged fibers has the lowest value, 35%, which is 65% lower than that for the reference fibers. Due to plasticized effect of moisture, RH98 fibers possess higher deformations that appear in higher (up to 23%) ε_0 compared to the unaged counterparts. These results correlate well with changes in the average failure strain (Table 2). Analogously to the discussions in Figure 11a for the strength data, a higher m_ε for UV-aged fibers is associated with a smoothened shape of the dependence $<\varepsilon>$ vs. LE. In other words, the gauge length effect on the failure strain of the fibers is mitigated after UV exposure. An opposite trend is observed for the moisture-saturated fibers.

4.5. Optical Microscopy and SEM Investigations

Optical micrographs of the reference wool fibers are shown in Figure 14. SEM pictures of the reference and UV-aged wool fibers are shown in Figure 15. Wool fibers are non-uniform and have a unique surface structure of overlapping scales called cuticle cells. A scaly surface of the fiber contributes to the wool's ability to felt and enhance adhesion when used as a reinforcement in composites.

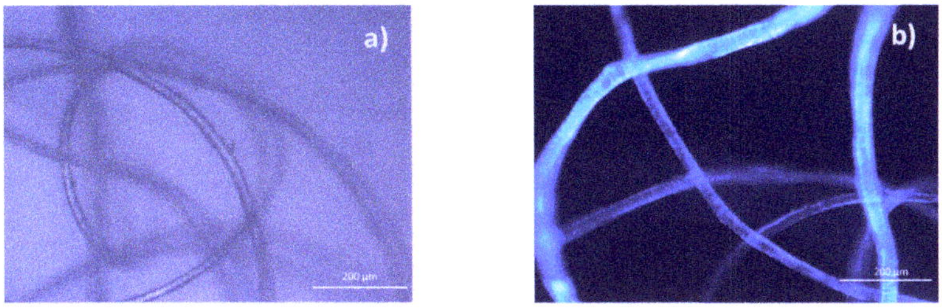

Figure 14. Optical micrographs of the reference wool fibers: original (**a**) and with a fluorescence filter set (**b**).

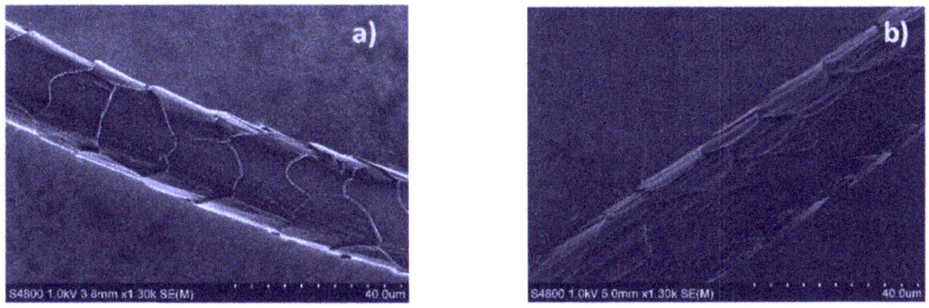

Figure 15. SEM images of the reference (**a**) and UV-aged wool fiber (**b**).

Fibers conditioned under a humid environment did not show any noticeable changes in their visual appearance that could be observed by an optical microscope. SEM is not efficient for the analysis of surface changes of moisture-saturated wool fibers since moisture is desorbed during sample preparation and investigation, and it greatly affects the quality of images. UV irradiation damaged the wool fibers and appears on their loose and non-smooth surface. This resulted in the brittleness of the fibers. In addition, yellowing of UV-aged fibers was noticed. Although yellowing is hardly to be distinguished when considering single fibers, this is well seen by the eye for wool balls (Figure 16). Further investigations are needed for quantified color analysis of wool yellowing. Despite degradation effects, diameters of all fibers before and after aging remained within the distribution range (Section 4.1).

Figure 16. Photo of wool balls of the reference (**left**) and UV-aged (**right**) fibers.

The wool cuticle is very resistant, which is due to the high degree of disulfide and isopeptide cross-linking. In the majority of cases, the amino acids in the cuticle are altered to a greater extent than in the cortex, as the outer layers of fibers receive more exposure to radiation. Since the cuticle protects the cortex, damage to this region usually occurs after extensive damage to the hair cuticle. These defects cause cystine degradation, but the exact mechanism is not precisely known [38].

The literature shows that photodegradation of cystine occurs via the C-S fission pathway and is unlike the chemical oxidation of cystine, which occurs mainly via the S-S fission pathway. Melanin provides a form of photochemical protection for hair proteins by absorbing and filtering falling radiation and then transferring this energy as heat. Its high absorption capacity can be attributed to an extensive system of conjugated carbonyl groups and double bonds. It not only traps a large part of the radiation but also immobilizes many of the free radicals, preventing the transfer of these free radicals into the keratin matrix. However, protecting the fiber proteins from light breaks down or bleaches the pigments. UV irradiation causes the formation of oxyradicals such as superoxide ($O2^{\bullet -}$) and hydroxyl ($OH^{\bullet}$). These compounds have a single unpaired electron in the outer orbital, which gives them a very strong ability to react, especially with molecules with a double bond structure, such as unsaturated lipids. These changes are believed to be caused by UV light-induced oxidation of sulfur-containing molecules in the fiber cortex [39].

In peptides and proteins, in particular, the hydrolytic reaction breaks the amide bond of the peptide and protein with a water molecule. This process results in the conversion of asparagine (Asn) to aspartic acid (Asp) (deamidation), the formation of protein fragments (peptide bond cleavage), or the cyclization of adjacent amino acid residues such as Arg-Pro and Lys-Pro [40]. Further investigations are needed to comprehend the degradation phenomena caused by environmental aging. Some results of the structural characterization of wool fibers will be highlighted in the next study.

5. Conclusions

Latvian dark-headed sheep wool fibers, those that are virgin, and those after environmental aging, were tested. The results of single fiber tensile tests showed a noticeable gauge length effect on the mechanical characteristics of the fibers. The strength and failure strain of long (50 mm) fibers are about 40% lower than those for short (10 mm) fibers. The elastic modulus decreased, but strength showed weak correlations with the growing diameter of fibers of each group. Environmental aging significantly affected the wool's mechanical performance. Aged fibers possessed lower strength and stiffness compared to their pristine counterparts. Long-term conditioning of fibers under a humid environment resulted in a decrease of the strength and elastic modulus of 34% and 43%, respectively. UV-aged fibers possessed up to a twofold reduction in the ultimate mechanical characteristics. The most severe degradation effect was observed under the coupled influence of moisture and UV irradiation.

Weibull distribution was utilized for statistical data analysis. The strength and failure strain was estimated by the two-parameter Weibull model, and their gauge length dependences were well-fitted by the model. The dispersion of the results is compatible with the nature of fibers and allows characterization of its mechanical behavior with reasonable confidence. Moisture-saturated and UV-aged fibers are characterized by less extensive strength dependencies on the fiber length. Environmental aging resulted in a horizontal shift of the strength and failure strain distributions to lower values.

The obtained results can be useful for material designers and engineers in selecting appropriate eco-friendly components for specific applications and predicting their environmental durability. Comprehensive characterization of wool fibers will promote their use in novel technical applications, contributing in this way to the effective use of local natural resources and waste management.

Author Contributions: Conceptualization, O.S. and V.V.; methodology, O.S., A.S. and V.V.; formal analysis, O.S., A.S., V.V. and A.O.; investigation, O.S., A.S. and V.V.; writing—original draft preparation, O.S., A.S. and V.V.; writing—review and editing, O.S., A.S., V.V. and A.O.; project administration, V.V.; funding acquisition, V.V. and A.O. All authors have read and agreed to the published version of the manuscript.

Funding: This research was funded by the European Regional Development Fund (Project No. KC-PI-2020/7, ("Sheep wool fibre processing technology for use in the development of multifunctional bio—filters").

Institutional Review Board Statement: Not applicable.

Informed Consent Statement: Not applicable.

Data Availability Statement: Not applicable.

Acknowledgments: This research was funded by the European Regional Development Fund (Project No. KC-PI-2020/7, ("Sheep wool fibre processing technology for use in the development of multifunctional bio—filters").

Conflicts of Interest: The authors declare no conflict of interest.

References

1. Cheung, H.Y.; Ho, M.P.; Lau, K.T.; Cardona, F.; Hui, D. Natural fibre-reinforced composites for bioengineering and environmental engineering applications. *Compos. Part B* **2009**, *40*, 655–663. [CrossRef]
2. Ramamoorthy, S.K.; Skrifvars, M.; Persson, A. A Review of Natural Fibers Used in Biocomposites: Plant, Animal and Regenerated Cellulose Fibers. *Polym. Rev.* **2015**, *55*, 107–162. [CrossRef]
3. Allafi, F.; Hossain, M.S.; Lalung, J.; Shaah, M.; Salehabadi, A.; Ahmad, M.I.; Shadi, A. Advancements in Applications of Natural Wool Fiber: Review. *J. Nat. Fibers* **2022**, *19*, 497–512. [CrossRef]
4. Bosia, D.; Savio, L.; Thiebat, F.; Patrucco, A.; Fantucci, S.; Piccablotto, G.; Marino, D. Sheep wool for sustainable architecture. *Energy Procedia* **2015**, *78*, 315–320. [CrossRef]
5. Fiore, V.; Di Bella, G.; Valenza, A. Effect of Sheep Wool Fibers on Thermal Insulation and Mechanical Properties of Cement-Based Composites. *J. Nat. Fibers* **2020**, *17*, 1532–1543. [CrossRef]
6. Voikiva, V.; Zarins, A.; Avotina, L.; Barmina, I.; Rudovica, V.; Kizane, G. Influence of Biomass Combustion Products on Element Content and Thermal Stability of Latvian Sheep Breed Wool Filter Fibres. *Mater. Sci.* **2020**, *26*, 438–443. [CrossRef]
7. Available online: https://appsso.eurostat.ec.europa.eu (accessed on 11 June 2022).
8. Available online: https://theworldnews.net/lv-news/ (accessed on 11 June 2022).
9. Jozwiak-Niedzwiedzka, D.; Fantilli, A.P. Wool-Reinforced Cement Based Composites. *Materials* **2020**, *13*, 3590. [CrossRef]
10. Guna, V.; Ilangovan, M.; Vighnesh, H.R.; Sreehari, B.R.; Abhijith, S.; Sachin, H.E.; Mohan, C.B.; Reddy, N. Engineering Sustainable Waste Wool Biocomposites with High Flame Resistance and Noise Insulation for Green Building and Automotive Applications. *J. Nat. Fibers* **2021**, *18*, 1871–1881. [CrossRef]
11. Beckman, I.P.; Lozano, C.; Freeman, E.; Riveros, G. Fiber Selection for Reinforced Additive Manufacturing. *Polymers* **2021**, *13*, 2231. [CrossRef]
12. Alyousef, R.; Mohammadhosseini, H.; Ebid, A.A.K.; Alabduljabbar, H. An Integrated Approach to Using Sheep Wool as a Fibrous Material for Enhancing Strength and Transport Properties of Concrete Composites. *Materials* **2022**, *15*, 1638. [CrossRef]
13. Parlato, M.C.M.; Cuomo, M.; Porto, S.M.C. Natural fibers reinforcement for earthen building components: Mechanical performances of a low quality sheep wool ("Valle del Belice" sheep). *Constr. Build. Mater.* **2020**, *326*, 126855. [CrossRef]
14. Kim, N.K.; Lin, R.J.T.; Bhattacharyya, D. Extruded short wool fibre composites: Mechanical and fire retardant properties. *Compos. Part B* **2014**, *67*, 472–480. [CrossRef]
15. Santulli, C.; Sarasini, F.; Tirillo, J.; Valente, T.; Valente, M.; Caruso, A.P.; Infantino, M.; Nisini, E.; Minak, G. Mechanical behaviour of jute cloth/wool felts hybrid laminates. *Mater. Des.* **2013**, *50*, 309–321. [CrossRef]
16. Conzatti, L.; Giunco, F.; Stagnaro, P.; Patrucco, A.; Marano, C.; Rink, M.; Marsano, E. Composites based on polypropylene and short wool fibres. *Compos. Part A* **2013**, *47*, 165–171. [CrossRef]
17. Manivannan, J.; Rajesh, S.; Mayandi, K.; Rajini, N.; Ismail, S.O.; Mohammad, F.; Kuzman, M.K.; Al-Lohedan, H.A. Animal fiber characterization and fiber loading effect on mechanical behaviors of sheep wool fiber reinforced polyester composites. *J. Nat. Fibers* **2020**, 1–17. [CrossRef]
18. Aida, H.J.; Nadlene, R.; Mastura, M.T.; Yusriah, L.; Sivakumar, D.; Ilyase, R.A. Natural fibre filament for Fused Deposition Modelling (FDM): A review. *Int. J. Sustain. Eng.* **2021**, *14*, 1988–2008. [CrossRef]
19. Bouagga, T.; Harizi, T.; Sakli, F.; Zoccola, M. Correlation between the mechanical behavior and chemical, physical and thermal characteristics of wool: A study on Tunisian wool. *J. Nat. Fibers* **2020**, *17*, 28–40. [CrossRef]
20. Rama Rao, D.; Gupta, V.B. Crystallite orientation in wool fibers. *J. Appl. Polym. Sci.* **1992**, *46*, 1109–1112. [CrossRef]
21. Andersons, J.; Sparnins, E.; Joffe, R.; Wallstrom, L. Strength distribution of elementary flax fibres. *Compos. Sci. Technol.* **2005**, *65*, 693–702. [CrossRef]
22. Zhang, Y.; Wang, X.; Pan, N. Weibull analysis of the tensile behavior of fibers with geometrical irregularities. *J. Mater. Sci.* **2002**, *37*, 1401–1406. [CrossRef]
23. Sia, C.V.; Fernando, L.; Joseph, A.; Chua, S.N. Modified Weibull analysis on banana fiber strength prediction. *J. Mech. Eng. Sci.* **2018**, *12*, 3461–3471. [CrossRef]

24. Andersons, J.; Poriķe, E.; Spārniņš, E. Modeling strength scatter of elementary flax fibers: The effect of mechanical damage and geometrical characteristics. *Compos. Part A Appl. Sci.* **2011**, *42*, 543–549. [CrossRef]
25. Guo, M.; Zhang, T.H.; Chen, B.W.; Cheng, L. Tensile strength analysis of palm leaf sheath fiber with Weibull distribution. *Compos. Part A* **2014**, *62*, 45–51. [CrossRef]
26. Joffe, R.; Andersons, J.; Spārniņš, E. Applicability of Weibull strength distribution for cellulose fibers with highly non-linear behavior. In Proceedings of the International Conference of Composite Materials (ICCM-17), Edinburgh, UK, 27–31 July 2009.
27. Shebanov, S.M. Extreme Values of the Strength of Single Filaments and Threads of a Paraaramide Fiber After Its Treatment by a Water Suspension of Carbon Nanoparticles. *Mech. Compos. Mater.* **2021**, *57*, 269–272. [CrossRef]
28. Starkova, O.; Gagani, A.I.; Karl, C.W.; Rocha, I.B.C.M.; Burlakovs, J.; Krauklis, A.E. Modelling of Environmental Ageing of Polymers and Polymer Composites—Durability Prediction Methods. *Polymers* **2022**, *14*, 907. [CrossRef] [PubMed]
29. Jiang, H.; Shi, J.; Zhou, N.; Tuo, X.; Gong, Y.; Guo, J. Study on the Relationship between Accelerated Aging, Color Characterization and Properties of Natural Fibers. *J. Nat. Fibers* **2021**, 1–11. [CrossRef]
30. Lu, D.; Yu, W.; Pan, N. Determination of the strength and elongation distribution of single wool through fiber bundle testing based on acoustic emissions. *Text. Res. J.* **2021**, *91*, 1263–1273. [CrossRef]
31. Lu, D.; Yu, W. Characterization and analysis of the tensile and acoustic emission parameter distributions of single wool fibers. *Text. Res. J.* **2020**, *90*, 2454–2466. [CrossRef]
32. Ntenga, R.; Saïdjo, S.; Beda, T.; Béakou, A. Estimation of the Effects of the Cross-Head Speed and Temperature on the Mechanical Strength of Kenaf Bast Fibers Using Weibull and Monte-Carlo Statistics. *Fibers* **2019**, *7*, 89. [CrossRef]
33. Naik, D.L.; Fronk, T.H. Weibull Distribution Analysis of the Tensile Strength of the Kenaf Bast Fiber. *Fibers Polym.* **2016**, *17*, 1696–1701. [CrossRef]
34. Gutans, J.A.; Tamuzh, V.P. Scale Effect of the Weibull Distribution of Fibre Strength. *Mech. Compos. Mater.* **1984**, *6*, 1107–1109.
35. Watson, A.S.; Smith, R.L. An examination of statistical theories for fibrous materials in the light of experimental data. *J. Mater. Sci.* **1985**, *20*, 3260–3270. [CrossRef]
36. Karimah, A.; Ridho, M.R.; Munawar, S.S.; Adi, D.S.; Ismadi; Damayanti, R.; Subiyanto, B.; Fatriasari, W.; Fudholi, A. A review on natural fibers for development of eco-friendly bio-composite: Characteristics, and utilizations. *J. Mater. Res. Technol.* **2021**, *13*, 2442–2458. [CrossRef]
37. Starkova, O.; Platnieks, O.; Sabalina, A.; Gaidukovs, S. Hydrothermal Ageing Effect on Reinforcement Efficiency of Nanofibrillated Cellulose/Biobased Poly(butylene succinate) Composites. *Polymers* **2022**, *14*, 221. [CrossRef]
38. Nogueira, A.C.S.; Dicelio, L.E.; Joekes, I. Photochem. About photo-damage of human hair. *Photobiol. Sci.* **2006**, *5*, 165–169. [CrossRef]
39. Dyer, J.M.; Plowman, J.E.; Krsinic, G.L.; Deb-Choudhury, S.; Koehn, H.; Millington, K.R.; Clerens, S. Proteomic evaluation and location of UVB-induced photo-oxidation in wool. *J. Photochem. Photobiol. B Biol.* **2010**, *98*, 118–127. [CrossRef]
40. Mozziconacci, O.; Schöneich, C. Chemical degradation of proteins in the solid state with a focus on photochemical reactions. *Adv. Drug Deliv. Rev.* **2015**, *93*, 2–13. [CrossRef]

polymers

MDPI

Review

Factors Affecting Mechanical Properties of Reinforced Bioplastics: A Review

Jet Yin Boey [1], Chee Keong Lee [2] and Guan Seng Tay [1,3,*]

[1] Bioresource Technology Division, School of Industrial Technology, Universiti Sains Malaysia, Penang 11800, Malaysia
[2] Bioprocess Technology Division, School of Industrial Technology, Universiti Sains Malaysia, Penang 11800, Malaysia
[3] Green Biopolymer, Coatings & Packaging Cluster, School of Industrial Technology, Universiti Sains Malaysia, Penang 11800, Malaysia
* Correspondence: taygs@usm.my; Tel.: +60-4-6532201

Abstract: The short life cycle and recalcitrant nature of petroleum-based plastics have been associated with plastic waste accumulation due to their composition rather than worldwide overproduction. The drive to replace single-use products has sparked a considerable amount of research work to discover sustainable options for petroleum-based plastics. Bioplastics open up a new horizon in plastics manufacturing operations and industrial sectors because of their low environmental impact, superior biodegradability, and contribution to sustainable goals. Their mechanical properties regarding tensile, flexural, hardness, and impact strength vary substantially. Various attempts have been made to augment their mechanical characteristics and capacities by incorporating reinforcement materials, such as inorganic and lignocellulosic fibres. This review summarizes the research on the properties of bioplastics modified by fibre reinforcement, with a focus on mechanical performance. The mechanical properties of reinforced bioplastics are significantly driven by parameters such as filler type, filler percentage, and aspect ratio. Fibre treatment aims to promote fibre–matrix adhesion by changing their physical, chemical, thermal, and mechanical properties. A general overview of how different filler treatments affect the mechanical properties of the composite is also presented. Lastly, the application of natural fibre-reinforced bioplastics in the automobile, construction, and packaging industries is discussed.

Keywords: mechanical properties; physical treatment; chemical treatment; biological treatment

Citation: Boey, J.Y.; Lee, C.K.; Tay, G.S. Factors Affecting Mechanical Properties of Reinforced Bioplastics: A Review. *Polymers* **2022**, *14*, 3737. https://doi.org/10.3390/polym14183737

Academic Editors: Domenico Acierno and Antonella Patti

Received: 31 July 2022
Accepted: 27 August 2022
Published: 7 September 2022

1. Introduction of Bioplastics

Plastics represent a broad category of polymer composites that constitute polymers as a building block. Polymeric materials can be divided into either thermoplastic (which softens when heated and stiffens again when cooled) or thermosetting polymers (which do not soften when they have been moulded). Most thermoplastic and thermosetting materials in present industrial use are petroleum-derived and non-renewable, posing a limitation to the polymer industry [1,2]. As a result, plastic waste has been primarily portrayed as a plastic composition issue rather than a global overproduction problem. Thankfully, plastics can be more sustainable through the convergence of technology improvements and consumer preferences, making them more achievable than ever. Bioplastic is the foundation of the principle of sustainable development, from exploiting more renewable content and recycling materials, to lowering manufacturing energy and returning material to nature at the end of its life. They reduce the pollutants induced by petroleum-derived plastics that remain solid for centuries, signalling a new era of packing technology and industry. The demand for bioplastics has gone through the roof in both industry and research. This is because people are worried about pollution in the environment more than ever, and local and international groups are passing stricter laws to protect the environment.

Bioplastics are a type of plastic material that is bio-based, biodegradable, or both, depending on the source from which they were created. To put it another way, the term bioplastic also refers to petroleum-based plastics that are biodegradable. It can be any combination of bio-based (partially, completely, or non-biobased), biodegradable, or compostable, provided that it is not both non-bio-based and non-biodegradable. "Bio-based" is termed as products created from biological material derived from biomass such as plants, bacteria, algae, etc. [1,3]. For instance, in conventional plastics (i.e., petroleum-derived and non-biodegradable), the traditional petrochemical resin is replaced by biopolymers extracted from animals or plants, while synthetic glass or carbon fibres are substituted by natural fibres like jute, bamboo, flax, and hemp [4–7]. The word "biodegradability" refers to a broad range of enzymatic and/or chemical reactions mediated by bacteria or biological organisms, the efficiencies of which are governed by the conditions in which these polymers biodegrade [8]. Microorganisms, industrial or home composting as an end-of-life option, as well as anaerobic digestion, may also decompose bioplastics, encouraging a more sustainable circular economy [9]. The substitution of petroleum-based feedstock with renewable feedstock provides an extra benefit as it relies less on fossil fuel as the carbon source. Nevertheless, it does not imply that the need for fossil fuels is eliminated. With this, the amount of greenhouse gas emissions associated with bioplastic production is reduced. Because the carbon dioxide (CO_2) taken from the air during photosynthesis compensates for the CO_2 released during biodegradation, it can be carbon neutral or even carbon negative [10,11].

With emerging innovation, it is now possible to design more sustainable plastics with distinct physical and aesthetic properties to compete with conventional plastics like polystyrene (PS), polypropylene (PP), and polyethylene terephthalate (PET). New bio-based materials have the increasingly popular ability to minimise environmental concerns while addressing the existing polymer and composite demand [1]. Bioplastic can be tailored to behave similarly to traditional plastics in the manufacturing phase but also excel from a performance standpoint. Currently, bioplastics account for only a relatively low proportion of global plastics production [10]. The cost of bioplastics is the main concern for the future extent of implementation as commercial manufacturing processes are expensive [8]. In this context, manufacturing costs can be cut down by integrating organic waste and residues, thus lowering the number of biodegradable polymers required to make bioplastics [2]. Metabolic and genetic engineering advancements have led to microbial and plant strains that may considerably boost yields and production capacities while being cost-effective [12]. When these considerations are paired with the conservation of scarce fossil fuels and the increment of environmental consciousness, it is foreseeable that the market dominance for bioplastics may develop in the future and replace petroleum-based single-use products, such as containers, straws, cups, and cutlery.

2. Types of Bioplastics and Process of Moulding Different Types of Bioplastics

Bioplastics, unlike conventional plastics, are mostly derived from renewable raw materials, including vegetable fats, oils, whey, starch, cellulose, and chitosan [1,3,10]. They are classified into two groups based on their backbone chemical composition: bio-based plastics and biodegradable plastics.

Bio-based plastics can be either biodegradable or non-biodegradable. Aliphatic polyesters like poly(lactic acid) (PLA), polyhydroxyalkanoates (PHA), starch, and cellulose are examples of both bio-based and biodegradable plastics. PLA is a thermoplastic biodegradable polyester that is produced through the polymerisation of bio-derived monomers, such as corn, potato, sugarcane, etc. It is recognised as one of the most popular "green" polymers in the polymer market, extensively used in food packaging applications and the biomedical sector [13]. In addition, PHA is an aliphatic bioplastic synthesised naturally by bacteria through the fermentation of lipids and sugar [14]. Besides utilising natural resources such as glucose, starch, and edible oils as the substrate for PHA production, several academics have investigated the potential of employing industrial, agricultural, and food waste, along with wastewater [12,15–17]. Starch is a biodegradable polysaccharide polymer that is widely used in food packaging applications owing to its abundance, low material cost, and food safety [18]. To be treated as a deformable thermoplastic polymer, a plasticiser (urea, glycerol, or sorbitol) with the addition of water to produce thermoplastic starch (TPS) under elevated temperature. TPS can then be extruded to make foam and solid moulded objects [19]. Cellulose, a polysaccharide composed of β-D-glucose subunits, is another biodegradable polymer [11].

As previously mentioned, the fact that they are bio-based does not imply that they are inherently biodegradable; that is, they contain renewable or fossil-fuel-based carbon [1]. For example, bio-polyethylene terephthalate (bio-PET), polyethylene-2,5-furandicarboxylate (PEF), and bio-polyethylene (bio-PE) are chemically identical to fossil-based PET and polyethylene (PE) [20]. This plastic type accounts for more than 42% of global bioplastic production capacities [3]. Polyamides (PA) represent another example of non-biodegradable bioplastics with high mechanical strength used in medical implants [21]. Over the years, attention has turned to PEF, a new polymer expected to hit the global market by 2023. PEF is like PET, but it is made entirely of bio-based materials and has better barrier properties, making it a great choice for bottles of drinks [10].

Aside from that, there are hydrocarbon thermoplastics that can be produced from renewable resources to replace a portion of the monomer, one of which is poly(1-butene) (PB). Cui et al. [22] synthesised isotactic poly(1-butene) (iPB) from eugenol, which is a phenol compound that can be extracted from different types of plant oil, such as clove oil, laurel oil, and camphor oil. The monomer of 1-butene and eugenol were copolymerized in the presence of Ziegler-Natta as a catalyst. The results indicated that the introduction of eugenol in the synthesis of poly(1-butene) has improved the thermal stability of the product and prevented the thermo-oxidative reaction of the polymer chain.

To specify whether the material is biodegradable or compostable, specific standards and protocols are required, which are standardised by the International Organization for Standardization (ISO), the American Society for Testing and Materials (ASTM International), government institutions, and other associations [1,23]. Their biodegradability is primarily determined by their physical and chemical structures but also by the environmental conditions in which they are placed [10,23]. Biodegradable polymers are typically derived from biological sources, but they can also be derived from petroleum resources. Biodegradable but non-biologically derived polymers include poly(butylene adipate-co-terephthalate) (PBAT), polycaprolactone (PCL), polyvinyl alcohol (PVA), and polybutylene succinate (PBS) [13].

There are different approaches for producing green composites, including injection moulding (IM), compression moulding (CM), extrusion, calendaring, thermoforming, and resin transfer moulding (RTM) [24–26]. Broadly, the selection of the processing techniques is a trade-off between the processing time, production cost, and final product design, shape, and size. The technique chosen affects the fibre dispersion, orientation, and aspect ratio, hence defining the mechanical properties [7,27]. Additionally, the processing conditions, such as temperature, pressure, and speed, vary from one technique to another. Factors, such as fibre length, content, type, and moisture content, can also affect the manufacturing process. For instance, there is a possibility that fibre would be thermally degraded if the temperature used is too high; therefore, it is preferable for those matrices that have melting points lower than the degradation temperature [26]. In view of this, it is essential to use the right methods and parameters to get the best results when making composites.

Injection moulding is a closed moulding process, which involves injecting a material into a mould under elevated temperatures and pressure. This technique is suitable for fabricating metal, glass, thermoplastic or thermosetting materials into assorted sizes and shapes of plastic products within a short time with high precision [21,28,29]. This process necessitates the use of an IM machine (equipped with a hopper, plunger, a heating unit, and a clamping system), raw plastic material, and a mould or die. The cycle starts with the injection stage, followed by the holding and plasticising stages, and finally the ejection of the moulded component. During the injection stage, the injection moulding compound (IMC) is introduced into the injection chamber via the feed hopper. After that, IMC will be conveyed by a screw-type plunger into a heated barrel, which transforms it into liquid form. It is then mechanically transferred into the closed mould cavity to cool and acquire the desired shape. During the holding and plasticising stages, the mould is kept at a constant temperature and pressure so that the melt can solidify rapidly after it is filled. Once the compound is hardened, the mould plates will open and eject the finished part via ejector pins. A new cycle can be started at this stage [28–30]. Regrettably, the significant expense of producing the mould is frequently a hurdle to IM technology. This method works best for the large-scale mass production of identical products [31].

Compression moulding is a traditional manufacturing technique that involves pressing thermoset and thermoplastic materials in the form of granules, sheets, or prepregs between two matched metal dies with huge presses [30]. The short cycle time and high production rate of CM make it ideal for applications in the automobile industry [7,27]. The common intermediate materials are sheet moulding compound (SMC), bulk moulding compound (BMC) prepregs, and glass mat thermoplastics (GMT). SMC and BMC are applicable to thermoset matrices, whereas GMT is normally used for thermoplastic matrices. CM can also be divided into two types, namely cold and hot CM. In cold CM, only pressure is applied, as it requires only room temperature for the curing process, whereas both heat and pressure are necessary for the latter [32]. The mould is preheated before transferring the heat to the composite and starting the curing process. Plastic materials are placed in between two preheated moulds, which are then pressed against each other and take the shape of the mould cavity with great dimensional accuracy. The process is carried out at a high temperature and pressure, depending on the requirements of the composite, for a set period until the moulding material is shaped [29,33]. It is also critical to keep the pressing time under control. Otherwise, there is a risk of cracking, scorching, or warping [28]. The advantages of the CM include a short cycle time and the potential to mould large, complex parts in a variety of forms and sizes. It helps to reduce waste material, which gives it a significant benefit when working with high-cost materials. In addition, it is one of the least expensive moulding techniques when compared to other production processes such as IM and RTM [31].

In extrusion, a thermoplastic resin is heated and plasticised through the action of the barrel of the extruder and the rotating screws. It is then extruded and driven out of the chamber via a die to form different cross-section products. This method has been adapted to fabricate short-fibre composites and may be utilised to make an IM precursor [30]. There are two types of extruder machines: single-screw and twin-screw. The former provides consistent mixing and distribution depending on the material's viscosity, whereas the latter is mostly preferred with short fibres to achieve more intense mixing [34]. A twin-screw system is applicable in the field of compounding and polymer blending, which gives better mechanical performance than a single-screw extruder [26]. Thermoforming is a unique method of transforming thermoplastic plastic sheets into functional plastic products. The sheet material is clamped and heated until it softens. Under the action of an external force (vacuum or air pressure), the softened sheet is forced against the shape of the mould. After cooling and shaping, the product is finished. Food packaging is the most prevalent use for thermoformed containers, trays, cups, and jars [11,35].

3. Mechanical Properties of Different Types of Bioplastics

There are various mechanical assessments performed to comprehend the composite with indicators including tensile, flexural, impact, and hardness tests. The study of mechanical characterisations from different tests reveals the mechanical behaviour of a polymer composite and provides information on the composite's suitability for its intended purpose [7,36]. The design criteria and specific application of the composite may be accessed by comparing multiple properties [37].

3.1. Tensile Properties

Tensile testing is among the most basic and intensively studied mechanical testing methods for polymeric materials, owing to its simplicity of testing and ease of interpretation of the results. It is often termed "tension testing" and is used to evaluate the stress-strain behaviour under tension [7]. During tensile testing, a tensile force (pulling force) is exerted on the material, and the specimens' response to the applied force (stress) is quantified. The samples are subjected to controlled tension until failure occurs. A stress-strain curve can be developed through this test to obtain the tensile strength, modulus of elasticity (Young's modulus), and elongation at break of the samples assessed [38]. The tensile strength indicates the highest stress that the specimen could sustain before it broke, as well as how much it stretches before it breaks [39].

3.2. Flexural Properties

The flexural test, also called the transverse beam test, is used to measure flexural strength and flexural modulus. It is a basic parameter to determine the feasibility of composite materials for structural applications [36]. Flexural properties are the outcome of the simultaneous effect of tensile, compressive, and shear stresses in the materials. Under flexural loading, a rectangular cross-section specimen is loaded in either a three-point bending or four-point bending mode [7]. The interfacial connection between fibre and matrix, and the extent of tension transfer between fibre and matrix, governed the flexural properties [5]. The flexural strength indicates the maximum breaking stress at failure, whereas the flexural modulus reflects the ratio of applied stress to deflection, computed from the starting slope of the stress-strain deflection curve [7,39,40]. These two values tell us the sample's resistance to flexure or bending forces.

3.3. Impact Properties

The impact test is used to evaluate structural materials' impact strength, toughness, and notch sensitivity, measured either by Charpy or Izod impact tests. The impact strength reflected the material's capacity to tolerate high-rate loading [7,39]. Toughness is a measure of the total energy absorbed per unit volume of material up to the point of rupture. Its value is quantified by measuring the total area under the stress-strain curve. Tough and brittle materials have high and low work-to-fracture values, respectively. Most polymer materials require an impact test because it concerns product performance, safety, liability, and service life [21,36,41].

3.4. Hardness

The hardness of a material signifies how resistant it is to abrasion, indentation, scratching, and plastic deformation under compressive load. These properties outline the material's wear and tear [42]. The surface hardness value can be determined by the Rockwell, Vickers, Brinell, and Shore hardness testers [43,44]. A greater hardness value suggests that the material is tougher and more resistant to penetration by other materials. From an engineering standpoint, a material's hardness is a benefit because it makes it less likely to wear down due to friction or erosion by water, oil, or steam [36].

4. Reinforcement Materials for Bioplastics

Bioplastics in the biocomposite category can be reinforced with reinforcing agents like inorganic fillers and natural fibres to increase their mechanical strength [1,3]. Fillers are often inexpensive, making the filler-reinforced biocomposites more economical. Importantly, these biocomposites unlock a new channel for the creation of innovative polymeric materials with more fascinating features [45]. In composite materials, a polymer matrix holds the fibres together, transfers the load to them, and distributes the load evenly. The fibres, on the other hand, serve as the primary load-bearing component because of their superior strength and modulus [46]. Figure 1 presents a schematic diagram of the filler–matrix interface, in which stress acting on the matrix is conveyed to the filler via the interface. For the composite to perform effectively, both phases must be well-bonded. When stress is transmitted, fibre–matrix adhesion is localised to the interphase region, which is a three-dimensional area between the phases, as illustrated in Figure 1b. With the formation of a "bridge" in the interface region, the efficiency of stress transmission can be enhanced. Likewise, the structure and properties of the interface are considered since they have a direct influence on the physical and mechanical characteristics of composite materials [47].

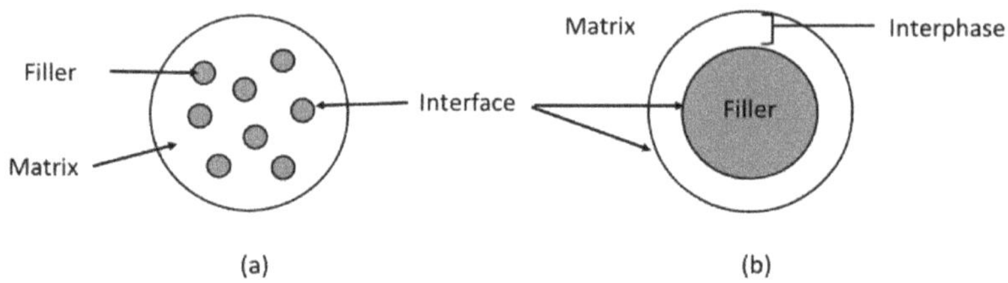

Figure 1. (**a**) Schematic of filler-matrix interface; (**b**) interphase region between the filler and matrix.

4.1. *Inorganic Filler*

Synthetic man-made fibres created by chemical synthesis are further categorised as organic or inorganic based on their composition. Glass fibre (GF), carbon fibre (CF), metal fibre, and ceramic fibre belong to the category of inorganic synthetic fibres [48]. These fibres are made from materials such as silica, alumina, aluminium silicate, zirconia, boron, boron carbide, boron nitride, graphite, silicon boride, silicon carbide (SiC), silicon nitride (SiN), etc. Inorganic fibre composites have been used to make lightweight structural materials with excellent strength and modulus values that can be tuned to specific loading requirements. They are heat-resistant, more robust, stiff, and have a higher melting point than conventional fibres [49].

Among these fibres, glass and carbon fibre are the most commonly used, with significant advances in plastic reinforcement applications. GF was the first continuous inorganic fibre, invented around 80 years ago [50]. They are abundant, available at cheap rates, and easy to use. GFs come in a variety of structures, including rambling, chopped strands, threads, fabrics, and mats, serving distinct functions for GFRP composite formation [31]. Numerous types of GFs are commonly used in polymer composites, depending on the raw materials employed and their quantities. Examples of GFs are A-Glass, C-Glass, D-Glass, E-Glass, R-Glass, etc. E-glass is the most universally used fibre in polymer composites [49,51,52]. GFs are usually employed in electronics, marine, aviation, and automotive applications. They have superior strength and durability, as well as thermal stability, impact resistance, electric insulators, and incombustibility.

Carbon fibres are fibres containing at least 92% carbon by weight, which first emerged in the market in the 1960s. The outstanding mechanical strength, stiffness, modulus of elasticity, high-temperature tolerance, and chemically inertness accurately describe these fibres [49]. Furthermore, alkaline materials and ultraviolet (UV) light do not affect CFs. These properties have made them very appealing in numerous engineering industries, including aerospace, civil engineering, sports, marine transportation, and the automobile industry. The main weakness of CFs is their high capital cost when compared to GFs, plastic fibres, or naturally occurring fibres. Moreover, their impact properties are comparatively weaker than those of GFs, but they are stiff and strong like steel [31].

4.2. *Lignocellulosic Materials*

The use of lignocellulosic fibre reinforcement has dominated research in recent years. A plant-derived natural fibre is known as lignocellulosic fibre (LCF). These are composed of cellulose, hemicellulose, lignin, pectin, waxes, and other water-soluble substances. The composition and percentages of these components differ depending on the type of biomass. Hence, they exhibit diverse mechanical behaviour [46]. Given their abundance, low density, non-abrasive, high specific mechanical strength, and modulus, these fibres are suitable for composite materials where the ideal property is centrally weight reduction. In addition to the enhancement of the properties of biopolymer composites, LCFs have also been reported to improve biodegradation at the end-use [53]. For modern uses, LCF-reinforced composites are better than synthetic fibre-reinforced composites for two reasons: they are cheaper and better for the environment.

The major framework component of lignocellulosic biomass is cellulose, which is a long-chain polysaccharide composed of D-glucopyranose units interlinked with β-1,4-glucosidic bonds [54]. Cellulose exists in both crystalline and amorphous regions, with crystalline cellulose consisting of chains with an orderly molecular arrangement and amorphous cellulose consisting of random arrangements [55]. Crystalline cellulose imparts strength and stability to the fibre. Because of the strong intramolecular hydrogen bonding, the hydroxyl groups (OH) within crystalline cellulose molecules are impermeable to chemicals, even water molecules. On the other hand, amorphous cellulose is soluble and more susceptible to enzyme degradation. The OH group in this region forms hydrogen bonds, allows water molecules to pass through, and gives polar fibres [46,56].

In contrast to cellulose, hemicellulose is a short-branched heteropolymer that exists in plants as an amorphous form, connected to cellulose microfibrils via hydrogen bonding, providing structural support to the fibre [57]. Hydrophilic hemicellulose is more prone to alkali and acid degradation [27]. Structurally, amorphous lignin synthesised by phenylpropane units that are arranged in a complex three-dimensional network structure is non-water-soluble and optically inactive. Lignin, coupled with cellulose and hemicellulose, provides additional strength to the hemicellulose–cellulose network [58,59].

4.3. Bacterial Cellulose

In addition to plant-based biomass, bacterial cellulose has become in demand due to its purity (without lignin, hemicellulose, and pectin as compared to plant cellulose), high aspect ratio, and high crystallinity [60]. Owing to its non-toxicity and mechanical stability characteristics, bacterial cellulose has received high demand for biomedical medical applications [61]. Wang et al. [62] have proven that bacterial cellulose could be used as a reinforcement to improve the mechanical properties of the composites as compared to the neat thermoplastic matrix. This improvement could be further enhanced when the bacterial cellulose was esterified, where a better bacterial cellulose distribution was indicated due to better interface compatibility [63]. Hence, based on the performance of bacterial cellulose-reinforced thermoplastic, it is believed to be an effective reinforcement for bioplastic.

5. Mechanical Properties of Reinforced Bioplastics

In the past decades, it has been a research hotspot to optimise the characteristics of polymers by integrating fillers, and research articles have been published to update the forefront of polymer-based composites for structural applications [4,37]. In this section, the focus is given to the mechanical behaviour of biocomposite materials. These mechanical studies are pivotal factors in correlating the changes to the bioplastic following reinforcement by fillers since their outcomes are highly related to the distribution and orientation of fillers within the matrix [28]. Figure 2 depicts the effect of the filler distribution in the composite system. The presence of uniformly distributed filler in the matrix allows the load to pass through without cracking or delamination. In contrast, poor filler dispersion and agglomeration create stress concentration points within the composite (Figure 2b). When the composite is loaded, cracks may initiate and propagate, resulting in a reduced load-bearing capacity in the fractured composites [46]. As previously stated, it is primarily driven by filler–matrix adhesion. The stress load can be effectively conveyed across the interface by strongly bound particles. In reverse, filler–matrix debonding causes physical discontinuities that cannot withstand mechanical forces [40,64]. The homogeneity of the filler in a composite system is also dependent on the mixing technique employed. The most popular mixing technique for preparing a reinforced thermoplastic composite is extrusion using a twin-screw extruder. The process parameters of a twin-screw extruder, i.e., the design of the screw and the co-rotating mechanism, determine the homogeneity level of a thermoplastic mixture [11,27]. Hence, the filler agglomeration issue can be resolved if a good mixing technique is used in which the stress concentration point formation can be avoided and the distribution of the load can be dispersed well, as depicted in Figure 2a.

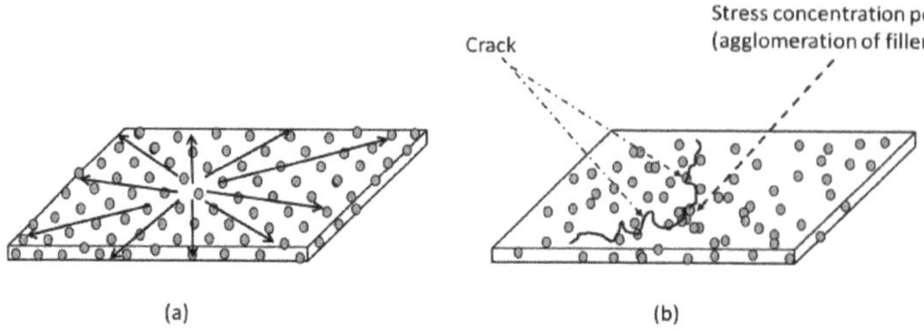

Figure 2. Load distribution for reinforced plastic (**a**) fillers are dispersed uniformly; (**b**) fillers are not well dispersed within the matrix.

In the composite system, the interaction between the thermoplastic and the filler can be classified into physical and chemical interactions. Mechanical interlocking is a form of physical force that holds two components of different interfaces together [65], whereas the formation of chemical bonding via functional groups between filler particles and matrix is classified as chemical interaction [43]. Without physical interaction, the filler may slip from the matrix when loaded, resulting in reduced stress transfer efficiency and lower composite strength. The composite strength can be further enhanced with the formation of chemical linkages between the filler and the matrix. For example, the functional group of isocyanates in the polyurethane matrix interacts with the OH groups of lignocellulosic filler for urethane linkage formation. This linkage may serve as a bridge where the load may be transferred efficiently from the matrix to the filler. With this, the strength of a composite can be enhanced [66].

Jiménez et al. [67] created a biocomposite using natural fibre reinforcement from sugarcane bagasse (SB) and a biodegradable starch-based matrix, Mater-bi® (PTA). SB fibres are prepared in the form of sawdust (WF), mechanical (MP), thermomechanical (TMP), and chemical-thermomechanical (CTMP) pulps. After incorporating 30% *w*/*w* of bagasse into the matrix, it turned out that the PTA/WF composite had a lower tensile strength than the neat PTA matrix. The major explanation for the decrease in tensile strength is that sawdust particles with a low aspect ratio are not perfectly adhered to the PTA matrix and have reduced reinforcing capabilities. The particles acted more as filler than as reinforcement. Conversely, all of the remaining fibres with higher aspect ratios behaved as reinforcements and produced composites with higher tensile strength. Moustafa et al. [53] identified the effect of reinforcement on the mechanical properties of the resulting composites. They incorporated coffee grounds (CG) into the PBAT matrix at varying levels of content, ranging from 10% to 50% in the presence or absence of PEG plasticiser. For PBAT/CG composites without PEG, large cavities are found in the rough fracture surface, translating into poor interfacial bonding between CG particles and the matrix. The tensile stress–strain curve demonstrated a significant loss in the mechanical properties as filler loading increased. In contrast, the PBAT/CG50 biocomposites had the highest elastic modulus values (777 MPa) of all the samples. However, the plasticization effect on the same PBAT/CG50 biocomposites made the values drop to 111 MPa.

Totaro et al. [68] highlighted the excellent mechanical properties of bioplastics by creating composites with silver skin coffee (SSK) as a filler and PLA and PBS as matrices. The incorporation of coffee by-product increased the value of the elastic modulus for both polymers, confirming optimum dispersion and wettability of the filler in the corresponding matrix. Aydemir and Gardner [24] explored the influence of cellulose nanofibrils (CNF) on the mechanical properties and discovered that the presence of CNF provides a mechanical property enhancement of 4–18% in the PHB/PLA blends. The introduction of filler to starch-based systems was investigated by Collazo-Bigliardi et al. [19]. They prepared coffee and rice husk cellulose fibres reinforced glycerol plasticised TPS films at 1, 5, and 10 wt%. For all composites, including both fillers, there was a strong tendency for the stiffness of the samples to increase. The obtained elastic modulus value reflects that even 1 wt% of filler enhanced the modulus of composites by approximately 60%, independent of the fibre type. Such an improvement might be attributed to the high purity of cellulose in the generated fibres, reflecting a higher degree of crystallinity of the material, which benefited the bonding between fibres and starch. Interestingly, adding 1 wt% of coffee fibre to TPS film did not change its ability to stretch, but adding the remaining 5 wt% and 10 wt% of both fillers made composites that were less stretchy.

Baek et al. [69] discovered that incorporating fibres into polymer matrices generates unstable interfaces and that the fibre's positive reinforcing effect is underutilised. The tensile strength of the green composites with PLA is negatively affected by the addition of natural fillers like bamboo flour (BF) and CG due to weak interface and obstructing stress concentration. A similar polymer matrix and filler combination were used by Kumar and Tumu [70], incorporating 5 wt% bamboo powder (BP) into the PLA matrix. When compared to neat PLA, the tensile properties of PLA/BP composites were reduced because of inadequate bamboo fibre dispersion and poor interfacial compatibility between the reinforced BP and matrix. When the adhesion is weak, the fibre does not play its role as a reinforcing material. Based on the data presented above, it can be assumed that the reinforcement of bamboo fillers weakens the mechanical properties of PLA/bamboo filler composites.

6. Factors Affecting the Mechanical Properties of Reinforced Bioplastics

The structure and final properties of composite materials, notably their mechanical properties, are dependent upon the reinforcement and the polymer matrix, and the interaction between the two constituents [26,71]. The type of filler, aspect ratio, filler loading, orientation, and many more are all important considerations. All these factors can be tuned to yield an optimum combination of mechanical strength and stiffness for future applications [52]. The contributions of surface treatments to the tensile, flexural, and impact properties of composites made from reinforced fibres will be discussed in the rest of this paper.

6.1. Types of Fillers

Strengthening polymers with fibres in varying ratios opens a world of options for developing materials with different attributes. These benefits have inevitably resulted in its widespread use in polymer applications, as it is observed that final properties are not feasible with a single polymer alone [43]. Fillers are widely used to improve the processability and mechanical properties of polymeric materials while at the same time lowering the material costs [47]. For reinforced composites, the dominant factor that governs their properties is the filler type and its properties. The behaviour of fillers should be studied to understand their actual contribution to the composite before incorporating them into the composite material [5,58]. Along with that, high-performance composite applications can only be met if there is a homogeneous dispersion of reinforcement inside the matrix and proper bonding between them to allow appropriate stress transmission from fibre to the matrix and vice versa. In other words, both components must be physically and chemically compatible [7,72,73]. In LCFs, mechanical properties vary subject to the composition and structure of fibres, which are listed as the following factors: fibre diameter, spiral angle of fibrils, degree of crystallinity, size of crystalline fibres and non-crystalline region, and chain orientation [7]. Moreover, the chemical composition of fibre represented by the percentage of cellulose, hemicellulose, lignin, and wax differs from plant to plant and throughout distinct areas of the same plant. Cellulose showed higher stiffening abilities than hemicellulose, while lignin is typically used as a coupling bonding agent between cellulose and hemicellulose. Therefore, the high cellulose content in the fibre explains the increase in mechanical strength [59]. Other factors to consider are the plant's growing conditions, such as its topography, climate, and age [46,74].

6.2. Aspect Ratio

The aspect ratio of fibre is a valid indicator of the reinforcing abilities of a certain fibre [11]. It is the length-to-diameter ratio of a fibre. This ratio depends on the extrusion process that breaks or shortens the fibre bundles. In general, when the aspect ratio increases, the stress may be transferred more effectively since there are more surface areas available for interaction. Interfaces are ruled by interfacial bonding, which is a key issue in composite science because it determines how stress can be transferred between the matrix and the fibres, compromising the mechanical characteristics of the whole material [34].

In a short-fibre composite, the tensile load applied is transferred to the fibre through shear loading at the interface. The tensile stress is zero at the fibre ends and increases along with the fibre length. Therefore, the ideal fibre length should be larger than the critical length (*Lc*) in order to properly convey the load during tensile loading [26]. When the fibre length is less than the critical length, debonding and pull-out of fibres will occur, indicating poor interfacial bonding in fibre-reinforced composite systems [75]. The value of the critical length can be determined as follows:

$$Lc = \sigma_f D/2\tau$$

where σ_f is the tensile strength of fibre, D is the diameter of the fibre, and τ is the fibre/matrix interfacial shear strength (IFSS) [7].

It is worth emphasising that the real reinforcing ability of fibre corresponds to the aspect ratio. Fibre with aspect ratios greater than 10 behaves as reinforcements, allowing preferential dispersion and distribution, which positively influences the mechanical behaviour due to particle alignment. On the other hand, lower aspect ratio fibres have less reinforcing capability and can even cause mechanical failure [67]. In the study of García-García et al. [40], the addition of spent coffee ground (SCG) particles (aspect ratios lower than 2) resulted in a deterioration in flexural strength. This is because no particle alignment can be accomplished during manufacturing. Karaduman et al. [76] reported that enzymatically treated jute fibre-reinforced polyester composites had a decrease in fibre diameter, which increases the aspect ratio. It created a large effective contact surface for resin impregnation and achieved great relevance to the final properties of the composites. When the fillers are compatible and have the right aspect ratio, they can intensify the material and allow it to be used in composites [5,11].

6.3. Percentage of Filler

The relative proportions of the filler materials in the formulation dictate the mechanical properties of the composites [68,77]. Various studies were undertaken on the effects of filler on the mechanical properties of biocomposites with different filler contents, and it can be said that aggregation phenomena are more evident for specimens at high filler loading [24]. In the study of calcium phosphate (CaP) and magnesium phosphate (MgP) nanoparticles' impact on pure PLA by Sahu et al. [77], the tensile strength of PLA nanocomposites increases linearly with the inclusion of CaP concentration up to 15 wt%, with a subsequence decrease in tensile strength at 20 wt% of fibre concentration. According to their team, this increase in tensile strength is related to the presence of a tensile stress-carrying filler in the polymer matrix. Similarly, the tensile strength of PLA/MgP composites confirms that beyond 2% MgP concentration, minimal improvement is observed. This behaviour is attributed to the presence of agglomerates on the surface at a concentration of 20% CaP nanoparticles and above 2% concentration of MgP nanoparticles, as observed by scanning electron micrographs. Ragoubi et al. [78] discovered that Young's modulus and deformation at break of PLA/miscanthus composites decreased at a higher ratio of filler. Indeed, signs of aggregation are present and heterogeneous dispersion of filler in the polymer matrix occurs, precluding the transmittance of stress from fibre to matrix. The results of X-ray tomography also showed that at this ratio, composites have larger holes and higher porosity. Due to the intrinsic stiffness of miscanthus fibres, the increment in reinforcing content reduces deformation at break significantly. The material becomes less plastic. This weak structure of the blend reversed the mechanical strength of PLA/miscanthus composites. Using coffee grounds and PBAT composites, Moustafa et al. [53] obtained equivalent results.

In summary, the mechanical properties of fibre-reinforced composites are found to improve linearly with increasing fibre content up to a certain optimum value. Further addition of fibre above that limit adversely affects the mechanical strength due to increased porosity and the formation of agglomerates. Likewise, there is insufficient resin to properly wet all of the fibres, preventing good filler dispersion within the matrix and limiting the two from sharing stress. Such an effect can be related to the degradation of mechanical properties [7]. Hence, future research should concentrate on the effect of filler loadings on the mechanical properties of composites, as composites with uniformly distributed particles yield a higher load capacity.

6.4. *Types of Treatment for Reinforcement Materials*

Researchers encountered several problems when fabricating biocomposites reinforced with natural fibres, most notably the hydrophilic nature of natural fibres, thermal instability of natural fibres, and a weak interface between the reinforcing phase and matrix phase, particularly for a matrix with hydrophobic behaviour [27,44,58,79]. Fibres and polymer matrices have distinct chemical structures. Fibre dispersion is dependent on compatibility [80]. The compatibility issues caused by the hydrophilic fibre and hydrophobic polymer matrix restrict the future application of composites, especially in an outdoor environment [7]. This is because the OH groups in the amorphous region of the fibre create new hydrogen bonds with water molecules from the atmosphere, resulting in extremely high moisture absorption. Consequently, fibre swells within the matrix, creating a poor linkage to the matrix that drives stress concentrated at the interface. In addition, micro-cracking also occurs between the swollen fibres and matrix, leading to dimensional instability with a detrimental effect on the mechanical performance of the resulting composite [7,51,81]. This confirms the need to remove hydrophilic OH groups and surface particles from the fibre surface via the surface modification process. To circumvent the heterogeneous interfacial problem, various fibre surface treatments have been proposed, namely physical, chemical, and biological treatments.

6.4.1. Physical Treatment

Physical treatment is aimed at increasing the mechanical bonding of the polymers by modifying the fibres' structural and surface properties without changing the chemical composition of the fibre extensively [7,51]. In other words, a stronger mechanical connection between the fibre and the matrix typically improves the interface. Physical approaches include mechanical comminution (chipping, grinding, milling) and electric discharge (plasma, corona, ultraviolet (UV) radiation, electron radiation) [6,58]. The objective of chipping and grinding is to disperse the particle size and facilitate the treatment process. This process is followed by milling methods, which can be ball milling, two-roll milling, hammer milling, etc., to become fine powder [54]. In the end, the crystallinity of the fibre is reduced, which affects the degree of polymerisation. The hydrolysis rate and mass transfer characteristics can be improved due to the reduction in crystallinity and particle size, respectively. The grinding conditions and intensity influence the final particle size of fibres, which in turn determines the energy requirement for mechanical comminution. This implies that mechanical processes are energy-intensive to achieve a high fermentable sugar yield, which is not economically feasible [82,83].

Plasma treatment is considered an environmentally friendly method for surface treatment using no chemical solvent. The plasma flows modified the fibre surface through ablation, etching, crosslinking, and surface activation. The fibre surface strength is enhanced after crosslinking the surface with free radicals [84,85]. Plasma etching generates hydrophobic surfaces by providing the desired roughness to the surface for physical adhesion and introducing new functional groups for higher polarity fibre surfaces [73,79]. The functional groups establish strong covalent bonds with the matrix, generating surface crosslinking to boost surface energy. In the end, the crosslinking process contributed substantially to an increase in mechanical strength [11]. The surface hydrophobicity can be altered by adopting different plasma parameters of exposure, i.e., nature of gas, exposure time, and applied power [85].

Another method of atmospheric plasma technique is corona treatment, which uses electric current to transmit changes in fibre properties and surface energy. Using corona discharge, chemical (surface oxidation) and physical (etching) effects are generated on treated fibres. Air plasma species bombardment increases surface roughness and coarseness, contributing to improved interactions between fillers and matrices [78,86]. To summarise, physical treatment is a non-polluting process with a short processing time and no specific conditions. During the treatment, a huge amount of material may be applied on a large scale, which benefits the manufacturing production of PF [58]. The fibre surface can be modified without affecting its integrity [79].

6.4.2. Chemical Treatment

Chemical treatment, which alters the chemical composition, surface topography, and morphology of natural fibres, is the most widely used method for strengthening fibre-matrix adhesion [5]. This treatment is described as the formation of a covalent bond between some reactive constituents of LCFs and chemical reagents, with or without the use of a catalyst [7]. The integration of hydrophilic fibre and hydrophobic matrix induces fibre swelling within the matrix and weakens bonding strength at the interface [56]. Chemical modifications destroy the fibre's hydroxyl groups and substitute them with hydrophobic chemical bonding. The seduction in the water absorption capability of the fibre is caused by the degradation of the OH group. In this context, fibre with lower hydrophilicity and a matrix with reduced cracking benefit the overall mechanical properties [87].

Acidic or alkaline treatments are the most commonly used and easiest treatments. These treatments usually focus on the fibre surface, where the soluble contents in fibres are dissolved using an acid (HCL) or alkali (NaOH) solution for hours, "washing" the surface from an uneven distribution layer of non-cellulosic components (lignin, hemicellulose, pectin, and impurities) that cover the fibre surface. These components are undesirable and may hinder the formation of physical, chemical, or both linkages between the matrix and the fibre (Figure 3). Alkali-treated fibres increased the ratio of exposed cellulose and experienced mass loss due to partial or complete elimination of non-cellulosic components, the majority of which are amorphous [45]. An easier fibrillation process is promoted after separating the fibre bundles into finer fibrils using a chemical solution to provide a larger surface area for interaction with the matrix [88,89]. Without impurities, fibre surfaces become rougher, providing additional sites for the polymer to anchor. Finally, potential mechanical anchorage and extra load-bearing capabilities at the interface can be accomplished as the fibres are surrounded by the matrix [79,90].

Figure 3. (**a**) The morphological change in the fibre surface before and (**b**) after surface treatment.

Coupling agents like silane, maleic anhydride, permanganates, and acetic acid function as the bonding agents to "bridge" the hydrophilic fibre and hydrophobic polymer through covalent bonding, hydrogen bonding, or polymer chain entanglement [90]. Chemically treated fibre has high moisture resistance properties via the removal of an OH group coating on the fibre surface. Grafting with compatibilisers (maleic anhydride) (MA) is a useful approach that allows the functional surface of fibre and matrix to interact efficiently. MA connects with OH groups in the fibre via covalent bonding and removes them from the fibre. The hydrophilic nature is reduced after a long polymer chain coating on the fibre surface. A maleated coupling agent creates a carbon–carbon connection between the OH groups of the fibre and the anhydride groups of MA. This covalent bond makes a bridge interface for efficient interlocking [43,56]. Silane treatment involves hydrolysis of alkoxy groups on silane to form silanol (Si-OH). During the condensation process, one end of silanol interacts with the OH group of cellulose (Si-O-cellulose), while the other end interacts with the functional group in a matrix (Si-matrix), forming a siloxane bridge between them. The number of OH groups of cellulose is reduced in the fibre cells, increasing the surface's hydrophobicity and ameliorating the interface's strength [6,26,91]. Acetylation substitutes the OH groups in fibres with acetyl groups, rendering the fibre surface more hydrophobic and rougher, providing stability to the composites [59]. Certainly, chemically treating fibre has significant benefits. However, there are some drawbacks. The well-recognised weakness is that these treatments provoke environmental issues attributable to the use of hazardous chemicals, inappropriate handling of chemical waste, and the generation of difficult-to-dispose-of by-products. This issue adds extra cost to the production process, making this treatment less widely adopted in manufacturing inexpensive products [44,92].

6.4.3. Biological Treatment

Given the environmental benefits, there has been an increasing interest in biological treatment. This treatment makes use of biological agents, either the microorganisms or enzymes secreted by the microorganisms, to fragment complex molecules of biomass into their constituents and change the structure and chemical composition of the fibre so that the treated fibre is more amenable to enzyme digestion [76]. Generally, this treatment is performed using different fungal species like white-, brown-, or soft-rot fungi and bacteria. White and soft-rot fungi specifically focus on both lignin and cellulose, whereas brown rot fungi depolymerise cellulose and hemicellulose. The specific extracellular enzymes secreted by these microorganisms increase the rate of enzymatic hydrolysis of the substrate through lignin degradation. White-rot fungi are reported to be the most efficient among these microorganisms, with *Phanerochaete chrysosporium* serving as the model strain for lignin breakdown [93]. Figure 4 illustrates the possible mechanical interlocking between filler and matrix in a composite system. The waxy layer coating the external surface is primarily responsible for the smooth native fibre surface [87]. Fungi produce hyphae during treatment, which create fine holes (pits) on the surface and provide roughness to the interface. It is believed that some of the filler components could be removed. A rougher fibre surface provides additional anchoring points, increasing the likelihood of mechanically interlocking with the matrix. Consequently, a high level of filler/matrix adhesion and good mechanical behaviour of the composite compared to the one with a smoother surface are expected [44,91,92,94].

Microorganisms are accountable for lignocellulosic materials degradation and demineralisation owing to the production of two types of extracellular enzymatic systems: the oxidative ligninolytic system, which acts on the phenyl rings in lignin, and the hydrolytic system, which attacks the cellulose and hemicellulose to liberate fermentable sugars [82]. Three major enzymes participate in the oxidative ligninolytic system: lignin peroxidase (LiP), manganese peroxidase (MnP), and laccase. The H_2O_2-dependent oxidation of lignin is catalysed by the LiP and MnP enzymes, while the demethylation of lignin components is catalysed by laccase, a copper-containing enzyme [93]. However, not all of these enzymes are secreted by fungal cultures. Bacterial laccase has also been identified in

Azospirillum lipoferum, *Bacillus subtilis*, and other organisms, but they are thought to have a minimal lignin degradation capability. Other than fungal treatment methods, several industrial enzymes, such as xylanase, cellulase, laccase, and pectinase, play a vital role in enzymatic hydrolysis. Non-cellulosic compounds conceal the external fibre surface and develop poor surface wetting, which impacts the interfacial adhesion between fibre and matrix [5,95]. Therefore, the enzyme catalyses biochemical reactions by binding a substrate at the active site specifically. Xylanase breaks down the hemicellulose, cellulase removes cellulose, laccase degrades the lignin structure, and pectinase is responsible for pectin degradation [76,96]. The modified fibre is less hydrophilic and has more exposed cellulosic fibrils, which improves wettability and mechanical interlocking between the fibre and matrix [87,97].

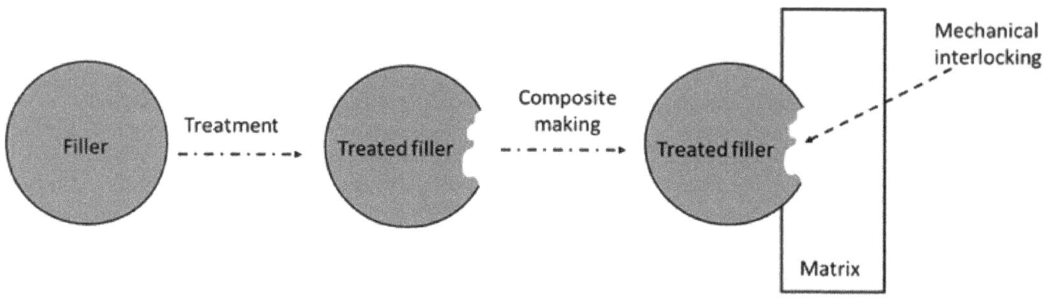

Figure 4. Formation of mechanical interlocking for a composite system.

Unlike physical and chemical treatment methods, biological treatment involves mild operating conditions and lower energy input and does not require acids, alkalis, or any reactive species. This process does not generate inhibitory substances or undesirable products. Another reason for embracing biological treatment is its cost-effectiveness, as no chemicals are employed and post-treatment washing and/or detoxification are not required [80,95]. However, there is a need to monitor the growth of microorganisms regularly and prolong treatment time for effective delignification, which restricts the industrial application of this method. Thus, the key parameters, such as the type of microorganisms involved, inoculum concentration, treatment time, pH, and temperature, should have optimum values to improve the techno-economic performance of the biological treatment. Despite that, biologically treating natural fibres is an innovative and emerging trend, considering the enzymes used are readily accessible and economical, and because the enzymes can be recycled, they produce little or no waste [71].

7. Effect of Filler Treatment on the Mechanical Properties

Prior to their incorporation into polymeric matrices, the morphological characteristics of fibre surfaces should be modified to minimise the shortcoming that comprises weak fibre-matrix interfacial attachment [91]. As previously discussed, the primary goal of surface treatment of filler is to achieve a high degree of fibre-matrix interlocking and stress transferability of the composites. Overall, the fibre surface treatment affects the physical, chemical, thermal, and mechanical properties of fibres and the resulting composite. The characterisation of mechanical properties, such as tensile strength, flexural strength, impact strength, and interfacial shear strength is studied to find out the effectiveness of various surface treatments on the performance of the resulting composite materials. The biocomposite after the application of different surface treatments to the fillers is summarised in Table 1.

Table 1. A summary of various surface treatments applied to fillers and their effects on the composite mechanical properties.

Treatment Type	Method	Filler	Matrix	Fabrication Method	Mechanical Performance	References
Physical treatment	Plasma treatment	Jute fibre	PLA	Injection moulding	Plasma polymerised jute fibre composites exhibited an increment in tensile strength, Young's modulus, and flexural strength up to 28, 17, and 20%, respectively. The IFSS of plasma polymerised jute fibre composites exhibited a significant increase of 90% more than untreated fibre, with a value of 6.84 MPa.	[79]
	Plasma treatment	Coir fibre	TPS	Compression moulding	Oxygen plasma with stronger etching was more influential in all conditions compared to air plasma, with the composite's tensile strength and elastic modulus increased by up to 300% and 2000%, respectively.	[73]
	Corona treatment	Miscanthus fibre	PLA	Extrusion-compression moulding	Lower fibre content (20% and 30%) increased Young's modulus more than the greater fibre content (40%).	[78]
	Corona treatment	Date palm fibre	PLA	Extrusion-compression moulding	Significant improvement in tensile strength and Young's modulus with 30% reinforcement of treated palm fibres in PLA, achieving the highest elastic modulus compared to untreated reinforcements and the PLA matrix.	[86]
	Electron beam irradiation	Bamboo powder	PLA	Injection moulding	The PLA/EBP5/ES 5phr composite demonstrated a 12% increase in notched impact strength over pure PLA.	[70]
Chemical treatment	NaOH alkali treatment	Alfa fibre	PLA	Injection moulding	The tensile strength and Young's modulus of the composite were strengthened by 17% and 45%, respectively, when 20 wt% NaOH-treated alfa fibres were included.	[89]
	NaOH alkali treatment	Rice husk	TPS	Compression moulding	The composites developed from alkaline-treated RH at a 20 wt% concentration gave the highest tensile strength by a factor of 220%.	[88]
	Acetylation	Sugarcane fibre	TPS	Extrusion	The addition of AcSF to the composite mixture increased the product's tensile strength while decreasing its water affinity.	[98]
	Acetylation and silanisation	Grape stalk powder	PBS	Injection moulding	Treated biocomposites showed better tensile properties than the control polymer. Acetylated GS powder gave the maximum improvement in Young's modulus from 616 MPa to 732 MPa.	[59]
	Maleic anhydride, NaOH alkali, and salinisation	Palm fibre (Macaíba)	PCL	Injection moulding	PCL composites with 15% and 20% MA treated MF showed the highest elastic modulus among all the samples. MA treatment presented the best mechanical performance, whereas NaOH treatment resulted in the worst.	[99]
	Silanisation	Coffee husk	PBAT	Melt extrusion	The addition of 40 wt% silane-treated CH increased the composite's mechanical properties (tensile strength, Young's modulus, and elongation at break) as compared to the 40 wt% untreated CH-reinforced PBAT composite.	[72]
	Silanisation	Silicon carbide	PBAT/PC	Solution casting and melt extrusion	The PBAT/PC composite with T-SiC showed a substantial enhancement in tensile strength and Young's modulus, with a reasonable drop in ductility.	[100]
	Maleic acid and silanisation	Coconut shell powder	PLA	Compression moulding	The treated composite's tensile strength and Young's modulus increased after the CS surface-treated with maleic acid and 3-APE coupling agent but had lower elongation at break.	[101]
	NaOH alkali treatment	Jute fibre	PLA	Injection moulding	Jute fibres treated with 5% NaOH concentration have good interaction with the PLA matrix, resulting in an improvement in tensile strength.	[79]

Table 1. *Cont.*

Treatment Type	Method	Filler	Matrix	Fabrication Method	Mechanical Performance	References
Biological treatment	Xylanase and pectinase enzymatic treatments	Alfa fibre	PLA	Injection moulding	The tensile strength of PLA/xylanase and PLA/pectinase composite samples is increased by ≈22% and ≈27%, respectively, when compared to that of unmodified samples.	[89]
	Xylanase and pectinase enzymatic treatments	Date palm fibre	PBS	Injection moulding	The combined action of two enzymes (xylanase and pectinase) gave the highest tensile modulus of reinforced composites (1600 MPa).	[45]
	Pectinase, laccase, and cellulase enzymatic treatments	Bamboo fibre	PHBV	Injection moulding	The values of tensile strength, impact strength, flexural strength, and flexural modulus were greatest for pectinase-treated bamboo fibre/PHBV composite.	[102]
	Cellulase enzymatic treatment	Ramie fibre	PBS	Compression moulding	The tensile and flexural strength of treated fibre reinforced biocomposites increased as the fibre concentration increased (0.5% to 1%).	[103]

7.1. Physical Treatment

Natural fibres have been physically modified to promote fibre-resin adhesion in fibre-reinforced composites, including plasma, corona discharge, and electron beam treatments [58]. Physical modification raises the IFSS of neat fibres with the polymer matrix, which was previously low before surface treatment. Gibeop et al. [79] studied the mechanical properties of jute fibre/PLA biocomposites by treating them with helium and acrylic acid as carrier gas and monomer, respectively, with a plasma power of 3 kV and 20 kHz for different exposure times (30, 60, and 120 s). Plasma polymerised fibre composites outperformed alkali-treated composites in terms of tensile strength, Young's modulus, and flexural strength by up to 28, 17, and 20%, respectively. Added to that, plasma-treated jute fibre composites aided in a pronounced improvement in the IFSS, which was determined by a micro-droplet test. The increment in IFSS value of 90% more than the untreated jute fibre/PLA composite was attributed to a rougher fibre surface indicated by an increased surface friction coefficient value. This is subjected to the heat and etching effect on the outer layer of fibre, which leaves more non-polar lignin on the surface. These results provide a great contact between jute fibres and the PLA matrix, which could be visualised by SEM micrographs. The significant improvement in the mechanical performance of the resulting composites suggests that plasma surface modification is capable of increasing the connectivity between hydrophobic matrix and hydrophilic fibre.

In a study performed on plasma treatment, de Farias et al. [73] treated coir fibres with oxygen and air before incorporating them into the TPS matrix. Their study demonstrated that plasma treatment (80 W, 7.2 min) was effective in improving both the tensile strength and elastic modulus of the composites when either oxygen or air was used. When compared to air plasma, oxygen plasma was more influential in all conditions, with the composite's tensile strength and elastic modulus achieved by up to 300% and 2000%, respectively. Stronger oxygen plasma etching removed more surface lignin, exposed the crystalline cellulose, and increased surface roughness and compatibility factor. The roughened surface points to fibre–matrix interlocking, which has a pronounced effect on the load transfer between them. The authors also pointed out that there was a correlation between plasma power and the resulting properties of the composites. Given this, these variables should be chosen wisely to reap the benefits of plasma treatment.

Miscanthus fibre was subjected to corona treatment at a discharge frequency of 50 Hz and a voltage of 15 kV for 15 min [78]. The fibres were blended with PLA granules containing 20–40 wt% fibre content, and the mixture was then extrusion-compression moulded to produce PLA/miscanthus composites. They experimented with both untreated and corona-treated fibres. Tensile measurements were used to determine the effect of fibres on the mechanical properties of PLA and composites. The effectiveness of corona-treated miscanthus fibres can be seen in the improvement of mechanical properties, including elastic modulus, stress, and strength at yield, in resultant composites when compared to PLA and composites containing unmodified fibre. Low fibre content (20% and 30%) showed better enhancement in Young's modulus than the higher one (40%) because good fibre dispersion is conducive to better stress transmission from matrix to fibre. The chemical (surface oxidation) and physical (etching) effects of corona treatment on fibres could explain the improvement in interfacial compatibility between PLA matrix and miscanthus fibres, observed using X-ray photoelectron spectroscopy (XPS) and SEM. At higher ratios of treated fibre, the composites display larger voids and higher porosity, while Young's modulus remained unchanged compared to composites with non-treated fibres. Amirou et al. [86] conducted another corona discharge treatment on date palm fibre (DPF) and PLA using the same corona discharge frequency and treatment time as the previous author. Extrusion-compression moulding techniques were used to create fibre mixtures with varying fibre content ranging from 30–40%. Before treatment, the inclusion of DPF did not show any improvement in the tensile strength, indicating inadequate adhesion between fibres and the PLA matrix. Through the corona treatment, there was a considerable improvement in tensile strength and Young's modulus, with the highest elastic modulus (2951 MPa) reached by 30% reinforcement of palm fibres in polylactic acid compared to untreated reinforcements (2708 MPa) and the PLA matrix (2396 MPa). This is attributed to the mechanical anchorage related to an etching effect caused by the bombardment of the air plasma species on the fibre surface. Indeed, the specimen surface became rougher and coarser. In both studies, it was found that higher mechanical anchorage helped improve the interfacial contact and compatibility between the two phases.

Kumar and Tumu [70] have utilised electron beam (E-beam) irradiation at various doses (30, 60, and 90 kGy) to achieve better interfacial adhesion of BP and PLA. E-beam irradiated bamboo powder (EBP) was melt blended with PLA at 5 wt% and 10 wt% concentrations, as well as the coupling agent epoxide silane (3-Glycidoxypropyltrimethoxy silane) (ES). They have asserted that the PLA/EBP5/ES 5phr with 5 wt% EBP and 5phr ES has better tensile properties than other PLA/BP composites. This could be because trapped free radicals in the EBP initiated the interaction with carboxylic terminal groups of PLA and epoxide groups of epoxide silane, forming PLA-g-ES copolymers. Because the silane alkoxy groups of PLA-g-ES are extremely reactive to the hydroxyl groups of bamboo powder, the copolymers function as an interface between the PLA matrix and the fillers to improve their miscibility. Besides, the composites have shown a noticeable improvement of 12% in the notched impact strength compared to pure PLA and rougher morphology with ideal distortions, indicating more impact energy was absorbed. The author points out that the incorporation of a higher percentage of EBP (10 wt%) leads to a decrement in the tensile properties because interfacial compatibility between matrix and filler decreases at a higher bamboo fibre content. Heterogeneous phase morphology, as corroborated by the SEM micrographs, which reflect a lack of adhesion between matrix and filler, may have contributed to lower mechanical properties. They also studied the effects of irradiation dose and concluded that a high-dose electron beam will generate excess free radicals that disrupt the intermolecular hydrogen bonding among the cellulose molecules.

7.2. Chemical Treatment

Most previous research identified that alkali-treated fibre improved the mechanical properties of the resulting polymer composite [51,88,89]. Boonsuk et al. [88] performed alkali treatment on rice husk (RH) using a high alkali concentration (11 wt% NaOH) and added it to the thermoplastic cassava starch (TPS) matrix at loadings of 5–20 wt%. The mechanical properties of untreated and alkali-treated RH/TPS biocomposite were studied and compared. The findings revealed that the addition of 20 wt% alkaline-treated RH/TPS biocomposites gave the highest tensile strength by 220% compared to the neat TPS but decreased elongation at break. The rough surfaces of treated RH and loss of hemicellulose after NaOH treatment recorded improved interface interaction and more effective fibre-matrix load transmission. Alkali treatment creates a smoother inner surface, splits fibres into fibrils (fibrillation), and makes OH-rich fibrils more accessible. After hemicellulose and lignin are removed, new hydrogen bonds can form between cellulose chains. Thus, from the above-reported finding, it can be extrapolated that the composites with high fibre content resulted in better tensile strength. In another study, alkali-treated alfa fibres were employed as reinforcement in PLA resin, and composites were prepared using IM with a fibre content of 20 wt% [89]. When 20 wt% NaOH-treated alfa fibres were included, the composite's tensile strength and Young's modulus were strengthened by 17% and 45%, respectively. At the surface of the fibres, it was seen that the fibre bundles were opening up and the cementing components (hemicellulose, lignin, waxes, and oils) were disappearing. This made the surface rougher and caused a high degree of fibrillation.

Aside from alkali treatment, acetylation is a popular fibre treatment method. Fitch-Vargas et al. [98] investigated thermoplastics made from acetylated corn starch composites reinforced with acetylated sugarcane fibre (AcSF). The AcSF-reinforced starch-based composite was prepared by extrusion. Through chemical modification and interactions between fibre-matrix, mechanical interlocking between the two phases was improved, as evidenced by an improvement in mechanical properties with AcSF of up to 12%. The water affinity property was reduced by the presence of hydrophobic acetyl groups in the biocomposite. Nanni et al. [59] applied two types of fibre surface treatments on grape stalk (GS) powder. Acetylation and silanisation, which were later reinforced in the PBS matrix. Acetylation reduced the polarity of GS and made its rougher and spongier, increasing the possibility of mechanically interlocking with polymer chains during melt compounding. AcGS had the best mechanical performance of all the samples tested, with Young's modulus increases from 616 MPa to 732 MPa. This trend is clarified by the degradation of hemicellulose under the harsh conditions of the acetylation process and is well interconnected between GS and the PBS matrix, supported by FTIR and SEM-FEG analysis. Moreover, acetylation worked well to minimise the moisture uptake of treated GS, showing that the surface of treated GS became less hydrophilic.

An investigation was carried out on the chemical treatments using (3-methacryloxypropyl) trimethoxysilane, MA, and NaOH on palm fibre (Macaíba fibre) (MF), which was subsequently melt extruded with polycaprolactone (PCL) [99]. Following that, the biocomposite with an MF concentration varying from 10–20% was then thermally, spectroscopically, mechanically, and morphologically characterised. For elastic modulus upon the addition of 10% treated fibre, silane treatment gave the best response among the treated samples and a neat PCL matrix, but NaOH treatment gave the lowest value, possibly due to excess delignification which weakens MF. Interestingly, biocomposites with 15% and 20% MA treated MF showed the highest elastic modulus among all the samples, most probably due to greater interaction between constituent components, namely PCL, fibre, and MA. Chemically treated biocomposites outperformed untreated ones in terms of flexural modulus. These enhancements are associated with enlarged contact points between fibre and matrix as a result of defibrillation. MA treatment also improved flexural modulus, which is thought to be related to the "anchoring" of succinic anhydride groups on the fibre surface and benefits the polar interaction between PCL and MF. Conversely, chemically treated MF biocomposites demonstrated lower impact strength than untreated MF biocomposites. This is owing to oil action in natural MF. The presence of oil in the pulp increases plasticization mechanisms, resulting in higher impact strength. The application of chemical treatment on MF and increased MF content lowers biocomposite elongation due to improved chemical interaction between MF and PCL, which restrains macromolecular movements, resulting in more stiff and brittle materials. Through the gathered findings in their work, the authors concluded that MA had the best mechanical performance and NaOH had the worst.

The creation of an interconnected network from silane treatment reduces the swelling property of fibre as a result of stable covalent bonds between fibre and matrix [104]. Lule and Kim [72] discussed coffee husk's (CH) mechanical properties against silanisation with a silane agent, 3-Glycidoxypropyl trimethoxysilane (GPTMS). When 40 wt% silane-treated

CH is reinforced in the PBAT matrix compared to the 40 wt% untreated CH-reinforced composite, mechanical parameters such as tensile strength, Young's modulus, and elongation at break are significantly improved. SEM micrographs also showed continuous phase morphologies with no gaps between their interfaces, achieving good interfacial interactions with the polymer matrix, which promoted greater physical and mechanical characteristics of the composites. Figure 5 outlines the stress transfer efficiency between filler and matrix. The absence of a gap between filler and matrix is attributed to the possible interaction, such as the development of covalent bonds. As a result, the stress transfer efficiency from matrix to filler is expected to be higher than that without interaction (Figure 5b). The stress could not be transferred due to the gap between filler and matrix, as depicted in Figure 5a. This demonstrated that silane treatment aided stress transmission between CH and the PBAT matrix by preventing the formation of voids and gaps. The same author studied the incorporation of surface-treated silicon carbide (T-SiC) particles in PBAT and polycarbonate (PC) matrices, which led to a substantial enhancement in tensile strength and Young's modulus, with a reasonable drop in ductility owing to greater SiC loadings [100]. According to Tanjung et al. [101], the inclusion of maleic acid-treated and silanated CS filler in the composite mixture has remarkably increased the PLA/CS composite's tensile strength and Young's modulus but reduced its elongation at break when compared to the untreated biocomposite. Wang et al. [105] studied the use of herb residue as a reinforcement material for PB. They found that the introduction of herb residue to PB improved its thermal stability, and this phenomenon was more obvious when the herb residue was treated with a silane coupling agent. This was attributed to the improvement of interfacial properties between the matrix and herb residue. The hydrophilicity of the reinforcement material decreased after it was treated with a silane coupling agent, and the compatibility between the treated reinforcement material and PB was improved [106].

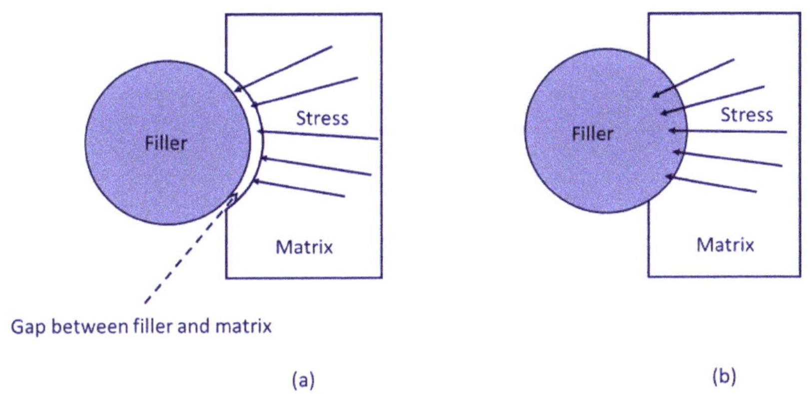

Figure 5. Stress transfer efficiency between filler (**a**) without interaction between filler and matrix (**b**) with interaction between filler and matrix.

The key operating parameters affecting the treatment, such as concentration of acid or alkali solutions, soaking time, and temperature, need to be optimised to have the most desired mechanical and physical properties. Increasing alkali concentrations have been linked to improved mechanical characteristics. However, exceeding the optimal concentration of chemical reagents may cause fibre degradation and have a detrimental impact on the tensile strength of composites. Gibeop et al. [79] revealed that alkali treatment with 3% NaOH concentration does not get rid of the amorphous material, with fibre pulling out holes in the PLA matrix, as shown by SEM images. On the other hand, jute fibres that have been treated with 5% NaOH concentration have good contact with the matrix, which makes the tensile strength better.

7.3. Biological Treatment

Biological treatment of fibres outcompetes chemical treatment without harsh chemicals or elevated temperatures. This treatment optimises the fibre surface for composite applications by using microorganisms, such as bacteria and fungi. These modifying agents are promising in developing composites with good mechanical properties that are both green and environmentally friendly. Enzymatic treatment is now gaining popularity, thanks to the high selectivity and specificity of enzymatic action that only targets the undesirable constituents without disrupting the structural modification of the important components [51,97].

Werchefani et al. [89] examined the impact of hemicellulases (cellulase-free xylanase) and pectinases on the alfa fibre surface, based on the hypothesis that hemicellulose and pectic components are accountable for moisture absorption and mechanical improvement. Their research demonstrated that these enzymes are excellent at improving the mechanical characteristics and water resistance of PLA composites. According to their findings, pectinase treatment was more effective than xylanase for eliminating undesirable materials, roughening the fibre surface, splitting alfa fibres into finer fibres, and enlarging surface accessibility for good polymer/filler interactions. As a result, an enhancement of tensile modulus and tensile strength was noticed when compared to that of unmodified samples. By getting rid of hemicellulosic and pectic components, enzyme treatments also make the surface less polar, which makes it less likely to absorb water.

The effects of three different enzymes (pectinase, laccase, and cellulase) on the reinforcing capability of bamboo fibres (BF) in poly(hydroxybutyrate-co-valerate) (PHBV) were studied by Zhuo et al. [102]. Melt blending of fibres and resin was followed by IM to fabricate the composites. All composites improved in mechanical properties following enzymatic treatment. However, the improvement was not significant. Pectinase had the best modifying impact of the three enzymes. The tensile strength, impact strength, flexural strength, and flexural modulus of PHBV composites with pectinase-treated BF increased by 4%, 7.1%, 6.2%, and 6.3%, respectively. They concluded that two factors contribute to the improvement of mechanical characteristics. The first is the surface roughness of BF, which is more favourable for stress transfer in composites. Second, the reduced polarity of BF after the removal of the OH group, lignin, and free cellulose on the surface. This feature is preferred for better compatibility with weakly polar PHBV and hence improves the interfacial compatibility of BF/PHBV composites. Werchefani et al. [89] reached the same conclusion: composites treated with pectinase had the best mechanical properties and the least amount of water absorption.

The combination treatment of xylanase and pectinase was conducted on DPF-reinforced PBS composites [45]. The highest tensile modulus (1600 MPa) was achieved with 20% of treated fibre reinforced composite, which was due to the synergistic effect of the two enzymes that impart the highest cellulose-rich fibre while degrading the amorphous components. The simultaneous action of both enzymes exposed more individual fibre bundles and cellulose microfibrils and reduced fibre diameters, which are believed to achieve the best mechanical properties. The efficacy of combined enzyme treatment was demonstrated by the depolymerisation of lignin, pectin, and hemicellulose during treatment with xylanase and pectinase, which verified an increase in the stiffness of the composites. Treated DPF has proven beneficial in a variety of applications where these mechanical properties are demanded. The study proved the combination of enzyme treatment benefits not only from the treatment efficiency but also from lowering the operational time. Another biological treatment using bacterial cellulase enzymes was applied to ramie fibres by Thakur and Kalia [103]. They used bacterial cellulases from two different bacterial strains (*Brevibacillus parabrevis* and *Streptomyces albaduncus*) to modify the surface properties of the PBS/ramie fibre biocomposites. The authors found that the ramie fibre surface is free from gums and polysaccharide layers and was cleaned and restructured to become more compatible with the hydrophobic PBS matrix. Therefore, there was better interlocking between the two phases, which helped the biocomposite to demonstrate satisfactory mechanical performance.

8. Advantages and Disadvantages of Reinforced Bioplastics and Its Treatment

Natural fibres derived from agricultural wastes serve as an ecological and cost-effective alternative to typical petroleum-based materials, since they substantially reduce the dependency on fossil fuels and greenhouse gas emissions. Depending on the plant source, the physical and mechanical characteristics of fibres can vary in terms of density, tensile strength, Young's modulus and elongation at break (Table 2) [26,27,107,108]. As shown in Table 3, these fibres have significantly higher strength and stiffness values than bioplastics and conventional plastics [3,14,109–111]. Given the properties of natural fibre, it is preferred to reinforce polymers with high-strength natural fibres to produce natural fibre-reinforced bioplastics (NFRP). The addition of fibres derived from renewable and infinite resources reduced the overall cost of the composite material while improving the waste management techniques in a sustainable manner.

Table 2. Properties of commonly used natural fibres to reinforce bioplastics.

Property/Natural Fibre	Bagasse	Bamboo	Cotton	Coir	Jute	Ramie	Oil Palm
Density (g/cm^3)	1.25	0.6–1.1	1.5–1.6	1.2–1.5	1.3–1.5	1.5	0.7–1.55
Tensile strength (MPa)	222–290	140–800	287–800	140–180	200–773	400–938	248
Young's modulus (GPa)	17	11–17	5.5–13	4–6	10–55	61.4–128	3.2
Elongation at break (%)	1.1	1.4	7–8	30	1.5–1.8	3.6–3.8	25

Table 3. Properties of bioplastics and conventional plastics.

Property/Polymer	PLA	PHA	PBS	PP	PET	PS
Density (g/cm^3)	1.24	1.25	1.26	0.91	1.3–1.4	1.05
Tensile strength (MPa)	37–66	20–40	30–35	15–27	55–79	24–60
Flexural modulus (MPa)	2392–4930	1280–3668	-	850–1050	1000–2300	2100–3000
Young's modulus (GPa)	2.7	2.95	0.27	0.95–1.77	2–4	3.4
Elongation at break (%)	0.5–9.2	1.4–5.5	8–13	100–600	15–165	1.6–2.5

Potential avenues for improving reinforced bioplastic properties for a variety of applications are being explored. Fibre treatment is a novel approach for improving interfacial adhesion between fibre and matrix. Bioplastics that contain surface-treated fibre as a reinforcing agent increase fibre–matrix binding. The modification with NaOH alkaline solution, for example, splits the fibre bundles into finer fibres. The smaller fibres were impregnated with polymer material, enhancing the interface between the fibres and the matrix. The opening of fibre bundles and partial removal of cementing constituents results in rougher fibre surfaces, which facilitates matrix penetration into fibres. This suggests that the fibre-resin integration can have a significant impact on the stress transmission at the interface via mechanical interlocking [89]. Without the mechanical locking or formation of linkages within this region, the efficiency of the stress transfer mechanism is thought to be low, and the composite could not withstand the load when loaded. This phenomenon will be worsened if the reinforcement material does not disperse well throughout the composite, causing uneven load distribution [46].

Higher compatibility between fibres and matrix was expected with the inclusion of reinforcement, and the reinforced bioplastics demonstrated superior mechanical performance with modified fibre compared to composites made with untreated fibre [70,73,79,86]. Reinforcing surface-treated fibres allows the mechanical performance to be improved without deconstruction. The high specificity of enzymatic treatment allows them to target the non-cellulosic fibre components while retaining the natural structure of cellulosic fibres [42]. Added to that, treating fibre with the combined action of two enzymes contributes to a more fibrillose structure and enhanced stiffness of the reinforced bioplastic [45]. The enzyme-treated fibre-reinforced bioplastics have the lowest moisture absorption properties after eliminating the hydrophilic components on the fibre surface. The ability to resist moisture is beneficial in the preparation of composites for construction, automotive, and packaging applications [89].

Disadvantageously, some studies found a considerable decline in the elongation at break upon reinforcement of surface-treated fibres [45,59,99]. This is most likely owing to the reinforcing action of treated fibres, which limits the molecular movements, leading to stiffer and less flexible bioplastics [89,99]. However, as the fibre loading increased, the Young's modulus decreased. This may be associated with the formation of aggregates at higher fibre content, leading to stress concentration zones and lower mechanical properties [78,86]. Another drawback of surface-treated bioplastics is the treatment parameters, which often deteriorate the mechanical performance of bioplastics when used in excess, as they damage the fibre surface [73,112,113]. As a result, optimal parameters and conditions should be carefully chosen to achieve the desired level of modification and boost the treatment efficiency.

9. Applications of Reinforced Bioplastics

Over decades, natural fibres have proved their excellence in substituting costly carbon and glass fibres. They have high specific tensile properties and lower density than synthetic fibres, making them lighter and more fuel efficient [43]. NFRP shows promise in a variety of areas, including automotives, aerospace, construction, consumer goods, protective equipment, packaging, and so forth. Because of their sensitivity to environmental degradation, NFRPs are currently limited to non-load-bearing interior components in civil engineering and automotive parts [91,114].

NFRPs have several advantages over conventional composites in the automotive industry, including increased acoustic insulation and mechanical properties, lower weight and manufacturing cost, recyclability, renewability, and eco-efficiency. They can be used to make door panels, seat cushions, armrests, and headrests [27,64]. Lighter composite parts used in vehicles to replace metal and heavier materials can lower transport weight, hence indirectly boosting fuel efficiency [115]. Werchefani et al. [89] fabricated biopolymer composites reinforced with alfa fibres from PLA. Mechanical testing shows that the composites have the required properties for interior automotive parts where composite strength is a necessity for performance.

By capitalising on its lower density, tool wear, and cost, natural fibre has surpassed synthetic fibre in many applications and is ideally suited for use as a reinforcement in polymer composites or cement matrices [58]. Indeed, natural fibre can be used to manufacture windows, doors, window frames, roof tiles, ceilings, and floor mats in the construction industry [27]. Sisal fibre and coir fibre have also been explored as roofing components instead of asbestos, which is carcinogenic [91,107]. Traditional composites have substantial pollution and disposal difficulties at the end of their useful life. As a result, there is a stronger desire to employ green products in order to leave a smaller environmental footprint.

When it comes to food applications, gas barrier properties (water vapour and oxygen permeability) are significant features to access the viability of materials because they affect the deterioration of moisture-sensitive products and their shelf-life [116]. In the PLA/PBS matrix, the presence of both cellulose nanocrystals (CNC) and surfactant-modified cellulose nanocrystals (s-CNC) provoked an improvement in oxygen and water vapour barrier properties [117]. In a prior study undertaken by Papadopoulou et al. [118], cocoa bean shells (CBS) as natural active additives in PLA composites were shown to represent a promising possibility for developing active food and biodegradable packaging materials by conferring antioxidant activity to the composites. Melt compounded PBAT/torrefied CG composites exhibited improved hydrophobicity, increased water contact angle values, and significant enhancement in the thermomechanical properties. Because of its high hydrophobicity, the biopolymer composite has the potential to be used in food packaging [119].

10. Conclusions

Bioplastics represent a new plastic generation, paving the way toward sustainability, renewability, and biodegradability. Their mechanical behaviour can be measured in terms of tensile, flexural, impact, and hardness. Reinforcing agents are added to bioplastics to strengthen their mechanical properties and expand their fields of application. For biocomposites, the choice of filler type, aspect ratio, filler loading, and surface treatment applied greatly influenced the final mechanical properties. To tailor the performance of final composites, uniform dispersion of reinforcement inside the matrix and a strong degree of interaction between them are required in composite materials. The hydrophilic fibre is modified for further compatibility enhancement with the hydrophobic behaviour of the polymer matrix. Treated fibres have a rough surface texture, which is critical for penetration into the matrix, enabling maximum stress transfer across the interface and a better mechanical interlocking system. Subject to the treatment strategies, most studies showed a better increment in fibre hydrophobicity, interfacial adhesion between fibre and matrix, and superior mechanical properties. The conditions and parameters used for surface treatments can cause changes in structure, morphology, and mechanical properties, consequently affecting the fibre-reinforced composites. Hence, proper fibre modifications enable better stimulation of their properties for usage as reinforcements in composites. A polymer composite with desired qualities that perfectly meets the requirements for a particular application can be fabricated by manipulating the fibre content, orientation, size, or manufacturing processes. Fibre-reinforced biocomposites find use in a variety of fields based on the qualities required. Further research on performance is needed to enlarge the domain of applications of biopolymer composites.

Author Contributions: Conceptualization, G.S.T.; investigation, J.Y.B., C.K.L. and G.S.T.; writing—original draft preparation, J.Y.B., C.K.L. and G.S.T.; writing—review and editing, J.Y.B., C.K.L. and G.S.T. All authors have read and agreed to the published version of the manuscript.

Funding: This research was funded by Universiti Sains Malaysia, grant number 1001/PTEKIND/8014123. The APC was funded by Universiti Sains Malaysia.

Institutional Review Board Statement: Not applicable.

Informed Consent Statement: Not applicable.

Data Availability Statement: Not applicable.

Conflicts of Interest: The authors declare no conflict of interest.

References

1. Garrison, T.F.; Murawski, A.; Quirino, R.L. Bio-Based Polymers with Potential for Biodegradability. *Polymers* **2016**, *8*, 262. [CrossRef] [PubMed]
2. Pudełko, A.; Postawa, P.; Stachowiak, T.; Malińska, K.; Dróżdż, D. Waste Derived Biochar as an Alternative Filler in Biocomposites-Mechanical, Thermal and Morphological Properties of Biochar Added Biocomposites. *J. Clean. Prod.* **2021**, *278*, 123850. [CrossRef]
3. Ibrahim, N.I.; Shahar, F.S.; Sultan, M.T.H.; Shah, A.U.M.; Safri, S.N.A.; Mat Yazik, M.H. Overview of Bioplastic Introduction and Its Applications in Product Packaging. *Coatings* **2021**, *11*, 1423. [CrossRef]
4. Burrola-Núñez, H.; Herrera-Franco, P.J.; Rodríguez-Félix, D.E.; Soto-Valdez, H.; Madera-Santana, T.J. Surface Modification and Performance of Jute Fibers as Reinforcement on Polymer Matrix: An Overview. *J. Nat. Fibers* **2019**, *16*, 944–960. [CrossRef]
5. Latif, R.; Wakeel, S.; Khan, N.Z.; Noor Siddiquee, A.; Lal Verma, S.; Akhtar Khan, Z. Surface Treatments of Plant Fibers and Their Effects on Mechanical Properties of Fiber-Reinforced Composites: A Review. *J. Reinf. Plast. Compos.* **2018**, *38*, 15–30. [CrossRef]
6. Sanivada, U.K.; Mármol, G.; Brito, F.P.; Fangueiro, R. PLA Composites Reinforced with Flax and Jute Fibers—A Review of Recent Trends, Processing Parameters and Mechanical Properties. *Polymers* **2020**, *12*, 2373. [CrossRef]
7. Shahzad, A. 11-Mechanical Properties of Lignocellulosic Fiber Composites. In *Lignocellulosic Fibre and Biomass-Based Composite Materials*; Jawaid, M., Md Tahir, P., Saba, N., Eds.; Woodhead Publishing: Sawston, UK, 2017; pp. 193–223. ISBN 978-0-08-100959-8.
8. Hubbe, M.A.; Lavoine, N.; Lucia, L.A.; Dou, C. Formulating Bioplastic Composites for Biodegradability, Recycling, and Performance: A Review. *BioResources* **2020**, *16*, 2021–2083. [CrossRef]
9. Narancic, T.; Cerrone, F.; Beagan, N.; O'Connor, K.E. Recent Advances in Bioplastics: Application and Biodegradation. *Polymers* **2020**, *12*, 920. [CrossRef]

10. Coppola, G.; Gaudio, M.T.; Lopresto, C.G.; Calabro, V.; Curcio, S.; Chakraborty, S. Bioplastic from Renewable Biomass: A Facile Solution for a Greener Environment. *Earth Syst. Environ.* **2021**, *5*, 231–251. [CrossRef]
11. Reichert, C.L.; Bugnicourt, E.; Coltelli, M.B.; Cinelli, P.; Lazzeri, A.; Canesi, I.; Braca, F.; Martínez, B.M.; Alonso, R.; Agostinis, L.; et al. Bio-Based Packaging: Materials, Modifications, Industrial Applications and Sustainability. *Polymers* **2020**, *12*, 1558. [CrossRef]
12. Nielsen, C.; Rahman, A.; Rehman, A.U.; Walsh, M.K.; Miller, C.D. Food Waste Conversion to Microbial Polyhydroxyalkanoates. *Microb. Biotechnol.* **2017**, *10*, 1338–1352. [CrossRef] [PubMed]
13. Luyt, A.S.; Malik, S.S. 16-Can Biodegradable Plastics Solve Plastic Solid Waste Accumulation? In *Plastics to Energy*; Al-Salem, S.M., Ed.; William Andrew Publishing: Norwich, NY, USA, 2019; pp. 403–423. ISBN 978-0-12-813140-4.
14. Loureiro, N.C.; Esteves, J.L.; Viana, J.C.; Ghosh, S. Mechanical Characterization of Polyhydroxyalkanoate and Poly(Lactic Acid) Blends. *J. Thermoplast. Compos. Mater.* **2013**, *28*, 195–213. [CrossRef]
15. Obruca, S.; Petrik, S.; Benesova, P.; Svoboda, Z.; Eremka, L.; Marova, I. Utilization of Oil Extracted from Spent Coffee Grounds for Sustainable Production of Polyhydroxyalkanoates. *Appl. Microbiol. Biotechnol.* **2014**, *98*, 5883–5890. [CrossRef] [PubMed]
16. Obruca, S.; Benesova, P.; Kucera, D.; Petrik, S.; Marova, I. Biotechnological Conversion of Spent Coffee Grounds into Polyhydroxyalkanoates and Carotenoids. *New Biotechnol.* **2015**, *32*, 569–574. [CrossRef]
17. Stavroula, K.; Simos, M.; Katherine-Joanne, H. Polyhydroxyalkanoates (PHAs) from Household Food Waste: Research Over the Last Decade. *Int. J. Biotechnol. Bioeng.* **2020**, *6*, 26–36.
18. Jiang, T.; Duan, Q.; Zhu, J.; Liu, H.; Yu, L. Starch-Based Biodegradable Materials: Challenges and Opportunities. *Adv. Ind. Eng. Polym. Res.* **2020**, *3*, 8–18. [CrossRef]
19. Collazo-Bigliardi, S.; Ortega-Toro, R.; Boix, A.C. Reinforcement of Thermoplastic Starch Films with Cellulose Fibres Obtained from Rice and Coffee Husks. *J. Renew. Mater.* **2018**, *6*, 599–610. [CrossRef]
20. Sisti, L.; Celli, A.; Totaro, G.; Cinelli, P.; Signori, F.; Lazzeri, A.; Bikaki, M.; Corvini, P.; Ferri, M.; Tassoni, A.; et al. Monomers, Materials and Energy from Coffee By-Products: A Review. *Sustainability* **2021**, *13*, 6921. [CrossRef]
21. Rajak, D.K.; Pagar, D.D.; Menezes, P.L.; Linul, E. Fiber-Reinforced Polymer Composites: Manufacturing, Properties, and Applications. *Polymers* **2019**, *11*, 1667. [CrossRef]
22. Cui, X.; Gu, G.; Li, C.; Liu, N.; Gong, Y.; Liu, B. Synthesis and Properties of Biomass Eugenol-Functionalized Isotactic Poly(1-Butene)s. *Polymer* **2020**, *202*, 122739. [CrossRef]
23. Su, S.; Kopitzky, R.; Tolga, S.; Kabasci, S. Polylactide (PLA) and Its Blends with Poly(Butylene Succinate) (PBS): A Brief Review. *Polymers* **2019**, *11*, 1193. [CrossRef] [PubMed]
24. Aydemir, D.; Gardner, D.J. Biopolymer Blends of Polyhydroxybutyrate and Polylactic Acid Reinforced with Cellulose Nanofibrils. *Carbohydr. Polym.* **2020**, 116867. [CrossRef] [PubMed]
25. Omar, L.; Zambrano-robledo, P.; Reyes-osorio, L.A.; Garza, C. Mechanical Characterisation of a Bamboo Fibre/Polylactic Acid Composite Produced by Fused Deposition Modelling. *J. Reinf. Plast. Compos.* **2020**, *39*, 932–944. [CrossRef]
26. Pickering, K.L.; Efendy, M.G.A.; Le, T.M. A Review of Recent Developments in Natural Fibre Composites and Their Mechanical Performance. *Compos. Part Appl. Sci. Manuf.* **2016**, *83*, 98–112. [CrossRef]
27. Zwawi, M. A Review on Natural Fiber Bio-Composites, Surface Modifications and Applications. *Molecules* **2021**, *26*, 404. [CrossRef] [PubMed]
28. Sandhu, S.S.; Singh, R. Comparative Analysis of Polymer Matrix Composites Processed by Injection and Compression Molding. *Int. J. Emerg. Technol. Innov. Res.* **2018**, *5*, 790–810.
29. Sapuan, S.M. Chapter 3-Composite Materials. In *Composite Materials*; Sapuan, S.M., Ed.; Butterworth-Heinemann: Boston, MA, USA, 2017; pp. 57–93. ISBN 978-0-12-802507-9.
30. Shubhra, Q.T.H.; Alam, A.K.M.M.; Quaiyyum, M.A. Mechanical Properties of Polypropylene Composites: A Review. *J. Thermoplast. Compos. Mater.* **2013**, *26*, 362–391. [CrossRef]
31. Rajak, D.K.; Wagh, P.H.; Linul, E. Manufacturing Technologies of Carbon/Glass Fiber-Reinforced Polymer Composites and Their Properties: A Review. *Polymers* **2021**, *13*, 3721. [CrossRef]
32. Balla, V.K.; Kate, K.H.; Satyavolu, J.; Singh, P.; Tadimeti, J.G.D. Additive Manufacturing of Natural Fiber Reinforced Polymer Composites: Processing and Prospects. *Compos. Part B Eng.* **2019**, *174*, 106956. [CrossRef]
33. Advani, S.G. 2.2 Role of Process Models in Composites Manufacturing. *Compr. Compos. Mater. II* **2018**, 24–41. [CrossRef]
34. Gallos, A.; Paës, G.; Allais, F.; Beaugrand, J. Lignocellulosic Fibers: A Critical Review of the Extrusion Process for Enhancement of the Properties of Natural Fiber Composites. *RSC Adv.* **2017**, *7*, 34638–34654. [CrossRef]
35. Jamshidian, M.; Tehrany, E.A.; Imran, M.; Jacquot, M.; Desobry, S. Poly-Lactic Acid: Production, Applications, Nanocomposites, and Release Studies. *Compr. Rev. Food Sci. Food Saf.* **2010**, *9*, 552–571. [CrossRef] [PubMed]
36. Shesan, O.J.; Stephen, A.C.; Chioma, A.G.; Neerish, R.; Rotimi, S.E. Improving the Mechanical Properties of Natural Fiber Composites for Structural and Biomedical Applications. In *Renewable and Sustainable Composites*; Agwuncha, C.S., Ed.; IntechOpen: Rijeka, Croatia, 2019; ISBN 978-1-78984-216-6.
37. Sadasivuni, K.K.; Saha, P.; Adhikari, J.; Deshmukh, K.; Ahamed, M.B.; Cabibihan, J.J. Recent Advances in Mechanical Properties of Biopolymer Composites: A Review. *Polym. Compos.* **2019**, *41*, 32–59. [CrossRef]
38. Long, S.; Zhong, L.; Lin, X.; Chang, X.; Wu, F.; Wu, R.; Xie, F. Preparation of Formyl Cellulose and Its Enhancement Effect on the Mechanical and Barrier Properties of Polylactic Acid Films. *Int. J. Biol. Macromol.* **2021**, *172*, 82–92. [CrossRef] [PubMed]

39. Zhang, K.; Chen, Z.; Smith, L.M.; Hong, G.; Song, W.; Zhang, S. Polypyrrole-Modified Bamboo Fiber/Polylactic Acid with Enhanced Mechanical, the Antistatic Properties and Thermal Stability. *Ind. Crops Prod.* **2021**, *162*, 113227. [CrossRef]
40. García-García, D.; Carbonell, A.; Samper, M.D.; García-Sanoguera, D.; Balart, R. Green Composites Based on Polypropylene Matrix and Hydrophobized Spend Coffee Ground (SCG) Powder. *Compos. Part B Eng.* **2015**, *78*, 256–265. [CrossRef]
41. Moon, R.J.; Martini, A.; Nairn, J.; Simonsen, J.; Youngblood, J. Cellulose Nanomaterials Review: Structure, Properties and Nanocomposites. *Chem. Soc. Rev.* **2011**, *40*, 3941–3994. [CrossRef]
42. Vishnu Vardhini, K.J.; Murugan, R.; Surjit, R. Effect of Alkali and Enzymatic Treatments of Banana Fibre on Properties of Banana/Polypropylene Composites. *J. Ind. Text.* **2018**, *47*, 1849–1864. [CrossRef]
43. Aaliya, B.; Sunooj, K.V.; Lackner, M. Biopolymer Composites: A Review. *Int. J. Biobased Plast.* **2021**, *3*, 40–84. [CrossRef]
44. Youssef, A.M.; Hasanin, M.S.; Abd El-Aziz, M.E.; Darwesh, O.M. Green, Economic, and Partially Biodegradable Wood Plastic Composites via Enzymatic Surface Modification of Lignocellulosic Fibers. *Heliyon* **2019**, *5*, e01332. [CrossRef]
45. Chaari, R.; Khlif, M.; Mallek, H.; Bradai, C.; Lacoste, C.; Belguith, H.; Tounsi, H.; Dony, P. Enzymatic Treatments Effect on the Poly (Butylene Succinate)/Date Palm Fibers Properties for Bio-Composite Applications. *Ind. Crops Prod.* **2020**, *148*, 112270. [CrossRef]
46. Kenned, J.J.; Sankaranarayanasamy, K.; Kumar, C.S. Chemical, Biological, and Nanoclay Treatments for Natural Plant Fiber-Reinforced Polymer Composites: A Review. *Polym. Polym. Compos.* **2021**, *29*, 1011–1038. [CrossRef]
47. Hemanth, R.; Sekar, M.; Suresha, B. Effects of Fibers and Fillers on Mechanical Properties of Thermoplastic Composites. *Indian J. Adv. Chem. Sci.* **2014**, *2*, 28–35.
48. Sathishkumar, T.P.; Naveen, J.; Satheeshkumar, S. Hybrid Fiber Reinforced Polymer Composites-A Review. *J. Reinf. Plast. Compos.* **2014**, *33*, 454–471. [CrossRef]
49. Liu, W.; Ma, J.; Yao, X.; Fang, R. Inorganic Fibers for Biomedical Engineering Applications. In *Materials for Biomedical Engineering: Bioactive Materials, Properties, and Applications*; Grumezescu, V., Grumezescu, A.M., Eds.; Elsevier: Oxford, UK, 2019; pp. 1–32. [CrossRef]
50. Ishikawa, T.; Oda, H. Heat-Resistant Inorganic Fibers. In *Advances in High Temperature Ceramic Matrix Compo sites and Materials for Sustainable Development*; Ceramic Transactions, Volume CCLXIII; John Wiley & Sons, Ltd.: Hoboken, NJ, USA, 2017; pp. 7–18. ISBN 978-1-119-40727-0.
51. Chihaoui, B.; Serra-Parareda, F.; Tarrés, Q.; Espinach, F.X.; Boufi, S.; Delgado-Aguilar, M. Effect of the Fiber Treatment on the Stiffness of Date Palm Fiber Reinforced PP Composites: Macro and Micromechanical Evaluation of the Young's Modulus. *Polymers* **2020**, *12*, 1693. [CrossRef]
52. Kahl, C.; Gemmeke, N.; Bagnucki, J.; Heim, H.P. Investigations on Fiber–Matrix Properties of Heat-Treated and UV-Treated Regenerated Cellulose Fibers. *Compos. Part Appl. Sci. Manuf.* **2022**, *152*, 106669. [CrossRef]
53. Moustafa, H.; Guizani, C.; Dufresne, A. Sustainable Biodegradable Coffee Grounds Filler and Its Effect on the Hydrophobicity, Mechanical and Thermal Properties of Biodegradable PBAT Composites. *J. Appl. Polym. Sci.* **2016**, *134*. [CrossRef]
54. Sun, S.; Sun, S.; Cao, X.; Sun, R. The Role of Pretreatment in Improving the Enzymatic Hydrolysis of Lignocellulosic Materials. *Bioresour. Technol.* **2016**, *199*, 49–58. [CrossRef]
55. Saha, P.; Chowdhury, S.; Roy, D.; Adhikari, B.; Kim, J.K.; Thomas, S. A Brief Review on the Chemical Modifications of Lignocellulosic Fibers for Durable Engineering Composites. *Polym. Bull.* **2016**, *73*, 587–620. [CrossRef]
56. Kabir, M.M.; Wang, H.; Lau, K.T.; Cardona, F. Chemical Treatments on Plant-Based Natural Fibre Reinforced Polymer Composites: An Overview. *Compos. Part B* **2012**, *43*, 2883–2892. [CrossRef]
57. Nwamba, M.C.; Sun, F.; Mukasekuru, M.R.; Song, G.; Harindintwali, J.D.; Boyi, S.A.; Sun, H. Trends and Hassles in the Microbial Production of Lactic Acid from Lignocellulosic Biomass. *Environ. Technol. Innov.* **2021**, *21*, 101337. [CrossRef]
58. Ahmad, R.; Hamid, R.; Osman, S.A. Physical and Chemical Modifications of Plant Fibres for Reinforcement in Cementitious Composites. *Adv. Civ. Eng.* **2019**, *2019*, 5185806. [CrossRef]
59. Nanni, A.; Cancelli, U.; Montevecchi, G.; Masino, F.; Messori, M.; Antonelli, A. Functionalization and Use of Grape Stalks as Poly(Butylene Succinate) (PBS) Reinforcing Fillers. *Waste Manag.* **2021**, *126*, 538–548. [CrossRef] [PubMed]
60. Azeredo, H.M.C.; Barud, H.; Farinas, C.S.; Vasconcellos, V.M.; Claro, A.M. Bacterial Cellulose as a Raw Material for Food and Food Packaging Applications. *Front. Sustain. Food Syst.* **2019**, *3*, 7. [CrossRef]
61. Swingler, S.; Gupta, A.; Gibson, H.; Kowalczuk, M.; Heaselgrave, W.; Radecka, I. Recent Advances and Applications of Bacterial Cellulose in Biomedicine. *Polymers* **2021**, *13*, 412. [CrossRef]
62. Wang, B.; Lin, F.; Li, X.; Ji, X.; Liu, S.; Han, X.; Yuan, Z.; Luo, J. Transcrystallization of Isotactic Polypropylene/Bacterial Cellulose Hamburger Composite. *Polymers* **2019**, *11*, 508. [CrossRef]
63. Wang, B.; Yang, D.; Zhang, H.; Huang, C.; Xiong, L.; Luo, J.; Chen, X. Preparation of Esterified Bacterial Cellulose for Improved Mechanical Properties and the Microstructure of Isotactic Polypropylene/Bacterial Cellulose Composites. *Polymers* **2016**, *8*, 129. [CrossRef]
64. Mohammed, L.; Ansari, M.N.M.; Pua, G.; Jawaid, M.; Islam, M.S. A Review on Natural Fiber Reinforced Polymer Composite and Its Applications. *Int. J. Polym. Sci.* **2015**, *2015*, 243947. [CrossRef]
65. Nasir, M.; Hashim, R.; Sulaiman, O.; Nordin, N.A.; Lamaming, J.; Asim, M. Laccase, an Emerging Tool to Fabricate Green Composites: A Review. *BioResources* **2015**, *10*, 6262–6284. [CrossRef]
66. Rozman, H.D.; Tay, G.S.; Abubakar, A.; Kumar, R.N. Tensile Properties of Oil Palm Empty Fruit Bunch–Polyurethane Composites. *Eur. Polym. J.* **2001**, *37*, 1759–1765. [CrossRef]

67. Jiménez, A.M.; Espinach, F.X.; Delgado-Aguilar, M.; Reixach, R.; Quintana, G.; Fullana-i-Palmer, P.; Mutjè, P. Starch-Based Biopolymer Reinforced with High Yield Fibers from Sugarcane Bagasse as a Technical and Environmentally Friendly Alternative to High Density Polyethylene. *BioResources* **2016**, *11*, 9856–9868. [CrossRef]
68. Totaro, G.; Sisti, L.; Fiorini, M.; Lancellotti, I.; Andreola, F.N.; Saccani, A. Formulation of Green Particulate Composites from PLA and PBS Matrix and Wastes Deriving from the Coffee Production. *J. Polym. Environ.* **2019**, *27*, 1488–1496. [CrossRef]
69. Baek, B.S.; Park, J.W.; Lee, B.H.; Kim, H.J. Development and Application of Green Composites: Using Coffee Ground and Bamboo Flour. *J. Polym. Environ.* **2013**, *21*, 702–709. [CrossRef]
70. Kumar, A.; Tumu, V.R. Physicochemical Properties of the Electron Beam Irradiated Bamboo Powder and Its Bio-Composites with PLA. *Compos. Part B Eng.* **2019**, *175*, 107098. [CrossRef]
71. George, M.; Mussone, P.G.; Alemaskin, K.; Chae, M.; Wolodko, J.; Bressler, D.C. Enzymatically Treated Natural Fibres as Reinforcing Agents for Biocomposite Material: Mechanical, Thermal, and Moisture Absorption Characterization. *J. Mater. Sci.* **2015**, *51*, 2677–2686. [CrossRef]
72. Lule, Z.C.; Kim, J. Properties of Economical and Eco-Friendly Polybutylene Adipate Terephthalate Composites Loaded with Surface Treated Coffee Husk. *Compos. Part Appl. Sci. Manuf.* **2021**, *140*, 106154. [CrossRef]
73. de Farias, J.G.G.; Cavalcante, R.C.; Canabarro, B.R.; Viana, H.M.; Scholz, S.; Simão, R.A. Surface Lignin Removal on Coir Fibers by Plasma Treatment for Improved Adhesion in Thermoplastic Starch Composites. *Carbohydr. Polym.* **2017**, *165*, 429–436. [CrossRef]
74. Murthy, P.S.; Madhava Naidu, M. Sustainable Management of Coffee Industry By-Products and Value Addition-A Review. *Resour. Conserv. Recycl.* **2012**, *66*, 45–58. [CrossRef]
75. Jia, W.; Gong, R.H.; Hogg, P.J. Poly (*Lactic Acid*) Fibre Reinforced Biodegradable Composites. *Compos. Part B Eng.* **2014**, *62*, 104–112. [CrossRef]
76. Karaduman, Y.; Gokcan, D.; Onal, L. Effect of Enzymatic Pretreatment on the Mechanical Properties of Jute Fiber-Reinforced Polyester Composites. *J. Compos. Mater.* **2012**, *47*, 1293–1302. [CrossRef]
77. Sahu, G.; Rajput, M.S.; Mahapatra, S.P. Polylactic Acid Nanocomposites for Biomedical Applications: Effects of Calcium Phosphate, and Magnesium Phosphate Nanoparticles Concentration. *Plast. Rubber Compos.* **2021**, *50*, 228–240. [CrossRef]
78. Ragoubi, M.; George, B.; Molina, S.; Bienaimé, D.; Merlin, A.; Hiver, J.M.; Dahoun, A. Effect of Corona Discharge Treatment on Mechanical and Thermal Properties of Composites Based on Miscanthus Fibres and Polylactic Acid or Polypropylene Matrix. *Compos. Part Appl. Sci. Manuf.* **2012**, *43*, 675–685. [CrossRef]
79. Gibeop, N.; Lee, D.W.; Prasad, C.V.; Toru, F.; Kim, B.S.; Song, J.I. Effect of Plasma Treatment on Mechanical Properties of Jute Fiber/Poly (*Lactic Acid*) Biodegradable Composites. *Adv. Compos. Mater.* **2013**, *22*, 389–399. [CrossRef]
80. De Prez, J.; Van Vuure, A.W.; Ivens, J.; Aerts, G.; Van de Voorde, I. Enzymatic Treatment of Flax for Use in Composites. *Biotechnol. Rep.* **2018**, *20*, e00294. [CrossRef]
81. Loureiro, N.C.; Esteves, J.L.; Viana, J.C.; Ghosh, S. Development of Polyhydroxyalkanoates/Poly(*Lactic Acid*) Composites Reinforced with Cellulosic Fibers. *Compos. Part B Eng.* **2014**, *60*, 603–611. [CrossRef]
82. Ravindran, R.; Jaiswal, A.K. A Comprehensive Review on Pre-Treatment Strategy for Lignocellulosic Food Industry Waste: Challenges and Opportunities. *Bioresour. Technol.* **2016**, *199*, 92–102. [CrossRef]
83. Saritha, M.; Arora, A. Lata Biological Pretreatment of Lignocellulosic Substrates for Enhanced Delignification and Enzymatic Digestibility. *Indian J. Microbiol.* **2012**, *52*, 122–130. [CrossRef]
84. Bozaci, E.; Sever, K.; Sarikanat, M.; Seki, Y.; Demir, A.; Ozdogan, E.; Tavman, I. Effects of the Atmospheric Plasma Treatments on Surface and Mechanical Properties of Flax Fiber and Adhesion between Fiber-Matrix for Composite Materials. *Compos. Part B Eng.* **2013**, *45*, 565–572. [CrossRef]
85. Ricciardi, M.R.; Papa, I.; Coppola, G.; Lopresto, V.; Sansone, L.; Antonucci, V. Effect of Plasma Treatment on the Impact Behavior of Epoxy/Basalt Fiber-Reinforced Composites: A Preliminary Study. *Polymers* **2021**, *13*, 1293. [CrossRef]
86. Amirou, S.; Zerizer, A.; Haddadou, I.; Merlin, A. Effects of Corona Discharge Treatment on the Mechanical Properties of Biocomposites from Polylactic Acid and Algerian Date Palm Fibres. *Sci. Res. Essays* **2013**, *8*, 946–952. [CrossRef]
87. Lee, C.H.; Khalina, A.; Lee, S.H. Importance of Interfacial Adhesion Condition on Characterization of Plant-Fiber-Reinforced Polymer Composites: A Review. *Polymers* **2021**, *13*, 438. [CrossRef] [PubMed]
88. Boonsuk, P.; Sukolrat, A.; Bourkaew, S.; Kaewtatip, K.; Chantarak, S.; Kelarakis, A.; Chaibundit, C. Structure-Properties Relationships in Alkaline Treated Rice Husk Reinforced Thermoplastic Cassava Starch Biocomposites. *Int. J. Biol. Macromol.* **2021**, *167*, 130–140. [CrossRef] [PubMed]
89. Werchefani, M.; Lacoste, C.; Belguith, H.; Gargouri, A.; Bradai, C. Effect of Chemical and Enzymatic Treatments of Alfa Fibers on Polylactic Acid Bio-Composites Properties. *J. Compos. Mater.* **2020**, *54*, 4959–4967. [CrossRef]
90. Essabir, H.; Raji, M.; Laaziz, S.A.; Rodrique, D.; Bouhfid, R.; Qaiss, A. el kacem Thermo-Mechanical Performances of Polypropylene Biocomposites Based on Untreated, Treated and Compatibilized Spent Coffee Grounds. *Compos. Part B Eng.* **2018**, *149*, 1–11. [CrossRef]
91. Geremew, A.; De Winne, P.; Demissie, T.A.; De Backer, H. Treatment of Natural Fiber for Application in Concrete Pavement. *Adv. Civ. Eng.* **2021**, *2021*. [CrossRef]
92. Khoshnava, S.M.; Rostami, R.; Ismai, M.; Valipour, A. The Using Fungi Treatment as Green and Environmentally Process for Surface Modification of Natural Fibres. *Appl. Mech. Mater.* **2014**, *554*, 116–122. [CrossRef]
93. Jayasekara, S.; Ratnayake, R. Microbial Cellulases: An Overview and Applications. In *Cellulose*; IntechOpen: London, UK, 2019.

94. Pickering, K.L.; Li, Y.; Farrell, R.L.; Lay, M. Interfacial Modification of Hemp Fiber Reinforced Composites Using Fungal and Alkali Treatment. *J. Biobased Mater. Bioenergy* **2007**, *1*, 109–117. [CrossRef]
95. Sindhu, R.; Binod, P.; Pandey, A. Biological Pretreatment of Lignocellulosic Biomass-An Overview. *Bioresour. Technol.* **2016**, *199*, 76–82. [CrossRef]
96. Dong, A.; Fan, X.; Wang, Q.; Yu, Y.; Cavaco-Paulo, A. Enzymatic Treatments to Improve Mechanical Properties and Surface Hydrophobicity of Jute Fiber Membranes. *BioResources* **2016**, *11*, 3289–3302. [CrossRef]
97. Bendourou, F.E.; Suresh, G.; Laadila, M.A.; Kumar, P.; Rouissi, T.; Dhillon, G.S.; Zied, K.; Brar, S.K.; Galvez, R. Feasibility of the Use of Different Types of Enzymatically Treated Cellulosic Fibres for Polylactic Acid (PLA) Recycling. *Waste Manag.* **2021**, *121*, 237–247. [CrossRef]
98. Fitch-Vargas, P.R.; Camacho-Hernández, I.L.; Martínez-Bustos, F.; Islas-Rubio, A.R.; Carrillo-Cañedo, K.I.; Calderón-Castro, A.; Jacobo-Valenzuela, N.; Carrillo-López, A.; Delgado-Nieblas, C.I.; Aguilar-Palazuelos, E. Mechanical, Physical and Microstructural Properties of Acetylated Starch-Based Biocomposites Reinforced with Acetylated Sugarcane Fiber. *Carbohydr. Polym.* **2019**, *219*, 378–386. [CrossRef] [PubMed]
99. Siqueira, D.D.; Luna, C.B.B.; Ferreira, E.S.B.; Araújo, E.M.; Wellen, R.M.R. Tailored PCL/Macaíba Fiber to Reach Sustainable Biocomposites. *J. Mater. Res. Technol.* **2020**, *9*, 9691–9708. [CrossRef]
100. Lule, Z.C.; Kim, J. Compatibilization Effect of Silanized SiC Particles on Polybutylene Adipate Terephthalate/Polycarbonate Blends. *Mater. Chem. Phys.* **2021**, *258*, 123879. [CrossRef]
101. Tanjung, F.A.; Arifin, Y.; Husseinsyah, S. Enzymatic Degradation of Coconut Shell Powder–Reinforced Polylactic Acid Biocomposites. *J. Thermoplast. Compos. Mater.* **2018**, *33*, 800–816. [CrossRef]
102. Zhuo, G.; Zhang, X.; Jin, X.; Wang, M.; Yang, X.; Li, S. Effect of Different Enzymatic Treatment on Mechanical, Water Absorption and Thermal Properties of Bamboo Fibers Reinforced Poly(Hydroxybutyrate-Co-Valerate) Biocomposites. *J. Polym. Environ.* **2020**, *28*, 2377–2385. [CrossRef]
103. Thakur, K.; Kalia, S. Enzymatic Modification of Ramie Fibers and Its Influence on the Performance of Ramie-Poly(Butylene Succinate) Biocomposites. *Int. J. Plast. Technol.* **2017**, *21*, 209–226. [CrossRef]
104. Jha, K.; Kataria, R.; Verma, J.; Pradhan, S. Potential Biodegradable Matrices and Fiber Treatment for Green Composites: A Review. *AIMS Mater. Sci.* **2019**, *6*, 119–138. [CrossRef]
105. Wang, B.; Mao, S.; Lin, F.; Zhang, M.; Zhao, Y.; Zheng, X.; Wang, H.; Luo, J. Interfacial Compatibility on the Crystal Transformation of Isotactic Poly (1-Butene)/Herb Residue Composite. *Polymers* **2021**, *13*, 1654. [CrossRef]
106. Wang, B.; Lin, F.; Zhao, Y.; Li, X.; Liu, Y.; Li, J.; Han, X.-J.; Liu, S.; Ji, X.; Luo, J.; et al. Isotactic Polybutene-1/Bamboo Powder Composites with Excellent Properties at Initial Stage of Molding. *Polymers* **2019**, *11*, 1981. [CrossRef]
107. Thyavihalli Girijappa, Y.G.; Mavinkere Rangappa, S.; Parameswaranpillai, J.; Siengchin, S. Natural Fibers as Sustainable and Renewable Resource for Development of Eco-Friendly Composites: A Comprehensive Review. *Front. Mater.* **2019**, *6*, 226. [CrossRef]
108. Faruk, O.; Bledzki, A.K.; Fink, H.-P.; Sain, M. Biocomposites Reinforced with Natural Fibers: 2000–2010. *Prog. Polym. Sci.* **2012**, *37*, 1552–1596. [CrossRef]
109. Zhao, X.; Cornish, K.; Vodovotz, Y. Narrowing the Gap for Bioplastic Use in Food Packaging: An Update. *Environ. Sci. Technol.* **2020**, *54*, 4712–4732. [CrossRef] [PubMed]
110. Zarrinbakhsh, N.; Wang, T.; Rodriguez-Uribe, A.; Misra, M.; Mohanty, A.K. Characterization of Wastes and Coproducts from the Coffee Industry for Composite Material Production. *BioResources* **2016**, *11*, 7637–7653. [CrossRef]
111. Chen, G.-Q. Introduction of Bacterial Plastics PHA, PLA, PBS, PE, PTT, and PPP. In *Plastics from Bacteria*; Chen, G.G.-Q., Ed.; Microbiology Monographs; Springer: Berlin/Heidelberg, Germany, 2010; Volume 14, pp. 1–16. ISBN 978-3-642-03286-8.
112. Jin, L.S.; Salimi, M.N.; Kamal, S.Z. Optimization of Pretreatment and Enzymatic Hydrolysis of Spent Coffee Ground for the Production of Fermentable Sugar. *IOP Conf. Ser. Mater. Sci. Eng.* **2020**, *743*, 012030. [CrossRef]
113. Kalia, S.; Vashistha, S. Surface Modification of Sisal Fibers (Agave Sisalana) Using Bacterial Cellulase and Methyl Methacrylate. *J. Polym. Environ.* **2012**, *20*, 142–151. [CrossRef]
114. Kopparthy, S.D.S.; Netravali, A.N. Review: Green Composites for Structural Applications. *Compos. Part C Open Access* **2021**, *6*. [CrossRef]
115. Nurazzi, N.M.; Harussani, M.M.; Aisyah, H.A.; Ilyas, R.A.; Norrrahim, M.N.F.; Khalina, A.; Abdullah, N. Treatments of Natural Fiber as Reinforcement in Polymer Composites—a Short Review. *Funct. Compos. Struct.* **2021**, *3*, 024002. [CrossRef]
116. Burgos, N.; Armentano, I.; Fortunati, E.; Dominici, F.; Luzi, F.; Fiori, S.; Cristofaro, F.; Visai, L.; Jiménez, A.; Kenny, J.M. Functional Properties of Plasticized Bio-Based Poly(Lactic Acid)_Poly(Hydroxybutyrate) (PLA_PHB) Films for Active Food Packaging. *Food Bioprocess Technol.* **2017**, *10*, 770–780. [CrossRef]
117. Luzi, F.; Fortunati, E.; Jiménez, A.; Puglia, D.; Pezzolla, D.; Gigliotti, G.; Kenny, J.M.; Chiralt, A.; Torre, L. Production and Characterization of PLA_PBS Biodegradable Blends Reinforced with Cellulose Nanocrystals Extracted from Hemp Fibres. *Ind. Crops Prod. Prod.* **2016**, *93*, 276–289. [CrossRef]

118. Papadopoulou, E.L.; Paul, U.C.; Tran, T.N.; Suarato, G.; Ceseracciu, L.; Marras, S.; D'arcy, R.; Athanassiou, A. Sustainable Active Food Packaging from Poly(*Lactic Acid*) and Cocoa Bean Shells. *ACS Appl. Mater. Interfaces* **2019**, *11*, 31317–31327. [CrossRef]
119. Moustafa, H.; Guizani, C.; Dupont, C.; Martin, V.; Jeguirim, M.; Dufresne, A. Utilization of Torrefied Coffee Grounds as Reinforcing Agent to Produce High-Quality Biodegradable PBAT Composites for Food Packaging Applications. *ACS Sustain. Chem. Eng.* **2017**, *5*, 1906–1916. [CrossRef]

polymers

Article

Properties of Wood–Plastic Composites Manufactured from Two Different Wood Feedstocks: Wood Flour and Wood Pellets

Geeta Pokhrel [1,*], Douglas J. Gardner [1,2,*] and Yousoo Han [1,2]

[1] School of Forest Resources, University of Maine, 5755 Nutting Hall, Orono, ME 04469, USA; yousoo.han@maine.edu
[2] Advanced Structures and Composites Center, University of Maine, 35 Flagstaff Road, Orono, ME 04469, USA
* Correspondence: geeta.pokhrel@maine.edu (G.P.); douglasg@maine.edu (D.J.G.)

Abstract: Driven by the motive of minimizing the transportation costs of raw materials to manufacture wood–plastic composites (WPCs), Part I and the current Part II of this paper series explore the utilization of an alternative wood feedstock, i.e., pellets. Part I of this study reported on the characteristics of wood flour and wood pellets manufactured from secondary processing mill residues. Part II reports on the physical and mechanical properties of polypropylene (PP)-based WPCs made using the two different wood feedstocks, i.e., wood flour and wood pellets. WPCs were made from 40-mesh wood flour and wood pellets from four different wood species (white cedar, white pine, spruce-fir and red maple) in the presence and absence of the coupling agent maleic anhydride polypropylene (MAPP). With MAPP, the weight percentage of wood filler was 20%, PP 78%, MAPP 2% and without MAPP, formulation by weight percentage of wood filler was 20% and PP 80%. Fluorescent images showed wood particles' distribution in the PP polymer matrix was similar for both wood flour and ground wood pellets. Dispersion of particles was higher with ground wood pellets in the PP matrix. On average, the density of composite products from wood pellets was higher, tensile strength, tensile modulus and impact strength were lower than the composites made from wood flour. Flexural properties of the control composites made with pellets were higher and with MAPP were lower than the composites made from wood flour. However, the overall mechanical property differences were low (0.5–10%) depending on the particular WPC formulations. Statistical analysis also showed there was no significant differences in the material property values of the composites made from wood flour and wood pellets. In some situations, WPC properties were better using wood pellets rather than using wood flour. We expect if the material properties of WPCs from wood flour versus wood pellets are similar and with a greater reduction in transportation costs for wood pellet feedstocks, this would be beneficial to WPC manufacturers and consumers.

Keywords: wood flour; wood pellets; wood–plastic composites; transportation costs; physical properties; mechanical properties

Citation: Pokhrel, G.; Gardner, D.J.; Han, Y. Properties of Wood–Plastic Composites Manufactured from Two Different Wood Feedstocks: Wood Flour and Wood Pellets. *Polymers* **2021**, *13*, 2769. https://doi.org/10.3390/polym13162769

Academic Editors: Domenico Acierno and Antonella Patti

Received: 29 July 2021
Accepted: 13 August 2021
Published: 18 August 2021

1. Introduction

With global awareness in addressing environmental impacts and minimizing the emission of harmful pollutants, the wood composites industry is seeking more environmentally friendly materials for their products. With the utilization of recycled plastics and waste wood-based fillers, wood plastic composites (WPCs) manufacturing can be considered a green technology [1]. The concept of WPCs is not new where its modern application began in the 1970s and since the 1990s, the popularity of WPCs in North America has increased in decking and railing production [2]. WPCs are any composite products manufactured using plant (wood or non-wood) fibers, thermoplastic or thermoset resins and a small number of additives. WPCs offer the advantages of enhancing mechanical properties with higher strength and stiffness, decreases in density and abrasion compared to inorganic filler composites [3–5] and compared to solid wood, higher water and decay resistance, better acoustic performance, reduced weight, lower production costs and biodegradability [6,7]. They have wide applications in the automotive, and construction industrial sectors.

In general, the processes followed in manufacturing WPCs are: extrusion, injection molding, and compression molding or thermoforming (pressing) [2]. Similarly, the newer production technologies are additive manufacturing processes based on extrusion processes and with laser sintering [2,8–11]. WPC compounding can be performed in an extruder (single-screw extruder, twin-screw extruder, conical twin-screw extruder or a combination type extruder) or high-speed mixers of the Henschel type. A Henschel mixer is a high intensity vertical mixer (causing the materials to move freely regardless of their size, density, friction coefficient, etc.) or melt type (used for making master batches with wax or materials with wax-like properties. Additive manufacturing or 3D printing consists of three general steps: usage of computer-aided design (CAD) to model the part, processing the model in 3D space through the slicing software, and with the G-codes printing and production of the part [12]. On an industrial scale, industries that manufacture WPCs either compound all the raw materials, i.e., wood filler, polymers, additives themselves or purchase pre-compounded WPC pellets. In 2019, the global market size of WPCs was USD $4.77 billion out of which North America (major application in decking) stood highest in market share with USD $2.12 billion [13]. Zion Market Research [14] has predicted that the WPCs global market in 2022 to reach to USD $8.76 billion in 2022. The expected mean annual growth rate is 12.3% between 2017 and 2022.

Thermosetting plastics cannot be melted repeatedly once cured whereas thermoplastics can be repeatedly melted. Thermosetting resins include epoxy and phenolics whereas polyethylene (PE), polypropylene (PP), polyvinyl chloride (PVC) and polystyrene (PS) are the thermoplastic resins. PP has advantages of cost-effectiveness, lower material weight, better processability, resistance to extreme environmental conditions and reusability. Further, PP-based composites have applications in automotive, packaging and building materials [15]. Additives such as colorants, coupling agents, stabilizers, blowing agents, reinforcing agents, foaming agents and lubricants can be used based on the target area of application. In North America, commercial WPCs are mostly manufactured from wood shavings, sawdust or wood chips produced from industrial processes [5,16–18]. Different wood species can be used to manufacture WPCs. To effectively utilize the wood-based fillers in the thermoplastics, a basic knowledge in morphological and chemical characteristics is vital [19]. Depending on the woody material, different properties of the composite material can be obtained. However, from a commercial standpoint, usually materials that are conveniently available from the supply side are utilized rather than based on the characteristics of the fibers [20]. In Southeast Asia, hemp, ramie, kenaf and so on are mostly used as fillers as they are abundant there. However, in the US, species mostly utilized as fillers are the softwoods such as white pine, spruce, hemlock; and hardwood such as aspen. In general, the size of wood flour for manufacturing WPCs ranges from 1 mm–100 μm (18 mesh–140 mesh) [20]. However, the particle size of wood flour most preferable for composite products is in the range of 180–425 μm (40–80 mesh size) [21]. Larger filler sizes offer an advantage in terms of cost of pulverization and higher filler content in composite but with the disadvantage of lack of water resistance and difficulties in fabrication during injection molding. Similarly, smaller filler sizes offer the advantages in terms of mechanical properties and durability but disadvantages in terms of comminution costs and filler contents. The risk of a dust explosion increases for fine powders as well [20].

Equally important is another aspect focusing on the logistics and supply chain of the product, which is often a major factor deciding the final cost of the end product. Most of the time the manufacturers of WPCs purchase the raw material of wood flour from the wood mills that can manufacture wood flour or other authorized manufacturers of wood flour. As of now, there is no advancement in technology that can compress fluffy wood flour to ship over long distances. Wood flour because of its lower bulk density has a larger volume and standard truck trailers are not able to reach the maximum weight load limits (40 tons). The low density and higher volume of the wood flour make longer distance transportation disadvantageous in terms of cost. Being a fluffy material, its shipment costs over longer distances exceed the actual material price. The higher price in purchasing raw material consequently has a negative impact on the production, distribution, trade, and/or retail sale on manufacturing the WPCs. Motivated by this problem, the current study aims at exploring the utilization of a compacted wood flour, i.e., wood pellets to manufacture WPCs. Then, the physical and mechanical properties of the resulting composite products are studied. Wood pellets are easier to transport and handle as well. Production of wood pellets might have slightly more costs than wood flour. However, a comparative analysis of bulk density of wood flour (190–220 kg/m^3) and wood pellets (700–750 kg/m^3) reveals that roughly three and half times more wood pellets can be transported in a truck trailer than wood flour by weight. This suggests it is economical to transport wood pellets over longer distances. Very limited articles such as [22] have reported the application of wood pellets to manufacture WPCs. However, no research has focused on transportation of raw materials for WPCs manufacturing in a cost-effective way. The first part of this study presented the results on the characterization of the properties of wood flour and wood pellets produced from secondary processing mill residues [23]. Wood flour and wood pellets were made from four different wood species: Northern White Cedar (*Thuja occidentalis,* Eastern White Pine (*Pinus strobus*), Eastern Spruce-Balsam Fir (*Picea rubens-Abies balsamea)* and Red Maple (*Acer rubrum*).

In this paper, discussion on WPCs manufactured utilizing wood filler as flour and pellets separately in a PP polymer matrix are presented. Utilizing wood pellets is a novel concept in the commercial manufacturing process of WPCs. Thus, a study of WPCs' physical and mechanical properties through the application of wood flour and wood pellets separately as raw material source of wood feedstock are highlighted in this study.

2. Materials and Methods

2.1. Materials

For the manufacturing of WPCs, the raw material based on wood, i.e., wood flour and wood pellets were prepared in the laboratory using local mill processing residues in Maine, USA. Part I of this paper series [23] has detailed information on equipment used and processing parameters followed during manufacturing of wood flour and wood pellets. The mill residues were clean (free from barks, adhesives, metals, etc.) and low in moisture content. They were grinded in a hammermill (Bliss Industries LLC, Ponca City, OK, USA) using a screen size of 0.5 mm. The produced wood flour was classified into different mesh sizes. Wood pellets made from 40-mesh wood flour was utilized as a wood feedstock in the manufacturing of WPCs. A pellet mill (Lawson Mills Biomass Solutions Ltd., North Wiltshire, PE, Canada) with a quarter inch thickness die was used for the production of wood pellets. Wood flour ground in hammermill had a relatively low moisture content. The low moisture of wood flour caused hindrance in the binding of particles to form pellets. Thus, water was added and mixed manually with the wood flour and the moisture content was maintained between 10–15% depending on the wood species. Under high temperature and pressure, pellets were formed from the wood flour. It should be noted that for each of the four wood species, the 40-mesh sized wood flour and the wood pellets pelletized from the 40 mesh-sized wood flour were utilized separately in the composite product manufacturing. Figure 1 below shows the 40-mesh wood flour and wood pellets made with 40-mesh flour for each of the wood species. Each scale bar is 2 cm in length. Similarly, Polypropylene (PP) was purchased from ExxonMobil Chemical Company ((Houston, TX, USA) with the melt flow index (MFI) of 20 g/min at 230 °C and a density of 0.900 g/cm^3. This nucleated, medium MFI rate homopolymer is suitable for general injection molding purposes. The coupling agent maleic anhydride polypropylene (MAPP) was purchased from the SI Group Inc. (Schenectady, NY, USA) with a MFI rate of 115 g/10 min at 190 °C and density of 0.6 g/cm^3. The application of MAPP offers advantage as a compatibilizer and adhesion promoter and thus improving the mechanical properties [24].

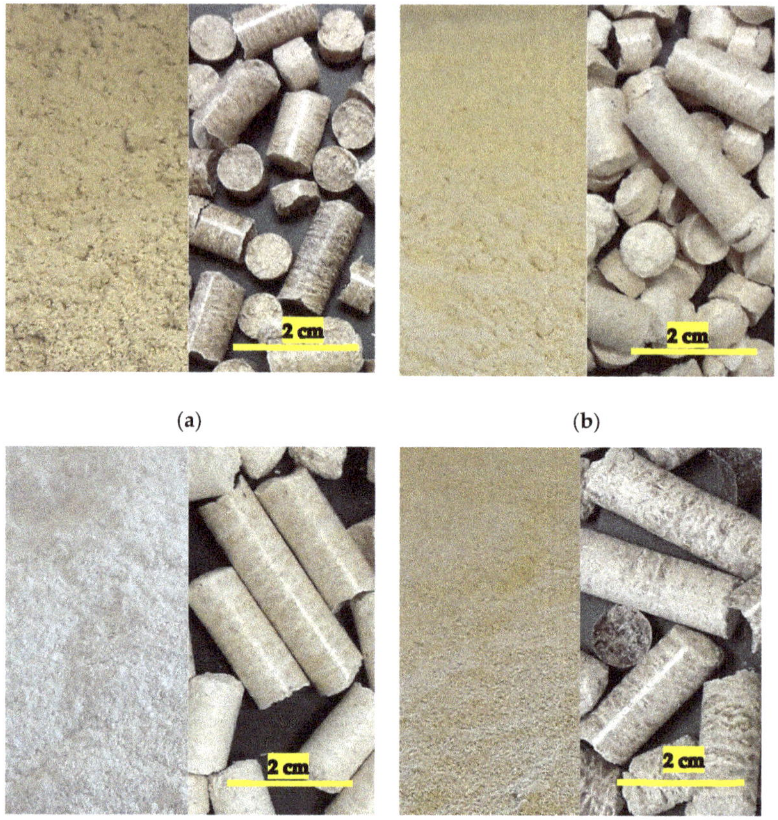

(**a**) (**b**) (**c**) (**d**)

Figure 1. Wood flour and wood pellets used as wood feedstock in WPCs (**a**) Cedar, (**b**) Pine, (**c**) Spruce-fir, (**d**) Maple. (Image source for wood pellets: [23]).

2.2. *Manufacturing Process of WPCs*

PP-based WPCs using wood fillers of wood flour and pellets separately were manufactured. For each of the wood species, 40-mesh wood flour and wood pellets pelletized from 40-mesh wood flour were utilized. As mentioned in the Introduction section, composites from 40–80 mesh wood fibers show better material properties. Similarly, the decking manufacturers mostly prefer using 40-mesh wood flour. Hence, wood feedstock of 40-mesh flour was utilized in this study. Initially, the wood flour and wood pellets of different wood species were dried in oven at 103 ± 2 °C for at least 24 h. The moisture content of the wood feedstocks were ensured to be below 1% during the manufacturing process. For each of the wood species, there were four different formulations with different percentages of wood filler, PP and MAPP additives by weight. A total of 16 different formulations were formulated to study the properties of WPCs. Table 1 below shows the four types of formulations for each of the wood species with the weight percentages of different raw materials:

Table 1. Four different raw materials' formulations of WPCs for each wood species.

Raw Materials	1st Formulation	2nd Formulation	3rd Formulation	4th Formulation
Wood flour	20%	0%	20%	0%
Ground wood pellets	0%	20%	0%	20%
PP	80%	80%	78%	78%
MAPP	0%	0%	2%	2%

Danyadi et al. [25] have reported larger wood content in the composite product leads to particle aggregation causing the lower strength of the WPCs so the wood filler of 20% was used. Similarly, Lu et al. [26] found the maximum values of tensile and flexural strength at 15% and 35% of wood particles by weight respectively.

For producing the WPCs using wood pellets, the wood pellets before extruding were ground into powder using a knife grinder (C.W. Brabender Instruments, Inc., South Hackensack, NJ, USA) using a screen size of 7.96 mm. It should be noted that the wood pellets discussed in this study represents the ground wood pellets using a knife grinder. The 2nd and 4th formulations, as mentioned in Table 1, were applicable on utilizing the wood pellets as feedstock for each of the wood species. After grinding the wood pellets, each component was first mixed manually for the equal mixing of the raw materials. A Brabender Twin Screw Extruder (Messrs. C.W. Brabender Instruments Inc., South Hackensack, NJ, USA) with a diameter of 20 mm, screw length L/D 40, flight depth of 3.75 mm, and screw speed of 150 min^{-1} was used for the extrusion process. The temperature profile of the extruder was set at 200 °C for the heating zones. The frequency of feeder was 15 Hz. Each formulation was fed to the extruders' feed throat. When the materials start to pass along the barrel, the plastic materials start to melt and results in compounding of the raw materials. The compounded materials are forced through a die to make strands of the composites. When the materials were falling from the feeder at a constant rate, the feed rate of the different formulations was also measured during extrusion. For this, the raw materials mixture from each formulation were collected in a small container and weighed after one minute. The WPC filaments after allowing them to cool were pelletized in the similar knife grinder (C.W. Brabender Instruments, Inc., South Hackensack, NJ, USA) used in grinding wood pellets using a screen size of 7.96 mm. Once the WPC pellets were ground, the pellets were then dried in oven at 103 $\pm$ 2 °C for at least 24 h before injection molding. Samples were injection molded using a Mini-Jector Injection Molder Model #55E (Miniature Plastic Molding, Solon, OH, USA) with a ram pressure of 17 MPa at 200 °C. They were then left in the molds for 10 s to cool. Molds used provided the samples with dimensions as specified in ASTM D638-14 (Type I) and ASTM D790-17 for testing of the properties. Before testing, the samples were conditioned for at least 40 h at 23 °C $\pm$ 2 °C and 50% $\pm$ 10% RH. For producing the WPCs using wood flour, the first step followed when using wood pellets filler, i.e., grinding of wood pellets using a knife grinder was not applicable. Besides this step, the other similar processes were followed. The 1st and 3rd formulations seen in Table 1 were applicable on utilizing the wood flour as wood feedstock for each of the wood species.

2.3. Testing of Physical and Mechanical Properties of WPCs

A Zeiss NVision 40 scanning electron microscope (SEM) (Carl Zeiss Microscopy, LLC, White Plains, NY, USA) and a capacity of up to 1.2 nm resolution was utilized to study the distribution of materials in the WPC samples. A Rotary Microtome (Warner-Lambert Technologies Inc., Buffalo, NY, USA) was used to make the flat and smooth cross-sections of the WPC samples. Coating of the samples was done with an Au/Pd conductive layer of 4 nm thickness before the SEM observations. The images were magnified 100×, with a surface area of 100 μm in a high vacuum 2.24 $\times 10^{-6}$ Torr, and the electron source voltage was 3 kV. The samples were also observed under a fluorescent microscope to observe the distribution more clearly. An Olympus BH2 fluorescent microscope (Olympus Scientific Solutions Americas Corp., Waltham, MA, USA) with a wide blue fluorescent filter: 450–480 nm excitation, mirror 500 nm, barrier filter 515 nm and a CHIU technical Corporation M-100 100W mercury illuminator was used. Samples for fluorescent microscopy were the thin wood polymer films prepared from the same rotary microtome used in making samples for the SEM. SHUR/Mount (Triangle Biomedical Sciences Inc., Durham, NC, USA) was used as a mounting medium and the samples were mounted properly on the coverslip. Thus, the WPC samples for observation in SEM were flat and smooth cross-sections and for fluorescent microscope were the thin films. Both the microscopes used a Zeiss Axiocam ERc 5 s camera and Zen Blue software (1.1.1.0, Carl Zeiss Microscopy, LLC, White Plains, NY, USA). Image J software (National Institutes of Health and the Laboratory for Optical and Computational Instrumentation, Madison, WI, USA) was run to study the wood fillers' particle dispersion in the polymer matrix. The area of each particle was calculated using the Image J software. Since the particles had variation in area, the number of particles for each area range (on every 50 square microns) was differentiated.

The American Society of Testing and Materials (ASTM) Standard D792-20 Standard Test Methods for Density and Specific Gravity (relative density) of Plastics by Displacement (ASTM International, West Conshohocken, PA, USA) was followed for determining the density of each sample after injection-molding. For each sample, five specimens were considered for the density testing. The temperature of the water was maintained at 23 °C ± 2 °C. Any bubbles observed were removed. Density was derived from the specific gravity calculation. The formulas to calculate specific gravity and density are:

$$\text{Specific gravity} = \frac{a}{(a+w-b)} \quad (1)$$

where, a is the apparent mass of specimen, without wire or sinker, in air, b is the apparent mass of specimen (and of sinker, if used) completely immersed and of the partially immersed wire in liquid, and w is the apparent mass of totally immersed sinker (if used) and of partially immersed wire.

$$\text{Density} = \text{Specific gravity} \cdot 997.5 \quad (2)$$

where, 997.5 is the density of water at 23 °C.

ASTM D 638-14 Standard Test Method for Tensile Properties of Plastics (ASTM International, West Conshohocken, PA, USA) was followed to determine the tensile properties of the WPCs. Tests were performed at room temperature of 23 ± 2 °C and 50 ± 10% RH. A universal testing machine Instron 5966 (Instron, Norwood, MA, USA) with a 10 kN load cell was used. A mounted extensometer in the Instron measured the elongation of the samples. The tensile test speed was set at 50 mm/min for breaking the specimen within 5 min. For each sample, 15 replicates were tested to report the average value.

ASTM D 790-17 Standard Test Methods for Flexural Properties of Unreinforced and Reinforced Plastics and Electrical Insulating Materials (ASTM International, West Conshohocken, PA, USA) was followed for measuring the flexural strength and modulus of elasticity. The room temperature of 23 ± 2 °C and 50 ± 10% RH was maintained for the testing. A universal testing machine Instron 5966 (Instron, Norwood, MA, USA) with a 10 kN load cell was used. The support span length was 52.8 mm with an average depth of the beam of 3.3 mm. The outer fiber strain rate was 0.01/min and the crosshead motion rate was 1.5 mm/min. For each sample, 15 replicates were tested and the average value was reported.

ASTM D 256-10 Standard Test Methods for Determining the Izod Pendulum Impact Resistance of Plastics (ASTM international, West Conshohocken, PA, USA) was followed to determine the impact strength of the samples. A Ceast Izod Pendulum Impact Tester (Ceast U.S.A., Inc., Charlotte, NC, USA) with hammer energy of 2.75 J was used for impact testing. Before testing, the samples were prepared following the recommended dimensions in the standard. The recommend length of the specimens was 63.5 ± 2 mm and depth of notch 10.16 ± 0.05 mm with the angle of the notch 45 ± 1°. Notches were prepared using a milling machine and the appropriate length was cut on a band saw. Fifteen replicates for each sample were tested to report the average value.

2.4. Statistical Analysis

A three-way Analysis of Variance (ANOVA) with a significance level of 0.05 was used for testing of the statistical significance in the means of the variables. The statistical association of type of wood filler, presence or absence of MAPP, and wood species with the density, tensile properties, flexural properties and impact properties of the WPC samples was analyzed. Here, type of wood filler, presence or absence of MAPP, and wood species are the three independent input variables and their two-way or three-way interaction on the output variables was studied. The dependent output variables are the physical and mechanical properties determined in the study.

3. Results and Discussions

For the WPCs manufactured using wood pellets and flour, the SEM images did not show significant differences in dispersion or distribution or uniformity of particles in the matrix of polymer. Lee et al. [27] mentioned that the quantitative determination of the fibers organization in the WPC profiles is difficult using SEM images. Thus, fluorescent images were also taken to study these parameters. Figure 2 below shows the fluorescent images of WPCs manufactured from four different species using wood flour and wood pellets in the presence of MAPP. Fluorescent images of WPCs for four different wood species with wood flour and wood pellets separately in controlled formulation are shown in Figure A1 of Appendix A. In the fluorescent images, the wood fillers can be clearly observed. The black portion represents the polymer matrix alone or the polymer matrix with MAPP. PP and MAPP could not be separated in the fluorescent images. These images showed that the particles are more uniform in size for wood flour utilized WPCs than the pellet samples. Similarly, the distribution of particles is similar for both WPCs made with flour and pellets but the dispersion was higher for those manufactured with pellets than flour. This could be attributable to the grinding action of the knife grinder to the pellets that created particles with less uniformity than the sieved 40-mesh wood flour. Analysis was done on the dimensions of wood pellet and wood flour particles in the polymer matrix. The major axis, i.e., length of the wood pellets particles ranged from finer particles of 4 microns to the larger particles of 427 microns and the minor axis, i.e., width ranged from 2 to 220 microns. The size range had a greater deviation. For the wood flour particles, the major axis ranged from finer sizes of 4 microns to larger 295 microns and the minor axis from 2 to 180 microns with lesser variation than the former case. Likewise, the average aspect ratio of the 40-mesh wood flour particles on the polymer matrix for cedar was 2.2, pine 2.1, spruce-fir 2.4 and of maple 2.3. For the ground wood pellets, the average aspect ratio of cedar was 2, pine was 2.2, spruce-fir 2.1 and the maple 2. The biggest wood particles appear more or less rectangle in shape whereas the smallest particles are more irregular in shape [28]. Because of high temperature and shear forces, wood fiber length is reduced but not the width during compounding and molding [29,30]. On comparing the hardwoods and softwoods, it can be clearly observed that the distribution and dispersion of particles is higher in softwoods than the hardwood maple. This might be because of the softwood fibers being flexible and hardwood fibers stiffer. Hardwood fibers are shorter (about 1 mm) and softwood fibers are longer than the hardwoods (about 3–8 mm). The discrepancies between hardwood and softwood fibers in terms of density, morphology and aspect ratio influences the dispersion of the fibers and consequently, the reinforcing for the polymer [31]. The color of WPC samples for each wood species was different correlating with the color of mill residues, wood flour and wood pellets. During the melt processing, the orientation and dispersion of fibers in the polymer is affected as well [30]. Mechanical behavior of the composites are greatly influenced by the consistency of the lignocellulosic materials concentration in the polymer [32,33]. Thus, on comparing the physical properties of wood pellets after pelletizing and then grounding operation with the original wood flour, the ground pellets had greater variation in sizes and were more bulky. The particle size variation of the ground pellets was smaller attributable to the usage of uniform size of the wood flour [34]. As the particles are smaller, the circularity of the particle sizes increases [35]. The same authors also suggested drier pellets produce finer particles. Studies have been conducted related to the particle shape of the ground pellets. Some authors have suggested the mill type influences [36] whereas others have suggested the material properties affect the particles' shape [37]. Jet mills produce particles with the highest aspect ratio. Likewise, in a dry condition, the chemical constituents of wood such as cellulose, hemicellulose and lignin usually do not change until reaching a temperature of around 200 °C [38]. However, surface energy of the particles might be reduced due to the thermal friction encountered by the particles during the milling operation. Grinding can change the surface properties of particles [39].

In this study, pellets were produced from the wood flour without any mechanical treatments. Durable pellets can be produced from the steam explosion mechanism of the raw materials [40]. Thermal pretreatment of the raw materials by torrefaction produce pellets with lower in properties such as density, hardness and energy yield, but higher in hydrophobicity [41]. The production factors such as additives, die temperature, pressure and raw materials can affect the properties of wood pellets [42]. With these changes in properties of pellets changes the material properties of WPCs. Butylina et al. [22] observed the physical and mechanical properties of PP-wood fibers produced from commercially manufactured wood pellets in between the values of wood flour and heat-treated wood fibers. Further detailed research on understanding the properties of wood pellets under different treatments and then its impact on composites manufacturing through different production and grinding treatments are suggested.

The feed rate of the raw materials under different formulations is shown in Table 2. Either in control or MAPP formulations, the feed rate using pellets was approximately 1.5 times greater than the feed rate using wood flour mixed with PP alone or with MAPP. This effect is slightly higher for formulations with MAPP than the controls as the weight of MAPP used was slightly more than the PP used. MAPP also acts as a processing aid, i.e., like a lubricant thus contributing to an enhanced production rate. Pellets after being processed in a grinder were bulkier than the wood flour fibers. These results suggest that time and energy are reduced for the processing of WPC pellets in an extruder using wood pellets rather than wood flour. Nevertheless, this is a case during extrusion only. The additional steps of pelletizing the wood flour and grinding operations still needs to be considered. The ground pellets possessed increased fragmentation in the polymer after the grinding.

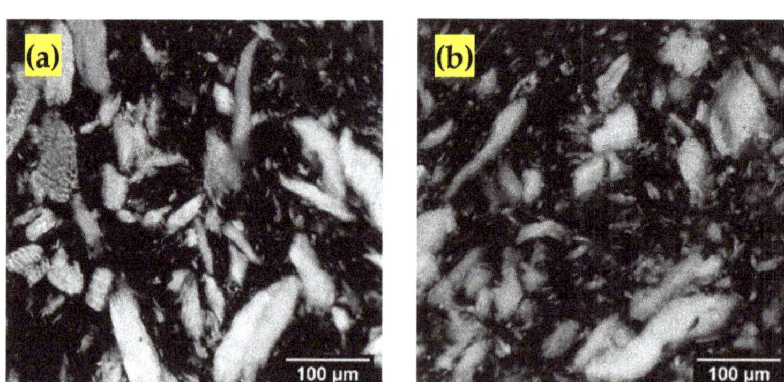

Figure 2. *Cont.*

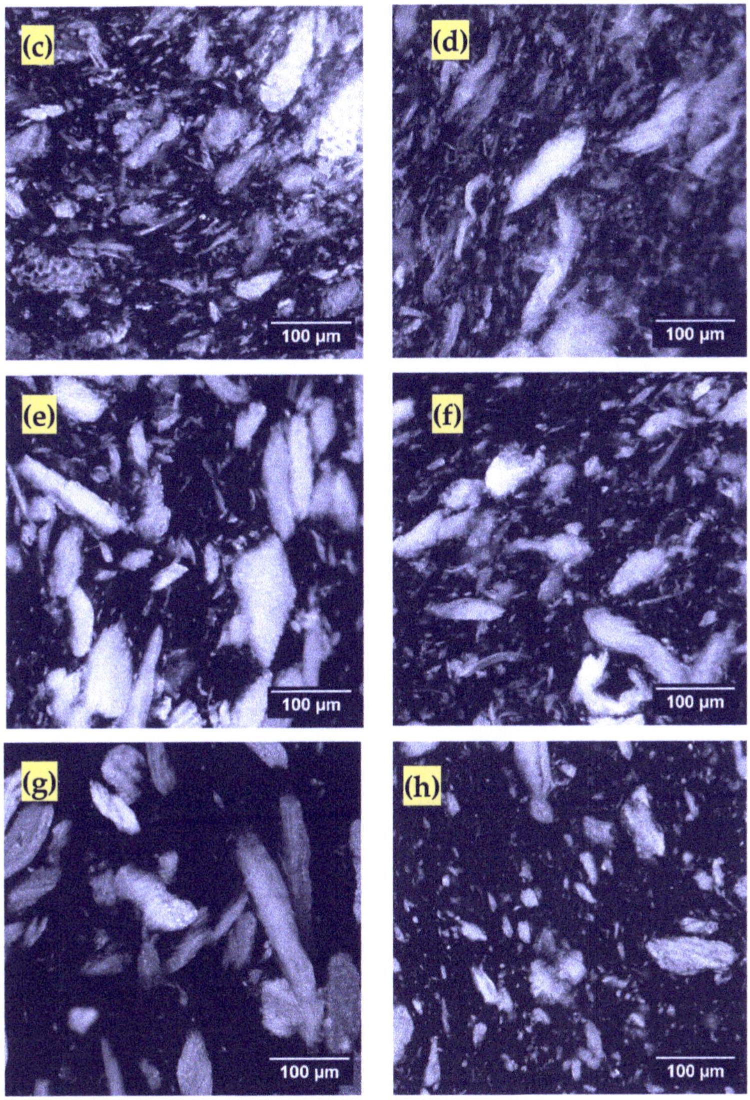

Figure 2. Fluorescent images of WPCs with MAPP (**a**) Cedar flour, (**b**) Cedar pellets, (**c**) Pine flour, (**d**) Pine pellets, (**e**) Spruce-fir flour, (**f**) Spruce-fir pellets, (**g**) Maple flour and (**h**) Maple pellets.

Table 2. Feed rate (g/min) from the feeder of twin-screw extruder.

Wood Species	Controlled Condition		MAPP Condition	
	Flour + PP	Pellets + PP	Flour + PP + MAPP	Pellets + PP + MAPP
White Cedar	26.40	40.81	26.42	40.84
White Pine	22.03	39.62	22.05	39.65
Spruce-Fir	24.51	39.17	24.53	39.20
Red Maple	31.03	43.50	31.05	43.53

1 g of PP = 1.04 g of MAPP.

The two graphs in Figure 3 below show the dispersion of particles in WPCs based on area of the particles. From both the graphs, it can be observed for both wood flour and wood pellets feedstock WPCs, most of the particles falls below 50 square microns followed by 50–100 square microns. However, the WPCs made from pellets have a steeper curve than that of utilizing wood flour. This suggests that within the same range of particles area, a higher number of the pellet particles fall than the flour particles. This then, affects the dispersion of particles in the polymer. A similar trend was observed through the visualization of fluorescent images in Figure 2. From the images, it can be inferred that WPCs made from wood pellets had better particles' dispersion than WPCs made from the wood flour. However, proper dispersion and distribution is always challenging for small particles in the polymer matrix [25,43]. The clustering of the particles usually can occur with particles having a smaller particle size and higher surface energy [44].

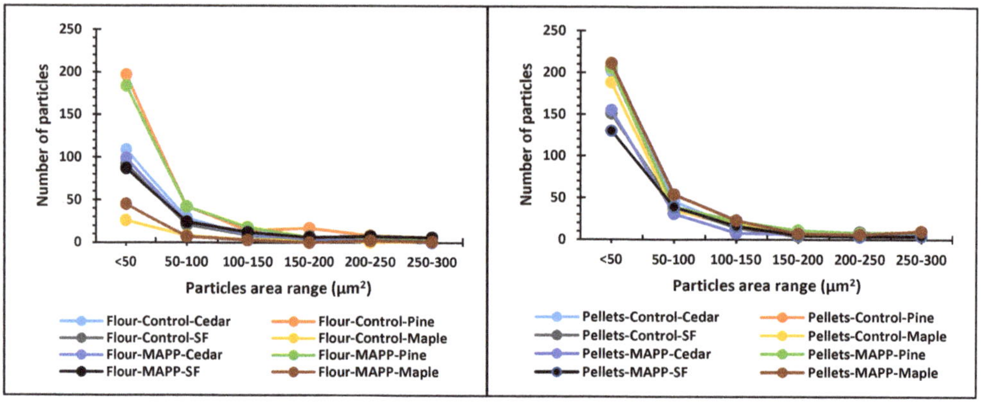

(a) (b)

Figure 3. Area of wood particles in the polymer matrix (**a**) wood flour and (**b**) wood pellets.

The average density of PP is 900 kg/m^3. Figure 4 below shows the density of WPC samples of wood flour and pellets in different formulations. For all sample formulations, the density observed was higher than 900 kg/m^3. For all cases, WPCs made from wood pellets with MAPP showed a higher density. The increase in density of WPCs made with wood pellets is because of ground wood pellets being denser than the wood flour used. Matuana and Stark [45] suggested that the compression of cell walls in the wood causes increase in density of the WPCs than the pure plastics. From our results, it can be indicated that the cell walls of pellets squeezed more than the flour. Similarly, in a microscopic level, the cell wall density of hardwoods and softwoods is almost similar that does not create significant differences in the density of composites. Results of three-way ANOVA showed that the p-value on the one-, or two- or three-way interaction of each variable (wood filler type, wood species, and additives condition) was greater than 0.05 in all conditions except the one-way interaction of wood filler. In the case of density, presence or absence of MAPP and wood species type or their interaction with the wood filler, does not make a notable difference. However, as explained before, filler type alone shows some significant differences as WPCs with pellets have a higher density than those with wood flour. On average, the density of WPCs for the wood flour controls was lower by 0.5% than the pellet formulations. Here, the density was 0.6% and 0.3% lower for the wood flour controls than the pellets for softwoods and hardwood, respectively. Likewise, in the wood flour–MAPP formulations, the average density was 0.6% lower than with pellet–MAPP formulations which was lower by 0.6% and 0.8% for softwoods and hardwoods respectively.

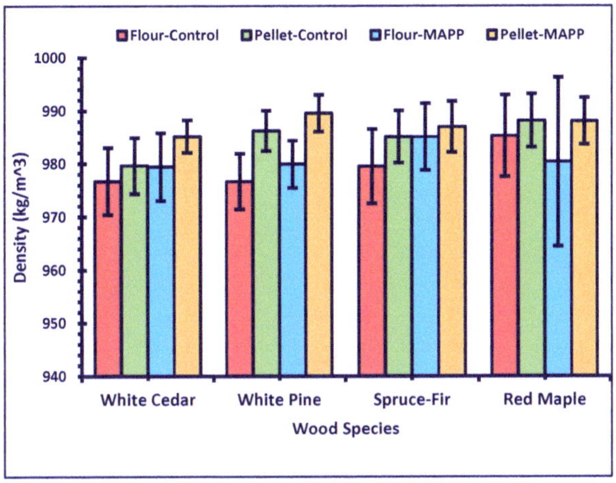

Figure 4. Density of WPCs.

The interplay of wood feedstock and the polymer impacts the mechanical properties of WPCs. The processes used in manufacturing WPCs also impact the composite products' mechanical behavior [46]. The physical and mechanical properties depend on the nature of wood filler such as the size and distribution of particles, the orientation of fibers, wood species and wood filler contents. Softwood pulps augment the tensile and flexural properties of WPCs as compared to the hardwood pulps [31]. However, Berger and Stark [47] reported PP composites from hardwoods performed better than the softwoods. Either with wood flour or pellets, the tensile and flexural properties of the samples containing MAPP were better than the PP controls. This is obvious as MAPP increases the adhesion between polymer and wood. Tensile flexural and impact properties are improved attributable to MAPP in composite of wood flour and PP [48].

Figure 5 conveys the results of tensile strength and modulus of the WPCs under different formulations. Similarly, values of the tensile strength at yield and break and % elongation at yield and break are tabulated in Table 3.

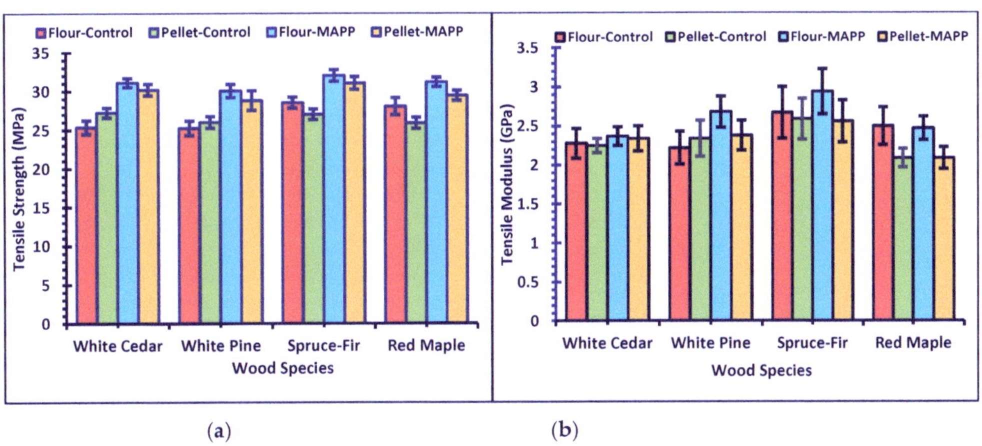

Figure 5. Graphs of (**a**) Tensile strength of WPCs and (**b**) Tensile modulus of WPCs.

Table 3. Tensile strength and % elongation at yield and break for different formulations of WPCs.

Wood Species	Formulations	Tensile Strength at Yield (MPa)		Tensile Strength at Break (MPa)		% Elongation at Yield		% Elongation at Break	
		Mean	S.D.	Mean	S.D.	Mean	S.D.	Mean	S.D.
White Cedar	Flour–Control	14.35	0.82	14.82	0.49	2.77	0.40	10.62	1.10
	Pellets–Control	16.56	0.73	16.12	0.32	2.54	0.18	9.35	0.56
	Flour–MAPP	20.22	0.97	18.32	0.47	3.48	0.30	9.21	0.64
	Pellets–MAPP	17.81	0.65	17.60	0.50	3.13	0.26	9.31	0.53
White Pine	Flour–Control	15.43	1.38	14.82	0.64	3.32	0.32	12.72	1.16
	Pellets–Control	16.23	0.88	15.14	0.48	3.07	0.29	11.27	0.82
	Flour–MAPP	18.96	1.19	17.60	0.61	3.16	0.21	9.66	0.66
	Pellets–MAPP	18.69	0.76	16.89	0.74	3.33	0.47	9.43	0.75
Spruce-Fir	Flour–Control	17.74	1.02	16.72	0.52	3.28	0.52	9.73	0.90
	Pellets–Control	15.95	0.85	15.85	0.42	3.03	0.30	11.06	0.73
	Flour–MAPP	19.35	1.48	18.84	0.56	3.25	0.28	9.37	0.69
	Pellets–MAPP	18.77	1.74	18.34	0.48	4.05	1.09	10.50	1.10
Red Maple	Flour–Control	16.48	1.12	16.47	0.64	2.76	0.31	9.16	0.53
	Pellets–Control	15.14	0.82	15.27	0.47	3.11	0.59	10.28	0.90
	Flour–MAPP	18.64	0.53	18.20	0.44	3.42	0.83	10.16	1.01
	Pellets–MAPP	17.37	1.32	17.18	0.39	2.49	0.61	8.69	1.01

Results of the three-way ANOVA showed the three variables alone, i.e., wood filler, additive and wood species or their two-way or three-way interactions indicate some statistical differences in the tensile properties. There was minute statistical difference in the tensile values attributable to the two-way interaction of wood species and the MAPP additive. Tensile properties with or without MAPP did not vary much among the different wood species. Similarly, tensile modulus was not influenced much based on the wood filler type and the three-way effect of filler type, species and MAPP additive. The graph in Figure 5 shows, WPCs manufactured using pellets or flour had no significant difference in the tensile values of the composite products. Either for the MAPP or control formulations, the wood pellet samples show similar or significantly lower differences in the tensile properties. On average, the tensile strength of the wood flour controls was 0.6% higher than the pellet controls. Tensile strength of the wood flour controls was lower by 1.7% for the softwoods and higher by 7.7% for the hardwood compared to the pellet controls. Similarly, the average tensile strength of the wood flour–MAPP samples was 3.9% higher than pellet MAPP samples. The softwoods and hardwood showed increases of 3.4% and 5.5% respectively for the wood flour–MAPP samples compared to pellet–MAPP samples. The average Modulus of Elasticity (MOE) for the wood flour controls was 3.8% higher than the pellet controls. In this case, the MOE of the wood flour control formulation compared to pellet controls were lower by 0.4% for the softwoods and higher by 16.4% for the hardwood sample. The average MOE of the wood flour–MAPP was higher by 10.2% than the pellet–MAPP formulation. The MOE was higher by 8.5% and 15.4% for softwoods and hardwood respectively for the wood flour–MAPP compared to the pellet–MAPP samples. These results show that the tensile property values differ based on the wood species. Rogers and Simonsen [49] suggested the type of wood species can influence the parameters such as: roughness, feasibility to grinding, and porosity which might regulate the bonding with WPCs. Compared to the other species, the spruce-fir WPCs exhibited the maximum tensile properties for all formulations. This is attributable to the higher fragmentation of fibers in the PP matrix. Spruce-fir fibers possess the highest aspect ratio among the wood species examined. Higher aspect ratio is crucial in influencing the mechanical performance of the WPCs than the length of the particle [16,17,50–53]. Damage to the wood fiber is induced by the grinding process [54]. WPCs made of hardwood, i.e., maple pellets have the lowest MOE values among the wood species made from pellets. Neagu et al. [55] reported the correlation between lignin content and stiffness of the composite materials where they observed maximum stiffness for softwood kraft fibers. These results also show similarities in the findings as maple pellets have less lignin content compared to the other softwood pellets.

Table 3 lists the tensile strength at yield and break and % elongation at yield and break for different formulations of the composites. The values show WPCs manufactured using pellets and wood flour did not have significant differences and any differences were small. Spruce-fir WPCs showed maximum strength compared to the other wood fillers on average. Likewise, the elasticity of the composites is indicated by the elongation at break. The percentage elongation at yield is in the range of 2–4% and a break in the range of 8–12% for the WPC samples. Between the pellets and wood flour WPCs, the values indicate minor differences. Compared to the maple and cedar, pine and spruce-fir have greater % elongation at yield and break. The tensile strength and modulus, and % elongation of the WPCs have a positive relation with the finer size of the particles [56,57].

In Figure 6, the flexural properties of WPCs made with wood flour and pellets in the presence or absence of MAPP are presented. From the three-way ANOVA, it was observed for the flexural properties the variables alone or their two-way or three-way interactions contribute to some significant differences in the bending properties except the one-way interaction of wood filler type. The wood filler either the wood flour or pellets did not influence the flexural strength. MOE was affected by the filler type by small statistical difference. MOE was not influenced by two-way effect of wood species and MAPP additive where, the MOE values with and without MAPP were similar among the different species. The average flexural strength for wood flour controls was lower by 1.3% than the pellet controls. Flexural strength of the wood flour controls was lower by 4% for softwoods and higher by 6.7% for the hardwood formulation compared to the pellet controls. Similarly, the average bending strength of wood flour–MAPP samples was 4.2% higher than the pellet MAPP samples. The softwoods and hardwood contributed to increase in bending strength by 4.1% and 4.35% respectively on flour–MAPP condition than the pellet MAPP samples. The average bending MOE for the flour–control case was 1.3% lower than the pellet-control case. In this scenario, the MOE for the flour–control formulation contrary to pellet-control formulation was lower by 7.5% for softwoods and higher by 17.5% for the hardwood. The average bending MOE for formulation with flour–MAPP was higher by 8.4% than the pellet–MAPP formulation. The MOE was higher by 8.7% and 7.6% for softwoods and hardwood respectively in flour–MAPP case than the pellet–MAPP case. Looking at the graphs of the bending tests in Figure 6, a similar trend to flexural properties existed for formulations with or without additives and with wood flour or pellets. Wood flour or wood pellets WPCs had similar or significantly fewer differences in flexural properties. In some cases, bending properties of WPCs made with pellets were higher than WPCs with wood flour. Wood species had a different impact to the bending properties than to the tensile properties. Compared to the rest of the wood species, red maple exhibited the maximum flexural strength using MAPP for both flour and pellets followed by spruce-fir. However, Pilarski and Matuana [58] mentioned flexural properties of WPCs produced from rigid PVC and HDPE matrices showed better performance for softwoods than the hardwoods. Higher values of tensile and flexural properties was shown by jack pine and black spruce than the white cedar for HDPE composite [59]. Similarly, the flexural modulus of elasticity of WPCs with red maple pellets was lower for both controls and MAPP formulations than the other species. This might be attributable to the correlation between lignin content and stiffness as concluded by Neagu et al. [55].

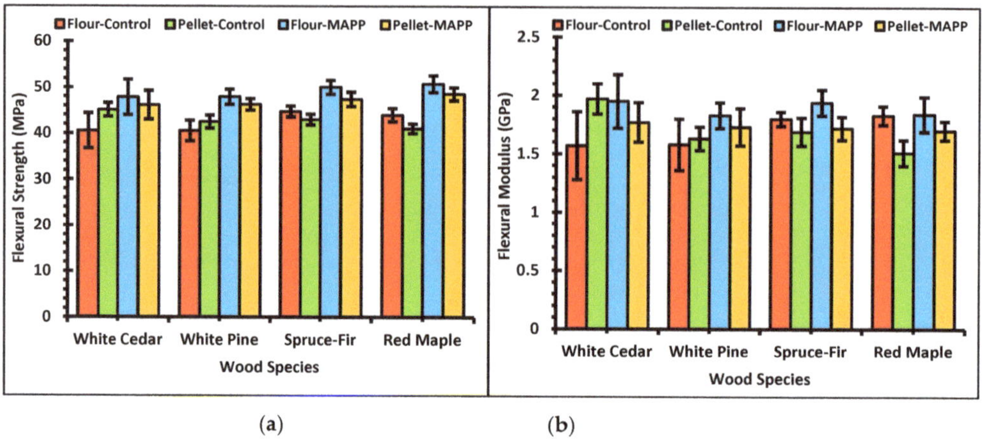

Figure 6. Graphs of (**a**) Flexural strength of WPCs and (**b**) Flexural modulus of WPCs.

Typically, the presence of wood fibers in the body of WPCs causes crack initiation and subsequent failure. The graph in Figure 7 show the impact properties of the different WPC samples. Unlike to the tensile and flexural properties, the presence of MAPP did not contribute to a higher impact strength than in the control formulations. Compared to the tensile and flexural properties, the impact strength values for different wood species had a greater variance from the mean value. On average, WPCs of red maple showed better impact strength than the other wood species. The study conducted by Bledzki and Faruk [60], on the effect of wood filler geometry on the physico-mechanical properties of PP/wood composites, reported hardwood flour reinforced PP composites had a better impact strength than the other PP composites of different fillers (softwood fiber, long wood fiber and wood chips). However, higher impact strength of PP based composites for ponderosa pine was greater than the hardwoods: oak and maple [47]. The composite samples made with fine wood particles have poorer impact strength than with coarse-wood particles [61]. In our study, the softwood particles were finer than the hardwood. Because of this, on average WPCs from softwoods had a lower impact strength than the hardwood. Similar to the tensile and flexural values, there was not a difference in the impact values due to the filler type. The average impact strength for the wood flour controls was 8.3% higher than the pellet samples. At flour–controls, it was higher by 14.3% for softwoods and lower by 9.6% for hardwood than the pellet-controls formulation. For the wood flour–MAPP formulations, the impact strength value was 2.3% higher than the pellet MAPP formulations. For softwoods it was higher by 3.7% and for hardwood lower by 1.8% at flour–MAPP formulation than pellet–MAPP formulation. Results of the three-way ANOVA on flexural properties showed the variables alone or their two-way or three-way interactions contribute to some significant differences in the impact properties. The statistical analysis also revealed that the interaction of wood filler and MAPP additive did not influence the impact strength with a greater difference. It means in each WPC formulation with presence or absence of MAPP, the impact strength for each filler type did not have much variation.

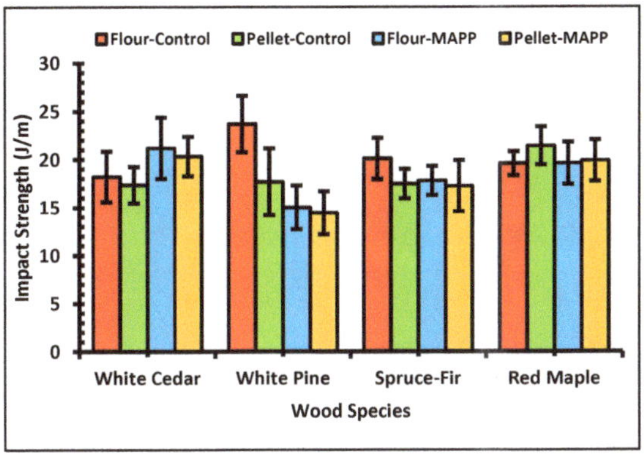

Figure 7. Impact strength of WPCs.

Future research work on using wood pellets as a wood feedstock in different polymers, different formulations, changing the properties of wood pellets during manufacturing, etc. are recommended to understand more about the material for its efficient application.

4. Conclusions

The following conclusions are highlighted from the present study:

- The physical and mechanical properties of WPCs made from either the wood flour or wood pellets were similar.
- Distribution of feedstock was similar for both wood flour and pellet formulated WPCs. However, dispersion was greater for WPCs with pellets than with wood flour.
- MAPP improved the physical and mechanical properties of WPCs for each wood feedstock and wood species.
- On average, WPC samples of spruce-fir species possessed the best properties.

Author Contributions: Conceptualization, D.J.G., Y.H. and G.P.; methodology, D.J.G., Y.H. and G.P.; software, G.P.; validation, D.J.G., Y.H. and G.P.; formal analysis, D.J.G. and G.P.; investigation, G.P.; resources, D.J.G.; data curation, G.P.; writing—original draft preparation, G.P.; writing—review and editing, D.J.G., Y.H. and G.P.; visualization, D.J.G. and G.P.; supervision, D.J.G.; project administration, D.J.G.; funding acquisition, D.J.G. and Y.H. All authors have read and agreed to the published version of the manuscript.

Funding: This research was funded by United States Department of Agriculture (USDA) Agriculture Research Service (ARS) with Agreement Number/FAIN: 58-0204-9-166 and Project Number: 0204-41510-001-90S.

Institutional Review Board Statement: Not applicable.

Informed Consent Statement: Not applicable.

Data Availability Statement: All the necessary data is included within the article. Upon request, the data in the article can be available from the corresponding authors.

Acknowledgments: The authors would like to acknowledge USDA ARS project for the financial support, School of Forest Resources at University of Maine for the availability of laboratory equipment and the associated lab assistants for the training and guidance. This project was supported by the USDA McIntire-Stennis Project Number ME0-41809 through the Maine Agricultural & Forest Experiment Station. Maine Agricultural and Forest Experiment Station Publication Number 3836.

Conflicts of Interest: The authors declare no conflict of interest.

Appendix A

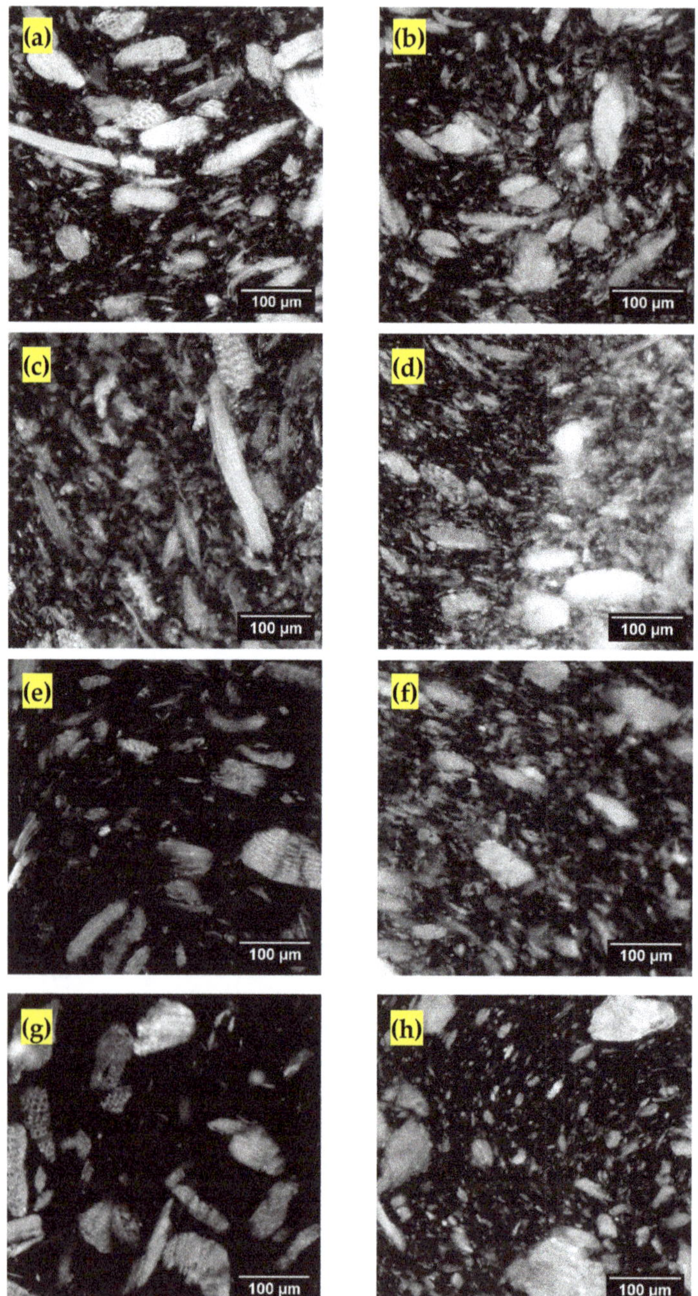

Figure A1. Fluorescent images of WPCs in controlled condition (**a**) Cedar flour, (**b**) Cedar pellets, (**c**) Pine flour, (**d**) Pine pellets, (**e**) Spruce-fir flour, (**f**) Spruce-fir pellets, (**g**) Maple flour and (**h**) Maple pellets.

References

1. Borah, J.S.; Kim, D.S. Recent development in thermoplastic/wood composites and nanocomposites: A review. *Korean J. Chem. Eng.* **2016**, *33*, 3035–3049. [CrossRef]
2. Gardner, D.J.; Han, Y.; Wang, L. Wood–Plastic Composite Technology. *Curr. For. Rep.* **2015**, *1*, 139–150. [CrossRef]
3. Bledzki, A.K.; Gassan, J. Composites reinforced with cellulose based fibres. *Prog. Polym. Sci.* **1999**, *24*, 221–274. [CrossRef]
4. Malkapuram, R.; Kumar, V.; Negi, Y.S. Recent development in natural fiber reinforced polypropylene composites. *J. Reinf. Plast. Compos.* **2009**, *28*, 1169–1189. [CrossRef]
5. Braghiroli, F.L.; Passarini, L. Valorization of biomass residues from forest operations and wood manufacturing presents a wide range of sustainable and innovative possibilities. *Curr. For. Rep.* **2020**, *6*, 172–183. [CrossRef]
6. Wolcott, M.; Muszynski, L. Materials and wood-based composites. In *Wood Engineering Challenges in the New-Millennium-Critical Research Needs, Proceedings of the Pre-Conference Workshop for ASCE Structures, Vancouver, BC, Canada, 23–24 April 2008*; American Society of Civil Engineers: Reston, VA, USA, 2008.
7. Thompson, D.W.; Hanse, E.N.; Knowles, C.; Muszynski, L. Opportunities for wood plastic composite products in the U.S. highway construction sector. *BioResources* **2010**, *5*, 1336–1552.
8. Wendel, B.; Rietzek, D.; Kuhnlein, F.; Feulner, R.; Hulder, G.; Schmachtenberg, E. Additive processing of polymers. *Macromol. Mater. Eng.* **2008**, *293*, 799–809. [CrossRef]
9. Gebhardt, A. *Understanding Additive Manufacturing*; Hanser Publications: Munich, Germany, 2011; pp. 1–164.
10. Jiang, K.; Guo, Y.; Bourell, D.L. Study on the microstructure and binding mechanisms of selective laser sintered wood plastic composite. In Proceedings of the 24th International SFF Symposium—An Additive Manufacturing Conference, Austin, TX, USA, 12–14 August 2013; pp. 497–504.
11. Ibrahim, M.; Badrishah, N.S.; Sa'ude, N.; Ibrahim, M.I. Sustainable natural bio composite for FDM feedstocks. *Appl. Mech. Mater.* **2014**, *607*, 65–69. [CrossRef]
12. Lamm, M.E.; Wang, L.; Kishore, V.; Tekinalp, H.; Kunc, V.; Wang, J.; Gardner, D.J.; Ozcan, S. Material extrusion additive manufacturing of wood and lignocellulosic filled composites. *Polymers* **2020**, *12*, 2115. [CrossRef]
13. Wood—Plastic Composite Market Size, Share & Industry Analysis, by Material (Polyethylene, Polypropylene, Polyvinyl Chloride and Others), by Application (Decking, Automotive, Sliding & Fencing, Technical Application, Furniture, Consumer Foods and Others), and Regional Forecast, 2020–2027. Available online: https://www.fortunebusinessinsights.com/wood-plastic-composite-market-102821 (accessed on 23 September 2020).
14. Global Wood Plastic Composites Market: By Type, Product, Application, Size, Share, Trends, Analysis, Segment and Forecast 2016–2022. Available online: https://www.zionmarketresearch.com/report/wood-plastic-composites-market (accessed on 7 February 2017).
15. Harper, C.A. *Modern Plastics Handbook*; The McGraw-Hill Companies, Inc.: New York, NY, USA, 2000. Available online: https://www.accessengineeringlibrary.com/content/book/9780070267145 (accessed on 24 July 2021).
16. Stark, N.M.; Rowlands, R.E. Effects of wood fiber characteristics on mechanical properties of wood/polypropylene composites. *Wood Fiber Sci.* **2003**, *35*, 167–174.
17. Migneault, S.; Koubaa, A.; Erchiqui, F.; Chaala, A.; Englund, K.; Wolcott, M.P. Effects of processing method and fiber size on the structure and properties of wood–plastic composites. *Compos. Part A Appl. Sci. Manuf.* **2009**, *40*, 80–85. [CrossRef]
18. Ashori, A. Wood–plastic composites as promising green-composites for automotive industries. *Bioresour. Technol.* **2008**, *99*, 4661–4667. [CrossRef]
19. Stokke, D.D.; Gardner, D.J. Fundamental aspects of wood as a component of thermoplastic composites. *J. Vinyl Addit. Technol.* **2003**, *9*, 96–104. [CrossRef]
20. Tanaka, T. Manufacturing and processing methods of biocomposites. In *Polymer Composites, Biocomposites*; Thomas, S., Joseph, K., Malhotra, S.K., Goda, K., Sreekala, M.S., Eds.; Wiley-VCH Verlag GmbH & Co.: Weihheim, Germany, 2014; Volume 3, pp. 179–211.
21. Clemons, C.M. Wood Flour. In *Functional Fillers for Plastics*; Wiley-VCH: Weinheim, Germany, 2010; pp. 269–290.
22. Butylina, S.; Martikka, O.; Kärki, T. Properties of wood fibre-polypropylene composites: Effect of wood fibre source. *Appl. Compos. Mater.* **2011**, *18*, 101–111. [CrossRef]
23. Pokhrel, G.; Han, Y.; Gardner, D.J. Comparative study of the properties of wood flour and wood pellets manufactured from secondary processing mill residues. *Polymers* **2021**, *13*, 2487. [CrossRef]
24. Kim, H.S.; Lee, B.H.; Choi, S.W.; Kim, S.; Kim, H.J. The effect of types of maleic anhydride-grafted polypropylene (MAPP) on the interfacial adhesion properties of bio-flour-filled polypropylene composites. *Compos. Part A Appl. Sci. Manuf.* **2007**, *38*, 1473–1482. [CrossRef]
25. Danyadi, L.; Renner, K.; Szabo, Z.; Nagy, G.; Moczo, J.; Pukanszky, B. Wood flour filled PP composites: Adhesion, deformation, failure. *Polym. Adv. Technol.* **2006**, *17*, 967–974. [CrossRef]
26. Lu, J.Z.; Wu, Q.; Negulescu, I.I. Wood-fiber/high-density-polyethylene composites: Coupling agent performance. *J. Appl. Polym. Sci.* **2005**, *96*, 93–102. [CrossRef]
27. Lee, C.-H.; Wu, T.-L.; Chen, Y.-L.; Wu, J.-H. Characteristics and discrimination of five types of wood-plastic composites by FTIR spectroscopy combined with principal component analysis. *Holzforschung* **2010**, *64*, 699–704. [CrossRef]
28. Yang, S.; Wei, B.; Wang, Q. Superior dispersion led excellent performance of wood-plastic composites via solid-state shear milling process. *Compos. Part B Eng.* **2020**, *200*, 108347. [CrossRef]

29. Bengtsson, M.; Baillif, M.L.; Oksman, K. Extrusion and mechanical properties of highly filled cellulose fibre–polypropylene composites. *Compos. Part A Appl. Sci. Manuf.* **2007**, *38*, 1922–1931. [CrossRef]
30. Peltola, H.; Paakkonen, E.; Jetsu, P.; Heinemann, S. Wood based PLA and PP composites: Effect of fibre type and matrix polymer on fibre morphology, dispersion and composite croperties. *Compos. Part A Appl. Sci. Manuf.* **2014**, *61*, 13–22. [CrossRef]
31. Maldas, D.; Kokta, B.V.; Daneault, C. Thermoplastic composties of polystyrene: Effect of different wood species on mechanical properties. *J. Appl. Polym. Sci.* **1989**, *38*, 413–439. [CrossRef]
32. Raj, R.G.; Kokta, B.V.; Maldas, D.; Daneault, C. Use of wood fibers in thermoplastics. VII. The effect of coupling agents in polyethylene–wood fiber composites. *J. Appl. Polym. Sci.* **1989**, *37*, 1089–1103. [CrossRef]
33. Chen, H.C.; Chen, T.Y.; Hsu, C.H. Effects of wood particle size and mixing ratios of HDPE on the properties of the composites. *Eur. J. Wood Wood Prod.* **2006**, *64*, 172–177. [CrossRef]
34. Manouchehrinejad, M.; Giesen, I.V.; Mani, S. Grindability of torrefied wood chips and wood pellets. *Fuel Process. Technol.* **2018**, *182*, 45–55. [CrossRef]
35. Masche, M.; Puig-Arnavat, M.; Jensen, P.A.; Holm, J.K.; Clausen, S.; Ahrenfeldt, J.; Henriksen, U.B. Influence of wood pellets properties on their grinding performance. *Biomass Conv. Bioref.* **2021**, 1–16. [CrossRef]
36. Rose, H.E. Particle shape and surface area. In Powders in Industry, Society of Chemical Industry. In Proceedings of the Symposium on Powders in Industry, London, UK, 29–30 September 1960.
37. Bond, F.C. Control particle shape and size. *Chem. Eng.* **1954**, *61*, 195–198.
38. Karinkanta, P.; Ammala, A.; Illikainen, M.; Niinimaki, J. Fine grinding of wood—Overview from wood breakage to applications. *Biomass Bioenergy* **2018**, *113*, 31–44. [CrossRef]
39. Galet, L.; Goalard, C.; Dodds, J.A. The importance of surface energy in the dispersion behaviour of talc particles in aqueous media. *Powder Technol.* **2009**, *190*, 242–246. [CrossRef]
40. Lam, P.S. Steam Explosion of Biomass to Produce Durable Wood Pellets. Ph.D. Thesis, University of British Columbia, Vancouver, BC, Canada, 2011.
41. Peng, J.; Wang, J.; Bi, X.T.; Lim, C.J.; Sokhansanj, S.; Peng, H.; Jia, D. Effects of thermal treatment on energy density and hardness of torrefied wood pellets. *Fuel Process. Technol.* **2015**, *129*, 168–173. [CrossRef]
42. Mostafa, M.E.; Hu, S.; Wang, Y.; Su, S.; Hu, X.; Elsayed, S.A.; Xian, J. The significance of pelletization operating conditions: An analysis of physical and mechanical characteristics as well as energy consumption of biomass pellets. *Renew. Sustain. Energy Rev.* **2019**, *105*, 332–348. [CrossRef]
43. Ess, J.W.; Hornsby, P.R. Characterization of distributive mixing in thermoplastics compositions. *Polym. Test* **1986**, *6*, 205–218. [CrossRef]
44. Móczó, J.; Pukánszky, B. Polymer micro and nanocomposites: Structure, interactions, properties. *J. Ind. Eng. Chem.* **2008**, *14*, 535–563. [CrossRef]
45. Matuana, L.M.; Stark, N.M. The use of wood fibers as reinforcements in composites. In *Biofiber Reinforcement in Composite Materials;* Elsevier: Amsterdam, The Netherlands, 2015; pp. 648–688. [CrossRef]
46. Soccalingame, L.; Bourmaud, A.; Perrin, D.; Benezer, J.C.; Bergeret, A. Reprocessing of wood flour reinforced polypropylene composites: Impact of particle size and coupling agent on composite and particle properties. *Polym. Degrad. Stab.* **2015**, *113*, 72–85. [CrossRef]
47. Berger, M.J.; Stark, N.M. Investigations of species effects in an injection-molding-grade, wood-filled polypropylene. In Proceedings of the Fourth International Conference on WoodFiber-Plastic Composites, Madison, WI, USA, 12–14 May 1997.
48. Kord, B.; Hemmasi, A.H.; Ghasemi, I. Properties of PP/wood flour/organomodified montmorillonite nanocomposites. *Wood Sci. Technol.* **2011**, *45*, 111–119. [CrossRef]
49. Rogers, J.; Simonsen, J. Interfacial shear strength of wood–plastic composites: A new pullout method using wooden dowels. *J. Adhes. Sci. Technol.* **2005**, *19*, 975–985. [CrossRef]
50. Zaini, M.J.; Fuad, M.Y.A.; Ismail, Z.; Mansor, M.S.; Mustafah, J. The effect of filler content and size on the mechanical properties of polypropylene/oil palm wood flour composites. *Polym. Int.* **1966**, *40*, 51–55. [CrossRef]
51. Liew, K.C.; Harum, J.; Tahir, P.M.; Yusoff, M.N.M.; Dahlan, K.Z.M. Properties of rubber wood flourpolypropylene composites blended at different fibre contents and fibre size fractions. *J. Trop. For. Prod.* **2000**, *6*, 21–27.
52. Lee, B.J.; McDonald, A.G.; James, B. Influence of fiber length on the mechanical properties of wood-fiber/polypropylene prepreg sheets. *Mater. Res. Innov.* **2001**, *4*, 97–103. [CrossRef]
53. Chaudemanche, S.; Perrot, A.; Pimbert, S.; Lecompte, T.; Faure, F. Properties of an industrial extruded HDPE-WPC: The effect of the size distribution of wood flour particles. In *Construction and Building Materials;* Elsevier B.V.: Amsterdam, The Netherlands, 2018; Volume 3, Available online: http://link.gale.com/apps/doc/A531285575/AONE?sid=summon&xid=1bdeae7e (accessed on 27 June 2021).
54. Thumm, A.; Dickson, A.R. The influence of fibre length and damage on the mechanical performance of polypropylene/wood pulp composites. *Compos. Part A Appl. Sci. Manuf.* **2013**, *46*, 45–52. [CrossRef]
55. Neagu, R.C.; Gamstedt, E.K.; Berthold, F. Stiffness contribution of various wood fibers to composite materials. *J. Compos. Mater.* **2006**, *40*, 663–699. [CrossRef]
56. Sain, M.M.; Imbert, C.; Kokta, B.V. Composites of surface-treated wood fiber and recycled polypropylene. *Die Angew. Makromol. Chemie* **1993**, *210*, 33–46. [CrossRef]

57. Siwek, S.; Oktaee, J.; Grasselt-Gille, S.; Wagenfuhr, A. Influence of different wood flour fractions on the mechanical properties of injection molded WPC with cellulose propionate. *Eur. J. Wood Wood Prod.* **2018**, *76*, 499–507. [CrossRef]
58. Pilarski, J.M.; Matuana, L.M. Durability of wood flour-plastic composites exposed to accelerated freeze–thaw cycling. II. high density polyethylene matrix. *J. Appl. Polym. Sci.* **2006**, *100*, 35–39. [CrossRef]
59. Bouafif, H.; Koubaa, A.; Perre, P.; Cloutier, A. Effects of fiber characteristics on the physical and mechanical properties of wood plastic composites. *Compos. Part A Appl. Sci. Manuf.* **2009**, *40*, 1975–1981. [CrossRef]
60. Bledzki, A.K.; Faruk, O. Wood fibre reinforced polypropylene composites: Effect of fibre geometry and coupling agent on physico-mechanical properties. *Appl. Compos. Mater.* **2003**, *10*, 365–379. [CrossRef]
61. Maiti, S.N.; Singh, K. Influence of wood flour on the mechanical properties of polyethylene. *J. Appl. Polym. Sci.* **1986**, *32*, 4285–4289. [CrossRef]

polymers

Article

Extraction and Characterization of Natural Cellulosic Fiber from *Pandanus amaryllifolius* Leaves

Z. N. Diyana [1], R. Jumaidin [2,*], M. Z. Selamat [1], R. H. Alamjuri [3,*] and Fahmi Asyadi Md Yusof [4]

1 Fakulti Kejuruteraan Mekanikal, Universiti Teknikal Malaysia Melaka, Hang Tuah Jaya, Durian Tunggal, Melaka 76100, Malacca, Malaysia; nurdiyanazakuan@gmail.com (Z.N.D.); zulkeflis@utem.edu.my (M.Z.S.)
2 Fakulti Teknologi Kejuruteraan Mekanikal dan Pembuatan, Universiti Teknikal Malaysia Melaka, Hang Tuah Jaya, Durian Tunggal, Melaka 76100, Malacca, Malaysia
3 Faculty of Tropical Forestry, Universiti Malaysia Sabah, Jalan UMS, Kota Kinabalu 88400, Sabah, Malaysia
4 Malaysian Institute of Chemical & Bioengineering Technology (UniKL MICET), Universiti Kuala Lumpur, Taboh Naning, Alor Gajah, Melaka 78000, Malacca, Malaysia; fahmiasyadi@unikl.edu.my
* Correspondence: ridhwan@utem.edu.my (R.J.); rhanim@ums.edu.my (R.H.A.)

Abstract: *Pandanus amaryllifolius* is a member of Pandanaceae family and is abundant in south-east Asian countries including Malaysia, Thailand, Indonesia and India. In this study, *Pandanus amaryllifolius* fibres were extracted via a water retting extraction process and were investigated as potential fibre reinforcement in polymer composite. Several tests were carried out to investigate the characterization of *Pandanus amaryllifolius* fibre such as chemical composition analysis which revealed *Pandanus amaryllifolius* fibre's cellulose, hemicellulose and lignin content of 48.79%, 19.95% and 18.64% respectively. Material functional groups were analysed by using Fourier transform infrared (FTIR) spectroscopy and X-ray diffraction analysis confirming the presence of cellulose and amorphous substances in the fibre. The morphology of extracted *Pandanus amaryllifolius* fibre was studied using a scanning electron microscope (SEM). Further mechanical behaviour of fibre was investigated using a single fibre test with 5 kN cell load and tensile strength was found to be 45.61 ± 16.09 MPa for an average fibre diameter of 368.57 ± 50.47 μm. Meanwhile, moisture content analysis indicated a 6.00% moisture absorption rate of *Pandanus amaryllifolius* fibre. The thermogravimetric analysis justified the thermal stability of *Pandanus amaryllifolius* fibre up to 210 °C, which is within polymerization process temperature conditions. Overall, the finding shows that *Pandanus amaryllifolius* fibre may be used as alternative reinforcement particularly for a bio-based polymer matrix.

Keywords: *Pandanus amaryllifolius* fibre; natural fibres; composite

Citation: Diyana, Z.N.; Jumaidin, R.; Selamat, M.Z.; Alamjuri, R.H.; Md Yusof, F.A. Extraction and Characterization of Natural Cellulosic Fiber from *Pandanus amaryllifolius* Leaves. *Polymers* **2021**, *13*, 4171. https://doi.org/10.3390/polym13234171

Academic Editors: Domenico Acierno and Swarup Roy

Received: 31 August 2021
Accepted: 11 November 2021
Published: 29 November 2021

Publisher's Note: MDPI stays neutral with regard to jurisdictional claims in published maps and institutional affiliations.

1. Introduction

Changes from the dominant usage of synthetic fibre to natural fibre indicate the rise of awareness among people around the world regarding the negative environmental impact that synthetic fibre brings which may damage human health. Synthetic polymers production utilized a large amount of energy which produces environmental pollutants during the production and recycling of synthetic composites [1]. The implementation of natural fibre-reinforced polymer (NFRP) composites has become an emerging trend driven by stringent environmental legislation that focuses on the development of more eco-friendly materials as an alternative to substitute synthetic composites. The incorporation of natural fibres in composite has presented many improvements in terms of favourable tensile properties, reduced health hazards, acceptable insulating properties, low density and decreased energy consumption [2]. Furthermore, cellulosic material has been widely utilized in many applications such as cellulose nanofibers in food packaging, drug delivery and biomedicine applications; chitosan widely used in biosensors and tissue engineering, membrane separation and treatment of water applications; and hybrid of bacterial cellulose and chitin nanofibers have produced nanocomposite film that provides an excellent barrier to act as the antioxidant and antibacterial film [3]. Besides that, many studies have been conducted with regards to the incorporation of natural fibre in composite applications which have proven that natural fibre has remarkable advantages and is a promising method to replace synthetic polymer [4–9].

Natural fibres can be simply defined as fibres that are not synthetic or man-made. They can be sourced from plants, animals or minerals where would be the common classification of natural fibres. In general, most common natural fibres come from plants and are composed of cellulose, thus making the fibre hydrophilic due to the presence of poly(1,4-β-D-anhydroglucopyranose) units which contain hydroxyl groups that enable them to form strong hydrogen bonds [10,11]. They are also called lignocellulosic fibres since their cellulose fibrils are embedded in the lignin matrix [12]. However, natural fibres properties and structural parameters differ among plants depending on the species, growing environment, topographical location and preparation sample fibre method. The plant cell wall is composed of cellulose, hemicellulose and lignin where cellulose is the fundamental composition of the fibre. [12]. Cellulose is an organic polysaccharide that is composed of thousands of D-glucose units resulting from condensation via β(1→4)-glycosidic bonds which permit cellulose to have physical and chemical properties that demonstrate high tensile strength, are environmentally friendly, non-toxic and totally renewable [13]. Hemicellulose is known as a branching polysaccharide polymer made up of glucose and other types of sugar groups including xylose, galactose, arabinos and mannose [11]. Lignin is composed largely of phenylpropane and is the second most abundant component after cellulose, responsible for cementing cellulose microfibrils as well as protecting cellulose and hemicellulose contributes rigidity and also carries out water transport [4]. Both lignin and hemicellulose are amorphous polymers and cellulose is a semicrystalline polymer.

Pandanus amaryllifolius is a tropical plant and a member of the Pandanaceae family. It is reported that there are more than 600 known species from the Pandanaceae family [13]. *Pandanus amaryllifolius* is one of the species that is known as an endemic plant to Malaysia and is famously called as pandan wangi. It is famous with regards to its unique fragrant leaves which are widely used for flavouring in the cuisines of the Southeast Asia region. Nevertheless, its species member called *Pandanus tectorius* has flowers that are scented but not the leaves [14]. The distinct flavour aroma from pandan wangi species is due to the presence of a high amount of 2-acetyl-1-pyrroline (2AP) [15]. Despite its unique aroma and colouring, pandan wangi also has been used traditionally as medicine for decades and is proven to enhance the immune system and anti-tumour agents [16]. Besides that, it is also reported that *Pandanus amaryllifolius* extract is able to reduce blood glucose levels as well as shows improvement in insulin resistance [17]. Table 1 shows *Pandanus* species and applications based on reported journals.

Table 1. Pandan species and potential applications.

Type of Pandan	Potential Application	References
Pandanus ceylanicus	Polymer composite	[13]
Pandanus tectorius	Natural fiber	[18]
Pandanus utilis	Low cost paper	[19]
Pandanus amaryllifolius	Polymer composite	[20,21]
Pandanus tectorius	Pharmaceutical, medicine	[16]
Pandanus amaryllifolius	Pharmaceutical, medicine	[17]
Pandanus odoratissimus	Traditional medicine	[22]
Pandanus amaryllifolius	Packaging application	[23]

Since *Pandanus amaryllifolius* plant has various applications and benefits, final products from extraction process or filtrate are taken out for further processes to leave *Pandanus amaryllifolius* fibre residue as a by-product. The remaining *Pandanus amaryllifolius* residues are usually unused and then discarded as waste. It is reported that about 1 kilogram of dried powder of pandan wangi can yield pandan extract of about 250 g [16]. Thus, if a huge scale of *Pandanus amaryllifolius* extract is produced this will contribute to the larger amount of pandan wangi waste that would be discarded. Moreover, since *Pandanus amaryllifolius* fibre is a source of lignocellulose which is abundant, can be obtained at low cost and is renewable, this would turn it into a suitable candidate for reinforcement agent in composite material application. Hence, it would be waste if the by-product is not fully utilized yet it has good potential to be recycled and able to produce to be another product.

A study reported by Adhamatika et al. [24], characterized physiochemical *Pandanus amaryllifolius* leaves in an attempt to convert the leaves to powder form for a variety of applications. Different parts of pandan leaves were taken; young and old leaves, and undergone three different drying methods; cabinet, vacuum and freeze-drying, prior turned into pandan powder. Although the investigation of physiochemical was conducted, the study has only focused on the non-cellulosic content in the pandan powder including ash content (%), chlorophyll content (mg/g), phenolic content (mg/g) and antioxidant (ppm) and no cellulosic information such as cellulose and hemicellulose content were demonstrated and no mechanical test conducted as the leaves were all converted into powder. Thus, the *Pandanus amaryllifolius* leaves characterization has not been well studied as source of fibre composite. Another study reported by Ooi et al. [25], focused on the potential development of new antiviral protein from purification and characterization of *Pandanus amaryllifolius* leaves. It was observed that the composition of lectin, designated Pandanin, possessed antiviral activities against human viruses namely herpes simplex virus type-1 (HSV-1). The extraction of lectin in the *Pandanus amaryllifolius* leaves was undertaken through a saline extract that underwent a few processes such as monium sulfate precipitation, affinity chromatography and gel filtration. It is interesting to acknowledge on the potential of Pandanin in *Pandanus amaryllifolius* could be a good candidate for a new class of anti-HIV or other antiviral agents. Therefore, it is clear that the reports of both Adhamatika et al. [24] and Ooi et al. [25] do not focus on the fibre characterization as the potential of fibre reinforcement.

Although, there are several studies reported on the potential of *Pandanus amaryllifolius* fibre as composite fillers application such as *Pandanus amaryllifolius* fibre in low-density polyethylene (LDPE) composite [23] and in epoxy resin composite [26]. However, to the best of our knowledge, there has been no research carried out on the characterization of *Pandanus amaryllifolius* fibre that comprehensively focused on its physicochemical and mechanical properties as a potential reinforcement agent. Hence, the study reported in this article focuses on the extraction and characterization of the physical, chemical and mechanical properties of *Pandanus amaryllifolius* fibre for possible utilization as fillers in polymer reinforcement matrix.

2. Materials and Methods

2.1. Materials

The *Pandanus amaryllifolius* (pandan wangi) leaves collected from rural area of Bahau (Negeri Sembilan, Malaysia) were used in this study. *Pandanus amaryllifolius* fibre (PAF) were extracted from fresh leaves. Samples were characterized in short length fibre form as shown in Figure 1h except for mechanical testing.

2.2. *Pandanus amaryllifolius* Fibre Extraction

The extraction of PAF involved in this study used the most common fibre extraction technique known as water retting. This technique was carried out by introducing moisture and chemical enzymatic reaction to extract the fibre in the pandan wangi leaves. Initially, bundles of pandan wangi leaves were cleaned and chopped into 12 cm × 3 cm pieces. Then the leaves were then placed in stagnant water to undergo water retting process for 8 weeks. Throughout the process, it was observed that the freshly collected pandan wangi leaves were green in colour at the beginning and changed to yellow-brownish colour after 8 weeks of water retting. The retted leaves were washed in running water and the pandan wangi fibres were removed by manual peeling. The extracted fibres were then cleaned and allowed to dry under direct sunlight for 1 day. After that, the extracted pandan wangi fibres were further dried in the oven because drying under the sunlight is not enough to remove all the moisture in extracted fibres. The fibres were then placed in the oven for 6 h at 80 °C. Some of the fibres were taken and ground using a disc mill to make the short fibre form. Both ground (short fibres) and unground fibres were kept in the zip-locked bag until further use. The images of different steps during the retting process are provided in Figure 1 below.

(**a**) (**b**) (**c**) (**d**) (**e**) (**f**) (**g**) (**h**)

Figure 1. Pictorial view of fibre extraction (**a**) *Pandanus amaryllifolius* plant; (**b**) cut bulk of leaves; (**c**) leaves cleaned and immersed in the water; (**d**) after 2 weeks; (**e**) skin degraded and removal of fibres; (**f**) extracted fibres collected and dried under sunlight; (**g**) ungrounded fibres after being further dried in oven; (**h**) ground (short fibres) PAF.

2.3. Characterization

2.3.1. Determination of Chemical Composition

The *Pandanus amaryllifolius* fibres (PAF) chemical composition was evaluated via neutral detergent fibre (NDF), acid detergent fibre (ADF) and acid detergent lignin (ADL), as well as ash content analysis. Using Equations (1) and (2), the cellulose and hemicellulose percentages can be determined respectively:

$$\text{Cellulose} = \text{ADF} - \text{ADL} \quad (1)$$

$$\text{Hemicellulose} = \text{NDF} - \text{ADF} \quad (2)$$

2.3.2. Fibre Length and Diameter

PAF fibre length was measured from the extracted fibre obtained after water retting and drying. Meanwhile, fibre diameter was evaluated under an optical microscope, Zeiss (Axiovert 200; Carl Zeiss Light Microscopy, Göttigen, Germany) that measured 15 individual fibre samples. The average diameter was randomly measured at three different positions of each image and the average value was determined.

2.3.3. Moisture Content

Five samples were prepared for the moisture content investigation. The weight of the samples before (W_i) was recorded prior to being heated in an oven for 24 h at 105 °C. The final weight after heating (W_f) was recorded after constant weight was obtained to ensure no moisture remained in the sample. The moisture content of the samples was determined using the following Equation (3):

$$\text{Moisture content } (\%) = \frac{W_i - W_f}{W_i} \times 100 \quad (3)$$

2.3.4. Scanning Electron Microscope (SEM)

Sample's morphology was observed by using a scanning electron microscope (SEM) machine, model Zeiss Evo 18 Research, (Jena, Germany) with an acceleration voltage of 10 kV.

2.3.5. Mechanical Testing

The ASTM D3379 (1998) guideline was adopted in the evaluation of the tensile properties of PAF by using an Instron universal testing machine (5556, Norwood, MA, USA) with 5 kN load cell capacity. Experimentation was performed with 30 mm gauge length and 1 mm/min crosshead speed. To ensure an adequate fibre fastening to the tensile machine, fibre was glued earlier to a rectangular structure of 20 mm in width and 50 cm in length as shown in Figure 2a and testing assembly is shown as Figure 2b.

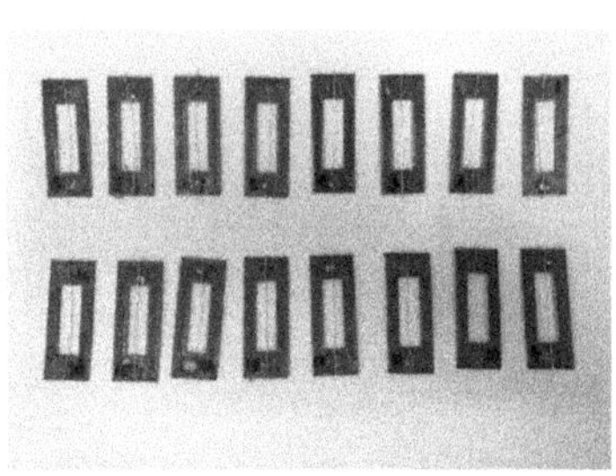

(a)

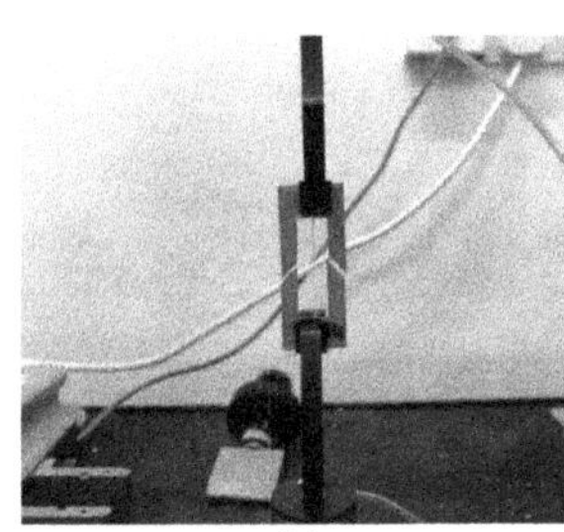

(b)

Figure 2. Tensile testing; (**a**) fastening to structure; (**b**) testing assembly.

2.3.6. Fourier-Transform Infrared (FTIR) Spectrometry

Fourier transform infrared (FT-IR) spectroscopy was used to investigate the intermolecular interaction and presence of functional group of PAF. Spectra of the material were obtained using JASCO FTIR-6100 Spectrometer (Japan). FT-IR spectra of the sample was collected in the range of 4000 to 500 cm^{-1} at resolution of 2 cm^{-1}.

2.3.7. Thermogravimetric Analysis (TGA)

Thermogravimetric analysis (TGA) was carried out using a Mettler-Toledo AG analyzer (Greifensee, Switzerland). The test was performed in a temperature range between 30 and 600 °C in inert (nitrogen) atmosphere heating rate of 20 °C/min with a flowrate 20 mL/min. A sample of 5–10 mg of the composite was heated in an alumina crucible pan. Before the thermal study, the samples were pre-conditioned for at least 48 h at 53%.

2.3.8. X-ray Diffraction (XRD) Analysis

The X-ray diffraction was conducted using a Rigaku D/max 2500 X-ray powder diffractometer (Rigaku, Tokyo, Japan) with Cu radiation run at 40 kV and 30 mA with 0.15406 nm light wavelength. The scanning rate of 2° min^{-1} in the range of diffraction angle 10° to 40° at room temperature was used to scan the samples. The crystallinity index (*CI*) was computed using the subsequent Segal expression, Equation (4):

$$CI = \frac{I_{002} - I_{am}}{I_{002}} \times 100\% \quad (4)$$

where I_{002} and I_{am} represent peak intensities of the crystalline and amorphous fractions, respectively. The crystallite size (*CS*) was determined using Scherrer's formula as shown in Equation (5),

$$CS = \frac{k\lambda}{\beta \cos\theta} \times 100\% \quad (5)$$

where k = 0.89 (Scherrer's constant), λ = 0.1541 nm is the radiation wavelength, β is the peak's full-width at half-maximum in radians, and θ is the corresponding Bragg angle.

3. Results and Discussion

3.1. Determination of Chemical Properties

Table 2 presents the chemical composition of *Pandanus amaryllifolius* fibre in comparison with other natural fibres. Mechanical properties and biodegradability of fibre depend on its chemical composition [27]. From Table 2, *Pandanus amaryllifolius* fibre shows value of cellulose content of 48.79% comparable to sugarcane fibre and *Arenga pinnata* fibre, 48% and 43.88% respectively, and shows lower cellulose content than other natural fibres such as *Calotropis gigantea, Ficus religiosa, Coccinia grandis,* bamboo, jute, sisal, kenaf and acacia tortilis with cellulose content of 64.47%, 55.58%, 63.22%, 73.83%, 66%, 65%, 72% and 61.89% respectively. Meanwhile, cassava bagasse, *Pandanus tectorius* and *Tridax procumbens* show lower cellulose value of 27%, 37.3% and 32% respectively. High cellulose content in fibre contributes to the enhancement of mechanical properties [11]. However, from the data tabulated in Table 2, *Pandanus amaryllifolius* fibre exhibits considerably low cellulose content that might be reflected in the low tensile value obtained in mechanical testing, besides the influence of fibre diameter. This also might be attributed to the presence of a high content of amorphous substances in *Pandanus amaryllifolius* fibre such as hemicellulose, lignin, ash and waxes [28].

Table 2. Comparative chemical composition of *Pandanus amaryllifolius* fibre and other natural fibres.

Fibre	Cellulose (%)	Hemicellulose (%)	Lignin (%)	Ash (%)	Ref.
Pandanus amaryllifolius	48.79	19.95	18.64	1.08	-
Cymbopogan citratus	37.56	29.29	11.14	4.28	[29]
Calotropis gigantea	64.47	9.64	13.56	3.13	[30]
Tridax procumbens	32.00	6.80	3.00	0.71	[31]
Ficus religiosa	55.58	13.86	10.13	4.86	[32]
Coccinia grandis	63.22	-	24.42	-	[33]
Bamboo	73.83	12.49	10.5	-	[34]
Sugarcane	48.00	14.60	12.10	12.10	[35]
Jute	66.00	17.00	12.50	-	[36]
Sisal	65.00	12.00	9.90	-	[36]
Kenaf	72.00	20.30	9.00	-	[36]
Cassava bagasse	27.00	30.00	2.70	-	[37]
Acacia tortilis	61.89	-	21.26	4.33	[27]
Arenga pinnata (sugar palm)	43.88	7.24	33.24	1.01	[38]
Pandanus tectorius	37.30	34.40	22.60	24.00	[11]

The biological role of hemicellulose is to strengthen plant cell walls by interaction with cellulose. *Pandanus amaryllifolius* fibre hemicellulose content of 19.95% is comparable to kenaf fibre, 20.3%, and jute fibre, 17%. It has relatively high hemicellulose content as compared to all other natural fibres except for kenaf, cassava bagasse and *Pandanus tectorius* with hemicellulose content of 20.3%, 30% and 34.4%, respectively. Besides that, the presence of hydrophobic substances such as lignin makes it less susceptible to moisture as well as providing better fibre stability and rigidity; however, it lowers the fibre tensile strength. The lignin content in *Pandanus amaryllifolius* fibre was 18.64% making it comparable to *Acacia tortilis*, 21.26% and *Pandanus tectorius*, 22.6%. It is also seen to have higher lignin content than bamboo, sugarcane, jute, sisal, kenaf and cassava bagasse at 10.5%, 12.1%, 12.5%, 9.9%, 9% and 2.7% respectively. Ash content in *Pandanus amaryllifolius* fibre is recorded at 1.08%, considered as among of the lowest in comparison to other natural fibres tabulated in Table 2 which determines it has less susceptibility to moisture when applied in a humid environment. The value is in good agreement with the result obtained from moisture content testing.

All chemical compositions exhibited in every plant natural fibre have their main functionality and role in maintaining the rigidity of the plant structure and they complement every constituent. However, it is worth noting that in filler composite application, the amount of cellulose in fibre becomes the main criteria that might influence the mechanical properties of the resulting composite.

3.2. Determination of Physical Properties

A total of 15 samples of *Pandanus amaryllifolius* fibre was picked randomly to determine the average fibre length. The extracted fibre were chosen in dried condition after undergoing the drying process as presented in Figure 1g. The average ultimate fibre length was found in this study to be 45.07 $\pm$ 4.81 mm long (coefficient of variation, CV = 0.11). The coefficient of variation value shows the variability in the individual reading is small meaning the sample is fairly homogenous. In general, fibre physical properties are varied depending on the climate, environmental condition, age and origin of fibre [11]. Besides, the *Pandanus amaryllifolius* fibres were collected from rural areas living alongside other wild plants as displayed in Figure 1a. Physically, it can be observed that the collected fibre is bigger and longer in size compared to *Pandanus amaryllifolius* fibres that are cultivated in residential areas.

Table 3 presents the diameter and moisture content of *Pandanus amaryllifolius* fibre compared to other natural fibres. Fibre diameter plays a vital role in determining fracture surface area thus reflecting the tensile value of the material in determining mechanical properties of the fiber. Thus, the diameter measurement of *Pandanus amaryllifolius* fibre is conducted under an optical microscope as displayed in Figure 3. Fibre average diameter was found to be 368.57 ± 50.47 μm which is comparable to the diameter of *Cissusquadrangularis* root (350 μm) and *Coccinia grandis* (543 ± 75 μm).

Table 3. Comparative physical properties of *Pandanus amaryllifolius* fibre and other natural fibres.

Fibre	Diameter (μm)	Moisture (%)	Ref.
Pandanus amaryllifolius	368.57 ± 50.47	6.00 ± 0.13	-
Cymbopogan citratus	326.67 ± 45.77	5.20 ± 2.28	[29]
Calotropis gigantea	32.70	7.27	[30]
Tridax procumbens	233.1 ± 9.9	11.20	[31]
Juncus effuses L.	280 ± 56	-	[39]
Ficus Religiosa	25.62	9.33	[32]
Date palm rachis	88 ± 12	-	[40]
Coccinia grandis	543 ± 75	9.14	[33]
Cissusquadrangularis root	350	7.30	[41]
Bamboo	-	7.00	[34]
Acacia tortilis	480	6.47	[27]
Arenga pinnata (sugar palm)	212.01 ± 2.17	8.36	[38]

(Note: Coefficient of variations of fibre diameter for *Pandanus amaryllifolius* fibre, CVFD = 0.14).

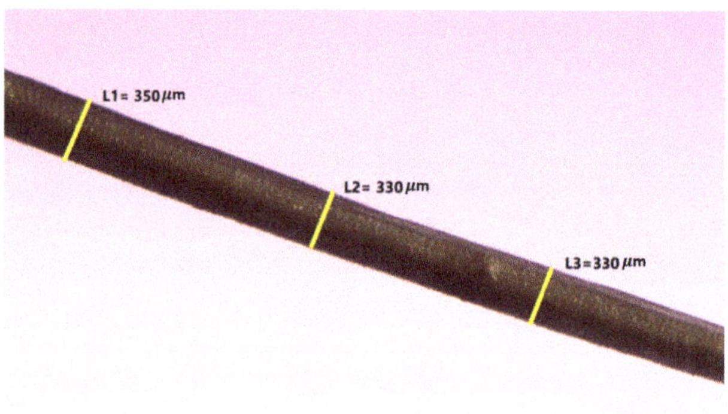

Figure 3. Optical microscopy image of the *Pandanus amaryllifolius* fibre.

Furthermore, moisture content of *Pandanus amaryllifolius* fibre was the lowest among natural fibres as shown in Table 3. However, the moisture content of *Pandanus amaryllifolius* fibre is seen to be the lowest among others and comparable to *Acacia tortilis* and bamboo at 6.47% and 7.00%, respectively. Meanwhile, the moisture content of *Tridax procumbens* fibre showed the highest which at 11.20%. Moisture content plays an important role in selecting the best properties of polymer products such as plastic packaging applications. Thus, the lower moisture content is preferable as it reflects the hydrophobicity of the fibre material.

3.3. Mechanical Testing

A total of 15 samples of fibre were chosen randomly to be tested under tensile test at a gauge length of 30 mm before obtaining the mechanical properties of the *Pandanus amaryllifolius* fibre. Figure 4 illustrates a typical stress-strain curve behaves quasi-linearly presented an initial elastic region with a strong slope, ($\varepsilon \leq 1.5\%$), continued with non-linear elastic region between 2% to 5% strain and reached final softening phase up to 8.3% maximum strain value. Besides, the elastic modulus was calculated from the elastic region values raging between 0.5% and 1.5% as shown in Figure 4. In general, the fibre demonstrated brittle properties due to the sudden drop in stress value in the stress-strain curve which is in good agreement with other reported studies on natural fibres such as *Tridax procumbens* [31], *Juncus effsus* L. [39] and *Agave americana* L. [42]. Typical yield point and breaking stress of *Pandanus amaryllifolius* fibre is labelled in Figure 4 and the elastic region is found to be smaller than the plastic region which is similar to the stress-strain curve obtained for *Juncus effsus* L. [39], *Agave americana* L. [42] and *Vernonia eleagnifolia* [43]. Unlike synthetic fibre, where the size and properties of the produced fibre can be controlled during the manufacturing process before being applied, natural fibre exhibits variation in diameter for every single fibre strand obtained. Therefore, the average value of mechanical tensile properties is displayed in Table 4 with standard deviation to show the variability in individual reading being small, and the material is fairly homogenous compared to tensile results extracted from different research studies.

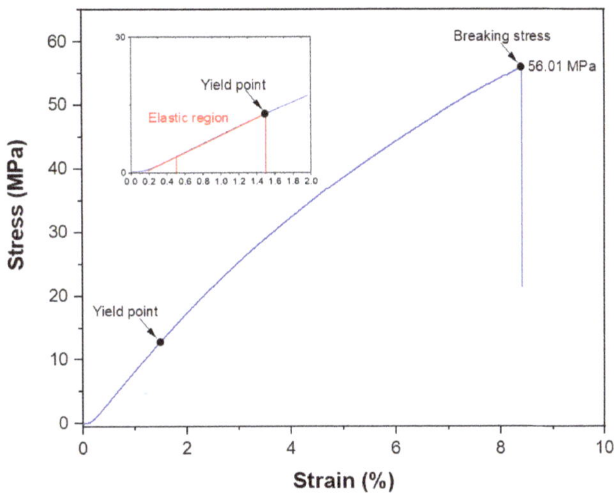

Figure 4. Typical stress-strain curve for *Pandanus amaryllifolius* fibre with magnification of elastic region.

Table 4. Comparison tensile properties of *Pandanus amaryllifolius* fibre with other natural fibre.

Fibre	Tensile Strength, σ (MPa)	Tensile Modulus, E (GPa)	Elongation at Break, ε (%)	Ref.
Pandanus amaryllifolius	45.61 ± 16.09	0.41 ± 0.18	8.17 ± 3.84	-
Cymbopogan citratus	43.81 ± 15.27	1.04 ± 0.33	0.84 ± 0.28	[29]
Tridax procumbens	25.75	0.94 ± 0.09	2.77 ± 0.27	[31]
Acacia tortilis	71.63	4.21	1.33	[27]
Juncus effuses L.	113 ± 36	4.38 ± 1.37	2.75 ± 0.6	[39]
Vernonia eleagnifolia	259.6	37.8	6.9	[43]
Ficus Religiosa	433.32 ± 44	5.42 ± 2.6	8.74 ± 1.8	[32]
Coccinia grandis	424.24	26.52	16.00	[33]
Date palm	530.5	21.9	3.6	[40]
Bamboo	583.0	25.5	2.1	[44]
Kenaf	393–773	26.5	1.6	[36]
Sisal	511–635	2.2–9.4	1.5	[36]
Kenaf	930.0	53.0	1.5	[36]

Sanyang et al. [12] found that the high cellulose content of fibre has a strong influence on their mechanical properties and different parts of the plant contribute to the different amounts of cellulose content from the fibre extracted. The study mentioned that sugar palm fibre extracted from the frond part contained the highest cellulose content (66.49%) as compared to bunch (61.76%) and trunk (40.56%) which leads to the highest fibre strength being found at the frond part (421.4 MPa) rather than the bunch (365.1 MPa) and trunk (198.3 MPa). Therefore, this could relate from data tabulated in Tables 2 and 4, since

Pandanus amaryllifolius is a leaf plant and tensile strength obtained is considerably lower than others which might be due to the cellulose content of fibre itself. A similar finding has also been reported on the enhancement of mechanical properties as the cellulose content increased after chemical treatment on fibre was conducted [27]. In general, from Table 4, the average tensile strength, elastic modulus, and elongation value of *Pandanus amaryllifolius* was found to be 45.61 ± 16.09 MPa, 0.41 ± 0.18 GPa, and 8.17 ± 3.84%, respectively. The tensile value found in the present work is higher than *Cymbopogan citratus*, 43.81 ± 15.27 MPa [29] and *Tridax procumbens*, 25.75 MPa [31] which might be attributed to higher cellulose content found in *Pandanus amaryllifolius* fibre (48.79%) as compared to *Cymbopogan citratus* (37.56%) and *Tridax procumbens* (32.00%). Meanwhile, Dawit et al. [27] reported better tensile strength of *Acacia tortilis*, 71.60 MPa than *Pandanus amaryllifolius* which also could relate to higher cellulose content found at 61.89%.

In the meantime, elastic modulus calculated from this study demonstrated an almost similar value to 0.94 ± 0.09 GPa obtained from fibre named *Tridax procumbens* as reported by Vijay et al. [31] and slightly lower than 1.04 ± 0.33 GPa reported for *Cymbopogan citratus* [29]. Moreover, *Pandanus amaryllifolius* fibre elongation at break is found to be an almost similar value of 8.74 ± 1.8% to *Ficus Religiosa* fibre studied by Arul et al. [32] while remaining higher than other fibres such as *Cymbopogan citratus* (0.84 ± 0.28%), *Tridax procumbens* (2.77 ± 0.27%), *Acacia tortilis* (1.30%), *Juncus effuses* L. (2.75 ± 0.6%), *Vernonia eleagnifolia* (6.9%) and date palm (3.6%). In general, although the overall mechanical properties of *Pandanus amaryllifolius* fibre as tabulated in Table 4 is averagely lower than other lignocellulosic fibres, it still can be considered as alternative reinforcement especially for a bio-based polymer matrix which is typically use for low strength application such as biodegradable packaging material.

3.4. Scanning Electron Microscope (SEM)

The SEM of PAF at magnification of 100× and 300× are shown in Figure 5a,b, respectively. In general, the surface micrograph of PAF is not smooth and is uneven with the deposition of other substances. From the presented longitudinal image at 100× magnificent, it is observed that the fibre structure exhibits vertical alignments in the same direction of the fibre axis. This is might be attributed to the structure of PAF that is composed of microfibrils, compactly aligned and bonded together by lignin, pectin and other non-cellulosic materials [4]. When the image is zoomed in at 300× magnification, Figure 5b, the structure of microfibrils can be seen and further prove the presence of lignin that is responsible for cementing cellulose microfibrils and protecting the structure. Similar results were reported by other authors [38,45]. Other than that, we also found shapes of regularly distributed semi-rectangular trays exist together along with PAF surface lines. The phenomenon might be exhibited in the deposition of amorphous substances including hemicellulose and lignin and might also be impurities that present during the binding process [40]. Nevertheless, the serration and uneven fibre surface somehow can help improve mechanical performance. It helps in enhancing the fibre-matrix interface in polymeric composite material, locking the fibre mechanically in the polymeric resin [45].

3.5. Fourier-Transform Infrared (FTIR) Spectrometry

The FTIR curve of *Pandanus amaryllifolius* fibre is displayed in Figure 6. The broadband and peak occurred at 3332 cm^{-1} reflecting water adsorption and the presence of the hydroxyl (O–H) group in the fibre which also indicates the existence of cellulose, lignin and water [45]. The presence of water in the fibre was revealed by moisture content tests. Meanwhile, the absorbance peak in the 2919 cm^{-1} and 1732 cm^{-1} represents the stretching of alkanes (C–H) and carboxyl group (C=O) of both cellulose and hemicellulose [38]. The presence of (C=C) stretch from aromatic lignin is observed at wave peaks of 1672 cm^{-1}. Besides that, the peak range of 1149–1414 cm^{-1} in the *Pandanus amaryllifolius* fibre spectrum is associated to the (C=O) stretching vibration of ester linkage of the carboxylic group of ferulic and p-coumaric acids of hemicellulose and lignin [46]. Constituents of polysaccharide in cellulose confirm the stretching vibration of (C–O) and (O–H) that occurred at the radiation of 1023 cm^{-1}. FTIR analysis verified the existence of cellulose and hemicellulose in the fibre's structural composition.

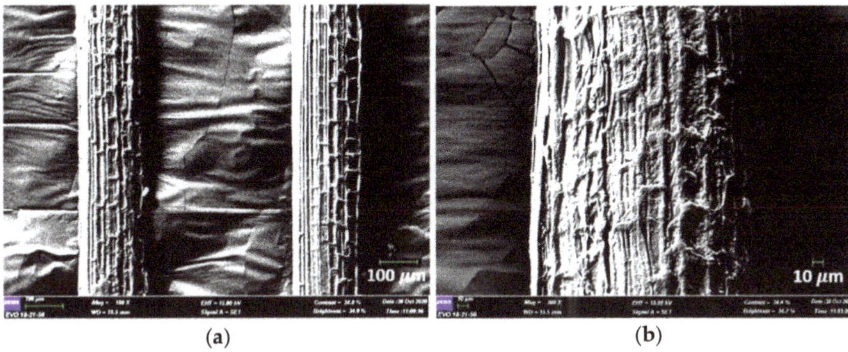

Figure 5. Longitudinal view of *Pandanus amaryllifolius* (pandan wangi) fibre by scanning electron microscope (SEM); (**a**) 100× magnification; (**b**) 300× magnification.

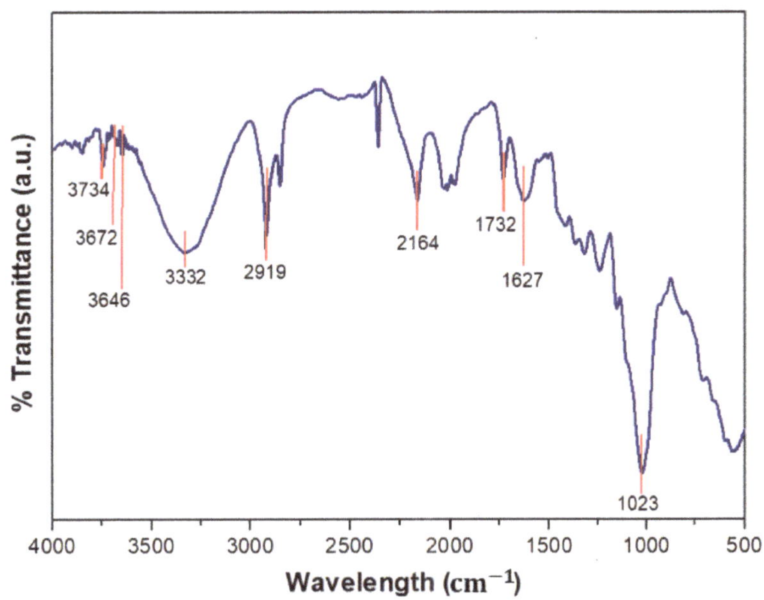

Figure 6. Fourier transform infrared (FTIR) spectra of *Pandanus amaryllifolius* fibre.

3.6. X-ray Diffraction Measurement

The diffraction pattern for *Pandanus amaryllifolius* fibre is shown in Figure 7 and the presence of three peaks can be observed. The highest intensity peak is found at $2\theta = 23.16°$, average density peak at $2\theta = 16.78°$ and the lowest intensity peak occurred at $2\theta = 34.69°$, which are denoted as the (002), (110) and (040) crystallographic planes, respectively [40]. The highest peak intensity (002) corresponds to the presence of semi-crystalline structure cellulose type I, associated to the parallel chain orientation with intra-sheet hydrogen bonds [47]. Meanwhile, the diffracted peak at intensity (110) proves the amorphous fraction in *Pandanus amaryllifolius* fibre [41]. This result is in corroboration with chemical and FTIR analysis results that confirm the presence of cellulose and amorphous substances.

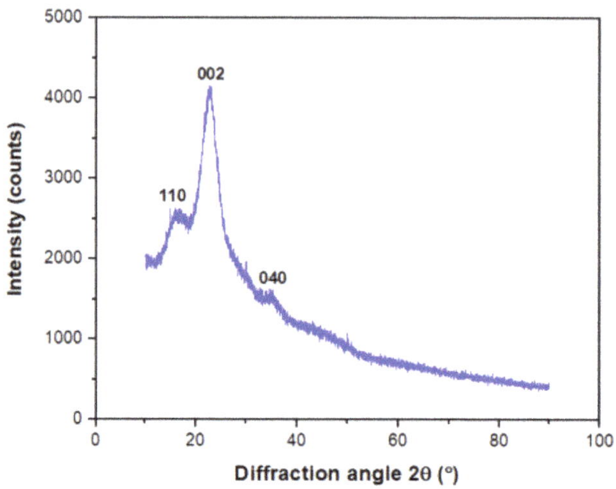

Figure 7. X-ray diffraction pattern of *Pandanus amaryllifolius* fibre.

High crystallinity values stipulate excellent mechanical properties of the fibre which can be identified from the sharpness of the diffraction peak where the sharper diffraction peak indicates a high crystallinity degree of the structure that might be influenced by the cellulose rigidity [48]. Segal's peak equation Equation (4) is employed to find the crystallinity index (CI) from the peaks found. The CI for *Pandanus amaryllifolius* fibre are calculated found to be 37.09% and 4.95 nm. This CI value is comparable to *Calotropis gigantea* (36.00%), higher than *Tridax procumbens* (34.46%) and *Juncus effuses* L. (33.40%), but lower than *Ficus religiosa* (42.92), *Pandanus tectorius* (55.10%), Cissusquadrangularis root (56.60%), pigeon pea plant (65.89%) and date palm rachis (69.77%).

Crystalline size (CS) for *Pandanus amaryllifolius* fibre is calculated using Segal's Equation (5) and found to be 4.95 nm. This value is within crystallite range size reported for other natural fibres such as *Ficus religiosa* (5.18 nm), *Tridax procumbens* (25.04 nm) and *Juncus effuses* L. (3.60 nm). The crystalline property of *Pandanus amaryllifolius* fibre is compared with the tested natural fibres shown in Table 5.

Table 5. Comparison table of crystalline properties of *Pandanus amaryllifolius* fibre with other natural fibres.

Fiber	C.I (%)	C.S (nm)	References
Pandanus amaryllifolius	37.09	4.95	-
Cymbopogan citratus	35.20	-	[29]
Calotropis gigantea	36.00	-	[30]
Tridax procumbens	34.46	25.04	[31]
Juncus effuses L.	33.40	3.60	[39]
Ficus Religiosa	42.92	5.18	[32]
Pandanus tectorius	55.10	-	[11]
Cissusquadrangularis root	56.60	7.04	[41]
Pigeon pea plant	65.89	11.60	[49]
Date palm rachis	69.77	5.63	[40]

3.7. Thermogravimetric Analysis

Figure 8 shows thermogravimetric analysis for *Pandanus amaryllifolius* fibre, both TGA and DTG curves were plotted as a function of temperature. This thermal analysis is used to measure weight degradation as the material is heated. In general, three degradation phases can be observed in Figure 8 which comprises small initial weight loss at the temperature range of 30–120 °C, major weight loss as the temperature rises from 220–420 °C, and final stage degradation afterwards until the maximum temperature was reached. The first stage with 7.84% weight loss corresponds to the evaporation of adsorbed moisture which is attributed to the hydrophilic nature of lignocellulosic materials [11,40]. After that, thermal stability is noted up to about 210 °C which reflects no appearance of significant peaks in both TGA and DTG curves showing that the work is in good agreement [39]. Then, major weight loss is observed, 67.70%, associated to the decomposition of glycosidic linkage cellulose and hemicellulose substances at the temperature range of 220–420 °C [41]. It is confirmed by the observation of the DTG peak curve at 370 °C which indicates the possible thermal decomposition of cellulose I and completion of α-cellulose decomposition [33]. Lastly, the final stage of degradation is associated to the fragmentation of waxen substances such as lignin and observed at temperatures 420–500 °C with small weight loss of 11.32%. Last but not least, char residue is spotted at 13.14% at the highest temperature of 600 °C. This char residue is mainly composed of carbon or carbonaceous material that cannot be further degraded into smaller volatile fragments and will remain in the combustion chamber until the process ends. It also might be attributed to the presence of inorganic matter that yields chars and may form the basis of quantitative estimation. Boumediri et al. [40] and Maache et al. [39] reported on char residue left at the end of combustion of an amount of 3.64% for *Juncus effuses* L. fibre and 18.29% for date palm fibre, respectively.

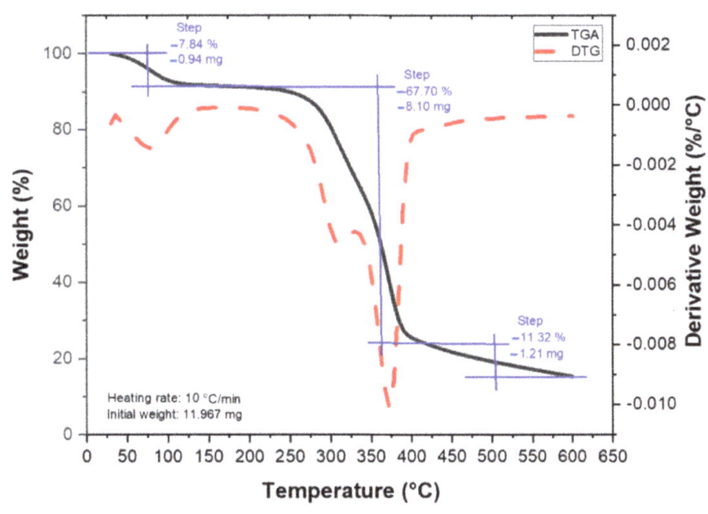

Figure 8. Thermogravimetric analysis (TGA) and DTG curves for *Pandanus amaryllifolius* fibre.

4. Conclusions

The *Pandanus amaryllifolius* fibres were successfully extracted from plants via a water retting extraction process. The *Pandanus amaryllifolius* fibres were characterized for physical, thermal, chemical and mechanical properties. The chemical composition analysis revealed that *Pandanus amaryllifolius* fibre has cellulose, hemicellulose, and lignin content of 48.79%, 19.95% and 18.64%, respectively. FT-IR and XRD analysis confirmed the presence of cellulose and amorphous substances in the fibre. The tensile strength of the fibre was 45.61 ± 16.09 MPa for an average fibre diameter of 368.57 ± 50.47 µm. Moisture content analysis indicated 6.00% moisture content of *Pandanus amaryllifolius* fibre. The thermogravimetric analysis showed the thermal stability of *Pandanus amaryllifolius* fibre up to 210 °C, which is within polymerization process temperature conditions. Overall, the finding shows that *Pandanus amaryllifolius* fibre may be used as alternative reinforcement particularly for a bio-based polymer matrix. Furthermore, utilization of *Pandanus amaryllifolius* fibre as the reinforcing agent will lead to greater development of more biopolymer products since this natural fibre can be reinforced with a synthetic and natural polymer matrix.

Author Contributions: Conceptualization, Z.N.D. and R.J.; methodology, Z.N.D. and F.A.M.Y.; formal analysis, Z.N.D.; R.J.; M.Z.S. and F.A.M.Y.; investigation, Z.N.D.; writing—original draft preparation, Z.N.D.; writing—review and editing, Z.N.D.; R.J.; R.H.A. and M.Z.S.; visualization, Z.N.D.; supervision, R.J.; project administration, Z.N.D.; funding acquisition, R.J. and R.H.A. All authors have read and agreed to the published version of the manuscript.

Funding: The research work was funded by FRGS RACER research grant RACER/2019/FTKMP-CARE/F00413 and the Article Processing Charge was funded by Universiti Malaysia Sabah.

Institutional Review Board Statement: Not applicable.

Data Availability Statement: The data presented in this study are available on request from the corresponding author.

Acknowledgments: The authors would like to thank Ministry of Higher Education Malaysia, Universiti Teknikal Malaysia Melaka for the financial support through grant number RACER/2019/FTKMP-CARE/F00413, Universiti Kuala Lumpur (UniKL MICET) for the experimental support and APC was funded by Universiti Malaysia Sabah.

Conflicts of Interest: The authors declare no conflict of interest.

References

1. Kumar, S.S.; Hiremath, S.S. *Natural Fiber Reinforced Composites in the Context of Biodegradability: A Review*; Elsevier Inc.: Amsterdam, The Netherlands, 2020; Volume 5, ISBN 9780128035818.
2. Satyanarayana, K.G.; Arizaga, G.G.C.; Wypych, F. Biodegradable composites based on lignocellulosic fibers—An overview. *Prog. Polym. Sci.* **2009**, *34*, 982–1021. [CrossRef]
3. Bertolino, V.; Cavallaro, G.; Milioto, S.; Lazzara, G. Polysaccharides/Halloysite nanotubes for smart bionanocomposite materials. *Carbohydr. Polym.* **2020**, *245*, 116502. [CrossRef]
4. Donaldson, L.; Nanayakkara, B.; Harrington, J.; Zealand, N. *Wood Growth and Development*, 2nd ed.; Elsevier: Amsterdam, The Netherlands, 2017; Volume 1, ISBN 9780123948076.
5. Izani, M.A.N.; Paridah, M.T.; Anwar, U.M.K.; Nor, M.Y.M.; H'ng, P.S. Effects of fiber treatment on morphology, tensile and thermogravimetric analysis of oil palm empty fruit bunches fibers. *Compos. Part B* **2013**, *1*, 1251–1257. [CrossRef]
6. Kaisangsri, N.; Kerdchoechuen, O.; Laohakunjit, N. Biodegradable foam tray from cassava starch blended with natural fiber and chitosan. *Ind. Crop. Prod.* **2012**, *37*, 542–546. [CrossRef]
7. Jumaidin, R.; Afif, M.; Khiruddin, A.; Asyul, Z.; Saidi, S.; Sapuan, S.M.; Ilyas, R.A. Effect of cogon grass fibre on the thermal, mechanical and biodegradation properties of thermoplastic cassava starch biocomposite. *Int. J. Biol. Macromol.* **2020**, *146*, 746–755. [CrossRef]
8. Ilyas, R.A.; Sapuan, S.M.; Ibrahim, R.; Syafri, E.; Asrofi, M.; Herlina, N.; Jumaidin, R. Effect of sugar palm nanofibrillated cellulose concentrations on morphological, mechanical and physical properties of biodegradable films based on agro-waste sugar palm (*Arenga pinnata* (Wurmb.) Merr) starch. *Integr. Med. Res.* **2019**, *8*, 4819–4830. [CrossRef]
9. Kaisangsri, N.; Kowalski, R.J.; Kerdchoechuen, O.; Laohakunjit, N.; Ganjyal, G.M. Cellulose fiber enhances the physical characteristics of extruded biodegradable cassava starch foams. *Ind. Crop. Prod.* **2019**, *142*, 111810. [CrossRef]
10. Yusof, F.M.; Wahab, N.A.; Rahman, N.L.A.; Kalam, A.; Jumahat, A.; Taib, C.F.M. Properties of treated bamboo fiber reinforced tapioca starch biodegradable composite. *Mater. Today Proc.* **2019**, *16*, 2367–2373. [CrossRef]
11. Sheltami, R.M.; Abdullah, I.; Ahmad, I.; Dufresne, A.; Kargarzadeh, H. Extraction of cellulose nanocrystals from mengkuang leaves (*Pandanus tectorius*). *Carbohydr. Polym.* **2012**, *88*, 772–779. [CrossRef]
12. Sanyang, M.L.; Sapuan, S.M.; Jawaid, M.; Ishak, M.R.; Sahari, J. Recent developments in sugar palm (*Arenga pinnata*) based biocomposites and their potential industrial applications: A review. *Renew. Sustain. Energy Rev.* **2016**, *54*, 533–549. [CrossRef]
13. Weerappuliarachchi, J.W.M.E.S.; Perera, I.C.; Gunathilake, S.S.; Thennakoon, S.K.S. Synthesis of cellulose microcrystals (CMC)/nylon 6, 10 composite by incorporating CMC isolated from *Pandanus ceylanicus*. *Carbohydr. Polym.* **2020**, *241*, 116227. [CrossRef]
14. Bhattacharjee, P.; Kshirsagar, A.; Singhal, R.S. Supercritical carbon dioxide extraction of 2-acetyl-1-pyrroline *from Pandanus amaryllifolius* Roxb. *Food Chem.* **2005**, *91*, 255–259. [CrossRef]
15. Watke, K.; Nadaf, A.; Thengane, R.; Jawali, N. *Pandanus amaryllifolius* Roxb. cultivated as a spice in coastal regions of India. *Genet. Resour. Crop Evol.* **2009**, *56*, 735–740. [CrossRef]
16. Awad, E.; Austin, D.; Lyndon, A.; Awaad, A. Possible effect of hala extract (*Pandanus tectorius*) on immune status, anti-tumour and resistance to Yersinia ruckeri infection in rainbow trout (*Oncorhynchus mykiss*). *Fish Shellfish Immunol.* **2019**, *87*, 620–626. [CrossRef]
17. Saenthaweesuk, S.; Naowaboot, J.; Somparn, N. *Pandanus amaryllifolius* leaf extract increases insulin sensitivity in high-fat diet-induced obese mice. *Asian Pac. J. Trop. Biomed.* **2016**, *6*, 866–871. [CrossRef]
18. Owolabi, L.; Sri, P.; Megat-yusoff, M.; Mohamad, Z.; Syahmi, M. Fabrication of pandanus tectorius (screw-pine) natural fiber using vacuum resin infusion for. *Integr. Med. Res.* **2019**, *8*, 3102–3113. [CrossRef]

19. Jaffur, N.; Jeetah, P. Production of low cost paper from Pandanus utilis fibres as a substitution to. *Sustain. Environ. Res.* **2019**, *6*, 20. [CrossRef]
20. Herlina, N.; Sanjay, M.R.; Arpitha, G.R.; Iulian, C.; Siengchin, S. Synthesis and properties of pandanwangi fiber reinforced polyethylene composites: Evaluation of dicumyl peroxide (DCP) effect. *Compos. Commun.* **2019**, *15*, 53–57. [CrossRef]
21. Johan, A.B.; Ratnawati, D.; Purnomo, S. Constructive learning through experiments utilization of pandan leaf fiber as a strengthening of composite material. *J. Phys. Conf. Ser.* **2019**, *1456*, 19. [CrossRef]
22. Adkar, P.P.; Bhaskar, V.H. *Pandanus odoratissimus* (Kewda): A Review on Ethnopharmacology, Phytochemistry, and Nutritional Aspects. *Adv. Pharmacol. Sci.* **2014**, *2014*, 19. [CrossRef]
23. Rahman, W.A.; Nur, S.; Sudin, A.; Din, S.N. Physical and mechanical properties of *Pandanus amaryllifolius* fiber reinforced low density polyethylene composite for packaging application. In Proceedings of the International Symposium on Humanities, Science & Engineering Research, Kuala Lumpur, Malaysia, 24 June 2012; pp. 345–349.
24. Adhamatika, A.; Murtini, E.S.; Sunarharum, W.B. The effect of leaf age and drying method on physico-chemical characteristics of pandan (*Pandanus amaryllifolius* Roxb.) leaves powder. In Proceedings of the International Conference on Green Agro-Industry and Bioeconomy, Malang, Indonesia, 25 August 2020.
25. Ooi, L.S.M.; Sun, S.S.M.; Ooi, V.E.C. Purification and characterization of a new antiviral protein from the leaves of *Pandanus amaryllifolius* (Pandanaceae). *Int. J. Biochem. Cell Biol.* **2004**, *36*, 1440–1446. [CrossRef]
26. Sembiring, T.; Sinuhaji, P.; Silitonga, R.Y. Manufacture and Characterization of Pandan Wangi (*Pandanus amaryllifolius* Roxb) Fiber-Based Composite Board with Epoxy Resin. *J. Technomater. Phys.* **2020**, *2*, 1–6. [CrossRef]
27. Dawit, J.B.; Regassa, Y.; Lemu, H.G. Property characterization of acacia tortilis for natural fi ber reinforced polymer composite. *Results Mater.* **2020**, *5*, 100054. [CrossRef]
28. Campos, A.; Sena Neto, A.R.; Rodrigues, V.B.; Luchesi, B.R.; Mattoso, L.H.C.; Marconcini, J.M. Effect of raw and chemically treated oil palm mesocarp fibers on thermoplastic cassava starch properties. *Ind. Crop. Prod.* **2018**, *124*, 149–154. [CrossRef]
29. Kamaruddin, Z.H.; Jumaidin, R.; Ilyas, R.A.; Selamat, M.Z.; Alamjuri, R.H. Characterization of Natural Cellulosic Fiber Isolated from Malaysian *Cymbopogan citratus* Leaves. *BioResources* **2021**, *16*, 7729–7750. [CrossRef]
30. Narayanasamy, P.; Balasundar, P.; Senthil, S.; Sanjay, M.R.; Siengchin, S.; Khan, A.; Asiri, A.M. Characterization of a novel natural cellulosic fiber from *Calotropis gigantea* fruit bunch for ecofriendly polymer composites. *Int. J. Biol. Macromol.* **2020**, *150*, 793–801. [CrossRef] [PubMed]
31. Vijay, R.; Singaravelu, D.L.; Vinod, A.; Sanjay, M.R.; Siengchin, S.; Jawaid, M.; Khan, A.; Parameswaranpillai, J. Characterization of raw and alkali treated new natural cellulosic fibers from *Tridax procumbens*. *Int. J. Biol. Macromol.* **2019**, *125*, 99–108. [CrossRef] [PubMed]
32. Moshi, A.A.M.; Ravindran, D.; Bharathi, S.R.S.; Indran, S.; Saravanakumar, S.S.; Liu, Y. Characterization of a new cellulosic natural fiber extracted from the root of *Ficus religiosa* tree. *Int. J. Biol. Macromol.* **2020**, *142*, 212–221. [CrossRef]
33. Demash, H.D.; Miyake, G. The effect of plasticizers on thermoplastic starch films developed from the indigenous Ethiopian tuber crop Anchote (*Coccinia abyssinica*) starch. *Int. J. Biol. Macromol.* **2020**, *155*, 581–587. [CrossRef]
34. Muhammad, A.; Rahman, M.R.; Hamdan, S.; Sanaullah, K. Recent developments in bamboo fiber-based composites: A review. *Polym. Bull.* **2019**, *76*, 2655–2682. [CrossRef]
35. Fitch-Vargas, P.R.; Camacho-Hernández, I.L.; Martínez-Bustos, F.; Islas-Rubio, A.R.; Carrillo-Cañedo, K.I.; Calderón-Castro, A.; Jacobo-Valenzuela, N.; Carrillo-López, A.; Delgado-Nieblas, C.I.; Aguilar-Palazuelos, E. Mechanical, physical and microstructural properties of acetylated starch-based biocomposites reinforced with acetylated sugarcane fiber. *Carbohydr. Polym.* **2019**, *219*, 378–386. [CrossRef]
36. Kumar, P.S.S.; Allamraju, K.V. A review of natural fiber composites [Jute, Sisal, Kenaf]. *Mater. Today Proc.* **2019**, *18*, 2556–2562. [CrossRef]
37. Paula, A.; Lamsal, B.; Luiz, W.; Magalhães, E.; Mottin, I. Cassava starch films reinforced with lignocellulose nano fi bers from cassava bagasse. *Int. J. Biol. Macromol.* **2019**, *139*, 1151–1161. [CrossRef]
38. Ilyas, R.A.; Sapuan, S.M.; Ibrahim, R.; Abral, H.; Ishak, M.R.; Zainudin, E.S.; Asrofi, M.; Atikah, M.S.N.; Huzaifah, M.R.M.; Radzi, A.M.; et al. Sugar palm (*Arenga pinnata* (Wurmb.) Merr) cellulosic fibre hierarchy: A comprehensive approach from macro to nano scale. *J. Mater. Res. Technol.* **2019**, *8*, 2753–2766. [CrossRef]
39. Maache, M.; Bezazi, A.; Amroune, S.; Scarpa, F.; Dufresne, A. Characterization of a novel natural cellulosic fiber from *Juncus effusus* L. *Carbohydr. Polym.* **2017**, *171*, 163–172. [CrossRef]
40. Boumediri, H.; Bezazi, A.; Garcia, G.; Pino, D.; Haddad, A. Extraction and characterization of vascular bundle and fiber strand from date palm rachis as potential bio-reinforcement in composite. *Carbohydr. Polym.* **2019**, *222*, 114997. [CrossRef] [PubMed]
41. Indran, S.; Raj, R.E.; Sreenivasan, V.S. Characterization of new natural cellulosic fiber from *Cissus quadrangularis* root. *Carbohydr. Polym.* **2014**, *110*, 423–429. [CrossRef]
42. Bezazi, A.; Belaadi, A.; Bourchak, M.; Scarpa, F.; Boba, K. Novel extraction techniques, chemical and mechanical characterisation of *Agave americana* L. natural fibres. *Compos. Part B Eng.* **2014**, *66*, 194–203. [CrossRef]
43. Wang, F.; Shao, J.; Keer, L.M.; Li, L.; Zhang, J. The effect of elementary fibre variability on bamboo fibre strength. *Mater. Des.* **2015**, *75*, 136–142. [CrossRef]
44. Jhon, C.; Martha, L.S.; Pati, W. Physical-mechanical properties of bamboo fibers-reinforced biocomposites: Influence of surface treatment of fibers. *J. Build. Eng.* **2020**, *28*, 101058. [CrossRef]

45. Shaker, K.; Muhammad, R.; Ullah, W.; Jabbar, M.; Umair, M.; Tariq, A.; Kashif, M.; Nawab, Y. Industrial Crops & Products Extraction and characterization of novel fi bers from *Vernonia elaeagnifolia* as a potential textile fi ber. *Ind. Crop. Prod.* **2020**, *152*, 112518. [CrossRef]
46. Hazrati, K.Z.; Sapuan, S.M.; Zuhri, M.Y.M.; Jumaidin, R. Extraction and Characterization of Potential Biodegradable Materials Based on Dioscorea hispida Tubers. *Polymers* **2021**, *13*, 584. [CrossRef] [PubMed]
47. Prado, K.S.; Spinacé, M.A.S. Isolation and characterization of cellulose nanocrystals from pineapple crown waste and their potential uses. *Int. J. Biol. Macromol.* **2019**, *122*, 410–416. [CrossRef]
48. Alemdar, A.; Sain, M. Isolation and characterization of nanofibers from agricultural residues—Wheat straw and soy hulls. *Bioresour. Technol.* **2008**, *99*, 1664–1671. [CrossRef] [PubMed]
49. Kulandaivel, N.; Muralikannan, R.; KalyanaSundaram, S. Extraction and Characterization of Novel Natural Cellulosic Fibers from Pigeon Pea Plant. *J. Nat. Fibers* **2020**, *17*, 769–779. [CrossRef]

polymers

MDPI

Article

Influence of Newly Organosolv Lignin-Based Interface Modifier on Mechanical and Thermal Properties, and Enzymatic Degradation of Polylactic Acid/Chitosan Biocomposites

Faisal Amri Tanjung [1,*], Yalun Arifin [2] and Retna Astuti Kuswardani [3]

[1] Faculty of Science and Technology, Universitas Medan Area, Medan 20223, North Sumatera, Indonesia
[2] Department of Food Business Technology, Universitas Prasetiya Mulya, BSD Raya Utama, Tangerang 15339, Banten, Indonesia; yalun.arifin@prasetiyamulya.ac.id
[3] Faculty of Agriculture, Universitas Medan Area, Medan 20223, North Sumatera, Indonesia; retna@staff.uma.ac.id
* Correspondence: faisalamri@staff.uma.ac.id or icalamri@gmail.com; Tel.: +61-7360-168

Abstract: This article aimed to study the effects of chitosan fiber and a newly modifying agent, based on organosolv lignin, on mechanical and thermal performances and the enzymatic degradation of PLA/chitosan biocomposites. A newly modifying agent based on polyacrylic acid-grafted organosolv lignin (PAA-g-OSL) was synthesized via free radical copolymerization using *t*-butyl peroxide as the initiator. The biocomposites were prepared using an internal mixer and the hot-pressed method at various fiber loadings. The results demonstrate that the addition of chitosan fiber into PLA biocomposites remarkably decreases tensile strength and elongation at break. However, it improves the Young's modulus. The modified biocomposites clearly demonstrat an improvement in tensile strength by approximately 20%, with respect to the unmodified ones, upon the presence of PAA-g-OSL. Moreover, the thermal stability of the modified biocomposites was enhanced significantly, indicating the effectiveness of the thermal protective barrier of the lignin's aromatic structure belonging to the modifying agent during pyrolysis. In addition, a slower biodegradation rate was exhibited by the modified biocomposites, relative to the unmodified ones, that confirms the positive effects of their improved interfacial interaction, resulting in a decreased area that was degraded through enzyme hydrolysis.

Keywords: polylactic acid; chitosan; organosolv lignin; modifying agent; biocomposites

Citation: Tanjung, F.A.; Arifin, Y.; Kuswardani, R.A. Influence of Newly Organosolv Lignin-Based Interface Modifier on Mechanical and Thermal Properties, and Enzymatic Degradation of Polylactic Acid/Chitosan Biocomposites. *Polymers* **2021**, *13*, 3355. https://doi.org/10.3390/polym13193355

Academic Editors: Domenico Acierno and Antonella Patti

Received: 15 September 2021
Accepted: 27 September 2021
Published: 30 September 2021

Publisher's Note: MDPI stays neutral with regard to jurisdictional claims in published maps and institutional affiliations.

1. Introduction

The manufacture of composite materials using ecologically friendly technologies is of great of interest to many academic and industrial practitioners in the areas of polymer science and engineering [1]. This trend is mainly driven by environmental considerations regarding the negative impact of petroleum-derived materials on the environment after their end-use, which are difficult to decompose in a landfill, and also by aiming to attain composite materials that possess the desired properties [2]. Composite materials often contain bio-based polymer. Among the bio-based polymers, polylactic acid (PLA) is the best biopolymer alternative for petro-polymers because of its renewability, biodegradability, biocompatibility, and good thermomechanical performance [3,4]. This biopolymer is derived from plants, such as corn and cassava, and it is known to have a relatively high melting point, strength, and versatility, with performance characteristics similar to synthetic polymers, such as polyethylene terephthalate, polyethylene, etc., [5–7].

PLA is regarded as a valuable and important biopolymer that can be utilized to replace synthetic counterparts in many applications, ranging from automotive parts to electronic devices [8–10]. However, PLA has some limitations that consequently restrict its widespread application, involving a low thermal resistance, heat distortion temperature, and crystallization rate, whereas other specific properties may be necessary in some different use sectors [11]. As a result, the incorporation of reinforcement agents, such as natural fibers and inorganic substances, represents a broad method for extending and improving the PLA's performance, particularly with regard to its tensile and thermal properties [12–16].

Many natural fibers have been studied in the design of PLA biocomposites that show promising results. Among existing natural fibers, chitosan has demonstrated excellent mechanical and thermal properties that are comparable to cellulose [17]. Chitosan is derived from chitin, which is obtained from the shells of crustaceans, such as crabs, shrimp, and prawns [18]. This natural fiber has been widely used in a variety of scientific applications, including medicine, biotechnology, the textile and food industries, as well as in fiber and plastic applications [19–21]. However, the main disadvantages of utilizing it as a reinforcing agent in polymer composites include low dispersion and poor interfacial adhesion, both of which are caused by incompatibility with the hydrophobic matrix polymer. This is demonstrated by the difficulties of the polar hydroxyl groups located on the chitosan surface in building a well-bonded interface with a nonpolar matrix polymer, as strength improvement is dependent on stress transfer at the composite's interface when an external force is applied [22]. If the interface is poor, the fiber-matrix adhesion will diminish with no enhancement in performance [23]. Consequently, this problem reduces the benefits of using potential reinforcements in polymer composites.

To address the issue of interaction, the interfacial adhesion in the composite material is chemically altered. Chemical modification has been frequently utilized to enhance the interfacial adhesion in composite systems, since it is an effective technique for reducing the hydrophilic properties of natural fibers. Previous research used treated chitosan fiber as a natural filler in polypropylene composites. The results show that the incorporation of chitosan fiber into the composites increased the tensile modulus and the impact strength, while decreasing the tensile strength significantly [24]. Therefore, the current research is focused on employing chitosan fibers to increase the performances of PLA biocomposites by filler surface modification using a newly developed modifying agent based on grafted organosolv lignin. To the best of our knowledge, investigations involving the utilization of grafted organosolv-lignin-based modifying agents in biocomposites have been less reported in the literature.

Lignin is well-known as a byproduct of the wood pulping process and is one of the abundant vegetal-derived compounds [25]. Because of its complex aromatic structure, which is linked by an ester-bridge, lignin is a highly stable polymer. It has a strong polarity that results from the existence of a huge number of hydroxyl groups, both aliphatic and aromatic [26–28]. The lignin used in this study was obtained from lignocellulosic fiber using an established organosolv method with organic solvent and water [29]. The procedure produces lignin with a low molecular weight and a huge number of reaction sites available, making this kind of lignin a suitable surface-modifying agent. However, because of the complexity of its structure, lignin is difficult to dissolve in conventional solvents, causing the limitation of its chemical reactivity [30]. Therefore, a simple copolymerization reaction with acrylic acid was used to improve the lignin reactivity and solubility. The lignin alteration produces a pendant carboxylic moiety, which provides a site for additional reactive reactions.

This research is aimed at studying the effects of chitosan fiber and a newly developed modifying agent based on grafted organosolv lignin on the mechanical and thermal performances of PLA/chitosan biocomposites. Furthermore, the weight loss of the PLA/chitosan biocomposites during enzymatic degradation is examined.

2. Materials and Method

2.1. Raw Materials

PLA (TT Biotechnology Sdn. Bhd, Penang, Malaysia) had a melt flow index of 5.6 g/10 min (180 °C/2160 g), and a density of 1.27 g·cm^{-3}. The chitosan (Hunza Nutriceuticals Sdn Bhd., Parit Buntar, Malaysia) had an average size of 80 µm and a 90% degree of deacetylation (DD). The characteristics of chitosan are listed in Table 1. Commercial-grade diastase (sourced from malt) was supplied by Sigma Aldrich (St. Louis, MO, USA). The ethanol (98%.*v*/*v*), hydrochloric acid, *t*-butyl peroxide, acrylic acid, acetic acid, NaOH, sodium acetate and the sulphuric acid (98%.*v*/*v*) were obtained from Sigma Aldrich and used without further purification.

Table 1. Physical and chemical characteristics of chitosan.

Item	Specification	Test Method
Appearance	Off-white powder	Visual
Particle size	80 μm	Malvern Particle Size Analyzer
Solubility of 1% chitosan in 1% acetic acid	>99.0%	Dissolution & Filtration
Viscosity	150–200 m Pa·s	Ubbelohde Viscometer
Moisture content	<10.0%	Infra-red drying
Ash content	<1.0%	Incineration

2.2. Extraction of Organosolv Lignin (OSL) from Lignocellulosic Fiber

The extraction procedure of organosolv lignin (OSL) from lignocellulosic fiber was conducted following the established organosolv method [31]. The fiber was first treated using a mixture of aqueous ethanol and a catalyst (sulphuric acid) at a temperature set of 190 °C for 1 h, with the solid to liquid ratio adjusted at 1:8. The pretreated fiber was then rinsed with aqueous ethanol. The washes were mixed, and 5 vol% of distilled water was added to precipitate the organosolv lignin. The OSL was centrifuged and then dried in an oven at a temperature of 80 °C for 24 h.

2.3. Preparation of PAA-grafted OSL

The chemical grafting reaction of OSL with acrylic acid was carried out by utilizing *t*-butyl peroxide as the initiator in a copolymerization reaction. Three grams of lignin sample and *t*-butyl peroxide were mixed with 60 ml of distilled water in the reactor. Acrylic acid was subsequently added slowly into the reaction mixture under constant stirring. The mixture pH was adjusted from around 8.5 to 9.0 by adding 2% NaOH solution. After the reaction was completed, the product was precipitated by adding 6 M HCl. The solid was allowed to stay overnight upon discarding the supernatant. The solid product was rinsed using acidified water (pH = 2–3) prior to drying in an oven.

2.4. Chemical Modification of Chitosan

PAA-g-OSL was used to chemically modify the chitosan surface in an ethanol medium. The amount of PAA-g-OSL used was 3% by weight of chitosan fiber. Chitosan fiber was added to the PAA-g-OSL solution, and then vigorously stirred for 4 h. Afterward, the chitosan suspension was filtered and dried for 24 h in an oven at a temperature of 80 °C.

2.5. Preparation of PLA/Chitosan Biocomposites

PLA/chitosan biocomposites were prepared in an internal mixer (MCN ELEC Co., Taichung, Taiwan) at 160 °C and 50 rpm. PLA was firstly fed into the mixer to initiate the melt mixing process for 12 minutes. Afterward, chitosan fiber was slowly added, and the compounding was carried out for another 3 min. Next, the compounds were removed and sheeted through a laboratory mill at a 2.0 mm nip setting. The compounds were compression molded at 120 kg·cm^{-2} in an electrically heated hydraulic press. The hot-press procedure consists of three-consecutive steps, involving preheating at 190 °C for 9 min, compression for 3 min at 190 °C, and then cooling pressure for 3 min. A similar procedure was carried out to synthesize the modified PLA/chitosan biocomposites. The formulation of unmodified and modified PLA/chitosan composites is exhibited in Table 2.

Table 2. Formulation of unmodified and modified PLA biocomposites at different chitosan contents.

Materials	Unmodified Biocomposites	Modified Biocomposites with PAAL
Polylactic acid (PLA) (php)	100	100
Chitosan (php)	0, 10, 20, 30, 40	10, 20, 30, 40
PAA-g-OSL (PAAL) (wt%)	-	3

php = part per hundred of polymer.

2.6. Characterization

Tensile tests were conducted using an Instron 5582 machine (Instron, Norwood, MA, USA)according to ASTM D 638-91. A Wallace die cutter was utilized to cut at least five dumbbell specimens of each composition, 1 mm thick, from the molded sheets. The test was conducted at 25 ± 3 °C with a crosshead speed of 20 mm/min. Thermogravimetric (TGA) and derivative thermogravimetric (DTG) analyses were carried out using a TGA Q500 (TA Instruments, New Castle, Germany). The samples were thermally scanned at a temperature range of 30 to 600 °C, at a heating rate of 20 °C/min, with a nitrogen flow of 50 mL/min. Differential scanning calorimetry (DSC) analysis was performed using a DSC Q 1000 (TA Instruments). Samples were scanned at a temperature range of 25 to 250 °C, at a heating rate of 20 °C/min, with a nitrogen flow of 50 ml/min. The instrument software was used to calculate the melting points (T_m), glass transition temperature (T_g), crystallization temperature (T_c), and the enthalpies of the PLA/chitosan biocomposites. The crystallinity of the composites (X_c) was manually calculated using Equation (1):

$$X_c\ (\%) = \Delta H_f \times 100/\Delta H_f^{\circ} \quad (1)$$

where, ΔH_f is the fusion enthalpy of the PLA and biocomposites, and ΔH_f° is the thermodynamic fusion enthalpy belonging to a fully crystalline PLA (93.6 J/g) [32]. Data from the second heating scan was taken for each sample. An average value from three samples of each specimen was recorded.

A morphological study of the fracture surfaces of PLA/chitosan biocomposites was performed using a scanning electron microscope (SEM), JEOL model JSM 6260 LE. The fracture ends of the specimens were mounted on aluminum stubs and sputter-coated with palladium to avoid electrostatic charging during analysis.

Fourier transform infrared spectroscopy (FTIR) was conducted on unmodified and modified chitosan in ATR mode (Perkin Elmer 1600 Series). Samples were scanned at a resolution of 4 cm^{-1} from 650 to 4000 cm^{-1}.

The number of the average molecular weight (M_n), and weight average molecular weight (M_w), of the lignin were measured using gel permeation chromatography (GPC) after the acetylation of the lignin to allow dissolution in THF [33]. GPC analysis was carried out using a Perkin Elmer instrument (Waltham, MA, USA) equipped with an interface (PE Series 900).

The biodegradation test was performed in a diastase-enzyme-containing solution with an activity of 480 KNU·g^{-1}. A buffer solution (pH = 7.3) was prepared in a beaker by mixing 5 mL of 0.2 M acetic acid with 45 mL of 0.2 M sodium acetate solution. The biocomposites were submerged in the solution for 30 days in an incubator (New Brunswich Scientific) at 37 °C. The samples were taken every 5 days and rinsed thoroughly with distilled water before a conditioning step in an oven at 60 °C for 24 h.

In the weight loss measurement, each specimen was taken periodically and wiped and dried to a constant weight at 60 °C in a vacuum oven. The weight loss percentage was measured using an analytical balance, and calculated using Equation (2):

$$\text{Weight loss} = \frac{W_{initial} - W_{final}}{W_{initial}} \times 100 \quad (2)$$

where, $W_{initial}$ and W_{final} stand for the weights of the biocomposites before and after the biodegradation test, respectively. The average value of the three measurements from each biocomposite was reported.

3. Results and Discussion

3.1. Structural Analysis of Modified Chitosan

The OSL had a weight-average molecular weight (M_w) of 3429, and a number-average molecular weight (Mn) of 2506. The polydispersity (M_w/M_n) of the OSL obtained was 1.37. These values were calculated from the chromatogram produced by the gel permeation chromatography. These results suggest that the extraction significantly ruptured the complex macromolecular structure of the lignin. A low polydispersity value indicates a narrow molecular weight distribution.

Figure 1 shows FTIR spectra for unmodified chitosan fiber, and modified chitosan with PAA-g-OSL. The FTIR analysis was aimed at investigating the possibility of a chemical bonding formation between the PAA-g-OSL modifying agent and the chitosan fiber. The IR spectra belonging to unmodified chitosan demonstrate the existence of characteristic peaks at 3358 cm^{-1} (O–H stretch), 2872 cm^{-1} (C–H stretch), 1675 cm^{-1} (N–H bend), and 1590 cm^{-1} (C=O stretch). The differences between the absorption spectra of the modified and unmodified chitosan indicates an alteration in the chemical structure of chitosan upon the esterification process. The distinct change in the absorption spectrum, occurring on the peak between 1760–1712 cm^{-1} and 3400–3200 cm^{-1}, was caused by the carboxyl groups (C–OH) of the modifying agent linked to the hydroxyl groups (–OH), chitosan-fiber-formed ester linkages (C–O–C), and the decrease of the content of chitosan's hydroxyl groups. The formation of ester bonds (C–O–C) resulted in a decrease in hydrophilic moieties on the chitosan surface. The modified chitosan had less hydrophilicity (24.2% lower) than the unmodified chitosan. It was shown by a relatively lower intensity of the peak at 3292 cm^{-1} of the modified chitosan with PAA-g-OSL than that of the peak at 3358 cm^{-1} of unmodified chitosan. Figure 2 shows a possible schematic of the chemical reaction between chitosan fiber and esterified lignin.

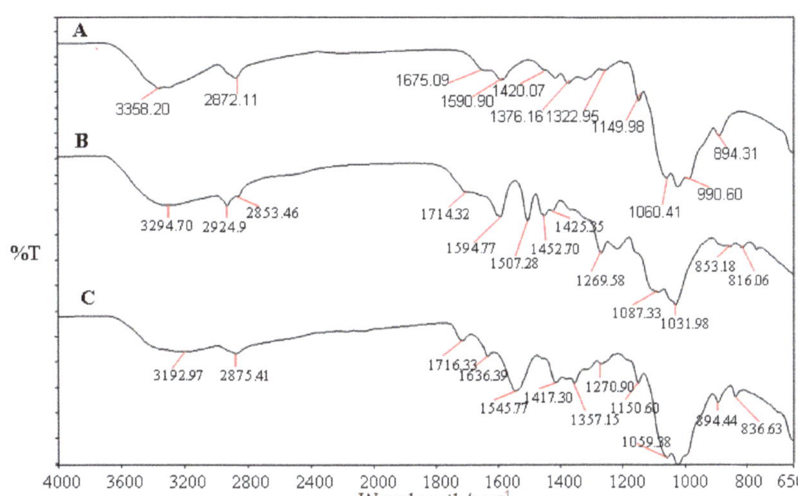

Figure 1. FTIR spectra of unmodified chitosan fiber and modified chitosan with PAA-g-OSL. (**A**) chitosan, (**B**) PAA-g-OSL modifying agent, (**C**) modified chitosan with PAA-g-OSL.

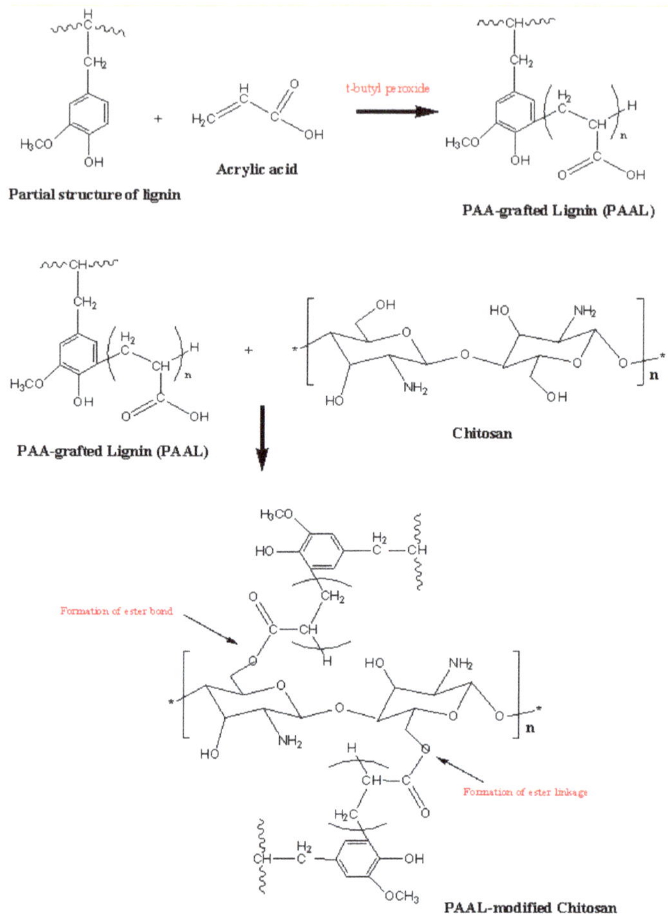

Figure 2. A possible schematic grafting reaction between OSL and acrylic acid, and the chemical modification of chitosan fiber.

3.2. Mechanical Properties of PLA/Chitosan Biocomposites

Figure 3 shows typical stress–vs–strain traces obtained from the tensile tests for neat PLA, unmodified PLA/chitosan biocomposites, and modified PLA/chitosan biocomposites, respectively. All of the PLA biocomposites contained 20 php of chitosan fiber. All curves clearly show the same trend, where the incorporation of chitosan decreased the strength and strain of the PLA biocomposites, corresponding to poor interface interaction between the PLA and chitosan. Chitosan is a natural polymer that possesses polar hydrophilic properties, whereas PLA is a nonpolar hydrophobic polymer. The differences in the affinities and polarity diminished the interaction on the filler-matrix interface [34]. The modified PLA/chitosan biocomposite with PAA-g-OSL, on the other hand, demonstrated a greater strength than those of the neat PLA and the unmodified biocomposites, which indicates the enhanced interfacial adhesion upon the existence of PAA-g-OSL on the chitosan surface. Interestingly, the strain increased by 24.57% when compared with the unmodified biocomposite, reflecting the enhanced chain mobility of the PLA matrix within the biocomposite. Overall, it can be observed that the interfacial compatibility improves the composites' strength. Although chemical modification had no effect on the stress–versus–strain trace character of the modified biocomposites, it did influence the measured strength at failure.

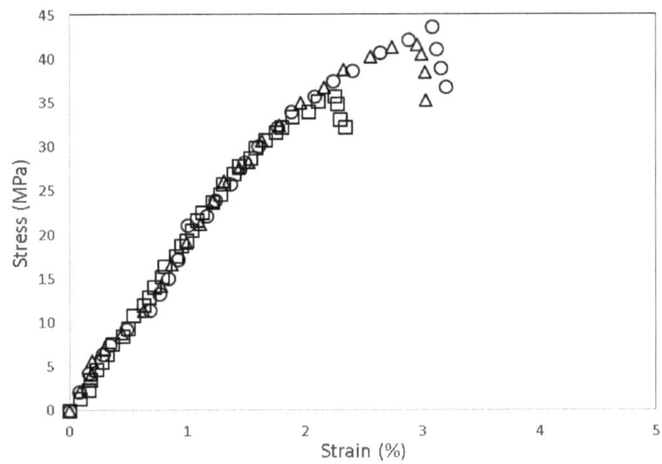

Figure 3. Stress–versus–strain traces of tensile testing for neat PLA (Δ), unmodified PLA/chitosan biocomposite (□), and modified PLA/chitosan biocomposite with PAA-g-OSL (o). Chitosan content: 20 php.

The tensile strength of unmodified and modified PLA/chitosan biocomposites at various fiber loading are exhibited in Figure 4. The tensile strength of the PLA biocomposites was inversely proportional to the chitosan content. However, the modified biocomposites showed a higher tensile strength than that of the unmodified biocomposites at a similar chitosan content. The tensile strength of the modified biocomposites improved by 36.43% as compared with the unmodified ones, indicating the enhanced filler-matrix interfacial interaction. This enhanced strength was clearly attributed to the high reactivity of PAA-g-OSL, which effectively facilitated surface modification, resulting in improved interfacial adhesion.

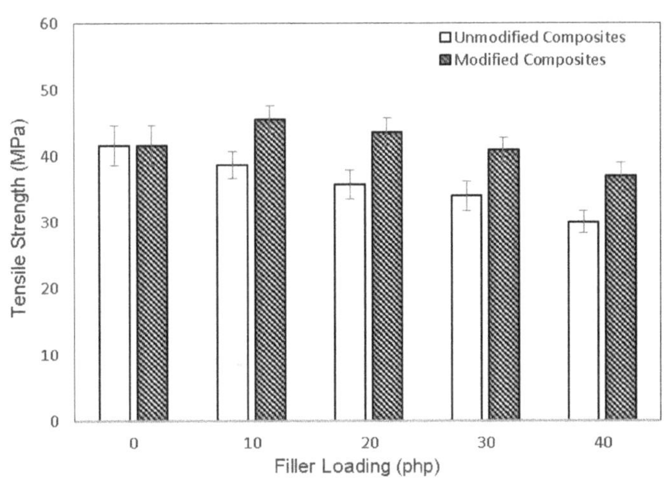

Figure 4. Effect of chitosan content and PAA-g-OSL on tensile strength of PLA/chitosan biocomposites.

Figure 5 shows the Young's modulus of unmodified and modified PLA biocomposites at different chitosan loading. The Young's modulus of the PLA/chitosan biocomposites increased as the chitosan content increased. The dispersion of chitosan fiber within the PLA matrix clearly improved the composite's stiffness, indicating the limited segmental molecular mobility of the PLA chain upon interacting with the fiber. The Young's modulus of the modified biocomposites was remarkably higher than the unmodified biocomposites. It increased by 55.33% as compared with the unmodified ones. The rigid nature of the filler, the high crystallinity index, and the presence of the modifying agent played an essential role in the improvement of the Young's modulus of the PLA biocomposites. The presence of esterified lignin improves the stiffness of the biocomposites by increasing the filler-matrix interfacial adhesion.

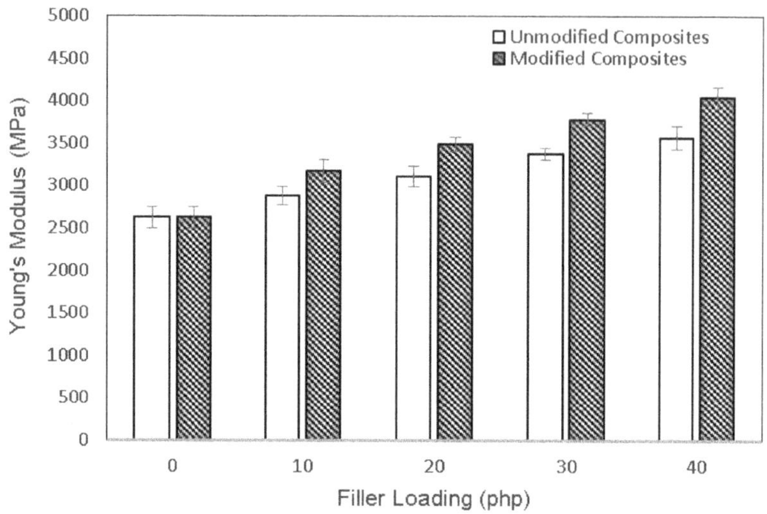

Figure 5. Young's modulus of unmodified and modified PLA biocomposites at various chitosan content.

3.3. SEM Image Analysis

Morphological images of the tensile fractured surfaces of the PLA matrix, unmodified PLA biocomposite containing 20 php and 40 php of chitosan fiber, and modified PLA composite filled with 40 php of chitosan fiber, are shown in Figure 6A–D, respectively. The surface of the PLA matrix was quite smooth, as seen in the SEM image (Figure 6A). A few voids, on the other hand, appeared on the fractured surface of the unmodified biocomposite (Figure 6B,C), exhibiting the fiber pulled out from the matrix phase. It pointed to the matrix's poor wetting of the filler, which could lead to the breakage at the interface of the chitosan and the PLA matrix. The difference in the affinities between both components has resulted in weak interaction at the interface. Meanwhile, a relatively smooth surface with fewer voids was seen on the fracture surface of the modified biocomposite (Figure 6D). This was clearly related to a good wetting of the filler by the matrix since the PAA-g-OSL modifying agents substantially modified the chitosan surface to become less-hydrophilic, leading to improved interaction with the polymer matrix. As a result, the modified composites had significantly fewer detached filler traces from the matrix phase.

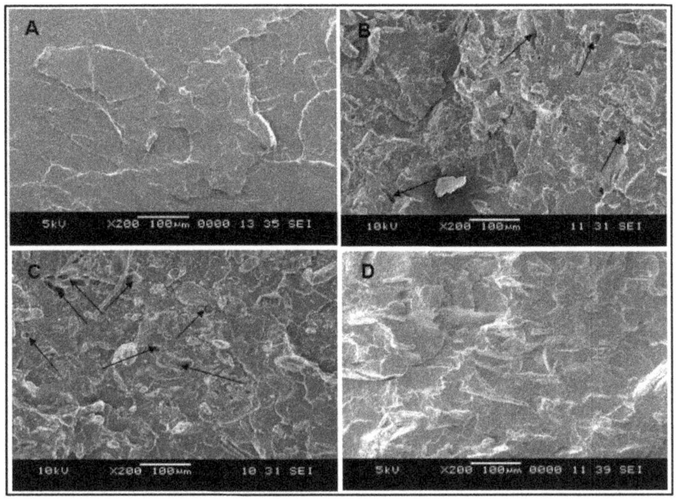

Figure 6. SEM micrographs of PLA/chitosan biocomposites, (**A**) neat PLA, (**B**) unmodified biocomposite (20 php), (**C**) unmodified biocomposite (40 php), (**D**) modified biocomposite (20 php). The arrows indicate the fiber pulled out traces.

3.4. Thermal Properties of PLA/Chitosan Biocomposites

Figure 7A,B show the TGA and DTG curves of neat PLA, unmodified PLA/chitosan biocomposites, and modified PLA/chitosan biocomposites, respectively. The thermal decomposition of neat PLA occurred over a one-stage process between 300 and 415 °C, leading to the formation of gaseous products. This degradation pathway was associated with the molecular structure of PLA that is composed of carbon–carbon linkages, where the degradation/depolymerisation occurs at the weak sites within the PLA backbone chain [35]. On the other hand, the PLA/chitosan biocomposites demonstrated three distinct stages of thermal decomposition. They were observed within the temperature ranges of 160–180 °C, 260–340 °C, and 400–490 °C, respectively. The similarity in the general patterns was exemplified by these typical thermograms. However, the biocomposites were indicated by their distinct temperature and weight losses. The first stage involved the release of typical strong hydrogen-bonded water and the evaporation of volatile compounds from the samples, while the second stage included the degradation and depolymerization of both the chitosan fiber and a modifying agent. The third stage could be attributed to the decomposition of char residue formed in the second stage.

The weight loss percentage for neat PLA, unmodified PLA/chitosan biocomposites, and modified PLA/chitosan biocomposites are tabulated in Table 3. An analysis of the weight loss data shows that 10% of the weight loss occurred at 328 °C for the unmodified biocomposites, and at 335 °C for the modified biocomposites. The temperature increased to 414 °C for the unmodified biocomposites, and to 440 °C for the modified biocomposites, when the weight loss was 50%. The obtained results clearly indicate that the modified biocomposites were more thermally stable than the unmodified ones. The attachment of the PAA-g-OSL modifying agent on the chitosan surface enabled it to perform as a thermal protective barrier for the biocomposites against thermo-decomposition.

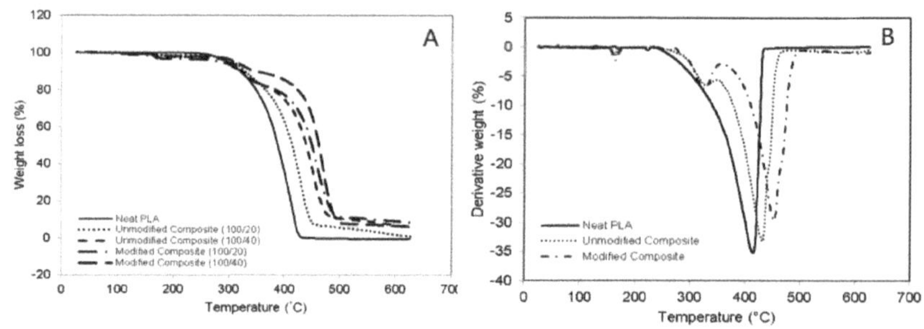

Figure 7. Effect of chitosan content and chemical modification on thermogravimetric properties of PLA/chitosan biocomposites. (**A**) weight loss versus temperature curves, (**B**) derivative thermogravimetry (DTG) curves.

Table 3. Weight loss percentage of unmodified and modified PLA biocomposites containing chitosan fiber at different temperatures.

Temperature (°C)	Neat PLA	Unmodified Biocomposites		Modified Biocomposites	
		20 php	40 php	20 php	40 php
100–150	0.002	0.039	0.277	0.342	0.600
150–200	0.016	0.271	0.555	0.464	0.773
200–250	0.150	1.017	1.894	0.943	1.561
250–300	0.476	0.157	0.212	0.428	0.673
300–350	4.993	0.495	1.262	2.102	3.419
350–400	14.813	4.540	5.080	3.988	2.005
400–450	43.150	13.539	13.105	10.197	9.663
450–500	36.400	40.499	36.512	38.738	31.997
500–550	0	33.153	32.855	32.834	29.461
550–600	0	1.494	1.428	1.359	9.302
600–630	0	1.886	1.692	0.348	0.674
Total	100	97.090	94.872	91.743	90.128

php = part per hundred of polymer.

Lignin is composed of a three-dimensional network of aromatic structures that can slow down the thermo-decomposition of polymeric materials. This, along with the complex heterogeneity of the chemical bonds in the cross-linked aromatic and aliphatic structures, is one of its primary defences against decomposition [36]. Furthermore, the data indicates that the modified biocomposites had higher thermal stability than the unmodified biocomposites. Despite the fact that chemical modification had no effect on the thermal decomposition mechanism of the PLA/chitosan biocomposites, the resulting different degradation profiles is indicative of the changing of the chemical structure and the thermal stability.

Figure 8 exhibits the DSC curves for neat PLA, unmodified PLA biocomposites, and modified PLA biocomposites, respectively. Table 4 provides the melting temperature (T_m), the glass transition temperature (T_g), crystallization temperature (T_c), and the degree of crystallinity (X_c) of all biocomposite categories. For all biocomposite categories, the endothermic peak was observed at a maximum temperature range of 150 to 155 °C. Neat PLA had a T_g of 57.6 °C, and a T_m of 152 °C. The addition of chitosan fiber increased both the T_g and T_m of neat PLA, while the crystallinity degree (X_c) value slightly declined. This decreased crystallinity indicated that the addition of chitosan in PLA biocomposites apparently became an obstacle for the nuclei crystal growth of the PLA chain, causing the crystallization process to be hindered. [37]. In addition, a crystallization peak (T_c) also appeared at a temperature of 115 °C, corresponding to the formation of the metastable crystalline phase within the biocomposites.

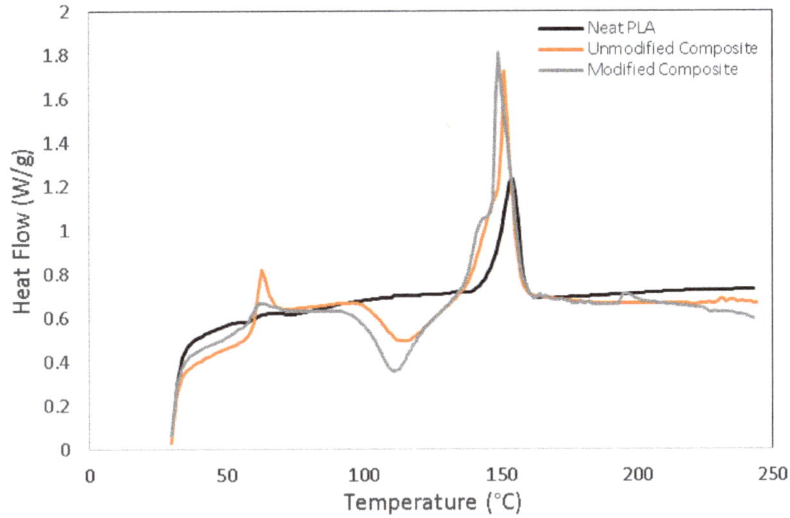

Figure 8. Differential scanning calorimetry (DSC) curves of unmodified and modified PLA/chitosan biocomposites at different fiber contents.

Table 4. DSC data obtained for unmodified and modified PLA/chitosan biocomposites.

Biocomposites	T_g (°C)	T_c(°C)	T_m (°C)	X_c (%)
Neat PLA	57.6	-	152.0(0.2)	32.74
PLA/chitosan: 100/20 php (unmodified)	63.4	117.7	151.3(0.3)	27.41
PLA/chitosan: 100/40 php (unmodified)	62.5	117.2	151.1(0.1)	24.25
PLA/chitosan: 100/20 php (modified)	61.3	115.3	150.4(0.5)	39.42
PLA/chitosan: 100/40 php (modified)	61.2	114.5	150.2(0.3)	33.80

T_g: glass transition temperature; T_c: crystallization temperature; T_m: melting temperature; X_c: crystallinity degree; php: parts per hundred of polymer.

On the other hand, a higher endothermic heat flow was observed for the modified biocomposites as compared to that for the unmodified biocomposites. Both the T_g and T_c of the modified biocomposites demonstrated a slight decrease; however, the X_c was obviously higher than that of the unmodified ones. It is thought that the presence of a PAA-g-OSL-based modifying agent improved the interfacial interaction between the chitosan and the PLA matrix, contributing significantly to the increased crystallinity of the modified biocomposites. It should be noted that the hydrophobic segment belonging to organosolv lignin may contribute to the effectiveness of the nucleation process, as well as the growth of the spherulitic structure in PLA biocomposites. The aromatic molecular structure and the chemical composition of organosolv lignin are expected as factors that can enhance the nucleation process. A previous study reported that some crystallization might occur when the fibers used contain sufficient quantities of lignin [38]. Nonetheless, the T_m did not show a significant change.

3.5. Weight Loss Analysis

Figure 9 exhibits the weight loss rate of neat PLA, unmodified PLA/chitosan biocomposites, and modified PLA/chitosan biocomposites as a function of the biodegradation days. As time passed, all the biocomposite categories degraded further, as evidenced by the increase in weight loss. The degradation of neat PLA was slower as compared to that of the two-biocomposite category. On the other hand, the addition of chitosan fiber apparently contributed to the increase in the biodegradation rate of the PLA biocomposites. The weight loss of the unmodified biocomposites was found to be about two-fold greater than that of the neat PLA, which confirmed the chitosan fiber dispersion within the PLA matrix and remarkably enhanced the degradation rate of the biocomposites. The chitosan fiber contributed to the high polarity of the biocomposites, resulting in decreased tensile strength. The improved biodegradation rate of the PLA/chitosan biocomposites was most likely because of the existence of the hydroxyl and amine groups belonging to chitosan, which, when dispersed in PLA, play a critical role in provoking the increasing hydrolytic degradation of the PLA main chain [39]. On the other hand, the biodegradation rates of the modified PLA/chitosan biocomposites were lower than that of the unmodified biocomposites, demonstrating that the chemical modification essentially enhanced the barrier performances of the composites and limited enzyme diffusion. The enhanced interfacial interaction between the PLA matrix and the chitosan fiber resulted from the PAA-g-OSL modifying effects, which decreased the area degraded through enzyme hydrolysis, leading to lower biodegradation rates.

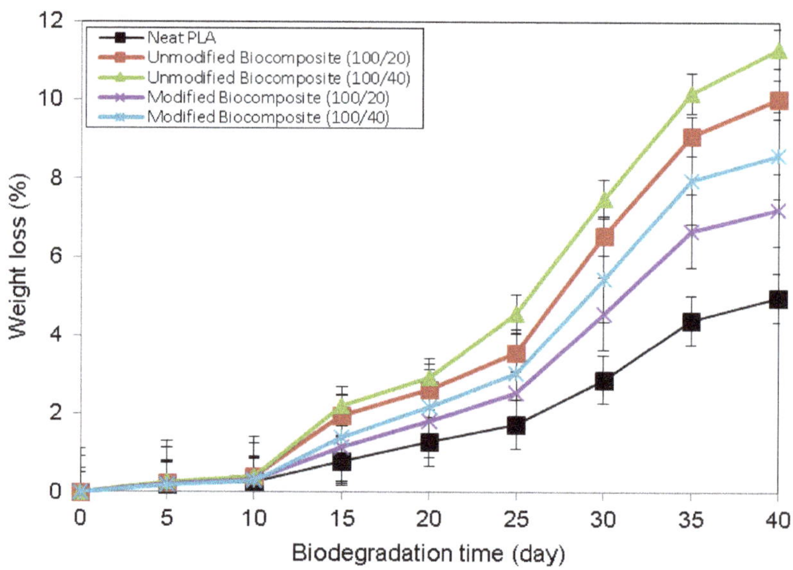

Figure 9. Percentage of weight loss of neat PLA, and unmodified and modified PLA/chitosan biocomposites, as a function biodegradation time.

One of the most important practical criteria for determining material degradation is the loss of mechanical properties. Table 5 provides tensile strength and the Young's modulus of neat PLA, unmodified PLA/chitosan biocomposite, and modified PLA/chitosan biocomposite at a 0- and 30-day biodegradation time. Before immersion in the enzyme-containing solution, the tensile strength of neat PLA was observed to have decreased along with the addition of 20 php of chitosan fiber. On the other hand, the Young's modulus improved. The presence of PAA-g-OSL on the chitosan surface contributed to the improvement in the tensile strength and the Young's modulus of the modified biocomposites as compared to the unmodified biocomposite. The results show a decrease in the tensile strength and the Young's modulus after 30 days of biodegradation. The action of the enzyme in the solution resulted in the apparent partial destruction of filler-matrix adhesion, void formation, and the subsequent random chain scission in the polymer chains, resulting in the decreased tensile properties.

Table 5. Tensile strength and Young's modulus of neat PLA, PLA/chitosan biocomposites, with and without treatment, at 0- and 30-day biodegradation times.

Biocomposites	0 Day		30 Day	
	Tensile Strength (MPa)	Young's Modulus (MPa)	Tensile Strength (MPa)	Young's Modulus (MPa)
Neat PLA	42.82	2674.68	35.67	2476.43
PLA/chitosan: 100/20 (unmodified)	37.74	3158.97	28.84	2847.91
PLA/chitosan: 100/40 (unmodified)	31.81	4125.86	25.65	3789.36
PLA/chitosan: 100/20 (modified)	45.64	3567.45	40.27	3254.29
PLA/chitosan: 100/40 (modified)	38.72	4180.87	32.85	3810.38

4. Conclusions

PLA-chitosan biocomposites modified with PAA-g-OSL were prepared. The properties of the modified biocomposites were investigated and are briefly discussed. The obtained organosolv lignin had a low molecular weight, which resulted in a higher efficiency of chitosan modification of up to 21.3%. The PAA-g-OLS-based modifying agent significantly altered the chitosan surface, resulting in improved interfacial adhesion within the biocomposites. The mechanical strength of the modified biocomposites was greater than the mechanical strength of the unmodified biocomposites. The SEM images clearly show that the modified biocomposites had fewer voids on their tensile surface. The TGA and DSC results show that the modified biocomposites had improved thermal stability and crystallinity. Moreover, the modified biocomposites had a higher crystallinity degree than the unmodified biocomposites. Overall, the obtained results clearly suggest the advantages in utilizing grafted organosolv lignin-based modifying agents, and other lignin derivatives, in the manufacturing of natural fiber-reinforced biocomposite materials, thereby opening up the "door" to the potential for a renewable low-cost chemical reagent alternative to synthetic reagents.

Author Contributions: Conceptualization and Investigation, F.A.T.; Methodology, Y.A.; Formal Analysis and Review, R.A.K. All authors have read and agreed to the published version of the manuscript.

Funding: This study received no external funding.

Institutional Review Board Statement: Not applicable.

Informed Consent Statement: Not applicable.

Data Availability Statement: Not applicable.

Acknowledgments: FA Tanjung is grateful to the Ministry of Education, Culture, Research and Technology of Indonesia for research funding under the World Class Research (WCR) grant scheme (1428/LP2M/03.7.1/VII/2021), and the Universitas Medan Area for the associated research facilities.

Conflicts of Interest: The authors declare no conflict of interest.

References

1. Akonda, M.A.; Shah, D.U.; Gong, R.H. Natural fiber thermoplastic tapes to enhance reinforcing effects in composite structures. *Compos. Part A* **2020**, *131*, 105822. [CrossRef]
2. Kun, D.; Kárpáti, Z.; Fekete, E.; Móczó, J. The role of interfacial adhesion in polymer composites engineered from lignocellulosic agricultural waste. *Polymers* **2021**, *13*, 3099. [CrossRef] [PubMed]
3. Alam, F.; Varadarajan, K.; Kumar, S. 3D-printed pol(ylactic acid) nanocomposite scaffolds for tissue engineering applications. *Polym. Test.* **2020**, *81*, 106203. [CrossRef]
4. Jiashen, L.; Yi, L.; Lin, L.; Frank, K.F.T.; Ling, Q. Fabrication and degradation of poly(l-lactic acid) scaffolds with wool keratin. *Compos. Part B Eng.* **2009**, *13*, 664–667.
5. Ramot, Y.; Haim-Zada, M.; Domb, A.J.; Nyska, A. Biocompatibility and safety of PLA and its copolymers. *Adv. Drug Deliv. Rev.* **2016**, *107*, 153–162. [CrossRef]

6. Sanchez-Olivares, G.; Sanchez-Solis, A.; Calderas, F.; Alongi, J. Keratin fibres derived from tannery industry wastes for flame retarded PLA composites. *Polym. Degrad. Stab.* **2017**, *140*, 42–54. [CrossRef]
7. Shah, A.A.; Hasan, F.; Hameed, A.; Ahmed, S. Biological degradation of plastics: A comprehensive review. *Biotechnol. Adv.* **2008**, *26*, 246–265. [CrossRef]
8. Spiridon, I.; Paduraru, O.M.; Zaltariov, M.F.; Darie, R.N. Influence of keratin on polylactic acid/chitosan composite properties. behavior upon accelerated weathering. *Ind. Eng. Chem. Res.* **2013**, *52*, 9822–9833. [CrossRef]
9. Suh, H.; Hwang, Y.S.; Lee, J.E.; Han, C.D.; Park, J.C. Behavior of osteoblasts on a type I atelocollagen grafted ozone oxidized poly-L-lactic acid membrane. *Biomaterials* **2001**, *22*, 219–230. [CrossRef]
10. Tanase, C.E.; Spiridon, I. PLA/chitosan/keratin composites for biomedical applications. *Mater. Sci. Eng. C* **2014**, *40*, 242–247. [CrossRef]
11. Bonilla, J.; Fortunati, E.; Vargas, M.; Chiralt, A.; Kenny, J.M. Effects of chitosan on the physicochemical and antimicrobial properties of PLA films. *J. Food Eng.* **2013**, *119*, 236–243. [CrossRef]
12. Quirk, R.A.; Chan, W.C.; Davies, M.C.; Tendler, S.J.B.; Shakesheff, K.M. Poly(L-lysine)-GRGDS as a biomimetic surface modifier for poly(lactic acid). *Biomaterials* **2001**, *22*, 865–872. [CrossRef]
13. Shimao, M. Biodegradation of plastics. *Curr. Opin. Biotechnol.* **2001**, *12*, 242–247. [CrossRef]
14. Olaiya, N.G.; Surya, I.; Oke, P.K.; Rizal, S.; Sadiku, E.R.; Ray, S.S.; Farayibi, P.K.; Hossain, M.S.; Khalil, H.P.S.A. Properties and characterization of a PLA-chitin-starch biodegradable polymer composite. *Polymers* **2019**, *11*, 1656. [CrossRef]
15. Rigotti, D.; Checchetto, R.; Tarter, S.; Caretti, D.; Rizzuto, M.; Fambri, L.; Pegoretti, A. Polylactic acid-lauryl functionalized nanocellulose nanocomposites: Microstructural, thermo-mechanical and gas transport properties. *Express Polym. Lett.* **2019**, *13*, 858–876. [CrossRef]
16. Mania, S.; Partyka, K.; Pilch, J.; Augustin, E.; Cieślik, M.; Ryl, J.; Jinn, J.-R.; Wang, Y.-J.; Michałowska, A.; Tylingo, R. Obtaining and characterization of the PLA/chitosan foams with antimicrobial properties achieved by the emulsification combined with the dissolution of chitosan by CO_2 saturation. *Molecules* **2019**, *24*, 4532. [CrossRef]
17. Chakraborty, H.; Bhowmik, N. Structure and stability analysis of biocompatible hydroxyapatite reinforced chitosan nanocomposite. *Polym. Compos.* **2018**, *39*, E573–E583. [CrossRef]
18. Tanjung, F.A.; Husseinsyah, S.; Hussin, K.; Hassan, A. Mechanical and thermal properties of organosolv lignin/sodium dodecyl sulphate binary agent-treated polypropylene/chitosan composites. *Polym. Bull.* **2016**, *73*, 1427–1445. [CrossRef]
19. Panariello, L.; Coltelli, M.B.; Buchignani, M.; Lazzeri, A. Chitosan and nano-structured chitin for biobased anti-microbial treatments onto cellulose based materials. *Eur. Polym. J.* **2019**, *113*, 328–339. [CrossRef]
20. Pillai, C.K.S.; Paul, W.; Sharma, C.P. Chitin and chitosan polymers: Chemistry, solubility and fiber formation. *Prog. Polym. Sci.* **2009**, *34*, 641–678. [CrossRef]
21. Sionkowska, A.; Walczak, M.; Sionkowska, M.M. Preparation and characterization of collagen/chitosan composites with silver nanoparticles. *Polym. Compos.* **2020**, *41*, 951–957. [CrossRef]
22. Mir, S.; Yasin, T.; Halley, P.J.; Siddiq, H.M.; Nicholson, T. Thermal, rheological, mechanical and morphological behavior of HDPE/chitosan blend. *Carbohydr. Polym.* **2011**, *83*, 414–421. [CrossRef]
23. Taufik, M.J.; Mansor, M.R.; Mustafa, Z. Characterization of wood plastic composite manufactured from kenaf fiber reinforced recycled-unused plastic blend. *Compos. Struct.* **2018**, *189*, 510–515. [CrossRef]
24. Tanjung, F.A.; Husseinsyah, S.; Hussin, K. Chitosan filled-polypropylene composites: The effect of filler loading and organosolv lignin on mechanical, morphological and thermal properties. *Fiber Polym.* **2014**, *15*, 800–808. [CrossRef]
25. Mishra, S.B.; Mishra, A.K.; Kaushik, N.K.; Khan, M.A. Study of performance properties of lignin-based polyblends with polyvinyl chloride. *J. Mater. Process. Technol.* **2007**, *183*, 273–276. [CrossRef]
26. Shikinaka, K.; Nakamura, M.; Otsuka, Y. Strong UV absorption by nanoparticulated lignin in polymer films with reinforcement of mechanical properties. *Polymer* **2020**, *190*, 122254. [CrossRef]
27. Cao, X.; Huang, J.; He, Y.; Hu, C.; Zhang, Q.; Yin, X.; Wu, W.; Li, R.K.Y. Biodegradable and renewable UV-shielding polylactide composites containing hierarchical structured POSS functionalized lignin. *Int. J. Biol. Macromol.* **2021**, *188*, 323–332. [CrossRef]
28. Wu, W.; Liu, T.; Deng, X.; Sun, Q.; Cao, X.; Feng, Y.; Wang, B.; Roy, V.A.L.; Li, R.K.Y. Ecofriendly UV-protective films based on poly(propylene carbonate) biocomposites filled with TiO_2 decorated lignin. *Int. J. Biol. Macromol.* **2019**, *126*, 1030–1036. [CrossRef]
29. Toledano, A.; Garcia, A.; Mondragon, I.; Labidi, J. Lignin separation and fractionation by ultrafiltration. *Sep. Purif. Technol.* **2010**, *71*, 38–43. [CrossRef]
30. Kabir, A.S.; Yuan, Z.S.; Kuboki, T.; Xu, C.B. Depolymerization of industrial lignins to improve the thermo-oxidative stability of polyolefins. *Ind. Crop. Prod.* **2018**, *120*, 238–249. [CrossRef]
31. Hage, R.E.; Brosse, N.; Chrusciel, L.; Sanchez, L.; Sannigrahi, P.; Ragauskas, A. Characterization of milled wood lignin and ethanol organosolv lignin from miscanthus. *Polym. Degrad. Stab.* **2009**, *94*, 1632–1638. [CrossRef]
32. Carrosco, F.; Pages, P.; Gamez, P.J.; Satana, O.O.; Maspoch, M.L. Processing of poly(lactic acid): Characterization of chemical structure, thermal stability and mechanical properties. *Polym. Degrad. Stab.* **2010**, *95*, 116–125. [CrossRef]
33. Huda, M.; Drzal, L.T.; Mohanty, A.K. Effect of fiber surface-treatments on the properties of laminated biocomposites from poly(lactic acid) (PLA) and kenaf fibers. *Compos. Sci. Technol.* **2008**, *68*, 424–432. [CrossRef]
34. Leluk, K.; Frąckowiak, S.; Ludwiczak, J.; Rydzkowski, T.; Thakur, V.K. The impact of filler geometry on polylactic acid-based sustainable polymer composites. *Molecules* **2021**, *26*, 149. [CrossRef]

35. Jang, W.; Shin, B.Y.; Lee, T.J. Thermal properties and morphology of biodegradable PLA/starch compatibilized blends. *J. Ind. Eng. Chem.* **2007**, *13*, 457–464.
36. Ibrahim, M.N.M.; Haras, M.R.A.; Sipaut, C.S.; Aboul-Enein, H.Y.; Mohamed, A.A. Preparation and characterization of a newly water-soluble lignin graft copolymer from oil palm lignocellulosic waste. *Carbohydr. Polym.* **2010**, *80*, 1102–1110. [CrossRef]
37. Perinovic, S.; Andricic, B.; Erceg, M. Thermal properties of poly(L-lactic)/olive stone flour composites. *Acta* **2010**, *510*, 97–102.
38. Gregorova, A.; Hrabalova, M.; Wimmer, R.; Saake, B.; Altaner, C. Poly(lactide acid) composites reinforced with fibers obtained from different tissue type of Picea sitchensis. *J. Appl. Polym. Sci.* **2009**, *114*, 2616–2623. [CrossRef]
39. Kijchavengkul, T.; Rafael, A.; Maria, R. Biodegradation and hydrolysis rate of aliphatic aromatic polyester. *Polym. Degrad. Stab.* **2010**, *95*, 2641–2647. [CrossRef]

polymers

MDPI

Article

Potency of Urea-Treated Halloysite Nanotubes for the Simultaneous Boosting of Mechanical Properties and Crystallization of Epoxidized Natural Rubber Composites

Indra Surya [1], Kamaruddin Waesateh [2], Sitisaiyidah Saiwari [3,4], Hanafi Ismail [5], Nadras Othman [5] and Nabil Hayeemasae [3,4,*]

1 Department of Chemical Engineering, Faculty of Engineering, Universitas Sumatera Utara, Medan 20155, Sumatera Utara, Indonesia; isurya@usu.ac.id
2 Islamic Sciences Demonstration School, Prince of Songkla University, Pattani Campus, Pattani 94000, Thailand; qamarud@hotmail.com
3 Research Unit of Advanced Elastomeric Materials and Innovations for BCG Economy (AEMI), Faculty of Science and Technology, Prince of Songkla University, Pattani Campus, Pattani 94000, Thailand; sitisaiyidah.s@psu.ac.th
4 Department of Rubber Technology and Polymer Science, Faculty of Science and Technology, Prince of Songkla University, Pattani Campus, Pattani 94000, Thailand
5 School of Materials and Mineral Resources Engineering, Engineering Campus, Universiti Sains Malaysia, Nibong Tebal, Penang 14300, Malaysia; ihanafi@usm.my (H.I.); srnadras@usm.my (N.O.)
* Correspondence: nabil.h@psu.ac.th; Tel.: +66-73-312-213; Fax: +66-73-331-099

Citation: Surya, I.; Waesateh, K.; Saiwari, S.; Ismail, H.; Othman, N.; Hayeemasae, N. Potency of Urea-Treated Halloysite Nanotubes for the Simultaneous Boosting of Mechanical Properties and Crystallization of Epoxidized Natural Rubber Composites. *Polymers* **2021**, *13*, 3068. https://doi.org/10.3390/polym13183068

Academic Editor: Domenico Acierno

Received: 12 August 2021
Accepted: 7 September 2021
Published: 11 September 2021

Abstract: Halloysite nanotubes (HNTs) are naturally occurring tubular clay made of aluminosilicate sheets rolled several times. HNT has been used to reinforce many rubbers. However, the narrow diameter of this configuration causes HNT to have poor interfacial contact with the rubber matrix. Therefore, increasing the distance between layers could improve interfacial contact with the matrix. In this work, Epoxidized Natural Rubber (ENR)/HNT was the focus. The HNT layer distance was successfully increased by a urea-mechanochemical process. Attachment of urea onto HNT was verified by FTIR, where new peaks appeared around 3505 cm^{-1} and 3396 cm^{-1}, corresponding to urea's functionalities. The intercalation of urea to the distance gallery of HNT was revealed by XRD. It was also found that the use of urea-treated HNT improved the modulus, tensile strength, and tear strength of the composites. This was clearly responsible for interactions between ENR and urea-treated HNT. It was further verified by observing the Payne effect. The value of the Payne effect was found to be reduced at 62.38% after using urea for treatment. As for the strain-induced crystallization (SIC) of the composites, the stress–strain curves correlated well with the results from synchrotron wide-angle X-ray scattering.

Keywords: epoxidized natural rubber; halloysite nanotubes; urea; tensile properties; wide-angle X-ray scattering

1. Introduction

Incorporation of fillers into rubber has been a major ingredient for rubber compounds. They are added to rubber for many purposes, such as improving durability, thermal stability, and even to cheapen the manufacturing costs [1]. So far, special awareness has been given to the addition of nano-fillers because they effectively boost the performance of rubber vulcanizate even at very low content. The factors affecting such enhancement are mainly due to the aspect ratio, degree of dispersion, and orientation of high aspect ratio filler particles [2]. Halloysite nanotubes (HNT) are a type of naturally occurring nano-filler composed of aluminum, silicon, oxygen, and hydrogen. Being an economically viable material, HNT has been available in many interesting research works; the novel properties and related applications of HNT have been successively reported over decades.

For instance, the use of HNT as a viable nanoscale container in drug delivery, their usage as controlled release [3], as well as other specific applications, namely an ion adsorbent, ceramic materials, especially biocompatible implants, and as template for synthesis of rod-like nanoparticles [4].

HNT have recently been added in various types of rubber matrices [5,6]. However, HNT are not compatible with non-polar rubbers due to its polarity in nature [7,8] and are especially not compatible with natural rubber (NR). Scientists have been trying to solve this problem and improve their compatibility using several methods. These cover using silane coupling agents, optimizing the preparation methods, and modifying the chemical structure of rubber matrix. The last approach might be of great practical relevance to ensure the compatibility at any location along the polymer molecular chain [9,10]. In this work, an Epoxidized Natural Rubber (ENR) was introduced for preparing the composite to assure filler–matrix compatibility. However, there are still concerns regarding the characteristics of HNT itself. HNT is made of aluminosilicate kaolin sheets rolled several times. The narrow layer distance in such rolled sheet gives HNT poor interfacial contact with the rubber matrix during processing. Therefore, increasing the layer separation distance could improve the interfacial contact.

Recently, increasing the basal distance of the HNT layers has been one of the major approaches to improve the interfacial adhesion between HNT and rubber matrix [11]. Nicolini et al. [12] observed that urea can increase the basal spacing through a mechano-chemical process. It was reported that the basal distance shifted from 7.4 Å to 10.7 Å when the urea percentage increased and stabilized when the amount of urea approached 18%. It is claimed that the layers of HNT were intercalated by urea, making a material with a basal distance of 10.68 Å and with an expansion of 3.5 Å [13]. By increasing interlayer separation in HNT, easier intercalation of a polymer inside the HNT lumen and/or in HNT interlayer is assumed. This study only focused on the treatment of HNT itself without any inclusion of a polymer matrix.

Therefore, the aim of this study is to use urea-treated HNT in the ENR matrix to improve the overall properties of the composites. With this idea, the interfacial adhesion between HNT and ENR may be improved, while good rubber-filler interactions are assured by the polarity of ENR. The filler modification is expected to improve the compatibility and homogeneity of filler dispersion in the composites and to thereby enhance the reinforcing efficiency of HNT in filled ENR composites. This study proposed certain methods of evaluating the reinforcing efficiency of the composites, namely mechanical properties, dynamic properties, and SIC. The latter method is considered an interesting route, and not many reports have been focused on it. This can only be correlated for certain types of rubber, such as NR [14–16]. This is because NR has a very long chain that has an ability in arrangement and orientation for crystallization under stretching [17]. This ability to crystallize under strain is explained by the high regularity of molecular structure that is almost 100% of cis-1,4-polyisoprene [18].

Many studies of NR have been carried out together with in-situ deformation through in-situ X-ray diffraction techniques. Tosaka et al. [19] studied the effect of the different crosslink density on the SIC of vulcanized rubber, and they found that faster development in crystallinity was observed in the sample with higher crosslink density but at limited amounts. Toki et al. [20] indicated that the crystallinity increased with strain. They considered that stretched rubber can be differentiated into three phases: (i) an un-oriented amorphous phase, (ii) an oriented amorphous phase, and (iii) a crystalline phase. In ENR, the SIC lattice parameters of ENR were found to be crosslink independent regardless of quantity and method of crosslink. Higher crosslink density may induce faster SIC, and the volume of unit cell is larger than the reported values for NR, which is simply due to the epoxide group at the main chain [21].

SIC of unfilled and filled NR has also been focused by Poompradub et al. [22]; they found out that the onset strain for SIC decreased after adding filler. The degree of lattice deformation decreased with filler content, especially in the carbon black (CB)-filled composite. Chenal et al. [23] further explained that onset strain for SIC is ruled by the strain amplification induced in the presence of filler. Furthermore, different fillers behave different characteristics associating to the formation of rubber-filler interaction/reactions. This can be either accelerate or slow down SIC depending on NR matrix–chemical crosslink density. Similar observation was observed in vulcanized NR containing CB particles by Candau et al. [24].

From the reports above, the rubber–filler interactions may speed up the crystallization process at the certain network chain density. In this report, we presented parallel wide-angle X-ray scattering and tensile measurement on urea-treated HNT-filled ENR composites. To date, no reports have been revealed of detailed investigations regarding the relationship between mechanical and dynamic properties with SIC of rubber composites. The results explored from this work aim to give a scientific understanding of how the role of urea affects the overall properties of ENR/HNT composites and will be useful for manufacturing of rubber products based on ENR/HNT composites

2. Experimental Details

2.1. Materials

High ammonia centrifuged latex (HA) with 60% of dry rubber content (DRC) was used to prepare ENR. This latex was centrifuged and supplied by Chalong Latex Industry Co., Ltd., Songkhla, Thailand. The chemicals involved in the synthesis of ENR were Teric N30 as non-ionic surfactant, Formic acid and hydrogen peroxide for performic acid reaction were purchased from Sigma Aldrich (Thailand) Co. Ltd., Bangkok, Thailand. The HNT was mined and supplied by Imerys Ceramics Limited, Matauri Bay, New Zealand. HNT consists of the following components: SiO_2 (49.5 wt%), Al_2O_3 (35.5 wt%), Fe_2O_3 (0.29 wt%), TiO_2 (0.09 wt%), as well as CaO, MgO, K_2O, and Na_2O as traces. Stearic acid was supplied by Imperial Industrial Chemicals (Thailand) Co., Ltd., Bangkok, Thailand. ZnO was supplied by Global Chemical Co., Ltd., Samut Prakan, Thailand. N-cyclohexyl-2-benzothiazole sulfenamide was obtained from Flexsys America L.P., Akron, OH, USA, and soluble sulfur was purchased from Siam Chemical Industry Co., Ltd., Samut Prakan, Thailand.

2.2. Preparation of Epoxidized Natural Rubber

The synthesis of ENR was done by diluting DRC of latex to 15%. Next, 1 phr of non-ionic stabilizer (10% Teric N30) was added while stirring for 30 min at ambient temperature to expel the ammonia dissolved in HA. The epoxidation was performed using formic acid and hydrogen peroxide at 50 °C in a 10-L glass container under a stirring rate of 30 rpm. The total reaction time was fixed to obtain ENR with 20 mol% epoxide. The resulting ENR was coagulated with methanol followed by washing with water. Finally, it was dried in a vacuum oven at 50 °C prior to use.

2.3. Preparation of Urea-Treated HNT

As for the urea-treated HNT, it was prepared according to Nicolini et al. [12]. The contents of urea varied among 10%, 14%, 18%, and 20% of HNT. Later, the samples were labeled as ENR20 for untreated HNT and as ENR20U10–ENR20U20 for the urea contents from 10% to 20%, respectively. The intercalation of urea into the interlayer of HNT was enabled mechano-chemically. Firstly, HNT was mixed with the urea and ground by ball milling in a ceramic container. The urea-treated HNT was purified by washing with isopropanol and dried in an oven at 70 °C for 12 h. The urea-treated HNT was finally ground in a mortar prior to use in compounding.

2.4. Preparation of ENR/HNT Composites

The recipe for the preparation of ENR/HNT composites is given in Table 1. ENR with 20 mole% epoxide was compounded with HNT (e.g., untreated and urea-treated HNT depending on the formulation) and other ingredients except for the curatives (CBS and sulfur) in a Brabender plasticorder. The initial mixing temperature was set at 50 °C with a rotor speed of 60 rpm. The compound was then sheeted on a two-roll mill while the curatives were incorporated. Finally, samples of the various compounds were later tested for curing characteristics.

Table 1. Formulation of ENR composites filled with untreated and urea-treated HNT.

Raw Material	Amount (phr)
ENR	100.0
Stearic acid	1.0
Zinc oxide	5.0
HNT *	5.0
CBS	2.0
Sulfur	2.0

Remark: * HNT was treated with various content of urea.

2.5. Measurement of Curing Characteristics

The curing properties of the composites were measured according to ASTM D5289 using a moving die rheometer (Rheoline, Mini MDR Lite, Prescott Instruments Ltd., Tewkesbury, UK). The operating temperature was set at 150 °C. The data in terms of torques, scorch time (t_{s2}), and curing time (t_{c90}) were recorded. The t_{s2} and t_{c90} were used in calculating the curing rate index (CRI) as follows:

$$CRI = \frac{100}{t_{c90} - t_{s2}} \quad (1)$$

2.6. Fourier Transform Infrared-Spectroscopic Analysis (FT-IR)

Attachment of urea onto HNT was confirmed by Fourier transform infrared-spectroscopy (FTIR) using FTIR spectroscope model TENSOR27 (Bruker Corporation, Billerica, Massachusetts, USA). The spectra were recorded in transmission mode with a 4 cm^{-1} resolution over 4000–550 cm^{-1}.

2.7. X-ray Diffraction Analysis (XRD)

The XRD analysis of pure HNT and urea-treated HNT was carried out by using PHILIPS X'Pert MPD (Eindhoven, Netherlands) with CuKα radiation (λ = 0.154 nm) at 40 kV and a current of 30 mA and Bruker D2 Phaser (Billerica, MA, USA) with CuKα radiation source (λ = 0.154 nm) and a current of 10 mA. The diffraction patterns were scanned in the diffraction angles 2θ of 5–30° with a step size of 0.05° and 3°/min scan speed. The d-spacing of HNT layers in particles was estimated from Bragg's equation.

2.8. Measurement of Mechanical Properties and Hardness

Measurement of tensile properties was done according to ASTM D412. The sample's dimension was based on Die C dumbbell shape. The test was performed using a universal testing machine (Tinius Olsen, H10KS, Tinius Olsen Ltd., Surrey, UK) at a crosshead speed of 500 mm/min. The determinations recorded were the modulus at 100% (M100) and 300% (M300) strains, tensile strength, and elongation at break. The tear strength of the respective composites was tested using the same machine and crosshead speed according ASTM D624. A type C (right angle) test piece was selected for experiment. The last measurement was the hardness property of the samples. It was performed according to ASTM D2240 using a Shore A type manual durometer.

2.9. Determination of Crosslink Density

The crosslink density of the composite was determined by equilibrium swelling method as described in ASTM D6814. The specimens were cut into a circular shape and weighed before and after immersing in toluene for 72 h. The modified Flory–Rehner equation was implemented for calculating the cross-link density (v) [25]:

$$\nu = \frac{1}{2M_c} \tag{2}$$

$$M_c = \frac{\rho \cdot V_0 \cdot \left(V_r^{\frac{1}{3}} - \frac{V_r}{2}\right)}{ln(1 - V_r) + V_r + \mu \cdot V_r^2} \tag{3}$$

where M_c is the number-average molecular weight of the rubber chains between crosslinks, μ is the parameter for rubber-toluene interactions (μ = 0.42), ρ is the bulk density of the specimen, V_0 is the molar volume of the toluene (V_0 = 106.2 cm^3/mol), and V_r is the volume fraction in the swollen specimen, defined as follows [26]:

$$V_r = \frac{(D - FT) \cdot \rho^{-1}}{(D - FT) \cdot \rho^{-1} + A_0 \cdot \rho_s^{-1}} \tag{4}$$

where T is the weight of the specimen, D is the weight of the de-swollen specimen, F is the weight fraction of the insoluble parts, A_0 is the weight of the toluene absorbed in swollen specimen, ρ is the density of the specimen, and ρ_s is the density of the toluene (0.886 g/cm^3).

2.10. Scanning Electron Microscopy

The freshly fractured surfaces of samples from tensile testing were used to observe the dispersion of untreated and urea-treated HNT in the rubber matrix. The morphology was captured using a scanning electron microscope (SEM; FEI Quanta FEG 400, Thermo Fisher Scientific, Waltham, MA, USA). A layer of gold/palladium coated the specimen to eliminate charge built during imaging.

2.11. Dynamic Properties

The dynamic properties of the composites were analyzed using a Rubber Process Analyzer (RPA), model D-RPA 3000 (MonTech Werkstoffprüfmaschinen GmbH, Buchen, Germany). At first, the tested samples were cured at 150 °C based on the t_{c90} tested by the same RPA. The samples were then cooled down to 60 °C. At this time, at 10-Hz fixed frequency, the strain was varied from 0.5 to 90%. This was to determine the storage modulus (G') as function of strain in the composites. The raw G' record was further used to study the filler–filler interactions via a so-called Payne effect. The Payne effect was calculated as follows:

$$\text{Payne effect} = G'_I - G'_f \tag{5}$$

where G'_i and G'_f were the G' at 0.5% and 90% strains, respectively. A larger Payne effect indicates weaker rubber–filler interactions.

2.12. Wide-Angle X-ray Scattering

The SIC of the composite was correlated with the stress-strain curves of the composite. SIC and other related results were obtained from a synchrotron wide-angle X-ray scattering (WAXS) analysis. The experiment was carried out using Beamline 1.3 W at the Siam Photon Laboratory, Synchrotron Light Research Institute (SLRI), Nakhon Ratchasima, Thailand. The distance between sample and detector was 115.34 mm, measured using a wavelength of 0.138 nm. A CCD detector (Rayonix, SX165, Rayonix, L.L.C., Evanston, IL, USA) with a diameter of 165 mm was equipped to capture the WAXS profile. The scattering angle was calibrated using 4-Bromobenzoic acid as a standard material.

Prior to the testing, a Die C type dumbbell specimen was held on the grips of a stretching apparatus. The sample was stretched at a crosshead speed of 50 mm/min to a given strain and was then relaxed in the deformed state for 30 s. WAXS was recorded and stretching then continued to the next predetermined strain until the characterization was complete. The degree of crystallinity (X_c) was calculated based on the data obtained from the WAXS profiles using the following equation:

$$\text{Degree of crystallinity } (X_c) = \left(\frac{A_c}{A_c + A_a}\right) \times 100 \tag{6}$$

where A_c is and A_a are the areas under the crystalline peak of interest and the amorphous halo, respectively.

The orientation parameter (OP) was also determined from the Hermann equation, as follows:

$$\text{OP} = \frac{3[\cos^2\varphi] - 1}{2} \tag{7}$$

where φ is the azimuthal angle related to the direction of strain. The mean value of $cos^2\ \varphi$ is calculated as follows:

$$[\cos^2\varphi] = \frac{\int_0^{\pi} I_c(\varphi) \cdot \cos^2\varphi \cdot \sin\varphi \cdot d\varphi}{\int_0^{\pi} I_c(\varphi) \cdot \sin\varphi \cdot d\varphi} \tag{8}$$

where I_c (φ) is the scattering intensity of the crystal at φ. I_c (φ) is normalized by subtracting the minimum scattering intensity of the amorphous component of the original WAXS intensity [27,28].

3. Results and Discussion

3.1. Curing Characteristics

Curing curves of the composites in the presence of untreated and urea-treated HNT are shown in Figure 1. The raw data are summarized in Table 2. It was found that the scorch time (t_{s2}) and curing time (t_{c90}) decreased, while CRI increased with urea content. Theoretically, an alkaline chemical substance relatively accelerates the vulcanization process, while an acidic compound would retard it. Urea is an amine substance with alkaline characteristics [29,30]. Therefore, increasing urea content contributed to the rate of vulcanization.

Table 2. Scorch time (t_{s2}), cure time (t_{c90}), minimum torque (M_L), maximum torque (M_H), delta torque (M_H–M_L), and CRI for the ENR/HNT composites made with untreated or urea-treated HNT.

Sample	t_{s2} (min)	t_{c90} (min)	M_L (dN.m)	M_H (dN.m)	M_H–M_L (dN.m)	CRI (min^{-1})
E20	2.29	4.75	0.76	8.40	7.64	40.65
E20U10	1.38	3.51	0.71	7.92	7.21	46.95
E20U14	1.42	3.52	0.68	8.35	7.67	47.62
E20U18	1.02	2.82	0.64	7.84	7.20	55.56
E20U20	1.18	3.06	0.63	7.74	7.11	53.19

It was observed that the use of urea-treated HNT slightly changed in M_L, while the M_H increased to its maximum at 14% of urea. This increase in M_H of the composites was attributed to the interactions of HNT and ENR with urea. Above 14% urea content, M_H decreased again, which can be attributed to the plasticizing effects of the urea. Similar observation was made regarding the delta torque (M_H–M_L), indicating that urea plays an important role in these composites.

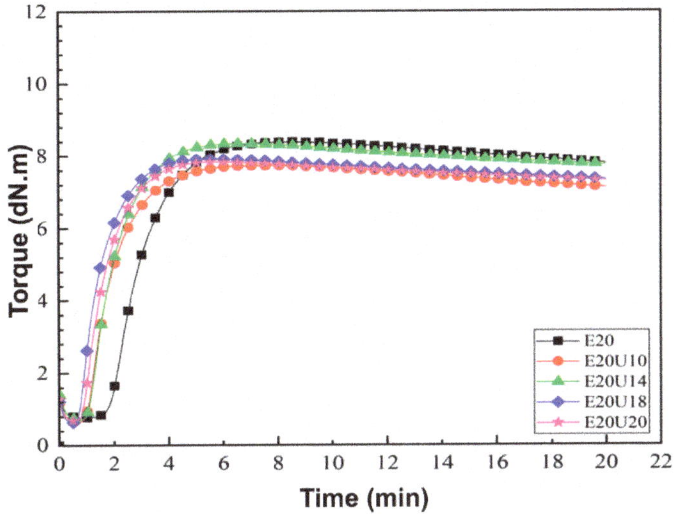

Figure 1. Curing characteristic curves of ENR/HNT composites filled with untreated and urea-treated HNT.

3.2. FT-IR Analysis

The FTIR spectra of urea, raw HNT, and urea-treated HNT are given in Figure 2. For the pure urea, peak absorption at 3432 cm^{-1} is assigned to N-H out-of-plane stretching vibrations and the bands at 3336 cm^{-1} and 3258 cm^{-1} are formed by N-H in-plane stretching vibrations. The N-H bending vibrations are also absorbing at 1619 cm^{-1} and 1592 cm^{-1}. The peak at 1460 cm^{-1} is assigned to the stretching vibrations of C-N, while the NH_2 rocking vibrations give absorption at 1152 cm^{-1} [31]. This clearly corresponds to the chemical structure of urea. The increased transmittance in the ranges 3500–3300 cm^{-1} and 1800–1400 cm^{-1} was clear as the level of urea was increased, corresponding to the reference band of pure urea. This corroborates that the urea was attached on the surfaces of HNT. The inner surface OH was represented at 3694 cm^{-1} and 3622 cm^{-1}.

To ensure that intercalation of urea to HNT was happening, the spectrum in the range 1800–1400 cm^{-1} was considered. Basically, the amine group available in urea can form hydrogen bonds with the hydroxyl groups on HNT. This could have caused the an amine band from 1618 cm^{-1} for pure urea to 1625 cm^{-1}, 1624 cm^{-1}, and 1624 cm^{-1} for urea-treated HNT. This happens together with the appearance of new peaks at around 3505 cm^{-1} and 3396 cm^{-1} [8] and also with shifting of carbonyl peak at 1680 cm^{-1} to 1674 cm^{-1}, 1670 cm^{-1}, and 1668 cm^{-1} [32]. Such findings are in good agreement with the report of Makó et al. [33], showing that the kaolinite–urea complex was formed after modification. Since HNT is chemically similar to clay, its interaction mechanisms with urea are expected to be similar.

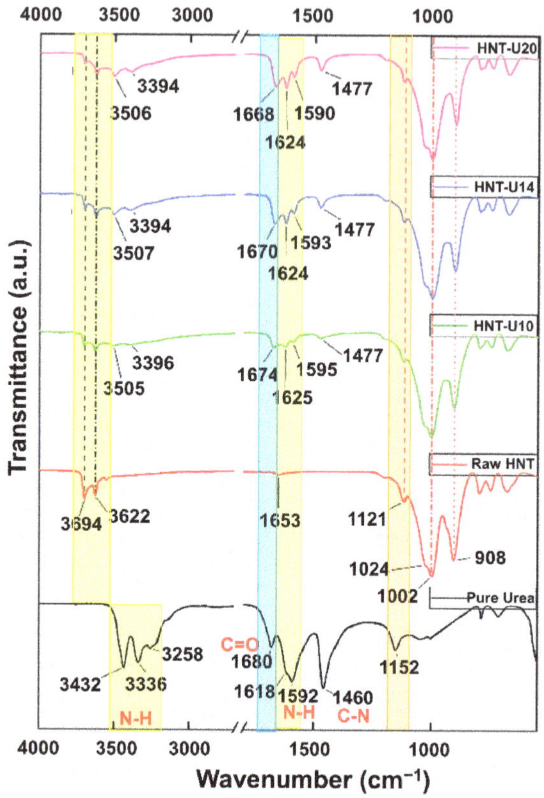

Figure 2. Infrared spectra of pure urea and untreated and urea-treated HNT.

The FTIR spectra of the ENR composites filled with untreated and urea-treated HNT are shown in Figure 3. The hydrocarbon characteristics absorption peaks at 1662 cm^{-1}, 1448 cm^{-1}, 1375 cm^{-1}, and 837 cm^{-1} relate to the stretching vibrations of C=C bonds, bending vibrations of CH_2 and CH_3 groups, and out-of-plane deformations of =C-H group, respectively. The epoxide ring absorption took place at 873 cm^{-1} and 1250 cm^{-1}. Interestingly, the absorption at 3694 cm^{-1} and 3622 cm^{-1} was assigned to inner surface OH, and the outer OH groups are reduced with increasing urea contents [34,35]. Broadened shoulder peak at 1152 cm^{-1} and weakened peak absorption at 3694 cm^{-1}, 3622 cm^{-1}, and 912 cm^{-1} were also found. This is a clear indication that the treatment of HNT with urea reduced transmittance of the inner surface hydroxyl groups on HNT, the interactions formed between HNT and urea. It is worth noting that the location of the Si–O stretching in the region of 1100–1000 cm^{-1} was shifted with the addition of urea. This suggests that ENR interacted with the surfaces of HNT through its epoxide groups, making hydrogen bonds with hydrogen attached to electronegative N atom in urea-treated HNT [36].

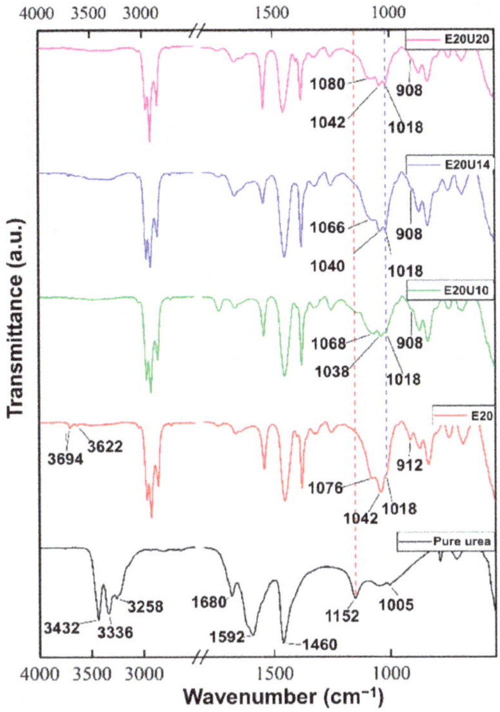

Figure 3. Infrared spectra of ENR/HNT composites in the presence of untreated and urea-treated HNT.

3.3. X-ray Diffraction Analysis

The XRD patterns of the untreated and urea-treated HNT are shown in Figure 4. The untreated HNT shows a 001 reflection at 2θ of 12.05°, indicating basal distance of 7.33 Å, which is characteristic of dehydrated Halloysite. This peak tended to disappear as the amount of urea was increased. All the HNT crystals were fully intercalated with urea, resulting an increase in basal distance from 7.33 Å (2θ = 12.05°) to 10.72 Å (2θ of 8.49° and 8.23°). This is verified by shifting of the 001 peak from 2θ of 12.05° to 8.49° and 8.23°. One important feature of intercalated HNT is the development of two peaks at 21.54° and 22.53° in the range with saw tooth diffraction peaks. These may be associated with decreased lattice strain and/or increased crystallite size [33]. A schematic model proposed for the interactions in urea-treated HNT is shown in Figure 5. This scheme demonstrates an increase of basal distance between layers of HNT due to the penetration of urea; hydrogen bonds are formed between the hydroxyl groups of HNT and amine contained in urea.

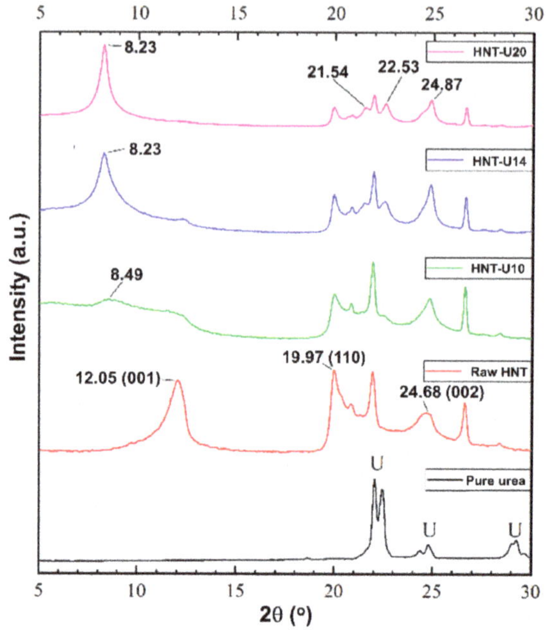

Figure 4. XRD scattering patterns of pure urea and untreated and urea-treated HNT.

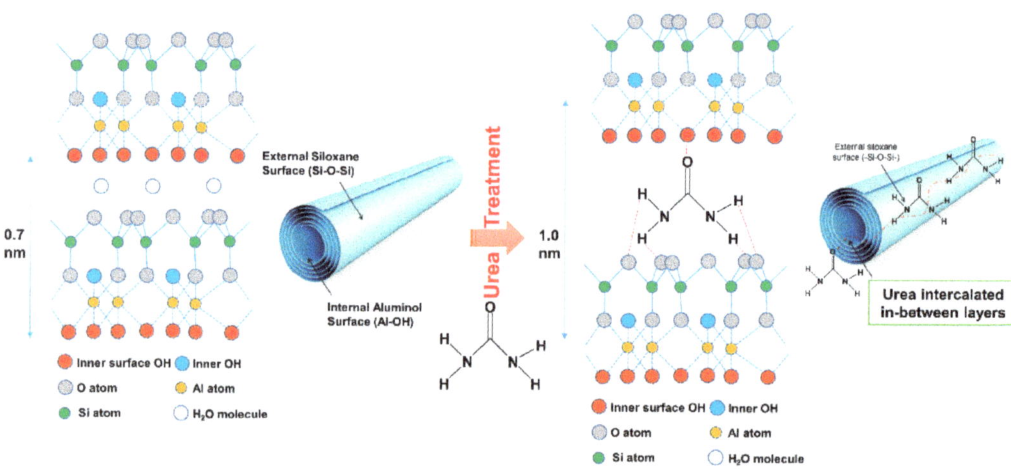

Figure 5. Proposed model of intercalation and interactions of urea with HNT via mechano-chemical processing modified from Yuan et al. [37].

Figure 6 shows the XRD patterns of the ENR composites filled with untreated and urea-treated HNT. The diffraction peak at 2θ of 12.05° (d is 7.33 Å) related to 001 plane and is absent after urea-treatment of HNT. This peak is shifted to the lower 2θ at 8.30°, 8.06°, 9.41°, or 8.60°, which correspond to basal gallery sizes of 10.64 Å, 10.95 Å, 9.38 Å, and 10.27 Å, respectively. The modification of HNT by mechano-chemical processing altered the crystal lattice of HNT. The modification with urea of HNT increased the basal distance, and hence, a lesser 2θ is presented due to the intercalation of ENR molecules to the gallery of HNT [38,39]. Once the intercalation occurred, possibly there were also further interactions between ENR and HNT in the system.

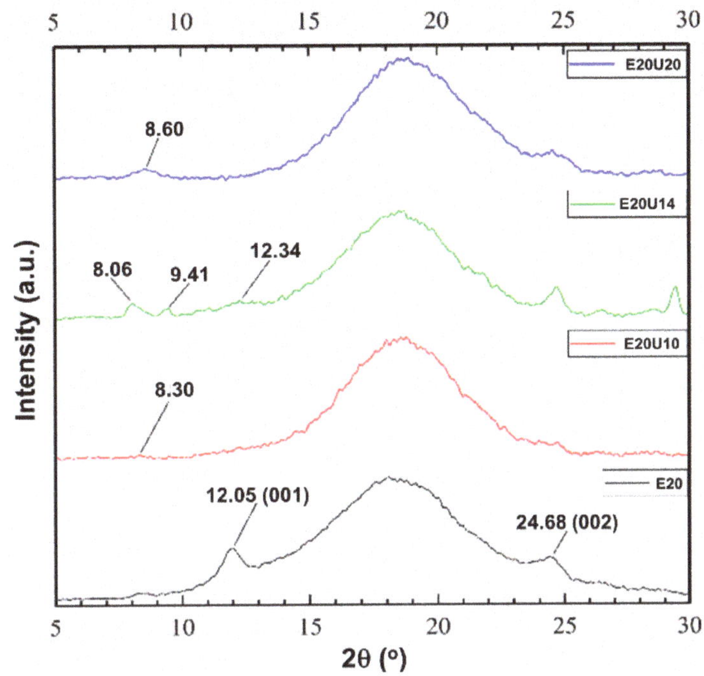

Figure 6. XRD scattering patterns of ENR/HNT composites filled with untreated and urea-treated HNT (un-deformed specimen).

3.4. Dynamic Properties

One approach to assessing possible interactions in a composite is by use of the dynamic properties. In this study, storage modulus and the Payne effect of the ENR/HNT composites made with untreated or urea-treated HNT filler were analyzed to assess the rubber–filler interactions. The results are presented in Figure 7. The storage modulus (G') of the composites was constant in the low strain region but slightly decreased with strains larger than 50%. This is common for a viscoelastic material and is due to the molecular stability of rubber. It is noticeable that the G' increased with urea content, indicating interactions between urea-treated HNT with ENR. There are two factors determining the increase in G', namely the dispersion of HNT facilitated by urea as well as improved interfacial interactions between HNT and ENR caused by urea-treatment of the HNT. The Payne effect relates to the rubber–filler interactions in the composite, and its measure here was the difference in G' between low and high strains [40]. A higher Payne effect indicates lesser rubber–filler interactions. It was found that the Payne effect decreased with urea content, where the value of Payne effect was found to be reduced at 62.38% after using urea for treatment. This might be due to stable rubber–filler interactions of urea-treated HNT filler in the ENR matrix. The results on the dynamic properties confirm interactions between urea-treated HNT and ENR, supporting the previous FTIR and XRD.

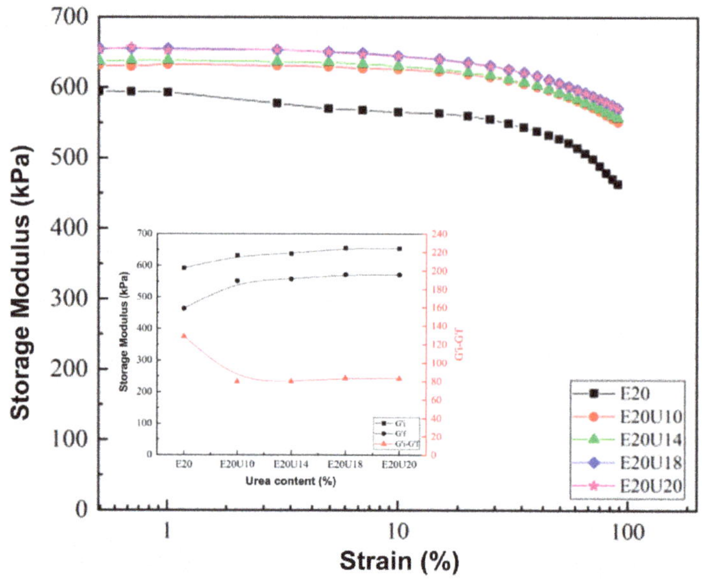

Figure 7. Strain dependence of storage modulus and Payne effect in ENR/HNT composites filled with untreated and urea-treated HNT.

3.5. Mechanical Properties

Figure 8 shows the stress–strain curves of the ENR composites made with untreated and urea-treated HNT. The stress–strain curves show the SIC. Higher stress response was found for the composites filled with urea-treated HNT, suggesting that the samples became stronger after urea-treated HNT was used. Higher compatibility between ENR and HNT in the presence of urea is responsible for these findings. Further, the area underneath the stress–strain curve was examined to confirm the compatibility of rubber and the filler. This indicates the toughness of a material [41]. Larger area underneath the curve corresponds to the greater toughness. The urea-treated HNT composites showed a greater area underneath the stress–strain curve than the untreated counterpart and therefore greater toughness. The curves shown are further discussed and regards crystallization behavior.

The mechanical properties, such as modulus, tensile strength, elongation at break, tear strength, and hardness, are shown and summarized in Table 3. It was found that the modification of HNT with urea via mechano-chemical process led to increase in the tensile and tear strength of the composites. The tensile strength with untreated HNT was 33.67 MPa and increased up to 35.15 MPa at 14% of urea content. The tear strength of reference sample increased from 38.29 N/mm to 38.36, 39.24, 37.42, and 35.60 N/mm. Using urea-treated HNT has evidently proven that an intercalation had taken place. The evidence for such boosting has already been shown by the previous sessions. The reduction of tensile and tear strength at urea content over 14% might be due to some destruction of the HNT structure that occurred during mechano-chemical process, as seen in SEM images. The significant change in the rubber–filler interaction of ENR and HNT can be also verified from the stresses at 100% (M100) and 300% (M300) strains (see Table 3). It can be seen that the M100 and M300 increased over the urea content. As the higher urea was introduced to the rubber, the greater interactions occurred, resulting in harder and stiffer composites. Such finding is clearer when examining the M300. The result obtained here corresponded well with the reduction of elongation at break of the composites. The reduction of elongation at break was found due to lower flexibility of molecular chain and contributed to the increase of interaction point in the composites. Similar observation for the modulus was also encountered for the hardness property.

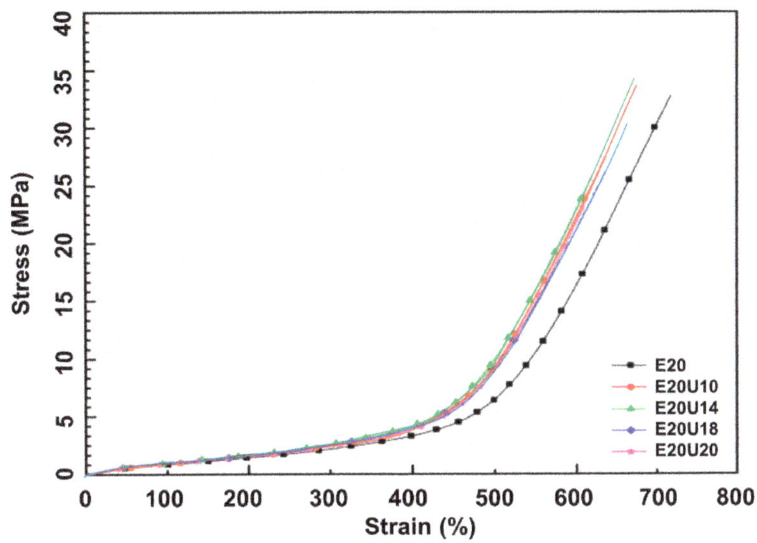

Figure 8. Stress–strain dependence of ENR/HNT composites filled with untreated and urea-treated HNT.

Table 3. Modulus at 100% (M100), 300% (M300), tensile strength (TS), elongation at break (EB), tear strength (Ts), and hardness of ENR/HNT composites filled with untreated and urea-treated HNT.

Sample	M100 (MPa)	M300 (MPa)	TS (MPa)	EB (%)	Ts (N/mm)	Hardness (Shore A)
E20	0.86 ± 0.03	2.25 ± 0.03	33.67 ± 1.61	717 ± 10	38.29 ± 0.94	39.3 ± 0.3
E20U10	0.89 ± 0.04	2.34 ± 0.23	34.95 ± 0.51	660 ± 49	38.36 ± 0.51	41.6 ± 0.4
E20U14	0.95 ± 0.05	2.58 ± 0.18	35.15 ± 0.42	627 ± 28	39.24 ± 0.54	42.2 ± 0.4
E20U18	0.96 ± 0.01	2.59 ± 0.07	30.59 ± 1.22	618 ± 20	37.42 ± 0.72	42.7 ± 0.3
E20U20	0.97 ± 0.03	2.63 ± 0.03	26.87 ± 1.11	615 ± 13	35.60 ± 0.50	43.9 ± 0.7

3.6. Morphological Properties

The dispersion of HNT within the rubber matrix can be assessed from the SEM images shown in Figure 9. Good dispersion of urea-treated HNT was observed with but shortened lengths of the dispersed HNT particles in the matrix are due to the destruction of HNT structure. This is in agreement with the decreased tensile strength and tear strength observed for comparatively high urea contents. Similar observations were previously reported for micro-fractured surfaces in NR composites having other fillers in the presence of a compatibilizer [42].

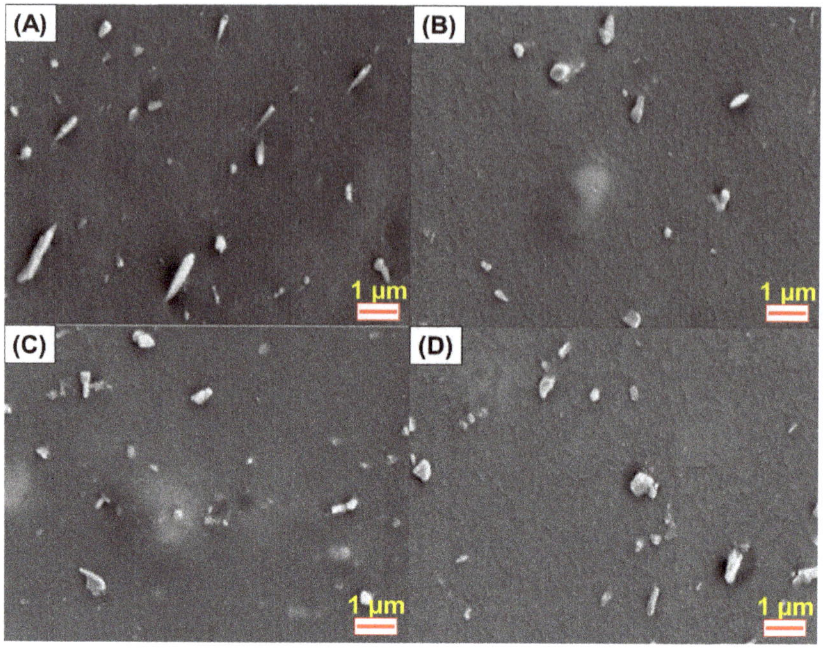

Figure 9. SEM photographs of ENR/HNT composites filled with untreated and urea-treated HNT; E20 (**A**), E20U10 (**B**), E20U14 (**C**), and E20U20 (**D**).

3.7. Wide-Angle X-ray Scattering

In the section on mechanical properties, the stress–strain behavior of the composites was associated with SIC. Since the nominal strain rate for tensile measurement and SIC is not similar (e.g., 0.42 s^{-1} and 0.042 s^{-1} for tensile test and WAXS, respectively). The correlation was made in the view of the stress versus crystallinity only. Previously, it was clear that the treatment of HNT by urea influenced the mechanical properties. The main factor is definitely the improved compatibility between ENR matrix and urea-treated HNT filler. The degree of crystallinity (X_C) versus strain deformation is shown in Figure 10. Crystallinity was estimated from the areas in diffraction pattern for 200 and 120 plane reflections [43,44]. The X_C increased with strain due to molecular chain orientation, as expected. The onset strain for SIC was determined from interception of a regression line for X_c as a function of strain (see Figure 11). The onset for SIC with urea-treated HNT filler was observed to be lower as a function of urea content. The interaction that takes place in the presence of urea can help to pull the surrounding molecular chains and speeds up the crystallization process. When considering the X_C and the stress propagation from tensile measurement (see Figure 8), it is clear that the X_C corresponded well with the stress, and the trend of the curves is also similar as urea content increased. From the stress–strain curves, it is obvious that the stress started to increase towards the urea content. This is responsible to the formation of interfacial contact, as discussed earlier.

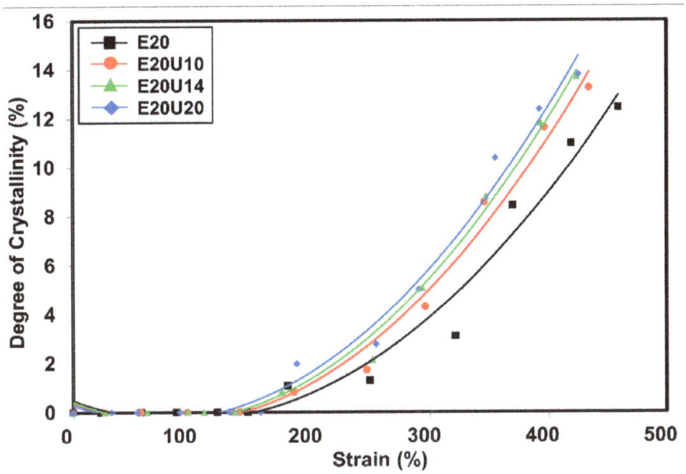

Figure 10. Strain dependence on degree of crystallinity of ENR/HNT composites filled with untreated and urea-treated HNT.

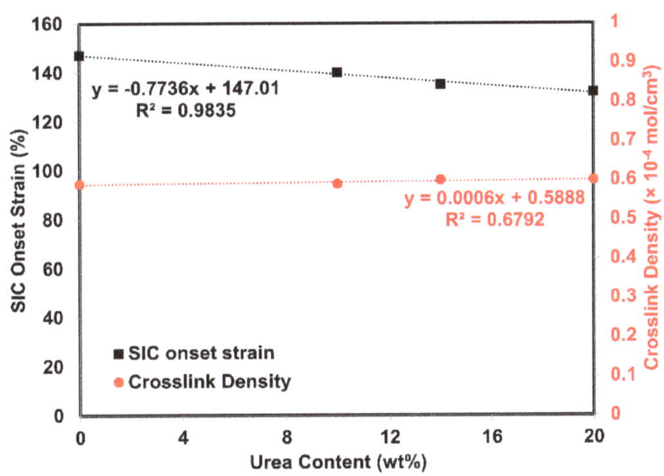

Figure 11. SIC onset strain (primary axis) and crosslink density of ENR/HNT composites filled with untreated and urea-treated HNT.

Earlier onset for SIC can be found usually for the filled composite. Poompradu et al. [22] reported that the lateral crystallite size was decreased, but the orientational fluctuation increased by the inclusion of filler. The lattice of the SIC changed almost linearly with the nominal stress. In addition, the degree of lattice deformation decreased with the filler content, especially in the CB-filled system. In addition to this, onset for SIC was differently observed depending on the filler's characteristics. Ozbas et al. [45] compared the SIC of graphene and CB-filled composites. They found that the onset of SIC occurs at significantly lower strains for graphene-filled NR samples compared with CB-filled NR even at low loadings. Chenal et al. [23] further explained that the onset of SIC is ruled by the strain amplification induced by the filler presence. Moreover, additional interaction in rubber network is also responsible for either accelerating or slowing down the crystallization rate depending on rubber matrix chemical crosslink density. Candau et al. [24] together with the report of Ozbas et al. [45] further emphasized that the rubber–filler interactions may fasten the SIC at low crosslink density. This is because high crosslinking may interfere the chain orientation and results in reduction of the SIC. Therefore, the crosslink density of this composite was also reported and shown in Figure 11. It can be seen that the crosslink density observed was more or less the same over the content of urea. This is a good indication that network chain density is unchangeably involved in the development of SIC regardless of urea content. As a consequence, it can be said that the change in SIC was promoted by rubber–filler interactions.

The orientation parameter (OP) indicates indirectly the molecular chain orientation and alignment and can be calculated from the Herman equation [27,28]. The OP for the composites is shown in Figure 12. Completely oriented molecular chains would give OP value 1 [46]. Here, the OP for the composites was lesser at low strains and grew with increasing strain, confirming that stretching oriented the molecular chains. The composites filled with urea-treated HNT showed higher OP values at low strains, indicating that the modification of HNT increases rubber–filler interactions, and accordingly, more molecular chain orientation was found for composites with urea-treated HNT.

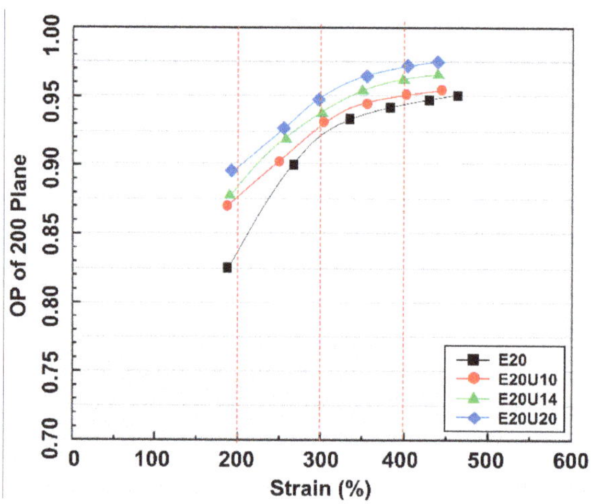

Figure 12. Strain dependence on orientation parameter of ENR/HNT composites filled with untreated and urea-treated HNT.

From these results, correlation between the SIC and corresponding interaction between ENR and HNT in the presence of urea is represented in a schematic model shown in Figure 13. Referring to the scheme, nothing happened at the un-stretched stage of the sample; the ENR matrix may be in contact with the HNT due to the interfacial interaction created by the unique characteristics of the HNT. When the strain was applied to the sample, crystallization of the ENR was then induced due to the stress concentration point on the HNT surfaces, and the crystallinity increased in association with the orientation of the HNT. HNT is orientated and aligned along the stretching direction. As a consequence, the ENR chains rearrange and crystallize. This always happens regardless of whether untreated or urea-treated HNT is concerned. This kind of phenomenon usually occurs in the filled composites, and it has been reported elsewhere [23,24,45]. However, it is interesting to highlight that the crystallinity of the ENR matrix increases steeply due to the collaborative crystallization of ENR and HNT in association with the contribution of the urea. Higher rubber–filler interactions as indicated by lower Payne effect is responsible for such change. The presence of the urea has played an important role in pulling the surrounding molecular chains. Thus, a significant increase in crystallization is observed at higher strains, and this is in agreement with results observed previously in the stress–strain behaviors and WAXS profiles.

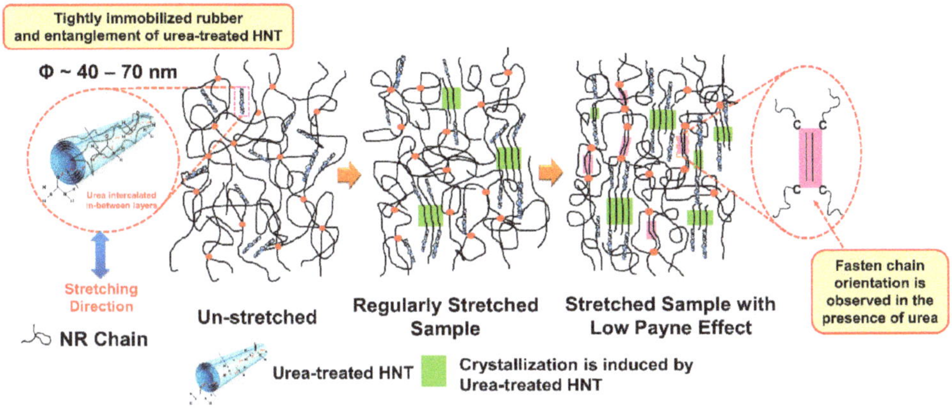

Figure 13. Schematic model representing the crystallization development of the ENR/HNT composites filled with urea-treated HNT.

4. Conclusions

Urea-treated HNT filler was successfully prepared via a mechano-chemical process. Attachment of urea on HNT was verified by FTIR, where new peaks appeared around 3505 cm^{-1} and 3396 cm^{-1}, corresponding to urea's functionalities. The intercalation of urea in the gallery of HNT was revealed by XRD results. The development of two peaks at 21.54° and 22.53° in the range of saw tooth diffraction peak and the shifting of the peak at 2θ of 12.05° to 8.23° correlated to the decrease of the lattice strain and/or to the increment of the crystallite size. The t_{s2} and t_{c90} decreased with urea content due to alkaline nature of urea. The M_H was optimal at 14% urea content. Increased tensile strength and tear strength were also observed due to the improved filler–matrix interfacial adhesions of HNT with the rubber matrix. This is confirmed by the dynamic properties of the composites, as the value of Payne effect was clearly reduced to 62.38% after using urea for treatment. The SIC in the composites exhibited a clear change, as the strain upturn was an earlier strain during stretching, indicating faster crystallization caused by better interfacial interactions within the composites. The X_c was directly observed with XRD during stretching, and it corresponded well with the tensile moduli of the composites. It is noted that correlation between SIC and mechanical strength may not be enough; an alternative testing, such as a fatigue test, is recommended in future work.

Based on the observations overall, it can be concluded that a urea content of about 14% in HNT filler is highly recommended for preparing composites with ENR matrix to ensure the compatibility and strength of the composite. On top of that, the use of urea can be the solution of choice for improving the interaction between ENR and HNT. It can offer improvements in processing behavior without the need to use complicated and costly silane coupling-agent systems.

Author Contributions: Investigation and writing—original draft preparation, I.S. and K.W.; methodology and validation, S.S., H.I. and N.O.; methodology, validation, writing—review and editing, supervision, N.H. All authors have read and agreed to the published version of the manuscript.

Funding: This research was funded by Faculty of Science and Technology, Prince of Songkla University, Pattani Campus, through a SAT-ASEAN research grant (Grant No. SAT590594S).

Institutional Review Board Statement: Not applicable.

Informed Consent Statement: Not applicable.

Data Availability Statement: The data presented in this study are available on request from the corresponding author.

Acknowledgments: We would like to thank the Research and Development Office (RDO), Prince of Songkla University, and Seppo Karrila for assistance with manuscript preparation.

Conflicts of Interest: The authors declare no conflict of interest in this publication.

References

1. Arrighi, V.; McEwen, I.; Qian, H.; Prieto, M.S. The glass transition and interfacial layer in styrene-butadiene rubber containing silica nanofiller. *Polymer* **2003**, *44*, 6259–6266. [CrossRef]
2. Ismail, H.; Pasbakhsh, P.; Fauzi, M.A.; Bakar, A.A. Morphological, thermal and tensile properties of halloysite nanotubes filled ethylene propylene diene monomer (EPDM) nanocomposites. *Polym. Test.* **2008**, *27*, 841–850. [CrossRef]
3. Price, R.R.; Gaber, B.P.; Lvov, Y. In-vitro release characteristics of tetracycline HCl, khellin and nicotinamide adenine dineculeotide from halloysite; a cylindrical mineral. *J. Microencapsul.* **2001**, *18*, 713–722. [PubMed]
4. Du, M.L.; Guo, B.C.; Jia, D.M. Thermal stability and flame retardant effects of halloysite nanotubes on poly(propylene). *Eur. Polym. J.* **2006**, *42*, 1362–1369. [CrossRef]
5. Jia, Z.; Luo, Y.; Guo, B.; Yang, B.; Du, M.; Jia, D. Reinforcing and flame-retardant effects of halloysite nanotubes on LLDPE. *Polym. Plast. Technol. Eng.* **2009**, *48*, 607–613. [CrossRef]
6. Vahedi, V.; Pasbakhsh, P.; Chai, S.-P. Toward high performance epoxy/halloysite nanocomposites: New insights based on rheological, curing, and impact properties. *Mater. Des.* **2015**, *68*, 42–53. [CrossRef]
7. Rooj, S.; Das, A.; Thakur, V.; Mahaling, R.; Bhowmick, A.K.; Heinrich, G. Preparation and properties of natural nanocomposites based on natural rubber and naturally occurring halloysite nanotubes. *Mater. Des.* **2010**, *31*, 2151–2156. [CrossRef]
8. Paran, S.; Naderi, G.; Ghoreishy, M. XNBR-grafted halloysite nanotube core-shell as a potential compatibilizer for immiscible polymer systems. *Appl. Surf. Sci.* **2016**, *382*, 63–72. [CrossRef]
9. Hayeemasae, N.; Sensem, Z.; Sahakaro, K.; Ismail, H. Maleated Natural Rubber/Halloysite Nanotubes Composites. *Processes* **2020**, *8*, 286. [CrossRef]
10. Hayeemasae, N.; Sensem, Z.; Surya, I.; Sahakaro, K.; Ismail, H. Synergistic Effect of Maleated Natural Rubber and Modified Palm Stearin as Dual Compatibilizers in Composites based on Natural Rubber and Halloysite Nanotubes. *Polymers* **2020**, *12*, 766. [CrossRef]
11. Khunova, V.; Kristóf, J.; Kelnar, I.; Dybal, J. The effect of halloysite modification combined with in situ matrix modifications on the structure and properties of polypropylene/halloysite nanocomposites. *Exp. Polym. Lett.* **2013**, *7*, 471–479. [CrossRef]
12. Nicolini, K.P.; Fukamachi, C.R.B.; Wypych, F.; Mangrich, A.S. Dehydrated halloysite intercalated mechanochemically with urea: Thermal behavior and structural aspects. *J. Colloid Interface Sci.* **2009**, *338*, 474–479. [CrossRef]
13. Fukamachi, C.R.B.; Wypych, F.; Mangrich, A. Use of Fe^{3+} ion probe to study the stability of urea-intercalated kaolinite by electron paramagnetic resonance. *J. Colloid Interface Sci.* **2007**, *313*, 537–541. [CrossRef] [PubMed]
14. Trabelsi, S.; Albouy, P.-A.; Rault, J. Stress-induced crystallization properties of natural and synthetic cis-polyisoprene. *Rubber Chem. Technol.* **2004**, *77*, 303–316. [CrossRef]
15. Toki, S.; Hsiao, B.S. Nature of strain-induced structures in natural and synthetic rubbers under stretching. *Macromolecules* **2003**, *36*, 5915–5917. [CrossRef]
16. Candau, N.; Chazeau, L.; Chenal, J.-M.; Gauthier, C.; Munch, E. A comparison of the abilities of natural rubber (NR) and synthetic polyisoprene cis-1, 4 rubber (IR) to crystallize under strain at high strain rates. *Phys. Chem. Chem. Phys.* **2016**, *18*, 3472–3481. [CrossRef]
17. Lake, G.J. Fatigue and Fracture of Elastomers. *Rubber Chem. Technol.* **1995**, *68*, 435–460. [CrossRef]
18. Huneau, B. Strain-induced crystallization of natural rubber: A review of x-ray diffraction investigations. *Rubber Chem. Technol.* **2011**, *84*, 425–452. [CrossRef]
19. Tosaka, M.; Murakami, S.; Poompradub, S.; Kohjiya, S.; Ikeda, Y.; Toki, S.; Sics, I.; Hsiao, B.S. Orientation and Crystallization of Natural Rubber Network As Revealed by WAXD Using Synchrotron Radiation. *Macromolecules* **2004**, *37*, 3299–3309. [CrossRef]
20. Toki, S.; Sics, I.; Ran, S.; Liu, L.; Hsiao, B.S.; Murakami, S.; Tosaka, M.; Kohjiya, S.; Poompradub, S.; Ikeda, Y.; et al. Strain-Induced Molecular Orientation and Crystallization in Natural and Synthetic Rubbers under Uniaxial Deformation by In-situ Synchrotron X-ray Study. *Rubber Chem. Technol.* **2004**, *77*, 317–335. [CrossRef]
21. Imbernon, L.; Pauchet, R.; Pire, M.; Albouy, P.-A.; Tencé-Girault, S.; Norvez, S. Strain-induced crystallization in sustainably crosslinked epoxidized natural rubber. *Polymer* **2016**, *93*, 189–197. [CrossRef]
22. Poompradub, S.; Tosaka, M.; Kohjiya, S.; Ikeda, Y.; Toki, S.; Sics, I.; Hsiao, B.S. Mechanism of strain-induced crystallization in filled and unfilled natural rubber vulcanizates. *J. Appl. Phys.* **2005**, *97*, 103529. [CrossRef]
23. Chenal, J.-M.; Gauthier, C.; Chazeau, L.; Guy, L.; Bomal, Y. Parameters governing strain induced crystallization in filled natural rubber. *Polymer* **2007**, *48*, 6893–6901. [CrossRef]
24. Candau, N.; Oguz, O.; Federico, C.E.; Stoclet, G.; Tahon, J.-F.; Maspoch, M.L. Strain induced crystallization in vulcanized natural rubber containing ground tire rubber particles with reinforcement and nucleation abilities. *Polym. Test.* **2021**, *101*, 107313. [CrossRef]
25. Flory, P.J.; Rehner, J., Jr. Statistical mechanics of cross-linked polymer networks I. Rubberlike elasticity. *J. Chem. Phys.* **1943**, *11*, 512–520. [CrossRef]

26. Marykutty, C.; Mathew, G.; Mathew, E.; Thomas, S. Studies on novel binary accelerator system in sulfur vulcanization of natural rubber. *J. Appl. Polym. Sci.* **2003**, *90*, 3173–3182. [CrossRef]
27. Osaka, N.; Kato, M.; Saito, H. Mechanical properties and network structure of phenol resin crosslinked hydrogenated acrylonitrile-butadiene rubber. *J. Appl. Polym. Sci.* **2013**, *129*, 3396–3403. [CrossRef]
28. Ran, S.; Zong, X.; Fang, D.; Hsiao, B.S.; Chu, B.; Phillips, R.A. Structural and morphological studies of isotactic polypropylene fibers during heat/draw deformation by in-situ synchrotron SAXS/WAXD. *Macromolecules* **2001**, *34*, 2569–2578. [CrossRef]
29. Surya, I.; Ismail, H.; Azura, A. Alkanolamide as an accelerator, filler-dispersant and a plasticizer in silica-filled natural rubber compounds. *Polym. Test.* **2013**, *32*, 1313–1321. [CrossRef]
30. Coran, A. Chemistry of the vulcanization and protection of elastomers: A review of the achievements. *J. Appl. Polym. Sci.* **2003**, *87*, 24–30. [CrossRef]
31. Piasek, Z.; Urbanski, T. The infra-red absorption spectrum and structure of urea. *Bull. L'Acadeie Pol. Sci. Sér. Sci. Chim.* **1962**, *10*, 113–120.
32. Horváth, E.; Kristóf, J.; Kurdi, R.; Makó, É.; Khunová, V. Study of urea intercalation into halloysite by thermoanalytical and spectroscopic techniques. *J. Therm. Anal. Calorim.* **2011**, *105*, 53–59. [CrossRef]
33. Makó, É.; Kristóf, J.; Horváth, E.; Vágvölgyi, V. Kaolinite–urea complexes obtained by mechanochemical and aqueous suspension techniques-a comparative study. *J. Colloid Interface Sci.* **2008**, *330*, 367–373. [CrossRef] [PubMed]
34. Jia, Z.; Luo, Y.; Yang, S.; Du, M.; Guo, B.; Jia, D. Styrene-butadiene rubber/halloysite nanotubes composites modified by epoxidized natural rubber. *J. Nanosci. Nanotechnol.* **2011**, *11*, 10958–10962.
35. Kadi, S.; Lellou, S.; Marouf-Khelifa, K.; Schott, J.; Batonneau-Gener, I.; Khelifa, A. Preparation, characterisation and application of thermally treated Algerian halloysite. *Microporous Mesoporous Mater.* **2012**, *158*, 47–54. [CrossRef]
36. Tan, W.L.; Salehabadi, A.; Mohd Isa, M.H.; Abu Bakar, M.; Abu Bakar, N.H.H. Synthesis and physicochemical characterization of organomodified halloysite/epoxidized natural rubber nanocomposites: A potential flame-resistant adhesive. *J. Mater. Sci.* **2016**, *51*, 1121–1132. [CrossRef]
37. Yuan, P.; Tan, D.; Annabi-Bergaya, F. Properties and applications of halloysite nanotubes: Recent research advances and future prospects. *Appl. Clay Sci.* **2015**, *112–113*, 75–93. [CrossRef]
38. Ismail, H.; Pasbakhsh, P.; Ahmad Fauzi, M.N.; Abu Bakar, A. The effect of halloysite nanotubes as a novel nanofiller on curing behaviour, mechanical and microstructural properties of ethylene propylene diene monomer (EPDM) nanocomposites. *Polym. Plast. Technol. Eng.* **2009**, *48*, 313–323. [CrossRef]
39. Pasbakhsh, P.; Ismail, H.; Ahmad Fauzi, M.N.; Abu Bakar, A. EPDM/modified halloysite nanocomposites. *Appl. Clay Sci.* **2010**, *48*, 405–413. [CrossRef]
40. Payne, A.R.; Whittaker, R.E. Low strain dynamic properties of filled rubbers. *Rubber Chem. Technol.* **1971**, *44*, 440–478. [CrossRef]
41. Nun-anan, P.; Wisunthorn, S.; Pichaiyut, S.; Nathaworn, C.D.; Nakason, C. Influence of nonrubber components on properties of unvulcanized natural rubber. *Polym. Adv. Technol.* **2020**, *31*, 44–59. [CrossRef]
42. Saramolee, P.; Lopattananon, N.; Sahakaro, K. Preparation and some properties of modified natural rubber bearing grafted poly (methyl methacrylate) and epoxide groups. *Eur. Polym. J.* **2014**, *56*, 1–10. [CrossRef]
43. Kuang, W.; Yang, Z.; Tang, Z.; Guo, B. Wrapping of polyrhodanine onto tubular clay and its prominent effects on the reinforcement of the clay for rubber. *Compos. Part A Appl. Sci. Manuf.* **2016**, *84*, 344–353. [CrossRef]
44. Hernández, M.; López-Manchado, M.A.; Sanz, A.; Nogales, A.; Ezquerra, T.A. Effects of strain-induced crystallization on the segmental dynamics of vulcanized natural rubber. *Macromolecules* **2011**, *44*, 6574–6580. [CrossRef]
45. Ozbas, B.; Toki, S.; Hsiao, B.S.; Chu, B.; Register, R.A.; Aksay, I.A.; Prud'homme, R.K.; Adamson, D.H. Strain-induced crystallization and mechanical properties of functionalized graphene sheet-filled natural rubber. *J. Polym. Sci. Part B Polym. Phys.* **2012**, *50*, 718–723. [CrossRef]
46. White, J.L.; Spruiell, J.E. The specification of orientation and its development in polymer processing. *Polym. Eng. Sci.* **1983**, *23*, 247–256. [CrossRef]

polymers

MDPI

Article

Improving Mechanical Properties of PLA/Starch Blends Using Masterbatch Containing Vegetable Oil Based Active Ingredients

Bianka Nagy, Norbert Miskolczi * and Zoltán Eller

Research Centre of Biochemical, Environmental and Chemical Engineering, MOL Department of Hydrocarbon & Coal Processing, Faculty of Engineering, University of Pannonia, H-8200 Veszprém, Hungary; nagy.bianka@mk.uni-pannon.hu (B.N.); ellerz@almos.uni-pannon.hu (Z.E.)
* Correspondence: mnorbert@almos.uni-pannon.hu

Abstract: The aim of this research was to increase the compatibility between PLA and starch with vegetable oil-based additives. Based on tensile results, it can be stated, that Charpy impact strength could be improved for 70/30 and 60/40 blends in both unconditioned and conditioned cases, regardless of vegetable oil, while no advantageous change in impact strength was obtained with PLA-g-MA. Considering sample with the highest starch concentration (50%), the flexural modulus was improved by using sunflower oil-based additive, Charpy impact strength and elongation at break was increased using rapeseed oil-based additive in both conditioned and unconditioned cases. SEM images confirmed the improvement of compatibility between components.

Keywords: PLA/starch; compatibilizer; vegetable oil-based additive; masterbatch

Citation: Nagy, B.; Miskolczi, N.; Eller, Z. Improving Mechanical Properties of PLA/Starch Blends Using Masterbatch Containing Vegetable Oil Based Active Ingredients. *Polymers* **2021**, *13*, 2981. https://doi.org/10.3390/polym13172981

Academic Editor: Domenico Acierno

Received: 9 August 2021
Accepted: 27 August 2021
Published: 2 September 2021

1. Introduction

The main challenges of the growing demand for petroleum-derived plastics are their long degradation periods, health risks, price volatility, waste disposal problems, and increasing demand for raw materials [1,2]. As a consequence, the development of biodegradable polymers from renewable sources has become increasingly conspicuous in recent years [2,3]. Biodegradable polymers can be converted to carbon dioxide, water, methane, and other products [4]. Many biodegradable polymers are known nowadays, such as polylactic acid (PLA), polycaprolactone (PCL), polybutylene adipate terephthalate (PBAT), polyhydroxybutyrate (PHB), and poly(hydroxyalkanoates) (PHA). Starch, known as a natural raw material, is also considered a promising alternative to biopolymers or their constituents [4].

Starch-containing polymers can be divided into four types: thermoplastic starch (TPS), starch/synthetic aliphatic polyester, starch/PBS or PBSA polyester, starch/PVOH [2]. Starch/biodegradable polymer blends are considered an auspicious method to improve the mechanical and thermal properties of native starch. Furthermore, due to their hydrophilic nature, the quality of starch-based blends also depend largely on their moisture content as well [4,5]. Blending the economically viable starch with PLA offers an attractive alternative. PLA generally has good mechanical properties, its strength and stiffness being comparable to, among others, polyethylene terephthalate (PET) and polystyrene (PS). However, its disadvantages include its fragility, its relatively slow rate of degradation in soil, and its higher production costs compared to petroleum-based polymers. By blending of PLA and starch, the good mechanical properties of PLA and the good biodegradability and low manufacturing cost of starch can be combined [6–8].

Most PLA-based plastic blends and composites show partial or complete incompatibility [9]. Hydrophobic PLA and hydrophilic starch are not thermodynamically miscible with each other resulting phase separation and weak interfacial adhesion in their blends. The lack of compatibility gives disadvantageous mechanical properties [7]. Thus, ensuring the compatibility of starch and PLA is essential to improve the mechanical properties [4,5].

During compatibilization, the agents are located at the interface reducing the interfacial tension and preventing the coalescence of the dispersed phase, improving the interfacial adhesion, and creating a thermodynamically stable structure [10]. There are basically four general methods for compatibilization: using copolymers, reactive compatibilization, using nanoparticles, and "radical" processing [10–12]. PLA-g-MA is a potential compatibilizer for PLA-based blends. In PLA/starch systems, interfacial adhesion can be improved by reducing the size of dispersed phase [13]. Another option is to use of bio-based agents. Vegetable oils provide a remarkable alternative instead of petroleum-derived additives with the result that their application is becoming more widespread [9,14]. Some fatty acids allow different chemical modifications due to their single or multiple unsaturation [14]. Modified vegetable oils can behave as a compatibilising agent in binary and ternary blends [9]. Now, epoxidized, MA-modified, and acrylated-epoxidized vegetable oils are known for industrial application [14].

The main motivation of this research is to focus on the production and application of additives based on vegetable oil (sunflower oil, rapeseed oil and castor oil) suitable for improving the miscibility properties of PLA/starch blends. The primary purpose was to produce a masterbatch with compatibilizing nature improving the mechanical properties of PLA/starch blends with different composition.

2. Materials and Methods

2.1. Materials

In this work corn starch (supplied by HungranaBioeconomy Company (Szabadegyháza, Hungary) was blended into commercial grade PLA (IngeoTM Biopolymer 4043D, Minnetonka, MN, USA) as matrix material. To achieve better interfacial properties of the PLA/starch blends, experimentally synthetized vegetable oil-based additives were tested. Three different types of technical grade vegetable oil (sunflower oil (Mw = 880 g/mol, Bunge PLC, Budapest, Hungary), rapeseed oil (Mw = 888 g/mol, Bunge PLC, Budapest, Hungary) and castor oil (Mw = 933 g/mol, Alfa Aeser, Haverhill, MA, USA) were used for additive synthesis.

2.2. Additive Synthesis

Vegetable oil-based additives were synthesized at the Department of MOL Hydrocarbon and Coal Processing, University of Pannonia. The synthesis of vegetable oil-based additives was carried out in a round-bottom flask equipped with a stirrer at temperature range of 130–150 °C in the presence of a hydrocarbon solvent. Stirring speed was set to be 120 rpm. The experiment was performed with three different types of vegetable oils: sunflower oil, rapeseed oil and castor oil. The molar ratio of vegetable oil to maleic anhydride was 1:1. Because of the radical initiated reactions, di-tert-butyl peroxide (supply from Merck KGaA, Darmstadt, Germany) was used. The volatiles and solvent were evaporated under vacuum at the end of reaction.

2.3. Sample Preparation

The additives prepared by the before mentioned method were tested in PLA/starch composites in the form of a masterbatch. Masterbatches containing the synthetized compatibilizer additives were produced by a two-roll mill (LabTech Engineering Ltd., LRM-100, Praksa, Muang, Samutprakarn 10280, Thailand) at temperature range of 150–165 °C using a friction ratio of 32.8:19.3. The matrix material of the masterbatches was PLA. Figure 1 shows the main steps of the sample preparation. The starch content of blends was between 10%–50%. Before processing, all of the polymers were conditioned at 80 °C for four hours to prevent hydrolytic degradation. After homogenization of PLA, starch and masterbatch by a two-roll mill, PLA/starch sheets with size of 170 mm × 170 mm × 2 mm were formed by a laboratory hot press (CARVER 3853-0, Carver, Inc., Savannah, GA, USA) at 170 °C for ten minutes. Then 10 mm wide and 50 mm long specimens were cut out from the sheets.

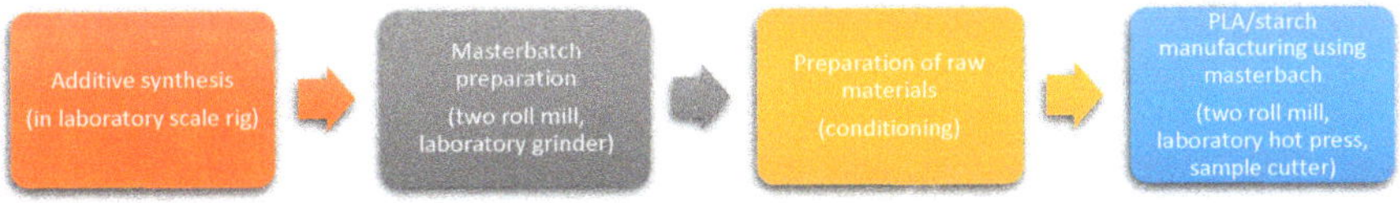

Figure 1. The main steps of the sample preparation.

2.4. Measurements

The main properties of vegetable oil-based additives were determined by standardized methods, by FTIR analysis (Bruker Tensor 27 instrument, USA, spectral range: 7500 to 370 cm^{-1}, with a standard KBr beam-splitter, resolution: better than 1 cm^{-1} (apodised), interferometer: RockSolid, permanent aligned, high stability, sample scan time: 16 scans, background scan time: 16 scans), and through their flow properties analyzed by rheological measurements (Anton Paar MCR301 dynamic shear rheometer, Graz, Austria).

The mechanical properties of PLA/starch blends were measured from the using of INSTRON 3345 universal tensile testing machine (USA), with 75 mm/min crosshead speed for tensile tests and 5 mm/min for flexural tests. The size of specimens were 10 mm wide and 50mm long, the clambing length was 30 mm. Three parallel measurements were carried out on unconditioned (at 20 °C) and conditioned samples (at 80 °C). A CEAST Resil Impactor machine (USA, 1J hammer) with "A" type notches in both unconditioned and conditioned cases was used to know the Charpy impact strength of the samples. Furthermore, the morphology of the samples was also followed via their SEM micrographs (SEM Apreo S LoVac, Waltham, MA, USA, HV: 5–10 kV, mag: 80–20,000×).

3. Results

3.1. Additive Characterization

The main properties of the synthesized additives are summarized in Table 1. Additives had Mn in the range of 6300–8280 g/mol, while the Mw changed between 8360 and 11,910 g/mol. Additive containing castor oil had the lowest polydispersity, while that of rapeseed oil-based additive was the highest. This result refers that additive containing rapeseed oil had the most components with a different structure.

Table 1. Acid number, iodine-bromide number and MA-content values of synthesized additives.

Properties	Sunflower Oil	Sunflower Oil-Based Additive	Rapeseed Oil	Rapeseed Oil-Based Additive	Castor Oil	Castor Oil-Based Additive
Mw, g/mol	-	8360	-	9680	-	11,910
Mn, g/mol	-	6300	-	6570	-	8280
Polydispersity	-	1.30	-	1.36	-	1.25
Acid number, mg KOH/g sample	3.3	54.1	8.6	46.2	2.9	45.4
Ionide-bromide number, I_2/100 g sample	106.7	84.6	101.2	75.5	93.3	75.5
MA-content, mg MA/g sample	-	1.4	-	1.5	-	1.6

The MA-content of the additives prepared based on the three different types of vegetable oils was nearly equal (1.4–1.6 mg MA/g sample). Regarding the acid number, the additive based on sunflower oil had the most carboxyl functional groups, therefore, this additive had the highest acid number (54.1 mg KOH/g sample). On the other hand, the acid numbers of rapeseed and castor oil-based additives were almost the same (46.2 and 45.4 mg KOH/g sample). Considering the degree of unsaturation, the additive most prone to saturation was the sunflower oil-based additive, and the additional properties of rapeseed and castor oil-based additives were almost the same as the acid number.

The FTIR spectra of the vegetable oil-based additives were determined using germanium ATR crystal. The spectra can be seen in Figures 2 and 3.

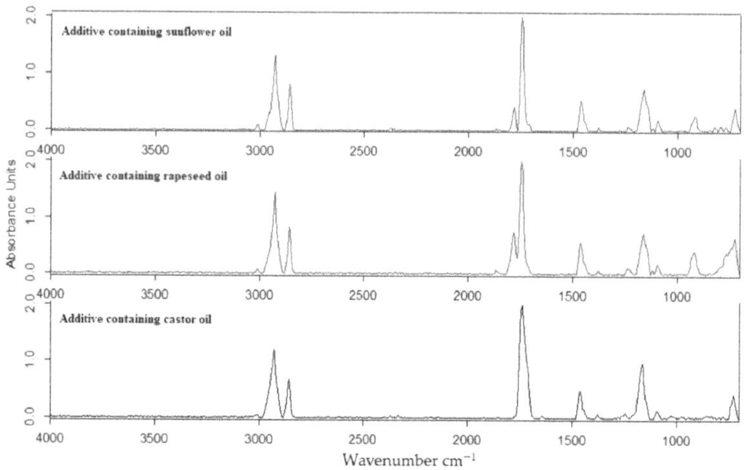

Figure 2. FT-IR spectra of vegetable oil-based additives in the wavenumber range 4000–600 cm^{-1}.

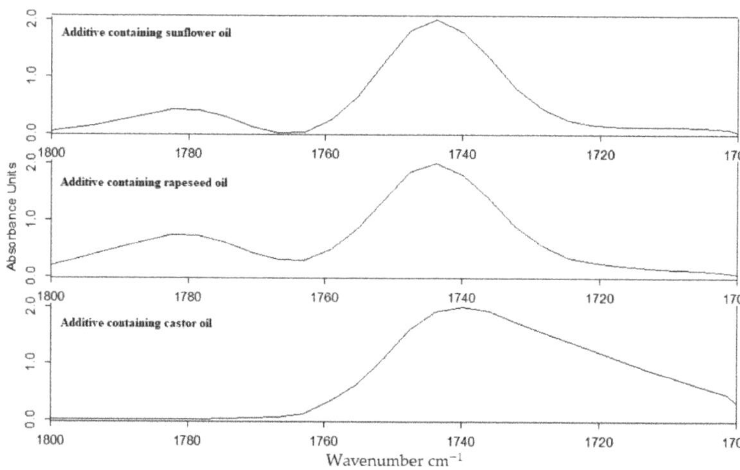

Figure 3. FT-IR spectra of vegetable oil-based additives in the wavenumber range 1800–1700 cm^{-1}.

In the wavenumber range of 3100–2800 cm^{-1}, asymmetric and symmetric stretching vibrations of methyl and methylene groups are observed [15]. The difference between the synthesized additives is manifested in the wavenumber range of 1900–1600 cm^{-1}. While in the wavenumber range of 1650–1630 cm^{-1} and 1750–1730 cm^{-1}, peaks (Figure 3) appeared for all three additives as well. In the range of 1790–1770 cm^{-1}, peaks appeared only in the case of rapeseed and sunflower oil-based additives. In case of the rapeseed oil, the ratio of the two peaks in the range of 1790–1740 cm^{-1} was 2.68, while in case of sunflower oil-based additive it was 3.71.

The dynamic viscosity of synthesized additives was measured at 25 °C (Anton Paar MCR301 instrument) in the shear rate range of 1–1000 $\frac{1}{s}$ as shown in Figure 4.

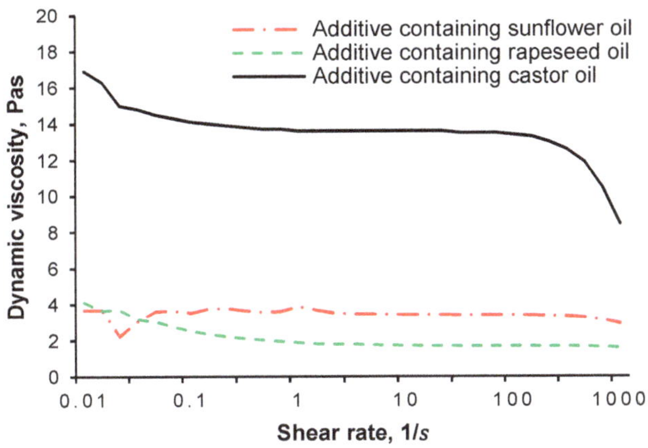

Figure 4. Rheological behaviour of synthesized additives in the range of 1–1000 $\frac{1}{s}$ (25 °C).

Based on the dynamic viscosity values, it can be stated that the additives containing castor oil had the highest dynamic viscosity, followed by the additive containing sunflower oil, and rapeseed oil. On one hand, it was caused by the effect of the length of hydrocarbon side chain of each vegetable oil. On the other hand, differences in viscosities were caused by the fact that the main components in vegetable oils are different with different structures. The main component of castor oil is ricinoleic acid which has a hydroxyl group in addition to the double bond [16]. The hydroxyl group allows maleic anhydride to be incorporated into the molecule, it allows the chemical modification of castor oil through esterification of this functional group to maleated half esters [17]. Most sunflower oils are basically linoleic acid, which have more double bonds than ricinoleic acid or oleic acid [18]. With 1 mole of unconjugated linoleic acid, maleic anhydride can form an "ene" adduct resulting in the formation of a conjugated diene, which can be further functionalized by Diels-Alder synthesis [17]. The main component of rapeseed oil is oleic acid, which has no hydroxyl group and contains less C-C double bonds than linoleic acid [19]. Oleic acid is reacted with maleic anhydride according to the "ene" reaction mechanism [17,20]. Since both rapeseed oil and sunflower oil can be found in both major components, the Diels-Alder and "ene" reaction mechanism may have occurred in both oils. This is probably the reason for the similar dynamic viscosity.

3.2. Masterbatch Characterization

Masterbatches were prepared by the mixing of the synthesized additives into PLA matrix. The main properties of the masterbatches are summarized in Table 2. In order to know the effect of vegetable oil-based additives to the starch-PLA composites, a masterbatch with PLA-g-MA was also prepared. The additive content of the masterbatches was 10%. Regarding the MFI values, it is clear, that PLA-g-MA had the lowest, while the sunflower oil-based masterbatch (SFO) had the highest values. However, there was no significant difference among the MFI values of the experimentally synthetized vegetable oil-based additives. Given the PLA granulates, 8.819 g/10 min MFI value was measured so it can be stated all of the vegetable oil components had softening effect, for PLA-g-MA it was not observed.

Table 2. The main properties of masterbatches (MB).

Properties	SFO-MB	RSO-MB	CO-MB	PLA-g-MA
Active agent	Sunflower oil-based additive	Rapeseed oil-based additive	Castor oil-based additive	PLA-g-MA
Additive content, %	10	10	10	10
MFI, g/10min (5.00 kg, 190 °C)	10.7	10.3	10.1	8.6

Figure 5 summarizes the rheological properties of the masterbatches. As it is well shown, the viscosity of masterbatches with PLA matrix did not change significantly up to the shear rate of $0.2\frac{1}{s}$. Reaching this shear rate, the dynamic viscosity started to decrease significantly. This change leads to the conclusion that molecular formation cannot occur. Regarding the results, the masterbatch containing rapeseed oil had the largest softening effect in the measurement range.

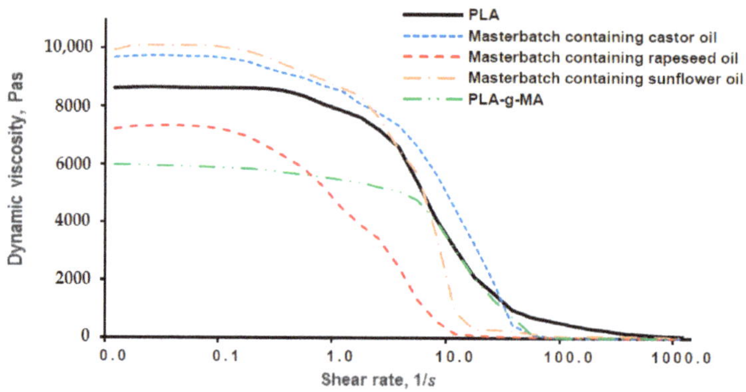

Figure 5. Rheological behaviour of PLA granulates and masterbatches in the range of 1–1000 $\frac{1}{s}$ (170 °C).

3.3. *Mechanical Properties of PLA/Starch Blends*

The purpose of the mechanical tests on PLA/starch blends was to determine the extent to which the starch content of the blends could be increased without deteriorating their mechanical performance.

3.3.1. Results

Results of tensile strength (Figure 6) well shows that the tensile strength decreased with increasing starch content in both unconditioned and conditioned samples for both additive-free and additive-containing blends. In the majority of both unconditioned and conditioned cases, the tensile strength decreased as function of starch content. Comparing the conditioned specimens containing PLA-g-MA (10%–40% starch content), the tensile strength also increased compared to the unconditioned samples. In the case of unconditioned samples, the tensile strength of blends containing vegetable oil-based additives was lower than that of additive free PLA/starch samples or PLA/starch/PLA-g-MA samples. This result was presumably due to the softening effect of vegetable oil-based additives. For all tested compositions, it was observed that the PLA/starch blends without masterbatches had the highest tensile strength (from 39 to 71 MPa). It is important to mention, that at additive free PLA/starch blends had lower tensile strength after the heat treatment in any cases, while the tensile strength of PLA/starch/PLA-g-MA samples could be increased by the heat treatment. Without any heat treatment, the additive free samples had higher tensile strength than the PLA/starch blend with additive. After conditioning, specimens prepared with PLA-g-MA had higher tensile strength (about 70 and 50 MPa of blends containing 10% and 30% starch (improvement was 5 MPa compared to additive free specimens)).

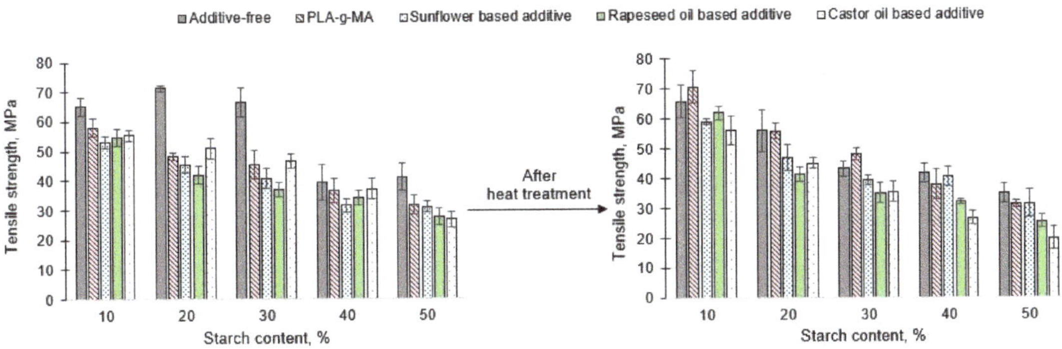

Figure 6. Tensile strength of unconditioned and conditioned PLA/starch blends.

Similar to tensile strength, it can be stated that the value of impact strength decreased with increasing starch content at both conditioning cases (Figure 7). The effect of heat treatment was more significant for samples with lower starch content (10% and 20%). To investigate the effect of masterbatches that did not contain vegetable oil-based additives, it can be concluded that the Charpy impact strength did not increase in either case. In general, the same trend was found for additive free samples, therefore, the impact strength decreased with increasing starch content. The PLA-g-MA was mostly able to compensate the negative effect of heat treatment, and no measurable difference between the two cases can be found. The positive effect of conditioning can also be observed in the case of specimens containing castor oil-based additives—in contrast to those found in the case of tensile strength—with the exception of the sample containing 50% starch, because the value of impact strength can be increased.

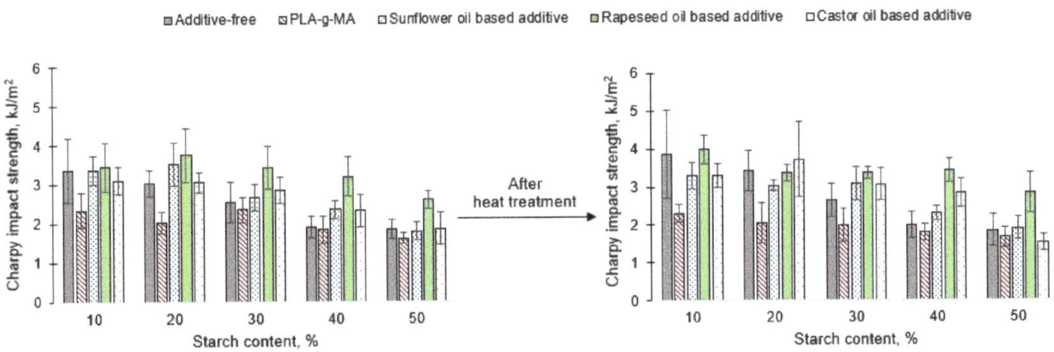

Figure 7. Charpy impact strength of unconditioned and conditioned PLA/starch blends.

Regarding the unconditioned samples, specimens containing 20%, 30%, and 40% starch, higher Charpy impact strength was found with additives than without that. Same phenomena can be stated in case of conditioned samples, thus, the conditioning did not affect the role of the tested additive. In the case of unconditioned samples, the tendency was the same: the value of impact strength decreased with increasing starch content. However, in case of conditioned samples, the starch content of 20% was an exception. Regarding the samples containing rapeseed oil-based additives, it can be stated that with the exception of the samples with 20% and 30% starch, the impact strength increased as a result of heat treatment. In case of both unconditioned and conditioned samples, in general, higher impact strength can be found by the using of vegetable oil-based additives (exception PLA with 10% starch). Regarding the effect of the sunflower oil-based additives, the impact strength values were improved for samples with 30%, 40% and 50% starch content after conditioning at 80 °C and without conditioning compared to the additive free blends. Thus, to investigate the effect of any additive in based on vegetable oil, it can be concluded that for certain compositions, an improvement can be obtained for unconditioned and 80 °C conditioned specimens. An exception to this phenomenon was the additive based on rapeseed oil, since in any composition. An improvement in impact strength can be obtained by its use in both unconditioned and conditioned cases. This is because the rapeseed oil has the longest hydrocarbon chains and consequently it has higher molecular weight.

Regarding the tensile modulus (Figure 8), no trend can be established in proportion to the increase in starch content. However, the highest tensile modulus was found in case of additive free PLA/starch composites without heat treatment (in exception of 70% PLA/30% starch composites). Comparing the results, it was found, that he 50/50 PLA/starch sample (without additives) had the highest tensile modulus among the unconditioned samples. As a result of conditioning, tensile modulus can be decreased, and the highest value (about 1500 MPa) was measured on the 90/10 PLA/starch specimen. Excluding tensile modulus, the effect of heat treatment on the mechanical properties could not be compensated in case of samples containing castor oil-based masterbatch. The tensile modulus of the 70/30 PLA/starch blend could be improved in both unconditioned and conditioned cases. For the samples containing masterbatch synthetized by the using of rapeseed oil-based additives, the tensile modulus could not be improved for any of the compositions compared to the additive free samples. The depreciation, due to 80 °C conditioning did not improve either. Although the masterbatch containing sunflower oil-based additive could not compensate for the negative effect of 80 °C conditioning in all cases. It was successful at 10% and 20% starch content, and a higher tensile modulus was measured compared to the unconditioned samples. However, a lower value was found compared that of to the additive free samples. For the 60/40 and 50/50 compositions, a higher value was measured compared to the additive-free blend after the heat treatment. Similar to the 30/70 PLA/Starch samples containing castor oil-based masterbatch, the tensile modulus of this composition can be improved at both temperatures with the sunflower oil-based masterbatch. By adding the PLA-g-MA, it can be found that for almost all compositions (10%–40% starch content) the values of the tensile modulus were higher after conditioning at 80 °C, which was not observed for either the additive free blends or the samples containing the different types of vegetable oil-based additives. For the unconditioned samples, no positive effect was concluded compared to the additive free blends, however, for the samples conditioned at 80 °C, a blend of some composition showed improvement.

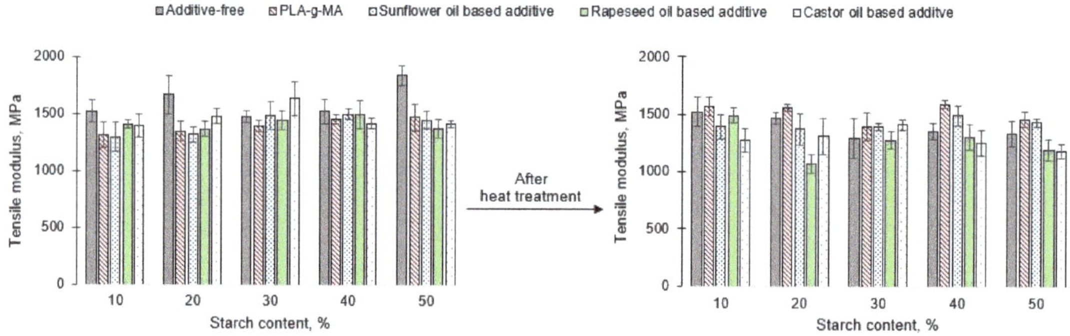

Figure 8. Tensile modulus of unconditioned and conditioned PLA/starch blends.

Regarding the flexural modulus (Figure 9), the unfavourable effect of conditioning also prevails, because in case of all tested compositions the values of flexural modulus after conditioning were lower.

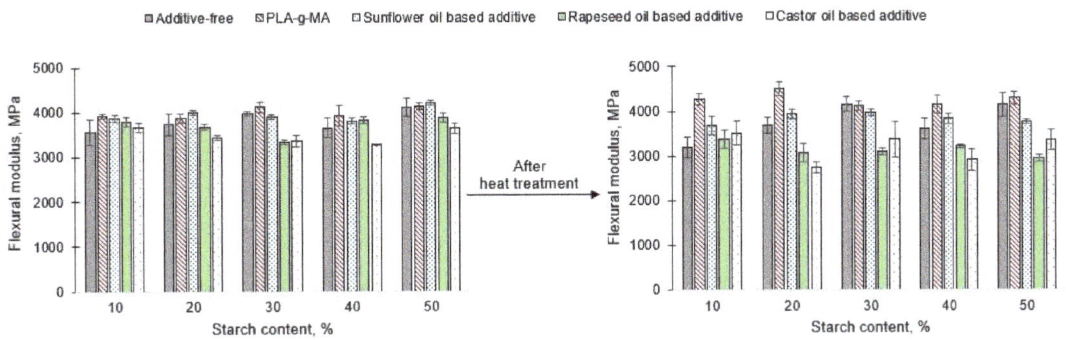

Figure 9. Flexural modulus of unconditioned and conditioned PLA/starch blends.

The effect of conditioning was eliminated with only PLA-g-MA at all ratios. In addition, unconditioned samples showed a significant improvement over the flexural modulus of the additive-free PLA/starch blends (for instance 260 MPa improvement was measured in case of 90/10 PLA/starch composition). Sunflower oil-based additives also caused a significant improvement in all blends of the examined composition and the disadvantage of the heat treatment can been reduced. Overall, for both conditioned and unconditioned samples, the flexural modulus could be improved with each masterbatch in case of 10% starch content.

Based on data in Figure 10, it was observed that increasing starch content causing a decrease in the elongation at break considering both the unconditioned and conditioned specimens. This tendency was more or less observed in the sample series containing the masterbatch. The effect of conditioning did not cause a large change in elongation at break. The largest change was a decrease from 5.6% elongation at break to 4.6% in case of additive free 30/70 PLA/starch blend. A decrease in the elongation at break was observed at both temperatures using PLA-g-MA. By using a rapeseed oil-based additive, the elongation at break could be increased regardless of composition and conditioning. Irrespective of conditioning, the castor oil and sunflower oil-based additives also resulted an increase in case of blends containing 10% and 20% starch as well.

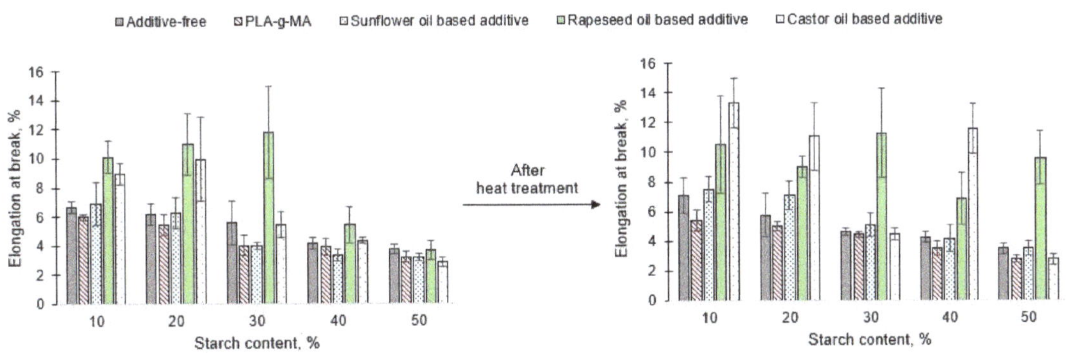

Figure 10. Elongation at break of unconditioned and conditioned PLA/starch blends.

3.3.2. Summary of Mechanical Properties

One of the purposes of mechanical tests was to track changes in the mechanical properties of specimens depending on their moisture content. For this reason, the specimens were conditioned at room temperature and at the drying temperature of the raw materials (80 °C) in order to minimize the moisture content of the specimens. Ke et al. [21,22] found that the thermal and crystallization properties of PLA or the compatibility between

PLA and starch were not modified significantly by moisture content. However, with increasing moisture content, the morphology of the mixtures became more uniform due to higher degree of gelatinization of the starch. Furthermore, it was found that as the starch content increased, both tensile strength and elongation at break decreased [22]. Given the mechanical properties, the type of starch has an influencing effect because different types of starch have different morphology and microstructure. The mechanical properties are significantly influenced by particle size and agglomeration of the particles, as well as a surface modification method influencing the agglomeration. [23] For example, in the experiment of Khalid et al. [23], PLA/starch composites containing large-sized starch granules showed higher strength.

The main conclusions of mechanical properties are summarized in Tables 3–6 indicating the measured improvements and depreciations compared to the additive free PLA/starch blends.

Regarding the results of the unconditioned blends, it can be concluded that the tensile strength was the only characteristic that could not be significantly increased by adding any of the vegetable oil-based additives. Only the PLA-g-MA could slightly increase the tensile strength of the conditioned samples (90/10 and 70/30 composition). However, the PLA-g-MA cannot increase the Charpy impact strength. In contrast, all of the additives containing vegetable oil can improve the Charpy impact strength, among other things, in case of blends containing 30% and 40% starch regardless of conditioning. Regarding the tensile modulus, the blends containing PLA-g-MA showed improvements for conditioned samples (each composition), while the additives made from sunflower oil and castor oil demonstrates improvement in case of 70/30 PLA/starch composition regardless of conditioning. In terms of the improvement of flexural modulus, the PLA-g-MA clearly proved to be the most effective regardless of conditioning. However, the sunflower oil-based additive can approach the effect of PLA-g-MA for all compositions. In terms of elongation at break, blends with rapeseed oil-based additive was the most effective for all cases. In the case of blends with 10% starch content, in addition to improvement of flexural modulus, the elongation at break could also be increased by using all types of the vegetable oil-based additives, and even by using the rapeseed oil-based additive, the impact strength can be also increased.

Regarding the blends containing 20% starch, the elongation at break was higher in the case of unconditioned and conditioned specimens. In addition to elongation at break, the flexural modulus was also increased by the sunflower-based additive. In case of 30% starch content, it was found that both the Charpy impact strength and the tensile modulus can be improved by using the castor oil or sunflower oil-based additives as well, while by using of rapeseed oil-based additive, the impact strength and elongation at break were increased in the temperature range of 25–80 °C. For 60/40 PLA/starch blend, the impact strength could be increased with all three vegetable oil-based additives, and the flexural modulus was improved by using of sunflower oil-based additives. Furthermore, the value of the elongation at break can be also increased due to the additive based on castor oil and rapeseed oil. Finally, regarding the samples containing 50% starch, no improvement could be obtained using castor oil-based additive in any of the properties, but the value of impact strength and elongation at break were better with the use of rapeseed oil-based additive, while the flexural modulus was increased as well using sunflower oil-based additive.

Table 3. Effects on mechanical properties of castor oil-based additive ("+": positive change relative to an additive-free blend of a suitable composition; "−": negative change relative to an additive-free blend of a suitable composition).

Starch Content, %	Unconditioned Blends					Conditioned Blends				
	Tensile Strength	Charpy Impact Strength	Tensile Modulus	Flexural Modulus	Elongation at Break	Tensile Strength	Charpy Impact Strength	Tensile Modulus	Flexural Modulus	Elongation at Break
10	−	−	−	+	+	−	−	−	+	+
20	−	−	−	−	+	−	+	−	−	+
30	−	+	+	−	−	−	+	+	−	+
40	−	+	−	−	+	−	+	−	−	+
50	−	−	−	−	−	−	−	−	−	−

Table 4. Effects on mechanical properties of rapeseed oil-based additive ("+": positive change relative to an additive-free blend of a suitable composition; "−": negative change relative to an additive-free blend of a suitable composition).

Starch Content, %	Unconditioned Blends					Conditioned Blends				
	Tensile Strength	Charpy Impact Strength	Tensile Modulus	Flexural Modulus	Elongation at Break	Tensile Strength	Charpy Impact Strength	Tensile Modulus	Flexural Modulus	Elongation at Break
10	−	+	−	+	+	−	+	−	+	+
20	−	+	−	−	+	−	−	−	−	+
30	−	+	−	−	+	−	+	−	−	+
40	−	+	−	+	+	−	+	−	−	+
50	−	+	−	−	+	−	+	−	−	+

Table 5. Effects on mechanical properties of sunflower oil-based additive ("+": positive change relative to an additive-free blend of a suitable composition; "−": negative change relative to an additive-free blend of a suitable composition).

Starch Content, %	Unconditioned Blends					Conditioned Blends				
	Tensile Strength	Charpy Impact Strength	Tensile Modulus	Flexural Modulus	Elongation at Break	Tensile Strength	Charpy Impact Strength	Tensile Modulus	Flexural Modulus	Elongation at Break
10	−	−	−	+	+	+	−	−	+	+
20	−	+	−	+	+	−	−	−	+	+
30	−	+	+	−	−	+	+	+	+	+
40	−	+	−	+	−	−	+	+	+	−
50	−	−	−	+	−	−	+	+	+	−

Table 6. Effects on mechanical properties of PLA-g-MA ("+": positive change relative to an additive-free blend of a suitable composition; "−": negative change relative to an additive-free blend of a suitable composition).

Starch Content, %	Unconditioned Blends					Conditioned Blends				
	Tensile Strength	Charpy Impact Strength	Tensile Modulus	Flexural Modulus	Elongation at Break	Tensile Strength	Charpy Impact Strength	Tensile Modulus	Flexural Modulus	Elongation at Break
10	−	−	−	+	−	+	−	+	+	−
20	−	−	−	+	−	−	−	+	+	−
30	−	−	−	+	−	+	−	+	+	−
40	−	−	−	+	−	−	−	+	+	−
50	−	−	−	+	−	−	−	+	+	−

3.4. Morphological Examination of the Structure

Morphological structures of samples containing 10% and 50% starch content were investigated by scanning electron microscopy (SEM) (Figure 11).

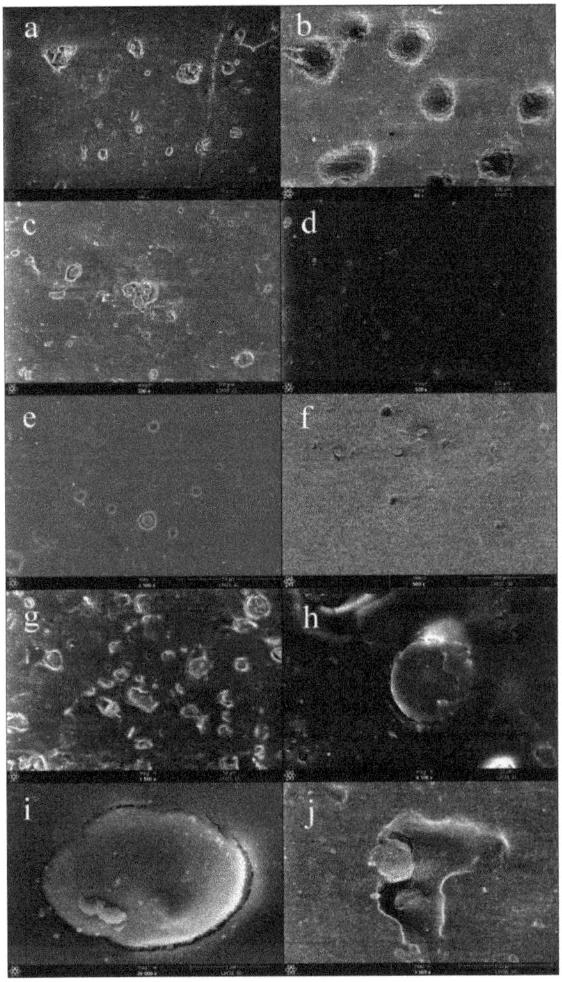

Figure 11. SEM images of compatibilized and uncompatibilised PLA/starch blends containing 10 and 50% starch ((**a**) 90/10 PLA/starch without compatibilization, (**b**) 50/50 PLA/starch without compatibilization, (**c**) 90/10 PLA/starch with castor oil-based masterbatch, (**d**) 90/10 PLA/starch with rapeseed oil-based masterbatch, (**e**) 90/10 PLA/starch with sunflower oil-based masterbatch, (**f**) 90/10 PLA/starch with PLA-g-MA, (**g**) 50/50 PLA/starch with castor oil-based masterbatch, (**h**) 50/50 PLA/starch with rapeseed oil-based masterbatch, (**i**) 50/50 PLA/starch with sunflower oil-based masterbatch, (**j**) 50/50 PLA/starch with PLA-g-MA).

The aim of this investigation was to determine whether the morphological structure was modified by adding masterbatch, to see if whether the compatibility was effective. Regarding the morphology of PLA/starch blends with composition of 90/10 and 50/50 without any additives (Figure 11a–c), it can be stated that the mutual miscibility of the two phases is only partial regardless of whether they contain 10% or 50% starch. Starch can be observed as a dispersed phase and PLA as a matrix. The images clearly show phase separation, in addition the starch particles were unevenly dispersed in the PLA matrix. The 50/50 PLA/starch blend has a more uneven structure than 90/10 blend. Regarding the morphology of the samples with 90/10 composition (Figure 11a,c–f), it was found that the most uniform surface is available in case of blend containing rapeseed oil-based additive. Given the blend containing castor oil-based additive, the separation of dispersed phases (starch) from the matrix was observed. The structure of the sample containing sunflower oil-based additive appears to be smoother than that of the sample containing castor oil, however, the separation of the dispersed starch phase is also evident in the SEM images. Although the structure is smoother for samples containing the PLA-g-MA masterbatch, the dispersed phase is also more or less separated from the PLA. Overall, compared to the additive-free samples, with the exception of the blend containing castor oil, a more uniform appearance of the morphology was observed in all cases, the distribution and the incorporation of the dispersed phase into the matrix could be improved. In case of sample containing 50% starch, the starch phases are completely separated from the matrix material resulting in a severe inhomogeneous morphology. Two important changes can be observed in the structure of the samples containing the additives: one is that the dispersed phases have smaller sizes and less agglomeration and are more evenly distributed. The other is that the matrix material can involve better than the dispersed phases. This positive effect is most pronounced for samples containing castor oil and least for blends containing PLA-g-MA masterbatch.

4. Conclusions

The aim of the experiment was to improve the compatibility between PLA and starch. To improve miscibility, PLA-g-MA and vegetable oil-based additives containing maleic anhydride were used in the form of masterbatches. The heat treatment at 80 °C has caused significant differences in mechanical properties, the effect of which could be eliminated in some samples by adding the synthesized additive. The effect of vegetable oil-based additives was most pronounced in Charpy impact strength. Using rapeseed oil-based additive, improvements were observed compared for several specimens to the additive-free specimens, for instance, in case of 50/50 PLA/starch blend, there was a nearly 40% improvement in the impact strength for the unconditioned sample, and a nearly 50% improvement after heat treatment compared to the additive-free blend. SEM images also confirmed the advantageous effect of the rapeseed oil-based additive. The effect of PLA-g-MA on the Charpy impact strength was not favourable. Taking into account the results of tensile tests, it was reduced in all cases, and the tensile modulus could be improved in case of 70/30 PLA/starch composition using castor oil and sunflower oil. The rate of improvement was about 10% both before and after the heat treatment. In terms of flexural modulus, PLA-g-MA proved to be more effective against vegetable oil-based additives.

Author Contributions: Conceptualization, B.N., N.M. and Z.E.; methodology, B.N. and N.M.; investigation, B.N.; writing—original draft preparation, B.N., N.M. and Z.E.; writing—review and editing, B.N., N.M. and Z.E.; visualization, B.N.; supervision, N.M. and Z.E. All authors have read and agreed to the published version of the manuscript.

Funding: This research received no external funding.

Institutional Review Board Statement: Not applicable.

Informed Consent Statement: Not applicable.

Data Availability Statement: Not applicable.

Acknowledgments: The authors would like to thank Nóra Lili Nagy, chemical technician for supporting our laboratory work. SEM images were performed at the electron microscopy laboratory of the University of Pannonia, established using grant no. GINOP-2.3.3-15-2016-0009 from the European Structural and Investments Funds and the Hungarian Government.

Conflicts of Interest: The authors declare no conflict of interest.

References

1. Méité, N.; Konan, L.K.; Tognonvi, M.T.; Doubi, B.I.H.G.; Gomina, M.; Oyetola, S. Properties of hydric and biodegradability of cassava starch-based bioplastics reinforced with thermally modified kaolin. *Carbohyd. Polym.* **2021**, *254*, 117322. [CrossRef] [PubMed]
2. Ashter, S.A. *Introduction to Bioplastics Engineering*, 1st ed.; William Andrew: Noewich, NY, USA, 2016; pp. 1–17. [CrossRef]
3. Rendón-Villalobos, R.; Ortíz-Sánchez, A.; Tovar-Sánchez, E.; Flores-Huicochea, E. The Role of Biopolymers in Obtaining Environmentally Friendly Materials. In *Composites from Renewable and Sustainable Materials*; IntechOpen: Rijeka, Croatia, 2016; pp. 151–159. [CrossRef]
4. Encalada, K.; Belén, M.; Proaño, E.; Valle, V. An overview of starch-based biopolymers and their biodegradability. *Artículo Investigación. Revista Ciencia e Ingeniería* **2018**, *39*, 245–258.
5. Muller, J.; González-Martínez, C.; Chiralt, A. Combination of Poly(lactic) Acid and Starch for Biodegradable Food Packaging. *Materials* **2017**, *10*, 952. [CrossRef] [PubMed]
6. Esmaeili, M.; Pircheraghi, G.; Bagheri, R.; Altstädt, V. Poly(lactic acid)/coplasticized thermoplastic starch blend: Effect of plasticizer migration on rheological and mechanical properties. *Polym. Adv. Technol.* **2018**, *30*, 1–13. [CrossRef]
7. Koh, J.J.; Zhang, X.; He, C. Fully biodegradable Poly(lactic acid)/Starch blends: A review of toughening strategies. *Int. J. Biol. Macromol.* **2018**, *109*, 99–113. [CrossRef] [PubMed]
8. Ohkita, T.; Lee, S.-H. Thermal degradation and biodegradability of poly (lactic acid)/corn starch biocomposites. *J. Appl. Polym. Sci.* **2006**, *100*, 3009–3017. [CrossRef]
9. Carbonell-Verdu, A.; Ferri, J.M.; Dominici, F.; Boronat, T.; Sanchez-Nacher, L.; Balart, R.; Torre, L. Manufacturing and compatibilization of PLA/PBAT binary blends by cottonseed oil-based derivatives. *Express Polym. Lett.* **2018**, *12*, 808–823. [CrossRef]
10. Ding, Y.; Feng, W.; Huang, D.; Lu, B.; Wang, P.; Wang, G.; Ji, J. Compatibilization of immiscible PLA-based biodegradable polymer blends using amphiphilic di-block copolymers. *Eur. Polym. J.* **2019**, *118*, 45–52. [CrossRef]
11. Ajitha, A.R.; Sabu, T. Introduction: Polymer blends, thermodynamics, miscibility, phase separation, and compatibilization. In *Compatibilization of Polymer Blends, Micro and Nano Scale Phase Morphologies, Interphase Characterization and Properties*; Elsevier Science: Amsterdam, The Netherlands, 2020; pp. 1–29. [CrossRef]
12. Seier, M.; Stanic, S.; Koch, T.; Archodoulaki, V.-M. Effect of Different Compatibilization Systems on the Rheological, Mechanical and Morphological Properties of Polypropylene/Polystyrene Blends. *Polymers* **2020**, *12*, 2335. [CrossRef] [PubMed]
13. Raj, A.; Samuel, C.; Prashantha, K. Role of Compatibilizer in Improving the Properties of PLA/PA12 Blends. *Front. Mater.* **2020**, *7*, 193. [CrossRef]
14. Garcia-Campo, M.; Quiles-Carrillo, L.; Masia, J.; Reig-Pérez, M.; Montanes, N.; Balart, R. Environmentally Friendly Compatibilizers from Soybean Oil for Ternary Blends of Poly(lactic acid)-PLA, Poly(ε-caprolactone)-PCL and Poly(3-hydroxybutyrate)-PHB. *Materials* **2017**, *10*, 1339. [CrossRef] [PubMed]
15. Stuart, B.H. *Infrared Spectroscopy: Fundamentals and Application*; John Wiley & Sons Ltd: Hoboken, NJ, USA, 2004; pp. 71–74. ISBN 978-0-470-85428-0.
16. Prakash, T.; Geo, V.E.; Martin, L.J.; Nagalingam, B. Effect of ternary blends of bio-ethanol, diesel and castor oil on performance, emission and combustion in a CI engine. *Renew. Energy* **2018**, *122*, 301–309. [CrossRef]
17. Wu, F.; Musa, O.M. Vegetable Oil–Maleic Anhydride and Maleimide Derivatives: Syntheses and Properties. In *Handbook of Maleic Anhydride Based Materials: Synthesis, Properties and Application*; Springer International Publishing Switzerland: Berlin/Heidelberg, Germany, 2016; pp. 151–210. [CrossRef]
18. Monfreda, M.; Gobbi, L.; Grippa, A. Blends of olive oil and sunflower oil: Characterisation and olive oil quantification using fatty acid composition and chemometric tools. *Food Chem.* **2012**, *134*, 2283–2290. [CrossRef] [PubMed]
19. Encinar, J.M.; Pardal, A.; Sánchez, N.; Nogales, S. Biodiesel by Transesterification of Rapeseed Oil Using Ultrasound: A Kinetic Study of Base-Catalysed Reactions. *Energies* **2018**, *11*, 2229. [CrossRef]
20. Stefanoiu, F.; Candy, L.; Vaca-Garcia, C.; Borredon, E. Kinetics and mechanism of the reaction between maleic anhydride and fatty acid esters and the structure of the products. *Eur. J. Lipid Sci. Technol.* **2008**, *110*, 441–447. [CrossRef]
21. Ke, T.; Sun, X. Effects of Moisture Content and Heat Treatment on the Physical Properties of Starch and Poly(lactic acid) Blends. *J. Appl. Polim. Sci.* **2001**, *81*, 3069–3082. [CrossRef]
22. Ke, T.; Sun, X. Physical Properties of Poly(Lactic Acid) and Starch Composites with Various Blending Ratios. *Cereal Chem.* **2000**, *77*, 761–768. [CrossRef]
23. Khalid, S.; Yu, L.; Meng, L.; Liu, H.; Ali, A.; Chen, L. Poly(lactic acid)/starch composites: Effect of microstructure and morphology of starch granules on performance. *J. Appl. Polim. Sci.* **2017**, *134*, 45504. [CrossRef]

MDPI
St. Alban-Anlage 66
4052 Basel
Switzerland
Tel. +41 61 683 77 34
Fax +41 61 302 89 18
www.mdpi.com

Polymers Editorial Office
E-mail: polymers@mdpi.com
www.mdpi.com/journal/polymers

www.ingramcontent.com/pod-product-compliance
Lightning Source LLC
Chambersburg PA
CBHW061455220926
51850CB00004B/95
9783036566856